19th International Conference on Aerospace Sciences & Aviation Technology (ASAT-19 2021)

AA003794

IOP Conference Series: Materials Science and Engineering
Volume 1172

Cairo, Egypt
6-8 April 2021

ISBN: 978-1-7138-4632-1
ISSN: 1757-8981

Printed from e-media with permission by:

Curran Associates, Inc.
57 Morehouse Lane
Red Hook, NY 12571

Some format issues inherent in the e-media version may also appear in this print version.

This work is licensed under a Creative Commons Attribution 3.0 International Licence.
Licence details: http://creativecommons.org/licenses/by/3.0/.

No changes have been made to the content of these proceedings. There may be changes to pagination and minor adjustments for aesthetics.

Printed with permission by Curran Associates, Inc. (2022)

For permission requests, please contact the Institute of Physics at the address below.

Institute of Physics
Dirac House, Temple Back
Bristol BS1 6BE UK

Phone: 44 1 17 929 7481
Fax: 44 1 17 920 0979

techtracking@iop.org

Additional copies of this publication are available from:

Curran Associates, Inc.
57 Morehouse Lane
Red Hook, NY 12571 USA
Phone: 845-758-0400
Fax: 845-758-2633
Email: curran@proceedings.com
Web: www.proceedings.com

TABLE OF CONTENTS

IOP Conf. Series: Materials Science and Engineering 1172 (2021) 011001 doi:10.1088/1757-899X/1172/1/011001

19th International Conference on Aerospace Sciences and Aviation Technology

6-8 APRIL 2021

CAIRO , EGYPT

asat.conferences.ekb.eg

www.mtc.edu.eg
asat@mtc.edu.eg

IOP Publishing

Egyptian Knowledge Bank
بنك المعرفة المصري

Content from this work may be used under the terms of the Creative Commons Attribution 3.0 licence. Any further distribution of this work must maintain attribution to the author(s) and the title of the work, journal citation and DOI.
Published under licence by IOP Publishing Ltd

IOP Conf. Series: Materials Science and Engineering 1172 (2021) 011001 doi:10.1088/1757-899X/1172/1/011001

ASAT-19

The 19th International Conference on Aerospace Sciences & Aviation Technology (ASAT-19)
April 6-8, 2021, Cairo, Egypt

Conference Program and Invited Talks

ASAT-19 2021 IOP Publishing
IOP Conf. Series: Materials Science and Engineering 1172 (2021) 011001 doi:10.1088/1757-899X/1172/1/011001

ASAT-19

The 19th International Conference on Aerospace Sciences & Aviation Technology (ASAT-19)
April 6-8, 2021, Cairo, Egypt

CONTENTS

IOP Conf. Series: Materials Science and Engineering 1172 (2021) 011001 doi:10.1088/1757-899X/1172/1/011001

ASAT-19

The 19th International Conference on Aerospace Sciences & Aviation Technology (ASAT-19)
April 6-8, 2021, Cairo, Egypt

PREFACE

The Military Technical College, Cairo, Egypt has the pleasure to organize the 19th International Conference on Aerospace Sciences & Aviation Technology (ASAT-19), APR 6 - 8, 2021, Cairo, Egypt, sponsored by the Egyptian Ministry of Defense.

The conference provides an academic platform for professionals and researchers from research, academia and, industrial sectors involved in Aerospace Science and Aviation Technology to exchange knowledge and gain an insight into the state of the art in the latest technology.

Confer-ence program includes 11th scientific sessions and 15th invited talks covering the following topics:

- Aerodynamics and Mechanics of flight
- Aerospace Combustion and Propulsion
- Communications and Networks
- Aerospace structures
- Aerospace Materials
- Remote sensing & image processing
- Renewable Energy
- Energetic Materials
- Aerospace Thermophysics
- Guidance, Navigation and Control
- Hydraulics and Fluid Power Systems
- Unmanned Systems
- Multidisciplinary Design Optimization
- Wind Energy

The (120) full manuscript submitted to the conference secretary, are re-viewed. Just (46) paper have been accepted for suitable publication in the conference.

These selected papers will be presented during the conference interval, APR 6 - 8, 2021, in different (11) scientific sessions. In addition to 15 invited talks.

IOP Conf. Series: Materials Science and Engineering 1172 (2021) 011001 doi:10.1088/1757-899X/1172/1/011001

ASAT-19

The 19th International Conference on Aerospace Sciences & Aviation Technology (ASAT-19)
April 6-8, 2021, Cairo, Egypt

PREFACE

Finally, the conference high committee hopes that the conference will achieve its planned mission and would like to acknowledge all contributors, members of the scientific committee and chairmen of the conference session.

ASAT-19

The 19th International Conference on Aerospace Sciences & Aviation Technology (ASAT-19)
April 6-8, 2021, Cairo, Egypt

LIST OF COMMITEES

Chairman

Maj. Gen. Prof. Ismail Mohamed Kamal Ismail

Vice Chairman

Maj. Gen. Prof. Moutaz Abo El-Noor

Conference Rapporteur

Mag. Gen. (R) Prof. Ahmed Elsaid Abdalla

Conference Co-Rapporteur

Prof. Mohamed Misbah Elkhatib

Rapporteur Assistants

- Brig. Gen. Assoc. Prof. Mohammad Yahya Mohammad
- Col. Dr. Mohamed Yahya Zakaria

Organizing Committee

- Maj. Gen. Prof. Yehia Z. Elhalwagy
- Maj. Gen. Prof. Hossam Abo Baker
- Brig. Gen. Assoc. Prof. Hazem Z. Faheem
- Brig. Gen. Assoc. Prof. Mohamed Ahmed Kamel
- Brig. Gen. Assoc. Prof. Sherif A El-Gamal
- Col. Dr. Ahmed Farid Namnam
- Lt. Col. Karim Abdelatty

Scientific Committee

- Maj. Gen. Prof. (R) Mahmoud Galal El-Din Mohamed Rabie
- Maj. Gen. Prof. (R) Ibrahim S. Mostafa
- Maj. Gen. Prof. (R) Ahmed El-Osmany Abdel-Fattah
- Maj. Gen. (R) Assoc. Prof. Nabil Gerges Michael
- Maj. Gen. (R) Assoc. Prof. M. Allam AlSanabawy
- Maj. Gen. (R) Assoc. Prof. Hamed M. Abdalla
- Maj. Gen. (R) Prof. Abdel Rahman H. Bardawiny

ASAT-19

The 19th International Conference on Aerospace Sciences & Aviation Technology (ASAT-19)
April 6-8, 2021, Cairo, Egypt

LIST OF COMMITEES

- Maj. Gen. (R) Prof. A. M. Reyad Mahmoud
- Maj. Gen. (R) Prof. Mostafa Adnan El-Shaffie
- Brig. Gen. (R) Prof. Ossama Morsy Moussa
- Maj. Gen. (R) Prof. Ahmed Elsaid Abdalla
- Maj. Gen. (R) Ossama M. Kamal
- Brig. Gen. Assoc. Prof. Ashraf Mahran
- Brig. Gen. Assoc. Prof. Mohamed A. Kamel
- Brig. Gen. Prof. Mohamed Misbah ElKhatib
- Brig.Gen. Dr. Ahmed Esmat
- Col. Dr. Assoc. Prof. Ahmed Ekhlas
- Brig. Gen. Assoc. Prof. M. Yahia Mohamed
- Col. Assoc. Prof. Bassem E. Sheta
- Col. Dr. Mohamed Yahia Zakaria

ASAT-19

The 19th International Conference on Aerospace Sciences & Aviation Technology (ASAT-19)
April 6-8, 2021, Cairo, Egypt

CONTRIBUTORS

- Department of Mechanical Engineering, College of Engineering, Taif University, Taif 21974, KSA
- Production and Mechanical Design Department, Faculty of Engineering, Mansoura University, Mansoura 35516, Egypt
- Civil Engineering Department, Military Technical College, Cairo, Egypt.
- Ships and Submarines Engineering Department, Military Technical College, Cairo, Egypt.
- Mechanical Engineering Department, Faculty of Engineering & Technology, Future University in Egypt, Cairo, Egypt
- Department of Aircraft Mechanics, Military Technical College, Cairo, Egypt.
- Mechanical Design and Production Department, Military Technical College, Cairo, Egypt.
- Rockets Department, Military Technical College, Cairo, Egypt
- Technical Research Centre, Cairo, Egypt
- Mechanical Engineering Department, Military Technical College, Cairo, Egypt.
- Computer Engineering, Pyramids Higher Institute for engineering & Technology
- (PHI), Giza, Egypt
- Computer Engineering, Arab Academy for Science, Technology & Maritime Transport (AASTMT) Cairo, Egypt
- Electronic Engineering Department, Military Technical College, Cairo, Egypt.
- Department of Communication Engineering, Military Technical College, Cairo, Egypt.
- NANO Master program, Faculty of Engineering. Cairo University, Giza, Egypt
- Avionics Department, Military Technical Collage, Cairo, Egypt
- Egyptian Academy for Engineering and Advanced Technology affiliated to ministry of Military Production, Cairo, Egypt.
- October 6 University, Cairo, Egypt.

ASAT-19

The 19th International Conference on Aerospace Sciences & Aviation Technology (ASAT-19)
April 6-8, 2021, Cairo, Egypt

CONTRIBUTORS

- Department of Civil Engineering, Military Technical College, Cairo, Egypt.
- Mechanical Engineering Department, Faculty of Engineering, Mataria-Helwan University in Cairo, Egypt.
- Mechanical Engineering Department, Faculty of Engineering and Technology, Badr University in Cairo (BUC), Cairo, Egypt.
- Key Laboratory of Power Station Energy Transfer Conversion and Sys-tem of Ministry of Education, School of Energy, Power and Mechani-cal Engineering, North China Electric Power University, Beijing 102206, China
- Mechanical Power Engineering Department, Faculty of Engineering – Mattaria, Helwan University, Cairo 11718, Egypt
- Centre for Renewable Energy Faculty of Engineering, British University in Egypt.
- Department of Chemical Engineering, Military Technical College, Cairo, Egypt.
- Department of Radar, Military Technical College, Cairo, Egypt.
- El-Nasr Company for Intermediate Chemicals, Industrial Area, Giza, Egypt.
- Radiation Chemistry Department, National Center for Radiation Research and Technology, P.O.Box 29, Nasr City, Cairo, Egypt.
- Engineering, Menoufia University, Egypt 32511
- Technical Research Center, Cairo, Egypt
- Mechanical Design and Production Engineering Department, Faculty of Engineering, Ain Shams University, Cairo, Egypt
- Chemistry Division, Faculty of Engineering, Ain Shams University, Cairo, Egypt
- Chemical Engineering Department, Military Technical College, Cairo, Egypt.

ASAT-19 2021 IOP Publishing
IOP Conf. Series: Materials Science and Engineering 1172 (2021) 011001 doi:10.1088/1757-899X/1172/1/011001

ASAT-19

The 19th International Conference on Aerospace Sciences & Aviation Technology (ASAT-19)
April 6-8, 2021, Cairo, Egypt

CONTRIBUTORS

- Drug Microbiology Laboratory, Drug Radiation Research Department, National Center for Radiation Research and Technology (NCRRT), Egyptian Atomic Energy Authority, Cairo, Egypt.
- Department of Mechanical Design and Production, Faculty of Engineering, Cairo University, Egypt
- Mechanical Engineering department, Faculty of Engineering at Shoubra, Benha University, Cairo 11629, Egypt.
- Department of Metallurgical Engineering, Faculty of Engineering, Cairo University, Cairo 12613, Egypt.
- Physics Department, Faculty of Science, Cairo University, Giza, Egypt
- Faculty of Engineering, Misr for Science and Technology, Cairo, Egypt.
- Space Technology Center (STC), Cairo, Egypt.
- Nuclear Engineering Department, Military Technical College, Kobry El-Koba, Cairo, Egypt
- Nuclear and Radiological Regularity Authority (ENRRA), Nasr City, Egypt
- Faculty of Engineering, 'Université Française d'Égypte, Cairo, Egypt
- Faculty of Energy Engineering, Aswan University, Aswan 81528, Egypt.
- School of Mechanical Engineering, Shanghai Jiao Tong University, Shanghai 200240, China.
- Science and Technology Center of Excellence, Military Production, Cairo, Egypt
- Naval Warning Engineering Department, Military Technical College, Cairo, Egypt
- Department of Computer Engineering and Artificial Intelligence, Military Technical College, Cairo, Egypt
- Guidance and Control Department, Military Technical College, Cairo, Egypt.
- Dept. of Computer Engineering, Military Technical College, Cairo, Egypt, 11766
- Aerospace Engineering Department, Military Technical College, Cairo, Egypt.

ASAT-19

The 19th International Conference on Aerospace Sciences & Aviation Technology (ASAT-19)
April 6-8, 2021, Cairo, Egypt

CONTRIBUTORS

- Department of Mechanical Power & Energy, Military technical college, Cairo, Egypt
- Department of Rockets, Military Technical College, Cairo, Egypt.
- Modern Academy for Engineering and Technology, Cairo, Egypt.
- Aircraft electrical equipment's, Military Technical College, Cairo, Egypt
- Department of Optoelectronics and Control, Military Technical College, Cairo, Egypt
- Aircraft Armament Department, Military Technical College,
- Nahda University, Egypt.
- Space Technology Centre, Cairo, Egypt.
- Faculty of Engineering, Zagazig University, Zagazig, Egypt.
- Rocket Department, Military Technical College, Egypt
- University of Science and Technology at Zewail City, Giza, Egypt
- National Centre for Combustion Research and Development & Department of Aerospace Engineering, Indian Institute of Technology Madras, Chennai, India

ASAT-19

The 19th International Conference on Aerospace Sciences & Aviation Technology (ASAT-19)
April 6-8, 2021, Cairo, Egypt

INVITED TALKS

Day	Room	9:00 - 10:00 / 10:15 - 11:15	11:30 - 12:30	12:45 - 13:45	14.00 - 15.00
Tue 6/4	C	The Fourth Industrial Revolution "Industry 4.0": Aerospace Technology and Skills Challenges Prof. Dr. Hany Moustapha • Fellow, ASME, CAE and CASI • Director of Innovation 4.0 Network, ÉTS, University of Québec • Siemens' Chair on Industry 4.0 Technology Integration • Senior Research Fellow, Pratt & Whitney Canada Canada	Egyptian Space Agency: National Space Program and the Related Space Activities Egyptian Space Agency Egypt	Trade-Off Analysis of Low Earth Orbit Spacecraft Power Supply System by Genetic Algorithm Space Technology Center Egypt	Mobile Mapping Technology: Recent Advances and Proliferation in GNSS-Challenging Environments Prof. Ayman F. Habib • Thomas A. Page Professor in Civil Engineering • Lyles School of Civil Engineering, Purdue University, USA USA
Chairman		Dr.Mohamed M. Elkhatib	Dr.Ahmed Esmat	Dr. Mahmoud Ashery	Dr.Mohamed Elkhatib

IOP Conf. Series: Materials Science and Engineering 1172 (2021) 011001 doi:10.1088/1757-899X/1172/1/011001

ASAT-19

The 19th International Conference on Aerospace Sciences & Aviation Technology (ASAT-19)
April 6-8, 2021, Cairo, Egypt

INVITED TALKS

Day	Room	9:00 - 10:00	10:15 - 11:15	11:30 - 12:30	12:45 - 13:45	14.00 - 15.00	15:15 - 16.15
Wed 7/4		Integrated Sensors: The synergistic combination of materials and microelectronics Prof. Khalid Nabil Salama Associate Dean, Computer, Electrical and Mathematical Science and Engineering KAUST, Kingdom of Saudi Arabia KSA	Low power approaches to communication and computing system design. Prof. Ahmed Eltawil University of California Irvine, USA USA	New Trends in Fractional-Order Circuits and Systems Prof. Ahmed Madian Nile University, IEEE secretary, Egypt section Prof. Lobna Said Faculty of Engineering and Applied Science, Nile University (NU) Egypt	AeroMACS: Overview of the Technology & Its Deployment Status Prof. Mahmoud T. Elhadidi Emeritus Professor of Computer Networks Department of Electronics & Electrical Communications Engineering Cairo University Egypt	Advancing Industry 4.0: Digital Transformation from Theory to Real-Life Applications Prof. Abdel-Moez Bayoumi • Professor and Associate Dean • Director, McNair Aerospace Center • University of South Carolina, Columbia South Carolina, USA USA	Revolutionizing Lyophilization As We Know It: A Chance to Lead Dr. Ahmed M. Abdelraheem • Ph.D. in Electrical & Computer Engineering from Purdue University • Asst. Prof. Department of Electronics Engineering at the Military Technical College Egypt
Chairman		Dr.Mohamed Elkhatib					Dr.Mahmoud Abdelrhaman

The 19th International Conference on Aerospace Sciences & Aviation Technology (ASAT-19)
April 6-8, 2021, Cairo, Egypt

ASAT-19

The 19th International Conference on Aerospace Sciences & Aviation Technology (ASAT-19)
April 6-8, 2021, Cairo, Egypt

INVITED TALKS

Day	Room	9:00 - 10:00	10:15 - 11:15	11:30 - 12:30	12:45 - 13:45	14.00 - 15.00
Thu 8/4	C	Introduction to The Egyptian Knowledge Bank (Ekb) Mahmoud Dawood Head EKB training Sector Egypt	How to get published Abdallah Mohmoud Asad IOP Egypt section UK	Geofacets: Helping Engineers & Geo-scientists everywhere to discover Dr. Yasmin AbdelRaouf Country Ambassador | Elsevier Engineering Solutions UK	Optical Rotation Sensors: Challenges and Opportunities Prof. Diaa Khalil Formal Acting Dean Faculty of Engineering Ain Shames University Egypt	On-Chip Dynamic Power Supplies for Mixed-Signal SoCs Prof.Ayman Fayed • Professor, Director, Power Management Research Lab (https://pmrl.osu.edu/), • Dept. of Electrical & Computer Engineering, • The Ohio State University, Columbus, Ohio, USA USA
Chairman		Dr.Mohamed Elkhatib		Dr.Nabil Nagy	Dr.Mohamed Elkhatib	

ASAT-19

The 19th International Conference on Aerospace Sciences & Aviation Technology (ASAT-19)
April 6-8, 2021, Cairo, Egypt

The Fourth Industrial Revolution "Industry 4.0"Aerospace Technology and Skills Challenges

Date: Tuesday, April 6, 2021 Time: 9:00 – 11:15

Abstract

The fourth industrial revolution "Industry 4.0" (Germany, 2011), with its automation, digitization and exponential technologies will change the landscape of the industry. It is characterized by optimization and management of assets, sharing and security of big data, tracking parts from cradle to grave: the "Digital Thread", data analytics, Internet of Things (IoT) and Artificial Intelligence (AI). It is the transformation from physical to digital: the "Digital Twin" and from clusters to cloud computing. It is the "Democratization of Technology" where humans, computers, machines and products collaborate digitally and communicate seamlessly through integrated and optimized processes across the total product value stream both within an enterprise, and upstream (suppliers) and downstream (customers) of an enterprise. Industry 4.0 will be affecting the total enterprise: marketing, engineering, manufacturing, etc. in order to achieve a "Digital Smart Enterprise". The four main research areas to achieve a Digital Smart Enterprise are: vertical networking, horizontal integration, through engineering and disruptive technologies. Compared to the previous three revolutions, Industry 4.0 is characterized by: real-time connectivity, speed of implementation and culture change. Industry 4.0 will drive a major change in tomorrow's workforce. Despite the greater use of robotics and automation, there will be a net increase in jobs. Some companies will suffer from "technology unemployment", unless they retrain their workforce on Industry 4.0 skills at a rate faster than the technology introduction. Some jobs will disappear; hence companies have to protect their workforce and not the jobs. Educational institutions need to respond to Industry 4.0 needs, provide broader skill sets and close the gap in Information and Communication Technology (ICT) skills. There will be a need for "Industrial Data

Scientists" with strong ICT and AI skills, user interface design, advanced analytics, root-cause-analysis skills and statistical programming. In addition, there will be a need for "Robot Coordinators" to oversee robots and respond to malfunction and emergency maintenance tasks. Embedded E-Learning tools to be used in real world situations and ICT skills need to be integrated in all technicians, engineering and business curricula.

Speaker — Prof. Dr. Hany Moustapha

Professor and Director, Innovation 4.0 Hub and Aerospace 4.0TM Siemens' Chair on Industry 4.0 Technology Integration Pratt & Whitney Canada (P&WC) Chair on Propulsion System Design École de technologie supérieure (ÉTS), Québec University P&WC Senior Research Fellow Chevalier de l'Ordre national du Québec (CQ), Fellow of American Society of Mechanical Engineers (ASME), Canadian Academy of Engineers (CAE), Canadian Aeronautics and Space Institute (CASI).

B.Sc. from Cairo University (1970), M.Eng. and Ph.D. from McMaster University, Canada (1978). With Pratt & Whitney Canada (P&WC) from 1978 to 2010, Dr. Moustapha was Senior Manager of Technology, Collaboration and Technical Training Programs for P&WC from 1999 to 2009 and was named P&WC Senior Research Fellow in 2011. Joined ÉTS in 2010 as Professor and Director of Aerospace Programs (AEROÉTS).

ASAT-19

The 19th International Conference on Aerospace Sciences & Aviation Technology (ASAT-19)
April 6-8, 2021, Cairo, Egypt

Mobile Mapping Technology: Recent Advances and Proliferation in GNSS-Challenging Environments

Date: Tuesday, April 6, 2021 Time: 14:00 – 15:00

Abstract

The continuous developments in direct geo-referencing technology (i.e., integrated Global Navigation Satellite Systems – GNSS – and Inertial Navigation Systems – INS) and remote sensing systems (i.e., passive and active imaging sensors in the visible and infrared range – RGB cameras, hyperspectral push-broom scanners, and laser scanning) are providing the professional geospatial community with ever-growing opportunities to generate accurate 3D information with rich set of attributes. These advances are also coupled with improvement in the sensors' performance, reduction in the associated cost, and miniaturization of such sensors. Aside from the sensing systems, we are also enjoying the emerging of promising platforms such as Unmanned Airborne Vehicles (UAVs) and Unmanned Ground Vehicles (UGVs). This talk will provide an overview of recent advances in mobile mapping technologies and their ability to address a wide range of application domains. Imposed challenges by the proliferation of mobile mapping technologies in GNSS-challenging environments will be also discussed while highlighting potential approaches for the mitigation of the impact of lower accuracy of GNSS/INS trajectory on the mapping products.

IOP Conf. Series: Materials Science and Engineering 1172 (2021) 011001 doi:10.1088/1757-899X/1172/1/011001

ASAT-19

The 19th International Conference on Aerospace Sciences & Aviation Technology (ASAT-19)
April 6-8, 2021, Cairo, Egypt

Speaker

Prof. Ayman F. Habib

Thomas A. Page Professor in Civil Engineering Lyles School of Civil Engineering, Purdue University, USA

Dr. Habib is the Thomas A. Page professor of Civil Engineering at Purdue University and the Co-Director of the Civil Engineering Center for Applications of UAS for a Sustainable Environment (CE-CAUSE). He is also the Associate Director of Purdue University's Joint Transportation Research Program (JTRP). Dr. Habib's research interests span the fields of terrestrial and aerial mobile mapping systems. In this regard, he has been focusing on modeling the perspective geometry of non-traditional imaging scanners, automatic matching and change detection, automatic calibration of low-cost digital cameras, object recognition in imagery and point cloud data, LiDAR mapping, and integrating photogrammetric data with other sensors/datasets (e.g., GNSS/INS, GIS databases, multi- and hyper-spectral sensors, and LiDAR).

ASAT-19
The 19th International Conference on Aerospace Sciences & Aviation Technology (ASAT-19)
April 6-8, 2021, Cairo, Egypt

On-Chip Dynamic Power Supplies for Mixed-Signal SoCs

Date: Thursday, April 8, 2021 Time: 14:00 – 15:00

Abstract

The diminishing returns of semiconductor technology scaling in terms of power reduction is forcing the electronics industry to shift from the traditional approach of using a few centralized static power supplies to power a System-on-Chip (SoC) to a numerous distributed dynamic power supplies approach. In this approach, the SoC is divided into sub-components, each with its own independent on-chip power supply that is dynamically adapted to the real-time demand of each sub-component. As a result, the overall power consumption of the SoC can be significantly reduced. However, such approach requires a large number of dynamic on-chip power supplies, and it becomes quite challenging to implement them in a size- and cost-effective manner while also maintaining high power conversion efficiency, particularly in nanometer CMOS technologies where the voltage rating of transistors is quite low. This presentation discusses techniques used to implement a large number of on-chip dynamic power supplies in mixed-signal SoCs, including high-frequency buck converters with on-chip passives, and it introduces the multi-frequency multi-phase SIMO buck converter topology with on-chip outputs as an attractive method for designing dynamic on-chip power grids.

IOP Conf. Series: Materials Science and Engineering 1172 (2021) 011001 doi:10.1088/1757-899X/1172/1/011001

ASAT-19

The 19th International Conference on Aerospace Sciences & Aviation Technology (ASAT-19)
April 6-8, 2021, Cairo, Egypt

Speaker

Prof. Ayman Fayed

Ayman Fayed received his B.Sc. degree in Electronics & Communications Engineering from Cairo University in 1998, and his M.Sc. and Ph.D. degrees in Electrical & Computer Engineering from The Ohio State University in 2000 and 2004, respectively. From 2000 to 2009, he held several technical positions in the area of analog and mixed-signal design at Texas Instruments Inc., where he contributed to many product lines for wire-line, wireless, and multi-media devices. From 2000 to 2005, he was with the Connectivity Solutions Dept. at TI, where he worked on the analog frontend design of high-speed wire-line transceivers such as USB, IEEE1394b, and HDMI, and on fully integrated switching/linear regulators and battery chargers for portable media players. From 2005 to 2009, he was a member of the technical staff with the wireless analog technology center at TI, where he worked on delta-sigma data converters for various wireless standards, and on the development of fully-integrated power management solutions for mixed-signal SoCs with multi-RF cores in nanometer CMOS. Dr. Fayed joined the Dept. of Electrical & Computer Engineering at Iowa State University in 2009, where he held the Northrop Grumman Assistant Professorship. He then joined the Dept. of Electrical & Computer Engineering at The Ohio State University in 2015, where he is currently a professor. He is the founder and director of the Power Management Research Lab (PMRL) and his current research interests include on-chip power supplies for dynamic energy distribution in VLSI systems, high-frequency switching regulators with on-chip and on-package passives for SoCs, low-noise power supplies and power supply modulators for analog and RF circuits, energy-harvesting circuits for power-restricted & remotely-deployed devices/systems, and analog/mixed-signal and power converter design in SiC and GaN technologies. He is currently an associate editor for IEEE TCAS-I and previously for TCAS-II, and serves in the technical program committee of RFIC, ISCAS, APEC, and the steering committee of MWSCAS. He is the author/co-author of many publications in the field and holds 8 US patents. Dr. Fayed is a recipient of NSF CAREER Award in 2013, and a co-recipient of the 2015 Darlington Best Transactions Paper Award from the IEEE Circuits and System Society. Dr. Fayed is a senior member of IEEE.

IOP Conf. Series: Materials Science and Engineering 1172 (2021) 011001 doi:10.1088/1757-899X/1172/1/011001

ASAT-19

The 19th International Conference on Aerospace Sciences & Aviation Technology (ASAT-19)
April 6-8, 2021, Cairo, Egypt

Optical Rotation Sensors: Challenges and Opportunities.

Date: Thursday, April 8, 2021 Time: 12:45 – 13:45

Abstract

Optical rotation sensors are key elements in many strategic applications due to their very high performance however, the practical implementation of these sensors are usually limited to specific countries because of the restriction and availabilities of components with high end performance. Recently there have been many research directions that targets to overcome these restrictions and to develop relatively high end optical gyroscopes using on the shelf components. The objective of this talk is to scan these directions to illustrate the potential directions in this field as well as the challenges to be faced on both the theoretical and practical levels. The techniques investigated in this direction will be discussed supported by the experiments developed for this purpose. The study will cover the different technologies in this domain including the Fiber optic Gyro, the Ring laser gyro and also the integrated optic gyroscope. The basic theory as well as the main features of the technology will be addressed emphasizing on their effects on the sensor performance and its compatibility with the target application.

Speaker **Prof. Diaa Khalil**

Prof. Diaa Khalil has over 35 years of experience in micro photonics systems, including integrated optoelectronics and optical MEMS technology. He obtained his PhD from INPG France in 1993.
Diaa Khalil is a Professor of photonics in the Faculty of Eng., Ain Shams University since 2004. In 2015 He was the Head of the ECE Dept., and in 2017 He

became the Vice dean of Research and Post Graduate. From 2019 to Feb. 2021, He was the Acting dean of the School of Engineering. He is leading a group of scientists working in the fields of Photonic microsystems. He supervised more than 60 MSc and PhD thesis in the fields of: Optics, Opto-electronics, Applied Spectroscopy, Optical MEMS and integrated Optics.

Diaa Khalil was also the CTO of the Optical MEMS Division in Si-Ware Systems from 2007 - 2020, leading a group of talented engineers developing an innovative FTIR MEMS spectrometer, a unique product that gained the Prism award in the Photonics West conference 2014 in SF USA. Prior to joining SWS, he worked with MEMScAP company as the head of the optical MEMS design group where he led the design and characterization of a 2×2 switch, and introduced a novel VOA (variable Optical Attenuator) achieving the world record of minimum PDL reported for wide dynamic range. Prof. Diaa Khalil is a holder of the Egyptian state incentive prize in engineering sciences in 1998. He is a senior member in the OSA, the SPIE, the IEEE Photonic Society, and URSI. He is currently the head of commission D in the National URSI committee, in Egypt. He is also a member in the editorial board of the journal, "Light: Science and Applications - Nature Publishing Group NPG. He is inventor of about 20 granted international patents and more than 10 other patent applications. He is author and co-author of more than 320 publications, in international journals and conferences, in addition to 3 book chapters and 1 ebook published by SPIE.

ASAT-19

The 19th International Conference on Aerospace Sciences & Aviation Technology (ASAT-19)
April 6-8, 2021, Cairo, Egypt

Egyptian Space Agency: National Space Program and the Related Space Activities

Date: Tuesday, April 6, 2021 Time: 11:30 - 12:30

Abstract

The Egyptian Space Agency was established in January 2018. According to the executive law no. 3 for the for 2018, The agency aims to create, transfer, localize, and develop space science and technology and possess the own capabilities to build satellites and launch them from the Egyptian lands in a manner that serves the country's strategy in the fields of development and achievement of the national security. The Agency is both legislative and executive body for space activities in Egypt. National Space Program was approved from the board of director in March 2020. The technical projects are carefully selected and planned to execute the national space program. The projects covers the six axes of the national space program which are capacity building, space systems, industrial development and infrastructure, space exploration, international cooperation, and legal and space standards. The present paper presents the current activities which include the technical and non-technical projects.

Speaker **Mohammed Khalil Ibrahim Iraqi**

Egyptian Space Agency

ASAT-19

The 19th International Conference on Aerospace Sciences & Aviation Technology (ASAT-19)
April 6-8, 2021, Cairo, Egypt

How to get published.

Date: Thursday, April 8, 2021 Time: 10:15 – 11:15

Abstract
It is targeted at mainly Early Career Researchers with the aim of educating them about all aspects of publishing in journals, right the way through from choosing the right journal to promoting a paper after it is published. The main sections of the presentation are:

- Introduction to IOP and IOP journals
- Why publish at all?
- Choosing your journal
- Writing your paper
- Peer review process
- Publication ethics
- Post-acceptance
- Post-publication.

Speaker **Abdallah Mahmoud Asad**

Abdallah Asad is currently working as "Regional Manager – Middle East and North Africa" for IOP Publishing, with more than 8 years of sales and training experience in academic publishing. He has conducted more than 100 platform, product and Guide to Getting Published sessions.

IOP Publishing

IOP Conf. Series: Materials Science and Engineering 1172 (2021) 011001 doi:10.1088/1757-899X/1172/1/011001

ASAT-19

The 19th International Conference on Aerospace Sciences & Aviation Technology (ASAT-19)
April 6-8, 2021, Cairo, Egypt

Geofacets: Helping Engineers & Geo-scientists everywhere to discover.

Date: Thursday, April 8, 2021 Time: 11:30 – 12:30

Abstract

- Agenda:
- What is Geofacets and how does it work?
- Search using Geofacets & Overlay Maps
- Reading Geoscience Papers
- Building Maps
- Use Case
- Practical Demo

ELSEVIER

Speaker **Dr. Yasmin AbdelRaouf**

Dr. Yasmin AbdelRaouf is the country ambassador for Egypt covering Elsevier Engineering and Geoscience solutions (Engineering Village – Geofacets – Knovel). Dr. Yasmin has organized on-site & online training workshops for universities and research institutes in Egypt as part of the Egyptian Knowledge Bank initiative. Dr. Yasmin AbdelRaouf is an assistant professor in Faculty of Engineering Cairo University where she received her bachelor and masters degrees in Chemical Engineering. She received her PhD in Civil Engineering from University of New South Wales (UNSW) in Australia in 2017.

ASAT-19

The 19th International Conference on Aerospace Sciences & Aviation Technology (ASAT-19)
April 6-8, 2021, Cairo, Egypt

Introduction to The Egyptian Knowledge Bank (Ekb)

Date: Thursday, April 8, 2021 Time: 9:00 – 10:00

Abstract
EKB is one of the largest projects initiated by the Egyptian presidency under the patronage of his Excellency president Abdel Fattah Al-Sisi. This project main concern is enhancing the Scientific Research, Education and General Literacy by availing content from the most well-known publishers all over the globe to all Egyptians in Egypt for free. It can be accessed at any time and from anywhere. It has many general resources trying to reach out for the general public with easy and attractive pieces of information that is well reviewed. It has also specialized resources targeting specific specialties giving them the moil. Updated research information that they need to excel in their field. Full text resources were difficult to be reached by most of researchers where now using EKB made this reachable even from the researcher home. Also, citation and bibliographic databases are available to help researchers reach the best journals and articles in their field of interest. Regarding Basic Education, EKB availed many resources in different formats to allow students to study and understand better through videos, images, animations and many other formats. The target is to link what Students team in school with the real world to get them excited about education. To help in achieving the best use of the above content, tools and software's are provided to EKB users to utilize the best benefit of the EKB. One of the boil outputs of the EKB is the "Local System of Egyptian Scientific Output" with its different platforms serving the multiple functions required by the scientific community. Many specialized workshops like "Essentials of Leadership ", "Writing a Successful Research Grant Proposal", "Writing and Publishing a Research Article" and many more are being undertaken to young researchers to help them build their future in the scientific world. A lot of specialized projects are being done under the umbrella of the EKB like the "Egyptian Universities International

Ranking Project", "Medical Education Project" and more. The EKB Learning Management System" is currently used in all governmental schools targeting Secondary one as a Mart and preparing to go for all grades to help students and engage them more by adding the modem technology in their education system to help them be more organized to get better results. One of the biggest achievements of the EKB is the Arabic Citation Index which will help researchers to reach the Arabic scientific output in a better and easier way.

Speaker — Mahmoud Dawoud

Mahmoud Dawoud has been involved in the Academic sector moil of his life from being a mailer's degree student to a certified trainer for many high prestigious academic and scientific publishers. He started his career in the academic field as a trainer for scientific and academic publishers to train their users in Egypt, Africa and some Arab world countries on how to effectively use these publishers' platforms to find the required scientific information and use them in their researches and has gained a big experience in how to communicate with researchers from being doing so for 24 years. He also trains on the best methods to write and arrange scientific papers using the moil known citation managers in the scientific field. He enjoys being able to help scientists in their scientific research to give us a better world.

ASAT-19

The 19th International Conference on Aerospace Sciences & Aviation Technology (ASAT-19)
April 6-8, 2021, Cairo, Egypt

New Trends in Fractional-Order Circuits and Systems.

Date: Wednesday, April 7, 2021 Time: 11:30 - 12:30

Abstract

Optical rotation sensors are key elements in many strategic applications due to their very high performance however, the practical implementation of these sensors are usually limited to specific countries because of the restriction and availabilities of components with high end performance. Recently there have been many research directions that targets to overcome these restrictions and to develop relatively high end optical gyroscopes using on the shelf components. The objective of this talk is to scan these directions to illustrate the potential directions in this field as well as the challenges to be faced on both the theoretical and practical levels. The techniques investigated in this direction will be discussed supported by the experiments developed for this purpose. The study will cover the different technologies in this domain including the Fiber optic Gyro, the Ring laser gyro and also the integrated optic gyroscope. The basic theory as well as the main features of the technology will be addressed emphasizing on their effects on the sensor performance and its compatibility with the target application. Fractional calculus has been proved through numerous research examples to be a superior tool for system description than the narrow integer order domain. This is achieved through the extra parameters introduced via allowing the differential or integral orders to take non-integer values. The promising capabilities of fractional-order devices challenge the research to find a way to simulate its behavior until its of-shelf appearance. Diferent integer order approximation techniques to the fractional-order transfer functions are investigated in the literature. The main idea behind approximating the fractance device is to achieve a nearly constant phase response of the impedance. There are many categories of fractional order system approximations in analog and digital domains. Research in circuits and systems using fractional-order elements is considered as an interdisciplinary topic. It serves many areas of applications such as electrical engineering, medicine, physics, control system, chaos theory signal processing, and bio-impedance. In bioengineering, conventional lumped element circuit models of tissue were extended as fractional-order generalization through modification of the defining current-voltage relationships. Such fractional-order models provide an improved description of observed bio-impedance behavior. Electrical impedance measurements have been widely used to estimate plant health, maturity of fruits, fruit damages, and structural cellular variation during fruit ripening, freeze or chill damages, and measurement of tree root growth. These analyses may be used in the field of importing and exporting fruits and vegetables. Generalizing the chaotic system to the fractional-order domain increases the rate of complexity level and provides more degrees of freedom in both the time and the frequency domain. In order to get higher performance, fractional-order has emerged in diferent chaotic systems. The fractional-order chaotic systems increase the chaotic behavior in new dimensions and add extra degrees of freedom, which increases system controllability. Fractional-order chaotic systems parameter identification is an essential issue in chaos control and synchronization process.

Speaker **Prof. Ahmed Madian**

Ahmed Madian (SM'12) received his Ph.D. and M.Sc. degrees from Cairo University, Egypt, in 2007 and 2002, respectively. He is currently Professor at the Department of Electronics and computer Engineering, Faculty of Engineering and applied science, NILE University, Giza, Egypt. He is the director of Microelectronics System Design Master Program since sept. 2015. Also, He is the director of Nanoelectronics Integrated System Design Research center (NISC) since 2016. He has published more than 150 papers in international conferences and journals. His H-index is currently 20. Also, he served in the many technical and organizing committee of many international conferences. He received many research grants as Principle Investigator (PI), CO-PI, or Consultant from different national/international organizations. He won the best researcher award (Dr. Hazem Ezzat award 2017) for his outstanding research profile. His research interests are in circuit theory; low-voltage analog CMOS circuit design, current-mode analog signal processing, Memristors, Fractional systems, VLSI, Encryption systems and mixed/ digital applications on field programmable gate arrays. Also, he is member of the national radio of science committee (NRSC) since 2018. Dr. Madian is actively serving as a reviewer in several journal and conference publications including IEEE conferences and journals. He served as guest associate editor for many international journals. He is the founder of IEEE Circuits and systems (CASS) Egypt technical chapter and co-founder of the IEEE Robotics and automations (RAS) Egypt technical chapter. He is currently the IEEE Egypt Section Secretary and member of Ex-COM.

Speaker — Assistant Prof./ Lobna A. Said

Lobna A. Said (Senior Member IEEE 2020) is a full-time Assistant Professor at the Faculty of Engineering and Applied Science, Nile University (NU). She received the B.Sc., the M.Sc., and the Ph.D. degrees in electronics and electrical communications from Cairo University, Egypt, in 2007, 2011, and 2016, respectively. She has over 100 publications distributed between high-impact journals, confer-

IOP Conf. Series: Materials Science and Engineering 1172 (2021) 011001 doi:10.1088/1757-899X/1172/1/011001

ences, and book chapters. Her research interests are interdisciplinary, including modeling, control, optimization techniques, and analog and digital integrated circuits, fractional-order circuits and systems, non-linear analysis, and chaos theory. She was involved in many research grants as a Senior Researcher/ Member such as JESOR-ASRT "wireless monitoring of fruit growth using an electrical bio-impedance sensor device," or STDF "Two Port Fractional-Order Oscillators and Filters Suitable for Tissue Modeling," or as a Co-PI as JESOR-ASRT "Smart Agriculture in the Internet Of Things Era,". She is the Vice-Chair of research activities at the IEEE Computational Intelligence Egypt Chapter 2018-present. She is the Counselor of the IEEE NU student branch 2018-present. She won the state encouragement award for the year 2019. She has received the Excellence Award from the Center for the Development of Higher education and Research in 2016. She is the winner of the Dr. Hazem Ezzat Prize for outstanding researcher, NU 2019, and 2020. She is one of the top 10 researchers at NU for 2018-2019 and 2019-2020. Her name was in the Top 2% Scientists According to Stanford Report for 2019, released in 2020. In 2019, she was selected as a member of the Egyptian Young Academy of Sciences (EYAS). In 2020, she was elected as the Co-Chair of EYAS. In 2020, she was selected to be an affiliate member of the African Academy of Science (AAS). In 2020, she was also chosen to be a Member of the Arab-German Young Academy of Sciences and Humanities (AGYA). Additionally, she is on the technical program committee for many International Conferences.

IOP Conf. Series: Materials Science and Engineering 1172 (2021) 011001 doi:10.1088/1757-899X/1172/1/011001

ASAT-19

The 19th International Conference on Aerospace Sciences & Aviation Technology (ASAT-19)
April 6-8, 2021, Cairo, Egypt

Low power approaches to communication and computing system design.

Date: Wednesday, April 7, 2021 Time: 10:15 – 11:15

Abstract

In this talk we discuss approaches to low power design for advanced communication and computing platforms. Specifically, we present the concept of cognitive power management, where contrary to common approaches that assume a 100% error free hardware, the algorithm is made aware of the statistical error performance of the underlying hardware platform. By accounting for hardware errors at the system level, the explorable power management design space is significantly expanded, leading to novel power saving schemes that deliver expected application performance at much lower power consumption. Sample case studies including LTE system design and in-memory computing platforms will be presented and discussed.

Speaker **Prof. Ahmed M. Eltawil**

Ahmed M. Eltawil has been a Professor at King Abdullah University of Science and Technology (KAUST) since 2019. Prior to that he was with the Electrical Engineering and Computer Science Department at the University of California, Irvine (UCI) since 2005. He is the founder and director of the Communication and Computing Systems Laboratory (CCSL) at KAUST and the inaugural founder and director of the Wireless Systems and Circuits Laboratory at UCI. His current research interests are in the general area of smart and connected systems with an emphasis on mobile systems. He received the Doctorate degree from the University of California, Los Angeles, in 2003 and the M.Sc. and B.Sc. degrees (with honors) from Cairo University, Giza, Egypt, in 1999 and 1997, respectively. Dr. Eltawil has been on the technical program committees and steering committees for numerous workshops, symposia, and conferences in the areas of low power computing and wireless communication system design. He received several awards, as well as distinguished grants, including the NSF CAREER grant supporting his research in low power computing and communication systems. He is a senior member of the IEEE and a senior member of the National Academy of Inventors, USA.

ASAT-19

The 19th International Conference on Aerospace Sciences & Aviation Technology (ASAT-19)
April 6-8, 2021, Cairo, Egypt

Integrated Sensors: The synergistic combination of materials and microelectronics

Date: Wednesday, April 7, 2021 Time: 9:00 – 10:00

Abstract
In this talk we will present the design and implementation of monolithic and hybrid sensors using integrated circuits. We will discuss the advantages and shortcomings of sensors built in silicon-based fabrication processes and examine, in detail, their integrated circuit topologies. Next, we will provide a comprehensive study of the design and analysis of CMOS integrated image sensors, integrated biosensors, and electronic backbone of MEMS hybrid sensors. We will conclude with a survey of advanced research topics in the area of integrated sensors such as smart sensors, RF-IDs, and nano sensors.

Speaker **Prof. Khaled Nabil Salama**

Prof. Khaled Nabil Salama is a full professor and associate dean of the Computer, Electrical, mathematical Science and Engineering Division at King Abdullah University of Science and Technology (KAUST). He received the B.S. degree from the Department Electronics and Communications, Cairo University, Cairo, Egypt, in 1997, and the M.S. and Ph.D. degrees from the Department of Electrical Engineering, Stanford University, Stanford, CA, USA, in 2000 and 2005, respectively. He was an Assistant Professor at Rensselaer Polytechnic Institute, NY, USA, between 2005 and 2009. He joined KAUST in January 2009 and was the founding Program Chair until August 2011. Currently, he is the director of the KAUST

SENSORS INITIATIVE a consortium of 9 universities (KAUST, MIT, UCLA, GATECH, Brown University, TU Delft, Swansea University, the University of Regensburg and the Australian Institute of Marine Science). In addition, he is the chair of winter enrichment program (WEP) for two consecutive years (2020-2021) at KAUST. His work on CMOS sensors for the National Institutes of Health (NIH) and the Defense Advanced Research Projects Agency (DARPA) was awarded the Stanford-Berkeley Innovators Challenge Award in biological sciences and was acquired by Ilumina Inc. He is the author of 300 publications and the inventor/co-inventor of 49 issued US patents on low-power mixed-signal circuits for intelligent fully integrated biosensors.

ASAT-19

The 19th International Conference on Aerospace Sciences & Aviation Technology (ASAT-19)
April 6-8, 2021, Cairo, Egypt

Advancing Industry 4.0: Digital Transformation from Theory to Real-Life Applications

Date: Wednesday, April 7, 2021 Time: 14:00 – 15:00

Abstract
Industries across the world are realizing the benefits of Industry 4.0 and the University of South Carolina has been using Digital Transformation to transform processes in multiple industries. The University of South Carolina has established a roadmap for implementing Digital Transformation and will introduce multiple use cases that can be used in real-world applications from an idea, to a design, to a manufacturing process, to operations and to maintenance and repairs.

Speaker **Prof. Abdel-Moez Bayoumi**

Dr. Bayoumi has about 40 years teaching, research and industry experience. He is currently the Director of the McNair Aerospace Center, Director of the Center for Predictive Maintenance, Associate Dean for Corporate Relations and Professor of Mechanical and Biomedical Engineering. Prior to joining USC, he was Professor of Mechanical and Aerospace Engineering at North Carolina State University, a project manager at Hewlett-Packard Company, and Professor of Mechanical and Materials Engineering at Washington State University. He has been actively involved in developing strong programs in mechanical, biomedical, nuclear and aerospace engineering. His research activities have been focused in mechanical behavior of materials, tribology, diagnosis and prognosis and life predictions of mechanical systems, manufacturing processes, and health monitoring systems, condition-based maintenance, and digital transformation. Dr. Bayoumi has received more than $40 M in funding from the US Department of Defense, NASA, US Department of Energy, US National Science Foundation and various US industries. He has published over 150 journal and conference papers.

ASAT-19
The 19th International Conference on Aerospace Sciences & Aviation Technology (ASAT-19)
April 6-8, 2021, Cairo, Egypt

Trade-Off Analysis of Low Earth Orbit Spacecraft Power Supply System by Genetic Algorithm

Date: Tuesday, April 6, 2021 Time: 12:45 – 13:45

Abstract

The spacecraft electrical power system design process is a very complex task, it depends on multiple issues, which will have a great effect on the trade-off analysis of designing Spacecraft Electrical Power System, it should also meet various design objectives under certain constraints. The main goal of designing the spacecraft power supply system is to reach the optimal solution from the available technologies, which is used in the power system configuration. Power supply subsystem components under study are the solar cells, configuration of solar arrays, technology of batteries, and bus voltages. This workshop discusses some important issues such as automation, intelligent design and trade-off processes of the Spacecraft Electrical Power System implemented by the practical design processes using GENETIC ALGORITHM. The new method reflects that simulation and optimization techniques can be effectively used for improving and automating the designing method. Multi-objective of optimal design solutions unnaturally evolved in the result of the interfacing with Genetic algorithm techniques with system sizing or examination tools under the supervision of the practical constraints.

Speakers

Ahmed Mokhtar

Received a B.S. in Engineering from Military Technical College, Cairo in 2005. Completed Master degree in 2017, currently a PhD candidate since 2021. Worked in the field of aeronautics since graduation and currently working in the field of space specialized in satellite power supply systems.

Ahmed M. Mahareek

Received a B.S. in Engineering from faculty of electronic engineering, alminofyia in 2013. Worked in the field of industrial electronics since graduation and currently working in the field of space specialized in electrical tests.

Hossam H. Kenawy

Received a B.S. in Engineering from faculty of electronic engineering, alminofyia in 2015. Worked in the field of industrial electronics since graduation and currently working in the field of space specialized in electrical tests.

IOP Conf. Series: Materials Science and Engineering 1172 (2021) 011001 doi:10.1088/1757-899X/1172/1/011001

ASAT-19

The 19th International Conference on Aerospace Sciences & Aviation Technology (ASAT-19)
April 6-8, 2021, Cairo, Egypt

AeroMACS: Overview of the Technology & Its Deployment Status

Date: Wednesday, April 7, 2021 Time: 12:45 – 13:45

Abstract

In 2007, a study by the International Civil Aviation Organization (ICAO) has recommended the use of WiMAX as the technology of choice for the future airport communications system. During the same year, the ITU-WRC-07 has allocated the frequency spectrum: 5091-5150 MHz for the new technology, what was named: Aeronautical Mobile Airport Communications System (AeroMACS). Since then, developments took place continuously and rapidly, paving the way for AeroMACS to become the predominant communication technology adopted by modern airports. In this presentation we highlight the issues that motivated the quest for a new airport communications system, the reasons for choosing WiMAX as the parent technology of AeroMACS, and the main technical features of AeroMACS. We also review the major benefits of using AeroMACS and the current status of its deployment worldwide.

Speaker **Prof. Mahmoud T. El-Hadidi**

Dr. Mahmoud T. El-Hadidi was born in Giza (Egypt) on 18 May 1952. He received his B.Sc. and M.Sc. in Electrical Engineering from Cairo University in July 1973, and October 1975, respectively. He pursued his studies at Princeton University (New Jersey – USA), from which he obtained his M.A. and Ph.D. in Electrical Engineering, in October 1977 and October 1979, respectively. During the period October 1979 till September 1980, Dr. El-Hadidi was a visiting researcher at C & C Central Research Laboratories of NEC (Kawasaki City, Japan). He joined the Teaching Staff at the Department of Electronics and Elec. Communications Engineering (EECE) in Cairo University since October 1980, where he assumed the positions of Assistant Professor (in 1980), Associate Professor (in 1985), and Professor (in 1992). During the period August 2009 till July 2012, Dr. EL-HADIDI was appointed as the Head of EECE Department. Dr. El-Hadidi's areas of interest cover a wide scope, including: Modeling and Analysis of Computer Networks, Modeling and Analysis of Telephone Networks, Wireless Networks, and Information Security. He has been the national representative of Egypt in Technical Committee 11 (TC11) of IFIP from 1995 till 2006. In 2002, Dr. El-Hadidi organized the 17th International Conference on Information Security, sponsored by IFIP, and served as its Secretary General. He has also helped organizing the series of International Conferences on Information and Communication Technology (ICICT) - sponsored by the Information Technology Institute (ITI) in Egypt - from 2003 till 2010, where he has served as its Secretary General and Technical Program Chair. Starting 2012 Dr. El-Hadidi was selected as a member of the National Radio Science Committee (NRSC) - affiliated with the Egyptian Academy for Scientific Research and Technology (ASR&T), where he currently serves as its Vice-Chair. He is also a member of International Union of Radio Science (URSI). Dr. El-Hadidi contributed more than 120 publications, ranging from books, technical reports, journal papers, conference papers, and short courses. In addition, Dr. El-Hadidi has been active as a consultant for the Egyptian Government and the Egyptian private sector, in relation to many computer network projects. Among these are: Ministry of Education (MOE), Ministry of Housing (MOH), Egyptian Universities Network (EUN), Housing and Building Research Center (HBRC), Egyptian Organization for Standards (EOS), National Telecommunication Institute (NTI), the Computer Network for Engineering Libraries (CNEL), Cairo Governorate, and others. Most recently, Dr. El-Hadidi has supervised the design and implementation of the information network for the New Central Library of Cairo University, the information network for the Egyptian National Railway (ENR), and the modernization of the information network for Al-Azhar Tunnel.

IOP Conf. Series: Materials Science and Engineering 1172 (2021) 011001 doi:10.1088/1757-899X/1172/1/011001

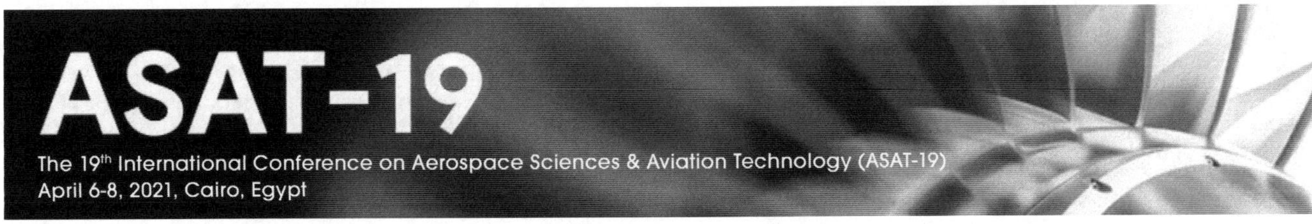

Revolutionizing Lyophilization as we know It: A Chance to Lead

Date: Wednesday, April 7, 2021 Time: 15:15 – 16:15

Abstract

The need for mass-produced injectable pharmaceutical products is growing massively especially with the most recent unfortunate pandemic of COVID-19. In 2019, the lyophilization equipment market exceeded USD 5 billion. Rising demand for lyophilized products and technological advancement in lyophilization methods are the major market driving forces. In its current form utilized in industry, lyophilization has not encountered a major change in the method since the 1960s. The process is still batch-based and has a very poor energy efficiency (~5%). In this talk, we present a key contribution that carries a lot of potential for the process of lyophilization, namely RF/RF-assisted lyophilization. A new method for conducting RF/RF-assisted lyophilization is presented that is directly applicable to the pharmaceutical and biological industry sectors. This would not only boost the process efficiency and reduce the time to market, but also enable the continuous process as opposed to the current batch-based process.

Speaker **Dr. Ahmed M. Abdelraheem**

Ahmed M. Abdelraheem is a faculty member at the Department of Electronics Engineering at the Military Technical College, Egypt. He received his B.Sc. and M.Sc. degrees in Electrical & Computer Engineering from the Military Technical College in 2009 and 2015, respectively, and his Ph.D. in Electrical & Computer Engineering from Purdue University in 2020. Dr. Abdelraheem is a recipient of the IEEE

IMS Flexible Energy Harvesters Design Competition Award in 2018. He is a co-recipient of the Egyptian Ministry of Defense (MoD) research grant. He authored and co-authored several research papers and contributed as a speaker/invited speaker in multiple international conferences. His current research is focused on bringing new electromagnetic techniques to the industry. He is the lead inventor in three US patents that brings new electromagnetic methods to enhance multiple industrial sectors that include, but not limited to, pharmaceutical and biological industries, wireless monitoring of industrial processes in harsh environments, and controlled industrial electromagnetic heating.

ASAT-19

The 19th International Conference on Aerospace Sciences & Aviation Technology (ASAT-19)
April 6-8, 2021, Cairo, Egypt

SCIENTIFIC SESSIONS

Session - Room	Subject	Day	Time	Paper ID	Paper Title	Session Board
S1-A	aerospace structures	06/04/2021	11:30 - 12:30	001_ASAT19	Dimensional structural mass optimization of forged steel connecting rod for aircraft piston engine	Brig. Prof. Mahmoud Yahia, MTC, Cairo, Egypt. Col. Assoc. Prof. Mohamed Zakria, MTC, Cairo, Egypt. Lt. Co. Assoc. Prof. Ashraf Kamal, MTC, Cairo, Egypt.
				002_ASAT19	Static and modal analysis of sine-wave-shaped spar	
				003_ASAT19	Vibration analysis of composite wing with geometric and material coupling	
				004_ASAT19	Numerical simulation of temperature field in a C/C composite multidisk brake during aircraft braking	
			12:45 - 1:45	005_ASAT19	Stress analysis of multi-layered composite cylinders subjected to various loadings.	
				006_ASAT19	Effect of damping material thickness on vibration analysis in pretension layer damping process	
				007_ASAT19	Vibration damping of aircraft propeller blades using shunted piezoelectric transducers	

ASAT-19

The 19th International Conference on Aerospace Sciences & Aviation Technology (ASAT-19)
April 6-8, 2021, Cairo, Egypt

SCIENTIFIC SESSIONS

Session - Room	Subject	Day	Time	Paper ID	Paper Title	Session Board
S2-A	Communications and Networks	06/04/2021	14:00 - 15:00	008_ASAT19	Design of a smart energy efficient IoT wireless node	Col. Assoc.Prof. Tamer Nabil MTC, Cairo, Egypt. Col. Assoc.Prof. Mahmoud Karm MTC, Cairo, Egypt. Assoc.Prof. Mohamed Hussein, MTC, Cairo, Egypt.
				009_ASAT19	Comparative study between cognitive radio techniques in FM broadcasting band	
				010_ASAT19	A satellite-based quantum key distribution using decoy-state protocol	

IOP Conf. Series: Materials Science and Engineering 1172 (2021) 011001 doi:10.1088/1757-899X/1172/1/011001

ASAT-19

The 19th International Conference on Aerospace Sciences & Aviation Technology (ASAT-19)
April 6-8, 2021, Cairo, Egypt

SCIENTIFIC SESSIONS

Session - Room	Subject	Day	Time	Paper ID	Paper Title	Session Board
S3-B	Image Processing/Remote sensing	06/04/2021	11:30 - 12:30	011_ASAT19	Enhanced image encryption algorithm by quantum key distribution	Maj. Gen. Prof. Fawzy Eltohamy MTC, Cairo, Egypt. Brig. Prof. Mohamed Samir MTC, Cairo, Egypt. Brig. Prof. Sherif El-gamel MTC, Cairo, Egypt
				012_ASAT19	SAR image formation enhancement using effective velocity estimation method	
				013_ASAT19	Frequency analysis of SAR system design for small satellite	
			12:45 - 1:45	014_ASAT19	Updating land use and land cover classes of Blue Nile basin using landsat-8 images	
				015_ASAT19	Assessment of top of atmosphere reflectance generation algorithm for world view-2 high resolution images	
				016_ASAT19	Hybrid fusion using Gram Schmidt and Curvelet transforms for satellite images	

ASAT-19

The 19th International Conference on Aerospace Sciences & Aviation Technology (ASAT-19)
April 6-8, 2021, Cairo, Egypt

SCIENTIFIC SESSIONS

Session - Room	Subject	Day	Time	Paper ID	Paper Title	Session Board
S5-A	Aerospace Materials	07/04/2021	9:00 - 10:00	020_ASAT19	Effective electromagnetic interference shielding using polyurethane / graphite composite foam materials	Brig. Prof. Ahmed Barka MTC, Cairo, Egypt Brig. Prof. Mohamed Gobara MTC, Cairo, Egypt. Col.Prof. Sherif El-basuney MTC, Cairo, Egypt.
				021_ASAT19	Parametric optimization of non-prismatic micro-plates to reduce stiffening and curling initiated during fabrication	
			10:15 - 11:15	022_ASAT19	On the effect of post-processing techniques of the additively manufactured aluminum alloy parts	
				023_ASAT19	Factors affecting Failure mode and load of rubber polyester composite joint	
				024_ASAT19	Water recycling system based on adsorption by activated carbon synthesised from c. verum for space exploration; an estimated design	

ASAT-19

The 19th International Conference on Aerospace Sciences & Aviation Technology (ASAT-19)
April 6-8, 2021, Cairo, Egypt

SCIENTIFIC SESSIONS

Session - Room	Subject	Day	Time	Paper ID	Paper Title	Session Board
S6-A	Aerospace Materials	07/041/2021	10:15 - 11:15	025_ASAT19	An investigation of plasma sprayed nickel-based and pure aluminum coatings on austenitic stainless steel AISI 304	Brig. Prof. Ahmed Barka MTC, Cairo, Egypt Brig. Prof. Mohamed Gobara MTC, Cairo, Egypt. Col.Prof. Sherif El-basuney MTC, Cairo, Egypt.
				026_ASAT19	Studying the distribution of hardness values in friction drilled 7075 Al-alloy sheets at different conditions	
				027_ASAT19	Low thermal emissivity coat-ing based on aluminium / acrylic composite coatings	
			12:45 - 1:45	028_ASAT19	Enhancement of carbon fiber/ epoxy composite electrical, optical and thermal proper-ties by using different types of nano-additives.	
				029_ASAT19	Non-destructive aerial and terrestrial analysis of urani-um isotopic activity by sen-sitive ?-spectrometer for El-Sella site	
				030_ASAT19	Investigation of foamcrete mechanical and physical properties	

ASAT-19

The 19th International Conference on Aerospace Sciences & Aviation Technology (ASAT-19)
April 6-8, 2021, Cairo, Egypt

SCIENTIFIC SESSIONS

Session - Room	Subject	Day	Time	Paper ID	Paper Title	Session Board
S7-A	Energetic Materials/Aerospace Thermophysics	07/04/2021	2:00 - 3:00	031_ASAT19	On the unsteady behaviours of the adiabatic endwall film cooling effectiveness	Maj. Gen. Prof. Hossam Elsaid Mostafa MTC, Cairo, Egypt; Brig. Prof. Ahmed Barka MTC, Cairo, Egypt; Brig. Prof. Chesham Ramzy Tantawy MTC, Cairo, Egypt
				032_ASAT19	Enhancing the propulsion characteristics of rockets by adding the energetic Ni-tro-hydroxyl-terminated polybutadiene	

ASAT-19

The 19th International Conference on Aerospace Sciences & Aviation Technology (ASAT-19)
April 6-8, 2021, Cairo, Egypt

SCIENTIFIC SESSIONS

Session - Room	Subject	Day	Time	Paper ID	Paper Title	Session Board
S4-B	Renewable Energy	06/04/2021	2:00 - 3:15	017_ASAT19	Output power boosting of a photovoltaic panel based on various back pipe structures: a computational study	Maj. Gen. Prof. Moahmed Amin MTC, Cairo, Egypt. Brig. Prof. Ali Elmahi MTC, Cairo, Egypt. Brig. Prof. Ayman Gomma MTC, Cairo, Egypt.
				018_ASAT19	Numerical investigation of four photovoltaic/thermal integrated structures from energetic point of view	
				019_ASAT19	Comparison of dust and high-temperature effects on mono and poly photovoltaic panels	

ASAT-19

The 19th International Conference on Aerospace Sciences & Aviation Technology (ASAT-19)
April 6-8, 2021, Cairo, Egypt

SCIENTIFIC SESSIONS

Session - Room	Subject	Day	Time	Paper ID	Paper Title	Session Board
S8-B	Guidance, Navigation and Control	07/04/2021	9:00 - 10:00	033_ASAT19	Acoustic threat detection and direction finding system	Brig. Prof. Mamoud Ashry MTC, Cairo, Egypt Col. Prof. Ahmed Tai-mour MTC, Cairo, Egypt Col.Prof. Mohamed Atef MTC, Cairo, Egypt
				034_ASAT19	Tactics Overview for Imple-menting High-Performance Computing on Embedded Platforms	
				035_ASAT19	Modelling and simulation of two axes gimbal fuzzy PI stabilization system in the presence of feedback sensors noise	
			10:15 - 11:15			

ASAT-19

The 19th International Conference on Aerospace Sciences & Aviation Technology (ASAT-19)
April 6-8, 2021, Cairo, Egypt

SCIENTIFIC SESSIONS

Session - Room	Subject	Day	Time	Paper ID	Paper Title	Session Board
S9-B	Guidance, Navigation and Control	07/04/2021	11:30 - 12:30	036_ASAT19	Performance comparison among popular implementations of H.264 encoders	Maj. Gen Prof. Ibrahim Elsherif MTC, Cairo, Egypt. Col. Assoc.Prof. Hossam Hendy MTC, Cairo, Egypt Lt. Col. Assoc. Prof. Mohammed Abozied MTC, Cairo, Egypt
				037_ASAT19	Modelling and transient response study of hydraulic servo actuator	
			12:45 - 1:45	038_ASAT19	Modeling, simulation and controller design for a typical bent axis electrohydraulic servo motor	
				039_ASAT19	Obstacle avoidance for multi-UAV path planning based on particle swarm optimization	

ASAT-19

The 19th International Conference on Aerospace Sciences & Aviation Technology (ASAT-19)
April 6-8, 2021, Cairo, Egypt

SCIENTIFIC SESSIONS

Session - Room	Subject	Day	Time	Paper ID	Paper Title	Session Board
S10-B	Unmanned Systems	07/04/2021	2:00 - 3:00	040_ASAT19	PID tuning approaches for quadrotors unmanned aerial vehicles	Brig. Prof. Ahmed Mohsen MTC, Cairo, Egypt. Col. Assoc. Prof. Basem Sheta MTC, Cairo, Egypt.
				041_ASAT19	Formation configuration of unmanned cooperative quadrotors via PID tuning approaches	

IOP Conf. Series: Materials Science and Engineering 1172 (2021) 011001 doi:10.1088/1757-899X/1172/1/011001

ASAT-19

The 19th International Conference on Aerospace Sciences & Aviation Technology (ASAT-19)
April 6-8, 2021, Cairo, Egypt

SCIENTIFIC SESSIONS

Session - Room	Subject	Day	Time	Paper ID	Paper Title	Session Board
S11-A	Aerodynamics/Aerospace Combustion and Propulsion	08/04/2021	9:00 - 10:00			Brig. Prof. .Mahmoud Yahia MTC, Cairo, Egypt. Col. Assoc. Prof. Ahmed Nemnem MTC, Cairo, Egypt. Col. Assoc. Prof. Mohamed Zakria, MTC, Cairo, Egypt.
				042_ASAT19	I Impact of nozzle profile on ballistic performance and structural loads	
				043_ASAT19	Design, test and build of a monopropellant thruster using 85% hydrogen peroxide	
			10:15 - 11:15	044_ASAT19	Parametric study of solid propellant slotted grain	
				045_ASAT19	Application of the level set method in solid propellant grain burnback with dogbone grain as a case study	
				046_ASAT19	Experimental Investigation of Recirculation Zone Characteristics of a Pre-filming Airblast Injector	

IOP Conf. Series: Materials Science and Engineering 1172 (2021) 011001 doi:10.1088/1757-899X/1172/1/011001

ASAT-19

The 19th International Conference on Aerospace Sciences & Aviation Technology (ASAT-19)
April 6-8, 2021, Cairo, Egypt

SCIENTIFIC SESSIONS

S1-A	Aerospace Structures
001_ASAT19	Dimensional structural mass optimization of forged steel connecting rod for aircraft piston engine
002_ASAT19	Static and modal analysis of sine-wave-shaped spar
003_ASAT19	Vibration analysis of composite wing with geometric and material coupling
004_ASAT19	Numerical simulation of temperature field in a C/C composite multi-disk brake during aircraft braking
005_ASAT19	Stress analysis of multi-layered composite cylinders subjected to various loadings.
006_ASAT19	Effect of damping material thickness on vibration analysis in pretension layer damping process
007_ASAT19	Vibration damping of aircraft propeller blades using shunted piezoelectric transducers

S2-A	Communications and Networks
008_ASAT19	Design of a smart energy efficient IoT wireless node
009_ASAT19	Comparative study between cognitive radio techniques in FM broadcasting band
010_ASAT19	A satellite-based quantum key distribution using decoy-state protocol

ASAT-19

The 19th International Conference on Aerospace Sciences & Aviation Technology (ASAT-19)
April 6-8, 2021, Cairo, Egypt

SCIENTIFIC SESSIONS

S3-B	Image Processing & Remote Sensing
011_ASAT19	Enhanced image encryption algorithm by quantum key distribution
012_ASAT19	SAR image formation enhancement using effective velocity estimation method
013_ASAT19	Frequency analysis of SAR system design for small satellite
014_ASAT19	Updating land use and land cover classes of Blue Nile basin using landsat-8 images
015_ASAT19	Assessment of top of atmosphere reflectance generation algorithm for world view-2 high resolution images
016_ASAT19	Hybrid fusion using Gram Schmidt and Curvelet transforms for satellite images

S4-B	Renewable Energy
017_ASAT19	Output power boosting of a photovoltaic panel based on various back pipe structures: a computational study
018_ASAT19	Numerical investigation of four photovoltaic/thermal integrated structures from energetic point of view
019_ASAT19	Comparison of dust and high-temperature effects on mono and poly photovoltaic panels

ASAT-19

The 19th International Conference on Aerospace Sciences & Aviation Technology (ASAT-19)
April 6-8, 2021, Cairo, Egypt

SCIENTIFIC SESSIONS

S5-A	Aerospace Materials
020_ASAT19	Effective electromagnetic interference shielding using polyurethane / graphite composite foam materials
021_ASAT19	Parametric optimization of non-prismatic micro-plates to reduce stiffening and curling initiated during fabrication
022_ASAT19	On the effect of post-processing techniques of the additively manufactured aluminum alloy parts
023_ASAT19	Factors affecting Failure mode and load of rubber polyester composite joint
024_ASAT19	Water recycling system based on adsorption by activated carbon synthesised from c. verum for space exploration; an estimated design

S6-A	Aerospace Materials
030_ASAT19	Investigation of foamcrete mechanical and physical properties
025_ASAT19	An investigation of plasma sprayed nickel-based and pure aluminum coatings on austenitic stainless steel AISI 304
027_ASAT19	Low thermal emissivity coating based on aluminium / acrylic composite coatings
026_ASAT19	Studying the distribution of hardness values in friction drilled 7075 Al-alloy sheets at different conditions
028_ASAT19	Enhancement of carbon fiber/epoxy composite electrical, optical and thermal properties by using different types of nano-additives.
029_ASAT19	Non-destructive aerial and terrestrial analysis of uranium isotopic activity by sensitive ■-spectrometer for El-Sella site

ASAT-19

The 19th International Conference on Aerospace Sciences & Aviation Technology (ASAT-19)
April 6-8, 2021, Cairo, Egypt

SCIENTIFIC SESSIONS

S7-A	Energetic Materials &Aerospace Thermo-physics
031_ASAT19	On the unsteady behaviours of the adiabatic endwall film cooling effectiveness
032_ASAT19	Enhancing the propulsion characteristics of rockets by adding the energetic Nitro-hydroxyl-terminated polybutadiene (NHTPB) in the propellant compositions

S8-B	Guidance, Navigation and Control
033_ASAT19	Acoustic threat detection and direction-finding system
034_ASAT19	Tactics Overview for Implementing High-Performance Computing on Embedded Platforms
035_ASAT19	Modelling and simulation of two axes gimbal fuzzy PI stabilization system in the presence of feedback sensors noise

IOP Conf. Series: Materials Science and Engineering 1172 (2021) 011001 doi:10.1088/1757-899X/1172/1/011001

ASAT-19

The 19th International Conference on Aerospace Sciences & Aviation Technology (ASAT-19)
April 6-8, 2021, Cairo, Egypt

SCIENTIFIC SESSIONS

S9-B	Guidance, Navigation and Control
036_ASAT19	Performance comparison among popular implementations of H.264 encoders
037_ASAT19	Modelling and transient response study of hydraulic servo actuator
038_ASAT19	Modeling, simulation and controller design for a typical bent axis electrohydraulic servo motor
039_ASAT19	Obstacle avoidance for multi-UAV path planning based on particle swarm optimization

S10-B	Unmanned Systems
040_ASAT19	PID tuning approaches for quadrotors unmanned aerial vehicles
041_ASAT19	Formation configuration of unmanned cooperative quadrotors via PID tuning approaches

ASAT-19

The 19th International Conference on Aerospace Sciences & Aviation Technology (ASAT-19)
April 6-8, 2021, Cairo, Egypt

SCIENTIFIC SESSIONS

S11-A	Aerodynamics & Aerospace Combustion and Propulsion
042_ASAT19	I Impact of nozzle profile on ballistic performance and structural loads
043_ASAT19	Design, test and build of a monopropellant thruster using 85% hydrogen peroxide
044_ASAT19	Parametric study of solid propellant slotted grain
045_ASAT19	Application of the level set method in solid propellant grain burnback with dogbone grain as a case study
046_ASAT19	Experimental Investigation of Recirculation Zone Characteristics of a Pre-filming Airblast Injector

ASAT-19
The 19th International Conference on Aerospace Sciences & Aviation Technology (ASAT-19)
April 6-8, 2021, Cairo, Egypt

ABSTRACTS

Dimensional structural mass optimization of forged steel connecting rod for aircraft piston engine

M Helal[1,2], B F Felemban1 , A Alharthi[1], S Almalki1, E Fathallah [3,4] , A A Elsherbiny[2] and H Eleashy[5]

[1]Department of Mechanical Engineering, College of Engineering, Taif University, Taif 21974

[2]Production and Mechanical Design Department, Faculty of Engineering, Mansoura University, Mansoura 35516, Egypt

[3]Civil Engineering Department, M.T.C. Kobry Elkobba - Cairo, Egypt

[4]Ships and Submarines Engineering Department, Military Technical College, Cairo, Egypt.

[5]Mechanical Engineering Department, Faculty of Engineering & Technology, Future University in Egypt, Cairo, Egypt

E-mail: saidhabib2000@hotmail.com

Abstract. The connecting rod is an important component of the engine. It conveys the kinetic energy from the piston to the crankshaft. All cars and aircraft engines contain at least one connecting rod, which differs from one motor to another in terms of length, size and shape. Hence, it is subjected to massive alternating load. This research aims to improve the connecting rod design by reducing its mass without sacrificing durability and safety especially for aircraft applications. Therefore, a static stress analysis is carried out on forged steel connecting rod using ANSYS APDL. Geometric modelling of the connecting rod was created using ANSYS APDL.Additionally, von-Mises stress and strain, principal stresses and strains, shear stress and the deflation results of the connecting rod are investigated. The results showed a great opportunity for mass weight reduction. Thus, a dimensional structural mass optimization was performed. The optimization results were promising, which reduced the mass by 55.13% (in the tensile case) and 56.7% (in the compression case) from the initial design. Therefore, the efficiency of aircraft engine can be maximized.

ASAT-19
The 19th International Conference on Aerospace Sciences & Aviation Technology (ASAT-19)
April 6-8, 2021, Cairo, Egypt

ABSTRACTS

Static and modal analysis of sine-wave-shaped spar

E Elshazly[1], M Kassem[1], M A Elshafei[1], and M Kamel[1]
[1] Department of Aircraft Mechanics, Military Technical College, EGYPT
E-mail: m_kassem@mtc.edu.eg

Abstract. Sine-Wave-shaped Spar (SWS) is recently involved in wing design as one of the corrugated shaped spars that highly improve the torsion and buckling resistance. SWS is an Ibeam made of sine wave shaped vertical web fixed to upper and lower flanges. The current research evaluates the behavior of SWS in comparison to the traditional straight-web spar in resisting two cases of loading; static bending loading and static torsion loading. For each case, deformation and stress are analyzed. Modal analysis is also performed for the two spar configurations. For this purpose, a finite element model is developed using Ansys Workbench software for the SWS and the straight web spar. To reach a fair comparison, similar weight is considered for the two spar structures. A parametric study is performed to investigate the effect of web thickness and flange thickness on the bending and torsion deformation and stresses. The effect of thickness variation on the natural frequencies and mode shapes is also considered. The current study can be considered a base to design wing spars that combine the merits of the two spar configurations.

IOP Conf. Series: Materials Science and Engineering 1172 (2021) 011001 doi:10.1088/1757-899X/1172/1/011001

ASAT-19

The 19th International Conference on Aerospace Sciences & Aviation Technology (ASAT-19)
April 6-8, 2021, Cairo, Egypt

ABSTRACTS

Vibration analysis of composite wing with geometric and material coupling

A Magdy[1], M. Kamel[1], M. A Elshafei[1] and M Kassem1
[1]Department of Aircraft Mechanics, Military Technical College, Egypt
E-mail: a.magdy@mtc.edu.eg

Abstract. Composite wing design is complicated but inevitable to enlighten modern airplanes while maintaining the required performance. Using the dynamic transfer method, this paper discusses intensively the dynamic characteristics of a cantilever composite wing with both torsion and bending coupling to represent both material and geometric coupling. The governing differential equations are obtained based upon the principle of Hamilton and are solved analytically using a harmonic oscillation assumption. For this purpose, a MATLAB code is developed and results are validated in comparison with published work. Such a comparison shows a good agreement between both results. Finally, a parametric study is carried out to show the influence of the variation of both geometric coupling and torsion bending coupling rigidity on the free vibration analysis of the composite wing. The study shows the crucial effect of both factors on the dynamic behavior of the composite wing. The current research can be considered as a base for aeroelasticians while designing composite structures.

ASAT-19

The 19th International Conference on Aerospace Sciences & Aviation Technology (ASAT-19)
April 6-8, 2021, Cairo, Egypt

ABSTRACTS

Numerical simulation of temperature field in a C/C composite multidisk brake during aircraft braking

E M Ramadan[1],*, A E Hussien [1], A M Youssef [1], and T M Abd El-Badia [1].
[1] Military technical college, Mechanical Design and Production Department, Cairo, Egypt.
E-mail: eramadan@mtc.edu.eg

Abstract. This article presents a finite element method for simulating the heat production during stopping the aircraft. Thermal analysis and simulation in the finite element model are based on the theory of energy transformation and transportation. A commercial software COMSOL Multiphysics 5.5a is used for simulating the braking operation. The internal temperatures of the brake disks were obtained and the variation in temperatures between disks were discussed. Thermomechanical behaviour is studied to show the effect of thermal energy on the contact mechanics for the friction surfaces between the brake disks. Aircraft mass, initial velocity and deceleration rate are responsible for heat generation and consequently the maximum reached temperature during braking. The friction surfaces between disks were the main heat energy source where the heat was concentrating on these layers of friction surfaces. For the selected braking operation, the maximum reached temperature was 1020K. The finite element odel was validated against historical data for a Boeing737-400 at constant deceleration for new disk brakes and for RTO brake energy.

ASAT-19
The 19th International Conference on Aerospace Sciences & Aviation Technology (ASAT-19)
April 6-8, 2021, Cairo, Egypt

ABSTRACTS

Stress analysis of multi-layered composite cylinders subjected to various loadings.

M H Elgohary[1], M I El-Geuchy[2], H M Abdallah[1] and S S Sayed[1]
[1] Rockets Department, Military Technical College, Cairo, Egypt
[2] PhD., Technical Research Centre, Cairo, Egypt
E-mail: Gohary1@live.com

Abstract: Composite cylinders are widely used in many applications, common examples for these applications are power drive shafts, chemical storage tanks, rocket motor cases and pressure vessels used in aerospace vehicles. The structure of the composite cylinder can be represented as a cylinder composed of coaxial orthotropic layers. In this paper, an analytical method is used to calculate stresses, strains and displacements through the wall thickness of multi-layered composite cylinder made of orthotropic material. The method is based on the theory of elasticity of bodies having cylindrical anisotropy. This method permits accurate stress analysis of thin and thick-walled composite cylinders subjected to axial load, torsional load and bending moment. The analytical method is modified to incorporate internal and external pressure loads beside the prescribed load cases. A numerical example is presented for a composite cylinder subjected to internal pressure load with a tensile axial load. The resulted stresses and strains are used to validate 3-dimensional finite element model. A parametric study has been performed using the analytical method to investigate the influencing parameters on stress distribution within the cylinder thickness and the results are found to be beneficial to look into during the preliminary design phase.

IOP Conf. Series: Materials Science and Engineering 1172 (2021) 011001 doi:10.1088/1757-899X/1172/1/011001

ASAT-19

The 19th International Conference on Aerospace Sciences & Aviation Technology (ASAT-19)
April 6-8, 2021, Cairo, Egypt

Effect of damping material thickness on vibration analysis in pre-tension layer damping process

A Mohamed[1], A A Omer[1] and A Hassan[1]

[1] Mechanical Engineering Department, Military Technical College, Cairo, Egypt.

E-mail: anwer_mohamed@mtc.edu.eg,

ABSTRACTS

Abstract. Vibration reduction is a critical necessity in many fields of engineering, technology, and industry. As a result, there is a need for vibration management. To restrict or adjust the system's vibration response, a variety of techniques are employed. In recent years, there has been a lot of enthusiasm for the easy implementation of these vibration-control structures. The damping properties of viscoelastic materials tend to be excellent. Damping is determined by the material's ability to dissipate energy. To minimise the vibration of vibrating surfaces, iscoelastic materials are commonly used. In this study, viscoelastic material (Dyad 606) is applied on the AL plate in the form of free layer damping and pretension layer damping. First aluminium structure using free layer damping and another one using pretension layer damping by applying tension load (two layers of damping plates, one on the top surface of the Al alloy and the other on the lower face). Passive vibration damping of the plate is achieved. These layers are influenced by axial uniform distributed load. The damping behaviour of the AL Plate is discussed in relation to the thickness variation of the pre-tension damping material. The results show that the loss factor and the attenuation percentage of the structure with pretension layer damping are increased as compared to free layer damping. It is also found that the most effective thickness of the damping material to increase the damping capacity of the framework is around half the base plate thickness

ASAT-19

The 19th International Conference on Aerospace Sciences & Aviation Technology (ASAT-19)
April 6-8, 2021, Cairo, Egypt

ABSTRACTS

Vibration damping of aircraft propeller blades using shunted piezoelectric transducers

A M Morad [1], M Kamel [2], and M K Khalil [2]
[1] Egyptian Armed Forces, Cairo, Egypt
[2] Aircraft Mechanical Engineering Department, MTC, Cairo, Egypt
E-mail: alaa_morad@yahoo.com

Abstract. Gas turbine engine blades experience vibrations due to the flow disturbances, these vibrations are critical to the engine durability and performance. Piezoelectric transducers (sensors and actuators) have been used for engine blade vibrations damping either through a passive or active vibration control. The propeller blades are part of turboprop engine and considered as one of the main source of turboprop engine vibrations. Piezoelectric blade damping ideas have been studied by other researchers for fan blades and compressor blades. In this research a vibration damping procedure using piezoelectric transducers applied to an unmanned aerial vehicle (UAV) composite propeller. Experimental investigation introduces an approach for the propeller vibration damping using piezoelectric transducers in conjunction with appropriate shunt circuit. Three thin piezoelectric transducers macro fiber composite (MFC) type PZT-5A are surface-mounted on the propeller, one at each blade. These transducers are placed at locations of high modal strain areas for the propeller first mode at each blade, where these locations are identified by finite element numerical simulation. Electronic resonance shunt circuit, resistor-inductor-capacitor type, for the piezoelectric transducers is designed and experimentally developed such that effective vibration suppression of the propeller is achieved. The experimental and numerical investigations in this research illustrate that piezoelectric transducers with appropriate shunt circuit reduces the aircraft propeller vibrations.

IOP Conf. Series: Materials Science and Engineering 1172 (2021) 011001 doi:10.1088/1757-899X/1172/1/011001

ASAT-19
The 19th International Conference on Aerospace Sciences & Aviation Technology (ASAT-19)
April 6-8, 2021, Cairo, Egypt

ABSTRACTS

Design of a smart energy efficient IoT wireless node

M S Nabil[1], A F Tammam[2] and M M ElKhatib[3]
[1] Computer Engineering, Pyramids Higher Institute for engineering & Technology (PHI), Giza, Egypt
[2] Computer Engineering, Arab Academy for Science, Technology & Maritime Transport (AASTMT) Cairo, Egypt
[3] Senior IEEE member, Electronic Engineering Department, Military Technical College (MTC) Cairo, Egypt
E-mail: mohamed.m.elkhatib@ieee.org

Abstract. The Internet of Things (IoT) is gaining greater effect in our life from day to day. Wireless sensor networks (WSN) are embedded into the "Internet of Things" and power saving is one of the problems that it faces. This study introduces a new circuit design for a DC-DC Buck-Boost converter for controlling the output voltage of Energy Harvester (EH) via adaptive Neuro-Fuzzy inference systems (ANFIS) controller. It has proven that there is an effective changes regarding settling time almost being zero, overshoot being negligible and steady state voltage. As an application for the proposed design, an implementation for Precision Agriculture (PA) self-sustainable Wireless Sensor-Node (WS-N) with low power consumption has been presented. A solar power management for Precision agricultural in internet of things system proved to enhance cost-efficiency and increased lifetime for the device. The simulation results are determined by means that of MATLAB/Simulink software. The cloud service of Blynk IoT has been implemented to monitor the collected data on real time. Data on solar cells, batteries and soil were regarded reliable and precise.

IOP Conf. Series: Materials Science and Engineering 1172 (2021) 011001 doi:10.1088/1757-899X/1172/1/011001

ASAT-19
The 19th International Conference on Aerospace Sciences & Aviation Technology (ASAT-19)
April 6-8, 2021, Cairo, Egypt

ABSTRACTS

Comparative study between cognitive radio techniques in FM-broadcasting band

I Hamed[1], M A Elmenyawi[2], and A Y Hassan[3]
[1] MSc student, Benha faculty of engineering, Benha University, Egypt
[2] Lecturer, Benha faculty of engineering, Benha University, Egypt
[3] Associate professor, Benha faculty of engineering, Benha University, Egypt
E-mail: ashraf.fahmy@bhit.bu.edu.eg

Abstract. Optimum usage of the existing frequency spectrum is a major requirement due to the large increase in the number of subscribers at the same frequency. The cognitive radio scheme has become an important application used to optimize spectrum utilization, detects spectrum bands that are not occupied by primary users to create communication links between secondary users in the same band. These non-occupied bands are described as white space. In this paper, a spectrum sensing method that is an important stage in cognitive radio communications is discussed. Various spectrum sensing methods such as energy detection (ED) and cyclostationary feature detection (CFD) are used to sense the spectrum in the FM broadcasting band. A comparison of energy detection and cyclostationary feature detection spectrum sensing methods is made to determine the empty bands in the FM broadcasting band. This enables secondary users to transfer information in these empty bands without affecting the primary users of the FM broadcasting system.

ASAT-19
The 19th International Conference on Aerospace Sciences & Aviation Technology (ASAT-19)
April 6-8, 2021, Cairo, Egypt

ABSTRACTS

A satellite-based quantum key distribution using decoy-state-protocol

A M Sofy[1], M Y Shalaby[2] and H M Dahshan[1]
[1] Military Technical College, Department of Communication Engineering, Cairo,
Egypt
[2] Ministry of Defence, Cairo, Egypt
E-mail: myousef73@hotmail.com

Abstract. An ideal QKD implementation that provides the promising unconditional security using single photon laser source is not practically easy, also sending these photons over a fiber channel affects the maximum distance of link up to few hundreds of kilometers because of the receiver's detectors losses. Therefore, the achievement of a global quantum network is not easy, and the need to use LEO satellites [1,2] for establishing quantum links makes the job easier as quantum free space link provides much less losses to the link than fiber links. Nowadays, a quantum communication link can be established by satellites through free space. in this paper, we propose an implementation [3] of a satellite-based quantum key distribution using decoystate protocol [4], we then compare this protocol with the BB-84 protocol [3] against security and the key length of the generated shared key.

ASAT-19
The 19th International Conference on Aerospace Sciences & Aviation Technology (ASAT-19)
April 6-8, 2021, Cairo, Egypt

ABSTRACTS

Enhanced image encryption algorithm by quantum key distribution

M A Mahmoud[1] ,T Mekkawy[2]and A D Elbayoumy[3]
[1]NANO Master program, Faculty of Engineering.Cairo University,Giza, Egypt
[2]Avionics Department, Military Technical Collage, Cairo, Egypt
[3]Communication Department, Military Technical Collage, Cairo, Egypt
E mail: Mohamed.khalifa84@eng1.cu.edu.eg

Abstract. Quantum Key Distribution (QKD) is one type of Quantum Cryptography (QC) which is based on quantum mechanics fundamentals such as Heisenberg's uncertainty principle and No-cloning theory. The usage of QKD warns the legitimate communicated parties to any attack attempt and this is the most interesting security parameter. Therefore, QKD provides unconditional secure communication method and supports a powerful encryption scheme. The combination between classical communication and QKD creates a new technique called semi quantum key distribution SQKD. Unfortunately, SQKD increases the schemes complexity and requires two steps for ciphering, scramble and encryption. In this paper an enhance image encryption algorithm is proposed based on QKD that eliminates most of the drawbacks of SQKD. The proposed algorithm is simpler than other encryption schemes as it exploits only one encryption step based on the power and the randomness of the generated secret key, which decreases the chance to be cracked. The correctness and efficiency of the proposed algorithm are validated by numerical simulations.

ASAT-19
The 19th International Conference on Aerospace Sciences & Aviation Technology (ASAT-19)
April 6-8, 2021, Cairo, Egypt

ABSTRACTS

SAR image formation enhancement using effective velocity estimation method

M A Fouad1, A Azouz[1], A A Mashaly[1] and A E Abdalla[1]
[1]Electrical Engineering, Military Technical College, Cairo, Egypt.
E-mail: mfouad10282@gmail.com

Abstract: Synthetic Aperture Radar (SAR) presents powerful tools for grounding mapping and remote sensing applications. Effective velocity is a vital index that controls the quality of SAR image formation. An accurate calculation of effective velocity provides a particular value of azimuth frequency modulated (FM) rate. The resultant azimuth FM rate could be used to produce a focused SAR image with sharp details. In this paper, SAR image formation enhancement is proposed using two guided methods based on precise effective velocity estimation. Firstly, effective velocity is estimated based on Sentinel-1data parameters extracted from selected image raw data. Secondly, an iteration method applies output image contrast, sharpness and entropy measurements to estimate the optimum value of the effective velocity based on the initial velocities calculated in the first method. Results are compared with extracted SAR images ignoring the effect of the effective velocity, to identify the performance of the proposed methods.

ASAT-19
The 19th International Conference on Aerospace Sciences & Aviation Technology (ASAT-19)
April 6-8, 2021, Cairo, Egypt

ABSTRACTS

Frequency analysis of SAR system design for small satellite

I Douidar[1], M Safy[2] and A Saleh[3]
[1] Egyptian Armed Forces, Cairo, Egypt.
[2] Egyptian Academy for Engineering and Advanced Technology affiliated to ministry of Military Production, Cairo, Egypt.
[3] October 6 University, Cairo, Egypt.
E-mail: Ibrahim.douidar@gmail.com

Abstract. Great interest is given in recent years to the small satellites as their main advantages are the low development cost and the reduced deployment time, which give attention to develop payloads that can be mounted on the platform of these satellite. the platform of small satellites imposes constraints on the sensor design to fit in the size, mass and power limitations, which leads to necessary trade-offs between the design parameters of the sensor. The selection of the operating frequency is dependent mainly on the mission requirements as each frequency band gives different information about the scene and has different penetration capabilities. However, the SAR sensors used on small satellites come with limitations, so other parameters should be considered during the design. After going through the SAR design process, a frequency analysis is presented at L, C and X bands, which is used to evaluate the effect on the size of the antenna (length and width) and the average power consumption that suit the constraints of small satellite.

ASAT-19
The 19th International Conference on Aerospace Sciences & Aviation Technology (ASAT-19)
April 6-8, 2021, Cairo, Egypt

ABSTRACTS

Updating land use and land cover classes of Blue Nile basin using landsat-8 images

A S Shobary[1], A S Elsharkawy[1], H E M El-Hanafy[1] and O M Moussa1
[1]Department of Civil Engineering, Military Technical College, Cairo, Egypt.
E-mail: a.s.shobary@mtc.edu.eg

Abstract. Land Use and Land Cover are considered as a main input layer for the estimation of surface runoff along with the soil maps, rainfall data, DEM. The main purpose of this research paper is to update the Land Cover and Land Use of the Blue Nile basin that can be used for further processing to estimate the surface runoff of the Blue Nile basin. Recent satellite images of Landsat 8, dated May and June, 2020, were used with less than 10% clouds covering the full scene of the study area. Six land cover classes were targeted namely; water, urban, barren land, forest, grass and agricultural crops. Supervised (maximum likelihood) classification method was used and accuracy assessment with the help of google earth as ground truth was performed.With the aid of 30-m DEM, the Blue Nile basin was divided into fourteen sub-watersheds. Classification scheme was applied to each of the fourteen sub-watersheds and the classification results were introduced as a percentage of the total area of each sub-watershed. Moreover, confusion matrices were formed for the subwatersheds.

ASAT-19
The 19th International Conference on Aerospace Sciences & Aviation Technology (ASAT-19)
April 6-8, 2021, Cairo, Egypt

ABSTRACTS

Assessment of top of atmosphere reflectance generation algorithm for world view-2 high resolution images

M Allam[1], T Mahmoud[1], A Elsharkawy[2] and B Sheta[1]
[1]Electrical Engineering, Military Technical College, Cairo, Egypt.
[2]Civil Engineering, Military Technical College, Cairo, Egypt.
E-mail: eng.ma.mo.allam@gmail.com

Abstract. Recently, there are many developments in remote sensing techniques using satellite imagery as a result of the variety of remotely sensed image sources with different spatial and spectral resolutions and from different sensors. Optical satellite products are affected by an error of payload, atmosphere (scattering and absorption), and variations relative to positions of Sun, Earth, and satellite during capturing data. Top of Atmosphere (TOA) correction is the process of removing the effects of variations relative to sensor error and positions of Sun, Earth, and satellite. In this paper, maximum benefits from high spectral and spatial resolution images are demonstrated and analyzed or compared with images from different. TOA correction algorithm, which was implemented using Python environment, is applied to high resolution images from WorldView-2 (WV-2) satellite. TOA reflectance is considered the first step in any algorithm dedicated to the change detection process. Reflectance conversion is also performed on the same images using Orfeo ToolBox (OTB), the open-source software. The performance and efficiency of the proposed algorithm are compared with that of the Orfeo ToolBox (OTB) TOA reflectance output. The achieved results show that the proposed algorithm, which is automatically performed, is faster and provides significant results for WV-2 images, and can be adapted to be applied on different optical satellite sensors.

ASAT-19
The 19th International Conference on Aerospace Sciences & Aviation Technology (ASAT-19)
April 6-8, 2021, Cairo, Egypt

Hybrid fusion using Gram Schmidt and Curvelet transforms for satellite images

M Hammad[1], T Ghoniemy[1], T Mahmoud[1] and A Amein[2]
[1]Electrical Engineering, Military Technical College, Cairo, Egypt.
[2]Electrical Engineering, Senior IEEE, FISCS, October 6 University, Cairo, Egypt.
E-mail: Radok2003@hotmail.com

ABSTRACTS

Abstract. Optical satellites generally provide high-resolution panchromatic but low-resolution multispectral images which provide structural details of features and spectral information respectively. Nowadays, fusion of the two types of resolutions, to have complementary information, becomes increasingly essential for many applications such as microscopic, astronomical and satellite imagery. In this paper, a novel hybrid pixel-level image fusion method is proposed for benefiting from both panchromatic (PAN) and multispectral (MUL) images. The proposed method integrates Gram Schmidt (GS) and curvelet transforms (CVT), by the aid of local energy and maximum fusion rules, for reducing individual method limitations and achieving both better spectral consistency and spatial details preservation. After a pre-processing stage, orthonormal bases are obtained for low spatial resolution images by using GS transform. Then, high-resolution and low-resolution images are fused using CVT by the aid of histogram matching. Finally, the fused image is obtained by applying both curvelet and GS inverse transforms. The performance of the proposed method is evaluated using publicly available Pleiades benchmark-datasets. Consequently, the spectral and spatial qualities of the fused images are assessed subjectively as well as objectively using different quality metrics. Moreover, the proposed method is compared with state-of-the-art fusion techniques and results show the robustness of the proposed method that has the best result in spatial and spectral evaluation metrics such as, Quality with No Reference (QNR), Peak Signal to Noise Ratio (PSNR), Standard Deviation (SD), Entropy (ENT) and Spectral Correlation Coefficient (SCC) metrics.

ASAT-19
The 19th International Conference on Aerospace Sciences & Aviation Technology (ASAT-19)
April 6-8, 2021, Cairo, Egypt

ABSTRACTS

Output power boosting of a photovoltaic panel based on various back pipe structures: a computational study

A Bayoumi[2], G M Abdo[2], and A A Emara[1,2]
[1] Mechanical Engineering Department, Faculty of Engineering, Mataria-Helwan University in Cairo, Egypt.
[2] Mechanical Engineering Department, Faculty of Engineering and Technology, Badr University in Cairo (BUC), Cairo, Egypt.
E-mail: gamal.abdo@buc.edu.eg

Abstract. The efficiency of photovoltaic panels (PV) drops due to the rise in temperature, which leads to a decrease in the PV output power. A PV/Thermal system is therefore used as a solution to increase the output power from the PV panels. Comsol Multiphysics software program, a simulator-based Finite Element Method (FEM) tool, and Matlab program simulations are used to perform this electro-thermal model. The simulation process for different back pipes follows a serpentine, and a new shape of pipes called (square shapes) which are attached under the PV module. These shapes were specifically chosen for higher conversion efficiency and increase the heat transfer of the system. Additionally, a comparison between electrical parameters and heat transfer characteristics of water and CuO/water nanofluid in a PV cooling system has been studied. The new shape leads to improve the photovoltaic (PV) module parameters, such as short circuit current ISC, open-circuit voltage VOC, and maximum power Pmax for a new shape more than the serpentine shape. These parameters are calculated under Air Mass 1.5 G (AM1.5G) with 1000 W/m2 of irradiance which is considered the average irradiance at the MENA (The Middle East and North Africa) region throughout the year. The results exhibit the PV module's total enhancement by using the new shape with CuO nano-fluid about 24.7 %.

IOP Conf. Series: Materials Science and Engineering 1172 (2021) 011001 doi:10.1088/1757-899X/1172/1/011001

ASAT-19
The 19th International Conference on Aerospace Sciences & Aviation Technology (ASAT-19)
April 6-8, 2021, Cairo, Egypt

ABSTRACTS

Numerical investigation of four photovoltaic/thermal integrated structures from energetic point of view

M. Abd El-Hamid[1, 2], Sherin. M[1], 2and Gaosheng Wei[1]
[1] Key Laboratory of Power Station Energy Transfer Conversion and System of Ministry of Education, School of Energy, Power and Mechanical Engineering, North China Electric Power University, Beijing 102206, China
[2] Mechanical Power Engineering Department, Faculty of Engineering – Mattaria, Helwan University, Cairo 11718, Egypt
E-mail: eng.mohamed.abdelhamed90@gmail.com

Abstract. Photovoltaic/Thermal (PV/T) module is considered to be one of the most recent technologies which offers harness and production of both electric and thermal energy. In the current study an energetic analysis is carried out to compare between four different configurations which are: the single pass single glazed (PV/T-I), the single pass double glazed with air gap (PV/T-II), the single pass double glazed with argon gap (PV/T-III) and the double pass double glazed (PV/T-IV) hybrid photovoltaic/Thermal air collector systems. A 3 dimensional numerical model is built up and validated with both the numerical and experimental results coming from the literature. The numerical simulations have been accomplished to investigate the energetic performance with a detailed thermal and electrical study taking in account an inlet coolant temperature the same as the ambient temperature for a coolant (air) mass flow rate equals to 0.025 kg/s of a typical day in August from 9:00 to 17:00 under the ambient conditions of Beijing, China. The results show that the single pass single glazed configuration has the highest electrical efficiency, whereas the double pass configuration has the greatest thermal and energy efficiencies among the proposed configurations. The average daily energy efficiencies are 53.14%, 75.92%, 77.63% and 82.19% for the (PV/T-I), (PV/T-II), (PV/T-III) and (PV/T-IV) configurations, respectively.

ASAT-19
The 19th International Conference on Aerospace Sciences & Aviation Technology (ASAT-19)
April 6-8, 2021, Cairo, Egypt

Comparison of dust and high-temperature effects on mono and poly photovoltaic panels

E Elshazly[1], A Abd El-Rehim[1] and I El-Mahallawi[1]
[1]Centre for Renewable Energy Faculty of Engineering, British University in Egypt (BUE),El-Sherouk 11837, Cairo, Egypt
E-mail: iemahallawi@bue.edu.eg

ABSTRACTS

Abstract. Solar Photovoltaic Panels are considered as one of the most powerful alternative renewable and sustainable energy sources. However, a major challenge is the effect of dust accumulation on the photovoltaic panels in natural outdoor environment as it reduces the transmissivity of the light on the surface of the solar panels. For many small communities, the decision of implementing mono or polycrystalline PVs should consider economic aspects. This study is a case study that is held at The British University in Egypt at El Sherouk city to study the effect of different parameters such as dust accumulation, water cooling and coating on the performance of both mono- and poly-crystalline panels at El-Sherouk City. The effects of high temperature and dust accumulation on different solar panels placed in natural outdoor conditions at El- Sherouk City are studied and the electrical performance of the solar panels is represented by measuring several characteristic parameters of dusty and cooled PV panels compared to cleaned and non-cooled panels. The effect of the tilt angle on the accumulation of dust on the surface of the solar panels is, also, studied. The mono-crystalline solar panels are installed at tilt angles 0°, 15°, 30°, 45°, and 60° for one month without cleaning, by any method. The results shows that the power reduction percentage is 17%,20%,25%,27% and 30% for tilt angles 60°,45°,30°,15° and 0°; respectively. Tilt angles 15° and 30° show to be optimal for the installation of the PV solar system, as they produce the highest amount of output power. It is found from the study that the accumulation of dust on the surface of different types of solar panels can reduce the efficiency by 30%. While the high temperature can reduce the efficiency by up to 10 %.

IOP Conf. Series: Materials Science and Engineering 1172 (2021) 011001 doi:10.1088/1757-899X/1172/1/011001

ASAT-19

The 19th International Conference on Aerospace Sciences & Aviation Technology (ASAT-19)
April 6-8, 2021, Cairo, Egypt

ABSTRACTS

Effective electromagnetic interference shielding using polyurethane / graphite composite foam materials

H Oraby[1], H R Tantawy[1], I Naeem[1], M Magdy[2], A A Ezz[3] and M Senna[4]
[1]Department of Chemical Engineering, Military Technical College, Cairo, Egypt.
[2]Department of Radar, Military Technical College, Cairo, Egypt.
[3]El-Nasr Company for Intermediate Chemicals, Industrial Area, Giza, Egypt.
[4]Radiation Chemistry Department, National Center for Radiation Research and Technology, P.O.Box 29, Nasr City, Cairo, Egypt.
E-mail: hussein.mohamed4544@yahoo.com

Abstract. Conversely, to metallic based electromagnetic interference (EMI) shielding materials; composite polymer-based foams are normally lightweight, cheaper than metals, and less sensitive to types of environmental degradation. In this work polyurethane - Graphite composite (PU-G) foam materials were prepared with different filler concentrations. Different characterization tools such as Fourier Transform Infra-Red (FTIR) and scanning electron microscope (SEM), were used to identify the structural and topological construction of the prepared composites. Further mechanical properties for the prepared samples were studied to elucidate the opportunity of utilizing these composites in applied applications, specifically for electromagnetic interference (EMI) shielding efficiency (SE) for aerospace applications. Also, in order to adjust this research in the area of aerospace EMI SE, the evaluations were executed in the X-band at (8-12) GHz. The obtained data indicated that the moreover in filler concentration enhanced the compressive strength and compressive modulus of the prepared samples. Moreover EMI SE reached -44 dB with 30 wt % graphite concentration. Finally polyurethane – graphite composite foam material can be taken into consideration a gratifying material to be utilised in EMI SE.

ASAT-19
The 19th International Conference on Aerospace Sciences & Aviation Technology (ASAT-19)
April 6-8, 2021, Cairo, Egypt

Parametric optimization of non-prismatic micro-plates to reduce stiffening and curling initiated during fabrication

B Bakeer[1], A Elsabbagh[1] and M Hedaya[1]
[1]Design and Production Engineering Department, Ain Shams University, Cairo, 11517, Egypt
Email: bahi.bakeer@eng.asu.edu.eg

ABSTRACTS

Abstract. Microelectromechanical devices such as accelerometers, gyroscopes, pressure sensors, and radiofrequency (RF) switches are widely used in aerospace applications. Reduction of stiffening and curling initiated during fabrication of these devices is one of the challenging issues in MEMS design. Reducing response time is also favorable in some applications such as RF MEMS switches. This paper aims at reducing stiffening, curling, and increasing the natural frequency for three well-known designs of micro-plates with fixed-fixed supports. To achieve these objectives, a parametric size optimization is carried out. For comparison purposes, same volume is set as a constraint for all three designs. Compared to conventional rectangular microplate, a reduction of 34% in stiffening in design 2, and 44% in curling in design 3. Design 1 showed the maximum fundamental natural frequency. Thus, it is predicted to have the lowest switching time. Moreover, design 2 showed the maximum critical buckling temperature, extending the operation range of the device. The effect of changing micro-plate material is also studied in this paper.

ASAT-19
The 19th International Conference on Aerospace Sciences & Aviation Technology (ASAT-19)
April 6-8, 2021, Cairo, Egypt

ABSTRACTS

On the effect of post-processing techniques of the additively manufactured aluminum alloy parts

A H Maamoun[1]
[1] Technical Research Center, Cairo, Egypt
E-mail: maamouna@mcmaster.ca

Abstract. Additive manufacturing (AM) or 3D printing of metals promises a significant impact on the upcoming industrial revolution "Industry 4.0" as it is considered one of its main pillars. However, some challenges related to high initial cost and the fabricated part's quality still existed. Selective laser melting (SLM) is one of the effective techniques used for additive manufacturing to fabricate metal products. This paper presents the impact of different postprocessing treatments on the microstructure and surface quality of the AlSi10Mg parts fabricated using SLM. This work illustrates the analytical view of the results obtained from two studies in a previous work by the author. A process map is presented for thermal post-processing treatment to customize the required quality and material characteristics of the AlSi10Mg parts. In addition, the shot peening results in a significant improvement for both surface roughness and hardness is illustrated. This work is a part of developing the manufacturing process of additively manufactured lightweight parts in some critical applications specifically for the metallic mirrors used in high power laser systems or wide view space telescopes.

ASAT-19
The 19th International Conference on Aerospace Sciences & Aviation Technology (ASAT-19)
April 6-8, 2021, Cairo, Egypt

ABSTRACTS

Factors affecting Failure mode and load of rubber polyester composite joint

B H Mousa[1], N A Azab[1], G Bassioni[2] and M H Abdellatif[1]
[1]Mechanical Design and Production Engineering Department, Faculty of Engineering, Ain Shams University, Cairo, Egypt
[2]Chemistry Division, Faculty of Engineering, Ain Shams University, Cairo, Egypt
E-mail: bassant.hany@eng.asu.edu.eg

Abstract. The joining of rubber polyester composite is affected by design and geometry parameters. These parameters affect the failure load of the composite joint and the failure mode. Catastrophic failure modes should be avoided, and the optimum parameters are selected to achieve a progressive failure that could be observed. Geometry parameters such as the edge distance and the center distance between holes are tested under tensile loading to get the optimum values. Design parameters such as tightening torque of bolt and washer size are also tested under tensile loading, and their effect on the failure load is obtained and studied. The failure modes are investigated under the digital microscope in each case. It is found that the failure load increases by increasing the edge distance, the tightening torque, and decreasing the center distance between holes. The optimum washer size is varied according to the used tightening torque. The optimum obtained failure mode in tests is a mixed - mode of bearing and net tension modes.

IOP Conf. Series: Materials Science and Engineering 1172 (2021) 011001 doi:10.1088/1757-899X/1172/1/011001

ASAT-19

The 19th International Conference on Aerospace Sciences & Aviation Technology (ASAT-19)
April 6-8, 2021, Cairo, Egypt

ABSTRACTS

Water recycling system based on adsorption by activated carbon synthesised from c. verum for space exploration; an estimated design

M N Ettish[1], O Abuzalat[1], G S El-Sayyad[2] and M A Elsayed[1]
1Chemical Engineering Department, Military Technical College, Kobry Elkobbah, Cairo, Egypt.
[2] Drug Microbiology Laboratory, Drug Radiation Research Department, National Center for Radiation Research and Technology (NCRRT), Egyptian Atomic Energy Authority, Cairo,Egypt.
E-mail: mnasr4810@gmail.com

Abstract. Water recycling is a crucial component of space flights. In this study, c.verum, a lowcost agricultural by-product abundant in Egypt, which was not utilized before for the preparation of porous carbons, and its ability for recycling water in space stations was estimated. The prepared samples show high porosity and surface area by physical activation. The influences of the pyrolysis temperature and activation hold-up time on the activated carbon's porosity were studied. The BET surface area and the total pore volume of the prepared carbon were used as the criteria for selecting the optimum preparation parameters. The optimum temperature for pyrolysis was found to be at a temperature of 900°C, hold-up time of two-hour, a nitrogen flow rate of 150 cm3/min, and a heating rate of 10°C/min. However, the optimum activation conditions were at a temperature of 900°C, a CO2 flow rate of 150 cm3/min, a heating rate of 20°C/min, and a hold-up time of 120 min. Equilibrium data is used for fitting to Freundlich, Langmuir, and Temkin isotherms models. The result revealed that the Langmuir model was the finest match for the equilibrium data, with an extreme monolayer adsorption capability of 12.37 mg/g at 25°C. The maximum monolayer adsorption capacity decreased with increasing temperature confirmed the exothermic character of the adsorption interaction.

ASAT-19
The 19th International Conference on Aerospace Sciences & Aviation Technology (ASAT-19)
April 6-8, 2021, Cairo, Egypt

ABSTRACTS

An investigation of plasma sprayed nickel-based and pure aluminum coatings on austenitic stainless steel AISI 304

K. Hamed[1], N. El-Mahallawy[2] and M.O.A. Mokhtar[3]
[1] M.Sc. Student, Design and Production Engineering Department, Faculty of Engineering, Ain Shams University, Egypt
[2] Professor Dr., Design and Production Engineering Department, Faculty of Engineering, Ain Shams University, Egypt
[3] Professor Dr., Department of Mechanical Design and Production, Faculty of Engineering, Cairo University, Egypt
E-mail: 13084@eng.asu.edu.eg

Abstract Atmospheric Plasma Spray (APS) is one of the most leading industrial techniques for protective coating, by improving the performance of parts in the thermal barrier, and wear resistance. Ni-Al alloys are very effective players in the field of design of protective coatings. Accordingly, mixed Al, Ni/Al, and Ni5Al powders were applied on 304stainless steel substrate to develop plasma sprayed coatings. The effect of different compositions on microstructure, microhardness, and porosity was measured. The microstructures of the as-deposited films were characterized utilizing X-ray diffraction (XRD), scanning electron microscopy (SEM), and microhardness measurements. The results showed the formation of two intermetallic compounds, namely: NiAl and Ni3Al. The existence of NiAl is inevitable in all samples, despite the amount of Ni-based alloys in mixtures, or even the atomic percentage of nickel, where the appearance of Ni3Al depends only on increasing the amount of the Ni-based alloy to 50 % percent in mixtures. As regards the steel substrate, the microhardness of the interdiffusion zone of the substrate has been significantly enhanced. Results have shown that the microhardness of the different tested coatings is increased directly with the increment of Ni-based percentage in the coating mixture. The average porosity of the plasma sprayed coatings has proven to be within the normal range.

ASAT-19
The 19th International Conference on Aerospace Sciences & Aviation Technology (ASAT-19)
April 6-8, 2021, Cairo, Egypt

ABSTRACTS

Studying the distribution of hardness values in friction drilled 7075 Al-alloy sheets at different conditions

N Hamzawy [1], MKhedr [1], T S Mahmoud [1], I El-Mahallawi [2] and T A Khalifa [1]
[1] Mechanical Engineering department, Faculty of Engineering at Shoubra, Benha University, Cairo 11629, Egypt.
[2] Department of Metallurgical Engineering, Faculty of Engineering, Cairo University, Cairo 12613, Egypt.
E-mail: Mahmoud.khedr.mk@gmail.com

Abstract. The thickness of sheet metal parts can be locally increased by friction drilling technology via forming of a hole with a bush by a special drilling tool. Here, a 7075 Al-alloy was drilled by friction using tool cone angles with values of 40, 45 and 50° under different feed rates (100, 200 and 315 mm/min) and rotational speeds (1000, 1250 and 1600 rpm). The present study investigates the hardness distribution in the thermally-formed bush and in the heat-affected zone around the bush. It was found that the hardness of the bush was slightly increased with increasing of the tool cone angle and reduction of the tool rotational speed. However, the hardness of the thermally-induced bush showed values lower than the parent metal. The hardness near the drilling surface was approximately 65±10 HV, while it recorded hardness values of 75±10 HV at 5 mm away from the drilling surface. In addition, the microstructure of the friction drilled specimens showed a very fine structure in the drilling zone due to crushing of the original structure during the friction drilling process.

ASAT-19
The 19th International Conference on Aerospace Sciences & Aviation Technology (ASAT-19)
April 6-8, 2021, Cairo, Egypt

ABSTRACTS

Low thermal emissivity coating based on aluminium / acrylic composite coatings

E E Ateia[1], M A Metwaly[1], H R Tantawy[2]
[1] Physics Department, Faculty of Science, Cairo University, Giza, Egypt
[2] Chemical Engineering Department, Military Technical College (M.T.C)
E-mail: cap.mahmoud2011@gmail.com

Abstract. Low emissivity coatings were synthesized by using fine flake Aluminium (Al) powder as a filler within acrylic resin to from the desired composite coatings. In the current work, ball milling was applied to prepare Al with different shapes and sizes. The technical parameters of prepared Al powders with respect to the initial raw Al are characterized by scanning electron microscope (SEM), X-Ray diffraction (XRD), and Energy Dispersive X-Ray Analysis (EDX). Moreover, parameters that affect the emissivity of the coating was investigated; such as coating thickness, particle size, spin coating, ball milling time and the content of coated Al powder. The thermal signature is highly affected by the variation in Al content (5%, 20%, 25%, 35%, 40% and 60%) at different temperatures (50ºC, 70ºC, 90ºC). The results indicate that the perfect percentage for the filler (Al) in the matrix within the range (35 wt. - 40 wt. %) and Al fine flake powder particle, which gives the lowest infrared emissivity of 0.385 μm and 0.412 μm for (3-5) μm, (8-12) μm, respectively.

ASAT-19

The 19th International Conference on Aerospace Sciences & Aviation Technology (ASAT-19)
April 6-8, 2021, Cairo, Egypt

ABSTRACTS

Enhancement of carbon fiber/epoxy composite electrical, optical and thermal properties by using different types of nano-additives.

L Bassiouny[1], T Samir[2], S Abdallah[2], H Ashour[1] and A Anwar[3]
[1] Faculty of Engineering, Misr for Science and Technology, Cairo, Egypt.
[2] Faculty of Engineering, Benha University, Shobra, Cairo, Egypt.
[3] Space Technology Center (STC), Cairo, Egypt.
E-mail: msc.a.anwar@gmail.com.

Abstract. Environmental space threats are becoming more critical as they affect the optical, thermal, and electrical properties of the reinforced fiber polymeric-based materials in spacecraft. Three different Nano-particles Alumina (Al)₂, MO₃ultiwall Carbon Nanotubes (MWCNT), and Reduced Graphene Oxide (RGO) were added to the epoxy matrix and then reinforced by bidirectional carbon fiber plain to form carbon fiber/epoxy by hand lay-up using autoclave curing technique to make three different reinforced materials. In this paper, the electrical, optical, and thermal properties of the carbon fiber/Epoxy Nanocomposite were studied. Fourier transform infrared (FTIR) was performed to evaluate the structural changes in the newly synthesized materials. The optical, thermal, and electrical properties were tested by UV-visible Spectroscopy, Photo-acoustic spectroscopy (PA), and Keithley 2635A respectively. The results showed an enhancement in the electrical, optical, and thermal properties of the epoxy matrix after the addition of Nano-particles. The optical test showed that the neat epoxy and epoxy/Nano-particles absorption spectra were in the infrared range. The thermal test indicated that the three thermal parameters diffusivity, effusivity, and conductivity showed the best enhancement after the addition of MWCNTs. The electrical test pointed out that after the addition of Nano-particles, neat epoxy changed from an insulating material to a semi-conductive material.

IOP Conf. Series: Materials Science and Engineering 1172 (2021) 011001 doi:10.1088/1757-899X/1172/1/011001

ASAT-19

The 19th International Conference on Aerospace Sciences & Aviation Technology (ASAT-19)
April 6-8, 2021, Cairo, Egypt

ABSTRACTS

Non-destructive aerial and terrestrial analysis of uranium isotopic activity by sensitive -spectrometer for El-Sella site

M Sabry[1], A Omar[1], M A Abdel-Rahman[1] and S A El-Mongy[2]
[1] Nuclear Engineering Department, Military Technical College, Kobry El-Koba, Cairo, Egypt
[2] Nuclear and Radiological Regularity Authority (ENRRA), Nasr City, Egypt
E-mail: mohy_sabry@mtc.edu.eg

Abstract. Any comprehensive radioactivity surveillance for a specific site should be carried out by a sensitive airborne system (e.g. carried on aircraft) together with terrestrial sampling and analysis. In the framework of ore, environmental and nuclear materials monitoring and investigation, the activity and isotopic compositions signature are considered crucial factors that must be considered, determined and evaluated. In this work and based on previous aerial monitoring, a very sensitive hyper pure germanium detector (Hp-Ge) of 50% efficiency with the Genie2000 software has been used for spectroscopic non-destructive assay (NDA) of radioactivity content for samples collected from the Egyptian eastern desert (El-Sella Site). The identification of specific signature isotopes through their characteristic gamma lines and a calculation of their specific activities, activity ratios of 235U/238U and 234U/238U and mathematically estimation of the natural enrichment percentage are very vital and nuclear forensic targets. The results of measurements, analysis and calculations show that the Site samples have high radioactivity, natural enrichment origin and high uranium concentration. The obtained results are given tabulated, depicted, discussed and compared with the recent nationally published work and the international levels and limits.

ASAT-19
The 19th International Conference on Aerospace Sciences & Aviation Technology (ASAT-19)
April 6-8, 2021, Cairo, Egypt

Investigation of foamcrete mechanical and physical properties

T. Youssef [1], Y Elsayed [2], T A Elsayed [2] and S M Lawaty [2,3],
[1] Faculty of Engineering, l'Université Française d'Égypte, Cairo, Egypt
[2] Faculty of Engineering (Mataria), Helwan University, Cairo, Egypt
[3] Faculty of Engineering & Technology, Badr University in Cairo, Egypt
E-mail: Tarik.youssef@ufe.edu.eg

ABSTRACTS

Abstract. Engineered Material Arresting Systems (EMAS) serve as a substitutional alternative/solution to airport runways when the Runway Safety Area (RSA) does not meet international Federal Aviation Administration (FAA) standards. The length of the runway can be shortened if an EMAS is installed on both ends of the runway. This paper provides experimental test results on foamcrete material used for such an important safety application. The objective of the paper is to present the required and measured properties of foamed concrete (density, compressive strength and water absorption) and results of an evaluation of a first phase of testing. In this study, a total of fourty eight mixes were conducted; yielding a range of densities, compressive strength(s) and water absorption characteristics that are: 554 to 1528 kg/m3, 1.1 to 21 MPa and 7.4 to 28.3 %, respectively. It is demonstrated herein – though the Analysis of Means statistical method – that foam volume is predominantly the main factor affecting the observed output characteristics. This is followed by Sand/Filler and Filler/Cement that yield marginal effect compare to the former foam volume ingredient.

ASAT-19
The 19th International Conference on Aerospace Sciences & Aviation Technology (ASAT-19)
April 6-8, 2021, Cairo, Egypt

ABSTRACTS

On the unsteady behaviours of the adiabatic endwall film cooling effectiveness

M Qenawy [1], Y Liu [2] andW Zhou [2]
[1] Faculty of Energy Engineering, Aswan University, Aswan 81528, Egypt.
[2] School of Mechanical Engineering, Shanghai Jiao Tong University, Shanghai 200240, China.
E-mail: m_qenawy@aswu.edu.eg.

Abstract. The film cooling technique is introduced in modern gas turbines to protect the blade from the high temperature of the incoming hot gases by forming a thin coolant blank over the blade surface. However, it is known as a jet in crossflow (JICF), where coolant and mainstream interact intensively and generate complex vortices leading to highly unsteady coolant coverage over the blades surface. In this study, a fast-response pressure-sensitive paint technique (fast-PSP) was used to measure the coolant unsteadiness with a high-resolution camera. The measurements were performed in a novel single-passage transonic wind tunnel to uncover the unsteady effectiveness of the endwall surface. Such effectiveness was dramatically influenced by the blowing ratios (M), showing attached flow at a low blowing ratio and lift-off at a high blowing ratio. The effectiveness was asymmetrically distributed due to the pressure gradients, jet compounding angle, and associated complex flows. The unsteady effectiveness was highly influenced by the energetic vortical structures, which interacted with the mainstream flow immediately behind the holes. It was featured by secondary structures (horseshoe, passage, and counter vortices) beside the JICF structures. Meanwhile, the unsteadiness was originated from the middle of the passage behind the holes. It is suggested to pay close attention to the locations of the holes for further optimization. This study could help the designers to understand the characteristics of unsteady effectiveness, promoting advanced cooling strategies for enhanced protection of future gas turbines.

ASAT-19
The 19th International Conference on Aerospace Sciences & Aviation Technology (ASAT-19)
April 6-8, 2021, Cairo, Egypt

ABSTRACTS

Enhancing the propulsion characteristics of rockets by adding the energetic Nitro-hydroxyl-terminated polybutadiene (NHTPB) in the propellant compositions

A Azazy[1], A Saleh[1], W F Aly[2], A Elbeih[2], A K Hussein[2], T Elshenawy[3], M Abdelhafiz[2], T Z Wafy[2], M G. Zaki[2] and H S Ahmed[1]
[1] Science and Technology Center of Excellence, Military Production, Cairo, Egypt
[2] Military Technical College, Kobry Elkobbah, Cairo, Egypt
[3] Technical Research Center, Cairo, Egypt.
E-mail: asaleh@stce-egypt.org

Abstract: Replacing the inert binder by an energetic one could increase the specific impulse of the propellants and enhance the propulsion characteristics of rockets. In this study, Nitrohydroxyl- terminated polybutadiene (NHTPB) was prepared by a simple method. The prepared NHTPB in addition to HTPB binder were characterized. FTIR spectra of both HTBP and NHTPB was determined and compared. The thermal behavior of the prepared NHTPB was studied using DSC technique at heating rate 5 degree/min. A composite propellant based on AP/NHTPB was prepared and the specific impulse was measured for AP/NHTPB using two inch motor. It was concluded that the energetic nitro-hydroxyl-terminated polybutadiene has a clear max. exothermic peak at 203 0C with heat release of 323 J/g. By comparing the results, the prepared propellant AP/NHTPB has specific impulse higher than the traditional AP/HTPB propellant. NHTPB is a promising binder for the application of rocket propellants and needs more tests for its approval.

IOP Conf. Series: Materials Science and Engineering 1172 (2021) 011001 doi:10.1088/1757-899X/1172/1/011001

ASAT-19
The 19th International Conference on Aerospace Sciences & Aviation Technology (ASAT-19)
April 6-8, 2021, Cairo, Egypt

ABSTRACTS

Acoustic threat detection and direction finding system

A Abdallah[1] and H Tork[1]
[1] Naval Warning Engineering Department, Military Technical College, Cairo,Egypt
Email: adel.abdallah@mtc.edu.eg

Abstract. In this paper, an acoustic threat detection system is investigated as preliminary study of one of the anti-terrorism systems. The proposed system is a prototype of acoustic threats detection, classification, and direction finding system. For this purpose, an experiment is performed to test the capability of a proposed warning system to perform the task accurately. The direction of the sound source is determined by means of two microphones. The angle to the acoustic threat is calculated based on time difference of arrival (TDOA). The limitations of the system are discussed and some possible solutions are introduced to improve the performance of the system.

ASAT-19
The 19th International Conference on Aerospace Sciences & Aviation Technology (ASAT-19)
April 6-8, 2021, Cairo, Egypt

ABSTRACTS

Tactics Overview for Implementing High-Performance Computing on Embedded Platforms

A Elshazly [1], A Elliethy [1] and M A Elshafey [1]
[1]Department of Computer Engineering and Articial Intelligence, Military Technical College, Cairo, Egypt
E-mail: ahmedelshazly@mtc.edu.eg

Abstract. Future space missions will rely on novel high-performance computing to support advanced intelligent on-board algorithms with substantial workloads that mandates rm realtime and power constraints requirements. Consequently, these advanced algorithms require signicantly faster processing beyond the conventional space-grade central processing unit capabilities. Moreover, they require careful selection of the target embedded platform from a diverse set of available architectures along with several implementation tactics to map the algorithms to the target architecture to fully unlock its capabilities. In this paper, we present a study of dierent architectures and embedded computing platforms for the satellite on-board computers. Moreover, we present a comprehensive overview of recent implementation tactics such as source code mapping and transformations. Additionally, we highlight some optimization techniques such as partitioning and co-designing using hardware accelerators. Finally, we discuss several implementation analysis methodologies to derive optimized code implementations. The top ranked YOLO-v3, as a deep learning based object detection algorithm, is selected as a case study model to be optimized using OpenVINO toolkit. The experimental results show an improvement ratios up to 73%, 41%, and 34% in terms of frames per second, CPU utilization, and cache memory, respectively. The study presented in this paper aims to guide the researchers in the eld of high performance embedded computing in terms of dierent hardware architectures along with several implementation tactics.

ASAT-19
The 19th International Conference on Aerospace Sciences & Aviation Technology (ASAT-19)
April 6-8, 2021, Cairo, Egypt

ABSTRACTS

Modelling and simulation of two axes gimbal fuzzy PI stabilization system in the presence of feedback sensors noise

A Fathi [1], A M Kamel[1], A O Elfarouk[1], Y Elhalwagy[1]
[1]Military Technical College, Cairo, Egypt.
E-mail: ahmedfathy111980@gmail.com

Abstract. The missile guidance in terminal phase is an optimization problem as the miss distance should be minimized. The optimization of missile miss distance is highly affected by the missile seeker performance. Design of gimbal control system suffers always from feedback sensors noise which leads to system instability. In this paper, a promising design of the fuzzy PID controller for a missile seeker gimbal is proposed considering the feedback sensor noise with practical gyro transfer function calculated based on real experimental measurements utilizing MATLB system identification toolbox. Also, the mathematical model of two stabilized axes gimbal. Stabilization is achieved considering the missile motion parameters such as rates, torques and coupling between yaw and roll channels. A Matlab simulation is carried out for evaluating the proposed system modelling and to test the robustness of the fuzzy PID controller in the presence of feedback sensor noise. A comparative analysis with PI based controller is conducted to evaluate the performance of the proposed controller which presents sufficient enhancement to the missile gimbal stability parameters.

ASAT-19
The 19th International Conference on Aerospace Sciences & Aviation Technology (ASAT-19)
April 6-8, 2021, Cairo, Egypt

ABSTRACTS

Performance comparison among popular implementations of H.264 encoders

H Y El-Arsh, A S Elliethy, A M Abdelaziz and H A Aly
Dept. of Computer Engineering, Military Technical College, Cairo, Egypt, 11766
E-mail: hassan.yakout@ymail.com

Abstract. Remote sensing videos captured by Unmanned Aerial Vehicle (UAV) air-born highresolution cameras require an ecient compression scheme that preserves the details of the visual contents of the videos while reducing the total size of the data to be managed in real-time. This paper presents a detailed comparison between dierent open-source implementations for the H.264 video compression scheme. While the high-resolution videos allow analysts to extract more descriptive interpretations and draw more conclusive results, the increase in the consequent data size consumes more storage, resulting in more channel bandwidth, more power, and encounters an extra delay in transmission time. An ecient implementation of video compression can alleviate these large data size eects. In this paper, we analyze and compare the JM-encoder, the X264, the FFmpeg, and Cisco's OpenH264 open-source implementations in terms of compression eciency, video quality, and computational load. Moreover, we present the rate-distortion curves in terms of PSNR as a quality metric against the bit-rate for a combination of 20 videos with various resolutions and dynamic contents. Albeit H.64 is superseded by H.265, till now H.264 is used in more than 65% of video coding applications. For example, YouTube only allows H.264 for live streaming.

IOP Conf. Series: Materials Science and Engineering 1172 (2021) 011001 doi:10.1088/1757-899X/1172/1/011001

ASAT-19
The 19th International Conference on Aerospace Sciences & Aviation Technology (ASAT-19)
April 6-8, 2021, Cairo, Egypt

Modelling and transient response study of hydraulic servo actuator

E E Ibrahim[1], T Elnady[1] ,I Saleh[1] and S Hassan[1]
[1]Department of Mechanical Power & Energy, Military technical college, Cairo, Egypt
E-mail: ezzbos@gmail.com

ABSTRACTS

Abstract. This paper presents a mathematical model of electro hydraulic servo system (EHS) using two-stage electrohydraulic servovalve (EHSV), by the aim of Matlab-simulink-simscape multibody to predict the flow characteristics (pressure/flow rate) with changing the parametric configuration of (EHSV), these parameters could be used to estimate the error at different operating conditions. It includes establishing mathematical model, controller model and validating the performance of the (EHSV) by experimental result of published work using the data of servovalve labelled B.31.210.12.1000.U2V (mechanical feedback) manufactured by PPT – Trstenik. The effect of changing servovalve orifice diameter (0.25, 0.30, and 0.35) mm, show that increasing the orifice diameter of the servovalve leads to increase the transient time but decrease the system overshoot. The pressure behaviour is plotted and shows that increasing the orifice diameter leads to increasing the pressure inside the system as to reach the relieve pressure.

ASAT-19
The 19th International Conference on Aerospace Sciences & Aviation Technology (ASAT-19)
April 6-8, 2021, Cairo, Egypt

Modeling, simulation and controller design for a typical bent axis electrohydraulic servo motor

M A Yousef[1], M El-Sanabawy[2], G Rabie[3] and R A Rateb[4]
[1] Department of Rockets, Military Technical College, Cairo, Egypt.
[2] Department of Rockets, Military Technical College, Cairo, Egypt.
[3] Modern Academy for Engineering and Technology, Cairo, Egypt.
[4] Egyptian Armed Forces, Cairo, Egypt.
E-mail: mohamed_abdelwahab@mtc.edu.eg

ABSTRACTS

Abstract. Bent axis electrohydraulic servo motors are one of the Electrohydraulic Servo Motors (EHSMs) family which are used in high frequency, speed, and precision applications such as aerospace applications as well as many military weapon system applications. Therefore, it is essential to understand how such motors affect the whole performance of the system which makes modeling of these motors is an important task to be achieved. Once an accurate model is obtained, an optimum controller can be applied. This paper introduces detailed mathematical modeling of a typical bent axis (EHSM). MATLAB SIMULINK package is used to simulate and control such (EHSM) using PID controller. The PID controller gains were tuned using PD-PI controller to obtain the precision response for the (EHSM). The validity of mathematical modeling was reviewed through some practical experiments.

ASAT-19
The 19th International Conference on Aerospace Sciences & Aviation Technology (ASAT-19)
April 6-8, 2021, Cairo, Egypt

ABSTRACTS

Obstacle avoidance for multi-UAV path planning based on particle swarm optimization

E N Mobarez [1], A Sarhan [2], and M M Ashry [1],
[1]Department of Optoelectronics and Control,
[2]Aircraft electrical equipment's, Military Technical College, Cairo, Egypt.
E-mail: mmaashry@mt c.edu.eg

Abstract. Traffic Collision Avoidance System aims to help aircraft to avoid collision with any object or other aircraft .One of the functions of this system is that it avoids threatening UAV to collide, it also addresses each threat separately with the best collision avoidance and the best suitable horizontal separation with other aircraft in the optimal path. In this paper the flight path planning for UAVs was designed to avoid obstacles depending on how the particle swarm was improved. Optimization problems are improved by using swarm dynamics (evolutionary computational technology). This is by describing avoiding obstacles and adapt the path planning for UAVs. The concept of concurrent restructuring has been integrated into path planning to stay away from both static obstacles. This optimization technique designed to decrease processing time and the shortest route of the path planning.

ASAT-19
The 19th International Conference on Aerospace Sciences & Aviation Technology (ASAT-19)
April 6-8, 2021, Cairo, Egypt

ABSTRACTS

PID tuning approaches for quadrotors unmanned aerial vehicles

M Mahfouz [1], A Taiomour [2], M M Ashry [1] and G Elnashar [1]
[1]Department of Optoelectronics and Control, Military Technical College,
[2]Aircraft Armament Department, Military Technical College,
E-mail: m.mahfouz.trc@gmail.com,

Abstract. In this paper, a comparative study between different PID tunning techniques is presented. The proposed techniques are applied to solve the formation configuration problem for a cooperative team of unmanned vehicles. The formation problem for the cooperative team is divided into two levels of control, one is the backstepping control technique for the stabilization of the team members positions as a higher controller. Simultaneously, PID controller receives the desired position to stabilize the attitude control as a lower controller to track the desired planning trajectories. The main contribution of this paper is the comparison between the different control approaches in tunning the PID gains to stabilize attitude control for the leader quadrotor. Simulation results present the assessment of the proposed PID control technique compared with different PID tuning approaches such as local optimal control, fraction order, Ziegler-Nichols and genetic algorithm. Moreover, disturbance rejection and white noise attenuation criterions are inspected to evaluate the ability of the proposed controllers to preserve the stability of the system.

IOP Conf. Series: Materials Science and Engineering 1172 (2021) 011001 doi:10.1088/1757-899X/1172/1/011001

ASAT-19
The 19th International Conference on Aerospace Sciences & Aviation Technology (ASAT-19)
April 6-8, 2021, Cairo, Egypt

Formation configuration of unmanned cooperative quadrotors via PID tuning approaches

M Mahfouz [1], A Taiomour [2], M M Ashry [1]and G Elnashar [1]
[1]Department of Optoelectronics and Control, Military Technical College,
[2]Aircraft Armament Department, Military Technical College,
E-mail: m.mahfouz.trc@gmail.com,

Abstract. Formation configuration is one of the major intrinsic strategies used in cooperative Unmanned Air Vehicles field. In this paper, Backstepping-PID control technique for cooperative quadrotors unmanned aerial vehicles are developed to solve the formation problem. The proposed controller is divided into couple of parts working together. Backstepping controller is used to stabilize the position control as a higher controller. Simultaneously, PID controller receives the desired position to stabilize the attitude control as a lower controller to track the desired planning trajectories. The main contribution of this paper is using Fraction Order Approach, and Local Optimal Approach to refine the PID lower controller gains. The tunning of the PID gains through the proposed PID tuning approaches guarantee the stabilization of the attitude control for all the team members. Simulation results present the success of the proposed PID tuning approaches in solving the formation problem for cooperative unmanned quadrotors tracking a desired path. Moreover, the simulation results present the ability of the proposed approaches to handle disturbance rejection and noise attenuation while preserving the stability of the system.

ASAT-19
The 19th International Conference on Aerospace Sciences & Aviation Technology (ASAT-19)
April 6-8, 2021, Cairo, Egypt

ABSTRACTS

Impact of nozzle profile on ballistic performance and structural loads

M Saleh[1], S Saleh[1], A Zayed[2] and M Y M Ahmed[1]
[1] Department of Rockets, Military Technical College, Cairo, Egypt.
[2] Nahda University, Egypt.
E-mail: saleh_uchiha@ymail.com

Abstract. A nozzle is a device that is designed to regulate the direction and characteristics of the combustion gas products of jet engines. So, the nozzle performance has a significant impact on the mission achievement. This paper is concerned with the internal ballistics of the nozzle aiming to estimate pressure and thermal loads on its walls. Computational fluid dynamics is applied to analyse the effect of changing nozzle internal profile on the resulting thrust, flow energy losses, and nozzle wall structure. Area ratios at the inlet, critical, and exit sections are considered as constraints for the examined design. Two different sets with 4 different profiles for each are investigated. The results show thrust, entropy losses across the nozzle and the static pressure and temperature at nozzle wall. Bell shape profiles produce better performance compared to other profiles. Changing the internal profile of the nozzle causes significant change in pressure and temperature loads acting on nozzle wall structure.

ASAT-19
The 19th International Conference on Aerospace Sciences & Aviation Technology (ASAT-19)
April 6-8, 2021, Cairo, Egypt

ABSTRACTS

Design, test and build of a monopropellant thruster using 85% hydrogen peroxide

Ah Deif[1], Ah El-S Makled[2], Mohamed K Khalil[3], and Y Elshaar[3]
[1] Space Technology Centre, Cairo, Egypt.
[2] Faculty of Engineering, Zagazig University, Zagazig, Egypt.
[3] Military Technical College, Cairo, Egypt.
E-mail: Deifmoh@yahoo.com

Abstract. The objective of this research is to design, build, and test about a 5N Hydrogen Peroxide(H2O2) monopropellant thruster (MPT). It is utilized in remote sensing satellites for attitude control and orbit manoeuvres. The MPT uses high test peroxide (HTP) of 85 % concentration. Firstly, H2O2 ≈85% concertation by weight is prepared in the laboratory. A distillation and filtration units are built. The distillation and filtration processes are performed. Next, the design of the monopropellant thruster is done based on the developed mathematical model using NASA CEA rocket performance code. The test facility is developed which consists of the thruster, the feeding system, static test stand and data acquisition system with measuring sensors. An experimental test stand is designed and fabricated with Pendulum thrust mechanism for measurements of thrust. Finally, the silver catalyst is prepared and packed inside the MPT chamber where silver screens of high purity 99.96 % are used. The 10-firing tests are conducted under atmospheric conditions. The firings performed without heating are not completely successful. The analysis of the results shows that the thruster has a thrust range from 3.8-4.2 N. The performance of thruster starts to decay after consuming 6 kg of stabilized H2O2. The specific impulse (Is) is evaluated to be ≈ 93-97s at decomposition pressure of ≈10 bars and mass flow rate ≈4.18 g/s. The performance evaluation is judged to be successful. However, using the whole potential of the 85 % concentrated H2O2, is expected to increase (Is) up to ≈111.5s.

ASAT-19
The 19th International Conference on Aerospace Sciences & Aviation Technology (ASAT-19)
April 6-8, 2021, Cairo, Egypt

Parametric study of solid propellant slotted grain

M El-Naggar[1], H Belal[2] and H M Abdalla[3]
[1] M.Sc. Student, Rocket Department, Military Technical College, Egypt
[2] Independent Researcher, AIAA Member, Egypt
[3] Associate Professor. Rocket Department, Military Technical College, Egypt
Email: khatab80@hotmail.com

ABSTRACTS

Abstract. One of the goals in solid rocket motor design is to have as large volumetric loading as possible keeping the basic requirements unaffected. Slotted grain can achieve this goal as it has the advantages of sliver-free and no stress-concentration regions that occur in other internal burning grains as star grain and wagon wheel grain. It has the disadvantage of exposing the motor wall to hot gases. In this paper, the geometry of slotted grain is discussed and the effect of design parameters (e.g., number of slots, dimensions of the slot, etc.) of slotted grain on grain burn back is explained. Also, a comparison between results and experimental data is performed.

ASAT-19 2021 IOP Publishing
IOP Conf. Series: Materials Science and Engineering 1172 (2021) 011001 doi:10.1088/1757-899X/1172/1/011001

ABSTRACTS

Experimental investigation of recirculation zone characteristics of a pre-filming airblast injector

M N Darwish[1], K P Shanmugadas[2] and S R Chakravarthy[2]
[1] Aerospace Engineering Department, University of Science and Technology at Zewail City, Giza, Egypt
[2] National Centre for Combustion Research and Development & Department of Aerospace Engineering, Indian Institute of Technology Madras, Chennai, India
Email: s-mohammednady@zewailcity.edu.eg

Abstract. The toroidal flow at the recirculation zone has a vital role in combustion process as it helps in mixing hot combustion products with incoming fresh air and fuel which increases combustion efficiency. In the present work, characteristics of recirculation zone are investigated using a pre-filming airblast injector. Particle Image Velocimetry is used to characterize the swirl flow field generated by the airblast injector. Moreover, olive oil is used as a tracer to be captured by a high-speed camera. Different flow rates are investigated in order for finding out the effect of varying flow rates on the characteristics of the recirculation zone. Results for recirculation zone shape, velocity field and shear strength at the primary zone of combustion are represented. Additionally, results show that recirculation zone is almost symmetrical with increasing trend in shear strength with increasing flow rates.

ASAT-19

The 19th International Conference on Aerospace Sciences & Aviation Technology (ASAT-19)
April 6-8, 2021, Cairo, Egypt

REVIEWERS LIST

Reviewer Name	Affiliation
Abalola, Adel	Prof. o Electric Power Engineering Department, Monufia University, Engineering College
Abdalla, Ahmed	Prof. of Electrical Eng., MTC, Cairo, Egypt.
Abdel Latif Soliman, Mohamed Samir	Assoc. Prof. of Electrical Eng., MTC, Cairo, Egypt.
Abdel-Rahman, Mohamed	Prof. Nuclear Engineering Department, MTC, Cairo, Egypt, Military Technical College
Abdelrahman, Mostafa	Egyptian atomic energy authority Cairo University
Abdo, Mohamed	Prof. of Physics Engineering Science, MTC
Aboazm, Ahmed	Prof. at Navy Engineering Branch, MTC, Cairo, Egypt
Aboelfotoh, Mohamed	Head of Chemical Warfare MTC, Dept.
Abozeid, Mohammed	Prof. Navigation, Guidance, and Control, MTC, Cairo, Egypt.
Ahmed, Hesham	Prof. of Elect. Eng. , MTC, Cairo, Egypt
Allam, Mohamed	Prof. of Rocket Mechanical, MTC
Almazy, Khaled	Prof. at Mechanical Eng. Branch, MTC, Cairo, Egypt
Arafa, Ibrahim	Prof. Guidance, Navigation and Control
Ashry, Amr	Prof. of Communications and Networks, MTC
Ashry, Mahmoud	Prof. Automatic Control , MTC, Cairo, Egypt.
Atef Kamel, Mohamed	Prof. of Mechanical Eng., MTC, Cairo, Egypt.
Atef, Mohamed	Automotive Engineering, MTC, Cairo, Egypt.

IOP Conf. Series: Materials Science and Engineering 1172 (2021) 011001 doi:10.1088/1757-899X/1172/1/011001

ASAT-19

The 19th International Conference on Aerospace Sciences & Aviation Technology (ASAT-19)
April 6-8, 2021, Cairo, Egypt

REVIEWERS LIST

Reviewer Name	Affiliation
Ateya, Ibrahim	Prof. at A/C Mechanical Dep. MTC, CA, Egy.
Atta, Mohamed	Prof. Communication Signal Analysis , MTC, Cairo, Egypt.
Azouz, Ahmed	Prof. SAR Digital Signal Processing , MTC, Cairo, Egypt.
Baraka, Ahmed	Head of Chemical Branch, MTC
Basuney, Sherif	Prof. Chemical Engineering, MTC
Bazan, Taher	Assoc. Prof. of Optoelectronics, MTC, CA, EG.
Bennay, Mohamed	Prof. of Mechanical Eng., MTC, CA, EG.
Dahshan, Hisham	Prof. Cryptography and Networks Security, MTC, Cai-ro, Egypt
Desouky, Ashraf	Prof. Navigation , MTC, Cairo, Egypt.
El_Khatib, Mohamed	Prof. of Electrical Eng., MTC, Cairo, Eg.
El-Dalil, Omar	Communication Engineering Alexandria University, Al-exandria, Egypt
Elmahihy, Ali	Prof. Mechanical Power, MTC, Cairo, Egypt
Elmahy, Wesam	Astronomy and space science Department – Faculty of science, Cairo University
Elmasry, Mohamed	Material science and technology , MTC, Cairo, Egypt.
Elmayyah, Wael	Mechatronics Engineering, MTC, Cairo, Egypt.
Elrewainy, Ahmed	Prof. of statistical Signal Processing, Avionics Depart-ment, MTC, Cairo, Egypt.
El-Sayyad, Gharieb	Egyptian Atomic Energy Authority
Elshafey, Mohamed	Faculty Member in Computer Engineering Dept., MTC,-Cairo, Egypt

IOP Conf. Series: Materials Science and Engineering 1172 (2021) 011001 doi:10.1088/1757-899X/1172/1/011001

ASAT-19

The 19th International Conference on Aerospace Sciences & Aviation Technology (ASAT-19)
April 6-8, 2021, Cairo, Egypt

REVIEWERS LIST

Reviewer Name	Affiliation
Elsherif, Ahmed	Engineering Mathematics, MTC, Cairo, Egypt.
Eltahlawy, Mohammed	Prof. Laser Physics
Eltalabawy, Waleed	Prof. material science and engineering, MTC, Cairo, Egypt.
Esmail, Ahmed	Engineering Mathematics, MTC, Cairo, Egypt.
Fathallah, Elsayed	Department of Civil Engineering, Military Technical College, Cairo, Egypt.
Foad, Mohamed	Prof. at Computer Eng. Dept., MTC, Cairo, Egypt.
Gobara, Mohamed	Prof. Material Properties, MTC, Cairo, Egypt
Hanna, Waleed	Material science and technology , MTC, Cairo, Egypt.
Hegazy, Shawky	Automotive Dept., MTC, Cairo, Egypt.
Hendy, Hossam	Prof. of Satellites, Attitude Determination and control system, MTC, Cairo, Egypt
Hussein, Ahmed	Prof. at Department of Mechanical Engineering, College of Engineering, University of Babylon, Babylon City, Hilla Republic of Iraq
Ibrahim, Mohamed	Dr., Electrical Eng. Dept., UAV Automation
Ismail, Gouda	Prof. of Computer Engineering
Kamal, Ashraf	PhD
Kamal, Osama	Prof. of Mechanical Engineering Rocket Applications
Kamel, Ahmed	Prof. Navigation, Guidance, and Control, MTC, Cairo, Egypt.
Kamel, Hazem	MTC Deputy Commandant for PGS & Researches
Kamel, Hisham	Automotive department, MTC
Kassem, Mohammed	Department of Aircraft Mechanics, Military Technical College, Cairo, Egypt

IOP Conf. Series: Materials Science and Engineering 1172 (2021) 011001 doi:10.1088/1757-899X/1172/1/011001

ASAT-19

The 19th International Conference on Aerospace Sciences & Aviation Technology (ASAT-19)
April 6-8, 2021, Cairo, Egypt

REVIEWERS LIST

Reviewer Name	Affiliation
Khalil, Mohamed	Prof. of Aerospace, MTC, Cairo, Egypt
Kotb, Wael	Automotive Engineering, MTC, Cairo, Egypt.
Magdy, Mohamed	Military Technical College
Mahr, Mahmoud	Prof. of Material Science, MTC, CA, EG.
Mashaly, Ahmed	Prof. Elect. Eng. ACB. Dept.
Mekkawy, Tamer	Prof. of Elect. Eng, MTC, Cairo, Egypt
Mokhtar, AYMAN	Prof. of Optoelectronics, MTC
Morsy, Sherif	Prof. of Computer Eng., MTC, CA, EG.
Mostafa, Hosamm	Prof. at Chemical Eng. Branch, MTC, Cairo, egypt
Nabiel, Tamer	Prof. Optical Wireless Communication , MTC, Cairo, Egypt.
Nagy, Amr	Cairo
nemnem, Ahmed	Prof. A/C Mechanical Dept., MTC, CA, Egy.
Ola, Mohamed	Mechatronics Engineering, MTC, Cairo, Egypt.
Omer, Adel	Prof. of dynamics, MTC, Cairo, Egypt.
Othman, Mohamed	Material science and technology , MTC, Cairo, Egypt.
Paravan, Christian	Space Propulsion Laboratory, Aerospace Science and Technology Dept., Politecnico di Milano
Ragheb, Hossam	Automotive Dept., MTC, Cairo, Egypt.
Raid, Ahmed	Prof. at Mechanical Engineering Branch, MTC, Cairo, Egypt
Rashad, Ayman	Dean of Faculty of Engineering at Shoubra
Rateb, Ragee	Assoc. Prof. Rocket Propulsion
Saaed, Tamer	Mechanical Engineering Branch, MTC, Cairo, Egypt
Safwat, Ehab	Head of Armament Navy Dept.

IOP Conf. Series: Materials Science and Engineering 1172 (2021) 011001 doi:10.1088/1757-899X/1172/1/011001

ASAT-19

The 19th International Conference on Aerospace Sciences & Aviation Technology (ASAT-19)
April 6-8, 2021, Cairo, Egypt

REVIEWERS LIST

Reviewer Name	Affiliation
Sanad, Ibrahim	Dr, EE, MTC, Cairo, Egypt
Shamekh, Mohamed	Prof. Science and Technology of Materials , MTC, Cairo, /Egypt
Sharaf, Alhossein	Prof., Automotive Dept., MTC, Cairo, Egypt.
shokry, khaled	PhD, Chemical Engineering,
Tork, HOSSAM	Head of Navy warning Dept.
Yacoub, Mustafa	Prof., Automotive Dept., MTC, Cairo, Egypt.
yahia, Mahmoud	Prof. of Rocket Mechanics
Zakria, Mohamed	Prof. Aerospace Mechanical Eng., MTC, Cairo, Egypt

IOP Conf. Series: Materials Science and Engineering 1172 (2021) 011001 doi:10.1088/1757-899X/1172/1/011001

ASAT-19

The 19th International Conference on Aerospace Sciences & Aviation Technology (ASAT-19)
April 6-8, 2021, Cairo, Egypt

AUTHORS INDEX

Author	Paper-ID	Author	Paper-ID
A Magdy	026_ASAT19	A F Tammam	008_ASAT19
A Mohamed	006_ASAT19	A Fathi	035_ASAT19
A A Emara	017_ASAT19		
A A Elsherbiny	001_ASAT19	A H Maamoun	022_ASAT19
A A Ezz	020_ASAT19	A Hassan	006_ASAT19
A A Mashaly	012_ASAT19		
A A Omer	006_ASAT19	A M Abdelaziz	036_ASAT19
A Abd El-Rehim	019_ASAT19		
A Abdallah	033_ASAT19		
A Alharthi	001_ASAT19		
A Amein	016_ASAT19	A M Kamel	035_ASAT19
A Anwar	028_ASAT19	A M Morad	007_ASAT19
A Azazy	032_ASAT19	A M Sofy	010_ASAT19
A Azouz	012_ASAT19	A M Youssef	004_ASAT19
A Bayoumi	017_ASAT19	A Magdy	003_ASAT19
A D Elbayoumy	011_ASAT19		
A Deif	043_ASAT19	A O Elfarouk	035_ASAT19
A E Abdalla	012_ASAT19	A Omar	029_ASAT19
A E Hussien	004_ASAT19		
		A S Elliethy	036_ASAT19
A Elbeih	032_ASAT19	A S Elsharkawy	014_ASAT19
A Elliethy_y	034_ASAT19	A S Shobary	014_ASAT19
A El-S Makled	043_ASAT19	A Saleh	013_ASAT19
A Elsabbagh	021_ASAT19	A Sarhan	039_ASAT19
A Elsharkawy	015_ASAT19	A Taiomour	040_ASAT19
A Elshazly	034_ASAT19		

ASAT-19

The 19th International Conference on Aerospace Sciences & Aviation Technology (ASAT-19)
April 6-8, 2021, Cairo, Egypt

AUTHORS INDEX

IOP Conf. Series: Materials Science and Engineering 1172 (2021) 011001 doi:10.1088/1757-899X/1172/1/011001

ASAT-19

The 19th International Conference on Aerospace Sciences & Aviation Technology (ASAT-19)
April 6-8, 2021, Cairo, Egypt

AUTHORS INDEX

ASAT-19

The 19th International Conference on Aerospace Sciences & Aviation Technology (ASAT-19)
April 6-8, 2021, Cairo, Egypt

AUTHORS INDEX

IOP Conf. Series: Materials Science and Engineering 1172 (2021) 011002 doi:10.1088/1757-899X/1172/1/011002

Peer review declaration

All conference organisers/editors are required to declare details about their peer review. Therefore, please provide the following information:

- **Type of peer review: Double-blind**

  ASAT 19 2021 uses double-blind review, the process is as follows:

  - In the title page the authors details (names and affiliations) are deleted.
  - Any refer to previous work should be in a passive way with replacement of authors names with "Anonymous" word.
  - Remove references to funding sources
  - Do not include acknowledgments

- **Conference submission management system:**

  Electronic Submission of a paper to ASAT-19 conference only requires the following steps:

  - ***Registration and/or login:*** The first time you use an online system "http://asat.conferences.ekb.eg/", you must register for an account. You will need your login information each time you return to the site.
  - *Entering manuscript information:* When submitting your paper in IOP style, the author must provide some basic information, such as title, authors, affiliations, abstract, etc.
  - ***Reviewing and submitting the paper in*** IOP style with a copy considering the peer-review instructions

- **Number of submissions received: 135**

- **Number of submissions sent for review: 120**

- **Number of submissions accepted: 46**

- **Acceptance Rate (Number of Submissions Accepted / Number of Submissions Received X 100): 38.5%**

- **Average number of reviews per paper: 2**

- **Total number of reviewers involved: 86**

Content from this work may be used under the terms of the Creative Commons Attribution 3.0 licence. Any further distribution of this work must maintain attribution to the author(s) and the title of the work, journal citation and DOI.
Published under licence by IOP Publishing Ltd

- **Any additional info on review process: NON**
- **Contact person for queries:**

Prof. Ahmed E. Abdalla
Conference rapporteur
Email: asat@mtc.edu.eg

Dimensional structural mass optimization of forged steel connecting rod for aircraft piston engine

M Helal[1,2], B F Felemban[1], A Alharthi[1], S Almalki[1], E Fathallah[3,4], A A Elsherbiny[2] and H Eleashy[5]

[1]Department of Mechanical Engineering, College of Engineering, Taif University, Taif 21974
[2]Production and Mechanical Design Department, Faculty of Engineering, Mansoura University, Mansoura 35516, Egypt
[3]Civil Engineering Department, M.T.C. Kobry Elkobba - Cairo, Egypt
[4]Ships and Submarines Engineering Department, Military Technical College, Cairo, Egypt.
[5]Mechanical Engineering Department, Faculty of Engineering & Technology, Future University in Egypt, Cairo, Egypt

E-mail: saidhabib2000@hotmail.com

Abstract. The connecting rod is an important component of the engine. It conveys the kinetic energy from the piston to the crankshaft. All cars and aircraft engines contain at least one connecting rod, which differs from one motor to another in terms of length, size and shape. Hence, it is subjected to massive alternating load. This research aims to improve the connecting rod design by reducing its mass without sacrificing durability and safety especially for aircraft applications. Therefore, a static stress analysis is carried out on forged steel connecting rod using ANSYS APDL. Geometric modelling of the connecting rod was created using ANSYS APDL. Additionally, von-Mises stress and strain, principal stresses and strains, shear stress and the deflation results of the connecting rod are investigated. The results showed a great opportunity for mass weight reduction. Thus, a dimensional structural mass optimization was performed. The optimization results were promising, which reduced the mass by 55.13% (in the tensile case) and 56.7% (in the compression case) from the initial design. Therefore, the efficiency of aircraft engine can be maximized.

1. Introduction:

One of the most critical parts of the internal combustion engines is the connecting rod which connects the piston to the crankshaft. In general, connecting rods are made of steel for vehicles and aircraft engines. But, it can be made from different materials such as Titanium for high-performance engines, or Aluminum (for the ability to absorb high impact and lightness at the expense of durability) or Cast iron for applications such as motor scooters. The connecting rod is under reciprocating load and due to this is under an incredible tensile and compression stress at every rotation. One of the most common engine failure cases is the connecting rod failure, which can result from a physical defect in the rod, lubrication, or bolt defects. Often these failures happen at competitive automobile events. But, not on production cars during normal daily driving, which rarely occurs, due to the greater safety factor in parts production, and quality control is often more systematic [1, 2]. The demand for robust and alternative materials has

Content from this work may be used under the terms of the Creative Commons Attribution 3.0 licence. Any further distribution of this work must maintain attribution to the author(s) and the title of the work, journal citation and DOI.
Published under licence by IOP Publishing Ltd

grown in the industry fields. A technique of hybridization involves a combination of two or more reinforcements to obtain better mechanical properties, such as strength, stiffness, higher strength to weight ratio, etc. And by combining two or more low-cost materials with different properties, the hybrid materials are formed. These advanced materials are used for high-strength; light-weight applications in automobiles, aerospace, etc. several techniques are used to create hybrid metal matrix composites such as stir cum squeeze, casting process, and stir casting process [3].

In the early years, connecting rod optimization has started. Consumers are looking for more robust and safer products at a reasonable price. However, the industry is looking for less time and cost production. The best values for these variables that achieve all these constraints can be determined by the optimization process, making it essential and primarily in the automotive industry. The car performance is affected by the design and weight of the connecting rod. Thus, it also affects car manufacture credibility. By considering the weight reduction structural factors during optimization, the component can be lighter and maintain higher strength, which will create striking success in the automotive and manufacturing industry. The benefits of connecting rod optimization go back to the consumers themselves. Among the key goals are to boost the engine's efficiency and enhance the product that guarantees human protection.[4]. The connecting rod can bear both static and dynamic loads. The compression and tensile stresses on the connecting rod are generated respectively by the pressures exerted by the combustion gases and the components of the inertial movement [5, 6]. In case of faulty connection or fatigue, connecting rods fail. Also, connecting rods may fail due to failure of loading, irregular adjustments of screws [7]. Connecting rod must resist the applied tensile stresses. consequently, some design techniques, material selection, and fatigue tests have been presented [8].

Failure of connecting rod is attributed to the availability of strength that is required to withstand the applied stresses. But by extending the life cycle by increasing the strength we can overcome it. Sathish et al.[1], performed static stress analysis based on the finite element method to AA2014, AA6061 and AA7075 aluminum alloy materials using ANSYS, and found that the AA2014 is the less weight and better stiffness which has the minimum values of equivalent stress, equivalent elastic strain, and the total deformation. Gopinath et al. [4], performed static analysis based on finite element analysis to forged steel, aluminum and titanium connecting rod. Further, a weight reduction by topology optimization technique on forged steel connecting rod is carried out. Rezvani et al. [9], studied the catastrophic joint deformation failure of the 645E3B diesel engine connecting rod. The findings confirm that the failure causes are assumed to be the hydro-lock condition due to the splash of water within the ignition chamber and the higher-pressure proportion. Abad et al. [10], performed dynamic load analysis as well as optimization of connecting rod. Lee et al. [11], investigated the buckling sensitivity by reducing the connecting rod weight. And found that in buckling, the stress sensitivity is more than or equal to fatigue and yield stress. Kreculj et al. [12], analyzed the impact load effect on the structure of aircraft. The accumulated impact damage and the impact damage effect are discussed. Additionally, Ranjan Pani et al. [13], performed a theoretical and a numerical buckling failure analysis and material selection for the 645E3b engine connecting rod. The theoretical part is taken to calculate the buckling load and stress using the merchant-Rankine approach and the slider-crank mechanism approach. The numerical analysis is done by ANSYS to validate the results. And conclude that, a proper buckling safety factor should be taken to avoid the buckling failure. The Al-2024-T6 is not ideal for heavy-duty diesel engines due to the lower buckling strength compared to 42CrMo4. Furthermore, Sharma et al. [14], carried out analysis and modeling of carbon steel and aluminum boron carbide connecting rod and found that the nearest working factory of safety to the theoretical factory of safety is the aluminum boron carbide connecting rod which has 48.55% more stiffness and 10.35% less stress. Pathade et al. [15], performed a finite element analysis (FEA) on the connecting rod using both ANSYS and pro-wild fire software and found that the small end stresses are greater than in the bigger end. In addition, concluded that the chance of the connecting rod failure may be at the fillet section of both ends. Bansal et al. [16], carried out a dynamic and a static FEA on aluminum alloy connecting rod. The static load analysis preformed for stress analysis, optimization and to determine the connecting rod von-Mises stresses. Ahmed et al. [17], carried out FEA on a connecting rod with three aluminum alloys materials (AA2014, AA7075, and AA6061) and carbon fiber. Bin et al. [18], performed FEA on a connecting rod to investigate the stress distribution and fatigue life cycle. Gautam et al. [19], carried out a static stress analysis based on FEA on SS304 connecting rod used in Cummins NTA885BC engine. Also, He et al. [20], carried out failure analysis of a diesel engine connecting rod. The reason for the connecting rod's failure was the high-stress concentration/high-stress level and lower yield strength. Londhe et al.[21], performed FEA and an experimental study of the connecting rod and found that both the experimental and the simulation results of the fatigue and static strain matched, thereby validating the FEA. Moreover, Rabb [22], studied the fatigue failure of the connecting rod. Similarly, Rakic et al. [23], predicted the failure of a connecting rod based on the FEA, which showed the concentration of maximum stress applied at position of original fracture. Witek et al. [24], conducted stress analysis on connecting rod using ANSYS. They observed that maximum principal stress can be found at zone near the bolt hole where the crack originated and it exceeded the fatigue limit of 42CrMo4 steel. Seralathan et al.[25], performed static stress analysis based on the FEA to different types of materials using ANSYS finite element code and found that, the A356-5%SiC-10% Flyash stir cum squeeze casting material has the minimum values of equivalent stress, equivalent elastic strain, and total deformation. A lot of research deals with weight loss [26, 27].

In this paper, the structural design optimization to reduce the weight of forged steel connecting rod in aircraft engines using a finite element method is performed. The connecting rod model is created using ANSYS APDL. The von-Mises stress, strain, shear stress and the deflation results of the connecting rod are investigated.

2. Finite element modeling and optimization

The FEA is widely used for solving different problems of engineering and predicting the behavior of structural elements as in [28-33]. FEA is the most significant applied technique for analyze engineering structures according to specific design considerations. It dives the real object into a large number small units called "elements" for dynamic and static analysis of simple to complex models in the presence of varied design constraints. Further checks could be done to improve a design for optimum performance and life about design failure [34]. The components of the connecting rod are illustrated in ***Figure 1***. Geometry and dimensions of the connecting rod are shown in ***Figure 2***. The model was crated and mapped in ANSYS software package by using ANSYS Parametric Design Language (APDL). The accuracy of the results is controlled by the element type and meshing of the model, which is responsible for dividing the model into a number of small elements. The smaller the elements size the more accurate the results. In the present paper, the model is meshed using Plane183 element as shown in ***Figure 3***. The total number of elements and nodes are (1157) and (3949), respectively. The mechanical properties for the Forged steel are as listed in **Table *1***.

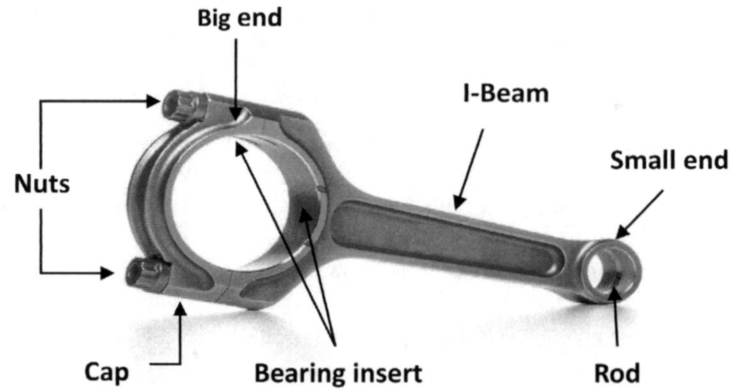

Figure 1: Connecting rod components.

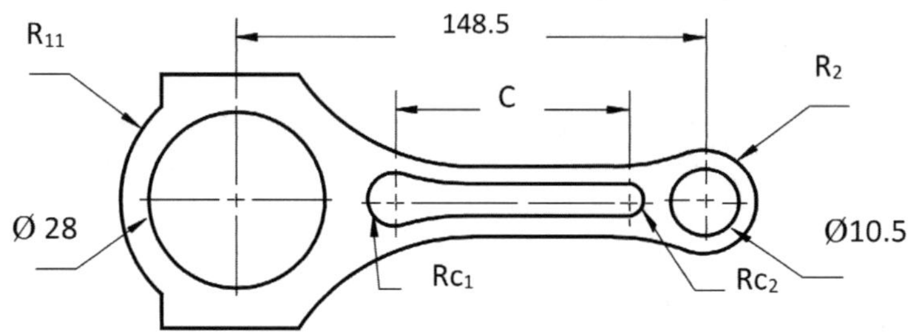

Figure 2: Connecting rod geometry and dimension (all dimensions are in mm).

Table 1: Mechanical properties of Forged Steel.

Young's Modulus, E	200GPa
Poisson's ratio, υ	0.3
Density, ρ	7850 Kg/m^3
Shear Modulus, G	78 GPa
Tensile Strength, σ_U	892.63 MPa
Yield Strength, σ_y	648.06 MPa

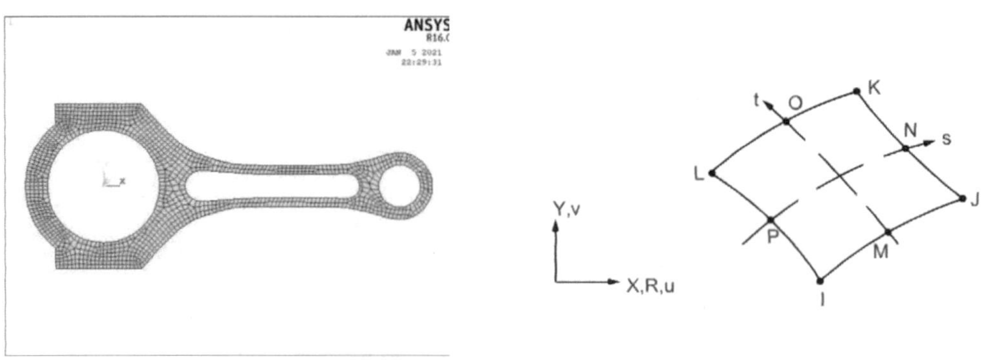

(a) (b)

Figure 3: Meshing and element type (a) Connecting rod mesh and (b) the geometry of plane183 element.

IOP Conf. Series: Materials Science and Engineering 1172 (2021) 012001 doi:10.1088/1757-899X/1172/1/012001

2.1 Boundary condition

The connecting rod is subject to two types of loads, the first being in compression and the other being in tension. In the tension case, the load is applied on the right half of small end at 120 degrees and the big end is fully fixed as shown in ***Figure 4*** (a). In the compression case: the load is applied on the left half of the small end at 120 degrees and the big end is fully fixed as shown in ***Figure 4*** (b). The chosen pressure value is 24.2 MPa [35].

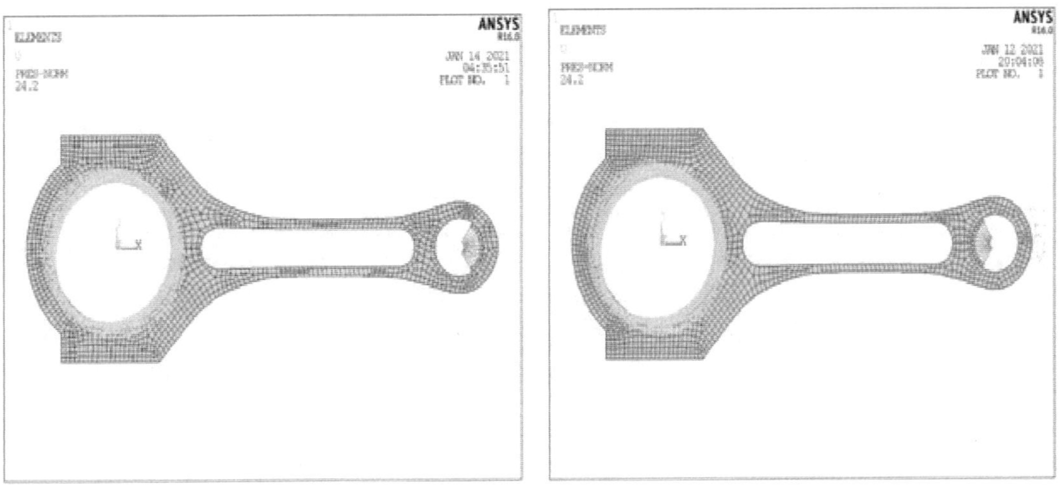

(a) (b)

Figure 4: Boundary condition of connecting rod (a) for tension (b) for compression.

2.2 Optimization statement

Design Optimization is a crucial mathematical method used in engineering and many other fields. Generally, it is a process of finding the optimal value of one or many design parameter that meets the system requirement and achieve the best performance [36]. Hence, the design optimization of the connecting rod is to minimize the mass of the connecting rod while bearing the compressive and the tension load acting on it [4]. The design optimization flow chart is shown in ***Figure 5***.

2.3 Objective function

The objective function f(x) is the parameter that desired to be minimized or maximized. In this paper the objective function is to minimize the connecting rod mass:

$$f(x) = minimaze\ the\ connecting\ rod\ mass(Kg)$$

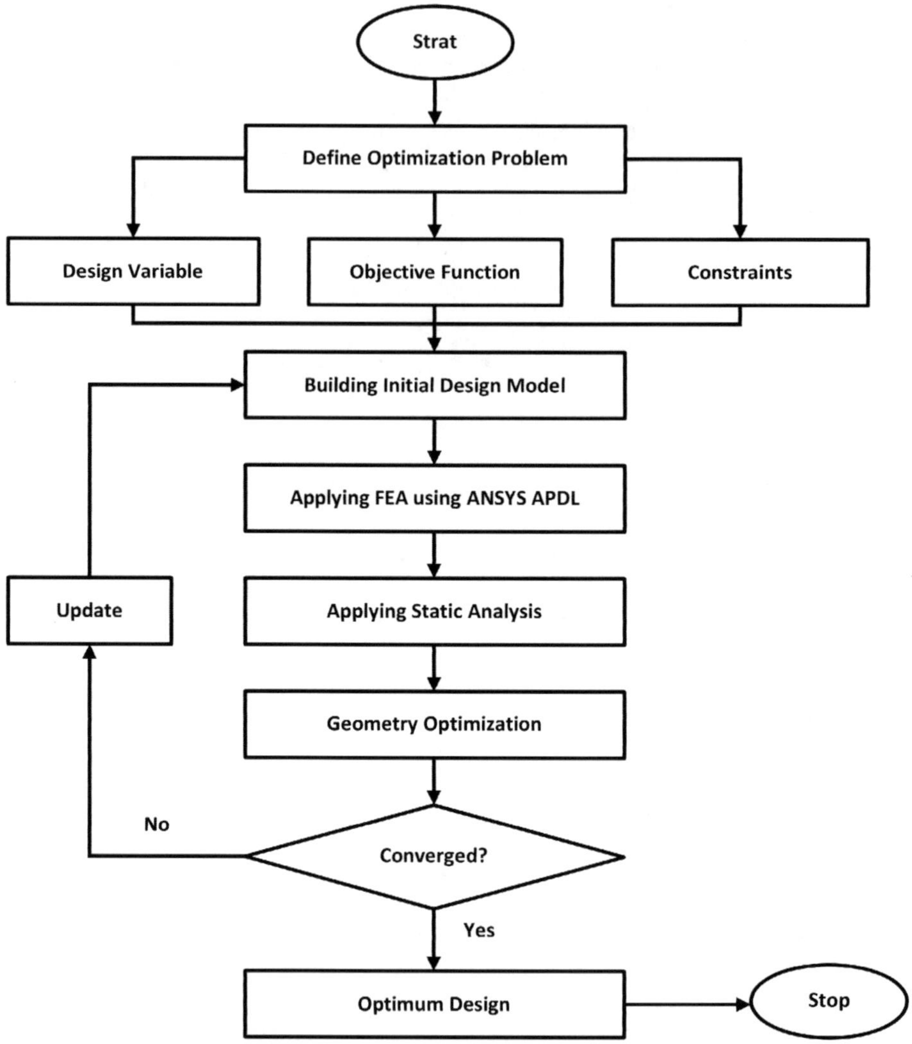

Figure 5: *Design optimization flow Chart*

2.4 Design constraints:

Design constraints are the range that parameters cannot go over or below it to avoid failure and meet the requirements. In this paper, constraints are characterized into two types: material and side constraints.

Material constraints:

To avoid connecting rod failure the following constraints must be satisfied:

1- The deflection $(\boldsymbol{\delta})$ of the connecting rod must not exceed the allowable deflection $(\boldsymbol{\delta_a})$ Mathematically:

$$\delta < \delta_a$$

2- The von-Mises stress $(\boldsymbol{\sigma})$ must not exceed the allowable stress $(\boldsymbol{\sigma_a})$ Mathematically:

$$\sigma < \sigma_a$$

IOP Conf. Series: Materials Science and Engineering 1172 (2021) 012001 doi:10.1088/1757-899X/1172/1/012001

Side constraints:
Side constraints are identified as the range limits of the dimensions parameter as listed in the Table 2.

Table 2: Side Constraints

	Lower limit	**Upper limit**
Outer big end (R_1)	${R_1}^L$	${R_1}^U$
Outer small end (R_2)	${R_2}^L$	${R_2}^U$
Big circle (Rc_1)	${Rc_1}^L$	${Rc_1}^U$
Small circle (Rc_2)	${Rc_2}^L$	${Rc_2}^U$
Center distant between big and small circle (C)	C^L	C^U
Connecting rod thickness (t)	t^L	t^U

3. Results and discussion

Based on Finite element analysis (FEA) and ANSYS Parametric Design Language (APDL), dimensional structural mass optimization is performed on forged steel connecting rod for aircraft engines under 24.2 MPa tension and compressive pressure.

3.1 Real model results

Numerical static stress analysis of forged steel connecting rod under 24.2 MPa tensile and compression pressure is obtained for a real model. Additionally, von-Mises stress, shear stress, total deformation, von-Mises strain and principal stresses and strains of the connecting rod are obtained in both tension and compression cases as shown in ***Figure 6*** and ***Figure 7***, respectively.

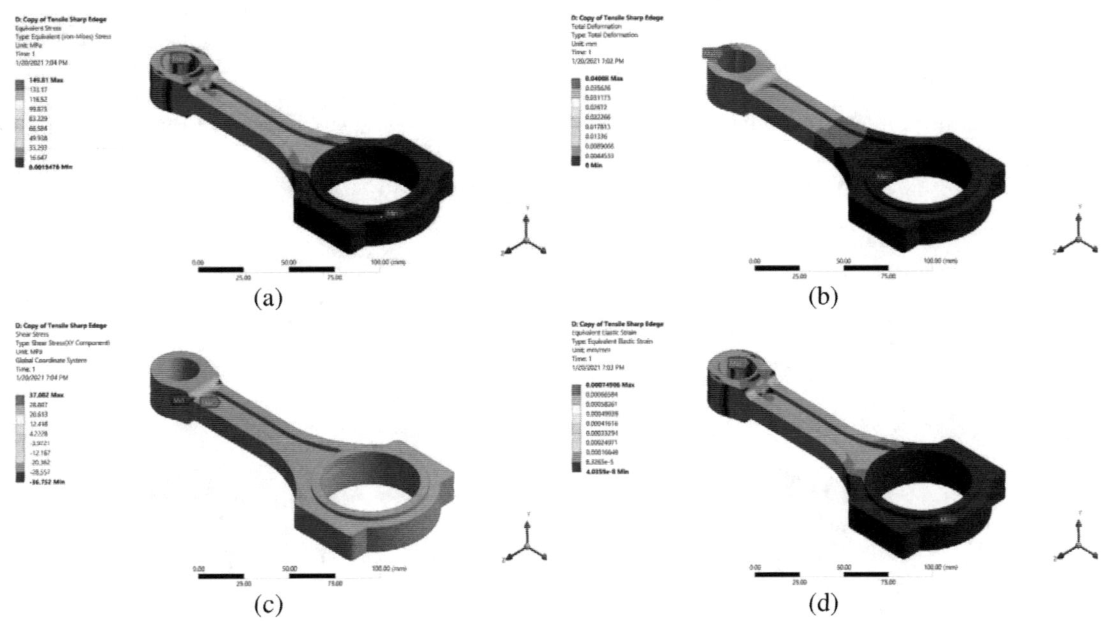

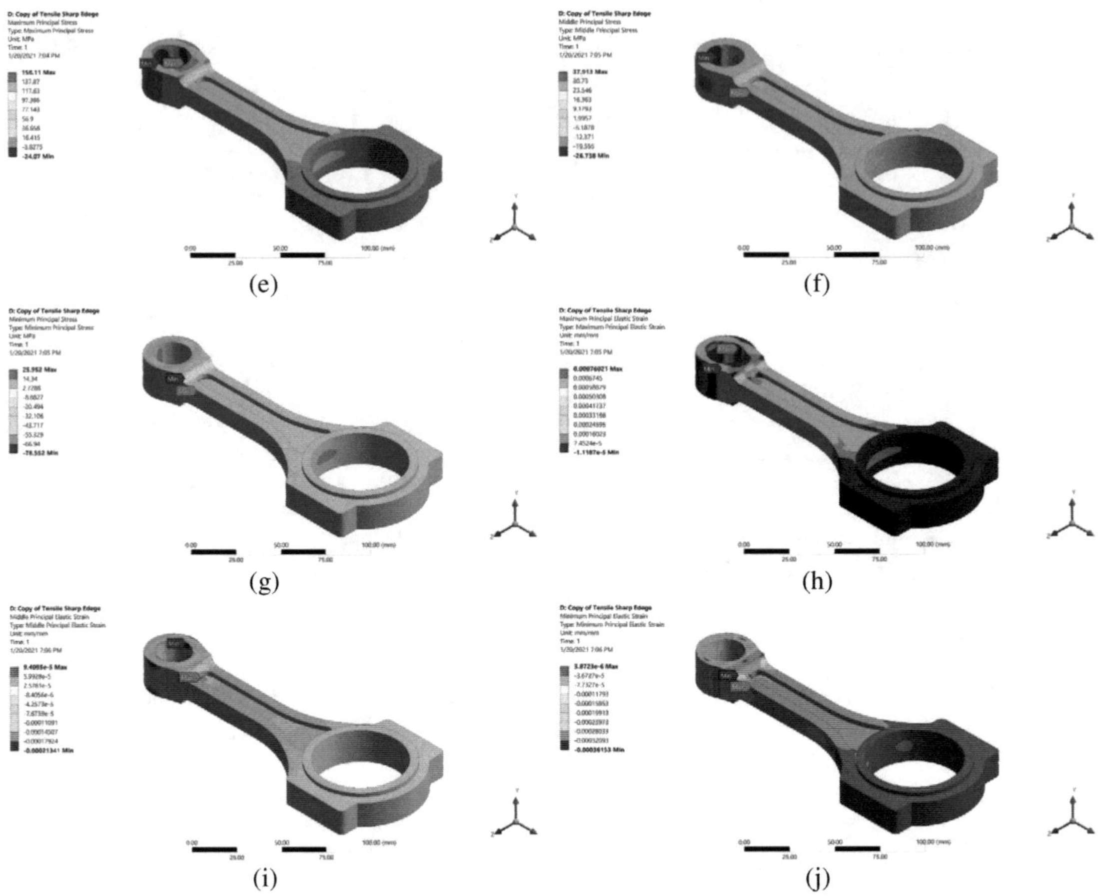

Figure 6: Connecting rod tensile case results of (a) von-Mises stress, (b) Total deformation (c) Shear stress, (d) von-Mises strain, (e) 1st Principal stress (f) 2nd Principal stress (g) 3rd Principal stress and (h-j) 1st, 2nd and 3rd Principal strain.

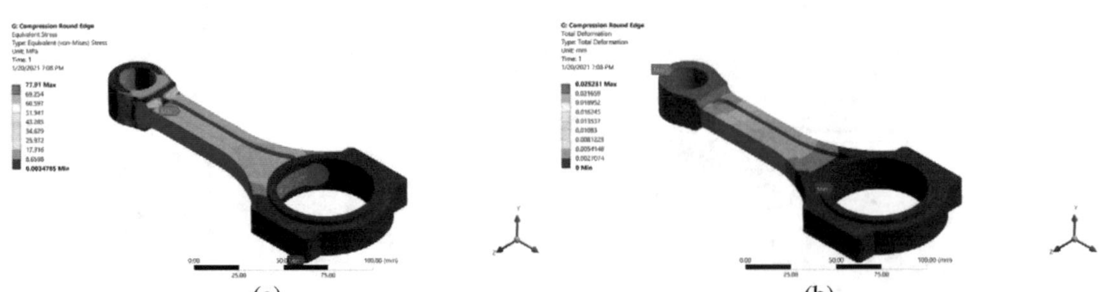

IOP Conf. Series: Materials Science and Engineering 1172 (2021) 012001 doi:10.1088/1757-899X/1172/1/012001

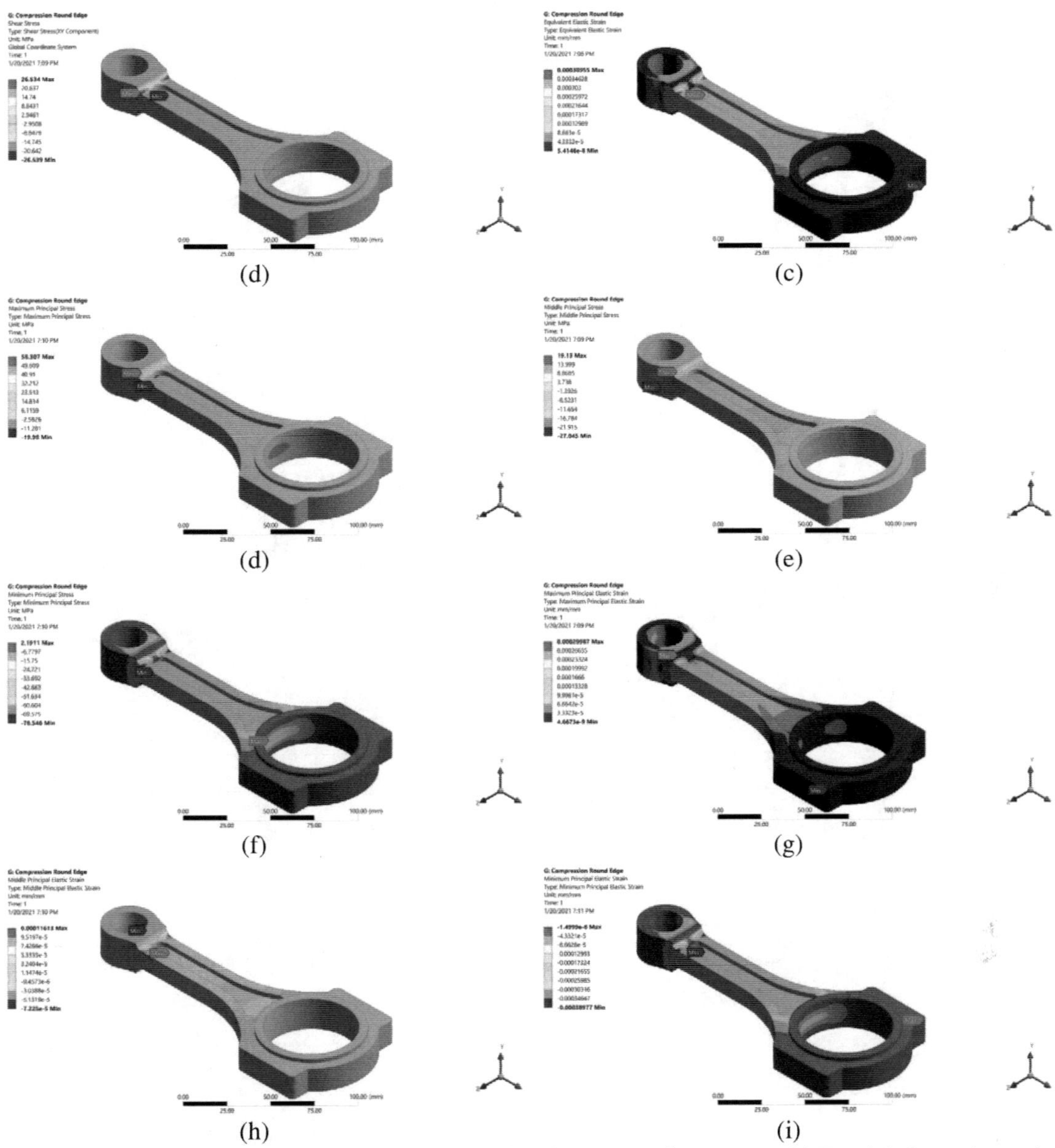

Figure 7: Connecting rod compression case results of (a) von-Mises stress, (b) Total deformation, (c) Shear stress, (d) von-Mises strain, (e) 1st Principal stress (f) 2nd Principal stress (g) 3rd Principal stress and (h-j) 1st, 2nd and 3rd Principal strain.

3.2 Optimized model results

The aim of the optimization is to reduce the connecting rod mass while meeting all the material and the side constraints. Numerical static stress analysis of forged steel connecting rod under 24.2 MPa tensile and compresion pressure is performed for the proposed optimized model. Also, von-Mises stress, shear stress, total deformation, von-Mises strain and prrincple stresses and strains of the connecting rod are obtained for the proposed optimized model in both tension and compression cases as demonstrated in ***Figure 8*** and ***Figure 9*** repectively. Detailed numerical results for real and optimized model can be summarized in ***Table 3***

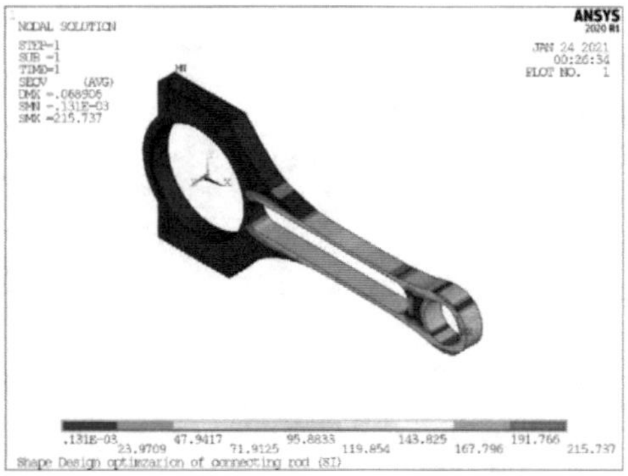
(a)

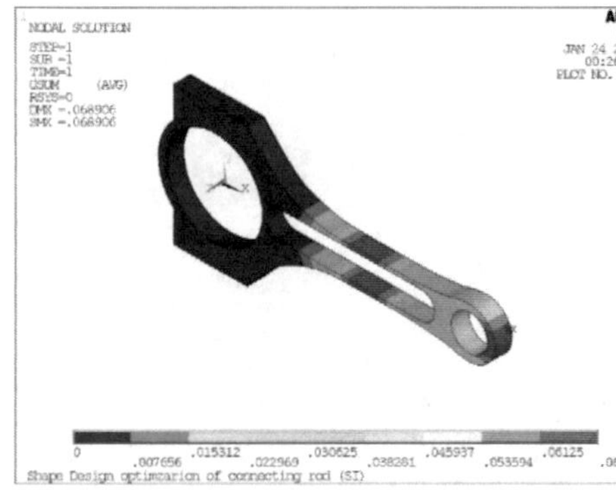
(b)

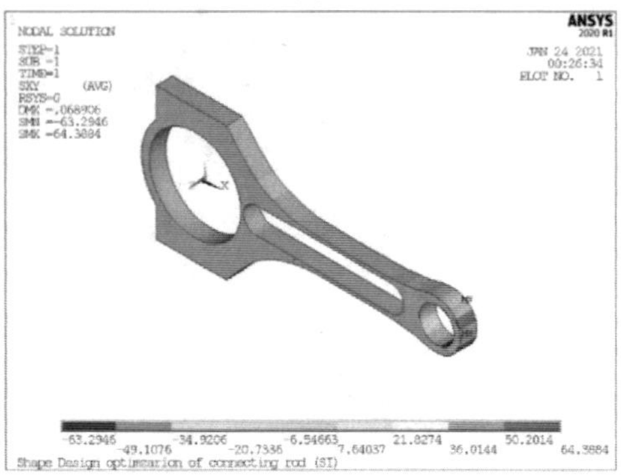
(d)

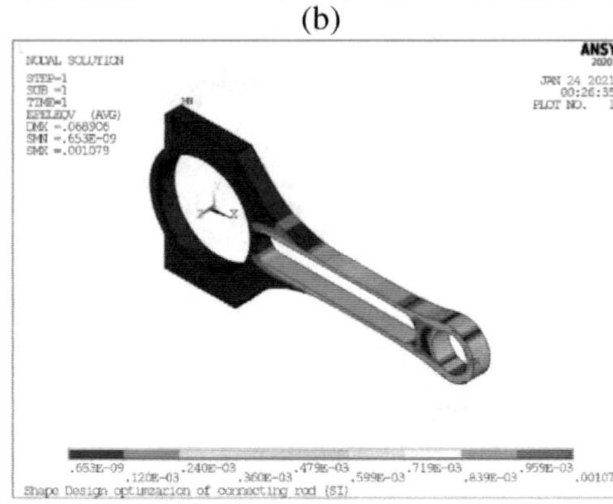
(c)

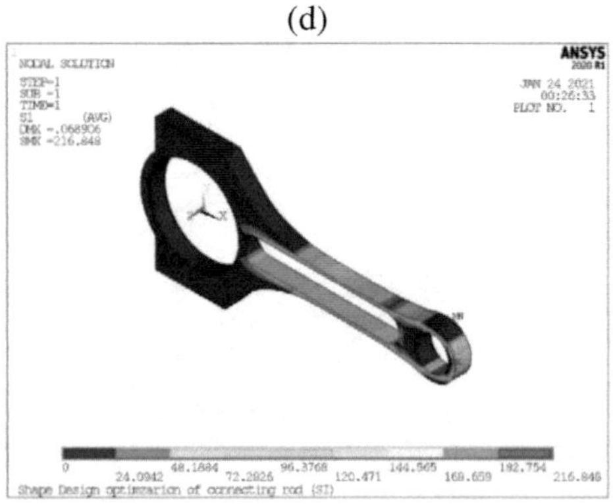
(d)

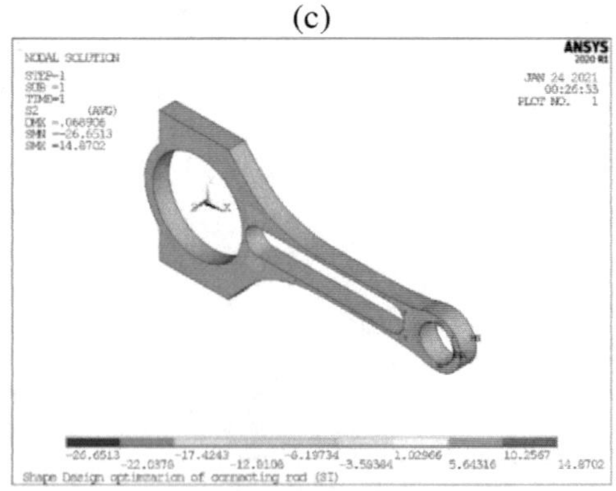
(e)

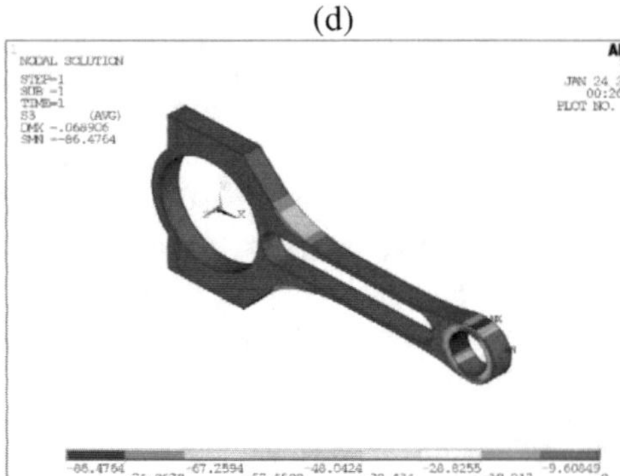
(f)

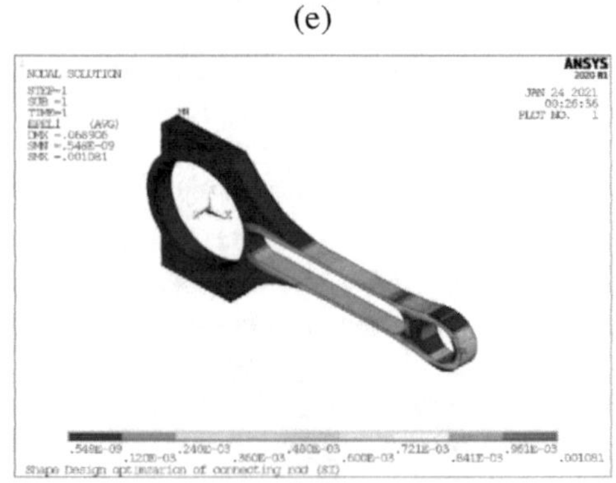
(g)

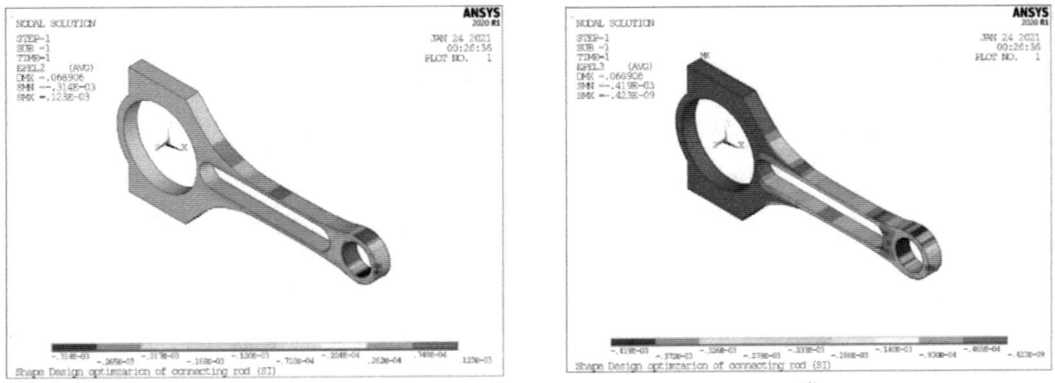

(h) (i)

Figure 8: Connecting rod optimized (tensile case) results of (a) von-Mises stress, (b) Total deformation, (c) Shear stress, (d) von-Mises strain, (e) 1st Principal stress (f) 2nd Principal stress (g) 3rd Principal stress and (h-j) 1st, 2nd and 3rd Principal strain.

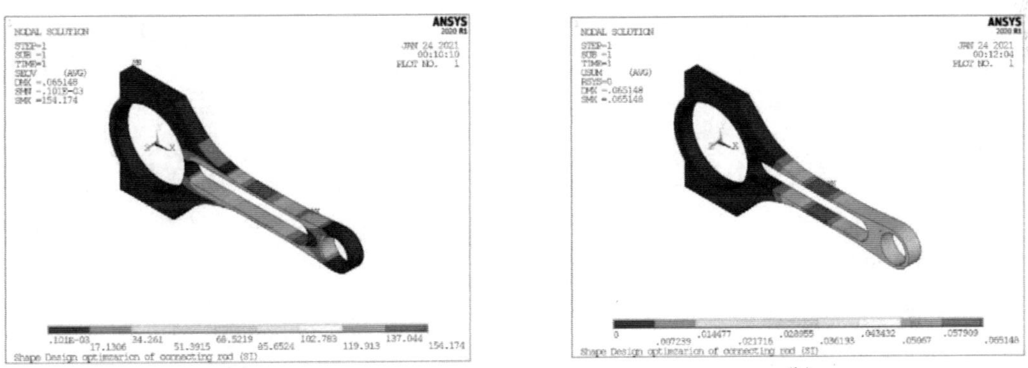

(a) (b)

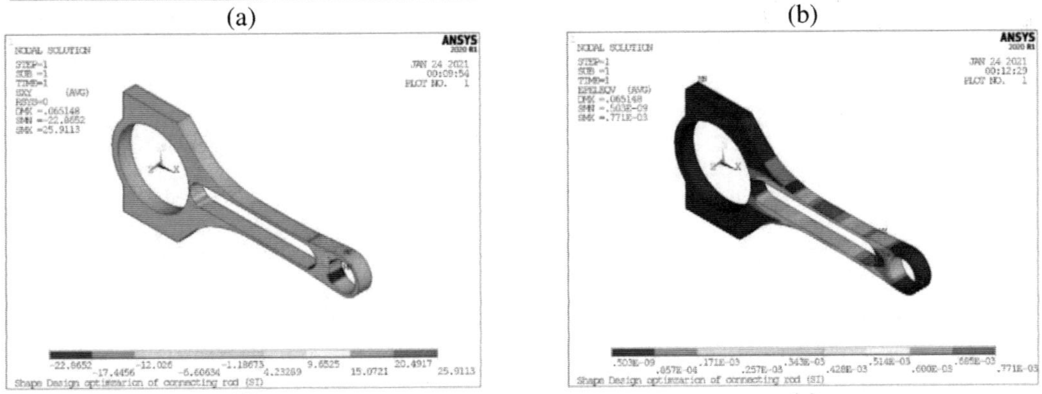

(d) (c)

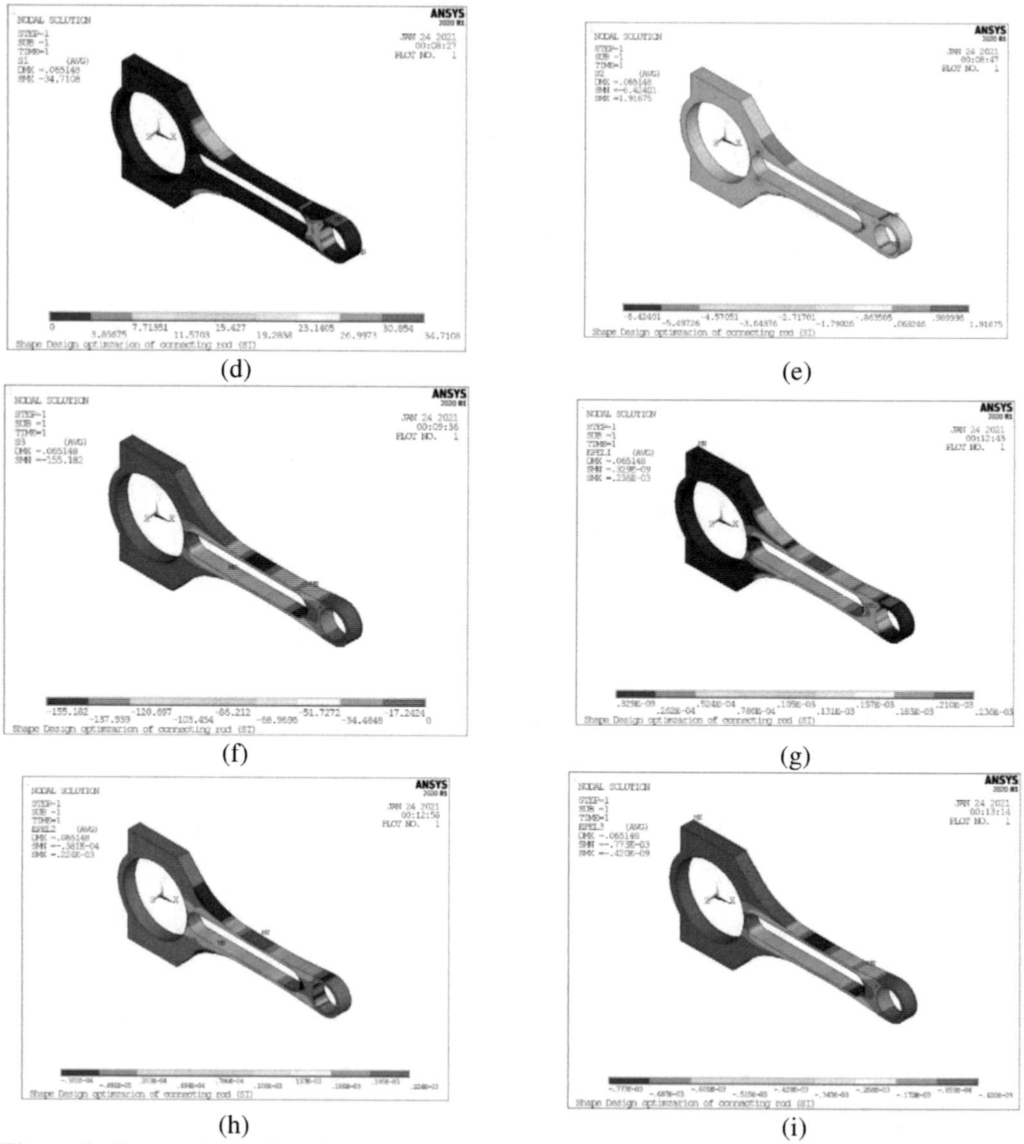

Figure 9: Connecting rod optimized (Compression case) results of (a) von-Mises stress, (b) Total deformation, (c) Shear stress, (d) von-Mises strain, (e) 1st Principal stress (f) 2nd Principal stress (g) 3rd Principal stress and (h-j) 1st, 2nd and 3rd Principal strain.

Table 3.Detailed numerical results for real and optimized model

			von-Mises stress (MPa)	Deflation (mm)	Shear stress (MPa)	von-Mises elastic strain	Principals stresses (MPa)	Principals strains
Real model	Tension case	*Max. Value*	148.81	0.04	37	7.5e-4	--	--
		Position	small end	small end	I-beam	small end	--	--
		Min. Value	2×10^{-3}	0	-36.75	4.04×10^{-8}	--	--

		Values	--	--	--	--	158.1, 37.91 and -78.55	7.6×10^{-4}, -2.13×10^{-4} and -3.6×10^{-4}
		Figure	6 (a)	6 (b)	6 (c)	6 (d)	6 (e-g)	6 (h-j)
	Compression case	*Max. Value*	77.91	0.03	26.53	3.9×10^{-4}	--	--
		Position	I-beam	small end	I-beam	I-beam	--	--
		Min. Value	3.5×10^{-3}	0	-26.53	5.4×10^{-8}	--	--
		Values	--	--			58.3, -27and -78.55	3×10^{-4}, 1.2×10^{-4} and -4×10^{-4}
		Figure	7 (a)	7 (b)	7 (c)	7 (d)	7 (e-g)	7 (h-j)
Optimized model	Tension case	*Max. Value*	215.74	0.07	64.4	1.1e-3	--	--
		Position	small end	small end	small end	small end	--	--
		Min. Value	1.3×10^{-4}	0	-63.3	6.53×10^{-10}	--	--
		Values	--	--	--	--	216.8 , -26.7 and -86.5	1.1×10^{-3}, 3.14×10^{-4} and -4.2×10^{-4}
		Figure	8 (a)	8 (b)	8 (c)	8 (d)	8 (e-g)	8 (h-j)
	Compression case	*Max. Value*	154.2	0.065	26	7.7e-4	--	--
		Position	I-beam	middle of connecting rod	small end	I-beam	--	--
		Min. Value	4	0	-23	5×10^{-10}	--	--
		Values	--	--	--	--	34.71, -6.42 and -155.2	2.4×10^{-4}, 2.24×10^{-4} and -7.73×10^{-4}
		Figure	9 (a)	9 (b)	9 (c)	9 (d)	9 (e-g)	9 (h-j)

4. Comparison between real and optimized model

From the previous results, structural optimization significantly minimizes the mass of the connecting rod by 55.13 % in the tensile case and 56.7 % in the compression case, which in turn reduces the production time and cost. Table 4 clarifies a comparison between the real and the optimized results of the connecting rod.

Table 4: Comparison between the real and optimized results.

Parameter	Real Model	Optimum Model

		Tensile	Compression	Tensile	Compression
1	von-Mises Stress (MPa)	148.81	77.91	215.7	154.2
2	Total Deformation (mm)	0.04	0.03	0.069	0.065
3	Factor of Safety (F.S)	4.35	8.32	3	4.2
3	Mass (Kg)	0.64		0.29	0.28
4	Volume (mm^3)	81112		36391	35233
5	Weight reduction (%)	-		55.13	56.7

5. Conclusion

In this paper, a static stress analysis was performed on forged steel connecting rod in aircraft engines under 24.2 MPa tensile and compression pressure using ANSYS APDL. The maximum and the minimum of von-Mises stress and strain, the principal stresses and strains, shear stress, and the deflection were investigated. The results show a significant opportunity for weight reduction optimization. Thus, a dimensional structural mass optimization was carried out to minimize the connecting rod's mass after performing static stress analysis. That, in turn, will enhance the performance of aircraft engines. The optimization results show notable weight reduction of the connecting rod by 55.13 % (in the tensile case) and 56.7 % (in the compression case) from the initial design. However, fatigue analysis for a connecting rod will be considered in further work.

Funding: This research received no external funding.
Acknowledgments: The authors are grateful to Taif University (Taif, KSA), Mansoura University (Mansoura, Egypt), Military Technical College (Cairo, Egypt) and Future University in Egypt (Cairo, Egypt) for providing all the required facilities to carry out the present research.
Conflicts of Interest: The authors declared no conflicts of interest.

Biography

Associate Professor Mahmoud M. K. Helal, received his B.Sc. and MSc. degrees from Production and Mechanical Design Department, faculty of Engineering, Mansoura University in 2000 and 2005, respectively. In 2011 he received his Ph.D. degree from Harbin Institute of Technology (HIT). He works as an assistant professor in the Production and Mechanical Design Department at Mansoura University. Dr. Helal's research interests lie in the structural optimization, the strength of the materials; MEMS design assembly, micromanipulation robot and solid mechanics.

Assistant Professor Bassem F. Felemban, revived his B.Sc. Degree from Production and Mechanical Design Department, King Abdelaziz University in 2009. Then he received the MSc. and Ph.D. degrees from Mechanical Engineering, University of Central Florida (UCF). He is working as an assistant professor in the Production and Mechanical Design Department at Taif University. Dr. Bassem F. Felemban works as a member of Mechanical of Materials Research Group at UCF.

Abdullah Alharthi, studied Mechanical Engineering at department of Mechanical Engineering, Faculty of Engineering, Taif University. Email: eng.abdullah.harthi@gmail.com.

Sari Almalki, is studied Mechanical Engineering at department of Mechanical Engineering, Faculty of Engineering, Taif University. Email: sariayed.3@gmail.com

Assistant Professor Elsayed Fathallah, received his bachelor and master degrees in Civil Engineering from Civil Engineering Department from Military Technical College in 1999 and 2007, respectively. In 2015 he received his Ph.D. degree from Harbin Engineering University, College of Aerospace and Civil Engineering. He works as an assistant professor at both Civil Engineering and Ships and Submarines

Engineering Department, Military Technical College. His research focuses on strength of the material and structural optimization.

Abdallah A. Elsherbiny, received his B.Sc., MSc. degrees from Production and Mechanical Design Department, faculty of Engineering, Mansoura University**.** He works as an assistant lecture in the same department.

Assistant Professor, Hassan M. S. Eleashy, received his B.Sc., MSc. and Ph.D degrees from Production and Mechanical Design Department, faculty of Engineering, Mansoura University in 2000, 2006, and 2014 respectively. He works as an assistant professor in the Mechanical Engineering Department at Future University in Egypt. Dr. Eleashy research interests lie in the mechanism design, structural synthesis of robots, planetary gear trains and stress analysis.

References

[1] Sathish, T., Kumar, S.D. and Karthick, S., 2020. Modelling and analysis of different connecting rod material through finite element route. *Materials Today: Proceedings*, ***21***, pp.971-975.

[2] Dinesh Kumar, S. and Purushothaman, K., 2019. Enhancement of thermal conductivity in a plate heat exchanger by using nanoparticles CNT, Al2O3, surfactant with de-ionised water as a coolant. *International Journal of Ambient Energy*, pp.1-4.

[3] Ranganathan, S., Kuppuraj, S., Soundarrajan, K. and Perumal, A., 2019. *Design and Analysis of Hybrid Metal Matrix Composite Connecting Rod via Stir/Squeeze Casting Route* (No. 2019-28-0113). SAE Technical Paper.

[4] Gopinath, D. and Sushma, C.V., 2015. Design and optimization of four wheeler connecting rod using finite element analysis. *Materials Today: Proceedings*, ***2*(4-5)**, pp.2291-2299.

[5] Heywood, J.B., 2018. Internal Combustion Engine Fundamentals (2nd edit). *McGrow-Hill Education.*

[6] Shenoy, P.S., 2004. *Dynamic load analysis and optimization of connecting rod* (Doctoral dissertation, University of Toledo).

[7] Ilman, M.N. and Barizy, R.A., 2015. Failure analysis and fatigue performance evaluation of a failed connecting rod of reciprocating air compressor. *Engineering Failure Analysis*, ***56***, pp.142-149.

[8] Ali, H.M.M. and Haneef, M., 2015. Analysis of fatigue stresses on connecting rod subjected to concentrated loads at the big end. *Materials today: proceedings*, ***2*(4-5)**, pp.2094-2103.

[9] Rezvani, M.A., Javanmardi, D. and Mostaghim, P., 2018. Diagnosis of EMD645 diesel engine connection rod failure through modal testing and finite element modeling. *Engineering Failure Analysis*, *92*, pp.50-60.

[10] Abad, M.R.A.A., Ranjbarkohan, M. and Dardashti, B.N., 2011. Dynamic load analysis and optimization of connecting rod of samand engine. *Australian Journal of Basic and Applied Sciences*, ***5*(12)**, pp.1830-1838.

[11] Lee, M.K., Lee, H., Lee, T.S. and Jang, H., 2010. Buckling sensitivity of a connecting rod to the shank sectional area reduction. *Materials & Design*, ***31*(6)**, pp.2796-2803.

[12] Kreculj, D. and Rasuo, B., 2018. Impact damage modeling in laminated composite aircraft structures. In *Sustainable Composites for Aerospace Applications* (pp. 125-153). Woodhead Publishing.

[13] Pani, A.R., Patel, R.K. and Ghosh, G.K., 2020. Buckling analysis and material selection of connecting rod to avoid hydro-lock failure. *Materials Today: Proceedings*, ***27***, pp.2121-2126.

[14] Sharma, P.K. and Borse Rajendra, R., 2012. Ftigue analysis and optimisation of connecting rod using finite element analysis. *International Journal of advance research in Science and Engineering*, ***1*(1)**, pp.3367-337.

[15] Pathade, V.C., Patle, B. and Ingale, A.N., 2012. Stress analysis of IC engine connecting rod by FEM. *International Journal of Engineering and Innovative Technology*, ***1*(3)**, pp.12-15.

[16] Bansal, R., 2013. Dynamic simulation of connecting rod made of aluminium alloy using finite element analysis approach. *IOSR Journal of Mechanical and Civil Engineering*, ***5*(2)**, pp.01-05.

[17] Ahmed, G.S., Khany, S.E. and Shareef, S.H., 2014. Design, Fabrication and Analysis of a Connecting rod with Aluminum alloys and carbon fiber. *International Journal of Innovative*

Research in Science, Engineering and Technology, **3(10)**, pp.16674-16687.

[18] Bin, Z., Lixia, J. and Yongqi, L., 2010, January. Finite element analysis and structural improvement of diesel engine connecting rod. In *2010 Second International Conference on Computer Modeling and Simulation* (**Vol. 1**, pp. 175-178). IEEE.

[19] AbhinavGautam, K. and Ajit, P., 2013. Static stress analysis of connecting rod using finite element approach. *IOSR Journal of Mechanical and Civil Engineering (IOSR-JMCE), 10*(1), pp.47-51.

[20] He, B.Y., Sun, J.B., Chen, S.Z. and Nie, R., 2013. Crack analysis on the toothed mating surfaces of a diesel engine connecting rod. *Engineering Failure Analysis*, ***34***, pp.443-450.

[21] Londhe, A., Yadav, V. and Sen, A., 2009. *Finite element analysis of connecting rod and correlation with test* (No. 2009-01-0816). SAE Technical Paper.

[22] Rabb, R., 1996. Fatigue failure of a connecting rod. *Engineering Failure Analysis*, **3(1)**, pp.13-28.

[23] Rakic, S., Bugaric, U., Radisavljevic, I. and Bulatovic, Z., 2017. Failure analysis of a special vehicle engine connecting rod. *Engineering failure analysis*, ***79***, pp.98-109.

[24] Witek, L. and Zelek, P., 2019. Stress and failure analysis of the connecting rod of diesel engine. *Engineering Failure Analysis*, ***97***, pp.374-382.

[25] Seralathan, S., Mitnala, S.V., Reddy, R.S.K., Venkat, I.G., Reddy, D.R.T., Hariram, V. and Premkumar, T.M., 2020. Stress analysis of the connecting rod of compression ignition engine. *Materials Today: Proceedings*.

[26] Gaikwad, S.S. and Aitavade, E.N., 2013. Static Analysis Of A Roller Of Gravity Roller Conveyor For Structural Strength & Weight Optimization. *PG student, Prof, Mechanical Engineering Department, Tatyasaheb Kore Institute of Engineering andTechnology, Warananagar, Tal.-Panhala, Dist. Kolhapur-416113, India, International Journal of Advanced Engineering Technology, 4*.

[27] Muhammad, A., Ali, M.A. and Shanono, I.H., 2020. Design optimization of a diesel connecting rod. *Materials Today: Proceedings*, ***22***, pp.1600-1609.

[28] Lotfy, M.N., Khalifa, Y.A., Dessouki, A.K. and Fathallah, E., 2020. Dynamic Behavior of Steel and Composite Ferry Subjected to Transverse Eccentric Moving Load Using Finite Element Analysis. *Applied Sciences*, **10(15)**, p.5367.

[29] Helal, M.M. and Elsayed, F., 2015. Dynamic behavior of stiffened plates under underwater shock loading. *Materials Testing*, **57(6)**, pp.506-517.

[30] Helal, M. and Fathallah, E., 2020. Finite element analysis and design optimization of a non-circular sandwich composite deep submarine pressure hull. *Materials Testing*, **62(10)**, pp.1025-1032.

[31] Helal, M., Huang, H., Wang, D. and Fathallah, E., 2019. Numerical Analysis of Sandwich Composite Deep Submarine Pressure Hull Considering Failure Criteria. *Journal of Marine Science and Engineering*, *7*(10), p.377.

[32] Fathallah, E., Qi, H., Tong, L. and Helal, M., 2015. Design optimization of lay-up and composite material system to achieve minimum buoyancy factor for composite elliptical submersible pressure hull. *Composite Structures*, ***121***, pp.16-26.

[33] Fathallah, E., Qi, H., Tong, L. and Helal, M., 2015. Numerical investigation of the dynamic response of optimized composite elliptical submersible pressure hull subjected to non-contact underwater explosion. *Composite Structures*, ***121,*** pp.121-133.

[34] Chen, X. and Liu, Y., 2018. *Finite element modeling and simulation with ANSYS Workbench*. CRC press.

[35] Alkalla, M.G., Helal, M. and Fouly, A., 2021. Revolutionary superposition layout method for topology optimization of nonconcurrent multiload models: Connecting-rod case study. *International Journal for Numerical Methods in Engineering*, **122(5)**, pp.1378-1400.

[36] Arora, J.S., 2017 Chapter 1 - Introduction to Design Optimization, in Introduction to Optimum Design (Fourth Edition), J.S. Arora, Editor, Academic Press: Boston. p. 3-18.

IOP Conf. Series: Materials Science and Engineering 1172 (2021) 012002 doi:10.1088/1757-899X/1172/1/012002

Static and modal analysis of sine-wave-shaped spar

E Elshazly[1], M Kassem[1], M A Elshafei[1], and M Kamel[1]
[1] Department of Aircraft Mechanics, Military Technical College, EGYPT

E-mail: m_kassem@mtc.edu.eg

Abstract. Sine-Wave-shaped Spar (SWS) is recently involved in wing design as one of the corrugated shaped spars that highly improve the torsion and buckling resistance. SWS is an I-beam made of sine wave shaped vertical web fixed to upper and lower flanges. The current research evaluates the behavior of SWS in comparison to the traditional straight-web spar in resisting two cases of loading; static bending loading and static torsion loading. For each case, deformation and stress are analyzed. Modal analysis is also performed for the two spar configurations. For this purpose, a finite element model is developed using Ansys Workbench software for the SWS and the straight web spar. To reach a fair comparison, similar weight is considered for the two spar structures. A parametric study is performed to investigate the effect of web thickness and flange thickness on the bending and torsion deformation and stresses. The effect of thickness variation on the natural frequencies and mode shapes is also considered. The current study can be considered a base to design wing spars that combine the merits of the two spar configurations.

1. Introduction

The spar is the principal structural element of the wing that supports an aircraft during take-off, cruise and landing. In fact, spar design tends to have several interferences with the aircraft structural design. Nowadays, the strength and weight of the spar has become an important factor. Efforts are being made to reduce the weight of the aircraft and consequently increase the payload. The spars may be made of metal, wood or composite materials depending on the design criteria of a specific aircraft. They are generally classified into four different types by their cross-sectional configuration; solid, partly hollows, formed an I-beam and box shaped. Conventionally, the spar web is made of a vertical straight plate. However, introducing a sine wave shaped webs has attracted many researchers at various applications.

R. N. Hadcock [1] introduced the advanced composite materials to produce the sine wave spars of the horizontal stabilizer torque box of the military aircraft F-14A for the lightweight and the safety of the aircraft structures. To improve the cost, weight and integrity of the Northrop F-5E Fighter, it included sine wave spars and sine wave ribs [2]. Robert. K. Perrons [3] discussed the details of involving the development and production of the sine wave spars for the wings of F-22 by the Boeing Defense and Space Group, as sine wave provides the spar with a higher degree of mechanical stiffness.

Ren Yiru et al. [4] integrated a design concept for Boeing 737-200 aircraft to improve the crashworthiness of the fuselage using sine-wave beam and strut to dissipate impact kinetic energy by energy absorption structure during impact process. Sine wave composite spars were used in the main wing box of AV-8B/G R Mk5 Harrier II airframe, with four composite ribs made of sine wave construction [5].

S. Hanagud et al. [6] analyzed the static energy absorption behavior of graphite epoxy composite corrugated sine wave webs loaded in axial compression where crashworthy design has become a standard feature of many vehicle design processes. Garyl Farley et al. [7] reported crushing characteristics of

Content from this work may be used under the terms of the Creative Commons Attribution 3.0 licence. Any further distribution of this work must maintain attribution to the author(s) and the title of the work, journal citation and DOI.
Published under licence by IOP Publishing Ltd

composite tubes with "Near-Elliptical" Cross Sections and showed that the energy absorbing beam webs evaluated in these studies consisted of sine-wave and integrally stiffened webs.

Harun Bayraktar, et al. [8] used three dimensional woven composites in sine-wave beam application to increase strength, stiffness, and thermal conductivity compared to conventional 2D laminated composites. Hartmut Pasternak and Dina Hannebauer [9] designed welded girders with sinusoidal corrugated webs. The sinusoidal corrugated web has a better buckling performance than classic welded sections and even comparable trapezoidally-shaped web. Krzysztof et al. [10] introduced numerical investigations of the behavior of a sinusoidal web loaded in shear due to buckling in the period from the onset of buckling until failure. William Rendall Johnson et al. [11] manufactured a sine wave beam web without bunching and distortion of the fibers of the composite web material and has a lightweight relative to strength and stiffness. The current research analyzes the sine-wave-shaped web spar compared with the straight-web spar from where static bending and static torsion are applied while keeping a similar weight of spars. A finite element analysis is performed using Ansys Workbench software. Moreover, a parametric study is conducted to find the effect of web thickness and flange thickness variations on the static response and modal analysis.

2. Spar configurations and modelling

The considered configurations of spar structures are shown in Figure 1. The spar consists of one vertical web and two horizontal flanges. The main difference between the two spar configurations is the vertical web. As shown in Figure 1(a), the straight-web spar is considered as an I-beam and is denoted by 'I' all through the paper. The straight-web thickness, length and height are denoted by twI, b, and LwI, respectively. The sine-web spar has a corrugated web in a sine-wave shape and is denoted by 's' all through the paper. The current research represents the sine-web as semi-circles connected in a zigzag pattern as shown in Figure 1(b). The sine-web thickness, length and height are denoted by tws, b, and Lws, respectively. Lws is the perimeter of the sine wave shaped web. The thicknesses of the upper and lower flanges are denoted by tfU and tfL, and the spar length is L for both spar configurations.

The static loading applied on the spar structure is calculated for a wing of span (L=3m), aspect ratio 6, length of chord (c= 0.5 m) and the area (S= 1.5m2). Schrenk method is used to calculate distributed pressure and torsion load.

$$S = L*c = 3 * 0.5 = 1.5m2. \tag{1}$$

$$Lift = \frac{1}{2}\rho V^2\ SC_l = \frac{1}{2} * 1.225 * 40^2 * 1.5 * 0.8 = 1176\ N. \tag{2}$$

$$P = \frac{L}{S} = \frac{1176}{1.5} = 784\ N/m^2. \quad \text{(Distributed pressure)} \tag{3}$$

$$T = L\left(\frac{c}{3} - \frac{c}{4}\right) = 1176 * \left(\frac{0.5}{3} - \frac{0.5}{4}\right) \quad \text{(Torsion load)} \tag{4}$$
$$= 1176 * 0.04166 = 4849000\ N/m$$

In the current research, the considered dimensions for the configurations of straight-web spar and sine-web spar are: the height of web is b=109 mm, the length of the straight-web spar is L_{W_I} =3000 mm, the length of the sine-web spar is L_{W_S}=4717mm as determined in equation (5). The radius of the sine-web is r=34.5mm, the number of the half cycles of the sine waves is n=43.5, the width of flange is a=70 mm, the web thickness (t_w) varies according to the loading. The flange thicknesses (t_{f_U},t_{f_L}) are respectively the upper and the lower flange thickness. In spar design, flange thicknesses may vary according to the loading, however for simplicity, the current research considers the upper and the lower flange thicknesses equal (t_{f_U}=t_{f_L} $= 3\ mm$), unless other dimensions are mentioned.

The circumference length of the sine-web spar:

$$L_{W_S} = \pi\, r\, n = \pi * 34.5 * 43.5 = 4717\ mm \tag{5}$$

To maintain the same weight of the two spar configurations, the web thickness of the straight-web spar is t_{W_I}=1 mm, while for the sine-web spar t_{W_S}=0.63542mm, unless other dimensions are mentioned.

Ansys workbench software is applied to establish the finite element model. SHELL181 element is used as it is suitable for analyzing thin to moderately-thick shell structures. It is a four-node element with six degrees of freedom at each node. Mesh selectivity analysis is conducted to achieve the convergence of deformation and stress results. Element size (9 mm), with a total number of elements (12691), is applied in the current research.

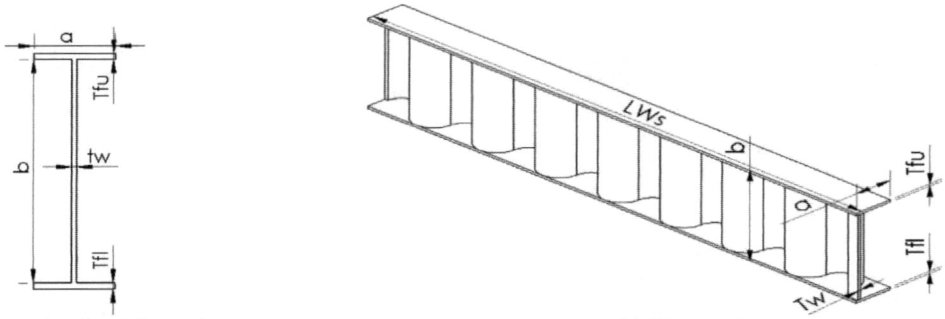

(a) Straight-web spar (b) Sine-web spar

Figure 1. Configuration of spar structures

3. Static and modal analysis

3.1 Stress analysis due to static bending.

The distributed pressure is applied on the upper flange of the straight-web spar and the sine-web spar. The spars are fixed as a cantilever beam. The distributed pressure leads to bending deformation as shown in Figure 2 (a) and (b) for straight-web and sine-web spars, respectively. The bending stresses are shown in Figure 3 (a) and (b).

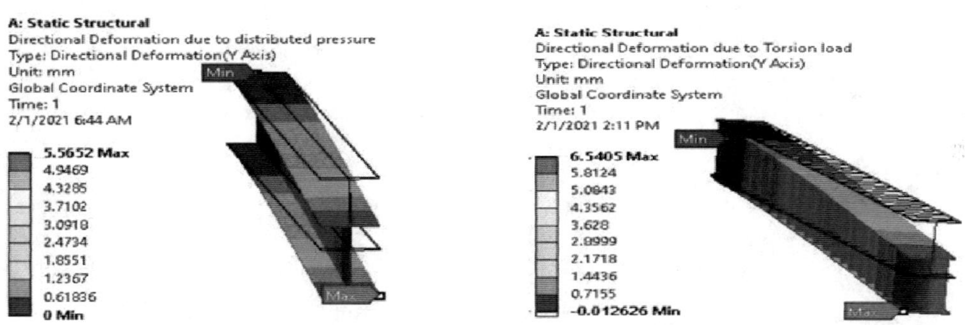

(a) Straight-web spar (b) Sine-web spar

Figure 2. Static Bending deformation.

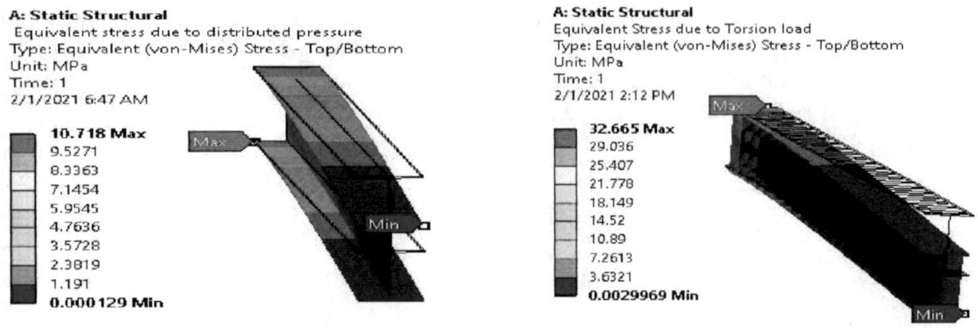

(a) Straight-web spar (b) Sine-web spar

Figure 3. Static Bending stress.

3.2. Stress analysis due to static torsion.

The spars are fixed as a cantilever beam and the torsion load is applied at the free-end of spars. The torsion load on the I-beam spars leads to torsion deformation as shown in the Figure 4 (a) and (b) for straight-web and sine-web, respectively, and torsion stress as shown in the Figure 5 (a) and (b).

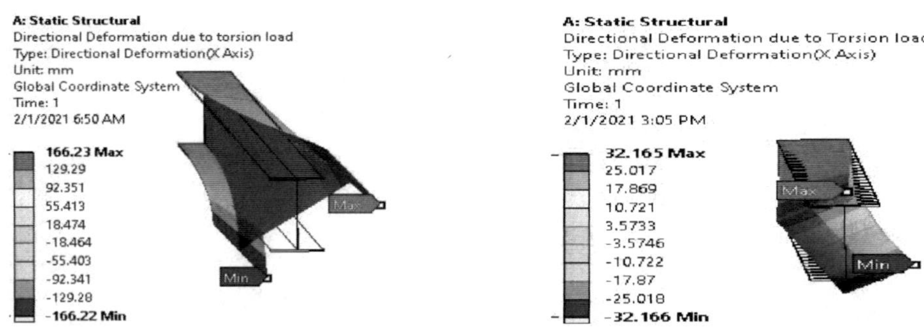

(a) Straight-web spar (b) Sine-web spar

Figure 4. Static torsion deformation.

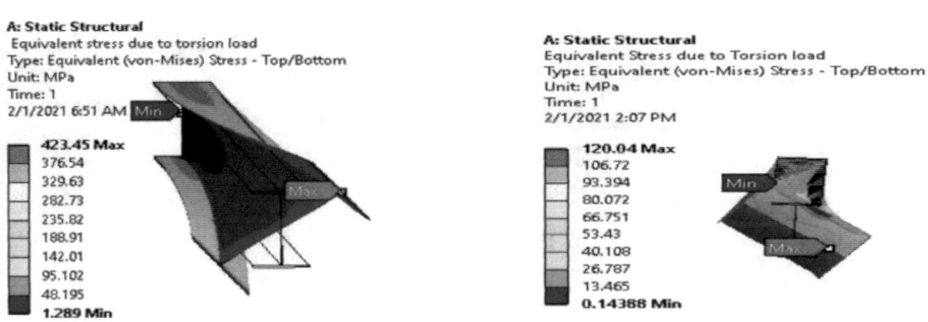

(a) Straight-web spar (b) Sine-web spar

Figure 5. Static torsion stress.

3.3. Modal analysis.

The mode shapes of the straight-web and the sine-web spars describe the deformation that the spars would show when vibrating at the natural frequencies as shown in Figure 6 to Figure 9.

Figure 6 (a) shows the first mode shape of the straight-web spar, the first in-plane bending mode where $\omega_{1I} = 5.6686\ Hz$. Figure 6 (b) shows the first mode shape of the sine-web spar, the first in-plane

IOP Conf. Series: Materials Science and Engineering 1172 (2021) 012002 doi:10.1088/1757-899X/1172/1/012002

bending mode where $\omega_{1S} = 5.731\ Hz$. There is a slight difference in the natural frequencies between the straight-web spar and the sine-web spar.

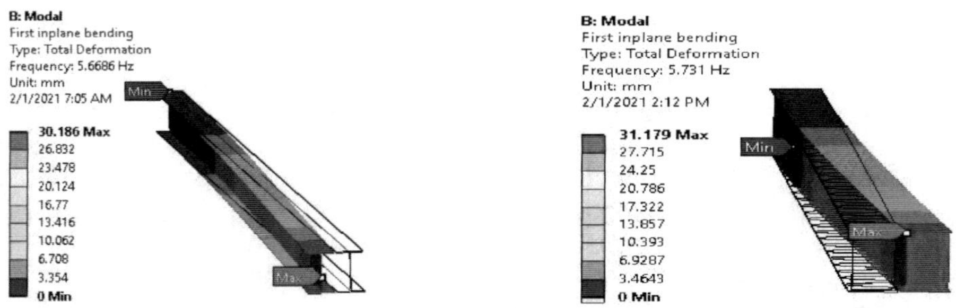

(a) Straight-web spar (b) Sine-web spar

Figure 6. First in-plane bending mode.

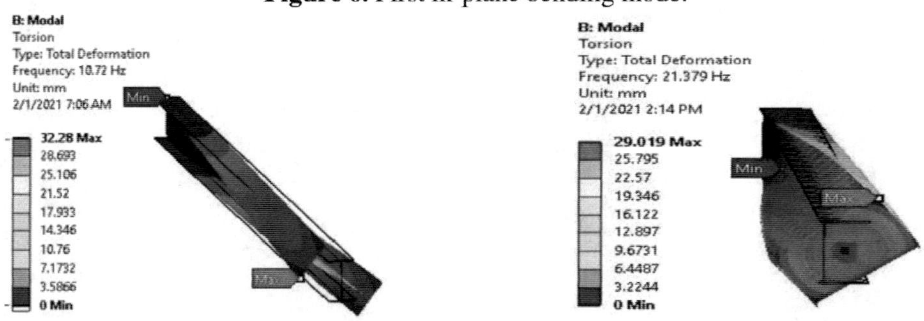

(a) Straight-web spar (b) Sine-web spar

Figure 7. First torsion mode.

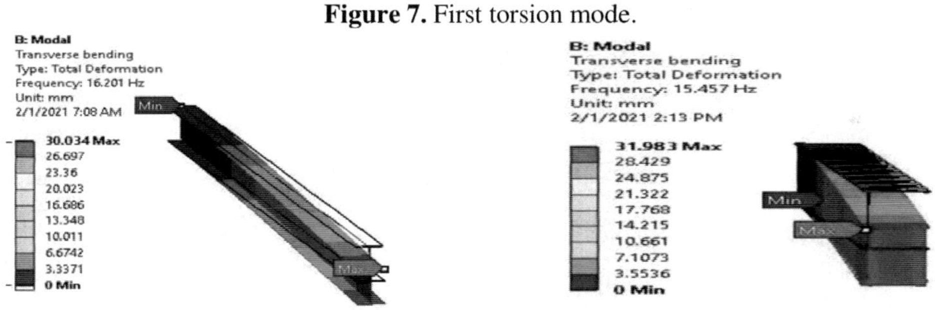

(a) Straight-web spar (b) Sine-web spar

Figure 8. First transverse bending mode.

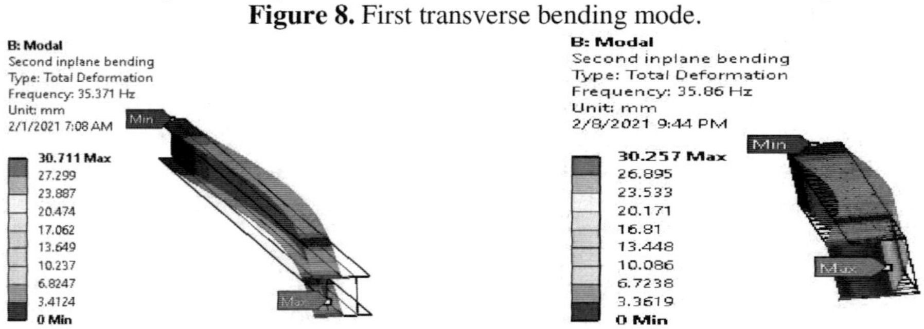

(a) Straight-web spar (b) Sine-web spar

Figure 9. Second in-plane bending mode.

4. Parametric Study

4.1 Web thickness effect.

To obtain the influence of web thickness variation on the bending and torsional rigidity and the modal analysis, a parametric study is conducted. The straight-web thickness is varied from 1 mm to 10 mm. To obtain a fair comparison, the equivalent sine-web thickness is determined to maintain the same weight of the corresponding straight-web spar. Hence, the sine-web thickness is varied from 0.635 mm to 6.354 mm. It is worth mentioning that in all the plots of the current sub-section, the x-axis presents the straight-web thickness. The flange thickness is constant (3 mm) in the current subsection, while the following sub-section investigates the effect of flange thickness variation.

Figure 10 shows the bending deformation due to a distributed pressure on the straight-web and the sine-web spars. The bending deformation decreases with increasing web thickness but the difference between the bending deformation of the straight-web and the sine-web spar increases starting at Δ w =17.529 % up to Δ w =75.832 %, according to the relation $\Delta w = \frac{w_S - w_I}{w_I}$ %.

Figure 11 shows the bending stress due to a distributed pressure on the straight-web and the sine-web spars. Generally, the bending stress decreases with increasing web thickness, but the difference between the bending stress of the straight-web and the sine-web spar varies starting at Δ σ = 204.777 % down to Δ σ = 97.199%, according to the relation $\Delta \sigma = \frac{\sigma_S - \sigma_I}{\sigma_I}$ %.

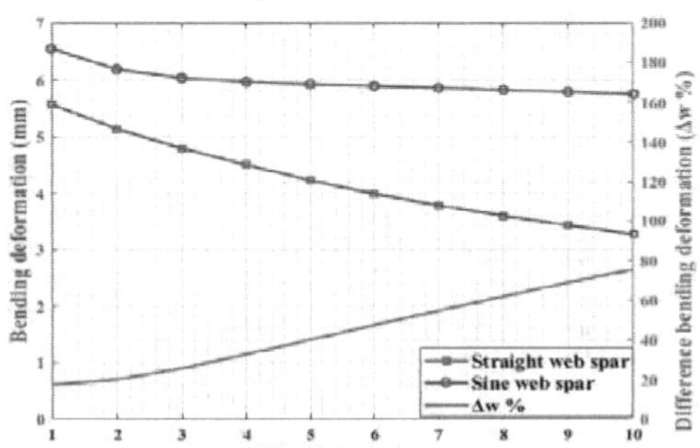

Figure 10. Bending deformation of straight and sine-web spar.

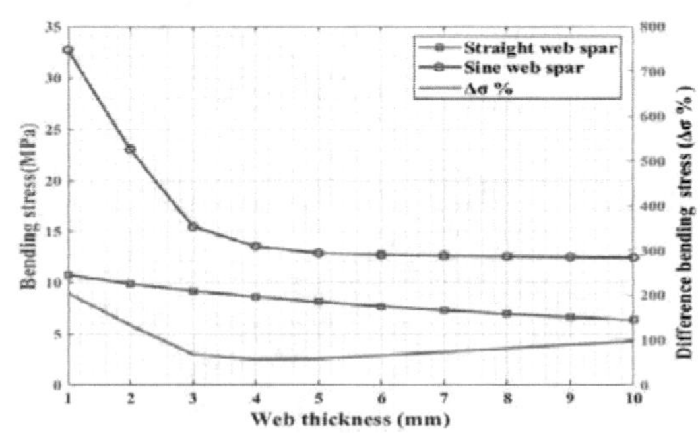

Figure 11. Bending stress of straight-web and sine-web spar.

Figure 12 and Figure 13 present the influence of web-thickness variation on the twist angle and the torsional stress, respectively, due to a static torsion on the free-end of the two spar configurations. As shown in Figure 12, the twist angle decreases with increasing the web thickness in both spar configurations. For small web thickness, the sine-web spar shows a torsional resistance greater than that of the straight-web spar, presented by a smaller twist angle. The difference of the twist angles starts at Δθ = -52.397 % at a straight web-thickness of 1 mm, then increases gradually to Δθ = 35.634 % at a straight-web thickness of 10 mm. The values of Δθ are determined according to the relation $\Delta\theta = \frac{\theta_S - \theta_I}{\theta_I}$ %.

IOP Conf. Series: Materials Science and Engineering 1172 (2021) 012002 doi:10.1088/1757-899X/1172/1/012002

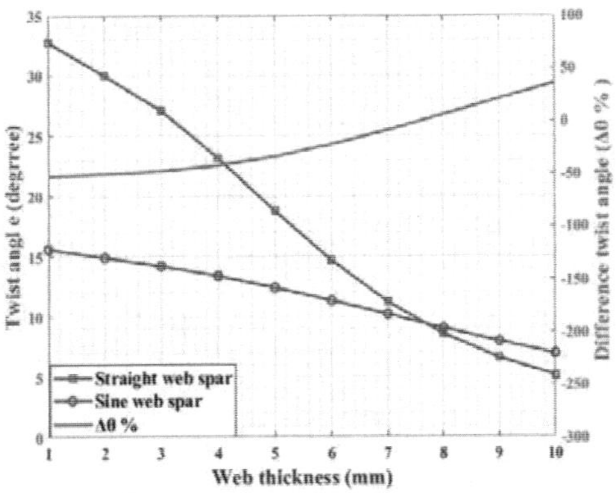

Figure 12. Twist angle due to torsion load on the straight-web and the sine-web.

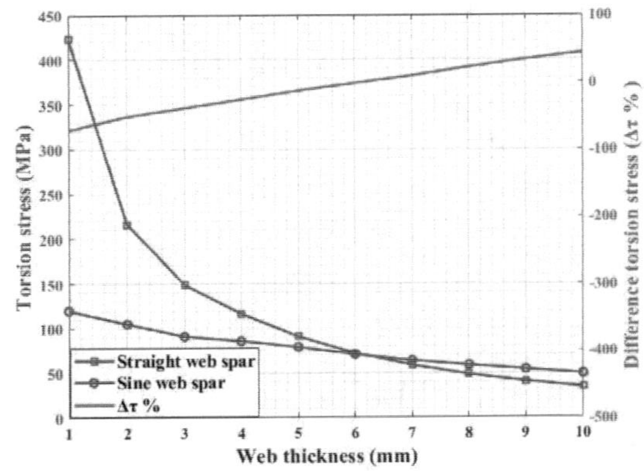

Figure 13. Torsion stress on the straight-web and the sine-web.

As shown in Figure 13, the torsion stress decreases with increasing web thickness but the difference of the torsion stress between the sine-web and the straight-web spars starts at Δ τ = -71.652 % at a straight web-thickness of 1 mm then increases to Δ τ = 43.119 % at a straight web-thickness of 10 mm. The values of Δτ are determined according to the relation $\Delta\tau = \frac{\tau_S - \tau_I}{\tau_{II}}$ %. Table 1 shows the results of the straight web spar and the sine web spar due to the distributed pressure and torsional load.

To compare the natural frequencies and mode shapes of the straight-web and sine-web spars of the same weight, the following parametric study is conducted by varying the web thickness. The results of the first four mode shapes are presented in Figure 14.

Figure 14 (a) shows the first mode shape of the straight-web and sine-web spars, the first in-plane bending mode. Generally, the first natural frequency of the sine-web spar is slightly greater than that of the straight-web spar. Figure 14 (b) shows the second mode shape of the straight-web spar, the first torsion mode and the third mode shape of the sine-web spar, the first torsion mode. Initially, the value of the first-torsion mode-shape of the sine-web spar is much greater than that of the straight-web spar at a straight-web thickness of 1 mm. Then, after a web-thickness of 6 mm, the value of the first-torsion mode-shape of the straight-web spar becomes greater. Figure 14 (c) shows the third mode shape of the straight-web spar, the first transverse bending mode and the second mode shape of the sine-web spar, the first transverse bending mode. Figure 14 (d) shows the fourth mode shape of the straight-web spar, the second in-plane bending mode and the fourth mode shape of the sine-web spar, the second in-plane bending mode, where there is a slight difference in the natural frequencies. Generally, the second in-plane bending mode natural frequency of the sine-web spar is slightly greater than that of the straight-web spar.

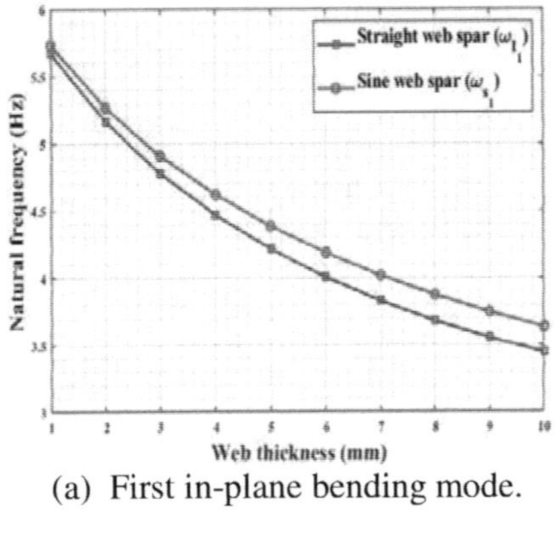

(a) First in-plane bending mode.

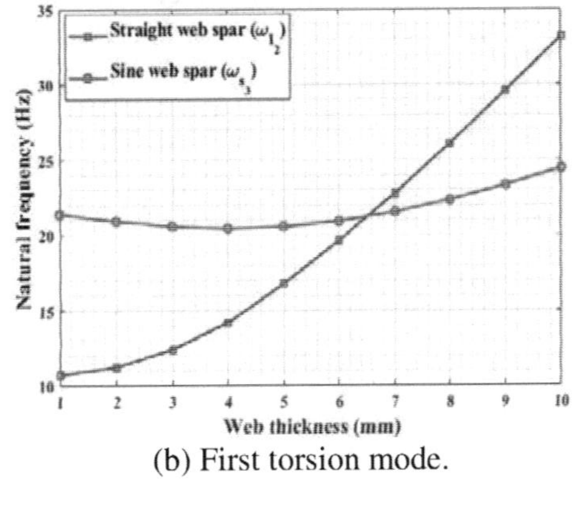

(b) First torsion mode.

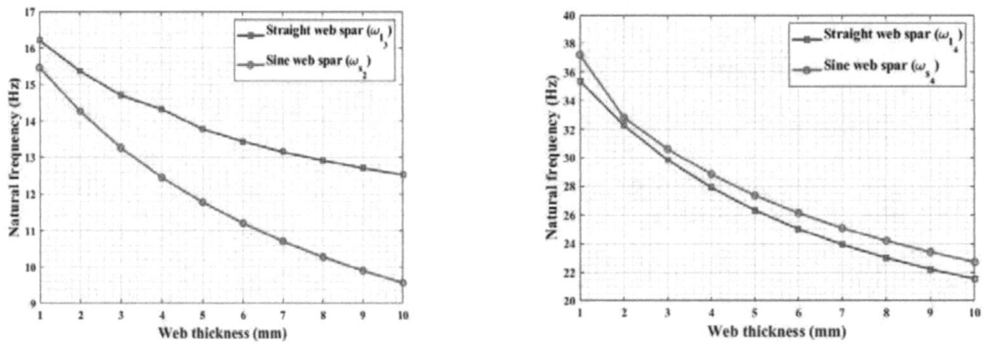

(c) First transverse bending mode (d) Second in-plane bending mode.

Figure 14. Mode shapes of straight and sine-web spars for variable web thickness (flange thickness is 3mm)

Table 1. Results of the straight web spar and the sine web spar due to the distributed pressure and torsional load.

		Distributed pressure				Torsion load			
		Straight web spar		Sine web spar		Straight web spar		Sine web spar	
Web thickness (mm)	Equivalent Web thickness (mm)	w_I (mm)	σ_I (MPa)	w_s (mm)	σ_s (MPa)	θ_I (deg.)	τ_I (MPa)	θ_s (deg.)	τ_s (MPa)
1	0.635	5.565	10.718	6.541	32.666	32.763	423.45	15.596	120.04
2	1.271	5.139	9.856	6.177	23.028	29.981	216.00	14.893	105.13
3	1.906	4.791	9.181	6.029	15.442	27.106	148.97	14.192	91.189
4	2.542	4.500	8.607	5.965	13.532	23.166	116.51	13.368	86.056
5	3.177	4.226	8.105	5.925	12.856	18.757	90.908	12.406	78.980
6	3.813	3.992	7.662	5.892	12.684	14.634	72.080	11.332	70.690
7	4.448	3.783	7.266	5.859	12.594	11.217	58.326	10.197	63.449
8	5.083	3.595	6.912	5.824	12.532	8.575	48.111	9.0581	58.544
9	5.719	3.425	6.589	5.788	12.476	6.598	40.349	7.9678	53.710
10	6.354	3.270	6.297	5.751	12.418	5.132	34.314	6.9605	49.110

4.2 Flange thickness effect.

A parametric study is conducted to investigate the effect of flange thickness variation on the static response and mode shapes of the straight-web and sine-web spars. The web thickness is maintained constant in the current sub-section. To keep the same weight of the two configurations, straight-web thickness is maintained at 1mm and the sine-web thickness is maintained at 0.63542 mm.

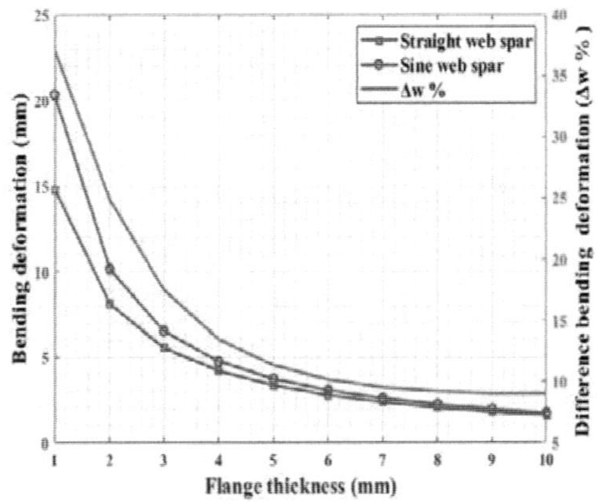

Figure 15. Bending deformation of straight and sine-web spars for variable flange thickness.

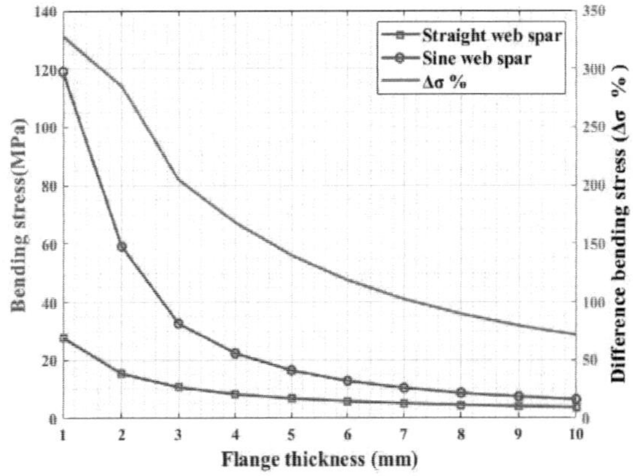

Figure 16. Bending stress on the straight-web and the sine-web for variable flange thickness.

Figure 15 shows the bending deformation due to distributed pressure on the straight-web and the sine-web spars at variable flange thicknesses. The bending deformation decreases with increasing the flange thickness but the difference between the bending deformation of the straight-web and the sine-web spar decreases starting at Δw =37.119 % down to Δw =9.041%. The values of Δw are obtained according to the relation $\Delta\,w = \frac{w_S - w_I}{w_I}\,\%$.

Figure 16 shows the bending stress due to distributed pressure on the straight-web and the sine-web spar at variable flange thicknesses. The bending stress decreases with increasing the flange thickness but the difference between the bending stress of the straight-web and the sine-web spar decreases starting at Δσ = 328.01 % down to Δ σ = 71.345 %. The values of Δσ are obtained according to the relation $\Delta\sigma = \frac{\sigma_S - \sigma_I}{\sigma_I}\,\%$.

Figure 17 shows the twist angle due to torsion load on the straight-web and the sine-web spars at variable flange thicknesses. The twist angle decreases with increasing flange thickness but difference of the twist angle has a small value for the straight-web at Δ θ = -5.320 % then this value decreases down to Δ θ= -85.532 %. The values of $\varDelta\theta$ are obtained according to the relation $\varDelta\,\theta = \frac{\theta_S - \theta_I}{\theta_I}\,\%$.

Figure 18 shows that the torsion stress decreases with increasing the flange thickness. The difference of the torsion stresses starts at Δ τ = -41.303 % at a straight-web thickness of 1 mm, then this value decreases down to Δ τ = -59.890 % at a straight-web thickness of 10 mm. The values of $\varDelta\,\tau$ are obtained according to the relation $\varDelta\,\tau = \frac{\tau_S - \tau_I}{\tau_{II}}\,\%$.

It is concluded that increasing flange thickness for the sine-web spar is better than the straight-web spar where the twist angle and the torsion stress due to the torsion load decrease with a higher rate than that of the straight-web spar.

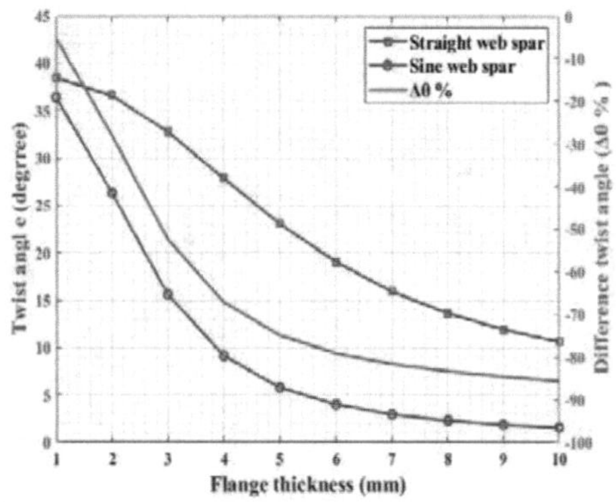

Figure 17. Twist angle due to torsion load on the straight-web and the sine-web for variable flange thickness.

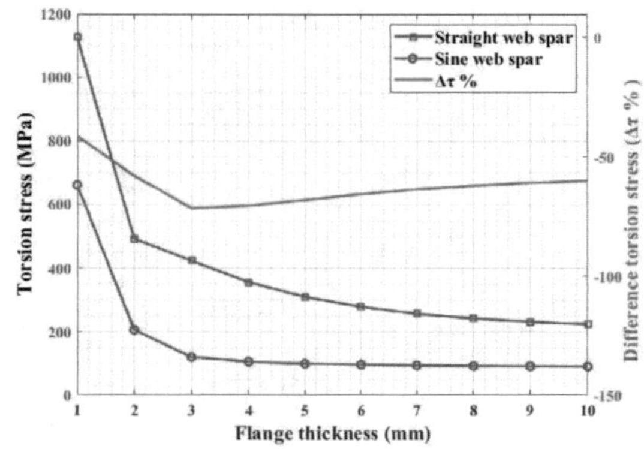

Figure 18. Torsion stress on the straight-web and the sine-web for variable flange thickness.

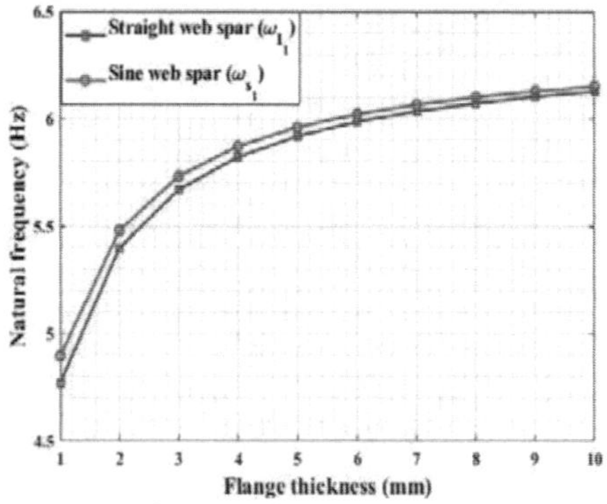

(a) First in-plane bending mode.

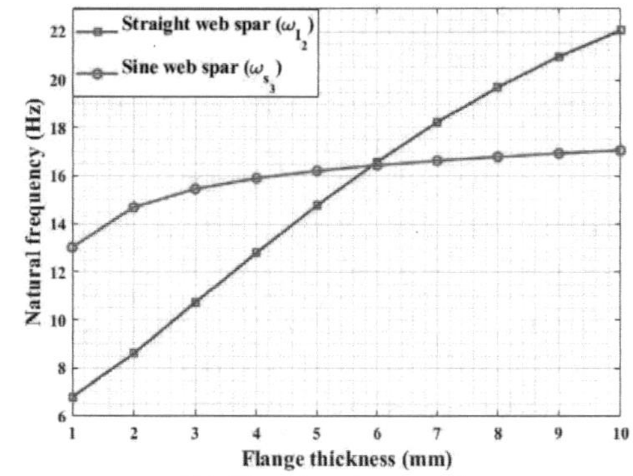

(b) First torsion mode.

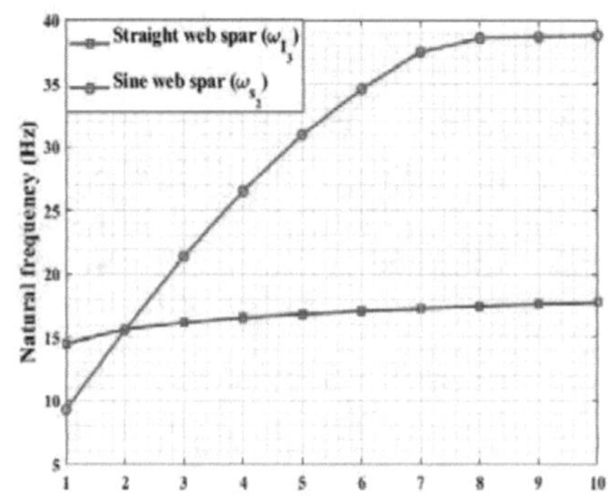

(c) First transverse bending mode

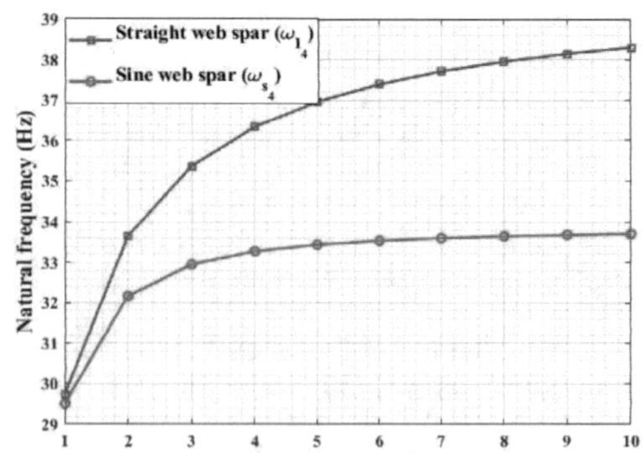

(d) Second in-plane bending mode.

Figure 19. Mode shapes of straight and sine-web spars for variable flange thickness (straight-web thickness is 1mm and the sine-web thickness is 0.63542mm)

The natural frequencies of the first four mode-shapes are obtained for the straight-web and sine-web spar to analyze the influence of flange thickness variation. Figure 19 (a) shows the first mode shape of the straight-web and the sine-web spars, the first in-plane bending mode, where there is a slight difference in the natural frequencies. Figure 19 (b) shows the second mode shape of the straight-web spar and the third mode shape of the sine-web spar, the first torsion mode. Figure 19 (c) shows the third mode shape of the straight-web spar and the second mode shape of the sine-web spar, the first transverse bending mode. Figure

19 (d) shows the fourth mode shape of the straight-web spar and the fourth mode shape of the sine-web spar, the second in-plane bending mode.

From the parametric study, it is concluded that varying the web thickness and the flange thickness greatly affects the static and dynamic performance of both straight-web and sine-web spar structures. Consequently, it is advisable to optimize the spar dimensions and configuration according to loading condition and application.

5. Conclusion

Static and modal analyses of sine-web spar are conducted and compared to the straight-web spar. To achieve a fair comparison, the sine-web thickness is selected to maintain the same weight of both configurations. Results show that although sine-web spar loses bending rigidity compared to the straight-web spar, it affords torsional resistance that exceeds that of the straight-web spar. It also retards the torsion mode by increasing its natural frequency. A parametric study is performed to investigate the effect of both web thickness and flange thickness variation on the static response and natural frequencies of the spar configurations. It is concluded that increasing the web thickness, while keeping the same weight of spars, increases the bending rigidity of the straight spar with a rate higher than that of the sine-web spar. It is clear that the web thickness needs to be optimized to obtain the optimum performance while keeping the weight as minimum as possible. This shows promising results to obtain an optimized spar structure that combines the merits of both spar configurations.

References

[1] Hadcok R 1977 The applications of advanced composites to military aircraft. I. p. 10.

[2] Figge F A and Bernhardt L 1975 Air superiority fighter wing structure design for improved cost, weight, and integrity. Journal of Aircraft, 12(8), pp.670-675.

[3] Perrons R K 1997 Make-buy decisions in the U.S aircraft industry,p.144.

[4] Ren Yiru X J 2016 Crashworthiness analysis of aircraft fuselage with sine-wave beam structure.Chinese Journal of Aeronautics, p. 8.

[5] Riley B L 1986 AV-SB/GR Mk 5 airframe composite applications. MechE, vol. 200, p. 17.

[6] Hanagud J I 1989 Energy absorption behavior of graphite epoxy composite sine webs. Journal of composite materials, vol. 23, p. 12.

[7] Gary R and Farley L 1992 Crushing characteristics of composite tubes with "near-elliptical" cross sections .Technomic Publishing Co.,Inc., p. 11.

[8] Bayraktar I 2012 Development of realistic simulation techniques to predict cure -induced microcracking in 3d woven composites.

[9] Pasternak D H Beams and Columns with sinusoidal corrugated webs. p. 8.

[10] Krzysztof S 2020 Finite element analysis of the stability of a sinusoidal web in steel and composite steel-concrete girders, p. 15.

[11] Johnson R R W 1978 Sine wave beam web and method of manufacture,united states patent 818,910.

[12] Jenkis J M 1979 Criteria for representing circular arc and sine wave spar webs by non-curved elements, national technical information service, p. 15

IOP Conf. Series: Materials Science and Engineering 1172 (2021) 012003 doi:10.1088/1757-899X/1172/1/012003

Vibration analysis of composite wing with geometric and material coupling

A Magdy[1], M Kamel[1], M A Elshafei[1] and M Kassem[1]

[1]Department of Aircraft Mechanics, Military Technical College, Egypt

E-mail: a.magdy@mtc.edu.eg

Abstract. Composite wing design is complicated but inevitable to enlighten modern airplanes while maintaining the required performance. Using the dynamic transfer method, this paper discusses intensively the dynamic characteristics of a cantilever composite wing with both torsion and bending coupling to represent both material and geometric coupling. The governing differential equations are obtained based upon the principle of Hamilton and are solved analytically using a harmonic oscillation assumption. For this purpose, a MATLAB code is developed and results are validated in comparison with published work. Such a comparison shows a good agreement between both results. Finally, a parametric study is carried out to show the influence of the variation of both geometric coupling and torsion bending coupling rigidity on the free vibration analysis of the composite wing. The study shows the crucial effect of both factors on the dynamic behavior of the composite wing. The current research can be considered as a base for aeroelasticians while designing composite structures.

1. Introduction

Free vibration analysis of composite beams has many practical engineering applications in mechanical and aerospace structural designs such as spacecraft, rotor blades of helicopters, some parts of robots and aircraft wings. It's a very important prerequisite when carrying out a response and aeroelastic analysis. Controlling and improving the dynamic effects of a structural element is a very important desirable target for all designers, this can be done in composite materials by changing the stacking sequence and ply orientation of fibers which in turns control the coupling between different modes of deformation due to the anisotropic properties of composite materials. Due to this coupling, composite and metallic structures have two different free vibration analyses [1].

The vibration analysis of a composite cantilever beam is chosen because of its important applications in the idealization of the structural elements such as the composite wing of high aspect ratio. The characteristics of the free vibration analysis of composite beams can be controlled favorably by selecting a suitable stacking sequence and ply orientation [2].

The coupling of composite materials occurs due to two main sources, the geometric coupling is one of these sources which comes from the distance between the shear center and the cross-sectional centroid of the beam. This kind of coupling occurs in the asymmetric cross-sections, of composite beams and metallic beams. Because of the inertial nature of this coupling bending and torsion motions under static loads are uncoupled.

Content from this work may be used under the terms of the Creative Commons Attribution 3.0 licence. Any further distribution of this work must maintain attribution to the author(s) and the title of the work, journal citation and DOI.
Published under licence by IOP Publishing Ltd

The other source of coupling occurs only in composite materials because its anisotropic properties arise from fiber orientations. This coupling depends only on material properties so, it is called material coupling, bending and torsion motions can occur under static and dynamic loads. For double symmetric cross sections, the coupling depends only on the material coupling and there is no effect for the geometric one [1]. Studying the free vibration analysis of composite beams has attracted the interest of many researchers due to their practical importance and potential advantages stated above. There are many analytical, numerical and experimental approaches are used for the vibration analysis of coupled bending and torsion in composite beams.

Hodges et al. [3] has proposed methods to predict composite beams' natural frequencies and mode shapes, two qualitatively different methods are used to evaluate the sectional elastic constants: simple analytical methods to calculate the stiffnesses that are given in a closed form and a detailed cross-sectional finite element method. They solved the equations of motion by an exact integration method and by a mixed finite element method. Eslimy-Isfany and Banerjee [4] developed a theory suitable for either open-section or closed-section composite beams of any cross-section, stacking sequence, and boundary conditions. They calculated the response of composite beams under the action of deterministic and random loads. Coupling between bending and torsion deformation, which comes from the anisotropic properties of the fibrous composite, investigated in their work is the material coupling.

RKaya and Ozgumus [5] used Hamilton's principle to derive the governing equations of motion of composite Timoshenko beams and solved them by using the differential transformation method to investigate the effects on natural frequencies due to bending-torsion coupling, the axial force and the slenderness ratio are studied. Mirtalaie et.al [6] presented a numerical solution using the method of differential quadrature to solve the coupling of torsional and lateral vibrations by modeling it with coupling rigidity of the bending twisting material. They took into account the effect of material coupling, shear deformation and rotary inertia.

Lottati [7] made an investigation for a swept forward composite cantilever rectangular wing carrying a pylon at the wingtip and a fuselage at its semi span assuming that the case of an unrestrained vehicle. He analyzed the variation of the divergence and flutter velocities due to the warping effect. He deduced that the aeroelastic characteristic changes with the warping effect. Lottati [8] made a similar analysis for the same wing without any load to obtain its divergence behavior and the aeroelastic flutter. He analyzed the variation in the influence of the torsion-bending stiffness coupling of that wing on the critical dynamic pressure of the flutter and divergence. He indicated that increasing the flutter velocity tends to decrease the divergence speed.

Kashani and Hashemi [9] made a free vibrations study for single delamination composite beams undergoing bending-torsion coupling by using the finite element method to analyze the delaminated beams subjected to tip moment and axial compressive load. He obtained a linear eigenvalue problem by discretizing the beam along its span to find the natural frequencies and mode shapes from "free mode" and "constrained mode".

In this paper, the dynamic transfer method is used to conduct free vibration analysis for a cantilever composite wing to calculate its natural frequencies and mode shapes and comparing them with a published work to validate this study then a parametric study is made to show the effect of geometric and material parameters on the natural frequency of the wing.

2. Theory

2.1. Formulation of the governing equations

In this part, the governing equations of a composite beam coupled in bending and torsional vibration are discussed as an idealized cross-section for the aircraft wing as shown in fig. 1. Y-axis represents the elastic axis of the wing, x_α is the distance between the elastic axis and the mass axis which represents the loci of the geometric mass centers of the wing cross-sections, EI is the bending rigidity, GJ is the torsion rigidity, K is the torsion bending coupling rigidity, m is the mass per unit length, I_α is mass moment of inertia per unit length about the Y-axis, L is the length of the wing, h(y , t) is the deflection out of the plane $\psi(y, t)$ is the angle of rotation about the Y-axis. The most important parameters are K and x_α because if x_α exists means there is no coincident between the shear center axis and the mass axis (geometric coupling) and K is an indication to the material coupling [1].

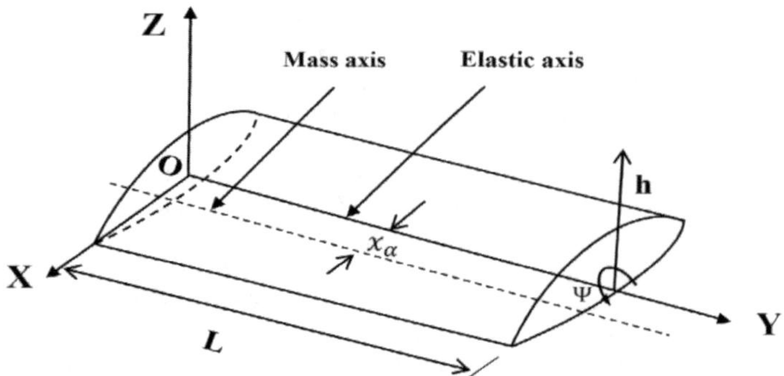

Figure 1. The coordinate system and notation for a bending–torsion coupled composite beam

It should be remembered that the presented theory does not involve the influence of rotatory inertia and shear deformation, it is applicable to analyze composite beams with high length to cross-sectional dimension ratio, for example, aircraft wing with high aspect ratio [10, 11]. The definition of kinetic energy and the potential energy for a composite beam subjected to torsion bending coupling are respectively given by [1].

$$T = \frac{1}{2}\int_0^L [m(\dot{h}) - 2mx_\alpha \dot{h}\dot{\psi} + I_\alpha (\dot{\psi})^2] dy \tag{1}$$

$$V = \frac{1}{2}\int_0^L [EI(h'') + 2Kh''\psi' + GJ(\psi')^2] dy \tag{2}$$

where an over dot and a prime denote partial differentiation with respect to time t and location y respectively. Using the definition of Hamilton's principal

$$\delta \int_{t1}^{t2} (T - V) dt = 0 \tag{3}$$

where the integration is made between the time interval from t_1 to t_2 and δ is the variational operator. Using the definitions of kinetic energy and potential energy to substitute into eq. (3), The governing equations in the free vibration analysis are given by:

$$EIh'''' + K\psi''' + m\ddot{h} - mx_\alpha \ddot{\psi} = 0 \tag{4}$$

$$GJ\psi'' + Kh''' - I_\alpha \ddot{\psi} + mx_\alpha \ddot{h} = 0 \tag{5}$$

Due to the existence of coupling, shearing force (F), bending moment (M_b) and torque moment (M_T) can be expressed by:

$$F = EIh''' + k\psi'' \tag{6a}$$
$$M_b = -EIh'' - K\psi' \tag{6b}$$
$$M_T = -Kh'' - GJ\psi' \tag{6c}$$

Assuming harmonic oscillation for h and ψ.

$$\begin{cases} h\,(y\,,t) = H(y)e^{i\omega t} \\ \psi\,(y\,,t) = \Psi(y)e^{i\omega t} \end{cases} \tag{7}$$

where ω represents the angular frequency, H is the amplitude of h and Ψ is the amplitude of ψ. By substituting Eq. (7) into Eqs. (4) and (5) the governing equations become:

$$EIH'''' + K\Psi''' - m\omega^2 H + m\omega^2 x_\alpha \Psi = 0 \tag{8}$$

$$GJ\Psi'' + KH''' + I_\alpha\omega^2\Psi - m\omega^2 x_\alpha H = 0 \tag{9}$$

Using the definition of the dimensionless length ζ and the differential operator D where

$$\zeta = \frac{y}{L} \qquad D = \frac{d}{d\zeta} \tag{10}$$

Putting Eqs. (8) and (9) into matrix form [12]

$$\begin{bmatrix} EI\frac{D^4}{L^4} - m\omega^2 & K\frac{D^3}{L^3} + m\omega^2 x_\alpha \\ K\frac{D^3}{L^3} - m\omega^2 x_\alpha & GJ\frac{D^2}{L^2} + I_\alpha\omega^2 \end{bmatrix} \begin{Bmatrix} H(\zeta) \\ \Psi(\zeta) \end{Bmatrix} = \begin{Bmatrix} 0 \\ 0 \end{Bmatrix} \tag{11}$$

Then the differential equations of the bending displacement $H(\zeta)$and the torsion rotation $\Psi(\zeta)$ are given by:

$$\left[\left(EI\frac{D^4}{L^4} - m\omega^2\right)\left(GJ\frac{D^2}{L^2} + I_\alpha\omega^2\right) - \left(K^2\frac{D^6}{L^6} - m^2\omega^4{x_\alpha}^2\right)\right] H(\zeta) = 0 \tag{12}$$

$$\left[\left(EI\frac{D^4}{L^4} - m\omega^2\right)\left(GJ\frac{D^2}{L^2} + I_\alpha\omega^2\right) - \left(K^2\frac{D^6}{L^6} - m^2\omega^4{x_\alpha}^2\right)\right] \Psi(\zeta) = 0 \tag{13}$$

Equations (12) and (13) can be simplified into sixth order as stated below.

$$\left[(EIGJ - K^2)\frac{D^6}{L^6} + (EII_\alpha\omega^2)\frac{D^4}{L^4} - (m\omega^2 GJ)\frac{D^2}{L^2} - m\omega^4 I_\alpha + m^2\omega^4{x_\alpha}^4\right] H(\zeta) = 0 \tag{14}$$

$$\left[(EIGJ - K^2)\frac{D^6}{L^6} + (EII_\alpha\omega^2)\frac{D^4}{L^4} - (m\omega^2 GJ)\frac{D^2}{L^2} - m\omega^4 I_\alpha + m^2\omega^4{x_\alpha}^4\right] \Psi(\zeta) = 0 \tag{15}$$

Representing the characteristic equations of the two previous equations [12]

$$(D^6 + aD^4 - bD^2 - abc)W = 0 \tag{16}$$

Where W refer to H or Ψ and

$$a = \frac{\bar{a}}{1-k_m} \quad , \quad b = \frac{\bar{b}}{1-k_m} \quad , \quad c = (1-k_g)(1-k_m) \tag{17}$$

$$\bar{a} = \frac{I_\alpha\omega^2 L^2}{GJ} \quad , \quad \bar{b} = \frac{m\omega^2 L^4}{EI} \quad , \quad k_g = \frac{m{x_\alpha}^2}{I_\alpha} \quad , \quad k_m = \frac{K^2}{EI\,GJ} \tag{18}$$

Using trial solution to solve eq. (16) [13] , let $W = e^{n\zeta}$ then the auxiliary equation is

$$N^6 + aN^4 - bN^2 - abc = 0 \tag{19}$$

let $\lambda = P^2$ then eq. (19) becomes

$$\lambda^3 + a\lambda^2 - b\lambda - abc = 0 \tag{20}$$

the previous equation can be reduced to the standard form

$$x^3 - qx - r = 0 \tag{21}$$

where

$$x = \lambda + \frac{a}{3} \quad , \quad q = b + \frac{a^2}{3} \quad , \quad r = a\left(bc - \frac{b}{3} - \frac{2a^2}{27}\right) \tag{22}$$

Let $\delta = 27r^2 - 4q^3$ then there are three cases for the three roots of eq. (21)

- $\delta > 0$ there are two conjugate imaginaries and one is real.
- $\delta = 0$ they are all real roots at least two of them are equal.
- $\delta < 0$ they are all real and unequal, one is positive and the other two are negative. $\delta < 0$ for all the physical meaningful values of a, b and c then [14]

$$\begin{cases} x_1 = 2(q/3)^{1/2}\cos(\phi/3) \\ x_2 = -2(q/3)^{1/2}\cos\{(\pi-\phi)/3\} \\ x_2 = -2(q/3)^{1/2}\cos\{(\pi+\phi)/3\} \end{cases} \quad (23)$$

where

$$\phi = cos^{-1}\frac{27abc-9ab-2a^3}{2(a^2+3b)^{3/2}} \quad (24)$$

Substituting eq. (23) into the definition of x given in eq. (22), the sixth roots are the plus and the minus sign of the square roots of the three values of λ, so if the sixth roots are given by α, $-\alpha$, $i\beta$, $-i\beta$, $i\gamma$ and $-i\gamma$ with α, β and γ are real. They are defined by [2]:

$$\left.\begin{aligned} \alpha &= \left[2(q/3)^{1/2}cos(\phi/3) - a/3\right]^{1/2} \\ \beta &= \left[2(q/3)^{1/2}cos\{(\pi-\phi)/3\} + a/3\right]^{1/2} \\ \gamma &= \left[2(q/3)^{1/2}cos\{(\pi+\phi)/3\} + a/3\right]^{1/2} \end{aligned}\right\} \quad (25)$$

Then the solution of eq. (16) is

$$W(\zeta) = a_1 cosh\alpha\zeta + a_2 sinh\alpha\zeta + a_3 cos\beta\zeta + a_4 sin\beta\zeta + a_5 cos\gamma\zeta + a_6 sin\gamma\zeta \quad (26)$$

Where $a_1 - a_2$ are constants, defining the expressions of H and Ψ.

$$H(\zeta) = A_1 cosh\alpha\zeta + A_2 sinh\alpha\zeta + A_3 cos\beta\zeta + A_4 sin\beta\zeta + A_5 cos\gamma\zeta + A_6 sin\gamma\zeta \quad (27)$$

$$\Psi(\zeta) = B_1 cosh\alpha\zeta + B_2 sinh\alpha\zeta + B_3 cos\beta\zeta + B_4 sin\beta\zeta + B_5 cos\gamma\zeta + B_6 sin\gamma\zeta \quad (28)$$

by substitution with the definitions of bending displacement and torsion rotation in eq. (8)

$$\begin{aligned} B_1 &= \tfrac{1}{L}(-A_1 e_\alpha g_\alpha + A_2 e_\alpha) \quad B_2 = \tfrac{1}{L}(A_1 e_\alpha - A_2 e_\alpha g_\alpha) \quad B_3 = \tfrac{1}{L}(A_3 e_\beta g_\beta + A_4 e_\beta) \\ B_4 &= \tfrac{1}{L}(-A_3 e_\beta + A_4 e_\beta g_\beta) \quad B_5 = \tfrac{1}{L}(A_5 e_\gamma g_\gamma + A_6 e_\gamma) \quad B_6 = \tfrac{1}{L}(-A_5 e_\gamma + A_6 e_\gamma g_\gamma) \end{aligned} \quad (29)$$

where

$$\begin{aligned} g_\alpha &= \bar{b}k_\delta/\alpha^3 \qquad g_\beta = \bar{b}k_\delta/\beta^3 \qquad g_\gamma = \bar{b}k_\delta/\gamma^3 \\ e_\alpha &= k_\alpha/(1-g_\alpha{}^2) \quad e_\beta = k_\beta/(1+g_\beta{}^2) \quad e_\gamma = k_\gamma/(1+g_\gamma{}^2) \end{aligned} \quad (30)$$

$$k_\alpha = \frac{EI}{K}\left(\frac{\bar{b}-\alpha^4}{\alpha^3}\right), \quad k_\beta = \frac{EI}{K}\left(\frac{\bar{b}-\beta^4}{\beta^3}\right), \quad k_\gamma = \frac{EI}{K}\left(\frac{\bar{b}-\gamma^4}{\gamma^3}\right), \quad k_\delta = \frac{EI}{K}\frac{x_\alpha}{L} \quad (31)$$

Substituting Eq. (29) into Eq. (28), the expression of Ψ will be:

$$\Psi(\zeta) = \frac{1}{L}\left(A_1 u_\alpha + A_2 v_\alpha + A_3 u_\beta + A_4 v_\beta + A_5 u_\gamma + A_6 v_\gamma\right) \quad (32)$$

where

$$\begin{aligned} u_\alpha &= e_\alpha(sinh\alpha\zeta - g_\alpha cosh\alpha\zeta) \quad v_\alpha = e_\alpha(cosh\alpha\zeta - g_\alpha sinh\alpha\zeta) \\ u_\beta &= e_\beta(-sin\beta\zeta + g_\beta cos\beta\zeta) \quad v_\beta = e_\beta(cos\beta\zeta + g_\beta sin\beta\zeta) \\ u_\gamma &= e_\gamma(-sin\gamma\zeta + g_\gamma cos\gamma\zeta) \quad v_\alpha = e_\gamma(cos\gamma\zeta + g_\gamma sin\gamma\zeta) \end{aligned} \quad (33)$$

The bending rotation $\Theta(\zeta)$ can be expressed by $\Theta(\zeta) = \frac{dH}{dy} = \frac{1}{L}\frac{dH}{d\zeta}$ then

$$\Theta(\zeta) = \frac{1}{L}(A_1\alpha sinh\alpha\zeta + A_2\alpha cosh\alpha\zeta - A_3\beta sin\beta\zeta + A_4 cos\beta\zeta - A_5\gamma sin\gamma\zeta + A_6\gamma cos\gamma\zeta) \quad (34)$$

Representing the expressions of shear force $S(\zeta)$, bending moment $M(\zeta)$ and torque moment $T(\zeta)$ mentioned in Eq. (6)

$$F(\zeta) = \frac{EI}{L^3}\frac{d^3H}{d\zeta^3} + \frac{K}{L^2}\frac{d^2\Psi}{d\zeta^2} = W_3\big[A_1(\alpha P_\alpha sinh\alpha\zeta - \alpha^2 e_\alpha g_\alpha k_b cosh\alpha\zeta) + A_2(\alpha P_\alpha cosh\alpha\zeta - \alpha^2 e_\alpha g_\alpha k_b sinh\alpha\zeta) + A_3(\beta P_\beta sin\beta\zeta - \beta^2 e_\beta g_\beta k_b cos\beta\zeta) - A_4(\beta P_\beta cos\beta\zeta + \beta^2 e_\beta g_\beta k_b cos\beta\zeta) + A_5(\gamma P_\gamma sin\gamma\zeta - \gamma^2 e_\gamma g_\gamma k_b cos\gamma\zeta) - A_6(\gamma P_\gamma cos\gamma\zeta + \gamma^2 e_\gamma g_\gamma k_b sin\gamma\zeta)\big] \tag{35}$$

$$M_b(\zeta) = -\frac{EI}{L^2}\frac{d^2H}{d\zeta^2} - \frac{K}{L}\frac{d\Psi}{d\zeta} = -W_2\big[A_1(P_\alpha cosh\alpha\zeta - \alpha e_\alpha g_\alpha k_b sinh\alpha\zeta) + A_2(sinh\alpha\zeta - \alpha e_\alpha g_\alpha k_b cosh\alpha\zeta) - A_3(P_\beta cos\beta\zeta + \beta e_\beta g_\beta k_b sin\beta\zeta) - A_4(P_\beta sin\beta\zeta - \beta e_\beta g_\beta k_b cos\beta\zeta) - A_5(P_\gamma cos\gamma\zeta + \gamma e_\gamma g_\gamma k_b sin\gamma\zeta) - A_6(P_\gamma sin\gamma\zeta - \gamma e_\gamma g_\gamma k_b cos\gamma\zeta)\big] \tag{36}$$

$$M_T(\zeta) = -\frac{K}{L^2}\frac{d^2H}{d\zeta^2} - \frac{GJ}{L}\frac{d\Psi}{d\zeta} = -\frac{W_1}{L}\big[A_1(q_\alpha cosh\alpha\zeta - \alpha e_\alpha g_\alpha sinh\alpha\zeta) + A_2(q_\alpha sinh\alpha\zeta - \alpha e_\alpha g_\alpha cosh\alpha\zeta) - A_3(q_\beta cos\beta\zeta + \beta e_\beta g_\beta sin\beta\zeta) - A_4(q_\beta sin\beta\zeta - \beta e_\beta g_\beta cos\beta\zeta) - A_5(q_\gamma cos\gamma\zeta + \gamma e_\gamma g_\gamma sin\gamma\zeta) - A_6(P_\gamma sin\gamma\zeta - \gamma e_\gamma g_\gamma cos\gamma\zeta)\big] \tag{37}$$

where

$$W_3 = \frac{EI}{L^3} \quad , \quad W_2 = \frac{EI}{L^2} \quad , \quad W_1 = \frac{GJ}{L} \tag{38}$$

$$\begin{matrix} P_\alpha = \alpha(\alpha + e_\alpha k_b) & P_\beta = \beta(\beta + e_\beta k_b) & P_\gamma = \gamma(\gamma + e_\gamma k_b) \\ q_\alpha = \alpha(e_\alpha + \alpha k_t) & q_\beta = \beta(e_\beta + \beta k_t) & q_\gamma = \gamma(\gamma + \gamma k_t) \end{matrix} \tag{39}$$

$$k_b = \frac{x_\alpha}{Lk_\delta} \quad , \quad k_t = \frac{k_m k_\delta L}{x_\alpha} \tag{40}$$

2.2. *Application of Dynamic transfer matrix method*

The boundary conditions as shown in Figs. (2) and (3)

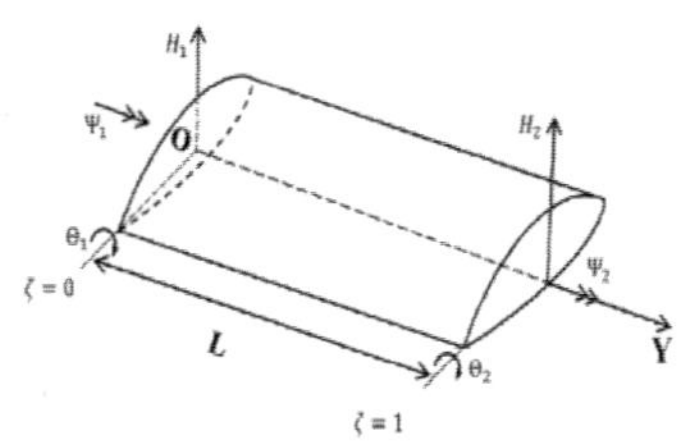

Figure 2. Boundary conditions for displacements.

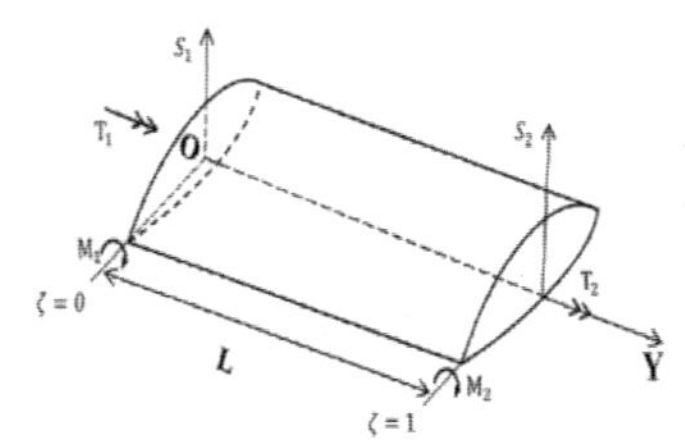

Figure 3. Boundary conditions for forces.

Displacements

$$\left.\begin{matrix} \text{at } y = 0\ (\xi = 0) & H = 0 \ , \Theta = 0 \ , \Psi = 0 \\ \text{at } y = L\ (\xi = 1) & H = H_2 \ , \Theta = \Theta_2 \ , \Psi = \Psi_2 \end{matrix}\right\} \tag{41}$$

Forces

$$\left.\begin{matrix} \text{at } y = 0\ (\xi = 0) & F = F_1, M_b = M_{b1}, M_T = M_{T1} \\ \text{at } y = L\ (\xi = 1) & F = 0 \ , M_b = 0 \ , M_T = 0 \end{matrix}\right\} \tag{42}$$

for a cantilever beam, fixed at the root ($\xi = 0$) and free at the tip ($\xi = 1$). Substituting with these boundary conditions into Eqs. (27), (32), (34), (35), (36) and (37)

$$\{Q\}_{F_x} = [R(0,\omega)]_{Fx}\{A\} \tag{43}$$

$$\{Q\}_{F_r} = [R(1,\omega)]_{Fr}\{A\} \tag{44}$$

IOP Conf. Series: Materials Science and Engineering 1172 (2021) 012003 doi:10.1088/1757-899X/1172/1/012003

where F_x and F_r refer to fixed end and free end, respectively and

$$\left.\begin{array}{r} \{Q\}_{F_x} = \{H_1\ \Theta_1\ \Psi_1\ S_1\ M_1\ T_1\}^T \\ \{Q\}_{F_r} = \{H_2\ \Theta_2\ \Psi_2\ S_2\ M_2\ T_2\}^T \\ \{A\} = \{A_1\ A_1\ A_1\ A_1\ A_1\ A_1\}^T \end{array}\right\} \tag{45}$$

From Eq. (43) and Eq. (44) the matrix relates the two ends with their boundary conditions will be

$$\{Q\}_{F_r} = [R]_{F_r}[R]_{F_x}{}^{-1}\{Q\}_{F_x} \tag{46}$$

Then the natural frequencies and mode shapes will be obtained by satisfying the boundary conditions and substituting in Eq. (46) with Eqs. (27), (32), (34), (35), (36) and (37), where $[R]_{F_r}[R]_{F_x}{}^{-1}$ is called the dynamic transfer matrix and the matrices $[R]_{F_x}$ and $[R]_{F_r}$ are defined as follows:

$$[R]_{F_x} = \begin{bmatrix} 1 & 0 & 1 & 0 & 1 & 0 \\ 0 & \frac{\alpha}{L} & 0 & \frac{\beta}{L} & 0 & \frac{\gamma}{L} \\ \frac{-e_\alpha g_\alpha}{L} & \frac{e_\alpha}{L} & \frac{e_\beta g_\beta}{L} & \frac{e_\beta}{L} & \frac{e_\gamma g_\gamma}{L} & \frac{e_\gamma}{L} \\ \alpha^2 e_\alpha g_\alpha k_b W_3 & \alpha P_\alpha W_3 & -B^2 e_\beta g_\beta k_b W_3 & -\beta P_\beta W_3 & -\gamma^2 e_\gamma g_\gamma k_b W_3 & -\gamma P_\gamma W_3 \\ -P_\alpha W_2 & \alpha e_\alpha g_\alpha k_b W_2 & P_\beta W_2 & -\beta e_\beta g_\beta k_b W_2 & P_\gamma W_2 & -\gamma e_\gamma g_\gamma k_b W_2 \\ -q_\alpha W_1/L & \alpha e_\alpha g_\alpha W_1/L & q_\beta W_1/L & -\beta e_\beta g_\beta W_1/L & q_\gamma W_1/L & -\gamma e_\gamma g_\gamma W_1/L \end{bmatrix} \tag{47}$$

$$\left[[R]_{F_r}\right]_{11} = \begin{bmatrix} C_{h\alpha} & S_{h\alpha} & C_\beta \\ \frac{\alpha}{L}S_{h\alpha} & \frac{\alpha}{L}C_{h\alpha} & -\frac{\beta}{L}S_\beta \\ \frac{u_\alpha{}^*}{L} & \frac{v_\alpha{}^*}{L} & \frac{u_\beta{}^*}{L} \end{bmatrix} \qquad \left[[R]_{F_r}\right]_{12} = \begin{bmatrix} S_\beta & C_\gamma & S_\gamma \\ \frac{\beta}{L}C_\beta & -\frac{\gamma}{L}S_\gamma & -\frac{\gamma}{L}C_\gamma \\ \frac{v_\beta{}^*}{L} & \frac{u_\gamma{}^*}{L} & \frac{v_\gamma{}^*}{L} \end{bmatrix} \tag{48}$$

where

$$\left.\begin{array}{lll} C_{h\alpha} = \cosh\alpha & C_\beta = \cos\beta & C_\gamma = \cos\gamma \\ S_{h\alpha} = \sinh\alpha & S_\beta = \sin\beta & S_\gamma = \sin\gamma \\ u_\alpha{}^* = e_\alpha(S_{h\alpha} - g_\alpha C_{h\alpha}) & & v_\alpha{}^* = e_\alpha(C_{h\alpha} - g_\alpha S_{h\alpha}) \\ u_\beta{}^* = e_\beta(-S_\beta + g_\beta C_\beta) & & v_\beta{}^* = e_\beta(C_\beta + g_\beta S_\beta) \\ u_\gamma{}^* = e_\gamma(-S_\gamma + g_\gamma C_\gamma) & & v_\gamma{}^* = e_\gamma(C_\gamma + g_\gamma S_\gamma) \end{array}\right\} \tag{49}$$

$$\left[[R]_{F_r}\right]_{21} = \begin{bmatrix} \alpha(\alpha e_\alpha g_\alpha k_b C_{h\alpha} - P_\alpha S_{h\alpha})W_3 & \alpha(\alpha e_\alpha g_\alpha k_b S_{h\alpha} - P_\alpha C_{h\alpha})W_3 & \beta(\beta e_\beta g_\beta k_b C_\beta - P_\beta S_\beta)W_3 \\ (P_\alpha C_{h\alpha} - \alpha e_\alpha g_\alpha k_b S_{h\alpha})W_2 & (P_\alpha S_{h\alpha} - \alpha e_\alpha g_\alpha k_b C_{h\alpha})W_2 & -(P_\beta C_\beta + \beta e_\beta g_\beta k_b S_\beta)W_2 \\ (q_\alpha C_{h\alpha} - \alpha e_\alpha g_\alpha S_{h\alpha})\frac{W_1}{L} & (q_\alpha S_{h\alpha} - \alpha e_\alpha g_\alpha C_{h\alpha})\frac{W_1}{L} & -(q_\beta C_\beta + \beta e_\beta g_\beta S_\beta)\frac{W_1}{L} \end{bmatrix} \tag{50}$$

$$\left[[R]_{F_r}\right]_{22} = \begin{bmatrix} \beta(P_\beta C_\beta + \beta e_\beta g_\beta k_b S_\beta)W_3 & \gamma(\gamma e_\gamma g_\gamma k_b C_\gamma - P_\gamma S_\gamma)W_3 & \gamma(P_\gamma C_\gamma + \gamma e_\gamma g_\gamma k_b S_\gamma)W_3 \\ (\beta e_\beta g_\beta k_b C_\beta - P_\beta S_\beta)W_2 & -(P_\gamma C_\gamma + \gamma e_\gamma g_\gamma k_b S_\gamma)W_2 & -(P_\gamma S_\gamma - \gamma e_\gamma g_\gamma k_b C_\gamma)W_2 \\ (\beta e_\beta g_\beta C_\beta - q_\beta S_\beta)\frac{W_1}{L} & -(q_\gamma C_\gamma + \gamma e_\gamma g_\gamma S_\gamma)\frac{W_1}{L} & -(q_\gamma S_\gamma - \gamma e_\gamma g_\gamma C_\gamma)\frac{W_1}{L} \end{bmatrix} \tag{51}$$

$$\left[[R]_{F_r}\right] = \begin{bmatrix} \left[[R]_{F_r}\right]_{11} & \left[[R]_{F_r}\right]_{12} \\ \left[[R]_{F_r}\right]_{21} & \left[[R]_{F_r}\right]_{22} \end{bmatrix} \tag{52}$$

2.3. Natural frequencies

substituting with the boundary conditions into Eq. (46), the following matrix equation is obtained

$$\begin{bmatrix} \bar{r}_{44} & \bar{r}_{45} & \bar{r}_{46} \\ \bar{r}_{54} & \bar{r}_{55} & \bar{r}_{56} \\ \bar{r}_{64} & \bar{r}_{65} & \bar{r}_{66} \end{bmatrix} \begin{Bmatrix} S_1 \\ M_1 \\ T_1 \end{Bmatrix} = \begin{Bmatrix} 0 \\ 0 \\ 0 \end{Bmatrix} \tag{53}$$

For a non-trivial solution and in order to obtain the frequency equation for the composite beam in the free vibration analysis, the determinant of the coefficient 3 × 3 matrix must be equal to zero.

2.4. Mode shapes

Taking the first three equations from Eq. (43) and the last three equations from Eq. (44) then substituting with the natural frequency ω_n, the following system of equations is obtained.

$$\begin{bmatrix} R_{Fx_{11}} & \cdots & R_{Fx_{16}} \\ \vdots & \ddots & \vdots \\ R_{Fr_{61}} & \cdots & R_{Fr_{66}} \end{bmatrix} \begin{Bmatrix} A_1 \\ \vdots \\ A_6 \end{Bmatrix} = \begin{Bmatrix} 0 \\ \vdots \\ 0 \end{Bmatrix} \quad (54)$$

To get these constants and to avoid the nontrivial solution, one row from the coefficient matrix will be deleted arbitrarily and its corresponding constant will be chosen arbitrarily, then expressing the remaining constants in terms of the chosen one.

3. Results and discussion

To validate the current work, a MATLAB code was used to obtain the dynamic equation which is used to get the natural frequencies and mode shapes of the analyzed composite beam. The data of this cantilever composite wing was taken from Banerjee.

Table 1. The characteristics used for the composite beam analysis

item	Value	units
EI	$9.75 * 10^6$	$N.m^2$
GJ	$9.88 * 10^5$	$N.m^2$
L	6	m
m	35.75	Kg/m
I_α	8.65	Kg. m
K	$9{:}75 * 10^6$	$N.m^2$
x_α	0.2	m

To study the change in the vibration analysis of a composite beam due to coupling in both material and geometric, some values of x_α and K have been analyzed. In the vibratory motion, there will be no coupling when x_α and K are both zero. The values of x_α and K must be taken equal to 10^{-5} or 10^{-6} to get the results with acceptable accuracy rather than putting them exactly zero to prevent any numerical overflow.

Table 2. The variation of the first natural frequencies for different values of x_α and K

x_α (m)	K($*10^6\ Nm^2$)		Natural frequencies (rad/s)		
			ω_1	ω_2	ω_3
0.1	0	Presented	50.539246	91.01975	258.42778
		Reference	50.539	91.02	258.43
	1.5	Presented	40251539	99.0718825	197.5697536
		Reference	40.252	99.072	197.57
	2	Presented	33.96224	100.4672	168.547189
		Reference	33.962	100.47	168.55
	2.5	Presented	25.44152	94.8172838	137.337642
		Reference	25.442	94.817	137.34
0.2	0	Presented	49.33140496	99.20276666	246.5961495
		Reference	49.331	99.202	246.6
	1.5	Presented	38.0708548	112.2243426	185.5736177
		Reference	38.071	112.22	185.57
	2	Presented	31.981323	114.7058	157.80644
		Reference	31.981	114.71	157.81
	2.5	Presented	23.88528092	104.3580622	133.8713782
		Reference	23.885	104.36	133.87

Banerjee made this analysis on the same cantilever beam using the assumption of harmonic oscillation and solved it with the dynamic stiffness method but in this paper with the same assumption of Banerjee, the dynamic transfer method has been used and a comparison between both methods has been made as shown in table (2)

The mode shapes of the two cases are shown in Figs. (4) and (5). In the first case as shown in fig. (4) the values of x_α and K are equal to zero to represent the uncoupled case, while the second case which represents the coupling between bending and torsion as shown in fig. (5) their values are 0.2m and $2.0{*}10^6 N.m^2$ respectively.

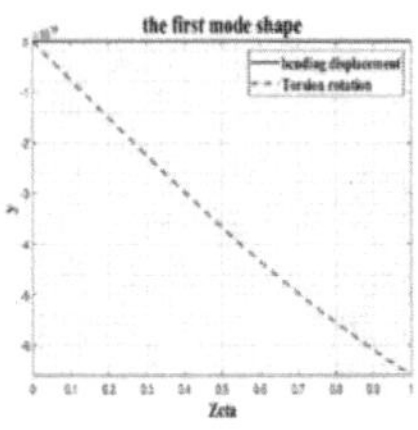

(a) ω_1=51.005

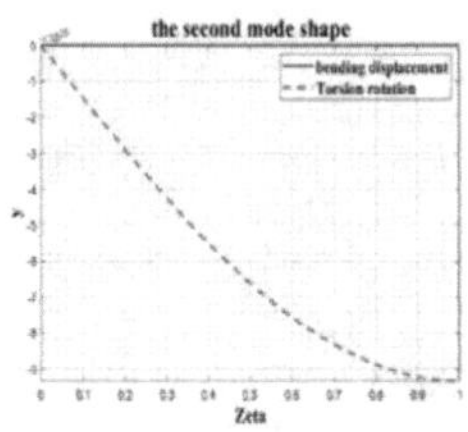

(b) ω_2=88.478

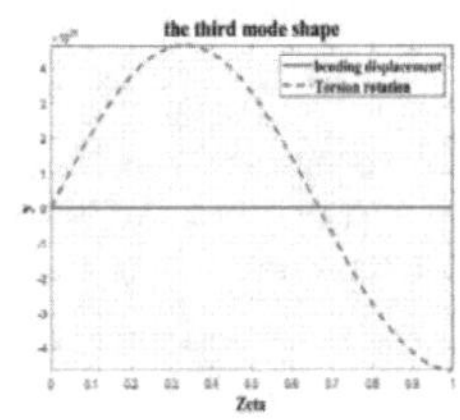

(c) ω_3=265.44

Figure 4. The uncoupled case

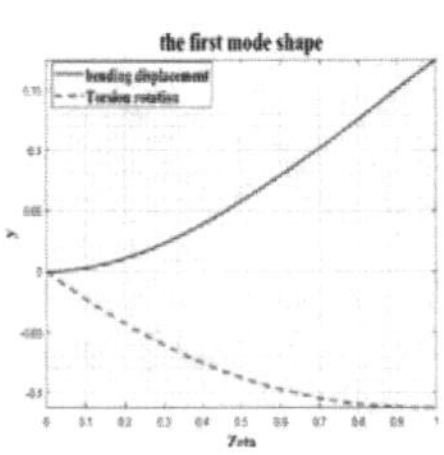

(a) ω_1=31.981

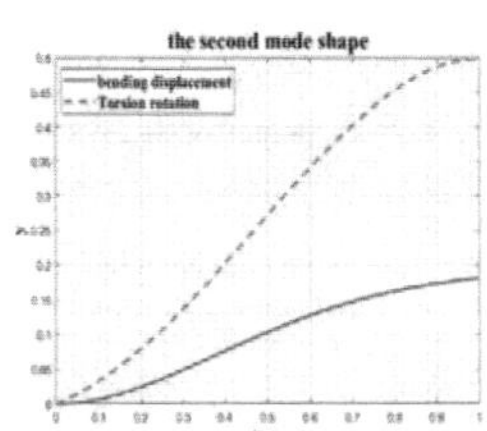

(b) ω_2=114.71

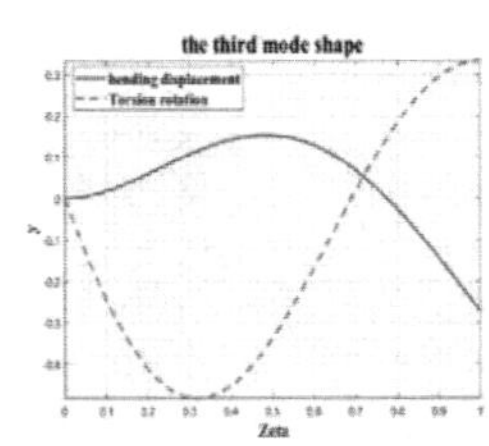

(c) ω_3=157.81

Figure 5. The coupled case

Note that natural frequencies and mode shapes are very different in the coupled case than the uncoupled one and this is important from the aeroelastic point of view.

After validation of the developed code, a parametric study was performed to show the effect of variation of K and x_α on the natural frequencies corresponding to the first three modes.

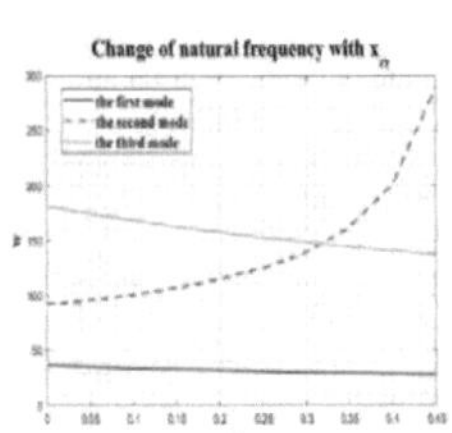

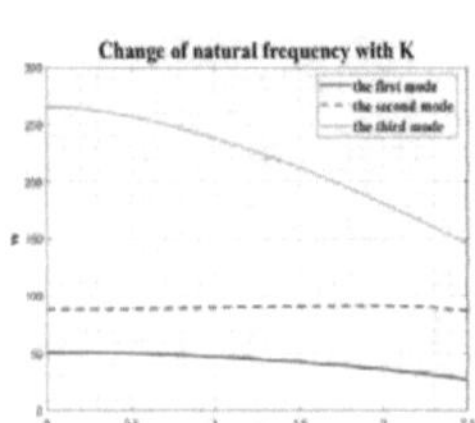

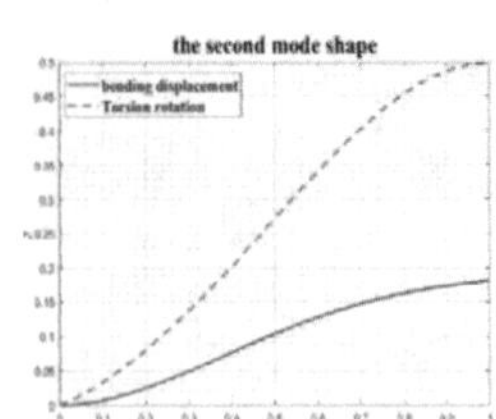

Figure 6. Change of natural frequency with x_α and K

Figure (6) shows the change in the first three mode shapes of the cantilever beam with the change in the values of K and x_α where as shown in Fig. 6 (a), ω_1 slightly decreases with x_α increasing, while ω_2 corresponding to the second mode is greatly increased. At $x_\alpha = 0.3169813$, the second and third modes intersect. Further increasing of x_α switches the order of the two modes (i.e., the natural frequency of the second mode becomes greater than that of the third mode. Figure 6 (b) represents the change of the natural frequency with the change in the value of the torsion bending coupling rigidity K when $x_\alpha = 0$ which means that the elastic axis and mass axis are coincided (i.e, the coupling depends only on the material coupling). As shown in this figure, ω_1 decreases with K increasing, while ω_2 increases very slowly with K increasing till approximately K = 2.0 *10^6 N. m^2 then decreases after that and ω_3 corresponding to the third mode is greatly decreased with K increasing. Figure 6 (c) represents the change of the natural frequency with the change of K at a constant value for x_α = 0.2 m (i.e., the coupling depending on both material and geometric coupling). In this figure with the increase in the value of K, ω_1 and ω_3 decrease but ω_2 increases till K = 2.0 *10^6 N. m^2 then decreases after that.

4. Conclusion

The present paper introduces an analytical solution for the free vibration analysis of composite beam with bending and torsional coupling representing material and geometric coupling for composite structures by a mathematical relation between both sides of the beam using the dynamic transfer method. This work was validated using a composite beam with fixed-free boundary conditions and the first three natural frequencies and mode shapes were calculated and compared with published results and a good agreement was found between them. The parametric study shows that x_α and K are very important parameters having a great effect on the free vibration analysis of composite structure, and this is an important benefit from the aeroelastic point of view that can be used to achieve desirable properties when design composite structures. In the future, this study will be conducted for composite wings under aeroelastic loading.

References

[1] Banerjee J R, Su H and Jayatunga C 2008 A dynamic stiffness element for free vibration analysis of composite beams and its application to aircraft wings. Computers & Structures, **86(6)**, pp.573-579.

[2] Banerjee J R 2001 Explicit analytical expressions for frequency equation and mode shapes of composite beams. International Journal of Solids and Structures, **38(14)**, pp.2415-2426.

[3] Hodges D H, Atilgan A R, Fulton M V and Rehfield L W 1991 Free-Vibration Analysis of Composite Beams. Journal of the American Helicopter Society, **36(3)**, pp.36-47.

[4] Eslimy-Isfahany S H R and Banerjee J R 1997 Dynamic response of composite beams with application to aircraft wings. Journal of aircraft, **34(6)**, pp.785-791.

[5] Kaya M O and Ozgumus O O 2007 Flexural–torsional-coupled vibration analysis of axially loaded closed-section composite Timoshenko beam by using DTM. Journal of Sound and Vibration, **306(3-5)**, pp.495-506.

[6] Mirtalaie S H, Mohammadi M, Hajabasi M A and Hejripour F 2012 Coupled lateral-torsional free vibrations analysis of laminated composite beam using differential quadrature method. International Journal of Mechanical and Mechatronics Engineering, **6(7)**, pp.1143-1148.

[7] Lottati I 1987 Aeroelastic stability characteristics of a composite swept wing withtip weights for an unrestrained vehicle. Journal of Aircraft, **24(11)**, pp.793-802.

[8] Lottati I 1985 Flutter and divergence aeroelastic characteristics for composite forward swept cantilevered wing. Journal of Aircraft, **22(11)**, pp.1001-1007.

[9] Kashani M T and Hashemi S M 2018 A finite element formulation for bending-torsion coupled vibration analysis of delaminated beams under combined axial load and end moment. Shock and Vibration, 2018.

[10] Murty A K 1970 Vibrations of short beams. AIAA Journal, **8(1)**, pp.34-38.

[11] Marur S R and Kant T 1998 A higher order finite element model for the vibration analysis of laminated beams. Journal of Vibration and Acoustics, **120(3)**, pp.822-824.

[12] Sarfaraz E, Afolabi J O and Hamidzadeh H R 2019 Coupled Flexural and Torsional Vibration Analysis of Composite Beams. In ASME International Mechanical Engineering Congress and Exposition, **Vol. 59414**, p. V004T05A076.

[13] Pipes L A and Harvill L R 2014 Applied mathematics for engineers and physicists. Courier Corporation.

[14] Dancila D S and Armanios E A 1998 The influence of coupling on the free vibration of anisotropic thin-walled closed-section beams. International Journal of Solids and Structures, **35(23)**, pp.3105-3119.

IOP Conf. Series: Materials Science and Engineering 1172 (2021) 012004 doi:10.1088/1757-899X/1172/1/012004

Numerical simulation of temperature field in a C/C composite multidisk brake during aircraft braking

E M Ramadan[1,*], A E Hussien [1], A M Youssef [1], and T M Abd El-Badia [1]

[1] Military technical college, Mechanical Design and Production Department, Cairo, Egypt.

E-mail: eramadan@mtc.edu.eg

Abstract. This article presents a finite element method for simulating the heat production during stopping the aircraft. Thermal analysis and simulation in the finite element model are based on the theory of energy transformation and transportation. A commercial software COMSOL Multiphysics 5.5a is used for simulating the braking operation. The internal temperatures of the brake disks were obtained and the variation in temperatures between disks were discussed. Thermomechanical behaviour is studied to show the effect of thermal energy on the contact mechanics for the friction surfaces between the brake disks. Aircraft mass, initial velocity and deceleration rate are responsible for heat generation and consequently the maximum reached temperature during braking. The friction surfaces between disks were the main heat energy source where the heat was concentrating on these layers of friction surfaces. For the selected braking operation, the maximum reached temperature was 1020K. The finite element model was validated against historical data for a Boeing737-400 at constant deceleration for new disk brakes and for RTO brake energy.

1. Introduction

C/C composite material is used over worldwide as friction brake material due to its good advantages such as light weight, it can absorb high energy during braking, low coefficient of thermal expansion, resist thermal shock, stable friction properties, good wear characteristics in heavy conditions, work in very high temperature ranges, high strength and it has no melting point at atmospheric pressure [1-5]. Aircraft friction material absorbs millions of joules during braking as the kinetic energy is transformed into thermal energy at the friction surfaces between the rotor and stator disks.

The temperature distribution between the disks and the reached temperature can affect the contact between the disks and thus will affect the tribological behaviour of the brake disks [6-8]. Therefore, it is essential to investigate the effect of temperature variation on the brake disks under landing conditions for developing the friction material.

The issue has been investigated for many years by different researchers to find an accurate method for measuring the disks temperature during braking while taking into consideration contact and noncontact measurement between friction surfaces [9-11]. None of these researchers reached to a completely successful method for measuring the disks temperature at the contacting surfaces because of the sever landing conditions the brakes are working at such as high speed and heavy load.

The finite element simulating software gives a great opportunity to simulate the braking operation to study the friction behaviour under the temperature and contact effects[12]. Kennedy and ling[13] studied the effect of thermal deformation on the contact pressure between the multidisk brakes and achieved a numerical simulation for thermomechanical behaviour for the disk brakes. Choi and lee [6] used

Content from this work may be used under the terms of the Creative Commons Attribution 3.0 licence. Any further distribution of this work must maintain attribution to the author(s) and the title of the work, journal citation and DOI.
Published under licence by IOP Publishing Ltd

IOP Conf. Series: Materials Science and Engineering 1172 (2021) 012004 doi:10.1088/1757-899X/1172/1/012004

ABAQUS finite element software for simulating an axisymmetric multidisk brake in aircraft including the thermoelastic contact problem and presenting the effect of temperature on the contact pressure between the contacting surfaces during braking.

However, in literature there are only a few investigations about the transient temperature field of C/C composite material during braking. For studying the transient temperature field changes and internal temperature gradient for C/C composite brake material a theoretical model was developed based on the phenomena of energy transformation during braking and transportation between disks. FEM was applied using the commercial software COMSOL 5.5a to solve the theoretical model.

2. Theoretical Analysis

2.1. Aircraft Brake Model

The aircraft brake assembly is shown in figure 1 introducing the brake system components. Aircraft brake is consisting of multiple disks stators, rotor disks, pressure plate and end plate. Number of pistons are used to produce pressure against the pressure plate in order to compress all the disks together to stop the aircraft. The torque plate or the end plate used as a constrain for the horizontal movement of the disks when pressure is applied. The torque plate and end plate are two stator disks but called with these names because of their function and position.

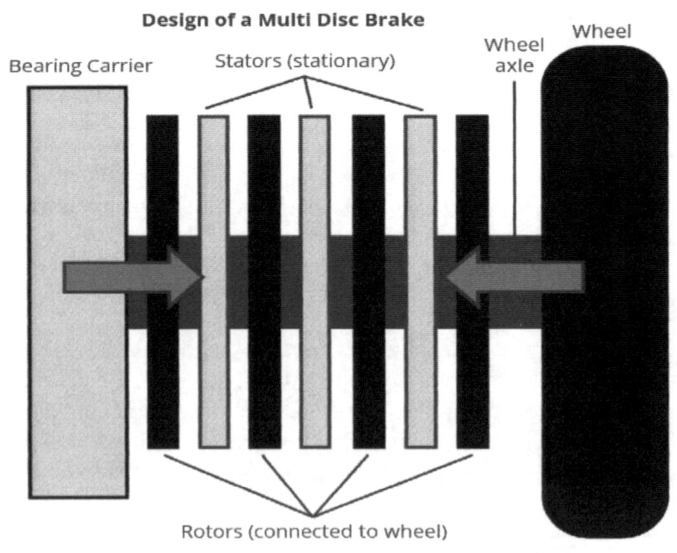

Figure 1. Aircraft brake assembly.

2.2. Theoretical Model

A 2D axisymmetric model is built to simulate the aircraft braking operation without scarifying the accuracy of the results as it reduce the computation time and reduce the data storage as well as shown in figure 2 . The aircraft brakes is consisting of seven disks three rotating disks which are connected with the rotating aircraft wheel and four stator disks which are consisting of the pressure plate and the end plate at the end of disks group. The disk brake dimensions are shown in table 1. The hydraulic pressure is uniformly distributed on the surface of the pressure plate as shown in figure 2. The displacement in z direction due to axial force applied on the pressure plate is constrained by the back plate. The pistons housing and torque tube has been neglected in the simulation model. There are six friction surfaces between the brake disks, which is used to produce friction to stop the aircraft, and these friction surfaces will be the heat source during braking.

During braking, disks temperature increase and transferred from disk to another depending on its thermal conductivity. In addition, disks dissipate heat by forced convection using air opposing the aircraft speed and heat radiation as heat diffuse from hot surfaces to the surroundings.

Depending on the law of energy transportation, the following differential equation is used to describe the heat flow between disks:

$$\rho\, c\, \frac{\partial T}{\partial t} + \nabla(-k\nabla T) = \rho\, c\, v.\Delta T \tag{1}$$

Where ρ the density of the disks material, c is the heat capacity of the disks material, k is the thermal conductivity of the disks material, v is the local velocity vector of the disks [7, 8, 14].

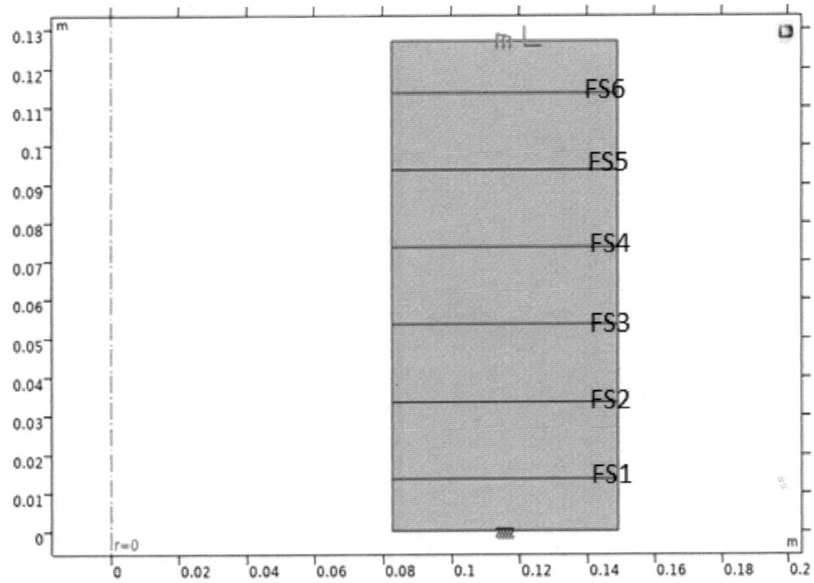

Figure 2. 2D axisymmetric aircraft brake model.

Table 1. Aircraft disk brake dimensions.

Disk type	Inner diameter (mm)	Outer diameter (mm)	Thickness (mm)
Stator and rotor disks	82.5	149	20
Back plate and end plate	82.5	149	13

2.3. Heat Power Generation

Heat power is generated due to the aircraft braking during landing with certain operating conditions; the speed of the aircraft was defined to decrease linearly with time in an explicit function. The heat power is obtained from the following equation:

$$p = -m\, v(t)\, a(t) \tag{2}$$

Where p is the heat power of one friction surface contact between two disks, v(t) is the velocity and a(t) is the deceleration of the aircraft, m represent the aircraft total mass [15-17].

IOP Conf. Series: Materials Science and Engineering 1172 (2021) 012004 doi:10.1088/1757-899X/1172/1/012004

2.4. Heat Transfer by Convection

Brake disks dissipate heat by forced convection during braking where the aircraft speed was assumed to be the same as the air speed facing the aircraft this will be the main reason for forced convection for the brake disks. Heat convection can be calculated from the following equation:

$$q = h\,(T_{ex} - T) \tag{3}$$

Where h is heat transfer coefficient, T_{ex} is the external disks temperature and T is the external air temperature [7, 8, 14]. Heat transfer coefficient is calculated from the following equation:

$$h = \frac{0.37\,k_{air}}{l}\,Re^{0.8}\,Pr^{0.33}$$

$$h = \frac{0.37\,k_{air}}{l}\left(\frac{\rho_{air}\,l\,v}{\eta_{air}}\right)^{0.8}\left(\frac{c_{air}\,\eta_{air}}{k_{air}}\right)^{0.33} \tag{4}$$

Where η_{air} is the air viscosity equals to 1.8e^{-5} N.S/m, l is the disk's diameter, $\rho_{air}$is the density of air; kair represents thermal conductivity of air; cair is specific heat capacity of air [7, 8, 14].

2.5. Heat transfer by radiation

The brake disks also dissipate heat by radiation to the surrounding air and can be calculated from the following equation:

$$q = \varepsilon\,\sigma\,(T_{sur}^4 - T^4) \tag{5}$$

Where ε is the emissivity, σ is the Stephan Boltzmann constant and T_{sur},T are temperature of the disks and surrounding [7, 8, 14].

3. Validation of Theoretical Model

The current model was validated against the results from Nihad et al. [18]. This paper is discussing modeling and computation of maximum braking speed of aircraft, brakes heat capacity and braking forces.

Nihad et al. build a model to indicate heat transfer and maximum temperature during braking conditions (landing and RTO) depending on historical data for Boeing737-400 series, which has the following data: The 737-400 aircraft has two main landing gears with four braked wheels. Every single wheel on axle is carrying five stators and four rotors with eight frictional contact surfaces.

Boeing 737-400 aircraft weight 65 tons to 68 tons and it has initial braking speeds 150 Knots, 160 knots and 170 knots, the aircraft initial temperature of disk brake is 400°K and the surrounding temperature is 300 °K. The disks initial high temperature is because of taxi out braking operation.

Steel alloys material is used for braking operation for the Boeing 737-400 and it was used for the model results and validations. Disk brake dimensions and steel alloy material property are shown in table 2.

In order to validate the model, Boeing 737-400 RTO condition and normal landing conditions have been simulated in the 2D axisymmetric model by using COMSOL Multiphysics software to compare the model results against the historical data for the aircraft.

Table 2. Boeing737-400 steel alloy brake material properties and disk dimensions [18].

Property	Rotor and stator

IOP Conf. Series: Materials Science and Engineering 1172 (2021) 012004 doi:10.1088/1757-899X/1172/1/012004

Density (kg/m³)	7780
Heat capacity C_p (J/kgK)	480
Thermal conductivity K (W/mK)	50
Disk outer radius (m)	0.25
Disk inner radius (m)	0.1443
Disk effective radius (m)	0.197
Disk thickness (m)	30
Wheel radius (m)	1.27

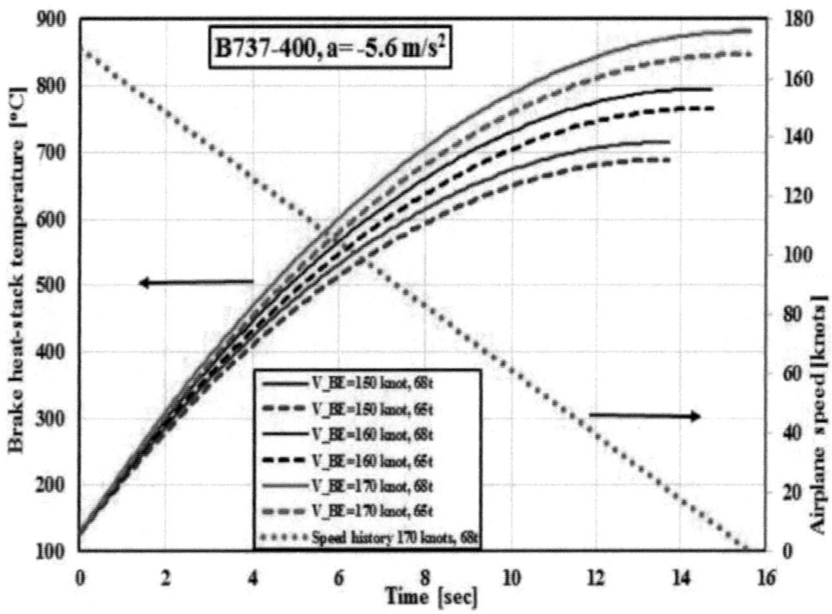

Figure 3. Temperature and speed histories for a Boeing737-400 at constant deceleration for new disk brakes and for RTO brake energy [18].

From the historical data for the selected aircraft, the RTO case has the following braking conditions initial speed 170knots, deceleration rate of 5.6 m/s^2 and 68 tons weight stopping at 15.76 sec where the disk peak temperature was 875°C. The model results showed that the maximum reached temperature was 842°c during the braking operation with error percentage 4.1% as shown in figure 4.

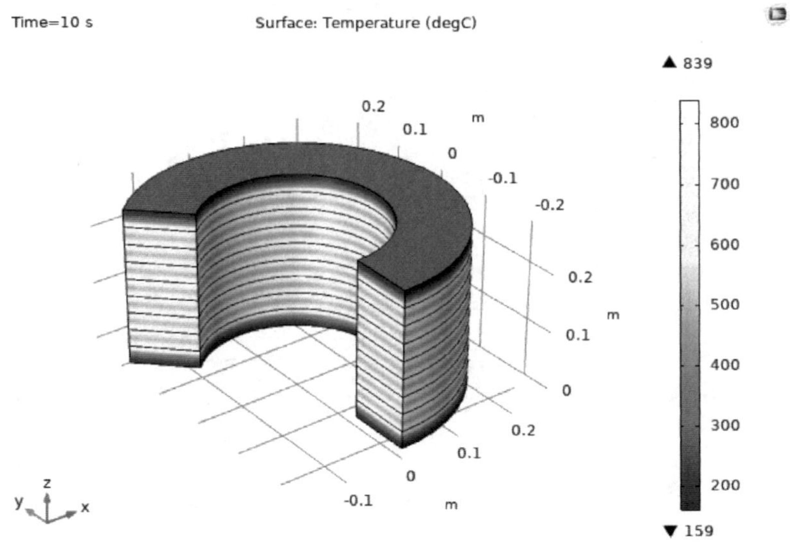

Figure 4. 3D temperature gradient for disk brakes during braking from v=170knots.

For further validating the model, another selected case from the historical data was simulated with weight of 65 tons, speed of 150 knots and deceleration rate of -5.6 m/s^2 stopping at t=13.77sec. The results showed that the maximum reached temperature was 710°c, which is in agreement with temperature history of the Boeing aircraft 737-400 with error margin of 1.2% as shown in figure 5.

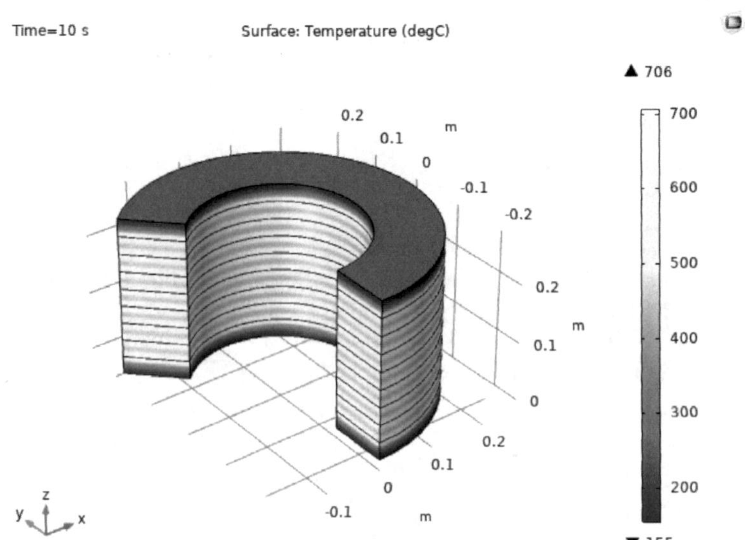

Figure 5. 3D temperature for disk brakes during braking for v=150knots.

4. Case study

C/C composite aircraft brake material is used in this model to simulate the aircraft braking operation with the following properties in table 3 [6]:

Table 3. C/C composite material properties.

Material property	Rotor and stator
Density, (kg/m^3)	1800
Modulus of elasticity, E_r (GPa)	50.9
Modulus of elasticity, E_z (GPa)	5.89
Poisons ratio, $\upsilon_{r\theta}$	0.3
Poisons ratio, υ_{rz}	0.33
Shear modulus, G_{rz} (GPa)	2.46
Thermal conductivity, k_r (W/mK)	50
Thermal conductivity, k_z (W/mK)	10
Heat capacity, c_p (J/kgK)	1420
Thermal expansion, α_r (10^{-6}/K)	0.31
Thermal expansion, α_z (10^{-6}/K)	0.29

The air used for dissipating heat by forced convection has the following properties in table 4 [7]:

Table 4. Air thermal properties.

Air property	Value
Density, (kg/m^3)	1.177
Thermal conductivity, k (W/mK)	0.026
Heat capacity, c_p (J/kgK)	1.1

An aircraft with the following landing conditions as shown in table 5 is simulated in the model:

Table 5. Aircraft landing conditions.

Parameter	Value	Definition
m_plan	9000[kg]	Aircraft mass
r_wheel	0.352[m]	Wheel radius
v0	260[km/h]	Initial aircraft touch down speed
a0	-3.611[m/s^2]	Initial vehicle acceleration
T_air	300[K]	Temperature, air
t_brake_start	1[s]	Braking time (start)
t_brake_end	21[s]	Braking time (end)
P	0.7[MPa]	Braking pressure
mu	0.32	Friction coefficient

Aircraft landing speed is assumed to decrease linearly with time during braking as shown in figure 6. Aircraft deceleration for stopping the aircraft is assumed to be constant and is defined in analytic function as a constant value during the whole braking operation as shown in figure 7.

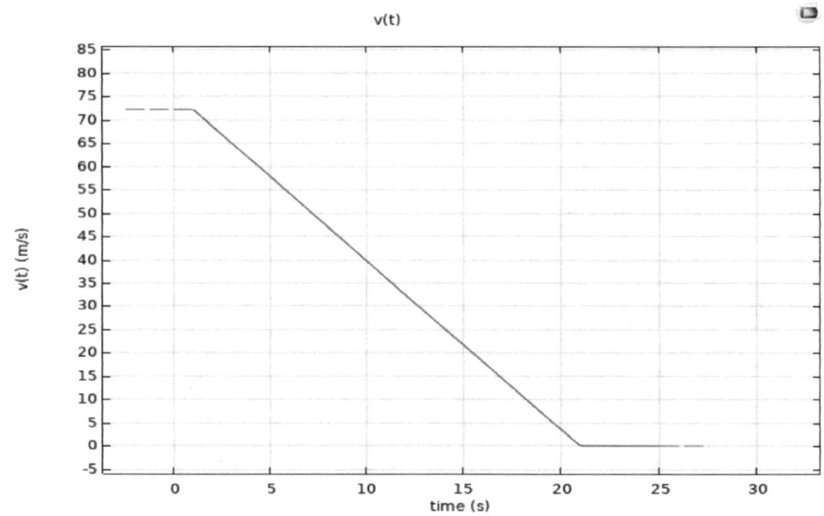

Figure 6. Aircraft velocity during braking.

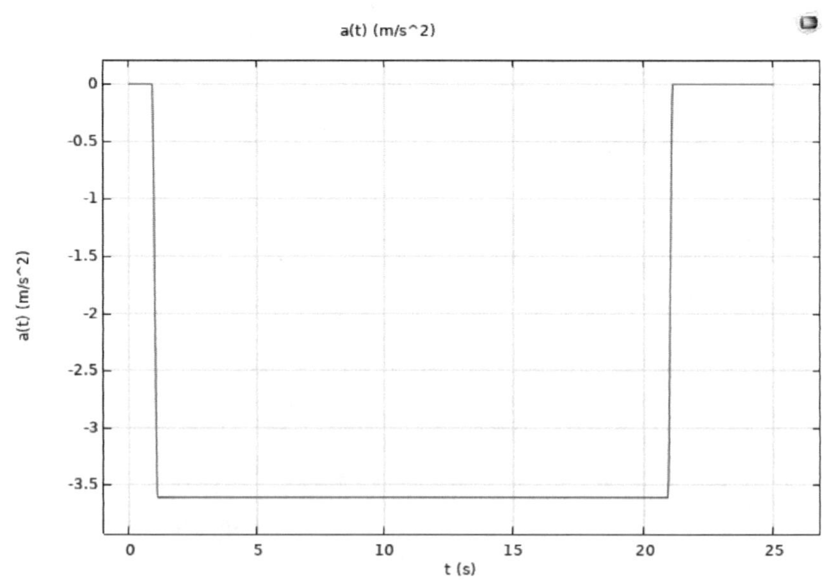

Figure 7. Aircraft deceleration during braking.

Mesh element size and number of elements has a significance effect on the accuracy of the results. Increasing the number of elements and reducing its size to certain size as fine or extra fine will give more accurate results, but also, it will increase time of computation and storage.

The model started as 2D axisymmetric model, mesh was build in free triangular elements by increasing the number of elements at the contact surfaces where the model results for contact pressure, and heat transfer is most relevant as shown in figure 8.

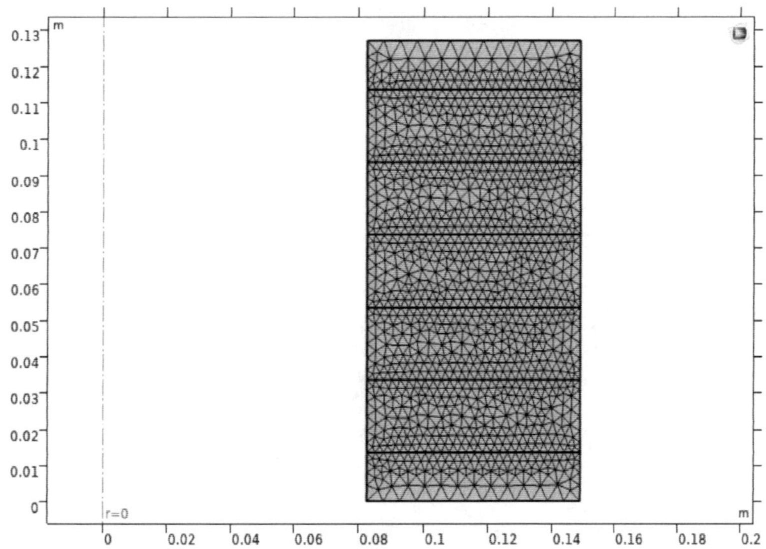

Figure 8. 2D axisymmetric model mesh build.

5. Results and Discussion

During braking, the kinetic energy is transferred into thermal energy. Using 2D axisymmetric model the temperature field simulation for braking operation was achieved. The highest temperature reached during braking at t=20sec was 1020 °K.

As shown in figure 9 at t=1sec the friction surfaces start producing friction between disks to stop the aircraft as the friction surfaces are the main sources for heat generation, this heat will start to dissipate and transport to the other disks and to the surrounding air.

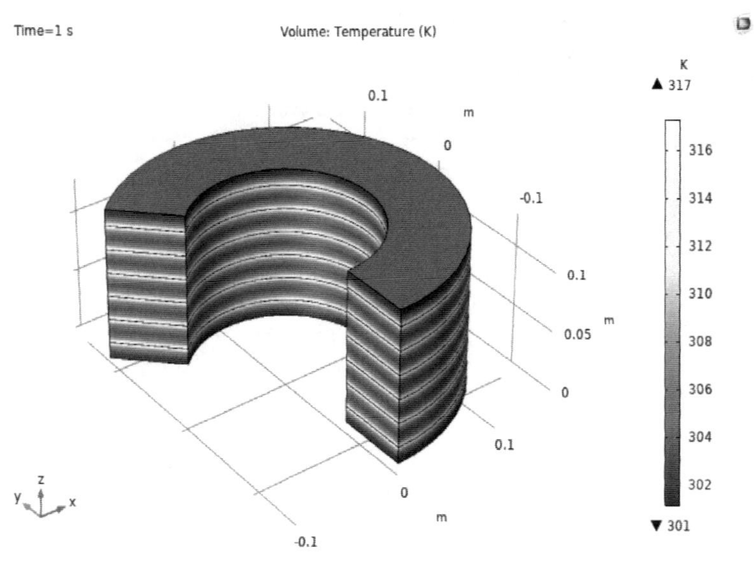

Figure 9. Disk brakes temperature while start braking at t=1sec.

Figure 10 shows how heat generation starts to propagate through the disks group as temperature increase with time during braking. It can be shown that the pressure plate and the end plate have lower temperature 374°k than the other brake disks, as these disks are the most affected surfaces by air forced convection. Depending on the thermal conductivity property of the friction material heat will start to propagate through the thickness of the disk brake and then it will be dissipated to the surrounding air.

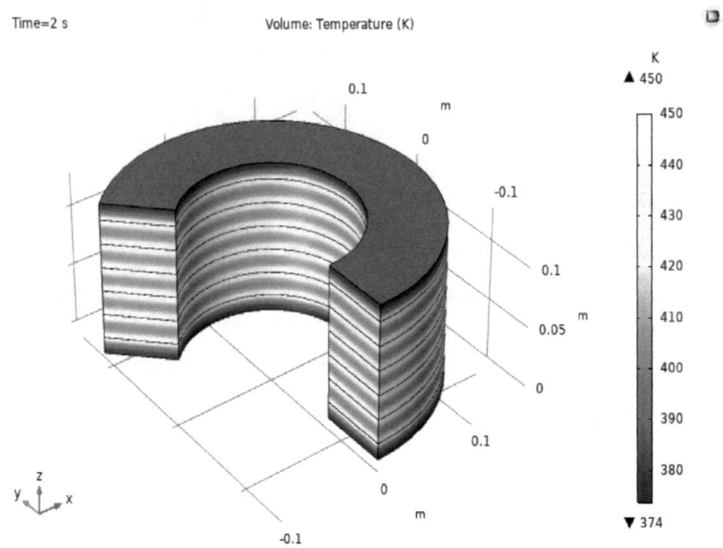

Figure 10. Disk brakes temperature during braking at t= 2sec.

Figure 11 presents the highest temperature the disks will reach in 3D and the hottest areas in the whole disks. The middle rotor disk (rotor disk that have fs3 and fs4) was found to have the highest temperature. This result explains that the temperature effect on the material thermal deformation and this affect the contact pressure between disks and the wear rate of the C/C composite disks. It can be deduced that the thermal effect has a clear effect on the tribological behaviour for C/C composite. It can be shown that minimum and maximum temperatures for the brake disks after braking are from 956K to 1033K at the end of braking the brakes thermal temperature will decrease and cooling starts.

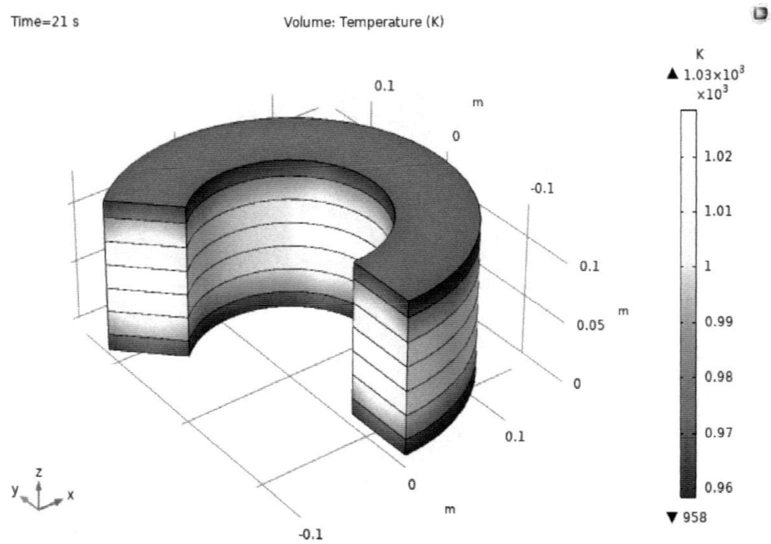

Figure 11. 3D disk brake temperature at t=21sec.

The contact pressure is affected by the heat dissipation during braking due to the material thermal deformation as shown in figure 12 where the contact pressure changes with the disk radius at different selected times. Figure 12 shows that for FS2 different points will get in contact with other points on the contacting friction surfaces or it may have loss of contact. This change in the contact pressure during braking will affect the wear rate and the surface morphology during braking.

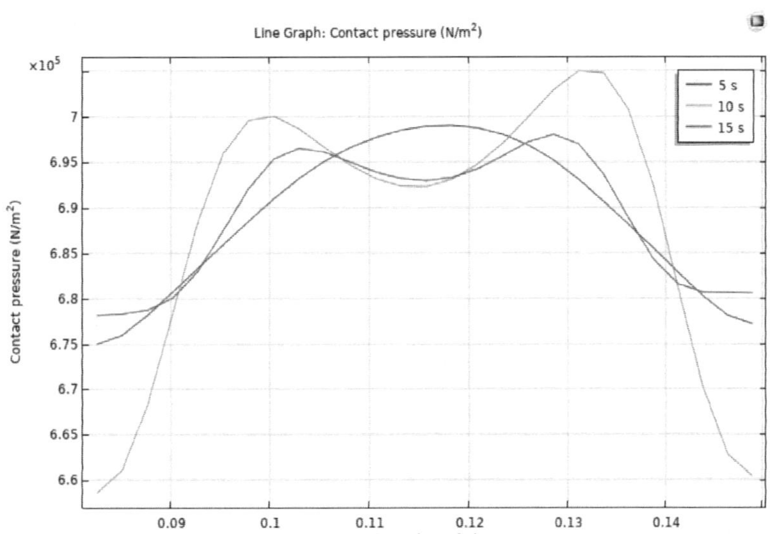

Figure 12. Fs2 contact pressure vs disk radius at various times during braking.

Figure 13 shows the brake disks temperature at t=15sec where the temperature is high at the middle rotor disk at fs3 and fs4, while it has a lower value at the pressure plate and end plate. Figure 13 also shows the 2D disk brake temperature at t=15sec where the temperature increases to about 1000K.

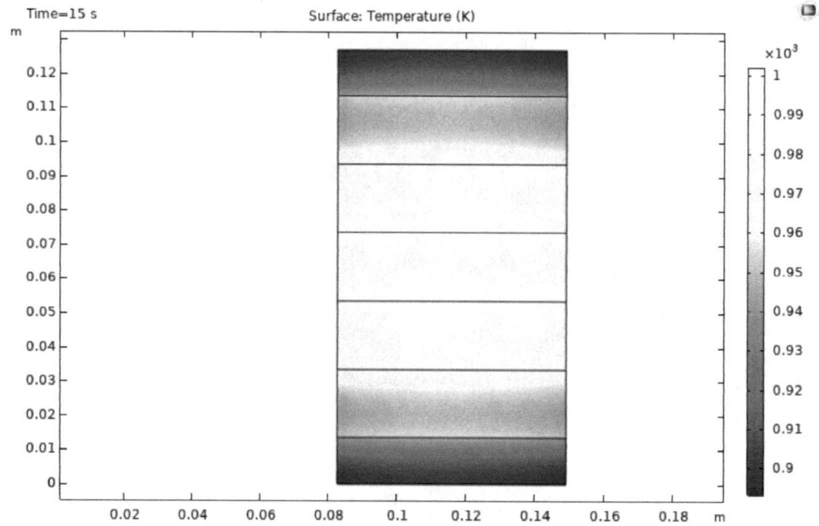

Figure 13. 2D disk brake temperature at t=15sec.

As shown in figure 14, temperature at friction surface number (3) at t=11sec at the middle of the disk is higher than at the ends because the ends of the disks are facing air much more. Also non-uniform contact pressure distribution between disks causes non-uniform temperature at the surfaces that will affect on deformation and so the wear of the contacting surfaces. It is noticed that heat dissipation at the end points of the disks near to the outer radius are cooler than any other points on the disk due to better cooling and better heat dissipation. Disk brakes temperature at the middle of friction surfaces are high and this also will affect the material properties and consequently thermal expansion resulting in some points will loss contact.

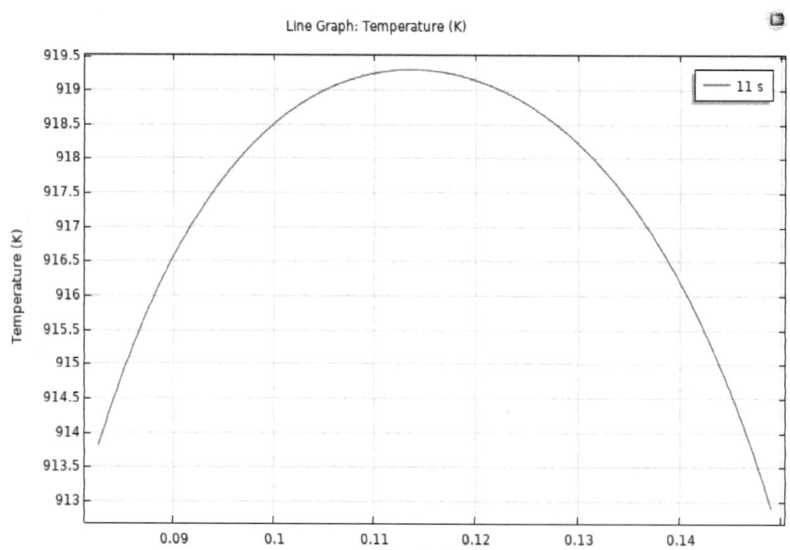

Figure 14. Temperature vs disk radius for fs3 at t=11sec.

Figure 16 shows the inner radius ends temperature numbered from the end plate to the pressure plate from one to eight as shown in figure 15. These results are in agreement with the experimental work and also literature for temperature of c-c composite [1, 4]. The reduction in curve slope is due to the reduction of the aircraft velocity during braking.

Figure 16 shows the points that have the highest temperature on the brake disks where points numbered as 4 and 5 have the highest temperature on the disks.

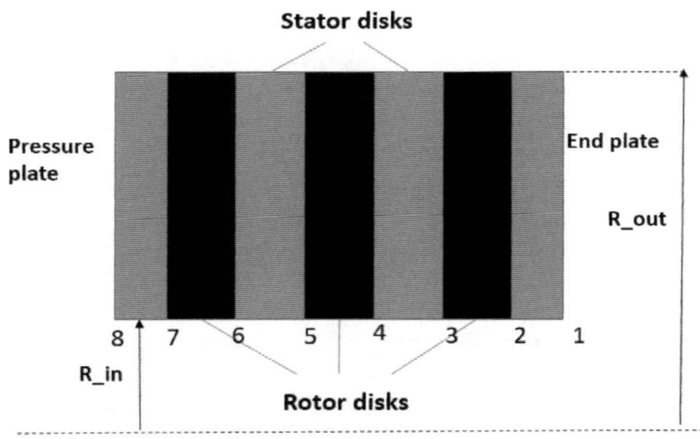

Figure 15. Disk brake numbering points for measuring temperature.

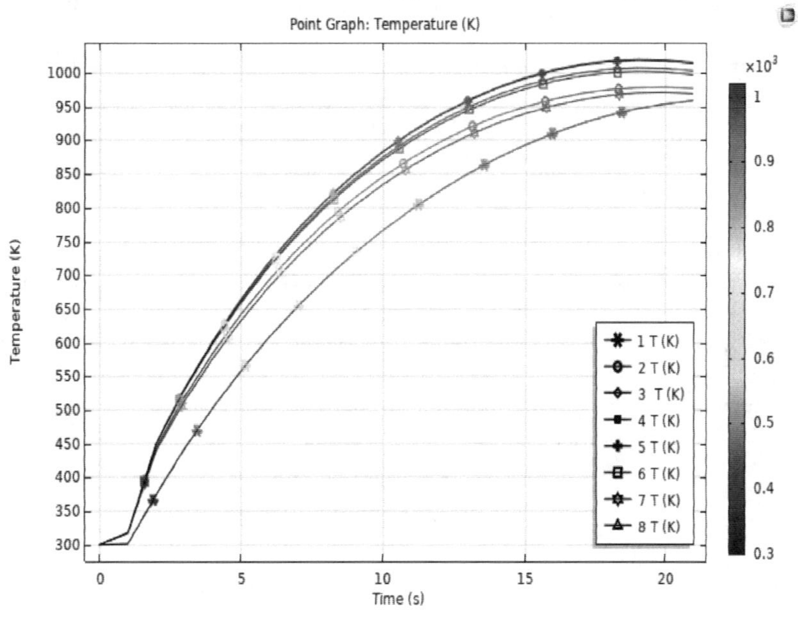

Figure 16. Temperature vs time for different points on the brake disks.

6. Conclusions

The finite element model can be used to simulate the aircraft braking conditions under all conditions. The friction surfaces are the main cause of heat production so it is actually the main heat source for the disks transforming the kinetic energy into heat energy.

The heat generation due to friction will affect the friction material thermal expansion, consequently material thermal deformation. This will affect the contacting points, some points will suffer from loss of contact, and thus all will affect the weight losses and so surface morphology. Heat dissipation at the ends of the disks were better than the middle area of the disk because of better cooling of air forced convection.
The middle rotor disk was found to have the highest temperature of the brake disks. During braking the aircraft velocity is decreasing and this will reduce the heat production due to friction between disks. The model was validated against historical data from literature for Boing 737-400 with cast steel material.

References:

[1] Blanco C, Bermejo J, Marsh H and Menendez R 1997 Chemical and physical properties of carbon as related to brake performance *Wear* **213** 1-12
[2] Orthwein W C 1986 *Clutches and brakes*: Dekker New York)
[3] Di Santo G 2001 Proper Operation of carbon brakes. In: *11th Airbus Performance and Operations Conference, Jalisco, Mexico,*
[4] Chen J and Ju C 1995 Effect of sliding speed on the tribological behavior of a PAN-pitch carbon-carbon composite *Materials chemistry and physics* **39** 174-9
[5] Chen J, Lin J C and Ju C 1996 Effect of load on tribological behaviour of carbon-carbon composites *Journal of materials science* **31** 1221-9
[6] Choi J-H, Han J-H and Lee I 2004 Transient analysis of thermoelastic contact behaviors in composite multidisk brakes *Journal of Thermal Stresses* **27** 1149-67
[7] Zhang C, Zhang L, Zeng Q, Fan S and Cheng L 2011 Simulated three-dimensional transient temperature field during aircraft braking for C/SiC composite brake disc *Materials & Design* **32** 2590-5
[8] Zhao S, Hilmas G E and Dharani L R 2008 Behavior of a composite multidisk clutch subjected to mechanical and frictionally excited thermal load *Wear* **264** 1059-68
[9] Buckley J D and Edie D D 1993 *Carbon-carbon materials and composites* vol 1254: William Andrew)
[10] Bijwe J, Majumdar N and Satapathy B 2005 Influence of modified phenolic resins on the fade and recovery behavior of friction materials *Wear* **259** 1068-78
[11] Belhocine A, Abu Bakar A and Bouchetara M 2014 Numerical Modeling of Disc Brake System in Frictional Contact *Tribology in Industry* **36**
[12] Deng D and Murakawa H 2006 Numerical simulation of temperature field and residual stress in multi-pass welds in stainless steel pipe and comparison with experimental measurements *Computational materials science* **37** 269-77
[13] Kennedy Jr F and Ling F 1974 A thermal, thermoelastic, and wear simulation of a high-energy sliding contact problem
[14] Zhao S, Hilmas G E and Dharani L R 2009 Numerical simulation of wear in a C/C composite multidisk clutch *Carbon* **47** 2219-25
[15] Fitzer E and Manocha L M 2012 *Carbon reinforcements and carbon/carbon composites*: Springer Science & Business Media)
[16] Petinrin M O and Oji J O 2012 Numerical simulation of thermoelastic contact problem of disc brake with frictional heat generation *New York Science Journal* **5** 39-43
[17] Singh H and Shergill H 2012 Thermal Analysis of Disc Brake Using Comsol *International Journal on Emerging Technologies* **3** 84-8
[18] Daidzic N E 2017 Modeling and Computation of the Maximum Braking Energy Speed for Transport Category Airplanes *Journal of Aviation Technology and Engineering* **6** 1

IOP Conf. Series: Materials Science and Engineering 1172 (2021) 012005 doi:10.1088/1757-899X/1172/1/012005

Stress analysis of multi-layered composite cylinders subjected to various loadings.

M H Elgohary[1], M I El-Geuchy[2], H M Abdallah[1] and S S Sayed[1]

[1] Rockets Department, Military Technical College, Cairo, Egypt
[2] PhD., Technical Research Centre, Cairo, Egypt

E-mail: Gohary1@live.com

Abstract: Composite cylinders are widely used in many applications, common examples for these applications are power drive shafts, chemical storage tanks, rocket motor cases and pressure vessels used in aerospace vehicles. The structure of the composite cylinder can be represented as a cylinder composed of coaxial orthotropic layers. In this paper, an analytical method is used to calculate stresses, strains and displacements through the wall thickness of multi-layered composite cylinder made of orthotropic material. The method is based on the theory of elasticity of bodies having cylindrical anisotropy. This method permits accurate stress analysis of thin and thick-walled composite cylinders subjected to axial load, torsional load and bending moment. The analytical method is modified to incorporate internal and external pressure loads beside the prescribed load cases. A numerical example is presented for a composite cylinder subjected to internal pressure load with a tensile axial load. The resulted stresses and strains are used to validate 3-dimensional finite element model. A parametric study has been performed using the analytical method to investigate the influencing parameters on stress distribution within the cylinder thickness and the results are found to be beneficial to look into during the preliminary design phase.

1. Introduction

Composite cylinders or pipes are laminated structures used in many applications such as pressurized fluid transportation or rocket motors manufactured using filament winding technology. With many advantages over conventional material, laminated composite cylinders are increasingly utilized due to their better stiffness to weight, strength to weight ratios and specific properties as corrosion resistance and chemical stability.

The first stumble for this type of industry and material is the many parameters involving the final design of the product. Due to design complexity parameters of interactions, trial and error in testing or manufacturing the composite products is very expensive and time consuming. Hence, there must be a method to decrease the number of experimentations to reach the final design, like implementing a mathematical model to perform a parametric study and finding out the effective parameters that control the design.

Mathematical analysis of laminated composite structures is used to predict their mechanical behavior using many approaches; one of the effective approaches is Lekhnitskii's approach [1], which used the three-dimensional theory of anisotropic elasticity to predict stresses and strains inside the thickness of composite laminates with fairly good accuracy.

Content from this work may be used under the terms of the Creative Commons Attribution 3.0 licence. Any further distribution of this work must maintain attribution to the author(s) and the title of the work, journal citation and DOI.
Published under licence by IOP Publishing Ltd

Panda and Natarajan [2],[3] performed a finite element analysis technique on laminated composite plates and shells to predict the stresses within the laminates of the structure and compared their findings with elasticity and exact solution finding high accuracy between their work and exact solution with thickness to radius ratio of (20.5) which was used in their example. Finite element analysis is very important as it is able to deal with more complex shapes and details than the analytical methods, also it is able to get slightly more accurate results when compared with same case conditions. The hindrance of such technique is the requirement of high computing devices and the large time of setting the conditions; for which the preliminary analysis and design of products require too much parameters changes until reaching a semifinal design. For example, the analysis of plies in finite element requires at least one element per ply in case of composite laminates to get acceptable outcomes and since the composite ply is usually less than 1 mm the meshing of such model would require large solving time while in theoretical analysis once the equation parameters are calculated, the parameters change and corresponding stress and strains calculations would just require a fraction of the time needed in finite element.

Joulicoeur and Cardou [4] derived the stress functions to calculate stress distribution within the cylinder thickness. They used Lekhnitskii's approach to solve a system of two coaxial hollow circular cylinders under bending, torsion and tensile loads. They found that there is no coupling between bending and tension-torsion loads, also the curvature caused by the bending load is in a plane that is perpendicular to the moment axis. Wild and Vickers [5] used Lekhnitskii's approach to predict stress distribution along the thickness including a combined loading of centrifugal, axial, and pressure loads. Bakaiyan et al. [6] also used the three-dimensional elasticity theory to inspect composite pipes under thermomechanical loads showing that hoop to axial stresses or strains tends increasing while increasing the winding angle and the opposite when decreasing the angle to axial direction. Sun et al. [7] who presented an analysis strategy to analyze stresses for composite cylindrical structures which resulted in finding that material anisotropy has great effect on the coupling stiffness coefficients for axial and torsional loads while having almost no effect on the bending stiffness coefficients as bending load is uncoupled with other load types.

Menshykova and Guz [8] made a comparative study on the stress analysis of composite cylinders under bending load and how the stress variation within the thickness of the composite cylinders is affected by the material and winding angle of the composite plies.

Pham et. al. [9] studied the usage of composite material in the oil transportation by composite tubes and their advantages to using conventional material in weight saving and how composites may have design and manufacturing complications according to their application. Guz et. al. [10] also studied different materials for long fiber reinforced filament wound composite cylinder under different loadings.

Mohanavel et. al. [11] modeled an aluminum matrix composite (AMC) pressure vessel using FEA to investigate the extreme usage of the material to estimate the life usage of the material to increase its service time.

Sharma et. al. [12] performed a theoretical analysis using ANSYS to simulate a 35 MPa hydrogen pressure vessel and verified the results experimentally within 10% error. A weight reduction has been achieved by changing the winding pattern design basis by at least 15%.

Moshir et. al. [13] discussed the behavior of composite thick tubes using 3D elasticity theory and compared it to different analysis methods to determine the stiffness and strains in the tube subjected to bending load. They concluded that while FEA can be accurate and its results agree with the experimental data, the 3D elasticity solution can be simpler and less time consuming.

The method used in this paper is an analytical method comparable with previous work by other authors[7],[14],[6]. The analytical solution used was derived by Joulicour and Cardou was used and extended to incorporate the static pressure loads on the internal surface of the inner cylinder and external surface of the outer cylinder beside axial load, torsion, and bending loading applied on a multi-layered composite cylinder. The work will be presented in the following section with validation by previous work and comparison to finite element model using the commercial software ANSYS ACP module.

2. Theoretical work

2.1. Objective and problem description

This study targets the derivation of a mathematical model of the laminated composite cylinder to produce a system of equations able to predict the stresses and strains within the cylinder thickness in a combined cylinder loaded under internal pressure, torsion, axial load and bending. The system at study is a combination of any number of concentric cylinders, figure 1. Each of them has an internal and external radius. The internal first cylinder is hollow and its external radius is the second cylinder's internal radius and so on. Figure 1.2 shows an example of a combined cylinder consists of n-concentric cylinders; each of different helix angle α_i, where i is the number of cylinders composing the composite cylinder.

As shown in figure 2.1 the combined cylinder is subjected to axisymmetric loadings; axial load (P), , torsion (C), internal and external pressure (P_{int}, P_{ext}) on surfaces (a) and (b) respectively denoted by the radii in figure 2.2. The cylinder could also be subjected to pure bending loads (Mx, My) in combination with the axisymmetric load. The resulting deformations due to these loads are global axial strain (ε), rotation per unit length (ϑ) and curvatures of the center line about the x and y axes (κ_x, κ_y). The combined cylinder has cylindrical anisotropy about the z axis. The coordinate system used here will be the cylindrical one, with cylinder axis (z) and (r, θ) as radial and tangential axes. All of the resultant stress and strains are functions of these coordinates only.

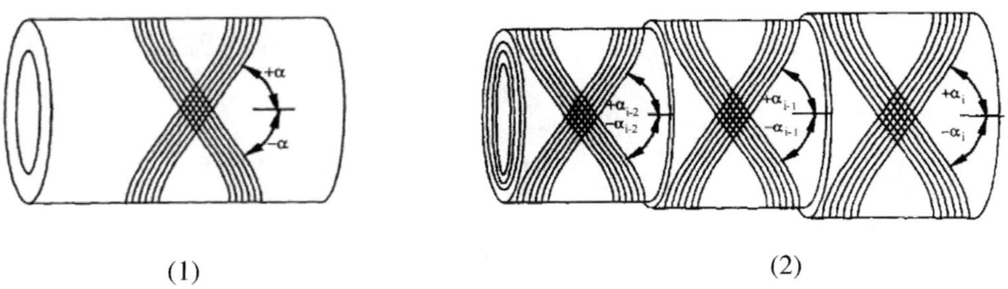

Figure 1. Scheme of composited laminated cylinder manufactured using filament winding

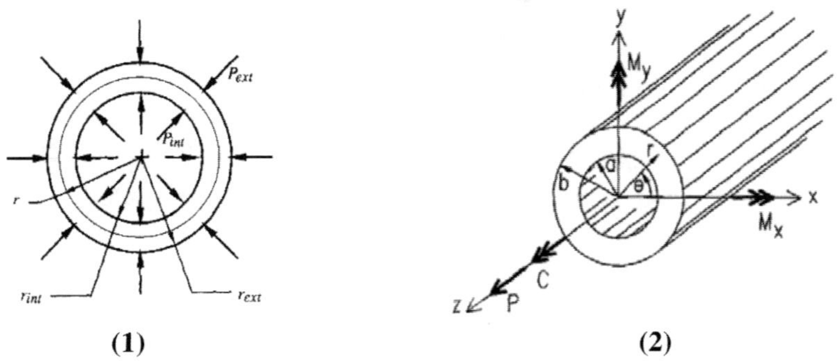

Figure 2. (1) Example of combined composite cylinder
(2) Loads and dimensions on the combined cylinder

The used material has orthotropic properties in which are defined by the fiber axis and the perpendicular directions, one in the same plane of the cylindrical ply (plane θ-z) and the other in the radial direction. The following assumptions are used in the analysis: the applied loads are constant along the z-axis, the body deformations are considered to be infinitesimal and elastic, resultant shear loads equal to zero, the resultant stresses and strains are independent of the z-axis, meaning that the curvature due to bending is constant.

2.2. Fundamental equations

Each cylinder has nine elastic independent constants in three principle directions (1, 2, 3) of its orthotropic material in its local coordinate system which has the fibre axis as the first axis of the material coordinate system. It is assumed that these constants are known from the material properties from mechanical testing or datasheet of the material. With these constants known, they are transformed into the compliance matrix for the global coordinate system of the composite cylinder. This transformation is done on the radial axis by the complement of the helix angle.

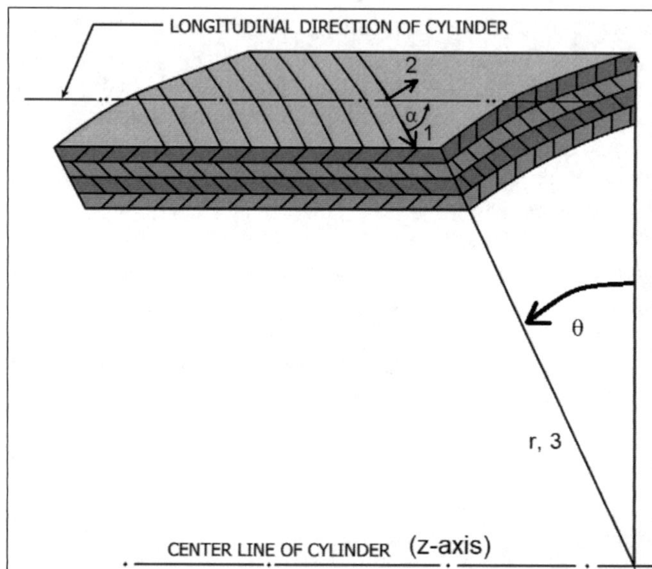

Figure 3. Coordinate Systems for composite material and cylinder

The transformed compliance matrix [Š] is calculated by:

$$[\check{S}] = [T]^{-1}[S][T] \quad (1)$$

The matrix [C] is used to calculate the reduced elastic constants [β] defined by Lekhnitskii by

$$\beta_{ij} = C_{ij} - \frac{C_{i3}C_{3j}}{C_{33}} \quad (2)$$

Noting that $\beta_{i3} = \beta_{3j} = 0$

2.3. Stress Functions

The stress components can be expressed using two functions F(r, θ) and Ψ(r, θ) derived by Lekhnitskii [1] using a system of partial differential equations that represent our problem. They are defined by

$$\sigma_r = \frac{\partial F}{r\partial r} + \frac{\partial^2 F}{r^2 \partial \theta^2}$$

$$\sigma_\theta = \frac{\partial^2 F}{\partial r^2}$$

(3)

$$\tau_{r\theta} = \frac{\partial F}{r^2 \partial \theta} - \frac{\partial^2 F}{r \partial r \partial \theta} \quad (4)$$

$$\tau_{rz} = \frac{\partial \Psi}{r \partial \theta} \quad (5)$$

(6)

$$\tau_{\theta z} = -\frac{\partial \Psi}{\partial r} \quad (7)$$

2.4. Separation of variables

To obtain a system of ordinary differential equations, Joulicoeur and Cardou [4] used separation of variables to acquire a solution in the form of

$$F = f_1(r)(\kappa_x \sin\theta - \kappa_y \cos\theta) + f_2(r) \quad (8)$$

$$\Psi = \psi_1(r)(\kappa_x \sin\theta - \kappa_y \cos\theta) + \psi_2(r) \quad (9)$$

the solution is divided into two systems, the first system in terms of f_1 and ψ_1 representing pure bending problem and the second system in terms of f_2 and ψ_2 representing axisymmetric problem.

2.4.1. Solution of Pure Bending system

The system mentioned is of the Cauchy-Euler type. This homogenous system is solved using the form

$$f_1 = K r^{m+1} \quad (10);$$

$$\psi_1 = K g r^m \quad (11)$$

Where K is an arbitrary constant

2.4.2. Solution of axisymmetric system

Lekhnitskii solved the general system as follows:

$$f_2 = \sum_{i=1}^{2} \frac{K_i'}{m_i'+1} r^{m_i'+1} + K_3' + K_4' r + \frac{K_5'}{2} r^2 + \frac{\mu_3}{3} \vartheta r^3 \quad (12)$$

$$\psi_2 = \sum_{i=1}^{2} \frac{K_i' g_i'}{m_i'} r^{m_i'} + K_4' \frac{\beta_{11}}{\beta_{14}} \ln r + K_5' \frac{\beta_{14}+\beta_{24}}{\beta_{44}} r + K_6' + \frac{C_{34}}{C_{33}\beta_{44}} \varepsilon r + \frac{\mu_4}{2} \vartheta r^2 \quad (13)$$

Where K_i' are six arbitrary constants

2.5. General complete solution

The complete general solution written using the previous results would be as follows:

$$F = (\kappa_x \sin\theta - \kappa_y \cos\theta)\left(\sum_{i=1}^{4} \frac{K_i}{m_i} r^{m_i'+1} + K_5 r + K_6 r \ln r + \frac{\mu_1}{2} r^3\right) + \sum_{i=1}^{2} \frac{K_i'}{m_i'+1} r^{m_i'+1} + K_3' + K_4' r + \frac{K_5'}{2} r^2 + \frac{\mu_3}{3} \vartheta r^3 \quad (14)$$

$$\Psi = (\kappa_x \sin\theta - \kappa_y \cos\theta)\left(\sum_{i=1}^{4} K_i g_i r^{m_i} + K_6 \frac{\beta_{56}}{\beta_{66}} + \mu_2 r^2\right) + \sum_{i=1}^{2} \frac{K_i' g_i'}{m_i'} r^{m_i'} + K_4' \frac{\beta_{11}}{\beta_{14}} \ln r + K_5' \frac{\beta_{14}+\beta_{24}}{\beta_{44}} r + K_6' + \frac{C_{34}}{C_{33}\beta_{44}} \varepsilon r + \frac{\mu_4}{2} \vartheta r^2 \quad (15)$$

2.6. Stresses and strains equations

$$\sigma_r = (\kappa_x \sin\theta - \kappa_y \cos\theta)\left(\sum_{i=1}^{4} K_i r^{m_i-1} + \mu_1 r\right) + \sum_{i=1}^{2} K_i' r^{m_i'-1} + \mu_3 \vartheta r + \mu_5 \varepsilon \quad (16)$$

$$\sigma_\theta = (\kappa_x \sin\theta - \kappa_y \cos\theta)\left(\sum_{i=1}^{4} K_i (m_i + 1) r^{m_i-1} + 3\mu_1 r\right) + \sum_{i=1}^{2} K_i' m_i' r^{m_i'-1} + \mu_3 \vartheta r + \mu_5 \varepsilon \quad (17)$$

$$\tau_{r\theta} = (\kappa_x \cos\theta + \kappa_y \sin\theta)\left(-\sum_{i=1}^{4} K_i r^{m_i-1} - \mu_1 r\right) \quad (18)$$

$$\tau_{rz} = (\kappa_x \cos\theta + \kappa_y \sin\theta)\left(\sum_{i=1}^{4} K_i g_i r^{m_i-1} + \mu_2 r\right) \quad (19)$$

$$\tau_{\theta z} = (\kappa_x \sin\theta - \kappa_y \cos\theta)\left(-\sum_{i=1}^{4} K_i g_i m_i r^{m_i-1} - 2\mu_2 r\right) - \sum_{i=1}^{2} K_i' g_i' r^{m_i'-1} - \mu_4 \vartheta r - \left(\mu_5 \frac{\beta_{14} + \beta_{24}}{\beta_{44}} + \frac{C_{34}}{C_{33}\beta_{44}}\right)\varepsilon \quad (20)$$

$$\sigma_z = \frac{1}{C_{33}}\left[\kappa_x r \sin\theta - \kappa_y r \cos\theta + \varepsilon - C_{13}\sigma_r - C_{23}\sigma_\theta - C_{34}\tau_{\theta z}\right] \quad (21)$$

2.7. Boundary conditions

To obtain the unknown constants K_i, K_i' and global deformations ε, ϑ, κ one should study the boundary conditions of case. Rigid-body displacements are also considered to maintain displacement compatibility of several coaxial cylinders; and it was found that assuming the rigid body displacement are equal to zero would overdetermine the problem in hand although its customary in this type to assume some of these displacements to be equal to zero.

From this point, the modification of the method is presented by adding the internal and external pressure in the boundary conditions to incorporate them in the solution. This will affect the part of axisymmetric loading case as shown in the next section. The theory could be used for cylinders that are bonded together or just coaxial cylinders with no friction between them maintaining no loss of contact between them. The concerned case is for the cylinders that are perfectly bonded together.

2.8. End conditions

$$P = \sum_{n=0}^{N} \int_0^{2\pi} \int_{r_a}^{r_b} \sigma_z r \, d\theta \, dr \quad (22)$$

$$C = \sum_{n=0}^{N} \int_0^{2\pi} \int_{r_a}^{r_b} \tau_{\theta z} r^2 \, d\theta \, dr \quad (23)$$

$$M_x = \sum_{n=0}^{N} \int_0^{2\pi} \int_{r_a}^{r_b} \sigma_z r^2 \sin\theta \, d\theta \, dr \quad (24)$$

$$M_y = -\sum_{n=0}^{N}\int_0^{2\pi}\int_{r_a}^{r_b}\sigma_z r^2\cos\theta\, d\theta\, dr \tag{25}$$

Such that P, C, M_x, M_y are external forces and moments.

Leading eventually to:

$$\begin{Bmatrix}\varepsilon\\ \vartheta\end{Bmatrix} = A\begin{Bmatrix}P_a\\ P_a\end{Bmatrix} + B\begin{Bmatrix}P\\ C\end{Bmatrix} \tag{26}$$

Where

$$A = (M3[M1]^{-1}[M2] + M4)^{-1}M3[M1]^{-1}(-M_p\,) \tag{27}$$

,

$$B = (M3[M1]^{-1}[M2] + M4)^{-1} \tag{28}$$

And Pa, Pb are the internal and external pressures of the inner and outer surfaces of the combined cylinder.

3. Case study with numerical examples

To get a clear comparison of results, all materials used in mathematical and finite element models are mentioned in the following table.

Table 1. Mechanical properties of the used composite materials in the study

	E_1 (GPa)	E_2 (GPa)	E_3 (GPa)	υ_{23}	υ_{13}	υ_{12}	G_{23} (GPa)	G_{13} (GPa)	G_{12} (GPa)
Graphite-polymer composite [15]	155	12.1	12.1	0.458	0.248	0.248	3.2	4.4	4.4
T300/934 [14]	141.6	10.7	10.7	0.495	0.268	0.268	3.88	3.88	3.58
Glass Reinforced Epoxy [15]	50	15.2	15.2	0.428	0.254	0.254	3.28	4.7	4.7
Kevlar49/Epoxy[16]	75	6	6	0.34	0.34	0.34	2	2	2

Table 2. Mechanical strength of the used composite materials in the study

	S_1^T (MPa)	S_2^T (MPa)	S_3^T (MPa)	S_1^C (MPa)	S_2^C (MPa)	S_3^C (MPa)	τ6 (MPa)	τ4 (MPa)	τ5 (MPa)
Graphite-polymer composite [15]	1500	50	50	1250	200	200	100	30	30
T300/934 [14]	2280	57	57	1725	228	228	76	60	76
E-Glass Epoxy [17]	1140	39	39	620	128	128	89	44	41

An implemented code ran by MATLAB using one or more of the materials mentioned in table 1 of four concentric cylinders each of 0.5 mm thickness with the composite cylinder inner radius of 50 mm under internal pressure of 10 MPa with closed ends (combined biaxial loads) is used to simulate stresses and strains with wall thickness of the cylinder.

4. Finite element analysis

Using ANSYS workbench and ACP module, a replica of the numerical case study example is done to compare results. The model used a custom material with the same mechanical properties used in the numerical example and in table 1. The finite element model shown in figure 5 shows a stress variation along the z-axis of the cylinder despite the numerical example assumption and this is due to the boundary conditions placed on the edges of the cylinder; one for the support and one for the axial force. And to overcome this variation the results are taken exactly at the middle of the cylinder within the thickness of the cylinder by making a construction path in the finite element model and comparing the results with the analytical model.

Starting with a shell 2D model then using the ACP module to create the solid model with the prescribed material plies thickness and orientation; meshed with a 3D solid element (8 nodded Hexahedron type Hex8) shown in figure 4 with nodal solution and maximum element size of 2 mm which is the thickness of one ply. SOLID185 Element which is used for 3-D modeling of solid structures is used as in this simulation. Defined by eight nodes the element has three degrees of freedom at each node: translations in the nodal x, y, and z directions, with large strain capabilities and other mechanical properties as hyper elasticity and plasticity.

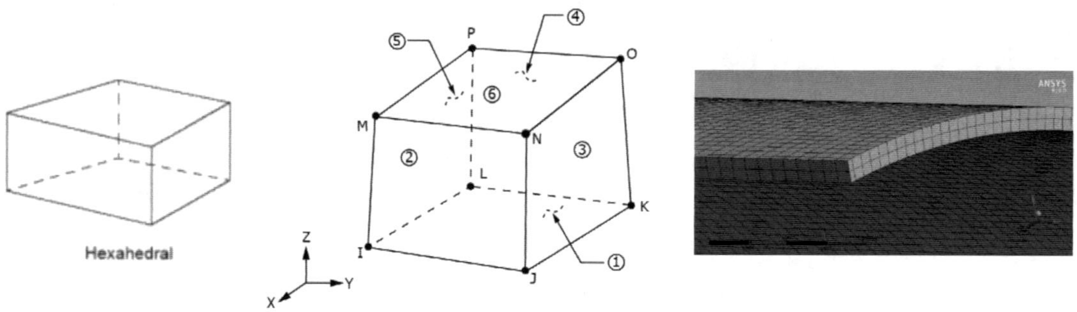

Figure 4. Solid model element used in the analysis (SOLID185 - 3-D 8-Node Structural Solid)

The mesh size element is chosen at least equal to one ply thickness as shown in figure 6. This was found in agreement with ref [7] as if the element size has increased more than one ply thickness, the output results would have a much increased error margin and low accuracy. And using less element size would require much more computational power with slight increase in output results accuracy.

To check the reliability of the finite element model agreement with the analytical model, the above-mentioned cylinder geometry has been checked with two of the most common used winding angles (±55°, ±30°) with different stacking sequences. The analytical model assumes constant stresses and strains through the axial direction while the finite element maybe affected by the boundary conditions necessary for the software calculations. The results of the finite element modeling show an agreement with the analytical results as shown in figure 7. By comparing the output results the error margin stays at less than 5% difference between the analytical model and the finite element model which is quite good. It's worth mentioning that the model implemented by the finite element software needs much more time to program into the software and each case would take quite the time to give results and take these results into other platform to compare; which is all fine and even versatile and accurate for

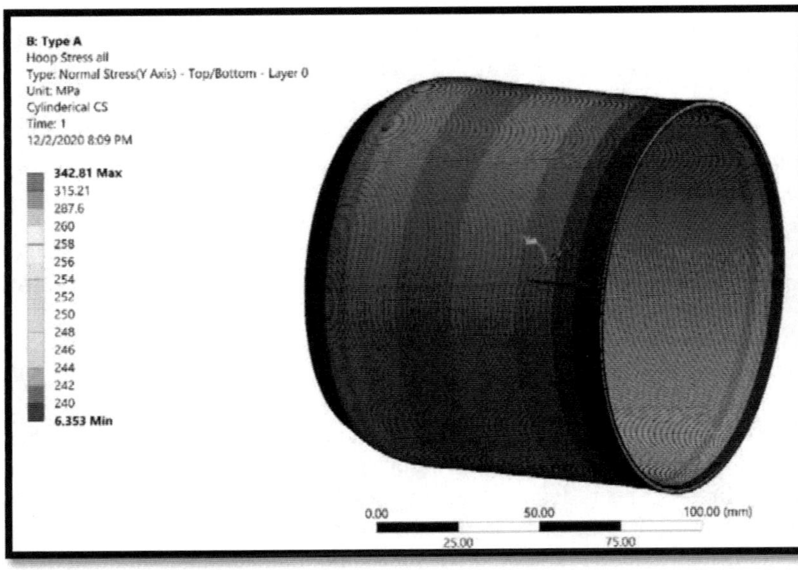

Figure 5. Hoop stresses on the numerical model

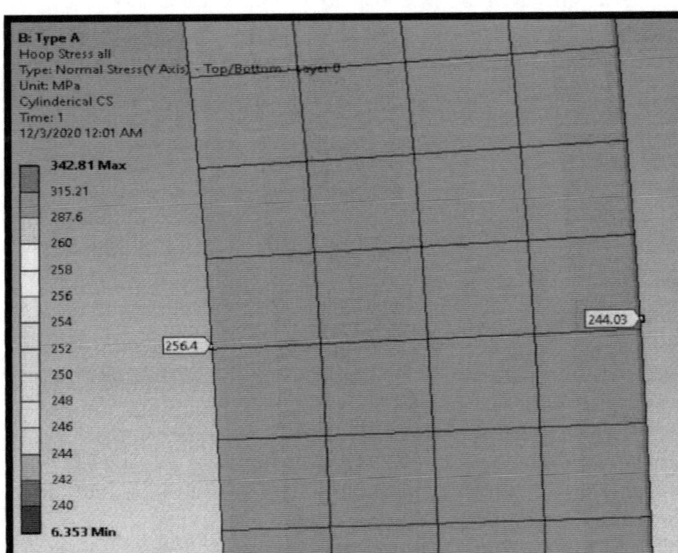

Figure 6. Numerical model elements within the thickness of the cylinder

unconventional geometry settings but may not be as suitable as analytical model that could study and give results for thousands of models during a parametric study or design optimization.

5. Parametric Study

The design parameters of the composite cylinder are further investigated to check their effect on the structure behavior under the same load and geometry. Using the previous example, the cylinder under consideration is consisted of 4 plies; all of which of the same material under internal pressure of 10 MPa with closed ends.

Starting with the winding angle as an important parameter in anisotropic material used in composite cylinders; a winding angle configuration ±α with range of 0° to 90° degrees as a value of α in a four-ply cylinder is studied and the results given in figure 7 show the stresses and strains within the wall thickness of the cylinder.

A general note on all stresses and strains; at the start and end of the range of the winding angle (i.e. at ±0° and ±90°) the results of all four plies have linear change in value without sudden or abrupt changes with ply migration within the thickness; this happens as the fibers tend to have the same direction which could mean that they may be considered as only one ply with the same winding angle. And to clarify that; at ±0° or ±90° the cylinder wall may be described as only one ply with 0° or 90° winding angle while at ±30° winding angle configuration; the cylinder wall may be described as (+30°/-30°/+30°/-30°).

5.1. Analytical and finite element comparison

Figures (7-1) and (7-2) show the hoop stresses and strains respectively. As mentioned before the start and end of the winding angle show linear change in stresses and strains within the thickness which is quite convenient to use as sudden stress change tend to create inter-ply high shear values that could create structure instability and even delamination way before the material ultimate strength values leading to structure failure. As it appears from figure (8-1) some comfort zones for the winding angle choice are worth mentioning. The first is (0°-30°) as the stress change is somehow smooth and without much difference between the stress at the inner and outer surface of the cylinder. The second is 90° as the stress is also linear and smooth but offers the highest stress difference between the inner and outer surface of the cylinder wall. The third is in the region around 68° as it may also offers somewhat smooth stress transition within thickness. Figure (8-2) is clear that more we get near the 90° winding angle, the less the strain within the wall thickness in the hoop direction meaning a stiffer structure, but it's also worth mentioning here that also around 68° winding angle is the local minimum of the strain's values even though it's not much different than 90° strain but it's an interesting phenomenon that may need further investigation to clarify.

5.2. Stresses and strains variation with winding angle

Figures (8-3) and (8-4) show the axial stresses and strains respectively. The comfort zones here are quite clear but by studying both figures simultaneously; the best of choice here would be (0°-10°) and around 53° winding angles as they offer the least strains and the smoothest stresses. The (0°-10°) range is quite clear to clarify as the axial stresses in those areas are mainly loaded on the fibers of the structure which has the stronger tensile stiffness strength. While the 53° range is explainable under the coupling between the pressure on the cylindrical wall and the closed ends and the usage of the netting analysis which could briefly described as the consideration of the cylinder as made of fibers only and the matrix strength is completely ignore and this angle offers the best resolution between hoop and axial stresses distribution according to pressurized cylindrical structure.

Figures (8-5) and (8-6) show the radial stresses and strains respectively. The strain figure offers the (50°±3°) range as the stiffest structure in the radial zone which could also be explained by the netting analysis as mentioned before. This could lead to that the netting analysis is worth considering greatly in the analysis of radial and axial stresses only which offer great agreement in the best winding angles range of ±53° configuration.

Figures (8-7) and (8-8) show the shear stresses and strains in the cylinder wall plane (Θ-z) direction respectively. Although these charts aren't quite cheerful in the point of smooth stress transition between plies but at least they offer a good insight into what to expect while choosing winding angles other than 0° and 90°, so that we could make sure the material and structure properties could with stand that kind of shear stresses values and transitions.

Figures (8-9) and (8-10) show the fiber and matrix tensile stresses respectively. And they may be considered a decision maker in the choice of winding angles configuration. It's well known that the tensile strength of fibers is much more of that of matrix strength, so it's only logical to choose the

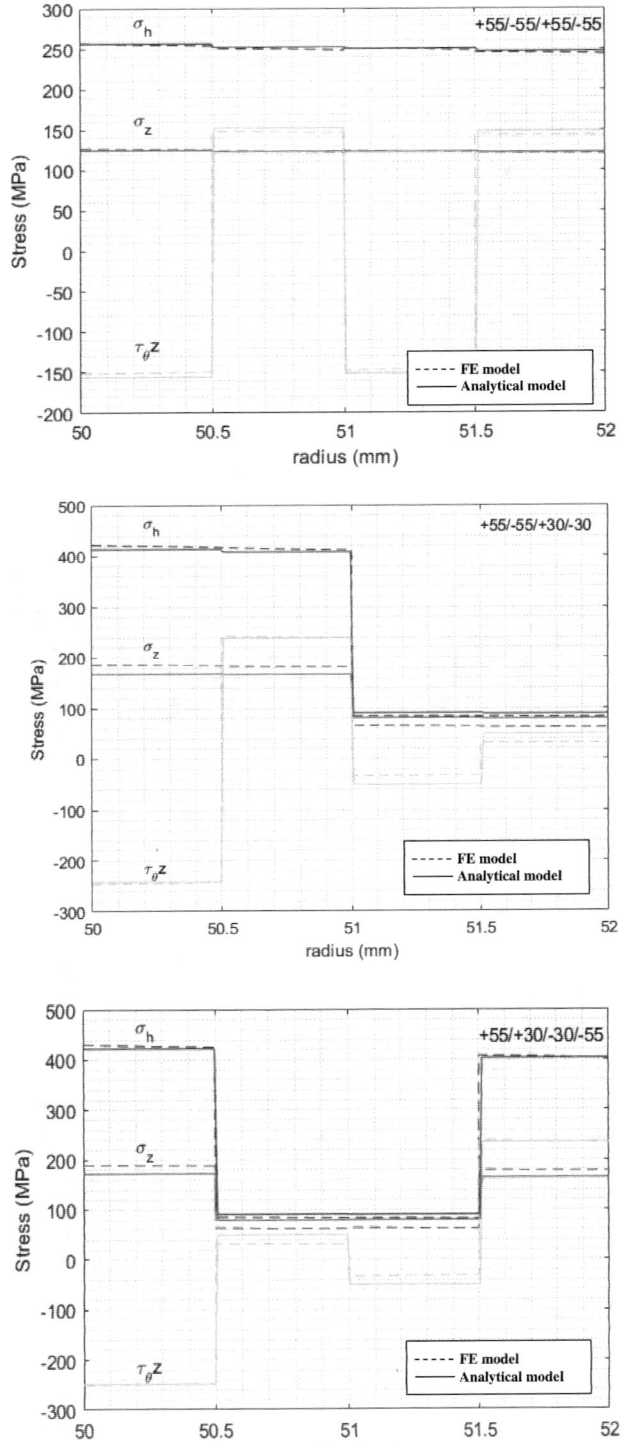

(1)

(2)

(3)

Figure 7. Stress distribution within thickness for a cylinder pressurized internally at 10 MPa using finite element analysis by ANSYS and analytical method for different stacking sequences

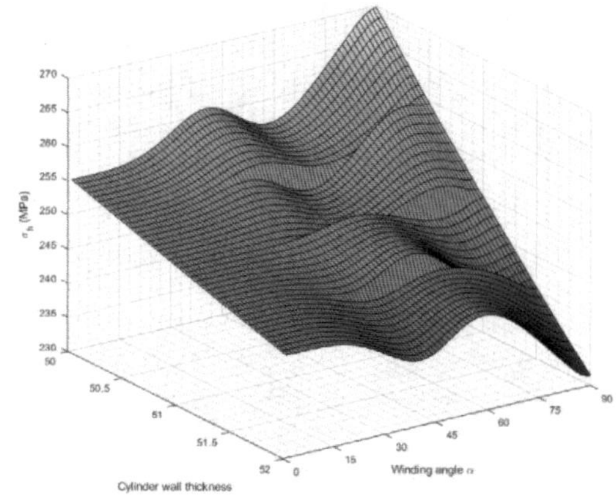

(1)

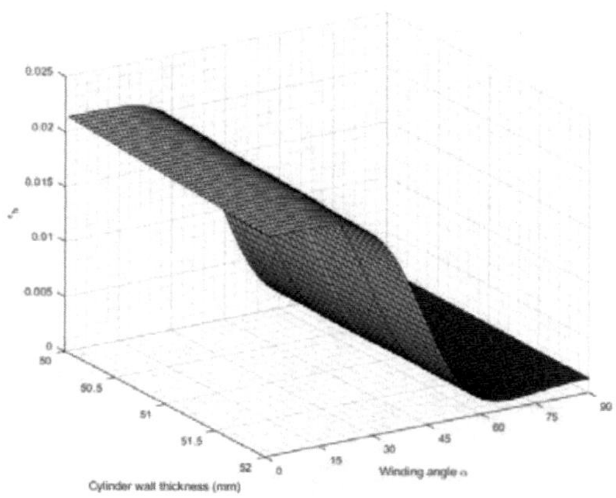

(2)

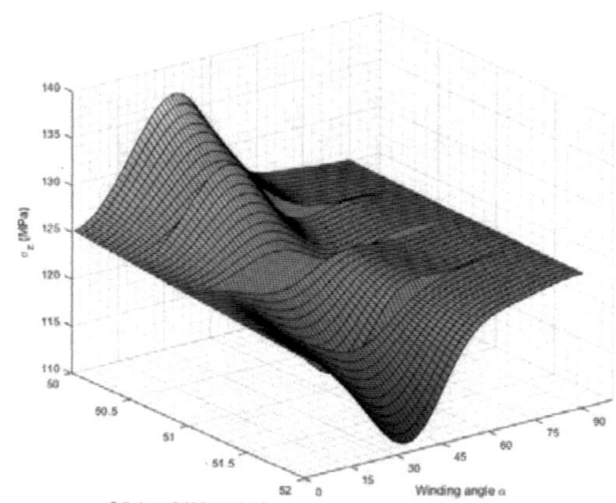

(3)

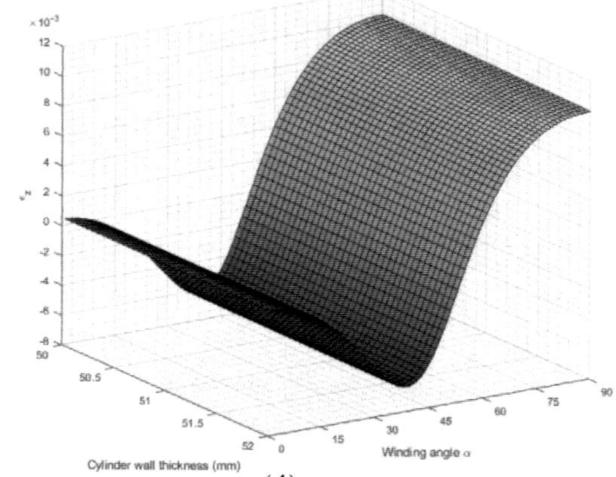

(4)

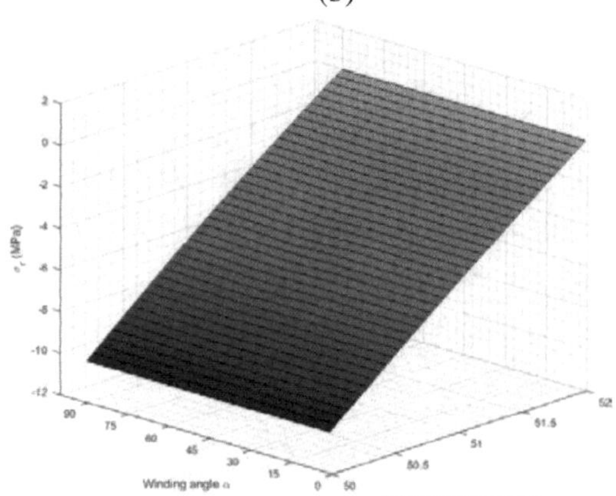

(5)

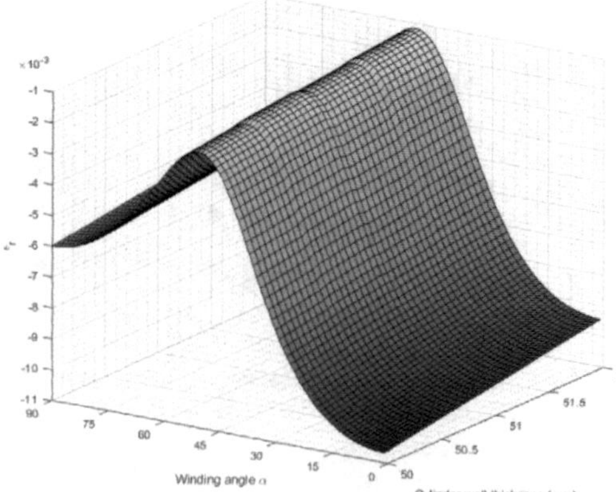

(6)

winding angle that offers the highest tensile stress on fibers and or the lowest tensile stress on matrix. And these two figures are actually complementary or very good to say that we could deduce one from

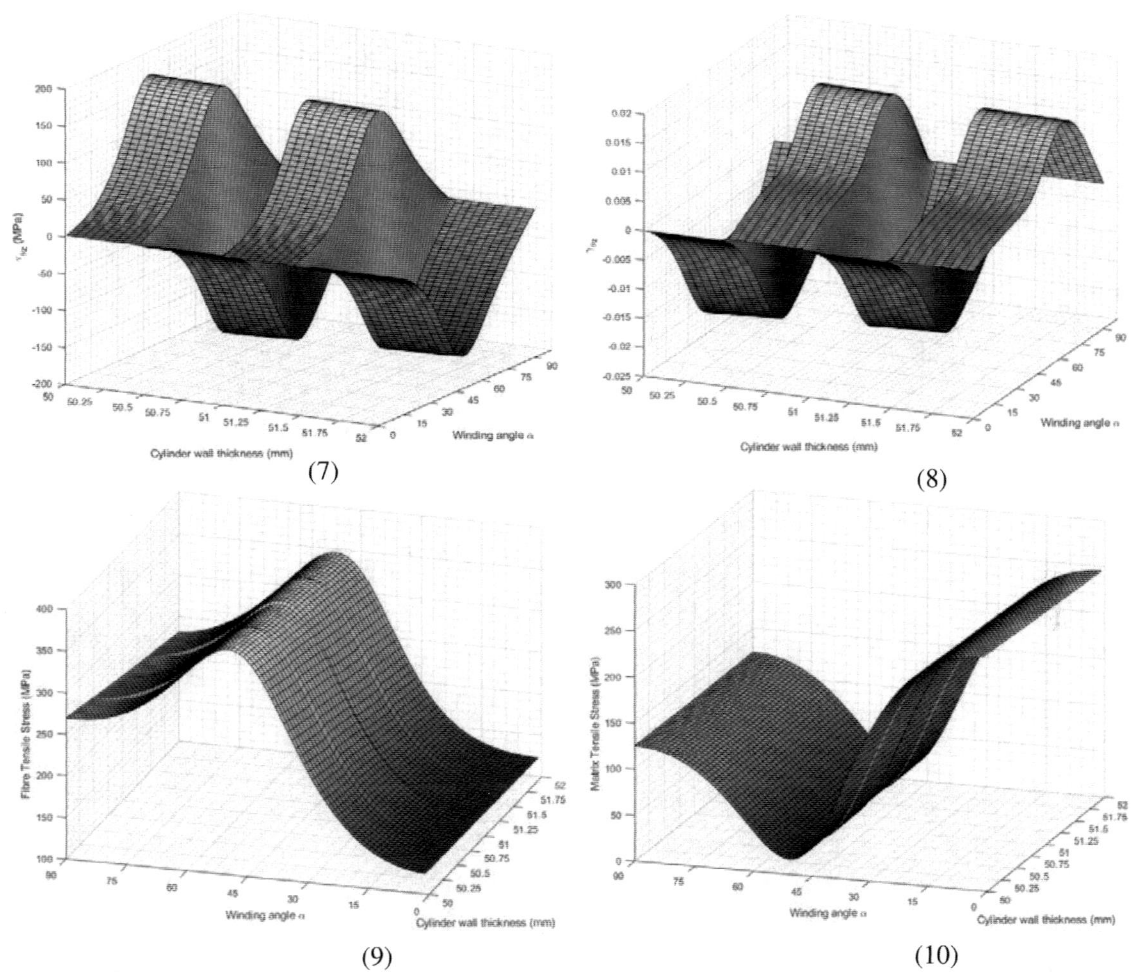

Figure 8. Stresses and strains distribution within the wall of Graphite-polymer composite cylinder (4plies) with closed ends pressurized internally under 10 MPa with winding angles ±α configuration

the other as the tensile stress is loaded one or both of them. The figures offer the (51°±3°) as the best choice to load most of the tensile stresses on the fibers as they have the best strength in the structure.

Although 3 cases are being investigated in the previous section; FEA is somehow time and resource consuming in the preliminary design phase where a parametric study of many inputs need to be studied to check the impact of every composite parameter for sensitivity of the final product to that parameter. Stacking sequence can vary a lot and is proven to be very effective in the safety of the final product; an optimization or a further investigation of many cases should be performed specially in the products with complex loading conditions (biaxial or triaxial loads) as these cases can require hundreds of checkups and geometry variations that can't be done efficiently on FEA or as easy as on mathematical modeling.

5.3. Ply thickness effect on stresses variation

Also, one of the affecting design parameters is the thickness of the single ply, figure 8 shows when increasing the number of plies within the same thickness, i.e. winding the same angle many times over until reaching specific thickness then changing the angle to the negative value, to study the effect of varying the winding angle over and over again until reaching the required thickness or is it better to stay on the same angle until a specific thickness and then change the angle. And the figure shows that staying on the same angle for large thickness without varying the angle periodically may create some kind of stress instability and larger maximum stresses in the overall thickness as the large variation between the maximum and the minimum value of the interlaminar stresses may create some kind of internal stresses that could lead to severe delamination. It is notable to say that the ability to increase the thickness may be only available via automated fibre placement method of winding as the conventional filament

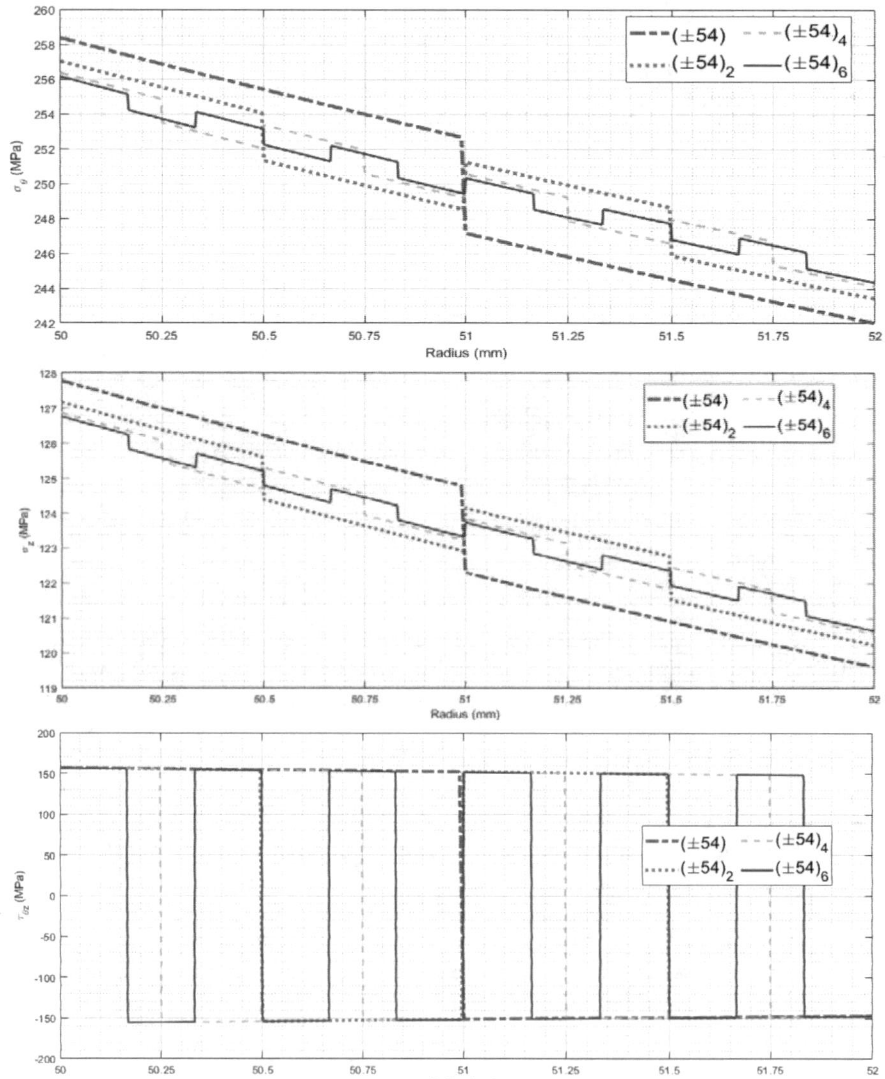

Figure 9. Stress distribution in UD Kevlar cylinder wall under internal pressure of 10 MPa for different ply thickness configurations

IOP Conf. Series: Materials Science and Engineering 1172 (2021) 012005 doi:10.1088/1757-899X/1172/1/012005

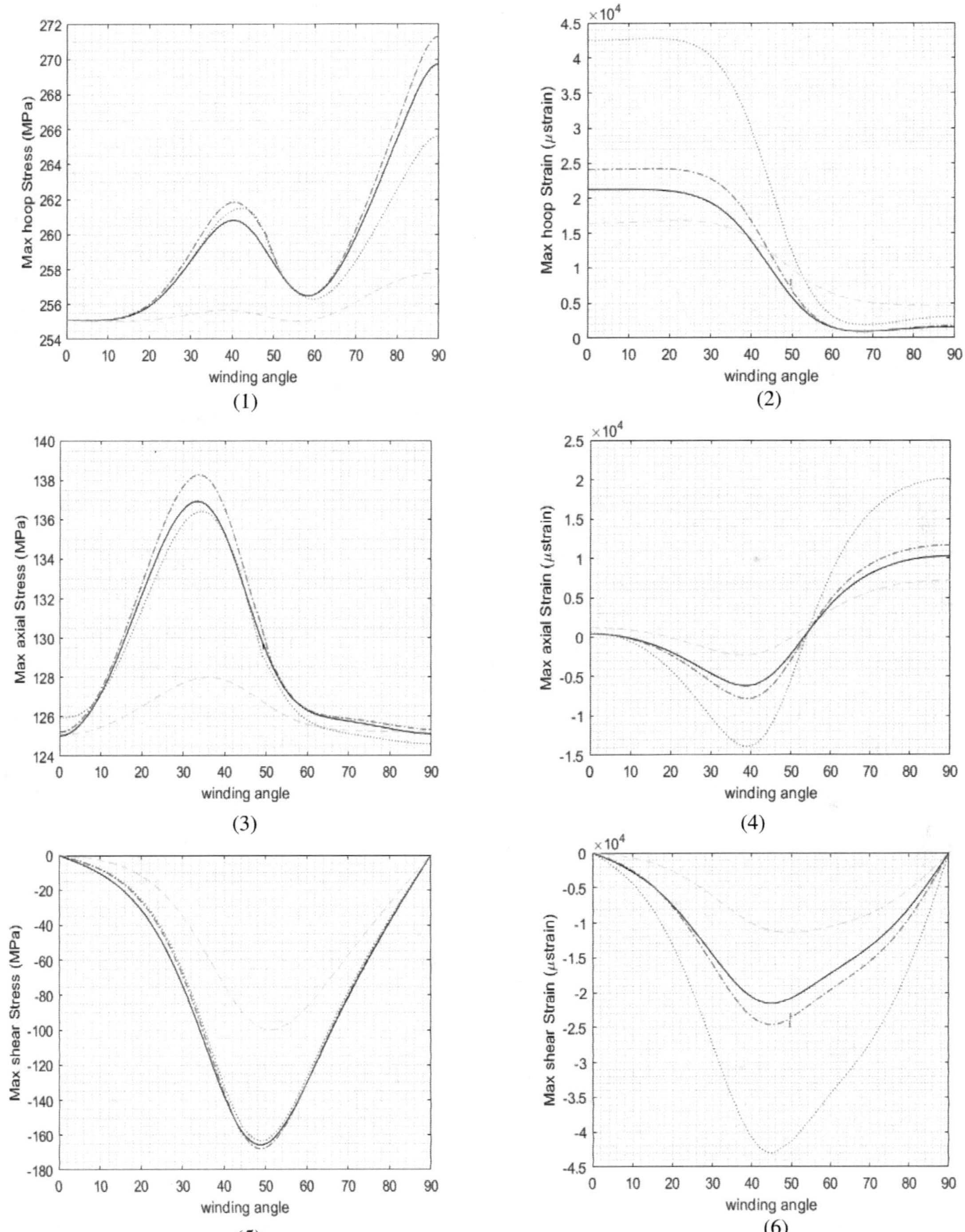

	T300/934	UD Kevlar 49	Glass Reinforced Composite	Graphite Reinforced Epoxy
E_1/E_2	13.23	12.5	3.29	12.81

Figure 10. Absolute maximum values of stresses and strains within wall thickness of different materials composite cylinder (4 plies) with winding angles ±α configuration

winding machine could only achieve reciprocating winding angle between positive and negative values of the defined winding angle by the designer.

5.4. Different material effect on stress variation

Figure (10) shows maximum absolute values of stresses and strains for different materials with different winding angles. Starting with 0° winding angle where the fiber is placed along the z-axis this angle is considered the worst choice for pressure vessels under internal pressure only as in this angle the resistant to the pressure is the matrix tensile strength which is the worst in comparable with the fiber strength.

Figure 10 shows the stresses and strains maximum values for different winding angles (0°-90°) for a cylinder of 4 plies with internal radius of 50 mm and thickness of 2 mm for all plies. As shown from figure 8 stresses vary along the thickness and the maximum stress within the thickness is chosen as a representing value for each winding angle for these charts. An internal pressure of 10 MPa load and normal forces due to closed ends on the cylinder ends are the principal loads on the charted cylinder. The stacking sequence is ±α for all plies starting from the inside ply.

Figures 10.1, 10.2 show the maximum hoop stresses and strains within the thickness of the cylinder wall for all winding angles in a stacking sequence of ±α. Although the stresses appear to be at a minimum value at near zero angles but if we look at the strain values at the same periods, we'll see that it has the maximum value which is logical as the winding angle zero means all the fibers are axial to the cylinder and the only resistance to the pressure load is the matrix tensile strength which has much lower modulus and strength. In terms of hoop strain the best choice for the winding angle would be angles more 45-50 degrees and up to 90 degrees, but the best choice according to the study for this case would be more than 60 degrees and for some materials as carbon fibers (T300 and Graphite) would be around 65 degrees as it presents the lowest hoop strains. In combination with stresses chart the best choices for winding angles would be less than 20 degrees and the period of 55-60 degrees; in which the first choice would be unacceptable due to large strains so the logical choice would be the latter.

Figures 10.3, 10.4 showing the axial stresses and strains offer us the green zones for choosing the winding angles at less than 10 degrees and more than 55 degrees for the stresses. While for strains the best choices would be less than 20 degrees and somewhere between 45-60 degrees. So, in terms of axial stresses and strains the best choice for winding angle would be either less than 10 degrees or 55-60 degrees.

Figures 10.5, 10.6 charting the shear stresses and strains offers the best winding angles at less than 20 degrees and more than 80 degrees for the stresses, and almost the same periods for the strains.

In general, the used mathematical model can be tailored to produce any required stresses or strains within the thickness of a composite laminated cylinder to use these results to investigate the impact of changing any one the production parameter or check the safety of the composite cylinder as explained in the next section.

6. Failure Criteria

Composite use many failure theories that can be compared with the output stresses of the analysis giving a safety factor or usability to the structure. In the following part a couple of the most commonly used composite failure will be demonstrated using the stress analysis results from the mathematical model.

6.1. Tsai Wu

$$F_1\sigma_1 + F_2\sigma_2 + F_{11}\sigma_1^2 + F_{22}\sigma_2^2 + F_{66}\tau_{12}^2 - \sqrt{F_{11}F_{22}}\,\sigma_1\sigma_2 \le 1 \qquad (29)$$

Where

$$F_1 = \left(\frac{1}{S_1^T} + \frac{1}{S_1^C}\right),\ F_{11} = -\frac{1}{S_1^T S_1^C}\ ,\qquad F_2 = \left(\frac{1}{S_2^T} + \frac{1}{S_2^C}\right),\quad F_{22} = -\frac{1}{S_2^T S_2^C},$$

$$F_{66} = \left(\frac{1}{S_{12}^F}\right)^2$$

Such that [15]:

- σ_i is normal or transverse stress according to the subscript (σ_1 is stress along the fiber direction and σ_1 is stress transverse to the fiber direction)
- S^i_j is the strength (i for compression or tension strength, j for direction of strength load)

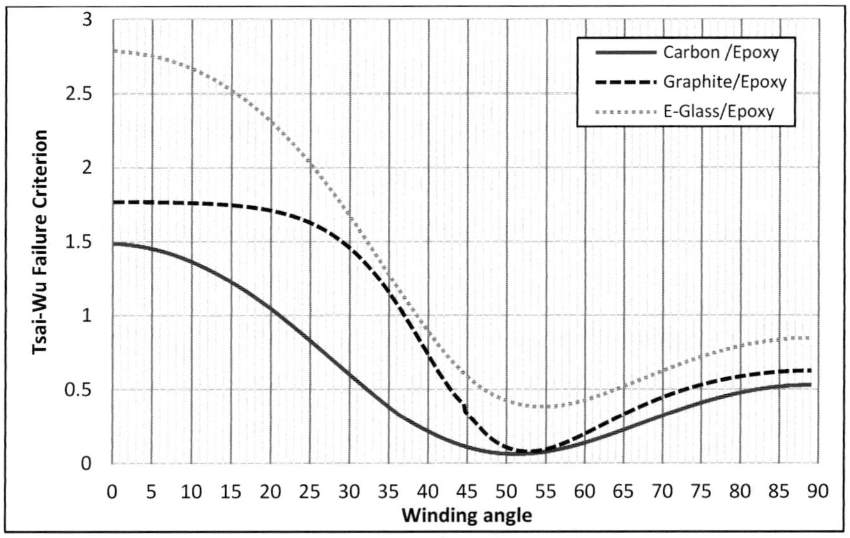

Figure 11. Tsai-Wu safety criterion variation with winding angle

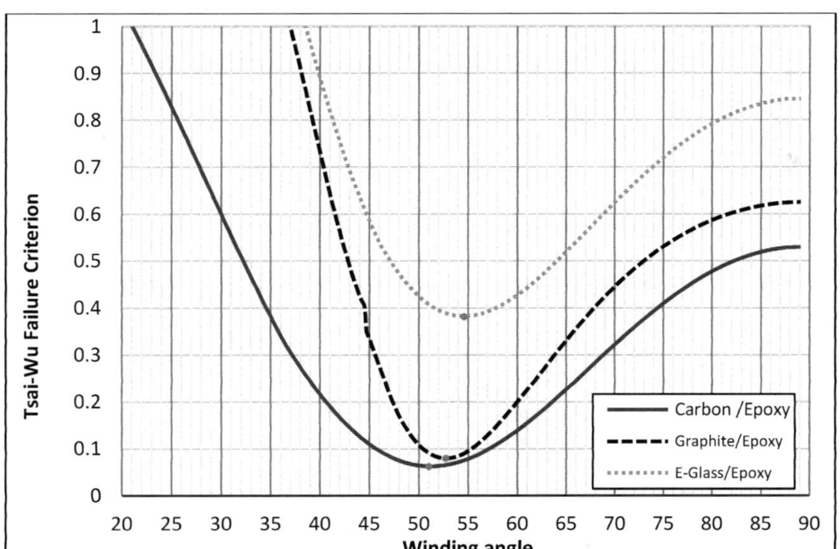

Figure 12. Winding angle safe zone (zoomed of figure 11)

Figures 11-12 show a cylinder of 50mm radius, thickness 4 mm manufactured with 4 plies of stacking ($\pm\alpha$) and internal pressure of 5MPa and closed ends (biaxial load); with the variation of the winding angle to check the first ply failure using the Tsai-Wu failure criterion figure 12 shows the safe zone of winding angle choice which varies according to the material type due to variation of strength and anisotropic ratio. The analysis shows that for carbon / epoxy the winding angle could be chosen between 21° and 90° with the safest choice around 51°, while for E-glass / epoxy the safe winding choice starts at 39° and the safest choice is around 54°. This change in optimum angle could be explained as the strength of the material increases the winding angle safe zone widens and the safest choice is dependable on the anisotropic nature of the material.

6.2. Tsai Hill

Based on the energy theory the Tsai-Hill criterion is

$$\left(\frac{\sigma_1}{S_1}\right)^2 - \left(\frac{\sigma_1\sigma_2}{S_1^2}\right) + \left(\frac{\sigma_2}{S_2}\right)^2 + \left(\frac{\tau_{12}}{S_{12}}\right)^2 \leq 1 \tag{30}$$

Analogous to the Von-mises criterion the Tsai-Hill failure criterion is considered one of the most used for orthotropic material as a failure criteria for design calculations [18].

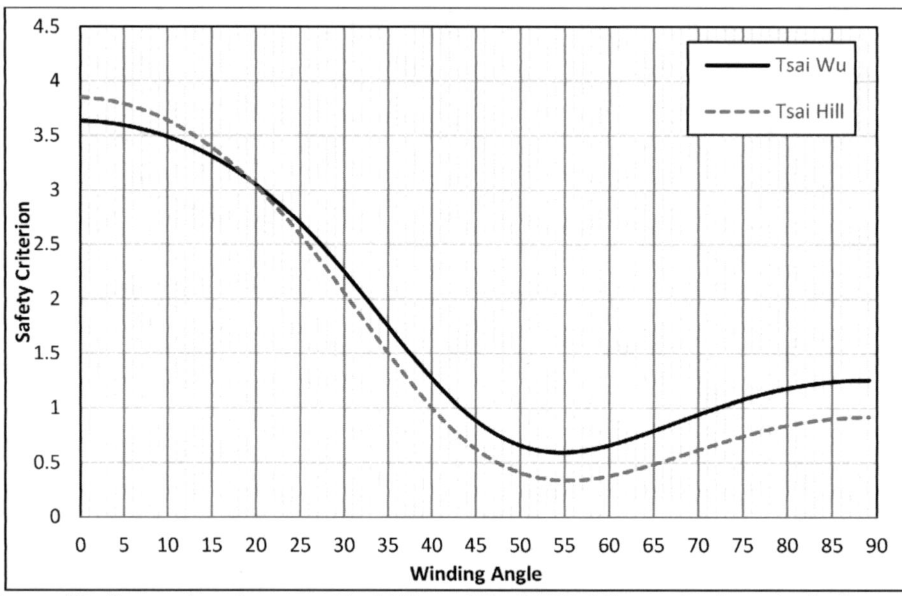

Figure 13. Comparison between Tsai-Wu and Tsai-Hill failure criteria and winding angle

Figure 13 shows that Tsai-Wu criterion is more conservative than Tsai-Hill in the safest winding angle zone as it gives less safe criterion number. An application of different failure criteria is beneficial to decrease the number of experiments required to test the final product or application.

Table 3. Optimum winding angles according to failure criteria

Optimum winding angle	**Carbon / Epoxy** E_1/E_2=14.27	**Graphite / Epoxy** E_1/E_2=12.81	**E-Glass / Epoxy** E_1/E_2=3.94
Netting Theory		54.74 °	
Tsai Wu	51.3 °	52.8 °	54.6 °
Tsai Hill	53.3 °	54.4 °	55.3 °

Table 3. shows the optimum winding angle according to different failure theories and from the output of the stress analysis by applying each stress value and comparing it to the material strengths. This shows us that according to the material properties and different loading conditions, the winding angle with the least chance to fail could change slightly or greatly, i.e. for custom material the material properties anisotropic factor E1/E2 could vary largely affecting the optimum winding greatly. As netting theory mainly depends on the existence of fibers only in the analysis only a single optimum winding angle is

provided for the same loading condition (for a pressure vessel with closed ends that optimum angle is 57.74°) no matter what the material properties or its anisotropic nature.

7. Conclusion

Lekhnitskii's formalism [1] is modified to incorporate the internal and external pressure loading beside the axial, torsional and bending loads for multilayered composite cylinder.

A numerical example for a composite cylinder subjected to internal pressure and axial tension is presented and the resulted stresses and strains within wall thickness are illustrated. These results are used to validate a built-up finite element model; the results are found to be comparable from both the analytical and the finite element model.

The modified analytical model is used to perform a parametric study for a composite cylinder subjected to internal pressure and axial tension, the results show the effectiveness of material anisotropy such that the higher E_1/E_2 the higher the variation of stresses with the winding angle.

Also, from the parametric study, one could conclude that the layer thickness is an effective parameter; as the layer thickness decreases, the obtained stresses through wall thickness is found to be smoother and have lower value.

The output of the stress analysis can be used in the safety check using any safety criteria, they are key to the design and geometry to perform such analysis is the design phase. The results of the safety criteria show that using the winding angle between 50-55 is the safest choice for biaxial load of pressure vessels and this agrees with the netting analysis with mild variation due to the matrix factor in strengthening the final product.

References

[1] S. G. Lekhnitskiĭ, Theory of elasticity of an anisotropic elastic body. Holden-day, 1963.
[2] S. C. Panda and R. Natarajan, "Finite element analysis of laminated composite plates," International Journal for Numerical Methods in Engineering, vol. 14, no. 1, pp. 69-79, 1979.
[3] S. Panda and R. Natarajan, "Analysis of laminated composite shell structures by finite element method," Computers & Structures, vol. 14, no. 3-4, pp. 225-230, 1981.
[4] C. Jolicoeur and A. Cardou, "Analytical solution for bending of coaxial orthotropic cylinders," Journal of engineering mechanics, vol. 120, no. 12, pp. 2556-2574, 1994.
[5] P. Wild and G. Vickers, "Analysis of filament-wound cylindrical shells under combined centrifugal, pressure and axial loading," Composites Part A: Applied Science and Manufacturing, vol. 28, no. 1, pp. 47-55, 1997.
[6] H. Bakaiyan, H. Hosseini, and E. Ameri, "Analysis of multi-layered filament-wound composite pipes under combined internal pressure and thermomechanical loading with thermal variations," Composite Structures, vol. 88, no. 4, pp. 532-541, 2009.
[7] X. Sun, V. Tan, Y. Chen, L. Tan, R. Jaiman, and T. Tay, "Stress analysis of multi-layered hollow anisotropic composite cylindrical structures using the homogenization method," Acta Mechanica, vol. 225, no. 6, pp. 1649-1672, 2014.
[8] M. Menshykova and I. Guz, "Stress analysis of layered thick-walled composite pipes subjected to bending loading," International Journal of Mechanical Sciences, vol. 88, pp. 289-299, 2014.
[9] D.-C. Pham, N. Sridhar, X. Qian, A. J. Sobey, M. Achintha, and A. Shenoi, "A review on design, manufacture and mechanics of composite risers," Ocean Engineering, vol. 112, pp. 82-96, 2016.
[10] I. A. Guz, M. Menshykova, and J. K. Paik, "Thick-walled composite tubes for offshore applications: an example of stress and failure analysis for filament-wound multi-layered pipes," Ships and Offshore Structures, vol. 12, no. 3, pp. 304-322, 2017.
[11] V. Mohanavel, S. Prasath, M. Arunkumar, G. Pradeep, and S. S. Babu, "Modeling and stress analysis of aluminium alloy based composite pressure vessel through ANSYS software,"

Materials Today: Proceedings, 2020.
[12] P. Sharma et al., "Theoretical analysis of design of filament wound type 3 composite cylinder for the storage of compressed hydrogen gas," International Journal of Hydrogen Energy, vol. 45, no. 46, pp. 25386-25397, 2020.
[13] S. K. Moshir, S. V. Hoa, F. Shadmehri, D. Rosca, and A. Ahmed, "Mechanical behavior of thick composite tubes under four-point bending," Composite Structures, vol. 242, p. 112097, 2020.
[14] M. Xia, H. Takayanagi, and K. Kemmochi, "Analysis of multi-layered filament-wound composite pipes under internal pressure," Composite structures, vol. 53, no. 4, pp. 483-491, 2001.
[15] M. W. Hyer and S. R. White, Stress analysis of fiber-reinforced composite materials. DEStech Publications, Inc, 2009.
[16] M. Jones, "The basic ply properties of a Kevlar 49/epoxy resin composite system," Royal Armament Research And Development Establishment Fort Halstead (England), 1983.
[17] I. M. Daniel, O. Ishai, I. M. Daniel, and I. Daniel, Engineering mechanics of composite materials. Oxford university press New York, 2006.
[18] D. Gay, Composite materials: design and applications. CRC press, 2014.

IOP Conf. Series: Materials Science and Engineering 1172 (2021) 012006 doi:10.1088/1757-899X/1172/1/012006

Effect of damping material thickness on vibration analysis in pretension layer damping process

A Mohamed[1], A A **Omer**[1] and **A Hassan**[1]

[1] Mechanical Engineering Department, Military Technical College, Cairo, Egypt.

E-mail: anwer_mohamed@mtc.edu.eg,

Abstract. Vibration reduction is a critical necessity in many fields of engineering, technology, and industry. As a result, there is a need for vibration management. To restrict or adjust the system's vibration response, a variety of techniques are employed. In recent years, there has been a lot of enthusiasm for the easy implementation of these vibration-control structures. The damping properties of viscoelastic materials tend to be excellent. Damping is determined by the material's ability to dissipate energy. To minimise the vibration of vibrating surfaces, viscoelastic materials are commonly used. In this study, viscoelastic material (Dyad 606) is applied on the AL plate in the form of free layer damping and pretension layer damping. First aluminium structure using free layer damping and another one using pretension layer damping by applying tension load (two layers of damping plates, one on the top surface of the Al alloy and the other on the lower face). Passive vibration damping of the plate is achieved. These layers are influenced by axial uniform distributed load. The damping behaviour of the AL Plate is discussed in relation to the thickness variation of the pre-tension damping material. The results show that the loss factor and the attenuation percentage of the structure with pretension layer damping are increased as compared to free layer damping. It is also found that the most effective thickness of the damping material to increase the damping capacity of the framework is around half the base plate thickness

1. Introduction

In different branches of engineering, technology, and industry, the issue of reducing the amount of vibration in a system arises. Vibrations that cause harm have a major effect on task execution, performance, and precision. [1]. As a result, vibration control is essential. The vibration response of the device is restricted or adjusted using a variety of techniques. In the last few years, there has been a lot of interest in bringing these vibration-control systems into operation. [2]. First and foremost, if an unwanted vibration problem needs to be regulated, it is desirable and always appropriate to comprehend its entire nature. It can then be determined that if an additional damping mechanism is to be successful, the increased damping must be much greater than the original damping. [3]. Then it's time to figure out if passive or active control strategies are the most effective way to solve the problem. To make the vibrational system less sensitive to its vibrational environment, passive control necessitates adjusting the rigidity, mass, and damping of the system. If one or more structural resonances overpower the unwanted vibration and acoustics, they can be effectively balanced by increasing the system's damping. For most applications, four types of noise and vibration management may be used: absorption, barriers and enclosures, structural vibration damping, and vibration insulation. [4]. the increased damping must be incomparably greater than the initial damping to be appropriate for an additional damping device. The most popular method of improving damping is to introduce highly damped polymeric material into

Content from this work may be used under the terms of the Creative Commons Attribution 3.0 licence. Any further distribution of this work must maintain attribution to the author(s) and the title of the work, journal citation and DOI.
Published under licence by IOP Publishing Ltd

the structure at key locations. The structure must interact with the polymer in such a way that the polymer dissipates as much energy as possible. In the common method, two forms of vibration damping and acoustic control treatments are provided [5]. The first one is referred to as thorough damping. This procedure is also known as the unregulated or free treatment of layer damping. When the structure is flex-mounted, the treatment is coated on one or both sides of the structure to cause tension-compression deformation of the damping material. But, due to the lower damping ratios, this technique cannot be used in a number of mechanical structures. [6]. The second type of damping is referred to as a shear type. The damping type is more effective for a given weight than the prolonged damping treatment. This efficiency is, however, balanced by more study and application complications. The therapy is identical to that of the unrestrained layer, with the exception of another layer limiting the viscoelastic content. Therefore, the extra layer will limit the viscoelastic material and deform the structure in shear when the structure is flexed [7]. Such damping decreases device performance and increases net costs.

Other damping systems with two layers of damping material, one on the top surface of the base vibrating component and the other on the bottom part, are investigated, to improve damping properties, reduce costs, simplify, and improve reliability [8].Damping material thickness is an effect of the vibration damping efficiency so that it is tested and the most effective thickness is clarified.

Damping is the dissipation of energy in the cyclical stress of a material or system. In any mechanical device, there are four major types of damping [9]:

- Viscous damping
- Magnetic damping
- material damping
- dry friction damping

2. Scope and objective

Structures that are both functional and lightweight are becoming increasingly necessary (Electric vehicles, engines, aircraft, boats, car engines, manufacturing equipment, and so on are examples of such products.). This criterion resulted in light structures shifting and/or vibrating at higher frequencies, producing higher temperatures and, as a result, increasing unwanted noise and oscillation levels. As a result, stronger vibration damping materials had to be found [10]. The aim of the optimum design of pretension layer damping is to maximise structural damping due to vibration modes' modal loss factors [11]. Here the maximization of damping is reported in the first free vibration mode. The damping intensification of pretension layer damping, and has been carried out as a thickness optimization of the polymeric sheet.

2.1. Damping material thickness

The vibration damping behaviour of the plate varies as the thickness of the damping material changes. However, it only increases dramatically near the damping material's glass transition temperature due to a substantial shift in the damping layers loss factor. [12]. If the damping layer is too thin or too thick, the damping effect is diminished. Maximize damping by increasing the thickness and/or rigidity of the material [13]. Since cost-effectiveness is often assumed when the damping film is too thin, the thickness of PTLD (which reflects the cost) is often considered a penalty. The following four types of plates are considered for analysis by taking into account the above references:

a) PTLD with VEM 0.3 mm.
b) PTLD with VEM 0.6 mm.
c) PTLD with VEM 0.9 mm.
d) PTLD with VEM 1.2 mm.

3. Numerical work

In ANSYS edition R15.0 [14], the method for building a finite element analysis begins in the 'Engineering data file.' In the second paragraphs, the materials used are discussed in depth, for example, by assigning material properties to all of the materials that were used in the FEM's design. Mechanical properties of FEM material. Base plate (AL 5051), and the damping layer (dyad 606) [8].

Table 1. The basic plate's mechanical and structural properties, as well as the damping material used in the (PTLD) procedure.

Layer	Thickness (mm)	Length (mm)	Width (mm)	Density (kg /m3)	Passion's Ratio	Young's Modulus (Pa)
AL	1.2	300	60	2660	0.33	70×109
DYAD 606	0.5	300	60	1120	0.49	20×106

3.1. Meshing of the plate FEM

The meshing method begins with the use of a fine mesh mesh tool on a shaped 3d geometry. After a mesh sensitivity analysis convergence study for the ANSYS model, the number of items is calculated. Convergence is done at a 5 mm element size.

The element's shape is "solid 186," which is a 20-node uniform structural solid. Figure 3 illustrates the solid 186 item's configuration. It's a polynomial displacement function with a 20-node higher-order 3-D solid feature. There are 20 nodes in the element, each with degrees of freedom. Descriptions of nodes in the x, y, and z directions. The factor supports plasticity, hyperelasticity, creep, stress stiffening, large deflection, and large strain capabilities. [15].This element's geometry, node locations, and element coordinate scheme are shown in Figure 2.

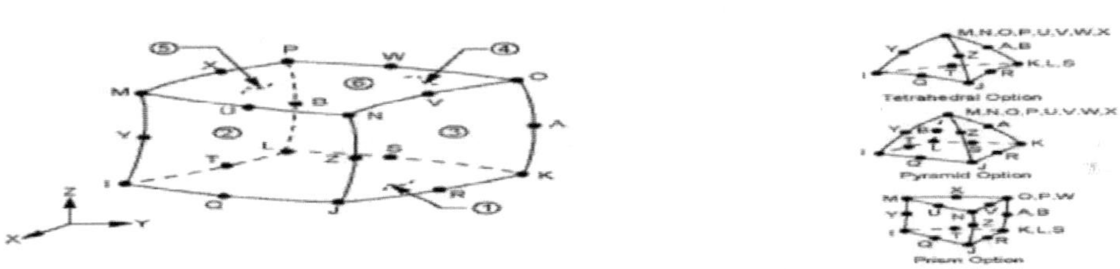

Figure 1. Solid 186 element configurations

Table 2.elements and nodes number used for the structure mesh.

model type	element number	nodes number
AL Alloy	722	5406
AL/UCLD	2161	16210

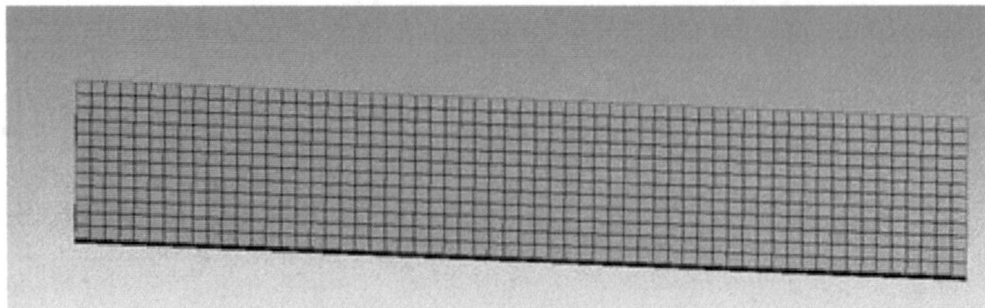

Figure 2. The final mesh of the plate FEM

3.2. Static analysis

In the application of continuous loads of inertia, such as plate gravity, the static structural analysis method is used to measure flat movements, stresses, pressures, and so on... Cantilever boundary conditions must be added to the FE model. Figure 3. Shows the plate tip displacement's gravitational strain.

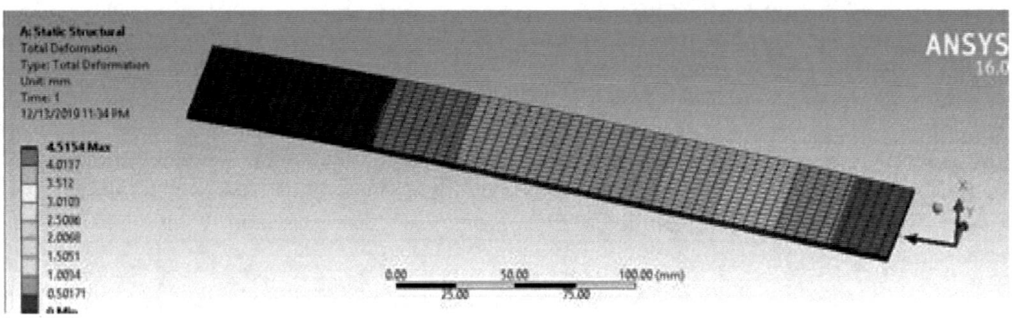

Figure 3. Gravitational load causes plate tip displacement.

3.3. Modal analysis

The vibrational characteristics of the platform and the platform using a polymer are classified using a modal analysis. (natural frequencies and modal shapes). It's been used as a jumping off point for more dynamic studies, such as transient, harmonic, or spectral dynamic analyses. Resonance occurs when dynamic load frequencies match one of the natural frequencies, which is the most important parameter in dynamic structure design. The modal analysis approach involves the following mode shapes, also known as modes, as well as the platform and polymer plate's Natural Frequencies (NF). The platform and the platform in polymer mode are depicted in the figures. The mode shape of the plate and the UCLD plate are shown in Table 3. The FRF of the current plate at various resonant frequencies is shown in Figure 7.

Table 3 .AL plate and the UCLD Plate mode shape.

Mode Shape	Type	AL Plate	UCLD Plate
1	Bending	9.7975 (Hz)	8.7 (Hz)
2	Bending	70.267 (Hz)	60.3 (Hz)

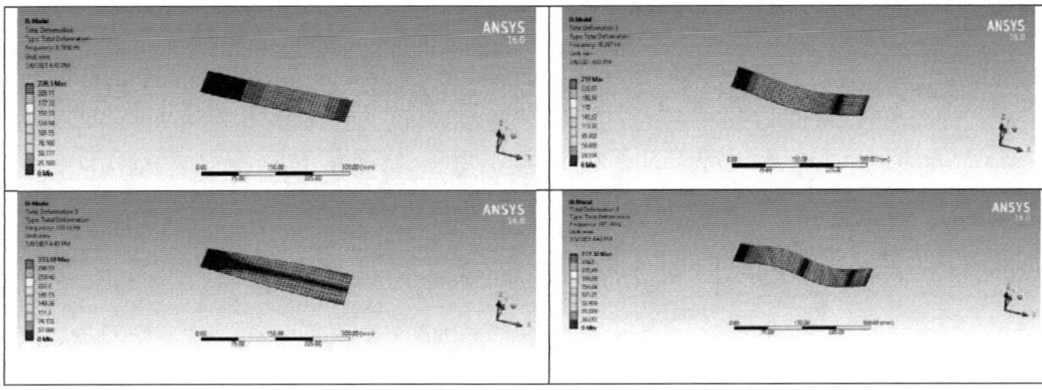

Figure 4.first four mode shape of AL plate

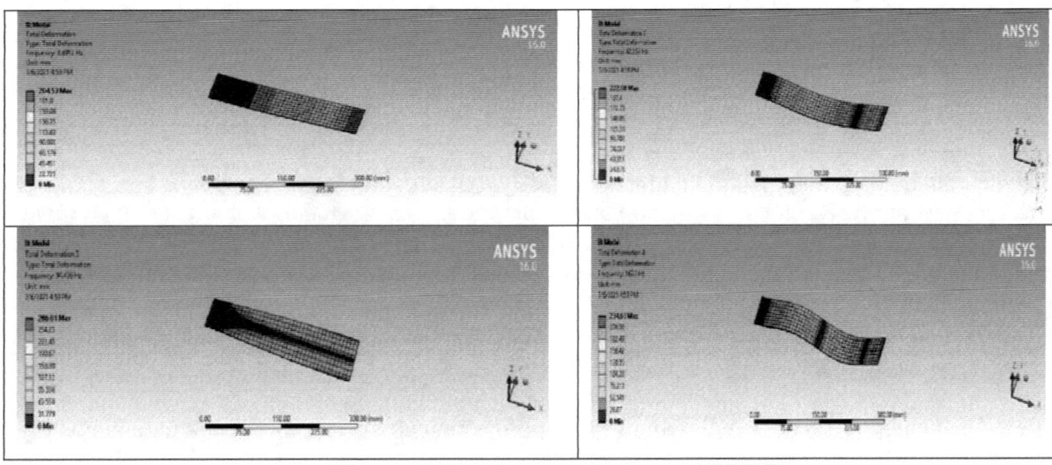

Figure 5. First four mode shape of UCLD

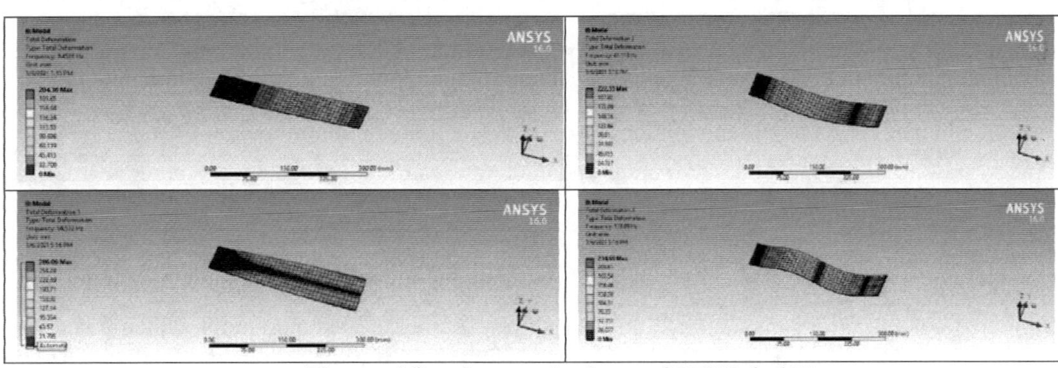

Figure 6.first four mode shape of PTLD f=5 N

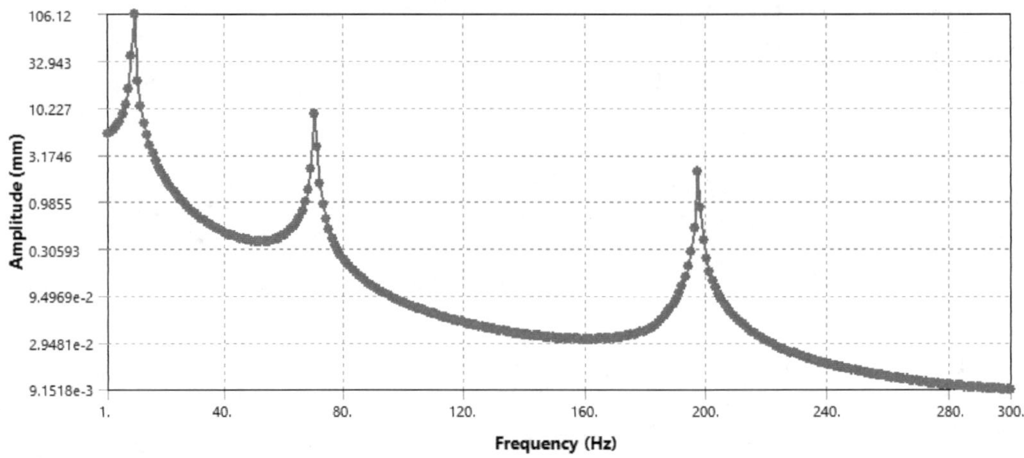

Figure7. Frequency response function (FRF) for the AL plate.

3.4. Verification of finite element modelling

AL plate was previously examined experimentally and analytically in comparison [8].

1. To account for weight due to gravitational acceleration (g = 9.81m/s2) and accelerometer mass modelled at the free-end mass level, the first step was to use the identical plate dimension as the reference by applying the same amount and distribution of the load applied obtained by the reference.
2. Table 4 compares the frequencies of the simulated results to the reference result, which can be referred to as the second step of testing.

Table 4. The comparison between the frequencies of the simulated work results and the reference results

No. of modes	Kind of mode shape	Resonant frequency (Hz)	
		Current model	Reference model
1	Bending	9.797	9.76
2	Bending	70.267	70.312

3.5. Parametric study:

3.5.1. Aluminum Plate with damping layer without pretension forces. The numerical analysis distinction between vibrational modes and FRF was shown in the previous paragraph, which had a good deal between the two models FE. The main goal of this study is to determine the smallest FRF amplitude that is proportional to the best damping layer thickness. We use two layers of damping material (DYAD 606) in this model, one on the top side of the base plate (AL 5056) and the other on the bottom side, to investigate the effect of damping material thickness on the FRF amplitude. Table 5 provides the results of the first natural frequency relation between the resonating frequencies and the loss factor for the numerical analysis. The comparison of various thicknesses of the vibration control layer effect of the plate is shown in Figure. The percentage attenuation is determined by the related equation.

$$\Delta_n^t(\omega_n) = \frac{|\alpha(\omega_n)| - |\alpha^t(\omega_n)|}{|\alpha(\omega_n)|} 100\%$$

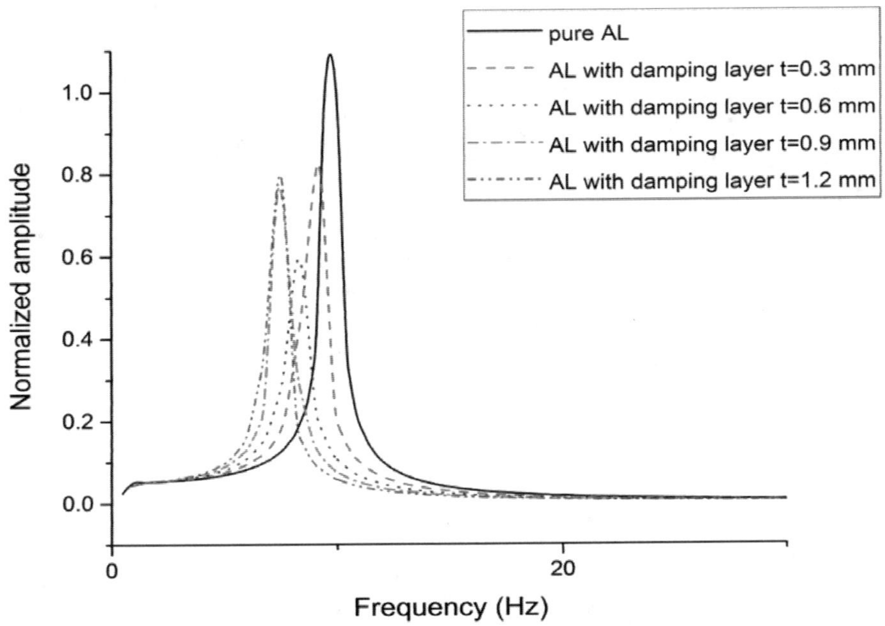

Figure 8. The comparison between various thicknesses of the damping layer effect without pretension force on the control of vibration of the plate

Table 5. The comparison between the first natural frequency and loss factor of the numerical simulation work results for the four models without pretension layer damping F=0 N.

No.	Plate type	Natural frequency (Hz)	Modal loss factor	Attenuation ratio[%]
1	UCLD PLATE -1	9.19	0.011	20%
2	UCLD PLATE -2	8.75	0.0204	41%
3	UCLD PLATE -3	8.112	0.0159	24.32%
4	UCLD PLATE -4	7.75	0.0125	22.1%

3.5.2. Aluminum Plate with damping layer with pretension forces. In this case, we demonstrate how to adjust the damping material thickness of the vibration control of the plate by using two layers of polymer plates (Dyad 606), which are influenced by an axial uniform distributed force when tension is applied, as shown in figure (11)[8]. With varying damping thickness, we apply a 5 N pretension force to achieve the smallest vibration amplitude. Figure 9 illustrates how the thickness of a damping material influences the vibration control of a flat. Table 6 compares the frequencies and the loss factor of the numerical simulation work's results for the first natural frequency.

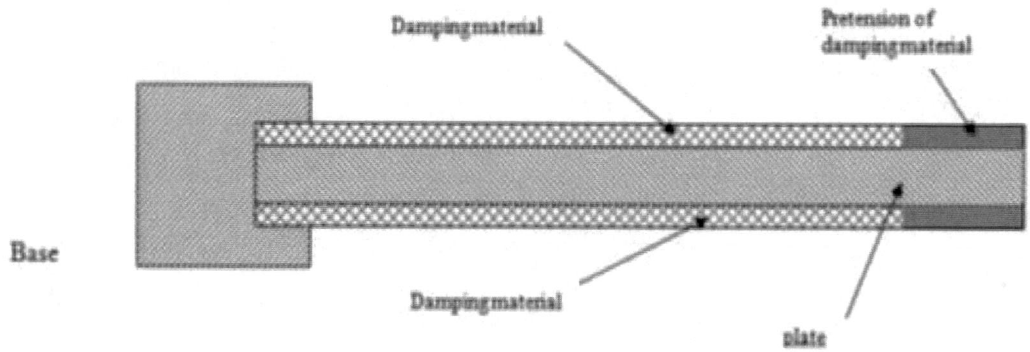

Figure 9. Base plate with pretension layer damping assembly

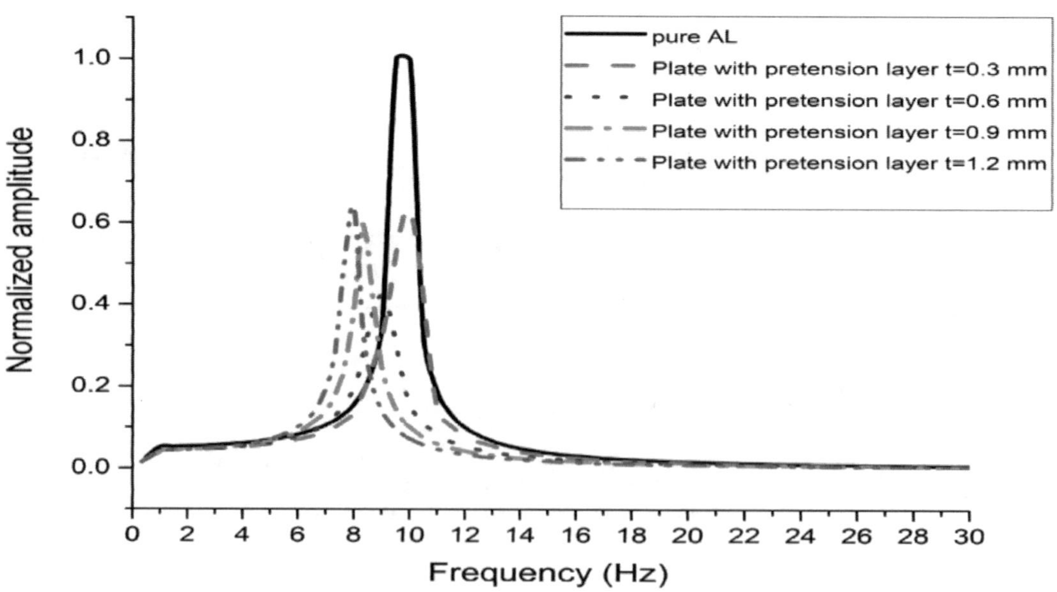

Figure10.the comparison between various thicknesses of the damping layer effect with pretension force F=5 N.

Table 6. The comparison between the first natural frequency and loss factor of the numerical simulation work results for the four models for pretension layer damping F= 5N

Sr. No.	Plate type	Natural frequency (Hz)	Modal loss factor	Attenuation ratio [%]
1	PTLD PLATE -1	9.85	0.021	41.2%
2	PTLD PLATE -2	9.2	0.0298	59.5%
3	PTLD PLATE -3	8.78	0.0195	38%
4	PTLD PLATE -4	8.5	0.0185	39.5%

4. Conclusion

The impact of a proper damping thickness on the loss factor and attenuation percentage of the PTLD plate is discussed in this paper. The loss factor results are obtained to the PTLD -1, PTLD-2, and PTLD-3, PTLD-4, from a numerical analysis using the ANSYS programme. It is notes that the most important increasing of loss factor in the PLTD-2. The amplitude of the vibration in plate-2 decreases as the modal loss factor increases. Vibration damping is enhanced when the damping layer thickness is about half of the plate thickness, whether pretension damping is used or not. The thickness ratio of the damping material layer to the plate must be calibrated to achieve the best damping in the vibrating plate.

References

[1] Hujare, P.P. and Sahasrabudhe, A.D., Abd-Elwahab, M. and Sherif, H.A., 2006. Pre-tensioned layer damping as a new approach for vibration control of elastic beams. 2014. Effect of thickness of damping material on vibration control of structural vibration in constrained layer damping treatment. In *Applied Mechanics and Materials* (**Vol. 592**, pp. 2031-2035). Trans Tech Publications Ltd.

[2] De Silva, C.W. ed., 2007. *Vibration damping, control, and design*. CRC Press.

[3] Cai, C., Zheng, H., Khan, M.S. and Hung, K.C., 2002, April. Modeling of material damping properties in ANSYS. In *CADFEM Users' Meeting & ANSYS Conference* (pp. 9-11).

[4] Koyasu, M., 1997. Recent activities in noise and vibration control engineering in Japan. *Noise News International*, *5*(3), pp.138-144.

[5] Casado, C.M., Díaz, I.M., de Sebastián, J., Poncela, A.V. and Lorenzana, A., 2013. Implementation of passive and active vibration control on an in-service footbridge. *Structural Control and Health Monitoring*, ***20*(1)**, pp.70-87.

[6] Sun, W., Yan, X. and Gao, F., 2018. Analysis of frequency-domain vibration response of thin plate attached with viscoelastic free layer damping. *Mechanics Based Design of Structures and Machines*, ***46*(2)**, pp.209-224.

[7] Zheng, H., Cai, C., Pau, G.S.H. and Liu, G.R., 2005. Minimizing vibration response of cylindrical shells through layout optimization of passive constrained layer damping treatments. *Journal of sound and vibration*, ***279*(3-5)**, pp.739-756.

[8] Mohamed, A., Hassan, A. and Omer, A.A., 2020, November. Passive vibration damping of a cantilever plate. In *IOP Conference Series: Materials Science and Engineering* (**Vol. 973**, No. 1, p. 012036). IOP Publishing.

[9] Finegan, I.C. and Gibson, R.F., 1999. Recent research on enhancement of damping in polymer composites. *Composite structures*, *44*(2-3), pp.89-98.

[10] Singh, R.V. and Pellny, M., 1998. *Lightweight, high-performance, constrained-layer sound dampers* (No. 980466). SAE Technical Paper.

[11] Moita, J.S., Araújo, A.L., Soares, C.M. and Soares, C.M., 2013. Finite element model for damping optimization of viscoelastic sandwich structures. *Advances in Engineering Software*, *66*, pp.34-39.

[12] Tudor, J., 2003. *Determination of Dynamic Properties and Modeling of Extensional Damping Materials* (No. 2003-01-1433). SAE Technical Paper.

[13] Mead, D.J., 2002. Structural damping and damped vibration. *Appl. Mech. Rev.*, *55*(6), pp.R45-R54.

[14] Chen, X. and Liu, Y., 2018. *Finite element modeling and simulation with ANSYS Workbench*. CRC press.

[15] Yadav, D., Sharma, A. and Shivhare, V., 2015. Free vibration analysis of isotropic plate with stiffeners using finite element method. *Engineering Solid Mechanics*, *3*(3), pp.167-176.

[16] Abd-Elwahab, M. and Sherif, H.A., 2006. Pre-tensioned layer damping as a new approach for vibration control of elastic beams.

IOP Conf. Series: Materials Science and Engineering 1172 (2021) 012007 doi:10.1088/1757-899X/1172/1/012007

Vibration damping of aircraft propeller blades using shunted piezoelectric transducers

A M Morad [1], M Kamel [2] and M K Khalil [2]

[1] Egyptian Armed Forces, Cairo, Egypt
[2] Aircraft Mechanical Engineering Department, MTC, Cairo, Egypt

E-mail: alaa_morad@yahoo.com

Abstract. Gas turbine engine blades experience vibrations due to the flow disturbances, these vibrations are critical to the engine durability and performance. Piezoelectric transducers (sensors and actuators) have been used for engine blade vibrations damping either through a passive or active vibration control. The propeller blades are part of turboprop engine and considered as one of the main source of turboprop engine vibrations. Piezoelectric blade damping ideas have been studied by other researchers for fan blades and compressor blades. In this research a vibration damping procedure using piezoelectric transducers applied to an unmanned aerial vehicle (UAV) composite propeller. Experimental investigation introduces an approach for the propeller vibration damping using piezoelectric transducers in conjunction with appropriate shunt circuit. Three thin piezoelectric transducers macro fiber composite (MFC) type PZT-5A are surface-mounted on the propeller, one at each blade. These transducers are placed at locations of high modal strain areas for the propeller first mode at each blade, where these locations are identified by finite element numerical simulation. Electronic resonance shunt circuit, resistor-inductor-capacitor type, for the piezoelectric transducers is designed and experimentally developed such that effective vibration suppression of the propeller is achieved. The experimental and numerical investigations in this research illustrate that piezoelectric transducers with appropriate shunt circuit reduces the aircraft propeller vibrations.

1. Introduction

Smart materials have been used in vibration damping. The main purpose is to reduce vibrations of a mechanical structure [1]. Piezoelectric transducers have been worked as sensors and actuators to damp vibrations. Piezoelectric transducers are able to change mechanical and electrical energy into each other's [2]. Piezoelectric materials are able to be work as passive or semi active vibration damping with shunt circuit [3]. Piezoelectric transducers with shunt circuit used for damping engineering structures is an identified method. This method depends on the piezoelectric result that changes the mechanical energy to the electrical energy [4]. In a structure using piezoelectric materials with shunt circuit for vibration damping. The shunt circuits may be used for active or passive damping. These circuits consist of capacitors, inductors and resistors [5]. The resistor of the shunt circuit dissipates the electrical energy transformed from mechanical energy due to the effect of piezoelectric transducer [6]. A resistor inductor shunted circuit faces a difficulty of large size of the inductor. Synthetic inductors can be used to decrease the volume and the weight of the circuit [7].

Many authors studied different procedures for blade vibration damping in gas turbine engines utilizing piezoelectric transducers. Duffy et al. [8] studied the result of shunted piezoelectric transducers put on a rectangular plates that loaded centrifugally. They stated that plate vibrations could be damped

Content from this work may be used under the terms of the Creative Commons Attribution 3.0 licence. Any further distribution of this work must maintain attribution to the author(s) and the title of the work, journal citation and DOI.
Published under licence by IOP Publishing Ltd

using shunted piezoelectric transducers. Duffy et al. [9] used flexible macro-fiber-composite piezoelectric patches applied to fan blades, they bonded the piezoelectric transducers on the blades. The piezoelectric transducers placed in a maximum modal strain area for the first mode that identified by finite element. They used spin test rig facility at for testing with magnetic bearing for excitation. The results showed that using active and shunted piezoelectric patches damp vibrations induced to the fan blades at different speeds. Min et al. [7] study experimentally and numerically rotating engine fan blades. The results are the numerical investigation is effective to the rotating engine fan blade. A finite element was studied to identify the best position for the placement of the piezoelectric transducers at the first bending mode high strain area. The results of the tests showed that piezoelectric patches can suppress vibrations of composite engine fan blades. Zhou et al. [10] used a shunt circuit with piezoelectric patches and then applied to a mistuned bladed disk. Piezoelectric patches with shunt circuit were attached to the blade to damp vibrations. Zhou and Zuo [11] used piezoelectric patches applied on cantilever beam to damp vibrations. They used numerical analysis, the numerical results showed that the vibrations are damped. Zakaria et al. [12] Study piezoelectric transducers oscillations to harvest energy. They used experimentally a fluttering composite piezoelectric beam positioned in an air flow at different directions. They show that the best amounts of electric power are got according to the direction of the air flow and its speed. Liu et al. [13] Used experimentally a carbon fiber L-shaped structure that consists of upright and level beams mounted on a shaker. They put two piezoelectric transducers onto each beam to get electrical energy from the beams vibrations. The results show that upright beam gets more electrical power than level one. Marques et al. [14] show experimentally the electrical power getting from vibrations of a an airfoil in wind tunnel. The results show in order to get more electrical power at low speeds the air foil axis moves proportional to the center of gravity. Santos et al. [15] did an experimental and numerical investigations for generating electrical energy using piezoelectric patches bonded to a composite cantilever beam. They used finite element approach for numerical investigations while used experimentally a wind tunnel.

In this paper, a finite element model of a composite propeller is introduced to determine the maximum modal strain position of the propeller for the best location of the piezoelectric transducers, and to obtain the propeller natural frequency and mode shape. Then applying piezoelectric transducers experimentally to the propeller blades with shunt circuit to damp vibrations, the results including natural frequencies and vibration damping performance.

2. Finite element modeling

A 3-blade fixed pitch propeller carbon fiber with diameter 24 in and pitch angle 12 in designed for unmanned aerial vehicles (UAV's) is used for numerical and experimental investigations. A finite element analysis of the propeller before bonding piezoelectric transducers is performed, using ANSYS. The propeller is fixed from the hub and the excitations are perpendicular to the propeller blades plane (Z axis direction). The FE model is used first to get the propeller maximum modal strain area that is the best location of the piezoelectric transducers in order to attain best damping. Figure 1 shows the equivalent elastic stain for first mode. In this figure the maximum strain locations are shown. Then modal analysis used to get natural frequencies and mode shapes as shown in figure 2 [16].

3. Theory of using shunt circuit for piezoelectric transducers to damp blade vibration

The illustration of blade damping using piezoelectric transducer with shunt circuit will be demonstrated in this section. Figure 3 explains schematically how the structure of a blade with shunt circuit damp vibrations. The vibration of the blade makes strains. A piezoelectric transducer produces an electric field. then electrical charges are collected piezoelectric patches. The shunt circuit resistor dissipates the electrical energy. The shunt circuit electric resonance ω_e is tuned to the desired resonance frequency ω_m, by shifting the current phase by 90º. The vibrations are damped in the structure [17].

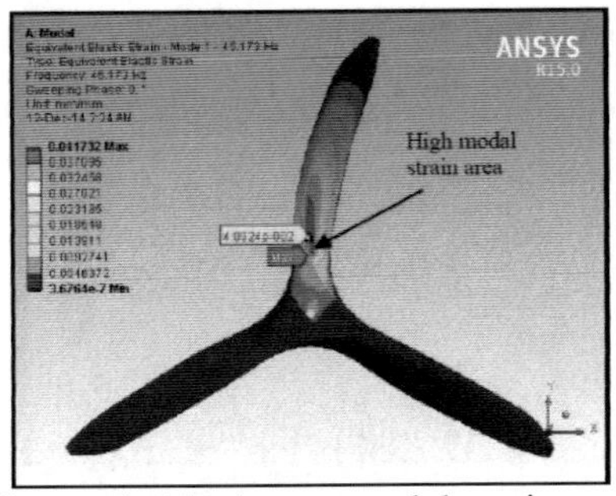

Figure 1. Maximum modal strain area

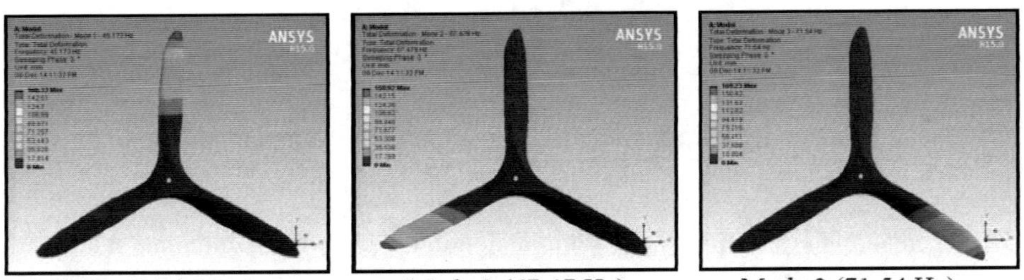

Mode 1(45.17 Hz) Mode 2 (67.47 Hz) Mode 3 (71.54 Hz)

Figure 2. Propeller modal displacement contours without shunted piezoelectric

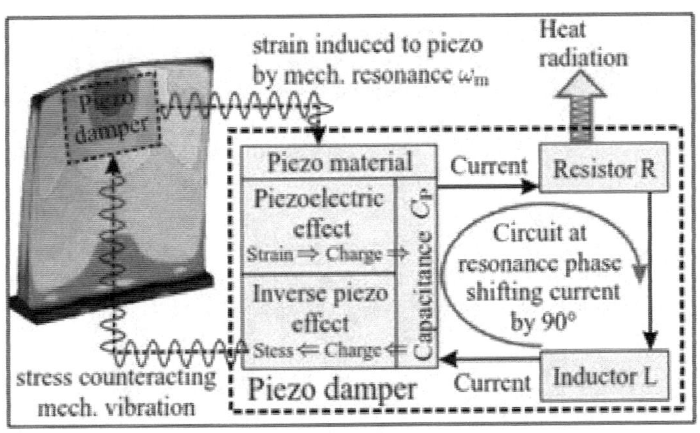

Figure 3. Design of blade vibration damping using shunt circuit [18]

4. Experimental investigation

The goal of this experiment is to design a vibration damping technique for the propeller using shunted piezoelectric transducers, the used transducers are (PZT) Lead Zirconate Titanate macro fiber composite (MFC). The experiments use piezoelectric transducers with shunt circuit to damp vibrations. Electrical resonance circuit is then tuned to the propeller resonance so that piezoelectric transducers act as vibrational absorber device.

4.1 Materials and approach

Piezoelectric transducers used in this experiment are DuraAct piezoelectric patch transducers PN: P-876.A15, ceramic type PIC255, (MFC) product of Physik Instrumente (PI) GmbH, their dimensions are 61×35×0.8 mm. Table 1 shows the specifications of the piezoelectric transducers.

Table 1. Specifications of DuraAct piezoelectric patch transducers

Operating voltage (V)	Min. lateral contraction ($\mu m/m$)	Blocking force (N)	Piezo-ceramic thickness (μm)	Electrical capacitance (nF)
- 250 to 1000	800	775	500	45

4.2 Shunt Circuit Design

The shunted piezoelectric transducers used for this study have passive electrical circuits. The components of this RLC shunt circuit: resistor, capacitor, and inductor as shown in figure 4.

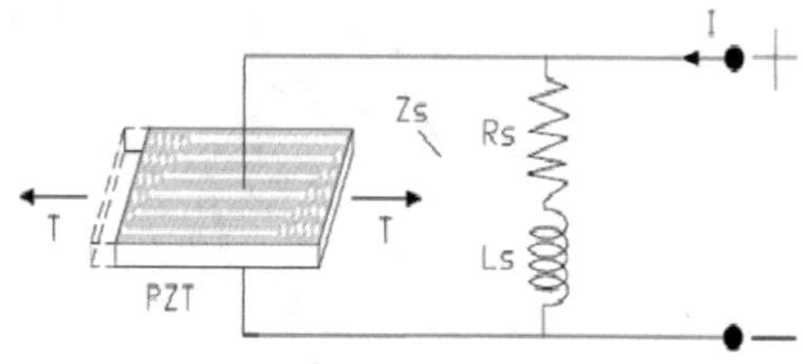

Figure 4. Shunt circuit design

The inductor and capacitor values of the shunt circuit are calculated as in the following equation:

$$\omega_e = \frac{1}{\sqrt{L_s C_P^S}} \tag{8}$$

Where ω_e is the electrical resonant frequency of the circuit, Ls is the inductance for the circuit, the capacitor C_P^S for the circuit is the PZT itself. Electric resonance ω_e is tuned to the propeller first mode ω_m which equal to 46 Hz. The inductor of the RLC circuit, Ls = 10501.99 H, is built electronically as synthetic inductor as shown in Figure 5 [18].

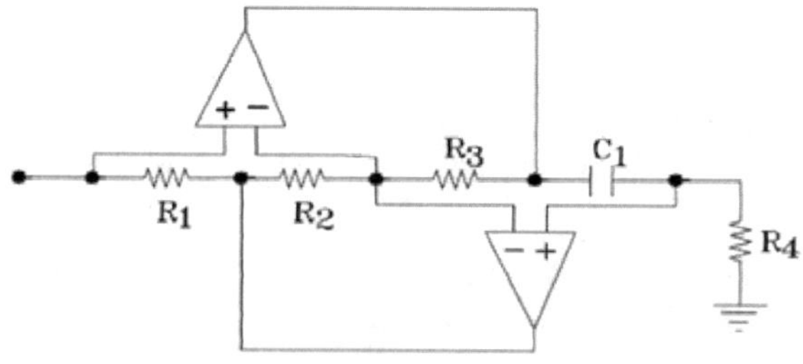

Figure 5. Synthetic inductor for variable inductor [18]

The resistor, R2, is a variable resistor to change the circuit inductance. The resistors R1, R3, and R4 are 10 kΩ. The shunt circuit uses a ±16 Volts, 1 Watt DC power supply.

4.3 PZT placement and application

Three piezoelectric transducers are applied to the propeller blades one at each blade and positioned at high modal strain areas of the first mode identified by finite element simulation as shown in Figure 6. Piezoelectric transducers are glued to the blades using epoxy adhesive.

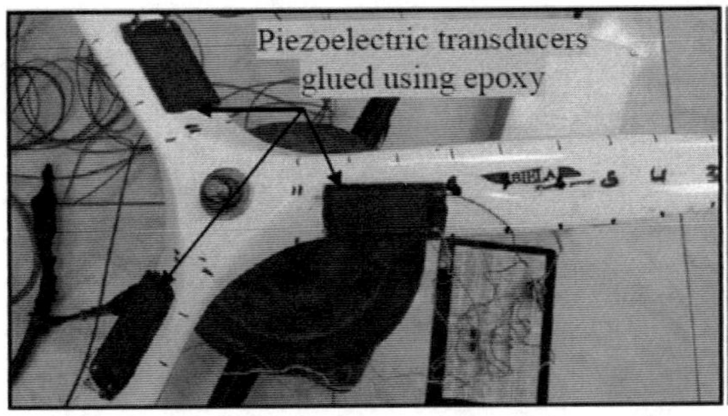

Figure 6. Piezoelectric transducers placement on the propeller

4.4 Experiment procedure

The digital signal processing unit used in the experiment is NI PXIe-1085 chassis product of NATIONAL INSTRUMENTS (NI), which includes all system controller modules. The software used is LabView. A NI rack mountable adapter BNC-2144 unit with a NI PXIe-4499 module is used to collect the output data through LabView. A NI rack mountable adapter BNC-2121 with a NI PXI-6602 module is used for transfer input signal from LabView to the shaker type 4808 manufactured by Bruel & Kjaer (B&K) through power amplifier type 2712 manufactured by B&K. Three PCB piezoelectric accelerometers attached to propeller blades tip and one accelerometer mounted on the shaker table are used to attain the frequency response functions (FRF) of the propeller. The shaker exited the propeller perpendicular to propeller blades plane. Experiment is performed with the test setup shown in figure 7. In this experiment the propeller is fixed from its hub onto the shaker. The output from the amplifier is connected to the shaker. The shunt circuit is connected to the piezoelectric transducers. The shunt circuit is powered by the power supply. The tests are taken from 40 to 80 Hz that contains the first 3 modes of the propeller.

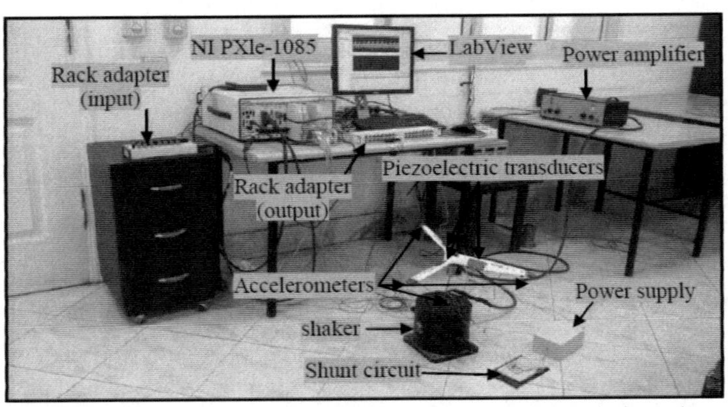

Figure 7. Experimental set up for propeller with shunted piezoelectric

5. Results and discussion

5.1 Propeller modal test and FE model verification

Experimental test is applied first to the propeller before bonding piezoelectric transducers. Figure 8 shows frequency response function (FRF) for the propeller before bonding piezoelectric transducers, this figure shows the first three modes for the propeller that identified by the phase angle change at the peaks appear from the FRF. The comparison of natural frequencies from numerical simulation for the propeller and the natural frequencies obtained from this experiment are shown in Table 2. The comparison results show good agreement with deviation from 0.7% to 2.2%, which validates the numerical simulation.

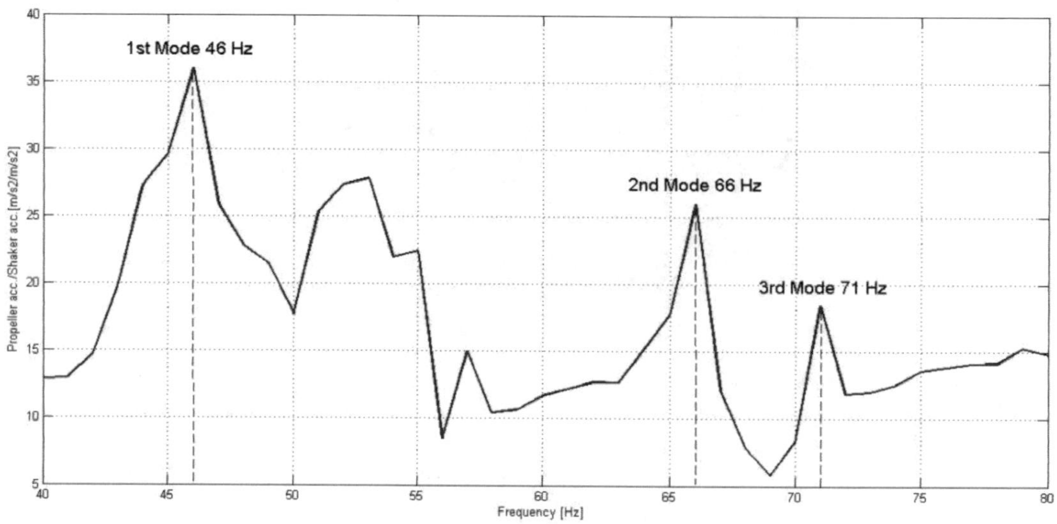

Figure 8. Propeller FRF before applying piezoelectric transducers

Table 2. The **c**omparison between propeller natural frequencies for numerical simulation and experimental investigations

Mode number	Frequency (Hz) Simulation	Frequency (Hz) Experimental	Deviation percent
1	45.173	46	1.7 %
2	67.479	66	2.2 %
3	71.54	71	0.7 %

5.2 Propeller with shunted piezoelectric test and FE model verification

Similar test is applied to the propeller after bonding shunted piezoelectric transducers. Figure 9 shows frequency response function (FRF) for the propeller after applying shunted piezoelectric transducers, in this figure the first 3 natural frequencies for the propeller are shifted up after applying shunted piezoelectric transducers due to increased stiffness of the propeller as a result of applying shunted piezoelectric transducers. Table 3 shows the comparison between the first 3 natural frequencies of the propeller before and after applying shunted piezoelectric transducers, this table shows that the increase in natural frequencies are from 4.2 % to 10.8 %.

A finite element simulation using ANSYS for the propeller with shunted piezoelectric transducers (sensors and actuators) is developed and analyzed as shown in figure 10, a perfect adhesive bonding between the transducers and blades is assumed.

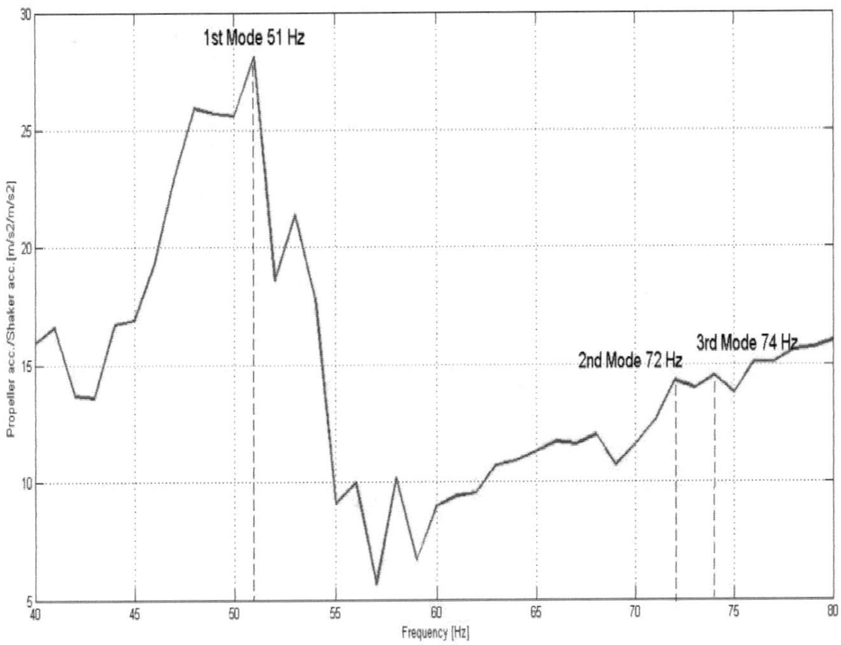

Figure 9. FRF for the propeller with shunted piezoelectric transducers

Table 3. Comparison between natural frequencies before and after applying shunted piezoelectric transducers to the propeller

Mode Number	Frequency (Hz) before applying piezoelectric	Frequency (Hz) with shunted piezoelectric transducers	Percent increase
1	46	51	10.8%
2	66	72	9.1 %
	71	74	4.2 %

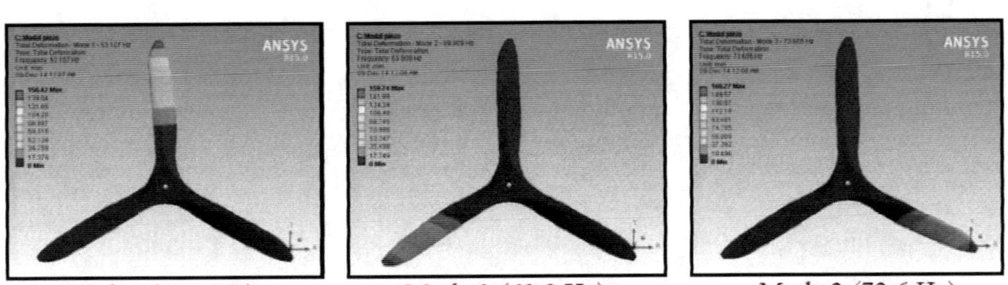

Mode 1(53.1 Hz) Mode 2 (69.8 Hz) Mode 3 (73.6 Hz)

Figure 10. Propeller modal displacement contours with shunted piezoelectric

Table 4 shows the comparison of the natural frequencies for the propeller with applying shunted piezoelectric transducers between experimental results and finite element (FE) simulation. The comparison results are in good agreement with deviation between 0.53% to 4.13%. These results validate the FE simulation.

Table 4. Comparison between natural frequencies of the experimental investigations and FE simulation for the propeller with applying piezoelectric transducers

Mode Number	Frequency (Hz) Experimental	Frequency (Hz) FE simulation transducers	Deviation percent
1	51	53.107	4.13 %
2	72	69.809	3.04 %
3	74	73.605	0.53 %

5.3 Comparison between the propeller for un-shunted and shunted piezoelectric

Tests are applied to the propeller after the piezoelectric transducers are bonded to the propeller with shunt circuit (shunted piezoelectric) and without shunt circuit (un-shunted piezoelectric) to study the effect of shunt circuit. The shunt circuit is then tuned to the first three modes of interest for the propeller to achieve best damping. Figure 11 demonstrates the effect of the shunt circuit on the propeller vibration response after applying shunted piezoelectric transducers. The first three modes accelerations are significantly reduced for the shunted piezoelectric propeller while the natural frequencies are the same values because the structure stiffness does not change.

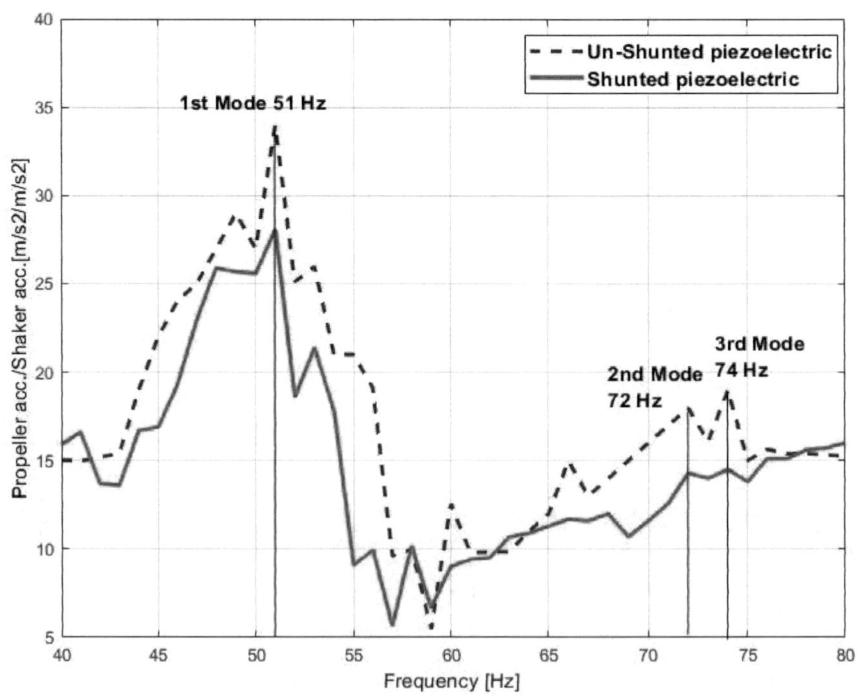

Figure 11. FRF for un-shunted and shunted piezoelectric propeller

Table 5 presents the decreases in the first three modes accelerations that were obtained using the tuned shunted piezoelectric transducers by comparing the values of peak accelerations results of propeller open loop (un-shunted) and closed loop (shunted). The results indicate that the shunted circuit significantly reduce the propeller vibrations by reducing acceleration peaks for the first three modes up to 20.99 %.

Table 5. Effect of shunt circuit on resonance acceleration peaks of the propeller

Mode Number	Un-Shunted PZT's $\left({}^{m/s^2}/_{m/s^2}\right)$	Shunted PZT's $\left({}^{m/s^2}/_{m/s^2}\right)$	Reduction percent
1	34.1	28.1	17.60%
2	18.1	14.3	20.99%
3	18.2	14.5	20.33%

5.4 Comparison between the propeller before and after applying shunted piezoelectric

Figure 12 illustrates the effect of the piezoelectric transducers with tuned shunt circuit on the propeller vibration response compared to the FRF of the propeller before bonding piezoelectric transducers (without piezoelectric). This figure shows reduction in the accelerations of first three modes for the shunted piezoelectric propeller and the increase in the first three natural frequencies indicated by (1), (2) and (3) for results without piezoelectric and with shunted piezoelectric.

The main goal of the experiments is to determine the vibration reduction by the application of piezoelectric transducers with shunt circuit to the propeller as shown in Figure 12. Table 6 shows the reductions in the resonance frequencies accelerations that are obtained using the shunted piezoelectric transducers by comparing the values of these acceleration of propeller before bonding piezoelectric transducers (without PZT's) and after applying piezoelectric transducers with shunt circuit (shunted PZT's). The results indicate that the shunted piezoelectric transducers significantly reduce the propeller vibrations by reducing acceleration peaks for the first three modes up to 45.04 %.

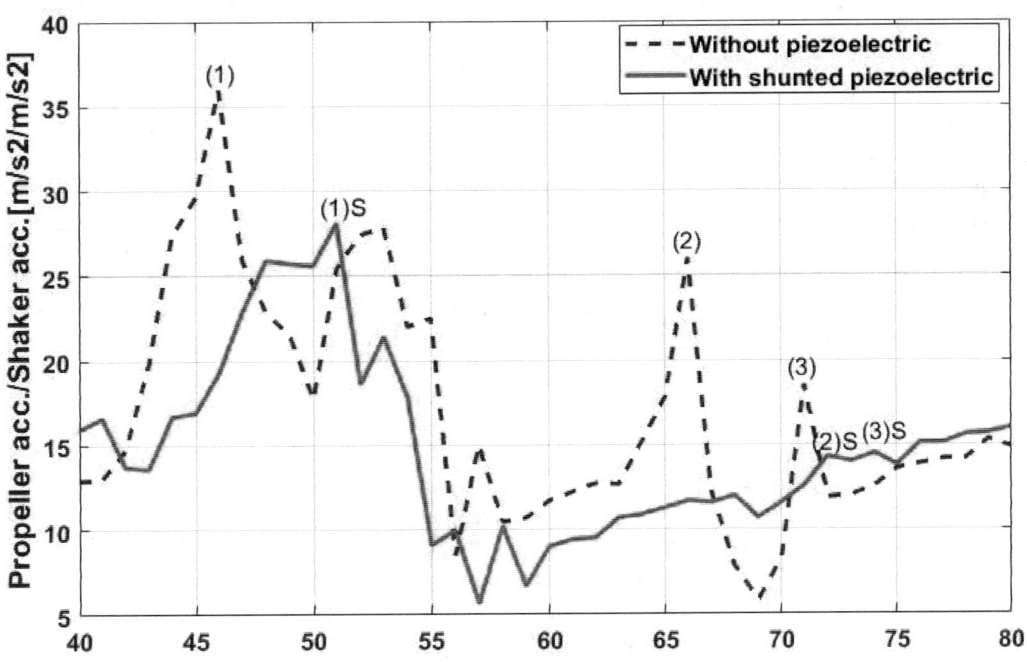

Figure 12. Effect of applying shunted piezoelectric transducers to the propeller

Table 6. Effect of shunted piezoelectric on vibration reduction

Mode Number	Without PZT's $\left(\frac{m/s^2}{m/s^2}\right)$	With shunted PZT's $\left(\frac{m/s^2}{m/s^2}\right)$	Reduction percent
1	35.97	28.1	21.87 %
2	26.02	14.3	45.04 %
3	18.51	14.5	21.66 %

6. Conclusions

This research used a numerical and experimental investigations to develop piezoelectric transducers with shunt circuit to damp vibrations for a real composite aircraft propeller blades.

Electronic resonance shunt circuit attached to piezoelectric transducers type PZT-5A is designed and experimentally examined such that vibration damping of the propeller is achieved. The shunt circuit used is resistor-inductor-capacitor type. The piezoelectric transducers are surface mounted on the propeller that attached to a shaker. The experimental vibration test and numerical simulations are conducted from 40 to 80 Hz, were the following results are concluded:

1- Applying shunted piezoelectric transducers cause the first 3 natural frequencies for the propeller to shift up from 4.2 % to 10.8 %. The cause of this increase is referred to the propeller increase structure stiffness.
2- The comparison of the propeller first 3 natural frequencies with applying shunted piezoelectric transducers between experimental results and finite element are in good agreement with deviation between 0.53% to 4.1%. These results validate the FE simulation.
3- Shunt piezoelectric transducers can effectively reduce the propeller vibrations throw reducing acceleration peaks of the first three modes up to 45.04 %.

The obtained results show that piezoelectric transducers with appropriate shunt circuit can be used to damp vibrations induced in aircraft propeller blades by applying these patches at propeller blades first mode maximum modal strain locations. However, for industrial application, piezoelectric transducers can be embedded to the propeller blades to avoid a strong air flow disturbances during operation.

References

[1] D'Assuncao D and Junior C 2014 Applied self-powered semi-passive control for a 2-degree-of-freedom aeroelastic typical section using shunted piezoelectric materials *J. Intelligent Material Systems and Structures* **26** pp 373-385.
[2] Zhou B *et al.* 2013 An adaptive control strategy based on passive piezoelectric shunt techniques applied to mistuned bladed disks *J. Computational and Applied Mathematics* **246** pp 289-300.
[3] Mayer D, Heuss H, Thiel O, Abele J and Melz T 2014 A*dditive manufacturing of active struts for piezoelectric shunt damping 25nd International Conference on Adaptive Structures and Technologies* (The Hague, The Netherlands)
[4] Neubauer M and Wallaschek J 2013 Vibration damping with shunted piezoceramics: fundamentals and technical applications *J. Mechanical Systems and Signal Processing* **36** pp 36-52.
[5] Pohl M and Rose M 2015 Piezoelectric shunt damping of a circular saw blade with autonomous power supply for noise and vibration reduction *J. Sound and Vibration* **361** pp 20-31.
[6] Hogsberg J and Krenk S 2015 Balanced calibration of resonant piezoelectric RL shunts with quasi-static background flexibility correction *J. Sound and Vibration* **341 pp** 16-30.
[7] Min J, Duffy K, Choi B, Provenza A and Kray N 2013 Numerical modeling methodology and experimental study for piezoelectric vibration damping control of rotating composite fan blades *J. Computers & Structures* **128** pp 230-42.
[8] Duffy K, Provenza A ,Trudell J and Min J 2009 *Passively shunted piezoelectric damping of centrifugally loaded plates Proc. of the 50th AIAA/ASME/ASCE/AHS/ASC Structures, Structural Dynamics, and Materials Conference* (Palm Springs, CA) p 2524.

[9] Duffy K, Choi B, Provenza J, Min B and Kray N 2012 *Active piezoelectric vibration control of subscale composite fan blades Proc. ASME Turbo Expo* (Copenhagen, Denmark) p 028.

[10] Zhou B, Thouverez F and Lenoir D 2014 Essentially nonlinear piezoelectric shunt circuits applied to mistuned bladed disks *J. Sound and Vibration,* **333** pp 2520-42.

[11] Zhou W and Zuo L 2015 A self-powered piezoelectric vibration control system with switch precharged inductor (SPCI) method J. *IEEE/ASME transactions on mechatronics* **20** pp 773-81.

[12] Zakaria M, Al-Haik M, and Hajj M 2015 Experimental analysis of energy harvesting from self-induced flutter of a composite beam *Applied Physics Letters* **107** (2) 023901.

[13] Liu D, Al-Haik M, Zakaria M and Hajj M 2017 Piezoelectric energy harvesting using L-shaped structures *Journal of Intelligent Material Systems and Structures* **29** (6) pp 1206-15

[14] Marques F, Pereira D, Zakaria M and Hajj M 2017 Power extraction from stall-induced oscillations of an airfoil *Journal of Intelligent Material Systems and Structures* **29** (7) pp 1407-17.

[15] Santos C, Pacheco D, Taha H, Zakaria M 2021 Nonlinear modeling of electro-aeroelastic dynamics of composite beams with piezoelectric coupling *Composite Structures* **255** 112968

[16] Morad A M, Elzahaby A, Abdallah S, Kamel M and Khalil M 2015 Application of piezoelectric materials for aircraft propeller blades vibration damping *Int. J. Scientific & Engineering Research* **6** pp 513-20.

[17] Sebastian M, Szwedowicz J, and Neubauer M 2012 *On blade damping technology using passive piezoelectric dampers Proc. ASME Turbo Expo* (Copenhagen, Denmark) p 68600.

[18] Horowitz P and Hill W 1994 *The art of electronics* (Cambridge: Cambridge University Press).

IOP Conf. Series: Materials Science and Engineering 1172 (2021) 012008 doi:10.1088/1757-899X/1172/1/012008

Design of a smart energy efficient IoT wireless node

M S Nabil[1], A F Tammam[2]and M M ElKhatib[3]

[1] Computer Engineering, Pyramids Higher Institute for engineering & Technology (PHI), Giza, Egypt
[2]Computer Engineering, Arab Academy for Science, Technology & Maritime Transport (AASTMT) Cairo, Egypt
[3]Senior IEEE member, Electronic Engineering Department, Military Technical College (MTC) Cairo, Egypt

E-mail: mohamed.m.elkhatib@ieee.org

Abstract. The Internet of Things (IoT) is gaining greater effect in our life from day to day. Wireless sensor networks (WSN) are embedded into the "Internet of Things" and power saving is one of the problems that it faces. This study introduces a new circuit design for a DC-DC Buck-Boost converter for controlling the output voltage of Energy Harvester (EH) via adaptive Neuro-Fuzzy inference systems (ANFIS) controller. It has proven that there is an effective changes regarding settling time almost being zero, overshoot being negligible and steady state voltage. As an application for the proposed design, an implementation for Precision Agriculture (PA) self-sustainable Wireless Sensor-Node (WS-N) with low power consumption has been presented. A solar power management for Precision agricultural in internet of things system proved to enhance cost-efficiency and increased lifetime for the device. The simulation results are determined by means that of MATLAB/Simulink software. The cloud service of Blynk IoT has been implemented to monitor the collected data on real time. Data on solar cells, batteries and soil were regarded reliable and precise.

1. Introduction

As an evolving concept, the IoT is seen as the next big thing that can have a major impact on the world's future. By implementing the recent IoT methods in the practice of agriculture, traditional farming methods can be essentially altered in every aspect. Modern wireless technology developments have revolutionized how farmers can interface with their farms and track their productivity. Advanced management ideas can be used by using latest technologies to supervise crops and by responding properly to their requirements. One strategy is known as Precision Agriculture (PA) [1], where technology is mixed with traditional farming methods. It is possible to achieve higher precision and control of crop growth and livestock farming through the use of PA in farming. By using latest technologies to support agriculture, farmers are able to boost effectiveness and decrease expenses as directly more exact remedies can be implemented by regulating the many elements of their farm management.

Fields are managed without the use of contemporary technology in traditional farming applications [14-16]. In order to retain adequate effectiveness, a higher quantity of experience is needed. When using traditional farming methods, choices such as planting, harvesting, and irrigation involve the use of

Content from this work may be used under the terms of the Creative Commons Attribution 3.0 licence. Any further distribution of this work must maintain attribution to the author(s) and the title of the work, journal citation and DOI.
Published under licence by IOP Publishing Ltd

IOP Conf. Series: Materials Science and Engineering 1172 (2021) 012008 doi:10.1088/1757-899X/1172/1/012008

present weather conditions and historical data to determine the optimal harvesting method. In contrast, PA utilizes techniques such as sensors, actuators, Global Positioning System (GPS), robots, and software for information analysis to restrict the quantity of labor applied by farmers and to boost the quantity of care provided to plants if necessary.

This paper's primary focus is on implementing a self-sustainable WS-N in IoT for PA, Since standard sensor nodes are loaded with independent batteries of limited capacity and durability. Due to harsh environments, the energy recharge will be expensive and complicated after energy drainage. A more sensible alternative is to provide EH such as solar photovoltaic (PV) within each node that allows the sensors to be loaded in their environments by the EH alone. There are two methods for handling the harvested energy; harvest-store-use and harvest-use-store. The first approach is to store all the energy collected from EH straight into the battery and then power the nodes. Though, the inefficiencies of the batteries only generate some of the harvested energy that can be stored, resulting in power waste. The second approach is to power the node directly from the EH and to store the remaining energy in the battery, from which to be drained if the energy harvested is not adequate. The approach of harvest-use-store will be used for power management. DC-DC Buck-Boost converters were suggested to regulate the EH output voltage. To deal with the temperamental consistent difference in collected energy to keep up settled balanced-out output energy while the input data may be incomplete or imprecise [2], Neuro-Fuzzy Controller (NFC) is utilized in DC-DC Buck-Boost converters [3-5] to manage and control with the output voltage. There are several controllers that can be used to manage the output voltage of DC-DC converters such as classical controllers (PID), Fuzzy Logic Controller (FLC) [3], and PID-like Fuzzy controller [5, 6], et cetera. Among these, NFC based method for controlling has been demonstrated more better looked at than every single other methods, it offers adaptability in taking linguistic data input sources and can adjust to any system [7-9]. Neuro-Fuzzy takes the benefits of accepting the linguistic data inputs from the FLC and the versatility from neural networks. The simulation results for the proposed method compared to other controllers, in this chapter between PID-like Fuzzy and NFC are presented in MATLAB/Simulink. In addition, through Blynk IOT cloud service the status of the solar cell system, battery and the moisture of the soil are being monitored.

2. Buck-Boost Converter (B-BC)

2.1. The new Proposed Buck-Boost Converter design

The proposed system depends on the structure of the NFC, adaptive Neuro-Fuzzy inference systems (ANFIS), this technique joins the learning capacities of artificial neural systems and the ability of fuzzy logic to deal with imprecise data [7-12]. Figure 1 describes the value of duty cycle by adding the value of ideal duty cycle (D_{cyci}) to the percentage of duty cycle error (D_{cyce}) for regulating the harvester output voltage. DCYCi 's value is provided in equation (2), where VOUT is substituted by SetPoint voltage (VSP) in reference to the Equation (1) of the duty cycle (D_{cyc}).

$$D_{cyc} = \frac{V_o}{V_o + V_i} \tag{1}$$

$$D_{cyci} = \frac{V_{SP}}{V_{SP} + V_i} \tag{2}$$

There are two inputs to the Neuro-Fuzzy controller; tracking error (E) and sequential change (ΔE) as shown in Figure 2, and the MATLAB simulation with ANFIS is presented in Figure 3. The value of E is obtained as the difference between set-point/reference while ΔE value is obtained as a difference between successive steps that is;

$$E = V_{SP} - V_{OUT} \tag{3}$$

$$\Delta E = E_{current} - E_{previous} = E(t) - E(t-1) \tag{4}$$

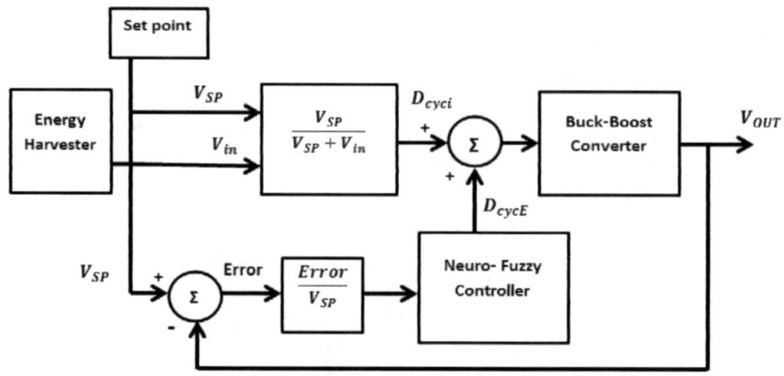

Figure 1. Buck Boost's proposed system architecture.

The value E is normalized then applied to the controller.

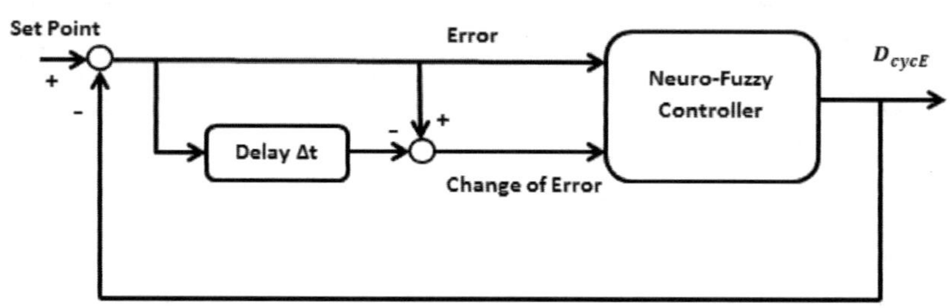

Figure 2. Neuro-Fuzzy control system's block diagram.

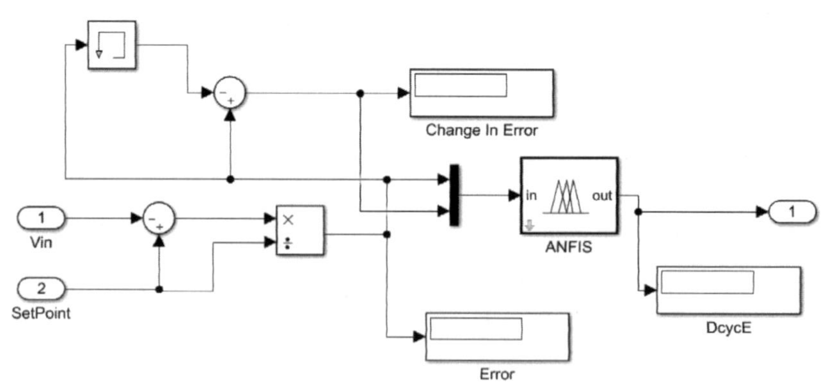

Figure 3. MATLAB Simulink's Neuro-Fuzzy control system.

2.2. Neuro-Fuzzy Implementation

The implementation of ANFIS in MATLAB/Simulink can be expressed by flowchart as shown in Figure 4, The training is done offline using ANFIS tool box in MATLAB, it can be summarized into four simple steps; Loading Data, Generating Fuzzy Inference System (FIS), Training FIS, and Testing FIS [7].

First step Loading Data; the input and output data is collected from PID-like Fuzzy controller, the total number of the data sets is 36079x3, the first two columns includes the value E and ΔE, while the third column includes the value of D_{CYCE}.

Second step Generating FIS; The FIS is generated via Grid partition with number of membership functions (MFs) equals to 3 (N, Z, P) for the MF of E, where N means Negative, Z means Zero and P means Positive.

Third step Training FIS; the FIS is trained for 100 epochs and with error tolerance equals to 0.1.

Fourth step Testing FIS; the FIS is tested according to the data loaded. The advantages of this method are its rapid tracking speed and high tracking accuracy.

2.3. The Proposed System Simulation and Results

In this paper; MATLAB/Simulink is used to design and simulate Neuro-Fuzzy Controller in comparison to PID-like Fuzzy Controller. The MATLAB/Simulink Neuro-Fuzzy Controller model for B-BC is shown in Figure 4 and Figure 5 shows the algorithm flow chart.

With the use of conventional methods such as PID-like Fuzzy Controller, it is found that the settling time and peak overshoot are high, on the other side, efficient adjustments have been observed with the use of Neuro-Fuzzy Controller, i.e. settling time is nearly zero and peak overshoot is negligible. The steady state error and rise time are found in a small range as compared to other methods. Hence, Neuro-Fuzzy Controller gives better controlling outcomes over PID-like Fuzzy Controller. The results of Neuro-Fuzzy controller and PID-like Fuzzy Controller of the B-BC are shown in Table 1, also in Figure 6 comparison between PID-like Fuzzy and Neuro-Fuzzy are shown within the duration of 0.5 seconds at voltage set point of 5v. It shows the implementation of different V_{in} values to B-BC. Neuro-Fuzzy Controller has achieved satisfying results in terms of less oscillation, less tracking error and very short rise time. There was no need for sensitivity analysis since the neural network was used to optimize the system and its performance to set the fuzzy system parameters.

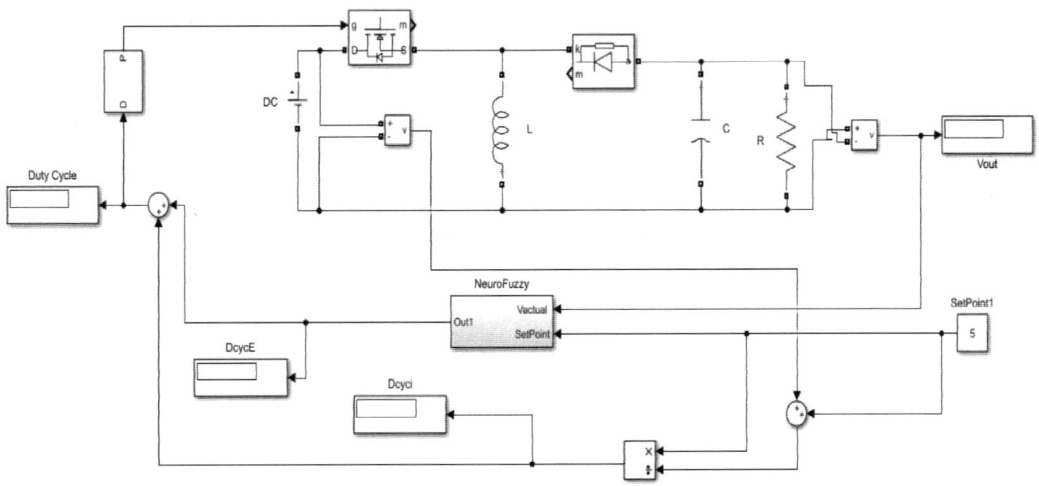

Figure 4. Neuro-Fuzzy Controller for B-BC

Table 1. Results from controllers of the B-BC at 5V Reference Voltage

Controller	PID-like Fuzzy			Neuro-Fuzzy		
	3v	5v	7v	3v	5v	7v

DcycE	0.0367 ~ 0.0898	0.0213 ~ 0.0898	0.0144 ~ 0.0898	0.0456	0.0423	0.0396
Dcyci	0.625	0.5	0.4167	0.625	0.5	0.4167
Steady state voltage (v)	5	4.99	4.99	5	5	5
Settling time (sec)	0.6	0.7	0.9	0.13	0.07	0.08
Maximum Overshoot (v)	NO	0.1	0.3	NO	NO	0.1

3. Proposed System Design

3.1. Hardware Design

In figure 7, it describes how the solar cell, Buck-Boost Converter, battery and water pump motor of PA operates; it is a circuit design of Solar Power Management for Precision Agriculture (SPM-PA) in internet of things system. The circuit diagram can be explained as follow:

- Arduino UNO input (UNO I/P)
  - Pin 1: R1 and R2 work as a voltage divider for the solar cell voltage coming from CELL_1 and CELL_2 directly to UNO I/P pin 1 for safety operation and not damaging the board.
  - Pin 2: R3 and R4 work as a voltage divider for the voltage coming from the BATTERY directly to UNO I/P pin 2 for safety operation and not damaging the board.
  - Pin 3: It is input pin from the soil sensor, to determine the moisture of the soil.
  - Pin 4: The output voltage (Vo) is used to get the value of tracking error (e) and its incremental change (Δe) to be used in Buck-Boost Converter Controller to determine the value of duty cycle (D_{cyc}) in order to control and manage the solar cell voltage and keep it 5v.

- Arduino UNO output (UNO O/P)
  - Pin 1: It is the value of duty cycle (D_{cyc}) of the controller for Buck-Boost Controller, to control and manage the solar cell voltage and keep it 5v.
  - Pin 2: When UNO O/P pin 2 is high, transistor Q1 will be switched ON and will lead to the power from CELL_1 and CELL_2 charging the battery through regulator U2.
  - Pin 3: When UNO O/P pin 3 is high, transistor Q2 will be switched ON and will tends to the power from BATTERY to the load.
  - Pin 4: When UNO O/P pin 4 is high, transistor Q3 will be switched ON will lead to pump motor turning ON.

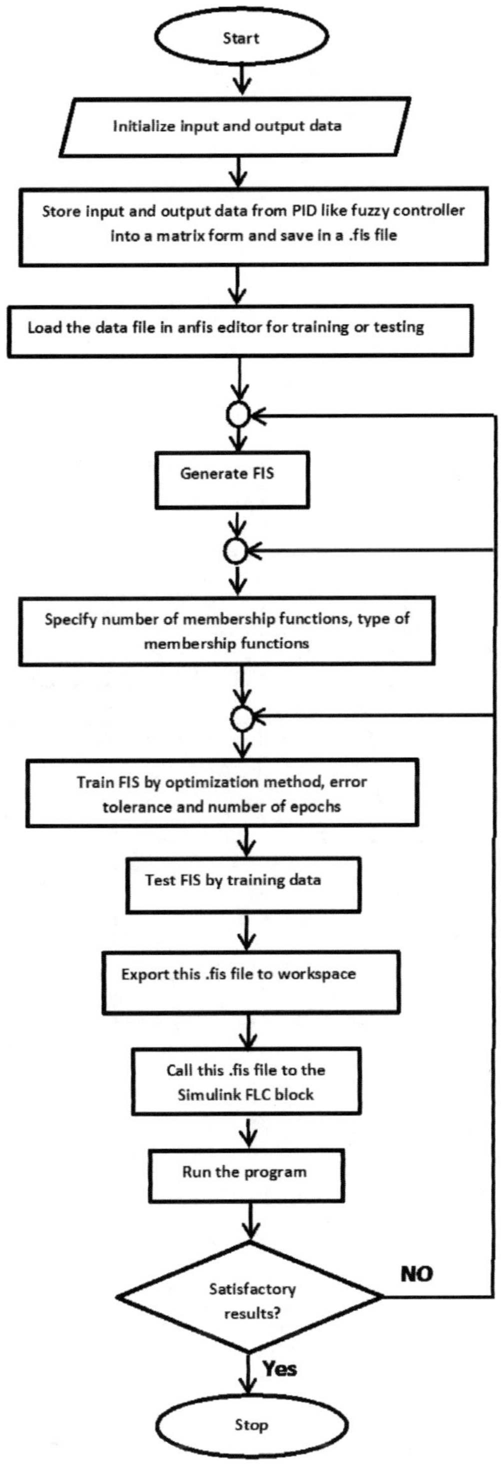

Figure 5. Flowchart for implementation of proposed Neuro-Fuzzy algorithm

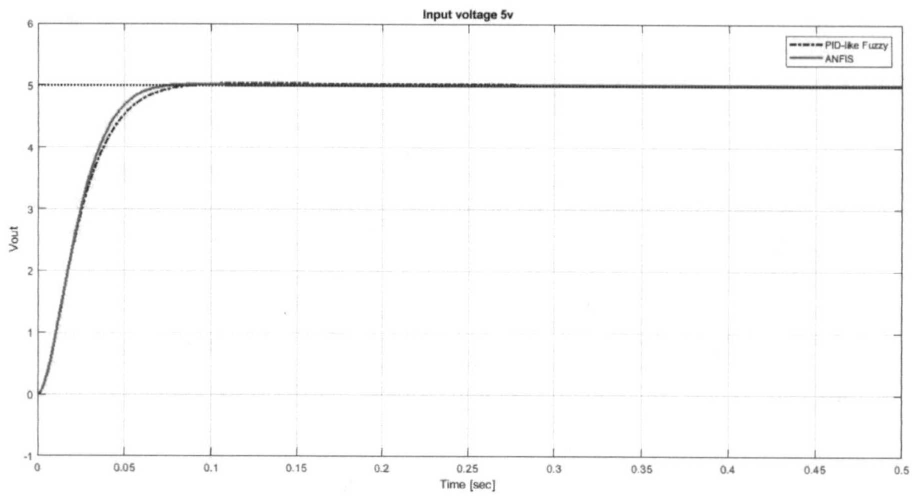

Figure 6. Output voltage comparison between PID-like Fuzzy and Neuro-Fuzzy

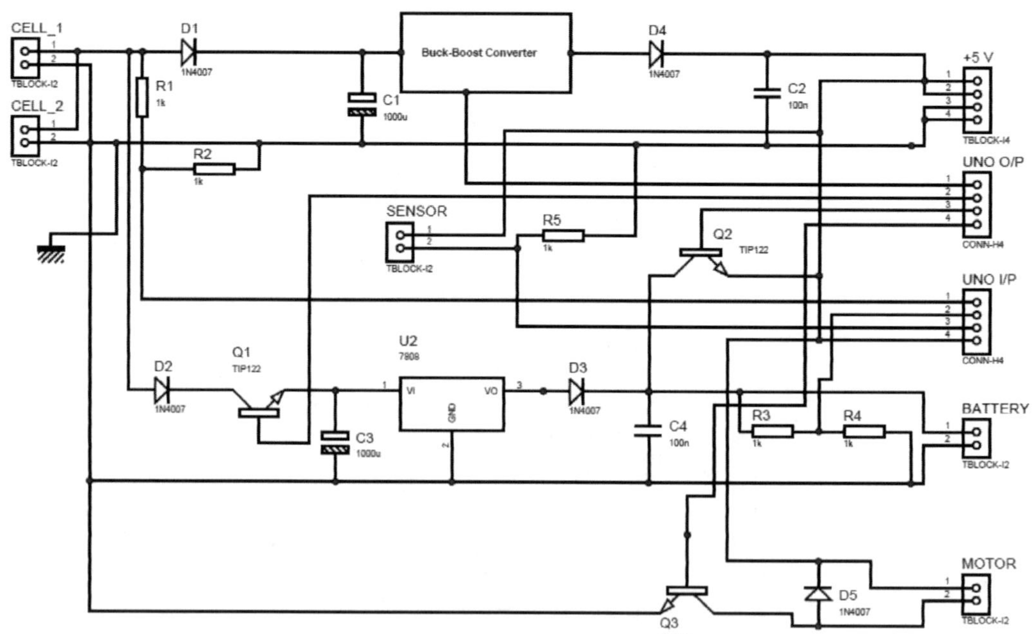

Figure 7. SPM-PA IoT circuit design.

3.2. Proposed Software Design

The software implementation can be divided into two parts; B-BC Controller and Solar Power Management for Precision Agriculture (SPM-PA). The first part, BBC controller has two inputs; tracking error (E) and its incremental change (ΔE). The value of E is normalized before being applied to the controller by diving it to set-pint voltage. The output of the B-BC Controller is percentage error duty cycle (D_{cyce}), where the ideal duty cycle (D_{cyci}) is added to percentage error duty cycle (D_{cyce}) for controlling the harvester output solar photovoltaic (PV), and the controlled voltage directly supply the water pump motor for irrigation.

The second part, software implementation of Energy Harvester Power Management for Precision Agriculture is expressed by flowchart in figure 8, where input/output pin of Arduino UNO is initialized and set, and then the values of Solar cell, battery and soil moisture data are read through (UNO I/P) as shown in figure 7. The solar cell voltage value that was read is then compared with minimum Cell voltage (MCV) which is the minimal amount of voltage that allows the motor to work efficiently with the sunlight. If Cell voltage is less than MCV then the solar cell will not charge the battery neither will load the Arduino. But if the Cell voltage is higher than MCV, then in case the battery arrived to minimum voltage level (MBV) it will load the battery till its full. And also while the solar cell level higher than MCV it will check the soil moisture level, if it the soil is dry according to Moisture data (M-data) that was collected then the water pump motor will run and irrigate the soil.

Finally all the data that is collected are sent through NodeMCU in real time which can be shown using mobile devices via Blynk services (app). NodeMCU is an IOT platform open source.

4. Proposed Hardware Implementation

4.1. The Requirements

This section focuses on hardware implementation of self-sustainable Wireless Sensor Node for PA, to save energy and prolong the lifetime of the battery. But at the same time the design of the hardware must be cost-effective and affordable. The hardware that is used and their purpose are discussed in Table 2. In Table 3 describes the software that is used for coding in Arduino UNO and NodeMCU. Table 4 illustrates Blynk [13], which is a platform service to connect the NodeMCU and the mobile app. it is used to monitor the data of solar cell, battery and soil moisture on the mobile app.

Table 2. Hardware and usage purpose

Hardware	Usage purpose
2 x Solar Cell panels 5Watts 9V	Transform light into energy
6v,4Ah Battery	Store solar energy
NodeMCU (ESP8266-12E)	Send Solar cell, Battery and Humidity data into internet via WIFI connection
Moisture Sensor	sensor is used to detect the amount of moisture present in the soil
5v Pump Motor	Used for water irrigation
Arduino UNO	Software code implementation for SPM-PA
SPM-PA system module	Buck boost, charger, motor driver

Table 3. Software and usage purpose

Software	Usage purpose
C++ on Arduino	Programming language on NodeMCU

Table 4. Service and protocol and usage purpose

Service and protocol	Usage purpose
Blynk App	IoT cloud service

4.2. *The proposed Concept Diagram*

In this research paper, the studies aim to expand their use as green energy resource and an alternative for a case study in the solar cell system for a smart energy efficient agriculture. The IOT is applied to measure and monitor the status of solar energy, battery and moisture. All data can be monitored in real time on mobile devices. Furthermore, as an IOT cloud service, all the data status of solar cell, battery and soil are preceded into a Blynk. All data on cell phones can be tracked in real time.

As explained in previous figure 7 for SPM-PA in IoT system module, it describes how the circuit design for the solar cell, Buck-Boost Converter, battery and water pump motor of PA operates and manages between them. In figure 9, it explains the whole concept diagram of SPM-PA in IoT.

The Battery will supply power and load Arduino via SPM-PA circuit, to be able to operate the rest of the system without shutting down. Among them controlling and managing solar voltage coming from solar cell via Buck-Boost Converter to stabilize the solar voltage to 5v to operate the water pump with efficiency. In order to increase the output current, two solar cells with 5 watts are connected in parallel.

Through the connection between Battery and SPM-PA, the battery level is determined by the code written in Arduino to charge the battery from solar cell.

The collected data of solar cell, battery and soil are sent from Arduino to NodeMCU serially, NodeMCU was used as a WIFI hardware to link the IOT to the access point of the router. This enables it to connect to the internet. Blynk was a service used to send data to an internet to show those information on smartphones in real time.

4.3. *Implementation and Results*

Arduino UNO, SPM-PA cicuit system and NodeMCU were put together in a small waterproof container as shown in figure 10.

All embedded devices are placed in containers that are waterproof. The solar panel, battery, device container and water pump are put in a handheld board as shown in figure 11 which describes the picture of the IOT solar system. The solar panels were designed to a 45 degree angle to the direction of sunrise.

The setup is designed to show the use IoT power management in precision Agriculture. From previous discussions, Blynk is known as a platform service to connect the NodeMCU and the mobile app. It is used to monitor the data of solar cell, battery and soil moisture on the mobile app that can be displayed in real time online.

The status of the current usages with the solar cell, battery and soil were displayed on Blynk app on mobile as shown in figure 12(a) when the sun at its peak, where solar cell voltage is high and battery level is high but the soil moisture level is low. On the other hand figure 12(b) shows the status at no sunlight, where solar cell level is low, battery level is high and soil moisture level is intermediate. The OFF button displayed in the figure is used to turn the light of the system box on and off to check for the connection.

When the soil moisture level is very low a notification was automatically displayed on mobile screens as shown in figure 13. Blynk's cloud service is an IOT app. Android and IOS support it. Solar-cell, battery and soil data can be presented every 2 senods in this study or when there are activities such as low or dry soil battery level.

IOP Conf. Series: Materials Science and Engineering 1172 (2021) 012008 doi:10.1088/1757-899X/1172/1/012008

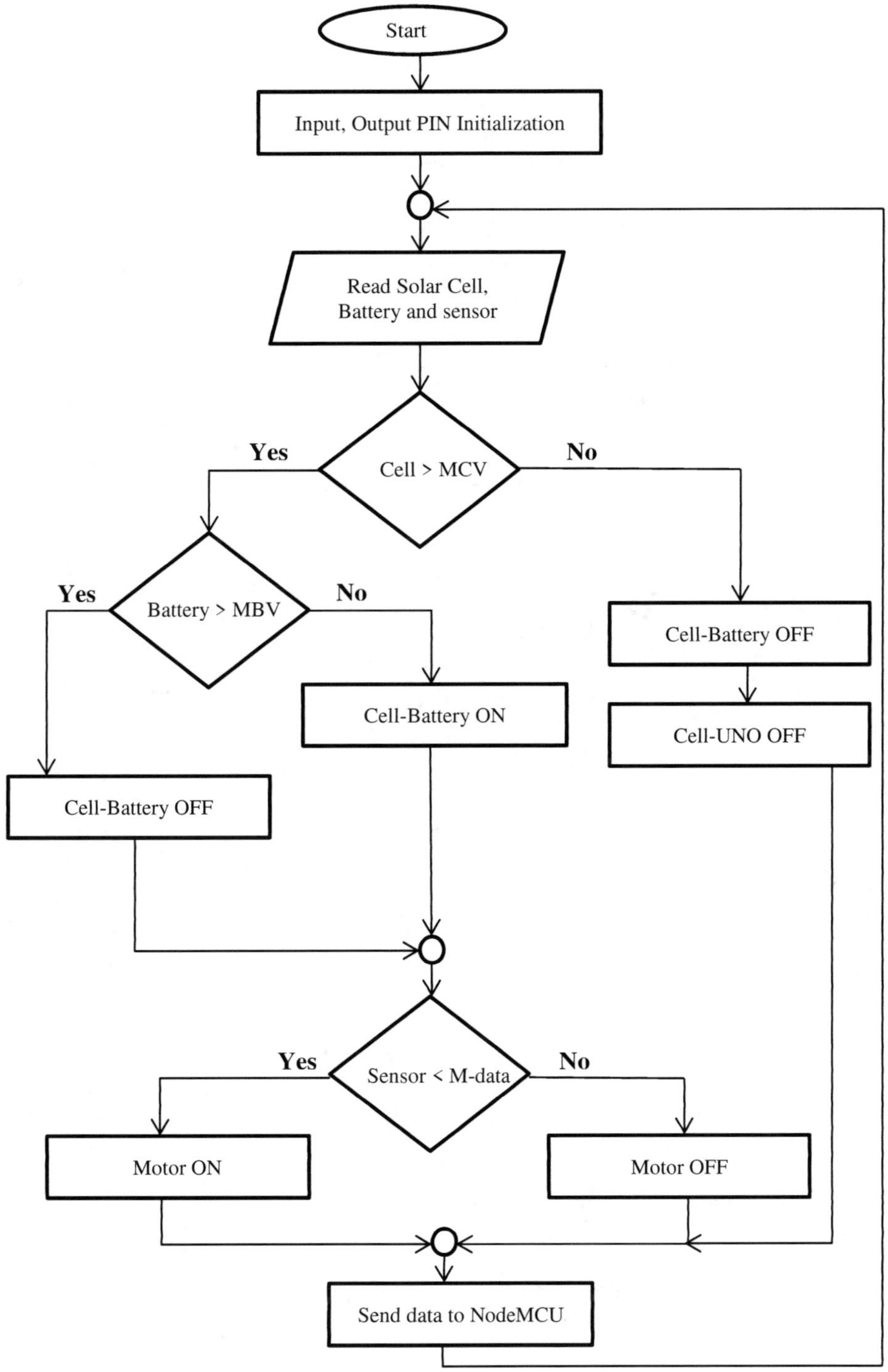

Figure 8. Flowchart of software implementation of SPM-PA.

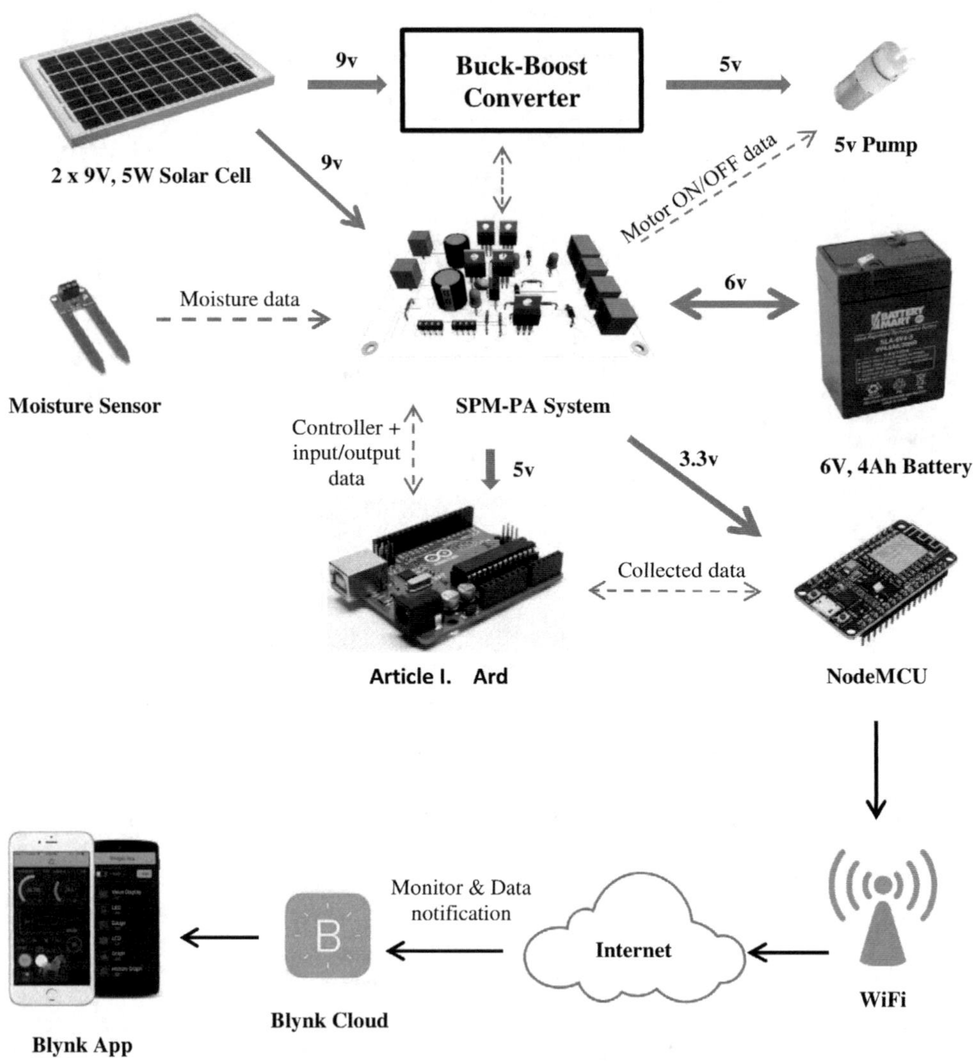

Figure 9. Concept diagram of Solar Power management System for Precision Agriculture in IoT

This work is a case study to incorporate as an alternative energy source for the PA model of a solar-cell device. Famers can therefore use this solar system in isolated areas. The status of solar-cell, battery and the soil were needed to be monitored on mobile devices. A total power usage was 19 watts as shown in Table 5.

Table 5. Power consumption per day

	Usage (Hours/Day)	Power rating (Watts)	Total daily used (Watts)
Water pump	0.4	2.5	1
Microcontrollers	24	0.75	18
			19

The daily usage of solar-cell system can be calculated as 19 watts, 5 hours for solar charging. Two panels of 5 watts were used which gives a total of 10 watts. The energy storing size of the battery can be calculated as Ah (Amp-hour) = total energy usage / battery voltage, Ah = 18/ 6 = 3 Ah, therefore the 4 Ah battery was applied to this case study. The average amount of time it takes to charge a solar-cell is approximately 5 hours. The battery was 4 Ah, which means that the battery was able to provide 6 volts X 4 Ah = 24 watts and the daily maximum system usage is 19 watts.

Figure 10. Electronic devices inside the box

Figure 11. Prototype Solar Power management System for Precision Agriculture in IoT

Application example supposes that it will be used for smart agriculture. Table 6 shows comparison of the average power consumed of the proposed system and the other discussed systems, the water pump is consuming 0.5 current as written in datasheet.

Table 6. Average power conception of the load

	Fuzzy	PID-like Fuzzy	Neuro-Fuzzy
Average mean voltage	5.0143	5.0095	5.0063
Power consumption	2.50715	2.50475	2.50315

IOP Conf. Series: Materials Science and Engineering 1172 (2021) 012008 doi:10.1088/1757-899X/1172/1/012008

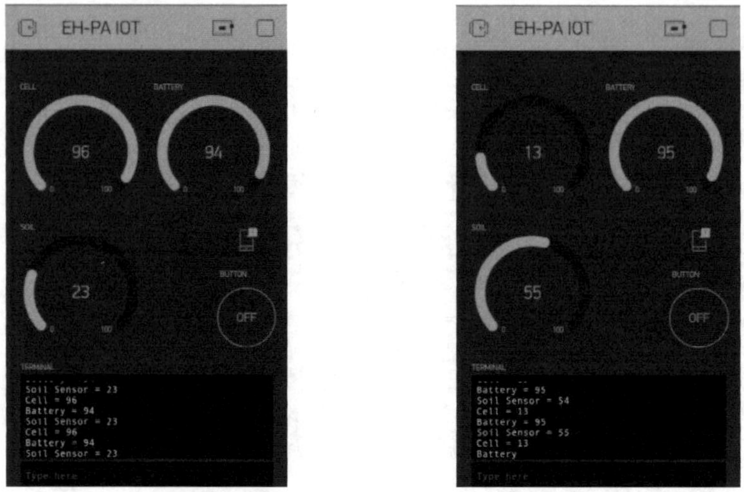

Figure 12. (a) System in the sun, (b) System in the shade.

Figure 13. Soil dryness notification on Blynk.

5. Conclusion

In this paper, a solar power management for Precision agricultural in internet of things system to enhance cost-efficiency and increased lifetime for the device was presented as an example and application of the proposed power management circuit. Neuro-Fuzzy Controller was proposed for the use in a new design of a DC-DC B-BC to regulate the voltage coming from Solar cell to the load. SPM-PA circuit was designed to manage between solar cell, battery, sensor and the motor with success. The water pump's functional status was performed properly. The cloud service of Blynk IoT has been implemented to monitor the collected data on real time. Data on solar cells, batteries and soil were regarded reliable and precise.

References

[1] Hassan, Q.F. ed., 2018. Internet *of things A to Z: technologies and applications.* John Wiley & Sons.

[2] Reznik, L., 1997. Fuzzy *controllers handbook: how to design them, how they work.* Elsevier.

[3] Blange, R., Mahanta, C. and Gogoi, A.K., 2015, June. MPPT of solar photovoltaic cell using perturb & observe and fuzzy logic controller algorithm for buck-boost DC-DC converter. In *2015 International Conference on Energy, Power and Environment: Towards Sustainable Growth (ICEPE)* (pp. 1-5). IEEE.

[4] X. Li and Z. Zhang, "Study on the output controllability of BUCK-BOOST Converter," in 2017

IEEE International Conference on Mechatronics and Automation (ICMA), 2017, pp. 107-112.
[5] Hussein, K.F., Abdel-Qader, I. and Hussain, M.K., 2015, May. Hybrid fuzzy PID controller for buck-boost converter in solar energy-battery systems. In *2015 IEEE International Conference on Electro/Information Technology (EIT)* (pp. 070-075). IEEE.
[6] Nabil, M.S., ElKhatib, M.M. and Tammam, A., 2019, April. Fuzzy Power Management for Internet of Things (IOT) Wireless Sensor Nodes. In *2019 36th National Radio Science Conference (NRSC)* (pp. 173-182). IEEE.
[7] Pedapenki, K.K., Gupta, S.P. and Pathak, M.K., 2015, December. Neuro Fuzzy Based Controller for Power Quality Improvement. In *2015 International Conference on Computational Intelligence and Communication Networks (CICN)* (pp. 1294-1298). IEEE.
[8] Tchoketch Kebir, G.F., Larbes, C., Ilinca, A., Obeidi, T. and Tchoketch Kebir, S., 2018. Study of the intelligent behavior of a maximum photovoltaic energy tracking fuzzy controller. *Energies*, *11*(12), p.3263.
[9] Mohan, T.K. and Mohammed, S.F., 2016, March. A neuro-fuzzy controller for multilevel renewable energy system. In *2016 International Conference on Electrical, Electronics, and Optimization Techniques (ICEEOT)* (pp. 4120-4123). IEEE.
[10] Sakunthala, S., Kiranmayi, R. and Mandadi, P.N., 2017, September. A study on fuzzy controller and neuro-fuzzy controller for speed control of PMSM motor. In *2017 IEEE International Conference on Power, Control, Signals and Instrumentation Engineering (ICPCSI)* (pp. 1409-1413). IEEE.
[11] Pedapenki, K.K., Gupta, S.P. and Pathak, M.K., 2015, December. Comparison of fuzzy logic and neuro fuzzy controller for shunt active power filter. In *2015 International Conference on Computational Intelligence and Communication Networks (CICN)* (pp. 1247-1250). IEEE.
[12] Gupta, R.K. and Chauhan, S., 2015, December. Comparision of PID controller & adaptive neuro fuzzy controller for robot manipulator. In *2015 IEEE International Conference on Computational Intelligence and Computing Research (ICCIC)* (pp. 1-4). IEEE.
[13] Saokaew, A., Chieochan, O. and Boonchieng, E., 2018. A smart photovoltaic system with Internet of Thing: A case study of the smart agricultural greenhouse. In *2018 10th International Conference on Knowledge and Smart Technology (KST)* (pp. 225-230). IEEE.
[14] Ramesh, B., Divya, M. and Revathi, G.P., 2020, November. Farm Easy-IoT based Automated Irrigation, Monitoring and Pest Detection using ThingSpeak for Analysis of Ladies Finger Plant. In *2020 International Conference on Recent Trends on Electronics, Information, Communication & Technology (RTEICT)* (pp. 237-241). IEEE.
[15] Sarma, B., Baruah, R. and Borah, A., 2020, October. Internet of Things based Smart Farming. In *2020 Fourth International Conference on I-SMAC (IoT in Social, Mobile, Analytics and Cloud)(I-SMAC)* (pp. 30-34). IEEE.
[16] Mageshkumar, C., 2020, January. IOT Based Smart Farming. In *2020 International Conference on Computer Communication and Informatics (ICCCI)* (pp. 1-6). IEEE.

IOP Conf. Series: Materials Science and Engineering 1172 (2021) 012009 doi:10.1088/1757-899X/1172/1/012009

Comparative study between cognitive radio techniques in FM broadcasting band

I Hamed[1], M A Elmenyawi[2], and A Y Hassan[3]

[1] MSc student, Benha faculty of engineering, Benha University, Egypt
[2] Lecturer, Benha faculty of engineering, Benha University, Egypt
[3] Associate professor, Benha faculty of engineering, Benha University, Egypt

E-mail: ashraf.fahmy@bhit.bu.edu.eg

Abstract. Optimum usage of the existing frequency spectrum is a major requirement due to the large increase in the number of subscribers at the same frequency. The cognitive radio scheme has become an important application used to optimize spectrum utilization, detects spectrum bands that are not occupied by primary users to create communication links between secondary users in the same band. These non-occupied bands are described as white space. In this paper, a spectrum sensing method that is an important stage in cognitive radio communications is discussed. Various spectrum sensing methods such as energy detection (ED) and cyclostationary feature detection (CFD) are used to sense the spectrum in the FM broadcasting band. A comparison of energy detection and cyclostationary feature detection spectrum sensing methods is made to determine the empty bands in the FM broadcasting band. This enables secondary users to transfer information in these empty bands without affecting the primary users of the FM broadcasting system.

1. Introduction

Recent years have witnessed an increase in the deployment of wireless devices and the need for a higher data rate transmission. Spectrum congestion and crowding have manifested as radio spectrum insufficiency. In most countries, national regulatory agencies such as the Federal Communications Commission (FCC) in the United State control the usage of the frequency spectrum. The FCC regulated the allocation of frequency bands and the granting of exclusive licenses to systems within a geographical region. At the same time, FFC prevents or at least coordinates other systems from working in these bands. According to the FCC report of the spectrum utilization, the utilization of licensed spectrum varies from 15% to 85% in the bands below 3 GHz [1]. FCC suggests that there is a substantial potential for improving the use of spectrum. Therefore, there is a need for new technologies to use empty spectrum bands and to increase the utilization of spectrum in dynamically changing environments.

Cognitive Radio (CR) [2] has emerged as the promising solution to the issue of spectrum crowding by allowing the opportunistic usage of frequency bands, which are not strongly used by licensed users. CR is an intelligent radio, which observes its surrounding environment and can learn and modify its actions and operation. It gives the best match between its surrounding environment and the needs of the users. CR can optimally adjust its operational parameters according to the surrounding radio environment activities. CR can identify white space holes in a frequency band that is not utilized by licensed users (primary users). It assigns these empty spectrum bands to a secondary user (SU) that uses them for transmitting its signal. However, when the primary user (PU) begins to reuse the spectrum, CR

Content from this work may be used under the terms of the Creative Commons Attribution 3.0 licence. Any further distribution of this work must maintain attribution to the author(s) and the title of the work, journal citation and DOI.
Published under licence by IOP Publishing Ltd

can recognize its activity via spectrum sensing techniques and hold the transmission produced by SU. There are various detection techniques available for spectrum sensing [3].

In this paper, two spectrum sensing techniques are used to scan the entire FM band (87.5 MHZ to 108 MHZ) and identify all unused spectrum holes. The first technique is energy detection, which is the most widely employed for spectrum sensing in CR systems because of its applicability and simplicity. The conventional energy detection approach [4] depends on a fixed threshold. It is sensitive to noise uncertainty that cannot be avoided in practical cases. In this paper, the calculation of the threshold of the energy-detection sensing method depends on some parameters that optimize the CR performance. The used method of threshold calculation optimizes the probability of detection and minimizes the probability of false alarms. Moreover, a cyclostationary feature detection based on a Fast Fourier Transform (FFT) filter with the spectral autocorrelation function is introduced [5]. Better detection of PU is accomplished through the use of the FFT filter in spectrum sensing at a very low value of SNR of the PU. It also allows the CR to track changes in the occupancy of the spectrum at a higher rate even at a low SNR value of PU and obtains more precise detection.

In this work, a proposed radio system is implemented based on the energy detection scheme and the cyclostationary feature detection. The proposed system detects the white space holes in the frequency band of the FM radio broadcasting. A simple adaptive method is also implemented to allocate the empty channels to the secondary user from the different holes identified by the previous two techniques.

2. Spectrum Sensing Techniques

Spectrum sensing allows CR to estimate, understand, and be aware of its operating environments like the availability of the spectrum and the state of the interference. If a frequency band is underused, by a PU at a specific time, a SU can use this frequency band in transmitting its signal. Spectrum sensing can also be carried out across the domains of frequency, time, and space. With the latest advancement of beam-forming technologies, several users can concurrently use the same channel/frequency in the same location [5], [3].

The receiver detection model of the SU uses a classical methodology known as binary hypothesis testing, where H0 is the null hypothesis (i.e., PU absence) and H1 is the alternative hypothesis (i.e., PU presence). Assume that the signal received by a cognitive radio user has a hypothesis model that is [3], [6].

$$y(t) = \begin{cases} x(t) + w(t) & for \quad H_1 \\ w(t) & for \quad H_0 \end{cases} \tag{1}$$

where y(t) is the received signal, x(t) is the transmitted signal of a PU, and w(t) represents the additive white Gaussian noise (AWGN). To differentiate between these two hypotheses, a binary hypothesis test is applied with a particular threshold calculated based on the probabilities of errors in the SU receiver. There are two error types in the binary hypothesis test that uses the model introduced in equation (1). The first error occurs when the receiver of the SU detects a PU signal while there is no PU signal is transmitted. The probability of this error is known as the false alarm probability (P_f). The false alarm probability in spectrum sensing is an important design parameter in specifying the threshold in the binary hypothesis test. The second error occurs when the receiver of the SU detects no PU signal while the PU signal exists. This probability of error is called the probability of missed detection (P_m) and it is equal to:

$$P_m = 1 - P_d \tag{2}$$

where P_d is the probability of detection. The missed detection results in collisions with primary user transmission. The error due to missed detection decreases the rates of data transmission for both the primary and secondary users. The probability of false alarm P_f (primary users exist in empty bands) and the probability of detection P_d (primary users exist in non-empty bands) defines the performance of the CR system. Generally, the constraints of the probability of false alarm as well as the probability of missed detection should be satisfied by a cognitive radio system. For the detection of the transmission of primary users, three schemes are typically used: matched filter detection, energy detection, and cyclostationary features detection.

2.1. Energy detection.

Energy detection [6] is a form of non-coherent detection technique which detects the primary signal depending on energy-sensing. Energy detection is considered to be the most common sensing technique in cooperative sensing since it's easy to use and doesn't need prior knowledge of PU signals. However, energy detection also comes by some drawbacks [7] such as:

- The sensing time needed to accomplish a given probability of detection can be high.
- The noise power uncertainties have an impact on the detection process efficiency.
- The differentiation between the PU and SU signals in the same CR system cannot be accomplished by energy detection technique.

Despite these drawbacks, in cooperative sensing, energy detection is the most common method of detection. Theoretical principle. Initially, the energy detector used a filter to remove the unwanted signals of the undesired frequency bands at the input of the energy detector. After that, the filter's output samples are squared. The summation of the squared samples is used to estimate the energy of the received signal. Finally, to compute whether a PU is present or not, the estimated energy is compared with a threshold value [3]. It is a difficult challenge to set the proper threshold since it must distinguish between the signal and noise.

It is important to have a priori knowledge of the amount of noise energy as its uncertainty degrades the performance of the detector. The probability of false alarm P_f and the probability of detection P_d must be measured in order to assess the energy detector's output. Moreover, the threshold value has to be specified to increase the probability of detection. The decision variable of an energy detector can be calculated as the mean of the energy of N samples as shown in equation (3) [8].

$$\mathbf{T} = \frac{1}{N}\sum_{n=1}^{N} |y(n)|^2 \tag{3}$$

where N is the number of samples used in the calculation of the average energy. The decision variable T in equation (3) is a chi-square random variable because the received samples y(n) are Gaussian random variables. The decision variable T is compared with a predetermined threshold λ to decide whether the spectrum is occupied by PU or not. The energy detector model for the cognitive radio can be defined as the binary hypothesis test shown in equation (4) [8].

$$d_{ED} = \begin{cases} H_1 & \mathrm{T} \geq \lambda \\ H_0 & \mathrm{T} < \lambda \end{cases} \tag{4}$$

Gaussian distribution is used to approximate T for a substantial number of samples using the central limit theorem as indicated in equation (5) [8].

$$\mathbf{T} \approx \begin{cases} \mathcal{N}\left(\mathrm{N}(\sigma_\omega^2 + \sigma_s^2), 2\mathrm{N}(\sigma_\omega^2 + \sigma_s^2)^2\right) & \text{for } H_1 \\ \mathcal{N}\left(\mathrm{N}\sigma_\omega^2, 2\mathrm{N}\sigma_\omega^4\right) & \text{for } H_0 \end{cases} \tag{5}$$

where σ_s^2 denotes the primary user signal variance, σ_ω^2 denotes the noise variance, and $\mathcal{N}$ denotes the normal distribution. For the binary hypothesis test in equation (4), the probability of detection P_d and the probability of false alarm P_f can be as shown in equations (6) and (7) [9].

$$P_d = \mathrm{Q}\left(\frac{\lambda - \mathrm{N}(\sigma_s^2 + \sigma_\omega^2)}{\sqrt{2\mathrm{N}(\sigma_s^2 + \sigma_\omega^2)^2}}\right) \tag{6}$$

$$P_f = \mathrm{Q}\left(\frac{\lambda - \mathrm{N}\sigma_\omega^2}{\sqrt{2\mathrm{N}\sigma_\omega^4}}\right) \tag{7}$$

where Q(.) is the Gaussian tail probability Q-function [10] and λ denotes the testing threshold. This threshold depends on the noise power and it is expressed from the probability of false alarm P_f as shown in equation (8).

$$\lambda = \left(\mathrm{Q}^{-1}(P_f)\sqrt{2\mathrm{N}} + \mathrm{N}\right)\sigma_\omega^2 \tag{8}$$

IOP Conf. Series: Materials Science and Engineering 1172 (2021) 012009 doi:10.1088/1757-899X/1172/1/012009

2.1.1. Implementation. In this work, an effective energy detector for FM broadcasting-band is introduced. In the used detector, some parameters are calculated and used to determine the system threshold λ. The FM broadcast band in Egypt is starting from 88 MHz to 108 MHz. This band is divided into 5 sub-bands. Five band-pass filters (PBFs) are used to filter the FM signal in these 5 sub-bands. One digital BPF has been used with a bandwidth of 5 MHz. The center frequency of this filter is changed according to the working sub-band. The working sub-bands and the center frequency of used BPF are shown in table (1).

Table 1. The frequencies of the used FM sub-bands

FM sub-band	Center frequency of the BPF
88 MHz – 92 MHz	90 MHz
92 MHz – 96 MHz	94 MHz
96 MHz – 100 MHz	98 MHz
100 MHz – 104 MHz	102 MHz
104 MHz – 108 MHz	106 MHz

The used digital BPF is a finite-impulse-response (FIR) filter whose coefficients are calculated according to the operation band. The filter coefficients for the different bands are stored in a look-up table (LUT) memory and they are recalled according to the working band. In the beginning, the detector selects the working FM band by selecting the coefficients of the corresponding BPF. The output signal from the BPF is sampled according to the second Nyquist theorem (under-sampling theory). The used sampling frequency is 20 MHz, which is 4 times the bandwidth of the used BPF. A squaring device is used to square the samples after the BPF. An accumulator is used to accumulate N samples from the output of the squaring device. The output of the accumulator is divided by N to get an estimate of the average energy of the received samples. The estimated average energy represents the decision variable to the threshold detector. This decision variable is compared with the appropriate threshold. The threshold value is set according to the number of samples N, the variance of the noise floor σ_ω^2, and the required probability of false alarm P_f as shown in equation (8). Finally, the energy detector decides whether the PU is present or not according to equation (4). The block diagram and the flowchart that illustrate the proposed technique are introduced in figure.1 and 2 respectively.

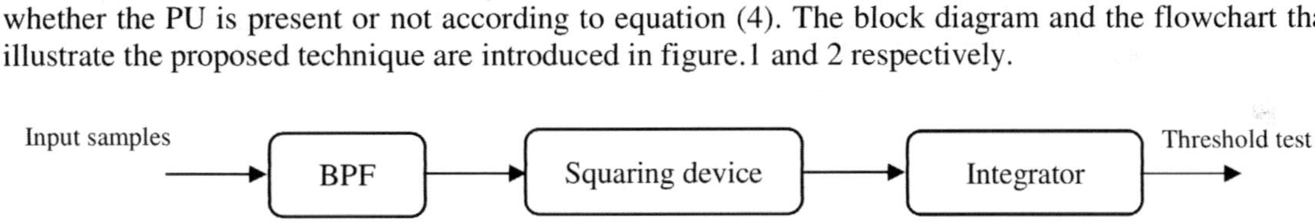

Figure 1. The block diagram of the energy detection scheme.

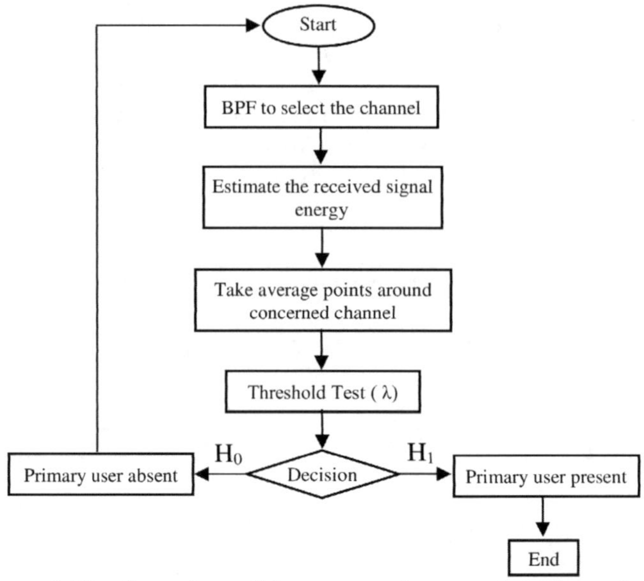

Figure 2. The flow chart of the proposed energy detection scheme.

2.2. *Cyclostationary feature detection.*

Cyclostationary feature detection [11] takes advantage of the periodicity feature in the received primary signal to detect the primary user presence. The periodicity is usually embedded in pulse trains, hopping sequences, sinusoidal carriers, or cyclic primary signals prefixes. The periodicity feature and the spectral correlation are only present in the useful signal and not in interferences signals and stationary noise. Therefore, the detection of cyclostationary features is immune to noise uncertainties and outperforms energy detection in low SNR regions. A signal is called cyclostationary if its autocorrelation and mean are periodic functions. Feature detection means obtaining features from the received signal and employing the detection task on the obtained features. This technique can differentiate PU noise and PU signal [12]. This technique is less common than energy detection due to some drawbacks such as:

- It is mandatory a priori awareness about the primary user.
- The observation time is long.
- The computational complexity is high.

2.2.1. Theoretical principle. This principle depends on the information extracted from the cyclic Spectral Correlation Function (SCF) and the PU signal periodicity. When the PU exists, this will generate a high value of the peak cyclic spectral correlation at the output. To obtain a more accurate decision about the correlation value, it is mandatory to use a large number of samples. The autocorrelation function for the signal x(t) is shown in equation (9) [13].

$$R_x(t,\tau) = E\left\{\chi\left(t+\frac{\tau}{2}\right)\chi^*\left(t-\frac{\tau}{2}\right)\right\} \quad (9)$$

Since the correlation function in equation (9) is periodic, it can be decomposed into Fourier series as.

$$R_x(t,\tau) = \sum_{k=1}^{M} R_x^{k\alpha_0}(\tau)e^{j2\pi k\alpha_0 t} \quad (10)$$

The fundamental cyclic frequency is $\alpha_0 = 1/T_0$, where T_0 is the hidden period and M represents the rank of the last harmonic. The Fourier coefficients of $R_x(t,\tau)$ are called the Cyclic Autocorrelation Function (CAF).

$$R_x^{k\alpha_0}(\tau) = lim_{T\to\infty}\frac{1}{T}\int_{-T/2}^{+T/2} R_x(t,\tau)e^{-j2\pi\alpha_0 t}\,dt \quad (11)$$

where T is the time duration that can be used to calculate the CAF. The Spectral Correlation Density function (SCD) is the Fourier transform of the CAF.

$$S_x^\alpha(f) = \int_{-\infty}^{+\infty} R_x^\alpha(\tau) e^{-j2\pi f\tau}\, d\tau \tag{12}$$

Equation (11) can be approximated as:

$$R_x^\alpha(\tau) = lim_{T\to\infty} \frac{1}{T}\int_{-T/2}^{T/2} x(t+\frac{\tau}{2}) x^*(t-\frac{\tau}{2})\, e^{-j2\pi\alpha t} dt$$

$$R_x^\alpha(\tau) = lim_{T\to\infty} \frac{1}{T}\int_{-T/2}^{T/2} x\left(t+\frac{\tau}{2}\right) e^{-j2\pi\frac{\alpha}{2}(t+\frac{\tau}{2})}\; x^*(t-\frac{\tau}{2})\, e^{-j2\pi\frac{\alpha}{2}(t-\frac{\tau}{2})} dt \tag{13}$$

The SCD is will be expressed after denoting $y(t) = x^*(t-\frac{\tau}{2}) e^{-2j\pi\frac{\alpha}{2}(t-\frac{\tau}{2})}$ as shown in equations (14 -16).

$$S_x^\alpha(f) = \mathbb{FT}\{S_x^\alpha(\tau)\}, \tag{14}$$

$$S_x^\alpha(f) = lim_{T\to\infty} \frac{1}{T} \mathbb{FT}\{y(\tau) * y^*(-\tau)\}, \tag{15}$$

$$S_x^\alpha(f) = lim_{T\to\infty} \{\frac{1}{T} X_T\left(f+\frac{\alpha}{2}\right) X_T^*\left(f-\frac{\alpha}{2}\right)\}, \tag{16}$$

In equation (16), $X_T(f)$ is the Fourier transform of the product of $x(t)$ and a rectangular window of width T.

$$X_T(f) = \int_{-T/2}^{T/2} x(t)\, e^{-j2\pi f t}\, dt \tag{17}$$

In this method, it's important to have advanced knowledge of the primary user. The probability of false alarm P_f and the probability of detection P_d must be calculated to evaluate the performance of the cyclostationary feature detection. It is common knowledge that to increase the probability of detection, the threshold value has to be specified. The probability of false alarm of cyclostationary feature detection (P_f) can be calculated from equation (18).

$$P_f = \mathrm{P_r(H1/H0)} = \exp\left(\frac{-\lambda^2}{2\sigma_A^2}\right) \tag{18}$$

where λ is the threshold value, $\exp(.)$ denotes the exponential function. $\sigma_A^2 = \frac{\sigma_\omega^2}{(2N+1)}$, and N is the number of received signal samples. The probability of detection (P_d) for cyclostationary feature detection can calculate from equation (19).

$$P_d = \mathrm{P_r(H1/H1)} = Q\left(\frac{\sqrt{2\gamma}}{\sigma_\omega}, \frac{\lambda}{\sigma_A}\right) \tag{19}$$

where γ is a signal-to-noise ratio, Q(.) is Generalized Marcum Q-function, and σ_ω^2 is noise variance. The threshold can be expressed from the probability of false alarm P_f as shown in equation (20).

$$\lambda = \sqrt{((-\log P_f) * 2 * \sigma_A^2)} \tag{20}$$

2.2.2. Implementation. In this section, the cyclostationary feature detection technique is implemented. The received signal is applied to digital BPF then the autocorrelation function is calculated. The autocorrelation function is calculated by multiplying the signal by shift versions of itself and averaging the result of the multiplication. The result is a function of the amount of shift. We then calculate the received signal's power spectrum density by getting the Fourier transform of the autocorrelation function. Finally, the cyclostationary detector determines the presence of the PU by comparing the PSD at the specified FM bands with a predetermined threshold for determining whether a signal is present or not. The block diagram and the flowchart that illustrate the proposed technique are introduced in figure.3 and 4 respectively.

Figure 3. The block diagram of the proposed cyclostationary feature detection scheme.

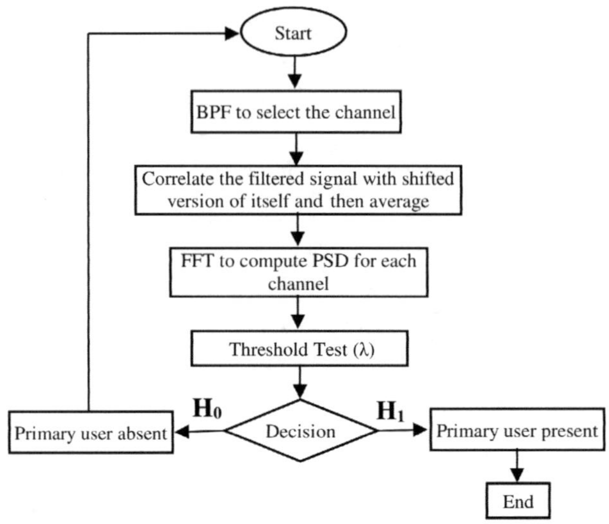

Figure 4. Flow diagram of the proposed cyclostationary feature detection scheme.

3. Results and Discussion

In this section, simulation results of the proposed system are represented. The proposed system is simulated using Matlab m-files. The primary users in the simulated system are 8 FM broadcasting channels that work in the FM band (88 MHz to 108 MHz). Two different detection methods are used to detect primary user's activity. The first detection method is the energy detection criterion represented in section (2.1), and the second detection method is the cyclostationary detection criterion represented in section (2.2). In the simulated system, 2 channels (users) out of the simulated 8 FM channels (users) are empty. These two channels are selected randomly each time the experiment is conducted.

In the simulated system, the channel output is sampled using an analog-to-digital (ADC) converter. The number of samples that are used as an input to the proposed detectors is equal to 10^6 samples. The received SNR at the input of the detector is changed from -20dB to 15dB. The simulated system's output is assessed by determining the obtained SNR value at which the detector can detect both empty and busy primary users with a probability of detection of 1.Three different probabilities of false alarms are used to calculate the threshold value in each detector according to equations (8) and (20), respectively. The used values of the probability of false alarm are $P_f = 0.1$, $P_f = 0.01$ and $P_f = 0.2$.

The decision variable at the output of each detector is calculated once for each received 10^6 samples from the ADC. The threshold in equations (8) and (20) is compared with this decision variable. If the decision variable is larger than the threshold, the detector will decide that the channel is busy and vice versa. The threshold may be compared with one decision variable or the average of more than one decision variable. The number "n" is the number of the samples of the decision variable that are used to calculate the average, which is compared with the threshold value.

Table 2a shows the SNR value at which the detector discriminates between the empty channel and the busy channel (primary user exists or not) with a probability of detection equal to 1. In our simulation, the detector indicator is the value of the SNR at which the detector can sense the primary user's presence with a probability of 1. Table 2b shows the SNR at which the detector discriminates between the empty channel and the busy channel when the threshold is compared with the average of 5 decision variables.

Table 2a. The threshold is compared with one decision variable (n=1)

Probability of false alarm	0.1	0.01	0.2
Cyclostationary Indicator	5dB	6dB	4dB
Energy Detection Indicator	10dB	12dB	9dB

Table 2b. The threshold is compared with the average of 5 decision variables (n=5)

Probability of false alarm	0.1	0.01	0.2
Cyclostationary Indicator	2dB	3dB	1dB
Energy Detection Indicator	8dB	9dB	7dB

Figure (5.a) and (5.b) show the probability of detection at the output of the energy detection detector (ED) and the cyclostationary feature detector (CFD) at different input SNR values. The probability of detection in figure (5.a) is calculated based on one value of the decision variable, however, the probability of detection in figure (5.b) is calculated based on the average of 5 values of the decision variable. The figures illustrate that the performance of the CFD detector is better than the performance of the ED detector. The CFD detector can detect the status of the channel (empty/busy) at lower received SNR than the ED detector. Moreover, using the average value of (n>1) samples of the decision variable gives better and accurate performance than using one sample of the decision variable.

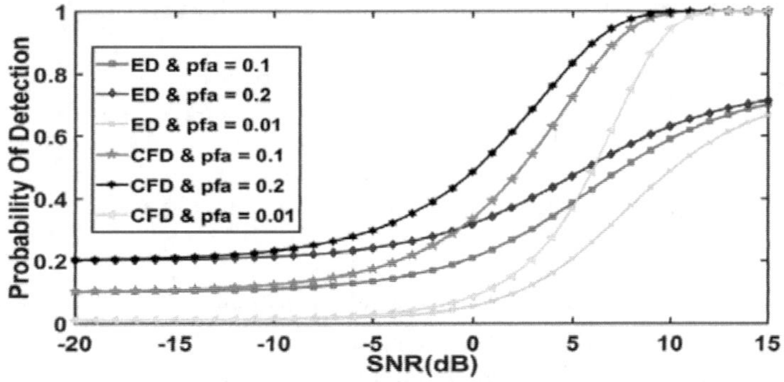

Figure 5a. performance of ED and CFD at n =1

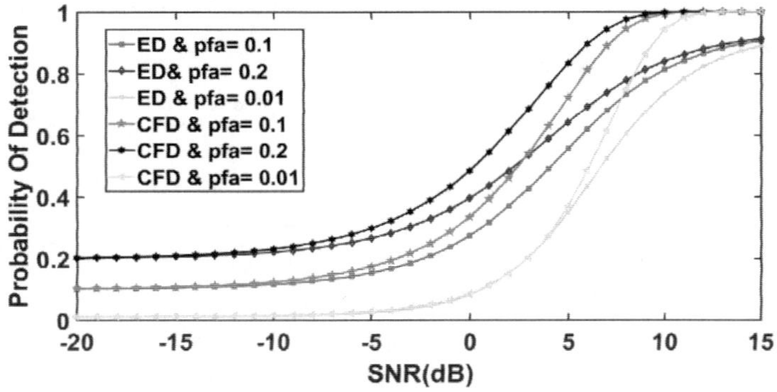

Figure 5b. performance of ED and CFD at n=5

Figures (6.a) and (6.b) show the channel indicator at the output of the energy detection detector (ED) and the cyclostationary feature detector (CFD) at different received SNR values. The figures illustrate that the channel indicator of the CFD detector will rise earlier at a lower SNR value than the channel indicator of the ED detector.

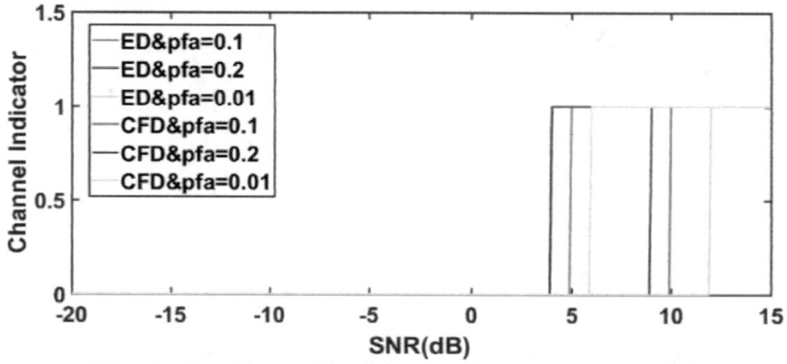

Figure 6a. Channel indicator for ED and CFD at n=1

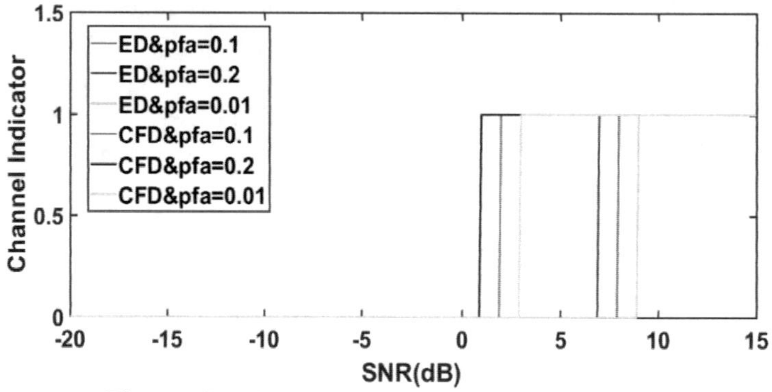

Figure 6b. Channel indicator for ED and CFD at n=5

4. Conclusion

In this paper, we use the energy detection method and the cyclostationary detection method for sensing spectrum in the FM broadcasting band. Each sensing technique has its own set of benefits and drawbacks. Energy Detection is the simplest technology and it is not necessary to have any prior knowledge about primary users. However, it performs poorly at low SNR. Cyclostationary detection technology is preferable to energy detection technology. It performs well at a low signal-to-noise ratio (SNR) as illustrated in all results, but it is a complex technique and needs prior knowledge of the primary users.

References

[1] Abdulsattar M 2012 Energy detection technique for spectrum sensing in cognitive radio: a survey *Int. J. of Comp. Net. & Comm.* **4** pp 223-242

[2] Mitola J and Maguire G Q 1999 Cognitive radio: Making software radios more personal *IEEE Pers. Comms.* **6** (4) pp 13-18

[3] Pandit S and Singh G 2017 Spectrum sharing in cognitive radio networks: medium access control protocol-based approach *Springer Int. Pub.* pp 35-75

[4] Pawełczak P 2011 Cognitive radio: Ten years of experimentation and development *IEEE Comm. Mag.* **49** (3) pp 90-100

[5] Yucek T and Arslan H 2009 A survey of spectrum sensing algorithms for cognitive radio applications *IEEE Comm. Surveys and Tutorials.* **11** pp 116-130

[6] Digham F *et al* 2007 On the energy detection of unknown signals over fading channels *IEEE*

Trans. on Comm. **55** pp 21-24

[7] Akyildiz A F *et al* 2011 Cooperative spectrum sensing in cognitive radio networks: A survey *Phys. Comm.* **4** pp 40-62.

[8] Xuping Z and Jianguo P 2007 Energy-detection based spectrum sensing for cognitive radio *Proc. IET Conf. on Wireless, Mobile and Sensor Net.* (*Shanghai*) 10194940

[9] Medina E and Barbin S 2018 Performance of spectrum sensing based on energy detection for cognitive radios *Proc. IEEE-APS Topical Conf. on Antennas and Propagation in Wireless Comm. (Cartagena des Indias)* 18196775

[10] Abramowitz M and Stegun C A 1972 *Orthogonal polynomials. Handbook of Mathematical Functions with Formulas, Graphs, and Mathematical Tables* 9th printing (New York: Dover) Chapter 22 pp 771-802.

[11] Gardner W A 1988 Signal interception: a unifying theoretical framework for feature detection *IEEE Trans. on Comm.* **36**, pp 897-906.

[12] Tripta V*et al* 2017 Techniques for energy detection in cognitive radio for spectrum sensing: a case study *Int. J. of Advanced Research in Computer* **8** pp 1027-1030.

[13] George G and Prema C 2019 Cyclostationary feature detection based blind approach for spectrum sensing and classification *Radioengineering* **27** (1) pp 289-303.

IOP Conf. Series: Materials Science and Engineering 1172 (2021) 012010 doi:10.1088/1757-899X/1172/1/012010

A satellite-based quantum key distribution using decoy-state protocol

A M Sofy[1], M Y Shalaby[2] and H M Dahshan[1]

[1] Military Technical College, Department of Communication Engineering, Cairo, Egypt
[2] Ministry of Defence, Cairo, Egypt

E-mail: myousef73@hotmail.com

Abstract. An ideal QKD implementation that provides the promising unconditional security using single photon laser source is not practically easy, also sending these photons over a fiber channel affects the maximum distance of link up to few hundreds of kilometers because of the receiver's detectors losses. Therefore, the achievement of a global quantum network is not easy, and the need to use LEO satellites [1,2] for establishing quantum links makes the job easier as quantum free space link provides much less losses to the link than fiber links. Nowadays, a quantum communication link can be established by satellites through free space. in this paper, we propose an implementation [3] of a satellite-based quantum key distribution using decoy-state protocol [4], we then compare this protocol with the BB-84 protocol [3] against security and the key length of the generated shared key.

1. Introduction

Quantum key distribution protocols are protocols for distributing secure key privately between two sides to establish a secure channel between them, This field of study has a quick progress since the first proposed protocol in 1984 (BB-84) named after its inventors (Bennett and Brassard) [5],QKD security doesn't depend on the computational difficulty of mathematical problems solving (mainly factorizing a large prime number as in most current encryption systems), but depends on the uncertainty principle of the quantum mechanics.

Another advantage of quantum cryptography represented in QKD protocols, a property called forward security' as if a generated QKD key is secure now so it will remain secure in the future against computing power advances, on the other hand, in public-key encryption, the keys and secured messages be stored and being subjected to different kinds of cryptanalysis and can be easily affected by computing power advances in the future.

In case of talking about the optical techniques for QKD implementation, it can be classified into two types, discrete-variable QKD (DV-QKD) in which each bit of private information is encoded into a discrete degree of freedom of optical signals. Another technique called continuous-variable QKD (CV-QKD) in which the private information is encoded using coherent communication techniques. Both techniques have consumed a lot of research effort to increase the rate of key generation and make it more compatible with nowadays communication infrastructure. but both techniques have a difficulty

Content from this work may be used under the terms of the Creative Commons Attribution 3.0 licence. Any further distribution of this work must maintain attribution to the author(s) and the title of the work, journal citation and DOI.
Published under licence by IOP Publishing Ltd

when being implemented in a wide-scale of QKD as there is big exponentially increasing transmission losses on the physical communication channels with distance, which makes a great limitation on the achievable key rates on these channels on long distances.

In this paper, we focus on DV-QKD protocols which can be categorized into prepare and measure protocols and entanglement-based protocols.

Entanglement based protocols Link quantum states of two far away (separated) objects together in a way that you can't describe one of them alone but only as a combined quantum state. In other words, measuring one will affect the other, so a pair of entangled objects is shared between two users, this is called entanglement. if any object was intercepted, it will be detected by affecting the other, and hence any unofficial third party will be detected.

Prepare and measure protocols where, Alice prepares (encodes) each classical bit individually as an optical signal then transmit it to Bob, who makes some measurements on the incoming signal individually, so he can get the classical data back. In this paper, we focus on prepare and measure protocols.

1.1. BB-84 Protocol

It is considered the pioneer work in quantum key distribution and it is the most commonly used prepare and measure protocol in which classical bits are encoded individually as polarization of a single photon, as Alice prepares the photons to be sent to Bob by choosing randomly the operating basis either horizontal/vertical basis (with polarization 0°/90°, respectively) basis or diagonal basis (with polarization 45°/135°, respectively) and then, the state with polarization 0° for horizontal/vertical basis or polarization 45° for diagonal basis is encoded as bit 0, and the state with polarization 90° for horizontal/vertical basis or polarization 45° for diagonal basis is encoded as bit 1. Then Bob receives these photons and makes his measurement on them one by one using randomly selected bases. After that, both sides announce in public the used basis every time, signals with different bases will be discarded and the others with the same basis will be kept, from which they create a secret key between them. In practical, applying such a protocol requires a true quantum random number generator such as (Quantis AIS 31) for random seeds production to make basis selection, and a single photon laser source which is not practically applicable as each laser pulse has some probability of containing more than a single photon.

1.2. Decoy-State Protocol

To overcome the problem of single photon sources, decoy-state protocol [4] uses weak coherent state laser source, where the mean of the number of photons can be controlled by adjusting the intensity of the laser beam, so decoy-state protocol, a one-way quantum key distribution protocol, actually works the same way as BB84, however it adds a precedent step which adds another degree of freedom. This adds another choice taken by Alice when sending every signal pulse to Bob, as well as another level of complexity for the eavesdropper to analyze.

In this case, Alice chooses randomly the intensity level of the sent state signal between two or more levels of intensity every time. Later, Alice announces publicly the used intensity level for sending every signal state (qubit). Since eavesdropper can't differentiate between signal and decoy pulses, any eavesdropping technique which depends on the number of photons (as Photon Number Splitting attack, PNS attack) has different impacts on the signal states and the decoy states and is therefore likely to be detected.

Decoy-state protocol has advantages over BB84 protocol, as in ideal BB84 protocol, only signals with single photon pulses sent from Alice are guaranteed to be secure, also using this would reduce losses in a comparison to BB84, moreover, using regular multi-photon sources leads to the ability of deploying these protocols using commercially available components.

1.3. Attacks and Security Proofs

Intercept and resend attack, Eve is going to quantify all the quantum states (photons) sent from Alice and then forward other states to Bob, prepared in the state she is estimating. It causes errors on the keys Alice and Bob share using BB84 protocol. Since, Eve has no idea the bases on which the states sent by Alice are encoded, she can only guess what bases to determine in the same way Bob does [6]. She estimates the correct state of photon polarization as sent by Alice when she chooses the same basis used by Alice, and sends the photon with the correct polarization to Bob. However, when she chooses the wrong basis, she estimates the state randomly and the state sent to Bob might not be the same as the state sent by Alice. If Bob then tests this state on the same basis that Alice sent, he gets a random outcome, with 50 percent he will get incorrect outcome (instead of the right outcome he would get without Eve being present).

Table I below gives an example of such an attack as the system uses two bases $|P\rangle$ and $|X\rangle$ of dimension d=2.

Table 1. Detecting Eve in Intercept and Resend Attack.

Randomly sent qubits	1	0	0	0	1	1	0	1	0
Qubits sent by Alice	$\|P;1\rangle$	$\|X;0\rangle$	$\|P;0\rangle$	$\|P;0\rangle$	$\|X;1\rangle$	$\|X;1\rangle$	$\|P;0\rangle$	$\|X;1\rangle$	$\|P;0\rangle$
Eve's measuring basis	P	P	P	X	X	P	X	X	X
Qubits sent by Eve	$\|P;1\rangle$	$\|P;1\rangle$	$\|P;0\rangle$	$\|X;1\rangle$	$\|X;1\rangle$	$\|P;1\rangle$	$\|X;0\rangle$	$\|X;1\rangle$	$\|X;1\rangle$
Bob's measuring basis	X	X	P	X	X	P	P	P	X
Qubits received by Bob	1	0	0	1	1	1	1	1	1
Shared key		0	0		1		**[1]**		

Photon number splitting attack, Alice in the BB84 protocol uses single photon to send the quantum state to Bob. In practical Implementations, laser pulses may have 2 or more Photons per pulse. When there are multiple photons in the pulse, then Eve can separate the excessive photons and Forward the remaining one photon to Bob. This is the concept of the splitting of photon number attack, where Eve uses a quantum memory to store those extra photons till Alice and Bob reveal the encoding basis. Eve then can calculate her photons in Right foundation and get crucial information without making detectable errors. this problem has many solutions. The most apparent is the use of a real single source of photons rather than an attenuated laser.

Although these sources are still in the developmental process, QKD was successfully carried out with them because current sources operate at low efficiency and frequency main levels and conveyance distances are restricted. Another option is replacing the BB84 Protocol with the Decoy states protocol, where Alice sends some randomly selected laser pulse's intensities with different average photon number of her laser pulses. You may use these decoy states to detect an PNS attack, since Eve has no way of knowing which pulses are signals and which are decoys. This concept was put into effect successfully, at the University of Toronto and in a number of QKD experiments, allowing high levels of security against known attacks.

Denial of service attack, which can be performed by simply cutting or blocking the line between two points connected by the quantum key distribution. This is one of the reasons why developing quantum key distribution networks must have alternative links in the case of interruption.

Security proofs, in case that Eve has unlimited resources, e.g. both classical and quantum computation abilities so she has a big chance to make a successful attack. BB84 protocol has been proved to be immune against any quantum mechanics-based attacks. Quantum mechanics allowed us to transmit information unconditionally safe using an ideal source of photons that emits just one photon at a time[7]. however, there are also other Necessary conditions to make the communication safe:

- Eve can't physically access Alice encoding equipment and Bob decoding equipment

IOP Conf. Series: Materials Science and Engineering 1172 (2021) 012010 doi:10.1088/1757-899X/1172/1/012010

- Using a true random generator for producing random seeds for both Alice's and Bob's selections.
- The classical channel of communication should be authenticated by the use of unconditionally secure Scheme for Authentication.
- One-time pad encryption scheme preferred to be used for encrypting the message.

2. Related Work

Nowadays usage of satellite QKD became applicable as it is moving so fast from the experimental phase to the practical one. An example, is the Chinese project carrying the name of (Micius) [8] a sun-synchronous satellite in a low-earth orbit at 500 KM altitude, which is a common project between Chinese Academy of Sciences (CAS), Austrian Academy of Sciences (AAS) and University of Vienna. It uses an optical transmitter to operate a Decoy-state QKD, as the Chinese quantum satellite works as a trusted relay between three nodes for key distribution.

Satellite optical links can be categorized into uplink and downlink. The downlink is more reliable than uplink as it is subjected to lower atmospheric loss than the uplink, because atmospheric properties like turbulence causes wandering of the optical beam which translates into a less accurate land transmitter than a space transmitter. But the uplink has an advantage that we don't have to place an accurate optical laser source in space which is somehow complex, instead we need to place a simple receiver onboard the satellite.

The most significant scenario of applying QKD in space is by using the satellite as a flying trusted node or it can be called (Key Relay Protocol), that the satellite makes QKD process with every ground station individually after all the satellite will store a shared secret key between it and each ground station. So for making a secure link between any two ground stations that have secret keys, (key 1) and (key 2) respectively, we need to provide each one with the key of the other, this is simply done by the satellite that sends the XOR combination of the two keys (key1, key2) to both stations, from which each station can retrieve the key of the other by XORing its own key to this combination, key1 $\oplus$ (key1$\oplus$ key2) = key2, key2 $\oplus$ (key1 $\oplus$ key2) = key1, this process is illustrated in Figure 1.

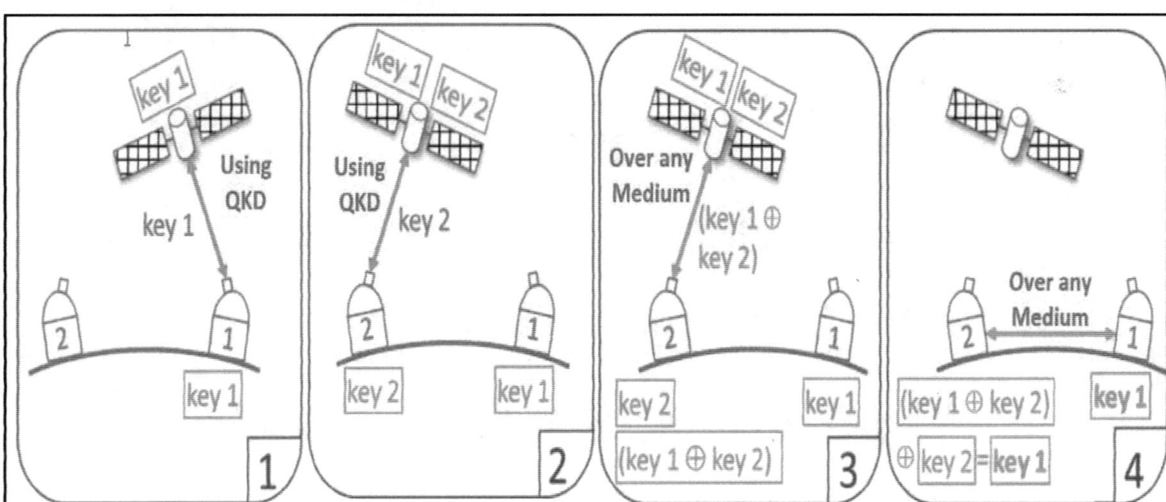

Figure 1. The use of a satellite for QKD.

The key relay protocol is implemented practically in the Chinese quantum satellite "Micius" that deploys a Decoy-state QKD protocol through a quantum downlink. The trusted node network principal

IOP Conf. Series: Materials Science and Engineering 1172 (2021) 012010 doi:10.1088/1757-899X/1172/1/012010

may be used in fiber network, however using a ground trusted node makes it easily be subjected to surveillance, probes and other kinds of attacks that can't affect a flying node (satellite) as it is a fast-moving target.

3. THE PROPOSED SIMULATION FRAMEWORK

We simulate the implementation of Decoy-state protocol using Optisystem v7.0 software package. It is an optical communication system simulator that provides a wide variety of QKD components. Most of the components used in simulation and their properties are exploited for the experimental use to setup QKD. In our experiment, a two-dimensional quantum system (d=2) is proposed with two orthogonal bases the rectilinear basis and the diagonal bases which are as presented in Table 2.

Table 2. Representation of Basis

Basis	Notation	Bit 0	Bit 1
Rectilinear Basis	+	0° →	90° ↑
Diagonal basis	X	45° ↗	135° ↖

3.1. Dataset

A hardware true quantum random numbers generator (Quantis AIS 31) shown in Figure 2 was used to generate the used random seeds that were used in the simulation work.

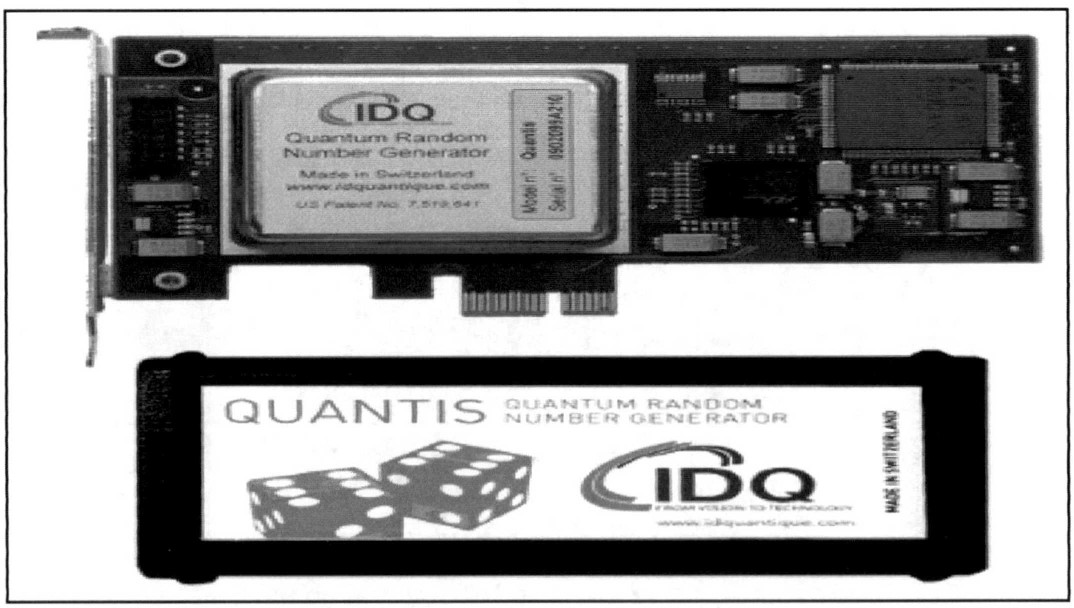

Figure 2. RNG used in simulation (Quantis AIS 31).

The generated random seeds were used for making the selection of the following:

- The basis of each iteration used by Alice
- Bob's basis for each iteration

IOP Conf. Series: Materials Science and Engineering 1172 (2021) 012010 doi:10.1088/1757-899X/1172/1/012010

- The basis for each iteration used by Eve
- Alice's intensity level of the sent state signal (from two levels of intensity)

A previous work was done by the Authors for modelling BB84 [3] on which this work is based. To make a fare comparison between the two protocols, this simulation is done with and without the presence of a third party (Eve).

3.2. IMPLEMENTATION SETUP FOR DECOY-STATE PROTOCOL

There are three basic functional blocks in this model as follows:

- Alice's side: where the laser source can emit a pulse with two intensities. The pulse is then encoded as a random state in random basis for each pulse
- Bob's side: where a random selection of the used basis for measurement is done with every received pulse from Alice
- Eve's side (if exist): where also a random measurement is done on the pulses passing through the quantum channel as Eve is supposed to have a full access to the quantum channel and can perform Intercept and resend attack. ,a more detailed description for the internal components of the model's blocks was previously illustrated in our previous work [3] of modelling the BB84 and The Two-Way QKD protocols.

The modelling diagram of Decoy state without Eve is shown in Figure 3, and the modelling diagram of Decoy state with Eve is shown in Figure 4.

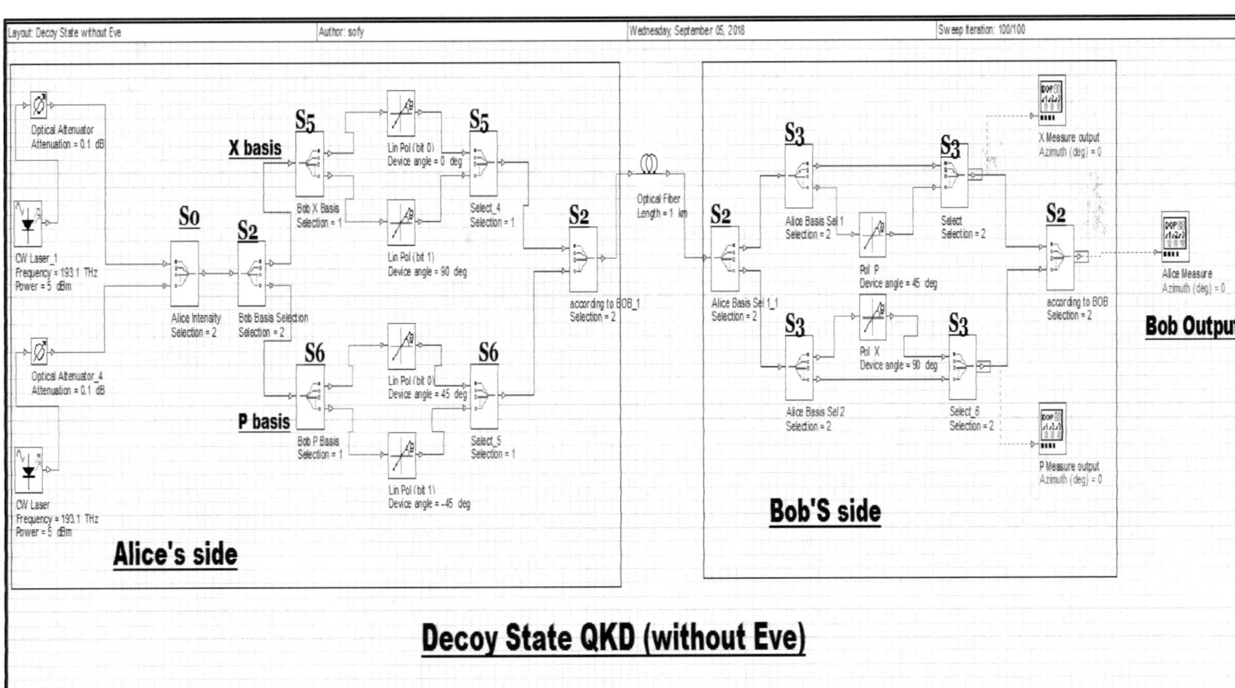

Figure 3. Model of Decoy-State QKD without Eve.

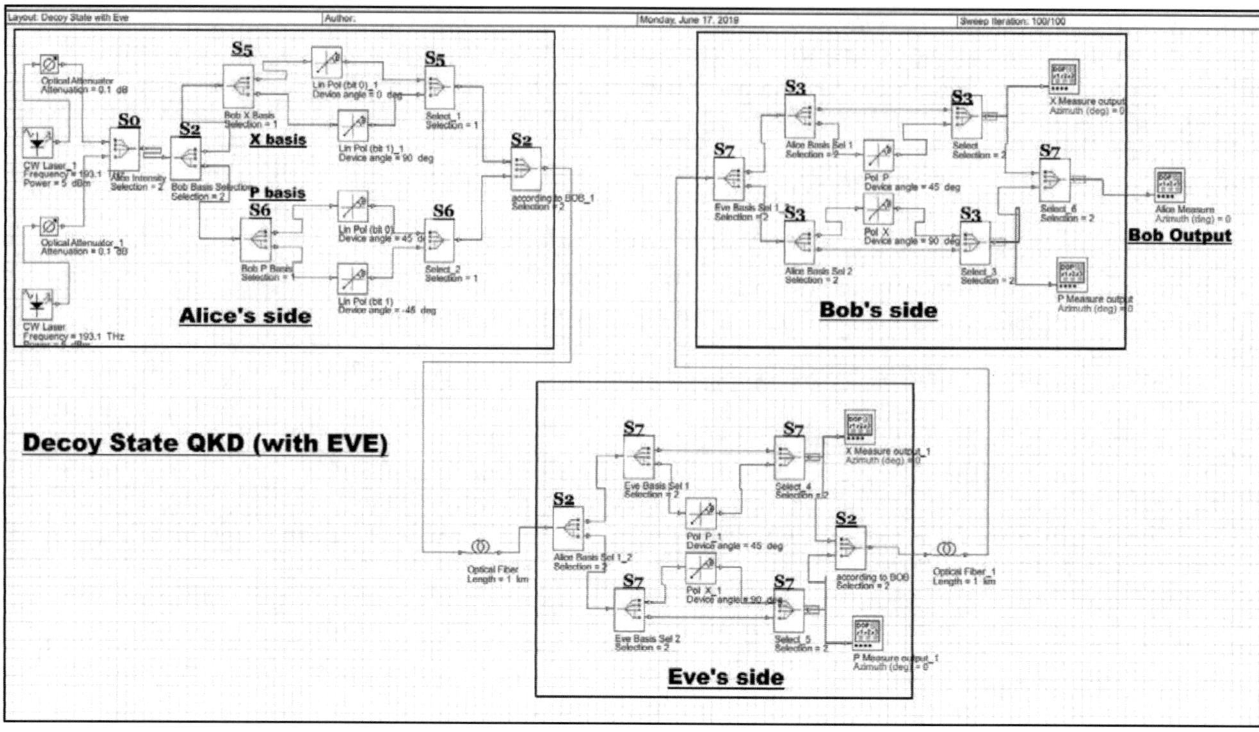

Figure 4. Model of Decoy-State QKD with Eve.

3.3. Results

In this model 1000 qubits were generated and sent from Alice to Bob. In the case that Eve is not present, the shared key size is 297 qubits, in the case of Eve presence, the shared key size is 213 qubits, as shown in Figure 6, taking into consideration that in both cases the same random seeds were used as various inputs for the modelling software for selecting Alice's encoding basis , Bob's measuring basis and also Eve's measuring basis, after analyzing the results and comparing the shared key size between the two cases of presence and absence of Eve we found that the presence of Eve affected 84 qubits which was detected as errors after the final announcement between the two parties out of 297 qubits (the shared key size in case of Eve's absence) leading to make the key size up to 213 qubits in case of exchanging 1000 qubits between the two parties.

To make a fare comparison between the proposed implementation of decoy-state protocol and the implementation of BB84 protocol[3,9], we select the same random seeds for both implementations at every iteration. The comparison results are shown below:

- without Eve's presence, the total shared key size decreased from 534 qubits in the case of BB84 [3] to 297 qubits in the case of our Decoy-state implementation.
- with Eve's presence, the total shared key size decreased from 412 qubits in the case of BB84 [3] to 213 qubits in the case of our Decoy-state implementation.

So, the probability of detecting Eve of BB-84 and Decoy state protocols can be calculated as shown in equation (1) and equation (2) sequentially.

$$P_d = \frac{\text{number of wrongly received qubits due to Eve's Measure}}{\text{Shared key size without Eve}} = \frac{122}{534} = 0.228 \quad (1)$$

$$P_d = \frac{\text{number of wrongly received qubits due to Eve's Measure}}{\text{Shared key size without Eve}} = \frac{84}{297} = 0.28 \quad (2)$$

Figure 5 shows the results of One-Way protocol that was previously obtained in our previous work [3], and Figure 7 shows the comparison between the two protocols.

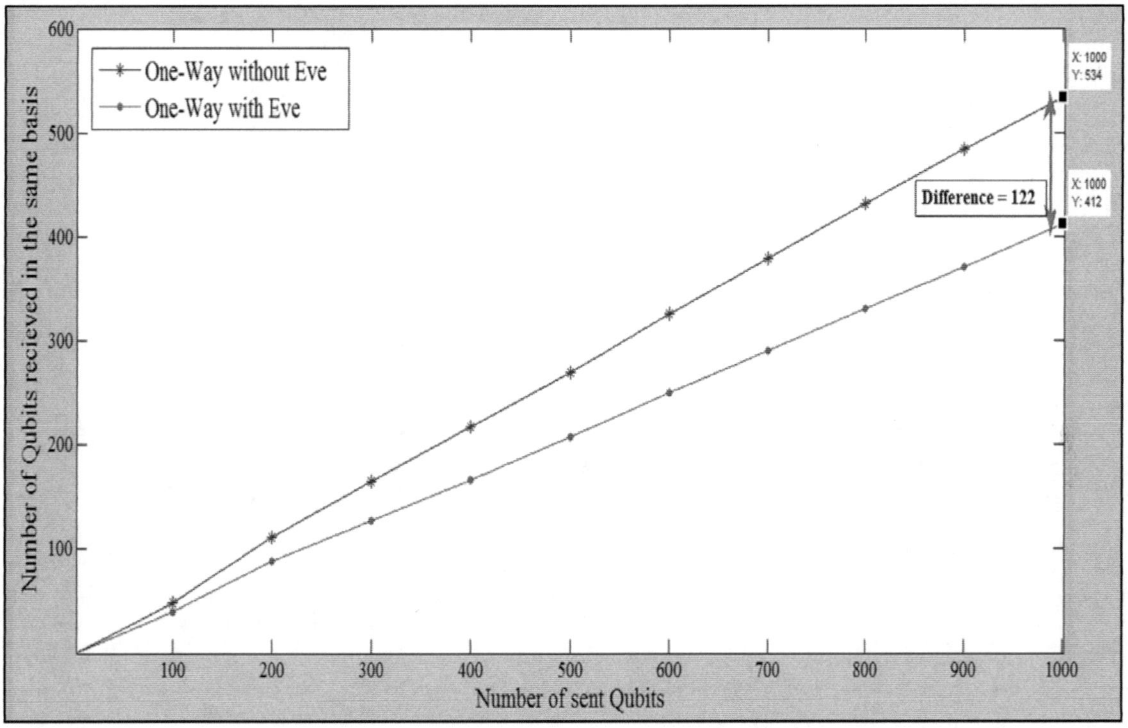

Figure 5. The number of qubits in the shared key between Alice and Bob in the presence and absence of EVE in One-Way protocol.

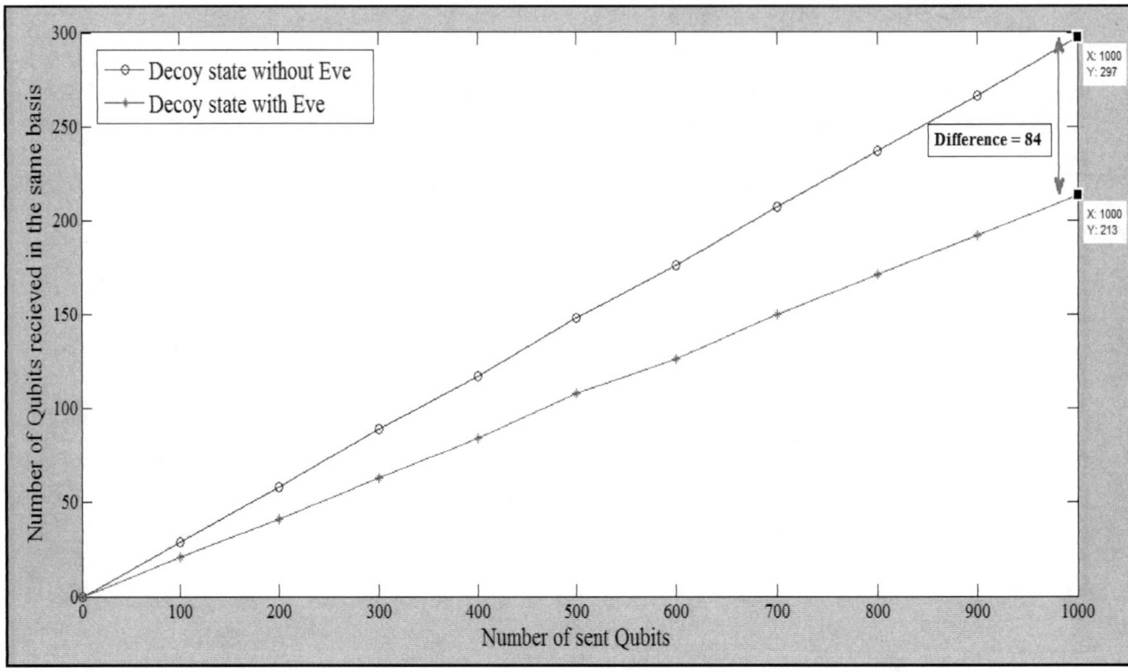

Figure 6. The number of qubits in the shared key between Alice and Bob in the presence and absence of EVE in Decoy-state protocol.

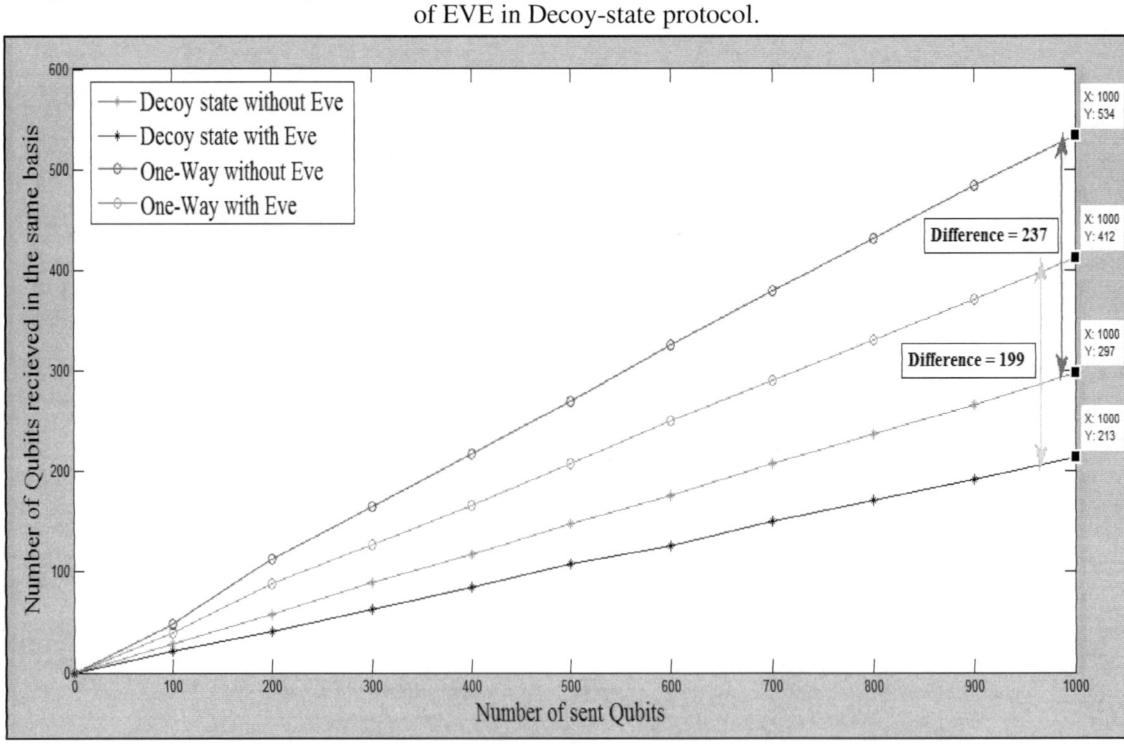

Figure 7. Comparison between the shared key size the presence and absence of EVE in One-Way protocol and Decoy-state protocol.

4. Conclusion

We have discussed the decoy-state protocol which is widely used for satellite quantum key distribution. Comparing the simulation of decoy-state protocol with BB84 protocol and according to the experimental results, the probability of detecting the presence of Eve, in the case of decoy-state implementation (Pd),equals 84/297 = 0.28, the probability of detecting the presence of Eve, in the case of implementing BB84 was 0.228 [3] in our previous work for a 1000 qubits sample.

we found out that decoy-state protocol is more secure against PNS attacks, and the experimental results of applying both protocol using Optiwave v7.0 software package shows that the probability of detecting Eve in the case of decoy-state protocol outperforms its counterpart in the case of BB84 protocol (0.28 versus 0.228) [10]. However, the number of shared qubits in the case of decoy-state protocol is less than its counterpart in the case of BB84 protocol.

References

[1] Liao S K, Cai W Q, Liu W Y, Zhang L, Li Y, Ren J G, Yin J, Shen Q, Cao Y, Li Z P, Li F Z, Chen X W, Sun L H, Jia J J, Wu J C, Jiang X J, Wang J F, Huang Y M, Wang Q, Zhou Y L, Deng L, Xi T, Ma L, Hu T, Zhang Q, Chen Y A, Liu N Le, Wang X Bin, Zhu Z C, Lu C Y, Shu R, Peng C Z, Wang J Y and Pan J W 2017 Satellite-to-ground quantum key distribution *Nature* **549** 43–7

[2] Scheidl T, Handsteiner J, Rauch D and Ursin R 2019 Space-to-ground quantum key distribution **11180** 67

[3] Sofy A M, Shalaby M, Dahshan H M and Rohiem A 2019 Modeling One-way and Two-way quantum key distribution protocols *Proc. - 2019 IEEE 9th Int. Conf. Intell. Comput. Inf. Syst. ICICIS 2019* 239–44

[4] Lo H K, Ma X and Chen K 2005 Decoy state quantum key distribution *Phys. Rev. Lett.* **94** 1–5

[5] Bedington R, Arrazola J M and Ling A 2017 Progress in satellite quantum key distribution *npj Quantum Inf.* **3** 1–12

[6] Fröhlich B, Lucamarini M, Dynes J F, Comandar L C, Tam W W-S, Plews A, Sharpe A W, Yuan Z and Shields A J 2017 Long-distance quantum key distribution secure against coherent attacks *Optica* **4** 163

[7] Alléaume R, Roueff F, Diamanti E and Lütkenhaus N 2009 Topological optimization of quantum key distribution networks *New J. Phys.* **11**

[8] Schmitt-Manderbach T, Weier H, Fürst M, Ursin R, Tiefenbacher F, Scheidl T, Perdigues J, Sodnik Z, Kurtsiefer C, Rarity J G, Zeilinger A and Weinfurter H 2007 Experimental demonstration of free-space decoy-state quantum key distribution over 144 km *Phys. Rev. Lett.* **98**

[9] Buhari A, Zukarnain Z A, Subramaniam S K, Zainuddin H and Saharudin S 2012 An efficient modeling and simulation of quantum key distribution protocols using OptiSystem™ *ISIEA 2012 - 2012 IEEE Symp. Ind. Electron. Appl.* 84–9

[10] Verma P K, El Rifai M and Chan K W C 2019 Quantum Key Distribution *Signals and Communication Technology* (Springer) pp 59–84

Enhanced image encryption algorithm by quantum key distribution

M A Mahmoud[1] ,T Mekkawy[2]and A D Elbayoumy[3]

[1] NANO Master program, Faculty of Engineering.Cairo University,Giza, Egypt
[2]Avionics Department, Military Technical Collage, Cairo, Egypt
[3]Communication Department, Military Technical Collage, Cairo, Egypt

E mail: Mohamed.khalifa84@eng1.cu.edu.eg

Abstract. Quantum Key Distribution (QKD) is one type of Quantum Cryptography (QC) which is based on quantum mechanics fundamentals such as Heisenberg's uncertainty principle and No-cloning theory. The usage of QKD warns the legitimate communicated parties to any attack attempt and this is the most interesting security parameter. Therefore, QKD provides unconditional secure communication method and supports a powerful encryption scheme. The combination between classical communication and QKD creates a new technique called semi quantum key distribution SQKD. Unfortunately, SQKD increases the schemes complexity and requires two steps for ciphering, scramble and encryption. In this paper an enhance image encryption algorithm is proposed based on QKD that eliminates most of the drawbacks of SQKD. The proposed algorithm is simpler than other encryption schemes as it exploits only one encryption step based on the power and the randomness of the generated secret key, which decreases the chance to be cracked. The correctness and efficiency of the proposed algorithm are validated by numerical simulations.

1. Introduction

Data encryption is one of the most important factors that used to determine the communication system quality and security. Communicated parties need their data to be ciphered to inhibit hackers from steal their data and personal information. Many schemes and algorithms were proposed to grantee data security among the last decades. Nowadays, online service takes place everywhere which increases data hacking and eavesdropping attacks attempts. Also, most of financial affairs are achieved through internet. This leads to the high requirement and the extremely need to powerful encryption schemes to save the mutual data between the service provider and the clients. Therefore, Quantum Key Distribution (QKD) [1] is adopted to improve the security performance due to its unconditional performance. The security of QKD is based on the principles of quantum mechanics, mainly, Heisenberg's uncertainty principle [2] and No-Cloning theory [3].

The quantum system sends notifications to the communicated parties for any attempt to sense or to hack the communication link. Because the measurement makes copies of the quantum states which conflicts with the No-Cloning theory, and similarly it warns the communicated parties to that eavesdropping. Quantum cryptography (QC) was introduced for the first time in 1970 [4]. It makes a great foundation for storing and transmitting the data by using linear or circular polarized photons. Moreover, when the Eavesdropper (Eve) tries to detect any information about the key that being used, the communicated parties (Alice and Bob) will notice that try immediately. In 1984, a secure communication protocol, BB84, was proposed [1]. It is based on conjugate observables and it describes photon polarized states to transmit the confidential information.

Content from this work may be used under the terms of the Creative Commons Attribution 3.0 licence. Any further distribution of this work must maintain attribution to the author(s) and the title of the work, journal citation and DOI.
Published under licence by IOP Publishing Ltd

Basically, BB84 uses two pairs of states, with each pair conjugate to the other pair, and the two states within a pair orthogonal to each other. These orthogonal pairs are the basis. The usual polarization state pairs are either the rectilinear basis of vertical (0°) or horizontal (90°), the diagonal basis of (45°) and (135°) or the circular basis of left- and right-handedness. Any two of these bases are conjugate to each other, and so any two can be used in the protocol. The usage of QKD in classical communication methods was discussed for the first time on 2007 in this technique one party is quantum (Alice) and the other party (Bob) is a classical one[5] so this algorithm named semi quantum key distribution SQKD. The first SQKD protocol [5] uses four different states. Those different states are randomly selected to be rectilinear or diagonal basis. The protocol was improved by two similar protocols were introduced in 2009[6]. The first protocol relayed on measurement and the second one was randomization dependent. A simplified SQKD protocol was employed in [7] with only three quantum states. Then, in 2011, another improved SQKD protocol was proposed [8] that was basing on entangled states, in which the qubit is more efficient by 50%, with respect to 25% of the protocol in [5]. Recently, in 2015 a new SQKD protocol was presented. The classical party in that protocol does not need the ability to measurement and needs only how to prepare, send and reorder qubits [9]. All the above SQKD protocols suppose the presence of an authenticated classical channel, which can be eliminated by pre-sharing a major secret key between two communicated parties. However, in the all mentioned SQKD protocols the classical parties are not really classical. This is because they need the ability of quantum preparing and measuring for the qubits in the computational basis, or quantum memory for the purpose of qubits reordering. In other words, the need of quantum devices for quantum operations is remaining exited. Then, Alice is supposed to be classical party and Bob is the quantum one [10].

From above literature, we observe SQKD algorithms consist of Quantum party, Classical party, Quantum channel, Classical channel, and Quantum memory. The main condition for SQKD algorithm is that the classical party can access the quantum channel otherwise no key will be shared with the communicated parties (quantum and classical). In 2016, a SQKD protocol based on a secure delegated quantum computation introduced [11], where Alice and Bob mutually transmit and receive data to third party (Charlie) to generate the secret key and it deeply requires the highest level of honesty to be in Charlie. Later in 2016, an algorithm proposed in which an integrated classical and QC scheme by using a three-party supposed to be authenticated [12]. In this scheme, three parties were involved and digital signature was added to ensure mutual authentication between the communicated parties. From the above preamble some drawbacks are concluded such as:

- Need for quantum memory.
- Need for quantum channel access for the classical party.
- The classical party is not completely classic one as he uses quantum devices for qubits measurement.
- Need for delegation of quantum computing for the third party quantum server.
- In many SQKD the key is one time pad OTP.
- Probability of the server adversary and dishonesty.
- Need for more added mutual authentication ways like digital signature.

On the other hand, image processing is a general term that represents any operation that applied to images for encryption, decryption, coding, decoding. Many researches were presented to secure images and hide their contained information between the communicated parties. On the opposite direction, attackers don't save efforts to hack communication systems and crack security keys. For example, in 2017 a scheme proposed a data transmission technique [13] by using QKD in which the data is firstly encoded by Huffman coding then the data is encrypted by QKD. This technique is limited in data size and also required a trusted third party to establish the secret key and need middle OTP coding step before data encryption and transmission.

Later in 2017, another scheme presented a robust encryption algorithm for quantum medical image [14]. In this algorithm they converted the normal medical images into quantum ones by a controlled-NOT gate then the converted images are scrambled by using quantum bit-plane and grey code and the last step is encrypting the scrambled quantum image. In 2020, an algorithm proposed for colour image encryption based on hyper chaotic map and mitochondrial DNA sequences [15] in which colour image encryption exploits an embarrassment operation that depending on a hybrid chaotic map. The first step is to separate each channel of colour images into number "n" of clusters. The second step is to create global shuffling over the total image.

The final step is to apply intra-pixel reconstructing in each cluster, which results in very non-ordered pixels in the encrypted image.

From above literature, we conclude that, QKD provides unconditional secure communication method and supports a powerful encryption scheme. The combination between classical communication and QKD creates a new technique called semi quantum key distribution SQKD. Unfortunately, SQKD increases the schemes complexity and requires two steps for ciphering, scramble and encryption. In this paper, a new enhancement for image processing algorithms is proposed. We applied QKD scheme to raise the security level to the confidential data to be transmitted over unsecure channel. The proposed scheme raises the difficulty for an attacker to hack the encrypted data. The proposed algorithm exploits the QC benefits to save images' information. Also there is no middle coding or scrambling step to reduce the algorithm complexity depending on the power of QKD. Also, the usage of the quantum channel and devices is limited to eliminate the SQKD drawbacks.

The rest of the paper is divided as follows. In section 2, an introduction to QC and OKD is presented. Section 3 explains the proposed algorithm. In section 4, results is showed and discussed. Also, a discussion is showed to compare the proposed algorithm to other algorithms [14] and [15] and finally, in section 5 the conclusion is presented.

2. Quantum cryptography and quantum key distribution

QC exploits quantum physics and quantum mechanics characteristics, which uses Heisenberg's uncertainty principle and the No-cloning theorem [2], [3]. Therefore, scientists adopted quantum mechanics to create a new branch of cryptography depending on the quantum properties. By QKD, any two communicated parties can encrypt and decrypt the messages with unique property of quantum mechanics, and the system can sense any eavesdropper. This is because the fact that the process of measuring any quantum system disturbs it. Therefore, the eavesdroppers' try will be detected easily. The first protocol of QKD is BB84 [1].

BB84 is related to its inventors "Charles H. Bennet and Gilles Brassard" in 1984. This protocol uses the polarization states of single photons to transmit a random key. Photons emerged from a light source often have a random polarization angle. To specify a selected polarization angle to the photon, a light filter is used to allow only photons that have the selected polarization angle to pass. The polarization angle is used to encode the transmitted bits.

The basis is chosen to substitute "0" and "1" without ambiguity. The first choice is rectilinear basis in which angles of 0° and 90° representing "0" and "1", respectively. The second choice is the diagonal basis; i.e., angle of 45° represents "0" and 135° represents "1". For simplicity, we denote 0°, 90°, 45° and 135° as H,V,D and A respectively. The polarization basis for H and V is denoted by + while the polarization basis for D and A is represented as ×. Table 1.shows the polarization angles for used basis in BB84 protocol.

Table 1 Polarization angles for BB84 protocol

Basis	0	1
+ (linear)	(0°) →	(90°) ↑
× (diagonal)	(45°) ↗	(135°) ↘

The data unit in QC protocols is qubit and it is defined as a physical structure defined by the quantum mechanics laws [16]. The qubit is mathematically represented by $|\psi\rangle$ [16] which is a unit vector in the two dimensional Hilbert space $|\psi\rangle \in \mathcal{H}^2$ [16]. A qubit $|\psi\rangle$ is expressed as [16]:

$$|\psi\rangle = \alpha|p\rangle + \beta|q\rangle \quad (1)$$

$$|\alpha|^2 + |\beta|^2 = 1 \quad (2)$$

Where $\alpha, \beta \in \mathbb{C}$, and $\{|p\rangle, |q\rangle\}$ is a random basis spanning $\mathcal{H}^2$[16]. The most important consequence of the vectorial nature of a qubit is that it is possible to write it in linear combination of elements of any basis. The choice of $\{|p\rangle, |q\rangle\}$ is often the orthonormal basis $\{|0\rangle, |1\rangle\}$ known as the computational basis.

IOP Conf. Series: Materials Science and Engineering 1172 (2021) 012011 doi:10.1088/1757-899X/1172/1/012011

In addition to the computational basis $\{|0\rangle, |1\rangle\}$, it is common in quantum cryptography protocols to use the diagonal basis $\{|+\rangle, |-\rangle\}$ defined as [16]:

$$|+\rangle = (|0\rangle + |1\rangle)/\sqrt{2} \tag{3}$$

$$|-\rangle = (|0\rangle - |1\rangle)/\sqrt{2} \tag{4}$$

3. Proposed algorithm

The proposed algorithm exploits QKD to generate a random key that used for image encryption operation. The key is totally random which provide a kind of difficulty to hack it. Also, the quantum mechanics principles will alarm the communicated parties for any hacking or eavesdropping attempt. The proposed algorithm is summarized as follows:

1- A QKD-based key is shared between Alice and Bob. That key has a unique property of quantum mechanics. By using BB84 protocol.
2- The confidential data is XORed with the generated key to encrypt and cipher it.
3- The encrypted data is sent through communication link.
4- At receiver, the received encrypted data is XORed again with the key to decrypt it.

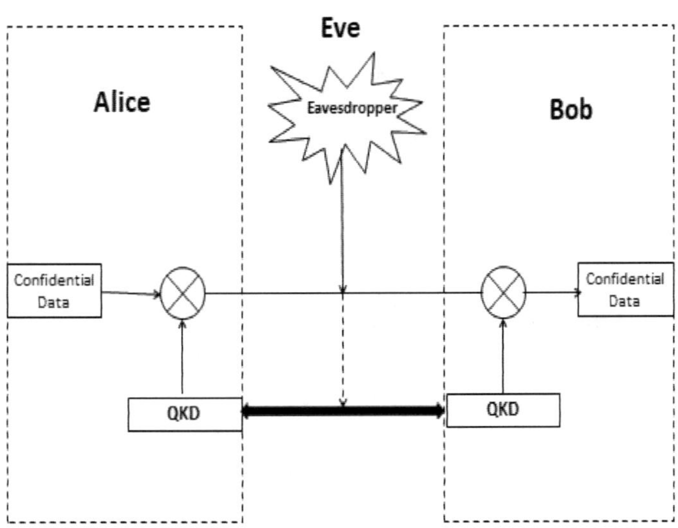

Fig.1 Schematic diagram for the proposed algorithm

The schematic diagram for the proposed algorithm is shown in Fig.1. Here Alice and Bob generate and share the quantum secret key before exchanging data. The quantum secret key and the confidential data are XORed together for data encrypting purpose before transmission. The encrypted data is transmitted through the communication link meanwhile Eve tries to hack both the communication link and QKD shared between Alice and Bob. This hacking try can be detected immediately due to the quantum mechanics properties as represented in Heisenberg's uncertainty principle and No-cloning theorem. As illustrated in Table 2, Alice and Bob use any selected QKD protocol (BB84) then the selected QKD protocol is applied for secret key establishment step.

Table 2 Random secret key extraction steps

Alice bits	1	0	1	0	1
Alice random basis	V	H	+	H	V
Random polarization	↑	→	↗	→	↑
Bob random selected polarization	V	-	V	H	V
Measuring basis	√	×	×	√	√
Bob measured basis	↑	↘	↑	→	↑
Result key	1	-	1	0	1

Where V, H, + and - represent state$|1>$, state$|0>$, state $|+>$ and state$|->$, respectively. While, the polarization angles 90°, 0°, 45° and 135° are represented as ↑ , ,↗, and , ↘espectively. Now the resultant key is generated as "1101" after sifting the different measured basis. Once the secret key is generated Alice and Bob use this key for image encryption operation. First, Alice applies XOR process to encrypt the image. On the other hands, Bob receives the encrypted image and uses XOR operation to extract the original image.

The communication here is normal with a new feature which is the secret key is created by QC which make the following advantages:

1- Unconditional security communication system based on QKD system.
2- Ease of eavesdropping detecting.
3- Ultimate secure key generated.
4- Totally random generated key.
5- No need for a quantum server or third authenticated party.
6- No need for adding a digital signature for mutual authentication.

The expecting scenario is as follows:

- First, Alice and Bob decide to exchange a confidential data between each other.
- Alice and Bob generate and share a completely secret random key based on QC. In other words a QKD.
- Alice prepares the confidential data to be XORed with the generated quantum key to encrypt it.
- Bob will receive the encrypted data. After eavesdropping check step he will apply XOR operation to encrypted data to decrypt it in order to extract the original confidential data.
- Any eavesdropping attempt will be discovered at once, alerts the communicated parties to that a hacker tries to steal their data then they decide to seize the communication.

4. Results and Discussions

In this section, we show the results of the proposed algorithm, by examining four datasets (Cameraman, Katrina, Aeroblk, and Honey badger images).

The key is created by QKD and then the image is XORed with the created key. As shown in Figs. 2, 3, 4, and 5, the original images are shown in Figs. 2a and 3a, and then the shared secret key which is created via QKD is XORed with the confidential dataset to encrypt it. The encrypted data is approximately hidden and has no information as shown in Figs. 2b and 3b.

On the other hand, at the receivers, the decrypted data is XORed with the shared secret key, as represented in Figs. 2c and 3c the decrypted image is rebuilt identical copy of the original dataset (image).

(a) Original image (b) Encrypted image (c) Decrypted image

Fig.2. the proposed algorithm is applied to "cameraman" dataset

(a) Original image (b) Encrypted image (c) Decrypted image

Fig.3. the proposed algorithm is applied to "Katrina" dataset

(a) Original image (b) Encrypted image (c) Decrypted image

Fig.4. the proposed algorithm is applied to "Aeroblk" dataset

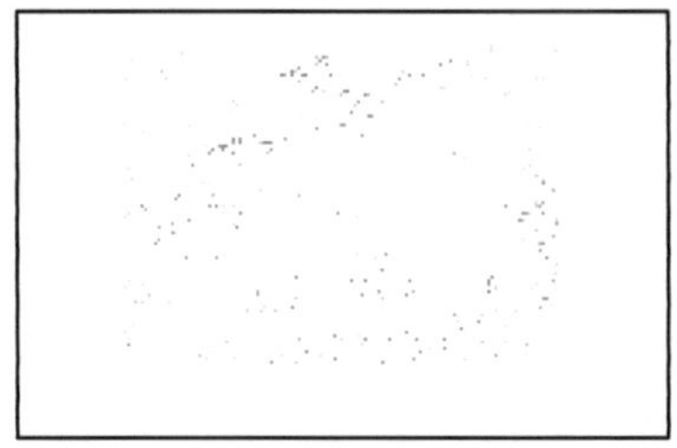

(a) Original image (b) Encrypted image (c) Decrypted image

Fig.5. the proposed algorithm is applied to "Honey badger" dataset

4.1 Simulation verification and statistical analysis

4.1.1 *Approximation quality metrics.* Approximation quality metrics are used to determine the image quality after encryption and decryption process. They consist of the Mean Square Error (MSE), Peak Signal-to-Noise Ratio (PSNR), Maximum Squared Error (MAXERR), and Ratio of Squared Norms (L2RAT), for an input image, X, and its approximation, Y. MSE is defined as the mean square error between an image, *X*, and an approximation, *Y*, is the squared norm of the difference divided by the number of elements in the signal or image N:

$$MSE = \frac{\|X-Y\|^2}{N} \quad (5)$$

Increasing MSE means the difference between the original image and the encrypted one is higher and is recommended for good encryption. PSNR is known as the peak signal-to-noise ratio in decibels (dB).

$$PSNR = 20 \log_{10} \frac{2^B-1}{\sqrt{MSE}} \quad (6)$$

It is recommended for PNSR to be low as possible for powerful encryption. MAXERR is the maximum absolute squared deviation of the data, X, from the approximation, Y. Higher MAXERR is better for encryption and represents good encryption algorithm. L2RAT is the ratio of the squared norm of the signal or image approximation, Y, to the input signal or image, X. The Structural Similarity (SSIM) Index quality assessment index is based on the computation of three terms, namely the luminance term, the contrast term and the structural term. The overall index is a multiplicative combination of the three terms. It measures the structural similarity between two images if it is the same, the result will be one and for different structure the better results is less than one. Naturalness Image Quality Evaluator (NIQE) no-reference image quality score calculates the no-reference image quality score for image X using the Naturalness Image Quality Evaluator NIQE. It compares X to a default model computed from images of natural scenes. Smaller score indicates better perceptual quality. If it is equality the structure of the two images is same and as it increases good encryption is achieved. The results of simulation for Cameraman and Katrina images are shown in table 3, table 4, table 5, and table 6 respectively.

Table 3.Cameraman image encryption/ decryption simulation results

	Original	Encrypted	Decrypted
PSNR	------	7.2463	Inf
MSE	------	1.2259×10^4	0
MAXERR	------	203	0
L2RAT	------	1.1642	1
SSIM	------	0.0822	1
NIQE	4.0994	11.2612	4.0994

Table 4.Katrina image encryption/ decryption simulation results

	Original	Encrypted	Decrypted
PSNR	------	5.4796	Inf
MSE	------	1.8413×10^4	0
MAXERR	------	203	0
L2RAT	------	1.3078	1
SSIM	------	-0.019843	1
NIQE	4.1943	9.1278	4.1943

Table 5.Aeroblk image encryption/ decryption simulation results

	Original	**Encrypted**	**Decrypted**
PSNR	------	3.8151e+00	Inf
MSE	------	2.7013×10^4	0
MAXERR	------	203	0
L2RAT	------	1.0649	1
SSIM	------	-0.016606	1
NIQE	3.2418	6.8626	3.2418

Table 6.Honeybadger image encryption/ decryption simulation results

	Original	**Encrypted**	**Decrypted**
PSNR	------	4.9242	Inf
MSE	------	2.0925×10^4	0
MAXERR	------	203	0
L2RAT	------	0.665546	1
SSIM	------	-0.028315	1
NIQE	4.2938	14.576	4.2938

As shown in Tables 3, 4, 5, and 6 the proposed algorithm fulfils the required conditions for a powerful algorithm. The obtained results show the powerful performance of the proposed. As shown, MSE is very high refers to a huge difference between the original data set and the encrypted one. PNSR is very low means a powerful encryption algorithm. MAXERR is extremely different between the original images and the encrypted ones points to approximately total difference between them and its value is zero compared to the decrypted images means that there is no difference between them which shows the power of the algorithm. Finally, L2RAT, SSIM, and NIQE focus how the encrypted images are different to the original ones and how the decrypted copies are the same to the original data sets.

4.1.2 Histogram analysis. Histogram is a type of bar plot for numeric data that group the data into bins. Figs.6, 7, 8, and 9 show the histogram of original images of Cameraman and Katrina and the corresponding histogram of the encrypted images and decrypted ones.

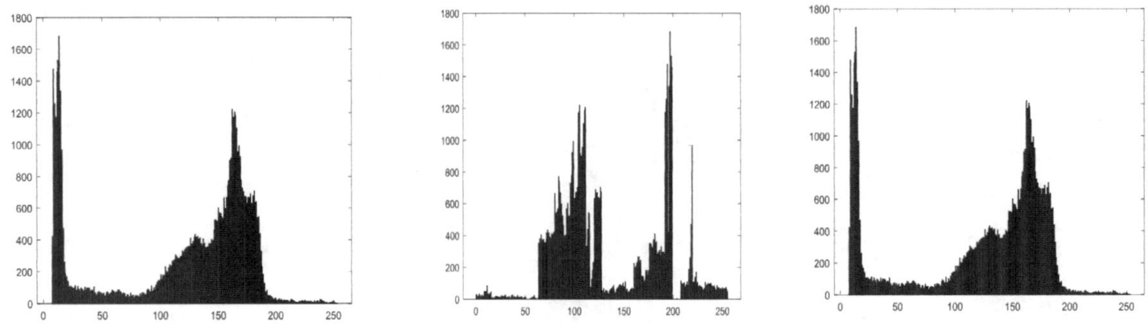

(a) Original image (b) Encrypted image (c) Decrypted image

Fig.6. the histogram analysis to "cameraman" dataset

In Fig. 6, the histogram of original cameraman image, encrypted image, and decrypted image are represented in Figs. 4a, 4b and 4c, respectively. As shown, the values of the numeric plots for the encrypted image are extremely different to the values of the original image because of the applying the powerful encryption process to the original image. On the other hand, the decrypted image is the same to original one without any small change. Similarly, the same results are found in Fig.7, Fig.8, and Fig.9 for "Katrina", "Aeroblk", and"Honey badger" datasets. Obviously, the histogram of the encrypted image is totally different to the original one, which proves the effectiveness and the powerfulness of the proposed algorithm with QKD.

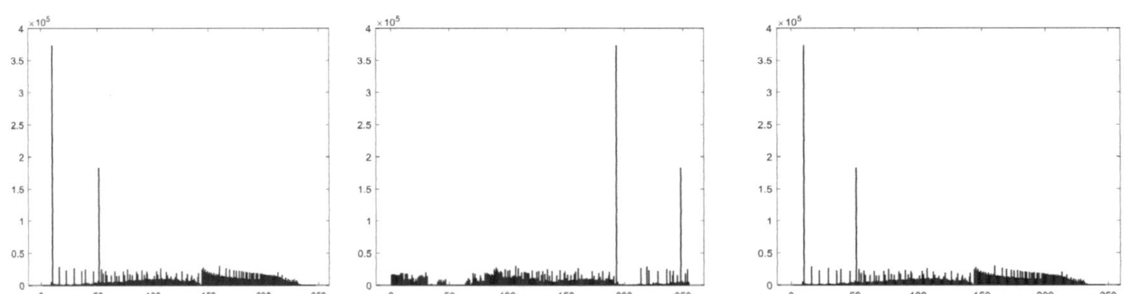

(a) Original image (b) Encrypted image (c) Decrypted image

Fig.7. the histogram analysis to "Katrina" dataset

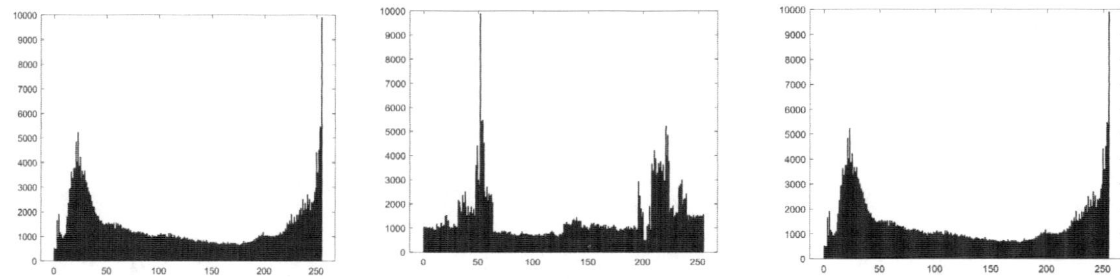

(a) Original image (b) Encrypted image (c) Decrypted image

Fig.8. the histogram analysis to "Aeroblk" dataset

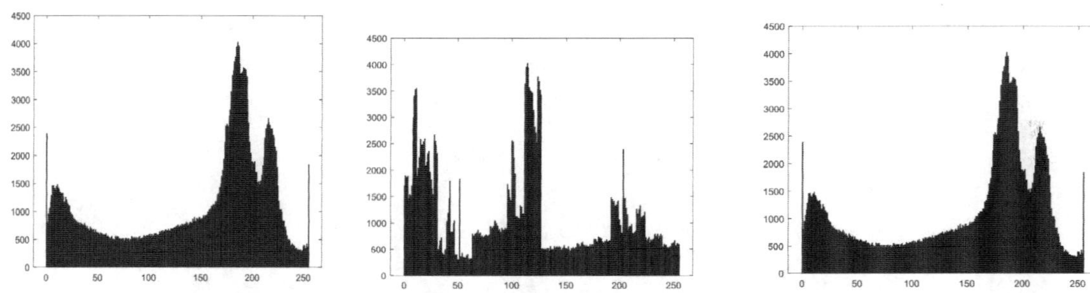

(a) Original image (b) Encrypted image (c) Decrypted image

Fig.9. the histogram analysis to "Honey badger" dataset

4.1.3 Correlation of adjacent pixels. The correlation between any pairs of two adjacent pixels is given by:

$$\text{coef} = \frac{\sum_{i=1}^{M}\left(k_i-\frac{1}{M}\sum_{j=1}^{M}k_j\right)\left(l_i-\frac{1}{M}\sum_{j=1}^{M}l_j\right)}{\sqrt{\sum_{i=1}^{M}\left(k_i-\frac{1}{M}\sum_{j=1}^{M}k_j\right)^2\sum_{i=1}^{M}\left(l_i-\frac{1}{M}\sum_{j=1}^{M}l_j\right)^2}} \quad (7)$$

Where M is the total number of adjacent pixels pairs in each direction and k_i, l_i are the values of adjacent pixels. Adjacent pixels in encrypted image should be de-correlated for powerful encryption algorithms. Table 7 shows the correlation of adjacent pixels for Cameraman, Katrina, Aeroblk, and Honey badger images.

Table7. Results of adjacent pixels of Cameraman, Katrina, Aeroblk, and Honey badger images

Image		**Direction**		
		Diagonal	Vertical	Horizontal
Cameraman	Original	0.908	0.959	0.939
	Encrypted	0.682	0.754	0.749
Katrina	Original	0.894	0.915	0.919
	Encrypted	0.771	0.798	0.803
Aeroblk	Original	0.968	0.98	0.982
	Encrypted	0.899	0.921	0.925
Honey badger	Original	0.911	0.919	0.944
	Encrypted	0.772	0.789	0.817

4.1.4 *Number of pixel changing rate and unified average changing intensity.* There are two quantities used to judge the power of any encryption algorithm. The first one is the number of pixel change rate (NPCR) and the second measure is unified average changing intensity (UACI). Suppose $U(i,j)$and $V(i,j)$be the (i,j)th pixel of two images U and V, respectively, Then NPCR and UACI can be expressed by using

$$\mathrm{NPCR} = \frac{1}{\mathrm{M}} \times \sum_{\mathrm{i:J}} \mathrm{D(i,j)} \times 100\% \tag{8}$$

$$\mathrm{UACI} = \frac{1}{\mathrm{M}} \left(\sum_{\mathrm{i:j}} \frac{|\mathrm{U(i,j)} - \mathrm{V(i,j)}|}{2^{\mathrm{N}} - 1} \right) \times 100\% \tag{9}$$

Where M is the total number of image pixels' and N is the number of bits required for pixels representation and $D(i,j)$is given by:

$$\mathrm{D(i,j)} = \begin{cases} 0 \text{ if } \mathrm{U(i,j)} = \mathrm{V(i,j)} \\ 1 \text{ if } \mathrm{U(i,j)} \neq \mathrm{V(i,j)} \end{cases} \tag{10}$$

Table 8 shows the simulation NPCR and UACI results.

Table8. NPCR and UACI simulation results

Image	NPCR%	UACI%
Cameraman	99.61	33.4
Katrina	99.61	33.4
Aeroblk	99.61	33.4
Honey badger	99.61	33.4

It is preferred for powerful encryption algorithm that NPCR to be about 99% and UACI to be about 33%. As shown in Table 6, the proposed algorithm fulfils the required conditions for a powerful algorithm.

4.1.5 *Entropy analysis.* Entropy is a scalar value representing the entropy of grey scale image I. Entropy is a statistical measure of randomness that can be used to characterize the texture of the input image. In other words, reflects the insecurity of image data, was denoted by E (m) of matrix m, and evaluated by:

$$\mathrm{E(m)} = -\sum_{\mathrm{i=1}}^{2^{\mathrm{L}}-1} \mathrm{p(u_i \log_2(p(u_i))} \tag{11}$$

Where $p(u_i)$ is the probability u_i. There are 2^8 probabilities for a gray scale image. The ideal entropy value is equal to 8-bit. So that the entropy value of the encrypted image should be close to 8 for efficient encryption algorithm. Table 9 shows the entropy of the proposed algorithm.

Table9. Entropy analysis

Image	Original	Encrypted
Cameraman	7.009716	7.016
Katrina	7.009716	7.016
Aeroblk	7.669	7.7273
Honey badger	7.669	7.7273

As shown in Table 9, the entropy is close to 8 which means that the proposed algorithm fulfils the required conditions for a powerful algorithm. The simulation operated with assumption of using AWGN channel. This looks like satellite communication link. So this new algorithm could be used as a secure communication link for image transmission and reception. The encrypted data is sent via Read Solomon Coded Channel (RSCC) with 255 as the total number of code symbols and 223 as the number of data symbols being encoded. This channel is equivalent to the satellite communication channel to discuss the channel performance and bit error rate BER after add the encrypted data by means of the proposed algorithm [17].

Table10. Comparison between proposed algorithm, [14], and [15]

factor	[15]	[14]	proposed
CF(av.) "image"	0.972	0.944	0.933
CF(av.)"enc. image"	0.00336	0.0089	0.728
NPCR	99.6	99.6	99.6
UACI	33.4	36.0075	33.4
Entropy	7.99	6.42	7.009716
SSIM	--------	--------	0.0822
NIOE	--------	--------	11.2612
MSE	--------	--------	$1.23x10^4$
PSNR	--------	--------	7.2463
MAXERR	--------	--------	203
L2RAT	--------	--------	1.1642

In this proposed algorithm most of the drawbacks of SQKD are eliminated. From another point of view, comparing the simulation results of the proposed algorithm to algorithms in [14] and [15] a kind of enhancement is achieved like simplicity as there is no middle scrambling or encoding stage beside that the proposed algorithm fulfills most of the required conditions for a powerful algorithm compared with [14] and [15] despite of being simpler as represented in Table 10. In Fig.10 the comparison between the proposed algorithm and both algorithms in [14], [15] results. As represented in the figure the results are approximately equal however the proposed algorithm is simpler and quietly powered by means of QKD.

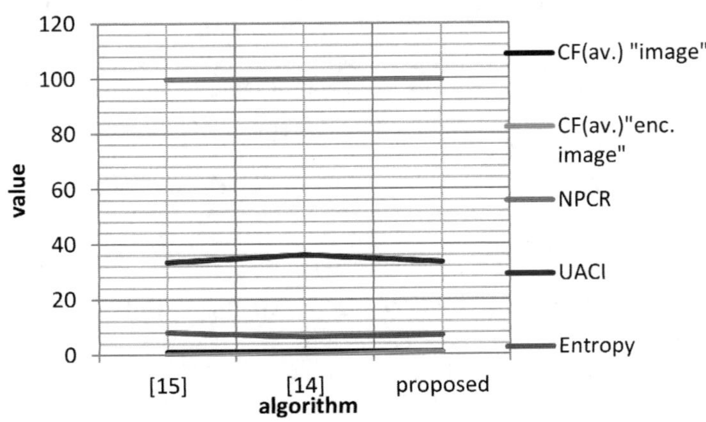

Fig.10 Comparison between proposed algorithms and [14] and [15]

5. Conclusion

In this paper a new image encryption algorithm is proposed. The proposed algorithm is based on quantum mechanics fundamentals that make it more secure as any hacking or eavesdropping try will be discovered at once. Also the generated secret key is totally random makes it very difficult to be cracked. It can be applied for any type of type of data that considered being confidential. Beside that it is a simple algorithm and does not require many complex processes and mutual authentication.

Acknowledgements

We would like to express our sincere appreciation to the reviewers for the valuable comments that have significantly enhanced the quality of this paper. The paper was introduced in the

19th International Conference on Aerospace Science and Aviation Technology ASAT-19, Military Technical Collage, Cairo, Egypt.

References

[1] Bennett, C.H. and Brassard, G., 1987. Quantum public key distribution reinvented. ACM SIGACT News, 18(4), pp.51-53.Wootters,
[2] Wootters, W.K. and Zurek, W.H., 1982. A single quantum - cannot be cloned. Nature, 299(5886), pp.802-803.
[3] D'Ariano, G.M. and Yuen, H.P., 1996. Impossibility of measuring the wave function of a single quantum system. Physical review letters, 76(16), p.2832.
[4] Wiesner, S., 1983. Conjugate coding. ACM Sigact News, 15(1), pp.78-88.
[5] Boyer, M., Kenigsberg, D. and Mor, T., 2007, January. Quantum key distribution with classical Bob. In 2007 First International Conference on Quantum, Nano, and Micro Technologies (ICQNM'07) (pp. 10-10). IEEE.
[6] Boyer, M., Gelles, R., Kenigsberg, D. and Mor, T., 2009. Semiquantum key distribution. Physical Review A, 79(3), p.032341.
[7] Zou, X., Qiu, D., Li, L., Wu, L. and Li, L., 2009. Semiquantum-key distribution using less than four quantum states. Physical Review A, 79(5), p.052312.
[8] Jian, W., Sheng, Z., Quan, Z. and Chao-Jing, T., 2011. Semiquantum key distribution using entangled states. Chinese Physics Letters, 28(10), p.100301.
[9] Zou, X., Qiu, D., Zhang, S. and Mateus, P., 2015. Semiquantum key distribution without invoking the classical party's measurement capability. Quantum Information Processing, 14(8), pp.2981-2996.
[10] Lu, H. and Cai, Q.Y., 2008. Quantum key distribution with classical Alice. International Journal of Quantum Information, 6(06), pp.1195-1202.
[11] Li, Q., Chan, W.H. and Zhang, S., 2016. Semiquantum key distribution with secure delegated quantum computation. Scientific reports, 6(1), pp.1-6.
[12] Renuka, D. and Reddy, P.C., 2018. Integrated Classical and Quantum Cryptography Scheme Using Three Party Authenticated Key Distribution Protocols. Materials Today: Proceedings, 5(1), pp.1017-1023.
[13] Armanuzzaman, M., Alam, K.M.R., Hassan, M.M. and Morimoto, Y., 2017, December. A secure and efficient data transmission technique using quantum key distribution. In 2017 4th International Conference on Networking, Systems and Security (NSysS) (pp. 1-5). IEEE.
[14] Abd El-Latif, A.A., Abd-El-Atty, B. and Talha, M., 2017. Robust encryption of quantum medical images. IEEE Access, 6, pp.1073-1081.
[15] Mohamed, H.G., ElKamchouchi, D.H. and Moussa, K.H., 2020. A novel color image encryption algorithm based on hyperchaotic maps and mitochondrial DNA sequences. Entropy, 22(2), p.158.
[16] Abd-El-Atty, B., Venegas-Andraca, S.E. and Abd El-Latif, A.A., 2018. Quantum information protocols for cryptography. In Quantum Computing: An Environment for Intelligent Large Scale Real Application (pp. 3-23). Springer, Cham.
[17] Samy, R. and Mahran, A., 2020. Low Complexity Viterbi Decoder for RSCC Concatenated Codes. In Journal of Physics: Conference Series (Vol. 1447, No. 1, p. 012040). IOP Publishing.

SAR image formation enhancement using effective velocity estimation method

M A Fouad[1], A Azouz[1], A A Mashaly[1] and A E Abdalla[1]

[1]Electrical Engineering, Military Technical College, Cairo, Egypt.

E-mail: mfouad10282@gmail.com

Abstract: Synthetic Aperture Radar (SAR) presents powerful tools for grounding mapping and remote sensing applications. Effective velocity is a vital index that controls the quality of SAR image formation. An accurate calculation of effective velocity provides a particular value of azimuth frequency modulated (FM) rate. The resultant azimuth FM rate could be used to produce a focused SAR image with sharp details. In this paper, SAR image formation enhancement is proposed using two guided methods based on precise effective velocity estimation. Firstly, effective velocity is estimated based on Sentinel-1data parameters extracted from selected image raw data. Secondly, an iteration method applies output image contrast, sharpness and entropy measurements to estimate the optimum value of the effective velocity based on the initial velocities calculated in the first method. Results are compared with extracted SAR images ignoring the effect of the effective velocity, to identify the performance of the proposed methods.

1. Introduction

The demand of high resolution SAR image in military and civil applications is considered as one of the major requirements in recent days [1]. There are many algorithms for SAR image formation like range Doppler algorithm, chirp scaling algorithm, omega-K algorithm, and back projection algorithm [2, 3]. These algorithms are continuously modified and upgraded to improve SAR image formation process and image quality, which differ due to on-hand application [2-5].

The range Doppler algorithm (RDA) was the first image formation algorithm used for SAR processing. Nevertheless, it is still the most known used algorithm due to its efficiency, simplicity and accuracy[6]. In other hand, RDA uses time domain interpolation, which needs intensive computations to overcome range cell migration (RCM) problem [2, 4, 5]. The Chirp Scaling Algorithm (CSA) is developed mainly to overcome the shortages of range Doppler algorithm (RDA). However, RDA is the principal algorithm for several SAR image formation applications. While, CSA has several advantages compared with RDA and the other used algorithms. These advantages of CSA could be summarized as computation efficiency and easy to model [3, 4, 7-9].

The Omega-k algorithms (WKA) shares RDA the same weakness point of intensive computation problem due to interpolation processing. Moreover, its assumption that SAR effective velocity is range independent, which reflects on its ability to work with large swaths.

Now, we are going to introduce in details the basic schematic block diagram of CSA as shown in figure 1 [4, 10]. The main concept of CSA is the ability to implement Range Cell Migration Correction (RCMC) through phase multipliers [4]. Firstly, azimuth FFT is done to transform the two-dimensional time domain raw data from range time domain to range Doppler domain. Then, the first phase

Content from this work may be used under the terms of the Creative Commons Attribution 3.0 licence. Any further distribution of this work must maintain attribution to the author(s) and the title of the work, journal citation and DOI.
Published under licence by IOP Publishing Ltd

multiplication is done to cancel the range migration for all received echoes. Next, a range FFT is done to convert the signal into two-dimensional frequency domains. Then, a second phase multiplication is performed to apply range compression, and RCMC. The next step is to perform range IFFT to convert the signal back to range Doppler domain. Then, third phase multiplication is performed to apply azimuth compression. Lastly, an azimuth IFFT is performed to transform the compressed data again to the two dimensional time domain, which presents SAR focused image [4, 8, 10].

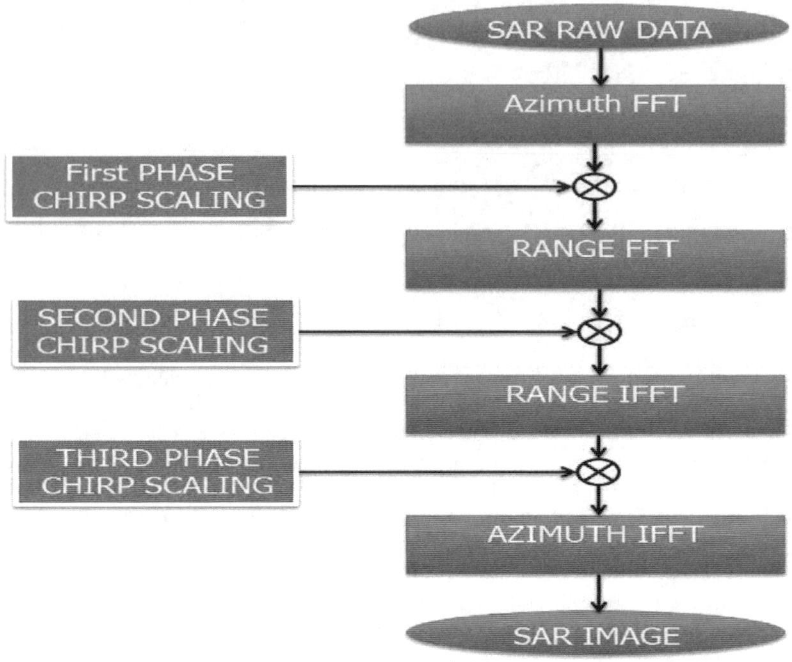

Figure 1. Basic block diagram of CSA

SAR image quality improvement is the fundamental objective for several SAR system designers [1, 3]. Conceptually, SAR image quality is guided by the precise calculation of the effective velocity [4, 11-13]. Theoretically, there are two system velocities [4]. One is the satellite/platform (radial) velocity, which presents the platform velocity along specific flight path. While, the other one is ground (beam footprint) velocity, which presents the sweep velocity of zero Doppler line along the ground as shown in figure 2 [4, 12, 13].

Based on radar slant range equation, the distance from SAR sensor to a target is calculated according to equation (1) [4, 12, 13]:

$$R(\eta) = \sqrt{R_o^2 + V_r^2\eta^2} \quad (1)$$

Where, R_o is the range of closest approach, η is the azimuth time (slow time), and V_r is the SAR effective velocity.

For spaceborne platforms, there is a difference between the radial and ground velocities, which equals approximately 12% [4]. However, these two velocities are almost the same for airborne platforms. Numerically, the velocity difference source for spaceborne case of study is the radius difference. Where, the satellite orbit is larger than the ground radius. Slant range equation supposes that the satellite moves in a straight line. Practically, this assumption is not true. In fact, the satellite rotates in a curvature orbit with a relative velocity, which is called as SAR effective velocity [4, 14].

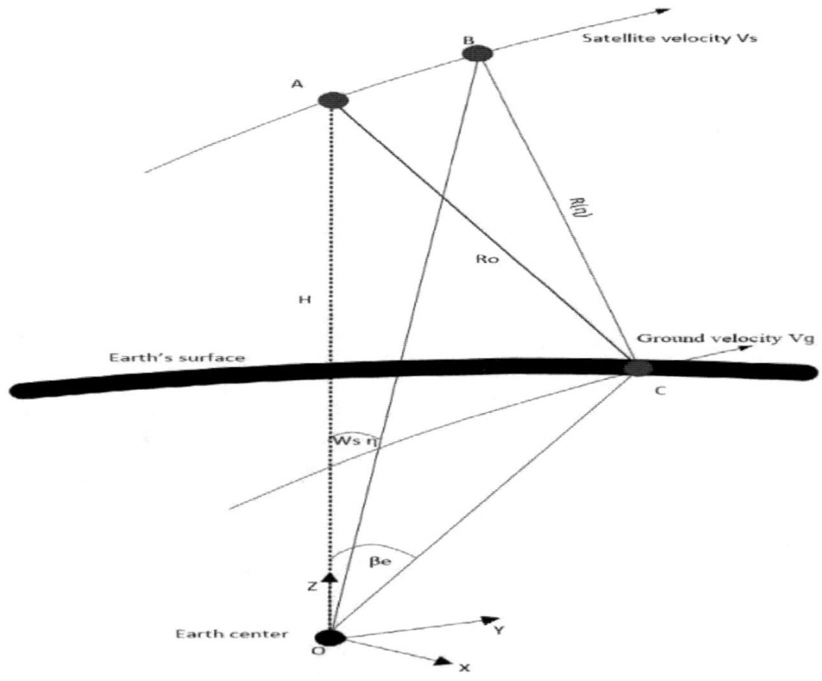

Figure 2. SAR Geometry for satellite velocity and beam footprint velocity presentation

The precise calculation of effective velocity is considered a critical parameter in producing a sharp focused image. Nevertheless, it is an effective parameter in calculating the azimuth FM rate, one of the powerful parameters that effects directly in image quality enhancement. Azimuth FM rate (K_a) could be estimated as follows:

$$K_a = \frac{2V_r^2 \cos^2(\theta_r)}{\lambda\, R(\eta)} \tag{2}$$

Where, θ_r is the squint angle, and λ is the antenna wavelength.
Based on the above equation of azimuth FM rate, it is obvious that, any error in estimating the effective velocity causes an error in azimuth FM rate. Moreover, it causes a matched filter phase error, which finally produces a low quality SAR image.
In this work, two methods are introduced to enhance SAR image quality based on the precise estimation of SAR effective velocity. In first method, the effective velocity is calculated numerically based on level-0 raw data[13]. While in the second method, an iteration algorithm is applied to precisely estimate the optimum effective velocity for optimal image enhancement. This method reflects a higher image quality calculated by its contrast, sharpness, and entropy [15].

The rest of this paper is organized as follows: in section 2, a numerical calculation for effective velocity estimation. Section 3 shows the iteration method used for the optimum effective velocity estimation. Finally, conclusion comes in section 4.

2. Numerical calculation of the effective velocity

Effective velocity (V_r) is numerically calculated based on mathematical derivation[13]. Firstly, satellite velocity (V_s) can be calculated by equation (3) guided by the extracted parameters after raw data treatment where, V_x, V_y, and V_z are the satellite velocities in the XYZ directions.

$$V_s = \sqrt{V_x^2 + V_y^2 + V_z^2} \quad (3)$$

Then, using the following equations in, the value of effective velocity could be numerically calculated.

$$H = R_e + h \quad (4)$$

$$W_s = \frac{V_s}{H} \quad (5)$$

$$\cos\beta_e = \frac{{R_e}^2 + H^2 - R_o^2}{2\, R_e\, H} \quad (6)$$

$$V_g = R_e * W_s * \cos\beta_e \quad (7)$$

$$V_r = \sqrt{V_g * V_s} \quad (8)$$

Where, V_g is the ground velocity, H is the local orbit radius, W_s is the radial velocity, R_e is the local earth radius, h is Sentinel-1 orbit altitude, β_e is the angle of center earth, and R_o is the range of the closest approach, which could be obtained from the extracted raw data. Firstly , calculate the satellite velocity after Sentinel-1A raw data level-0 treatment , secondly, calculate the ground velocity by applying for example, Sentinel-1A parameters listed in table 1 in equations(4-7), finally, the effective velocity is calculated from equation (8). The calculated values of the three velocities are listed in table 2.

Table 1. Sentinel-1 parameters [16]

Parameter	Value	Unit
Local earth radius R_e	6378	km
Angle of centre earth β_e	0.0435	Rad
Sentinel-1 orbit altitude h	693	km
Local orbit radius H	7071	km

Table 2. Calculated velocities using numerical method

Parameter	Value	Unit
Satellite velocity V_s	7589	m/s
Ground velocity V_g	6839	m/s
Effective velocity V_r	7200	m/s

3. Iteration method for optimum effective velocity estimation based on measuring the image contrast, entropy and, sharpness.

Conceptually, the satellite has a fixed altitude, and moves in a defined orbit with, approximately constant speed. From this point of view, the effective velocity could be also used as constant values during the whole flight path without degrading the image quality. The main advantageous of this concept is to reduce the processing time, which is a big issue in SAR systems but with a sensible reduction in SAR image quality [2].

To overcome on this problem, we need to find criteria that calculate the effective velocity precisely. It was found from previous research works that contrast, sharpness and entropy are widely used for image quality measurements. The iteration calculation of contrast, sharpness, and entropy as given in figure 3 are obtained based on results of the previous section.

As a result, the image quality is improved as the values of the contrast and sharpness increase[17, 18]. In the contrary, the image quality increases by minimizing the calculated entropy value [19]. Finally, the best output results from the three metrics lead to the precise value of the effective velocity at which, the optimal image resolution is obtained. The definitions of the three metrics are presented as follows: The ratio of the standard deviation to the mean of the image intensity named as image contrast [15]:

$$\mathrm{Co} = \frac{\sqrt{\mathrm{E}\{[\mathrm{I}^2(\mathrm{m,n}) - \mathrm{E}\{\mathrm{I}^2(\mathrm{m,n})\}]^2\}}}{\mathrm{E}\{\mathrm{I}^2(\mathrm{m,n})\}} \tag{9}$$

where, $\mathrm{I}(\mathrm{m,n})$ is the pixel intensity m=1,..M, n=1,..N, and E [•] is the mean value of the arithmetic mean of the samples and is defined by:

$$\mathrm{E}\{\mathrm{I}^2(\mathrm{m,n})\} = \frac{1}{MN}\sum_{m=1}^{M}\sum_{n=1}^{N} I^2(m,n) \tag{10}$$

The image sharpness could be obtained from convolving an image with Sobel operator, and sums the square of the gradient vector components as follows [15]:

$$\mathrm{SH} = \sum_{m} Height \sum_{n} Width\ (S_x(m,n)^2 + S_y(m,n)^2) \tag{11}$$

Where, $\mathrm{S_x(m,n)}$ and $\mathrm{S_y(m,n)}$ are the resultant image from convolution with Sobel summation.

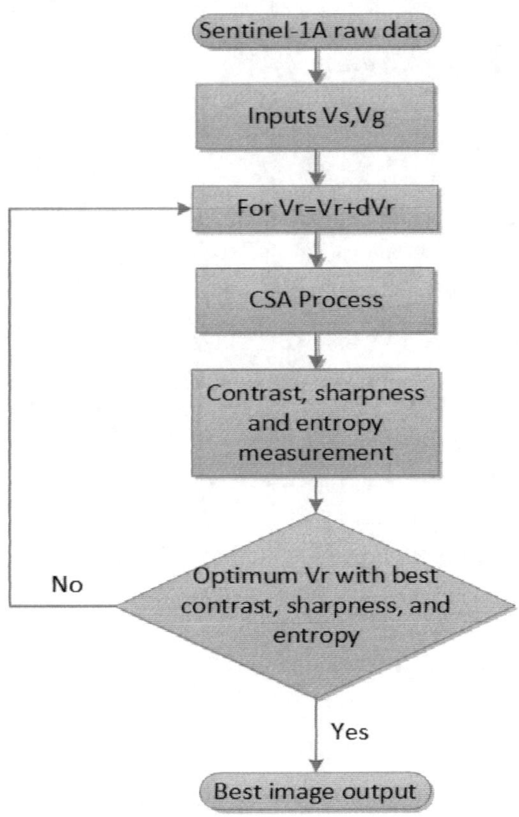

Figure 3. Effective velocity estimation flowchart based on iteration calculation of contrast, sharpness, and entropy

Finally, the image entropy could be expressed as follows [15]:

$$\text{Ent} = -\sum_{m}\sum_{n}\left(\frac{I^2(m,n)}{S}\right)\ln\left(\frac{I^2(m,n)}{S}\right) \tag{12}$$

Where,

$$S = -\sum_{m}\sum_{n} I^2(m,n) \tag{13}$$

The simulator parameters used in the modelled CSA are listed in table 3 [20] , and Sentinel-1A real raw data for a selected area shown in figure 4 are applied as an input raw data. The iteration process evaluates the values of contrast, sharpness, and entropy. This process continues for all the values of effective velocity between ground and satellite velocities calculated in previous section. The optimal obtained image resolution based on the metrics calculations leads to the precise effective velocity.

The results from numerical and iteration methods are compared with that obtained by ignoring effective velocity calculation is shown in Table 4. It is obvious that, the best-obtained result, which gives the better image quality, is obtained from using the iteration method.

Figure 4. Sentinel-1 preprocessed image raw data used in the proposed CSA

Table 3. Sentinel-1A SAR simulator parameters

Parameter	Value	Unit
Carrier frequency	5.504	GHz
PRF Range	1000-3000	Hz
RF peak power	4.141	Kw
Maximum Range Bandwidth	100	MHz
Antenna length	12.3	M

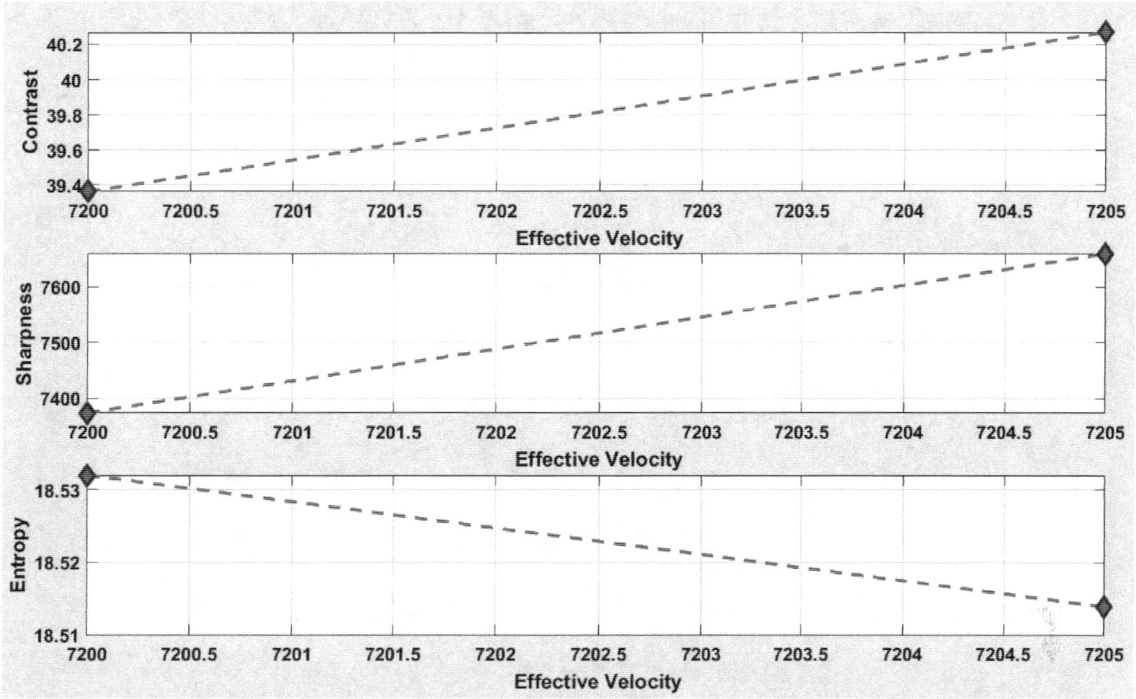

Figure 5. Iteration results (V_r= 7205 m/s) versus numerical results (V_r=7200 m/s) for image (a) contrast, (b) sharpness, and (c) entropy.

Table 4. Calculation of contrast, entropy and sharpness for the obtained values of effective velocity compared with ignoring effective velocity.

Metric parameter	Without effective velocity estimation $V_r = V_s$	With numerical calculation $V_r = 7200\ m/s$	With optimum iteration output $V_r = 7205$ m/s
Contrast	10.53	39.36	40.27
Sharpness	346	7374	7660
Entropy	18.96	18.53	18.51

Finally, the output images from applying the Sentinel-1A raw data in the modelled CSA at the three values V_r in table 4 is shown in figure 6. It is clear that, there is an improvement in the image quality. Where, image (a) is produced with ignoring effective velocity ($V_r = V_S$), image (b) is produced through using the numerical method result (V_r= 7200 m/s), while, image (c) is produced through using the iteration method result (V_r= 7205 m/s).

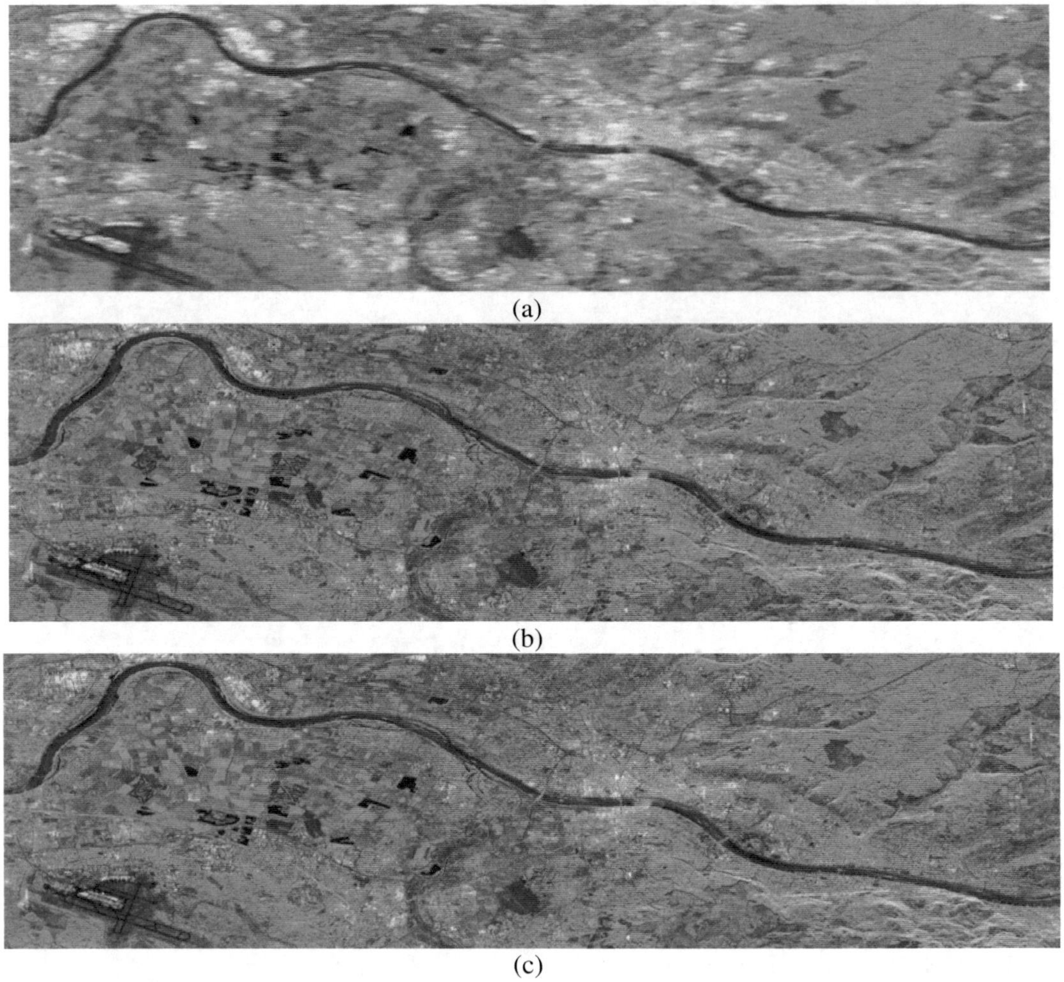

Figure 6. The output image from modelled CSA, (a) without effective velocity estimation, (b) with effective velocity enhancement using numerical calculation (V_r = 7200), and (c) with effective velocity enhancement using the proposed iteration method (V_r=7205 m/s).

4. Conclusion

The work in this paper investigates the effectiveness of the accurate calculation of effective velocity on SAR image formation enhancement. An Iteration method for optimum effective velocity estimation based on measuring the image contrast, entropy and, sharpness is developed. Simulation of the proposed algorithm is applied on Sentinel-1A real raw data of a selected area. The study explores that the accurate calculation of effective velocity provides a particular value of azimuth frequency modulated (FM) rate which lead to produce a focused SAR image with sharp details. An additional study is needed on the optimization to restrain the high computational cost.

References

[1] Sastri P, Pal T and Prasad B 2020 pp 19-28
[2] Kim C-K, Lee J-S, Chae J-S and Park S-O 2019 A Modified Stripmap SAR Processing for Vector Velocity Compensation Using the Cross-Correlation Estimation Method *Journal of Electromagnetic Engineering and Science* **19** 159-65
[3] Yang, R., Li, H., Li, S., Zhang, P., Tan, L., Gao, X. and Kang, X., 2018. *High-Resolution*

Microwave Imaging. Springer Singapore
[4] Cumming I G and Wong F H 2005 Digital Processing of Synthetic Aperture Radar Data: Algorithms and Implementation Artech House Print on Demand, p 660
[5] Kulpa J, Malanowski M, Gromek D, Samczynski P, Kulpa K and Gromek A 2013 Experimental Results of High-Resolution ISAR Imaging of Ground-Moving Vehicles with a Stationary FMCW Radar International *Journal of Electronics and Telecommunications* 59
[6] Ran L, Liu Z, Xie R and Zhang L 2019 Focusing High-Squint Synthetic Aperture Radar Data Based on Factorized Back-Projection and Precise Spectrum Fusion Remote Sensing 11
[7] Sastri P P, Pal T K, Prasad B S V and Kumar T K 2018 Analysis of SAR Imagery with different Data Processing Algorithms and Comparison thereof. In: 2018 *IEEE MTT-S International Microwave and RF Conference (IMaRC)*, pp 1-4
[8] Yuan Y, Chen S, Zhang S and Zhao H 2017 A Chirp Scaling Algorithm for Forward-Looking Linear-Array SAR With Constant Acceleration *IEEE Geoscience and Remote Sensing Letters PP 1-4*
[9] Zeng, T., Wang, Z., Liu, F. and Wang, C., 2020. An Improved Frequency-Domain Image Formation Algorithm for Mini-UAV-Based Forward-Looking Spotlight BiSAR Systems. *Remote Sensing*, ***12*(17)**, p.2680.
[10] Denham M, Areta J and Tinetti F G 2016 Synthetic aperture radar signal processing in parallel using GPGPU The Journal of Supercomputing **72** 451-67
[11] Tao Z, Wei Y, Liangbo Z and Zegang D 2013 An improved velocity calculation model in GEO SAR system. In: *IET International Radar Conference 2013,* pp 1-5
[12] Wong F H, Ngee Leng T and Tat Soon Y 2000 Effective velocity estimation for space-borne SAR. In: *IGARSS 2000. IEEE 2000 International Geoscience and Remote Sensing Symposium. Taking the Pulse of the Planet: The Role of Remote Sensing in Managing the Environment. Proceedings (Cat. No.00CH37120),* pp 90-2 vol.1
[13] Xiaojin S, Yunhua Z and Wenshuai Z 2007 Effective velocity estimation for space-borne SAR based on chirp scaling algorithm. In: *2007 1st Asian and Pacific Conference on Synthetic Aperture Radar,* pp 427-30
[14] Satellites C L 07 June 2019 Sentinel-1 Level 1 Detailed Algorithm Definition. ed T E S AGENCY: Sentinel-1 Document Library) p 158
[15] Azouz A A E and Li Z 2015 Improved phase gradient autofocus algorithm based on segments of variable lengths and minimum-entropy phase correction. In: *IET Radar, Sonar & Navigation*: Institution of Engineering and Technology) pp 467-79
[16] Agency E S 2013 ESA'S RADAR observation mission for copernicus operational services. ed E S Agency
[17] Fienup J and Miller J 2003 Aberration correction by maximizing generalized sharpness metrics *Journal of the Optical Society of America. A, Optics, image science, and vision* **20** 609-20
[18] Martorella M, Berizzi F and Bruscoli S 2006 Use of Genetic Algorithms for Contrast and Entropy Optimization in ISAR Autofocusing *EURASIP J Appl Signal Process* **2006**
[19] Zhang L, Sheng J-l, Duan J, Xing M-d, Qiao Z-j and Bao Z 2013 Translational motion compensation for ISAR imaging under low SNR by minimum entropy *EURASIP Journal on Advances in Signal Processing* **2013** 33
[20] Satellites C L 2016 Sentinel-1 Product Definition. ed T E S AGENCY: Sentinel-1 Document Library) p 129

Frequency analysis of SAR system design for small satellite

I Douidar[1], M Safy[2] and A Saleh[3]

[1] Egyptian Armed Forces, Cairo, Egypt.
[2] Egyptian Academy for Engineering and Advanced Technology affiliated to ministry of Military Production, Cairo, Egypt.
[3] October 6 University, Cairo, Egypt.

E-mail: Ibrahim.douidar@gmail.com

Abstract. Great interest is given in recent years to the small satellites as their main advantages are the low development cost and the reduced deployment time, which give attention to develop payloads that can be mounted on the platform of these satellite. the platform of small satellites imposes constraints on the sensor design to fit in the size, mass and power limitations, which leads to necessary trade-offs between the design parameters of the sensor. The selection of the operating frequency is dependent mainly on the mission requirements as each frequency band gives different information about the scene and has different penetration capabilities. However, the SAR sensors used on small satellites come with limitations, so other parameters should be considered during the design. After going through the SAR design process, a frequency analysis is presented at L, C and X bands, which is used to evaluate the effect on the size of the antenna (length and width) and the average power consumption that suit the constraints of small satellite.

1. Introduction

SAR sensor design process for small satellites comes with limitation in compactness, power consumption and mass to fit within these platforms, which affects the design parameters selection of the sensor system specially the operating frequency. However, the operating frequency selection is dependent mainly on the mission requirements like the satellite application, the required resolution and the technological aspects [1], [2].

A design analysis between three operating frequencies at L, C and X band and their effect on the other satellite system parameters like dimensions and mass is discussed. Higher frequencies – higher than X-band- has been excluded from this study because of the technological issues that come with them, also the lower frequencies -lower than L-band- have large dimensions for the antenna which make them out of small satellite scope.

From mission perspective the L-band, suits the missions that require a higher degree of penetration as it has longer wavelength [3,4]. on the other hand, for higher resolution the X-band is preferred due to wide available band widths and the technology maturity of the components, while for a compromise between the two bands the C-band provides a rational performance between the surface penetration and resolution [5], [6].

SAR system Design parameters equations including the minimum antenna area required, PRF selection, the ambiguity and the average transmitted power is presented is section 2. The design methodology used for evaluating the design parameters is given is section 3. The design methodology

Content from this work may be used under the terms of the Creative Commons Attribution 3.0 licence. Any further distribution of this work must maintain attribution to the author(s) and the title of the work, journal citation and DOI.
Published under licence by IOP Publishing Ltd

validation is in section 4. The frequency analysis is given in section 5. The results of the different frequencies on the design parameters are provided in section 6. The last section is the conclusion.

2. SAR parameters

2.1. Minimum antenna area

A typical antenna of SAR is one antenna with high gain used for both transmission and reception. One of the design parameters of SAR system is the dimensions of the antenna, and it is constraint by the antenna minimum area constraint for usable design. It is for the treatment of a special SAR design case, which has the design goals of achieving the widest possible swath and the best possible resolution [7]. Thus, SAR system should every time transmit when it advances a $L_{az}/2$ distance, at least one pulse, therefore the PRF minimum is:

$$PRF_{min} = \frac{2.V_{st}}{L_{az}} \tag{1}$$

So, to avoided the Doppler spectrum of azimuth from being overlapped with each other, the antenna length Laz should achieve:

$$L_{az} > \frac{2.V_{st}}{PRF} \tag{2}$$

The maximum swath Wg_{max} is the projection on the ground of the unambiguous slant-range swath determined by the minimum PRF:

$$Wg_{max} = \frac{c}{2.PRF_{min}.\sin\theta_i} = \frac{c.L_{az}}{4.V_{st}.\sin\theta_i} \tag{3}$$

To avoid echoes in range from being overlapped, the antenna pattern is used to suppress them in elevation direction. So, the antenna footprint in elevation must be narrower than the maximum width of swath Wg_{max}.

$$\text{footprint}_{elevation} = \theta_e.R = \frac{\lambda.R}{W_a} < Wg_{max} \tag{4}$$

Therefore, the width of the antenna Wa should be:

$$W_a > \frac{2.\lambda.R.PRF.\tan\theta_i}{c} \tag{5}$$

Using the equations (2) and (5), the minimum area of antenna that fits the ambiguity limitation is:

$$A_{min} = \frac{4.\lambda.R.V_{st}.\tan\theta_i}{c} \tag{6}$$

But at the case of SAR system designing which doesn't require simultaneously the possible wider swath width and the possible best resolution, a smaller antenna can be selected to achieve these goals [8].

PRF for a spaceborne SAR is an important design parameter, as it influences most of the other system parameters. PRF value affects many system parameters like peak transmitter power, raw data rate and pulse duty factor. while, its availability is affected by many factors including SAR altitude, incidence angle, swath width, transmitter pulse length and antenna length [9].

2.1.1 PRF constraints. For acquiring a high azimuth resolution, SAR sends a pulse for every time SAR platform travels half of the length of the antenna in azimuth direction, to achieve the azimuth beam width Nyquist sampling. The following condition presents the lower bound of the PRF:

$$PRF_{min} = \frac{2.V_{st}}{L_{az}} \tag{7}$$

Thus, for improving azimuth resolution, the antenna length in the direction of motion is critical which require faster pulsing from the SAR. Which reduces the inter-pulse period that is used to collect data in the range direction. Unfortunately, for given SAR frequency if there is more than one pulse exist in the scan area simultaneously produces the range-ambiguity, i.e., the PRF maximum is defined by reception of sequential pulses.

The upper limit for the PRF is related to the swath width by the following condition:

$$PRF_{max} = \frac{c}{2.Wg} \tag{8}$$

Using the equations (7) and (8) the usable PRF range can be defined as:

$$\frac{2.V_{st}}{L_{az}} < PRF < \frac{c}{2.Wg} \tag{9}$$

2.1.2 *Eclipsing Avoidance (Blind ranges).* Unfortunately, the range of PRFs values between the lower and upper limits of PRFmin and PRFmax is not all available. The receiver becomes blind during transmission because of the inherent in SAR systems isolation problems. The signal received is said to be eclipsed, when the echo returned from the scene coincides with a transmit event. So, avoiding PRFs that results in eclipsing is necessary by achieve the inequality:

$$\frac{(N-1)}{(\tau_{near} - \tau_p)} < PRF < \frac{N}{(\tau_{far} + \tau_p)} \tag{10}$$

Where:

- N Whole numbers (1,2,3,..) representing pulses.
- τ_p Transmitted pulse width.
- τ_{near} Delay time between transmission and reception of the pulse to the near edge of the swath.
- τ_{far} Delay time between transmission and reception of the pulse to the far edge of the swath.

$$\tau_{near} = \frac{2.R_n}{c} \tag{11}$$

$$\tau_{far} = \frac{2.R_f}{c} \tag{12}$$

Where:

- R_n Slant range from the antenna to the near edge of the swath.
- R_f Slant range from the antenna to the far edge of the swath.

PRFs which causes eclipsing depends on the transmitted pulse-width and slant range from the antenna to the swath far and near edges [10].

2.1.3 Avoiding Nadir Returns. If PRF value is very large, the returns will be overlapped of Nadir (near range) and far range. These happens at a time τ_{nadir} from the transmit-pulse start:

$$\tau_{near} = \frac{2.H}{c} \tag{13}$$

Where:

- H SAR height over the nadir point

The echoes that come from near range are much stronger than that come from far range as the small incidence angles has large backscattering coefficient. The echo duration is equal to the width of the transmit pulse at least and will become longer for the terrain at near range [16].

One approach to overcome this limitation is to avoid PRFs that causes the echo from near range to reach the receiver with the echo of far range simultaneously. Like in the case of the far range echo transmitter-pulse eclipsing. The PRFs values which make the avoidance of collapsing of echoes from near range with that of the far range are expressed by the following inequality:

$$\frac{(M-1)}{(\tau_{near}-\tau_p-\tau_{nadir})} < PRF < \frac{M}{(\tau_{far}+\tau_p-\tau_{nadir})} \quad (14)$$

Where:

M Whole numbers (1,2,3,...) corresponding to pulses

During calculating the minimum antenna area, a zero-order ambiguity analysis is presented. For the rejection of out of swath target ambiguous radar echoes, the footprint on the ground of the antenna pattern must be larger than the swath. This footprint is given by the one-way 3 dB beam width.
Ambiguous targets which are located at the edges of the footprint would only have 6 dB suppression capability. The big dynamic-range of σ0, makes this insufficient for the suppression. The range-ambiguities resulting from side-lobes of the antenna elevation pattern are illustrated in Figure 1.
The azimuth direction antenna beam-width is used to determine the Doppler bandwidth. As, the minimum PRF which is used to sample the azimuth Doppler spectrum with at the least the Nyquist rate. the Nyquist rate of the 6 dB Doppler bandwidth, which allow antenna pattern side-lobes spectral components to fold back into the spectrum main area, resulting in azimuth ambiguities, which is illustrated in Figure 2 [11].

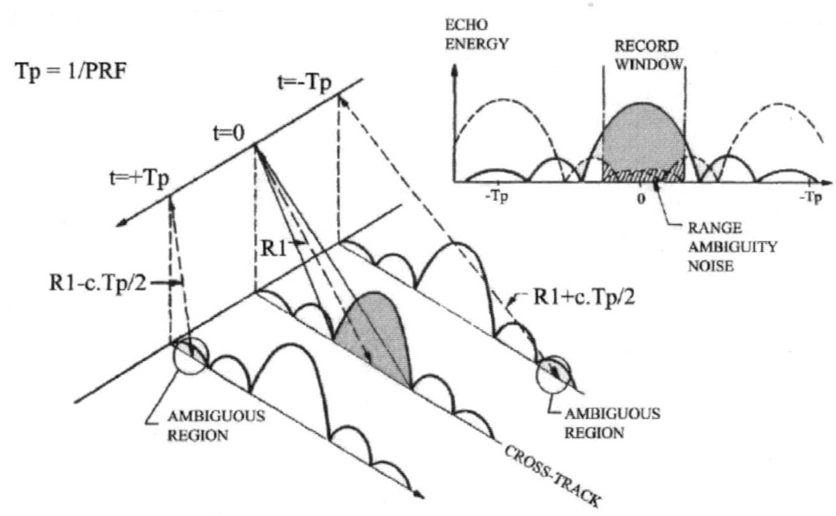

Figure 1. Illustration of SAR range ambiguities

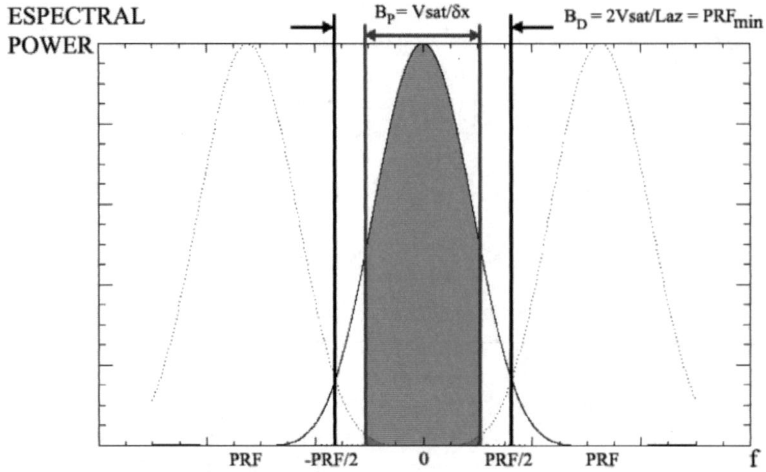

Figure 2. Doppler spectrum spectral folding for PRF=B_D

For a given antenna pattern in azimuth and range, the PRF value is chosen so the whole ambiguity noise relative to the signal is very small. On the other hand, for certain PRF value or group of values, the antenna dimensions must achieve the ASR specification.
The ambiguous signal power at certain Doppler frequency f0 and sometime delay τ0 is expressed as:

$$S_a(f_0,\tau_0) = \sum_{m,n\neq 0} G^2(f_0 + mPRF, \tau_0 + n/PRF).\sigma_0(f_0 + mPRF, \tau_0\, n/PRF) \quad (15)$$

Where

m	Whole numbers (1,2,3,...) corresponding to pulses
n	Whole numbers (1,2,3,...) corresponding to pulses
$G^2(f,\tau)$	Two-way far field antenna power pattern
σ_0	Radar reflectivity

Hence, the integrated ambiguity to signal ratio (ASR) is given by

$$ASR(\tau) = \frac{\sum_{m,n\neq 0}\int_{-B_p/2}^{B_p/2} G^2(f_0 + mPRF, \tau_0 + n/PRF).\sigma_0(f_0 + mPRF, \tau_0\, n/PRF)df}{\int_{-B_p/2}^{B_p/2} G^2(f_0,\tau_0).\sigma_0(f_0,\tau_0)df} \quad (16)$$

Where

B_p Processing bandwidth

The PRF is upper bounding this processing bandwidth. However, abortion from this bandwidth can be used to exclude a fragment of the folded Doppler spectrum into the main Nyquist window, so in the processor, the ambiguity ratio can be enhanced at the price of azimuth resolution worsen.
The ASR is expressed with dependence on τ, which indicates the place at cross-track in the image. Knowledge of the antenna two-dimensional pattern is required by this ASR expression, and the time delay and the Doppler frequency are used to formulate the reflectivity of the target. Also, these quantities derivation from the data measured requires relations. Typically, using local incidence angle antenna pattern can be given. The separation of the range and the azimuth components for design purposes becomes more practical [11].

2.1.4 *Range ambiguity to signal ratio (RASR).* Ambiguities appear in Range results from previous and following echoes received simultaneously at the antenna with the desired echo. For airborne SAR data this phenomenon is typically not significant, since the echo spread is relatively small to the interpulse period. Spaceborne SARs contrary, between transmitting and receiving a pulse several inter-pulse periods are elapsed, which make range ambiguities to become significant. to deduce the exact value of RASR, at a certain time (t_i) in the window of data record, ambiguous signals came from the succeeding ranges [12]:

$$R_{i,j} = \frac{c}{2}\left(t_i + \frac{j}{PRF}\right) \tag{17}$$

Where

j number is positive for preceding interfering pulses (j=0 for the desired pulse), and negative for succeeding ones

After that the integrated RASR is calculated by taking the sum of all signal components that come from preceding and succeeding echoes in the data window to the wanted pulse integrated signal ratio.
The ambiguous signals power in any given resolution cell, is given by:

$$S_{i,a} = \sum_{j \neq 0} \frac{\sigma_{i,j} G_{i,j}^2}{R_{i,j}^3 \sin(\theta_{i,j})} \tag{18}$$

Where

$\sigma_{i,j}$	Normalized backscatter coefficient at a given incidence angle
$\theta_{i,j}$	A cross-track antenna pattern at that incidence angle
$G_{i,j}$	A cross-track antenna gain at that incidence angle
σ_0	Radar reflectivity

The desired signal is given by:

$$S_i = \frac{\sigma_{i,0} G_{i,0}^2}{R_{i,0}^3 \sin(\theta_{i,0})} \tag{19}$$

Finally, the RASR is given by:

$$RASR = \frac{\sum_i S_i}{\sum_i S_{i,a}} \tag{20}$$

Representing the ratio between useful signal and ambiguous signal levels average [17].

2.1.5 *Azimuth ambiguity to signal ratio (AASR).* It is commonly referred to the ratio of the ambiguous signal to the wanted signal, within the azimuth processing bandwidth of the SAR correlator and it can be calculated using the following equation:

$$AASR = \frac{\sum_{m \neq 0} \int_{-B_p/2}^{B_p/2} G^2(f_0 + mPRF)df}{\int_{-B_p/2}^{B_p/2} G^2(f)df} \tag{21}$$

Assuming a uniform target reflectivity across the scene in azimuth and in range and separating the two-dimensional pattern of the antenna as the product of the elevation and azimuth patterns, which to most significant side-lobes is a reasonable approximation. The typical range of AASR is about -20dB. Nevertheless, even at this value high bright targets next to dark targets can have ambiguous signals observed in images [13].

The radar range equation is the initial start point for any radar design process, as it connects a number of system parameters with the receiver signal to noise (SNR) and the target's radar cross-section and its distance to the radar. The radar range equation can be altered in a number of forms and used in many ways. In a SAR system, a useful form is the single look signal to noise [14].
The sensitivity of the system is usually expressed by the noise equivalent σ_0 (NESZ), which results from setting SNR=1, so:

$$P_{avg} = \frac{2.(4.\pi)^3 R^3.K.F.T_s.V_{st}.a_w.L_{tot}}{NESZ.\lambda^3.G^2.\rho_{rg}} \quad (22)$$

3. Design methodology

The target application has a significant effect on the design process of the SAR system. Which determines the desired specification that the design should achieve and include Desired swath width, Ground azimuth and Range resolution, Incidence angle, Frequency, Polarization, Radiometric accuracy, SNR and Sensitivity represented by NESZ.

During mission design the available platform resources imposes more constraints such as the allowable mass for the payload, available generated power, allowable size and dimensions, working altitude, attitude control, accuracy of attitude determination, downlink data rate, and so on. SAR system achievable performance is constraints by these limitations. After conflicting requirements trading, a final optimal design can be achieved.

The design flow shown in Figure 3, at the beginning converts the mission requirements (resolution, swath width, AASR, RASR, NESZ) into the required range and azimuth bandwidths. The required resolution in azimuth interprets to an antenna length (Laz). The minimum antenna width depends on the minimum antenna effective area (A_{eff}), but this minimum antenna dimension usually results in range ambiguities above the requirements. Therefore, an iterative design process is performed to optimize the antenna width and its illumination.

The minimum required PRF can be determined, when the azimuth dimension of the antenna is determined. After that, the valid PRF values map definition by applying the transmit interference and the nadir interference constraint.

Then the AASR and RASR can be calculated and also selection of the PRF value that optimizes the ASR at each incidence angle. If the optimized ASR does not meet the requirements, the antenna width is increased and the process is repeated.

After setting antenna parameters and the bandwidth, comes the determination of the power transmission necessary to satisfy the sensitivity requirement.

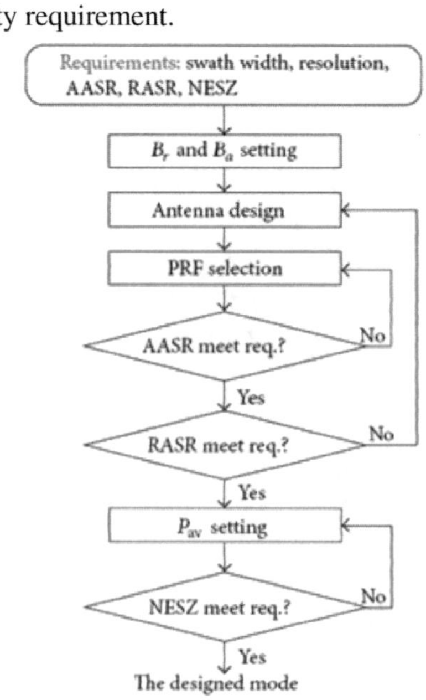

Figure 3. Flow chart of strip-map SAR system design

4. Design methodology validation

The validation is performed by deriving the system parameters from the mission requirements of a real system and comparing the obtained parameters. Therefore, for process validation, the TerraSAR-X main mission parameters have been set as constraints for obtaining the final mission specifications.

The main objective is to use the mission parameters of **TerraSAR-X** which are listed in the following table to derive the SAR system parameters using the suggested methodology and software tool and comparing the obtained results with that of the actual mission.

Table 1. TerraSAR-X mission parameters

Parameter	Value
Frequency f_0	9.65 GHz (X-band)
Orbital altitude	514 Km
Look angle	Between 20° and 45°
Resolutions	δR_g=1,7m δx=3m
Swath width	30Km
σ_0	-20 dB
ASR	20 dB
Mode	Strip-map

The variation of the minimum antenna area with the incidence angle is calculated and shown in Figure 1. As indicated at 20°, 45° incidence angles the minimum areas are 0.69 m^2 and 2.95 m^2 respectively. The lower bound for the required antenna area, is indicated by this figure without considering the ambiguity requirements [15].

optimum for this satellite. But the antenna length that TerraSAR-X satellite actually flown with is about 4,8m, which can be explained as despite decreasing the antenna length to improve the resolution the ambiguity requirements is still met. So, depending on this minimum antenna area restriction and this provided length the minimum antenna width should be no less than 0.5m.

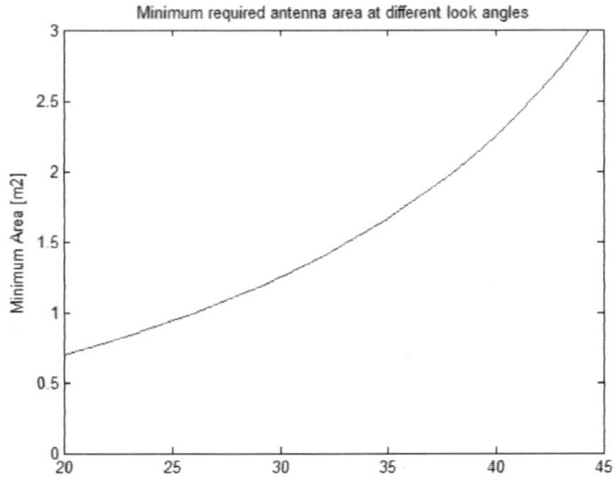

Figure 4. Minimum Antenna Area

After that the non-interference PRF values are calculated according to the design methodology and indicated as with areas in Figure 5.

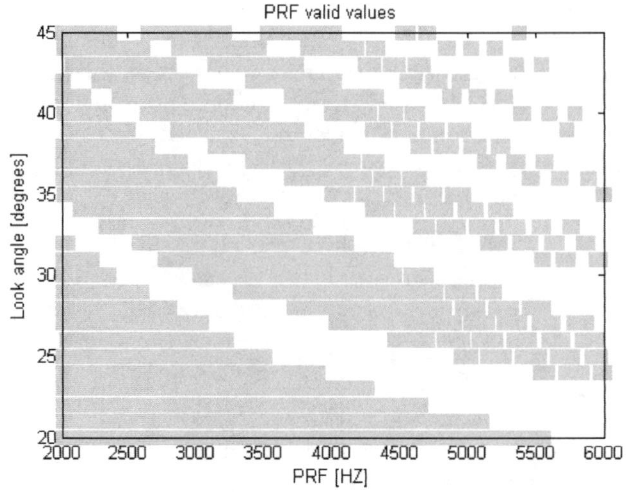

Figure 5. Non_interference PRF values

After that in the design methodology at every incidence angle in the required range of values a valid PRF value is chosen, such that the required ASR requirement is achieved at the required 30 Km swath. But if the ASR requirement cannot be achieved the process is repeated after increasing the antenna width. For the TerraSAR-X mission an antenna width of about 0.8 m fulfilled the ASR requirements at the all-incidence angles range of values. At each incidence angle the calculated ASR values fulfils the condition and are less than 20dB as shown in Figure 6. As indicated, all the values are less than 20dB.

A uniform antenna illumination pattern in azimuth assumption is used like in most orbital missions, for calculating the AASR. The AASR levels in practice are not enhanced by azimuth tapering which reduces the antenna gain. While, tapering in elevation pattern has a great impact on improving RASR. Which can be optimized for each system operating mode regarding SNR and ASR using Hanning tapering which is conventionally used in SAR systems.

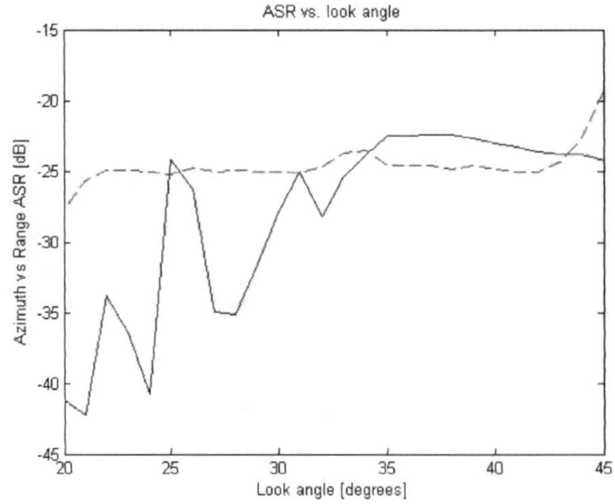

Figure 6. AASR and RASR levels with Hanning tapering in elevation

After that for achieving the required sensitivity the required average transmitted power is calculated as the last step, and it is calculated with the dependency on the range resolution for indicating the range resolution dramatic effect on it as shown in Figure 7. to achieve. For range resolution value of 1,7 m with -20 dB of NESZ requires a transmitted power around 300W.

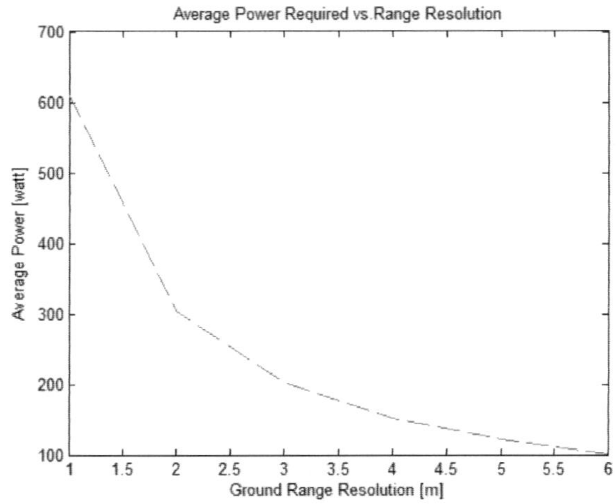

Figure 7. Average transmission power vs. Range Resolution

The results of the system design parameters values obtained achieving the mission requirements and that of TerraSAR-X are indicated in Table 2.

Table 2. TerraSAR design comparison

parameters	TerraSAR-X	Results
Antenna Length	4.8	4.8
Antenna Width	0.7	0.8
Average Transmit Power	360	359

So, based on the comparison results the design methodology produces valid and realistic results.

5. Frequency analysis

At this stage of the design of the system design process comes the frequency selection, which is mainly dependant on the application, but it has significant effect on other system parameters. A frequency analysis is presented at L, C and X frequency bands with the same mission parameters.

The required mission and quality parameters are listed in the following table with minimum acceptable sensor quality values of σ_{NE} and ASR values. The values stated in the table are from actual SAR missions like TerraSAR, RadarSAT or Envisat. Ambiguity problems can arise from the required lower values of ASR. For improving the sensitivity in case of more is required, decreasing the final range resolution is a choice.

Table 3. Quality and Mission Requirements

Quality requirements	
ASR_{min}	≈ 20dB
σ_{NE}	≈ -20dB
Ground Swath Width	30 km
Mission Requirements	
Frequency bands	9.65 GHz (X-band) 5.3 GHz (C-band) 1.3 GHz (L-band)
Altitude	510km
Look angle	20° to 45°
Polarization	single
Laz	≤ 5 m (if the quality requirements are not achieved it may be increased)

As a design objective, SAR antenna size is minimized as possible with considering the width of the swath to fit the required application. So, 5 m is the starting antenna length value of and it will be increased if the quality requirements are not met.

The application intended determines mainly the width of swath required, taking into consideration that small swaths are undesired for most application where the large swaths are not appropriate for small satellites because of the size of the antenna needed and for analysis simplification a 30 km swath is considered, also only a single polarization is considered.

At the different frequency bands the required minimum antenna area is illustrated in Figure 8, indicating that at the lower bands of frequency larger antenna sizes are needed.

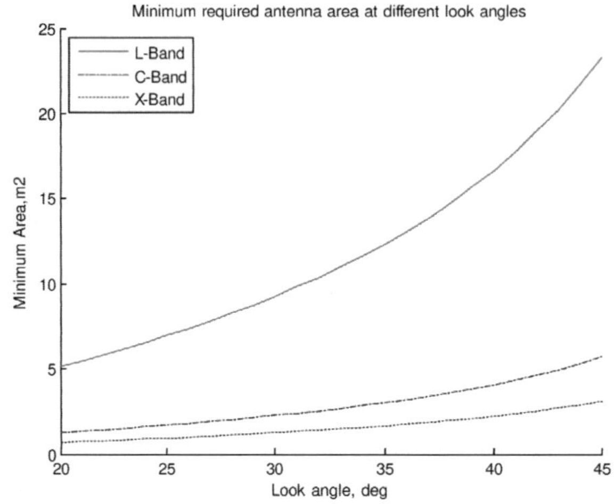

Figure 8. Initial antenna areas at L, C and X frequency bands

There is dependency between the required azimuth resolution and the needed antenna length, also long antenna lengths are required for the lower frequency bands but with the price of resolution reduction. Each frequency band is characterized by its penetration capability, but mainly frequency band selection is determined by the application.

Within the technological limitations, increasing the pulse width increases the range resolution at the price of sensitivity reduction.

As the look angle is increased the available PRF values is decreased as illustrated in Figure 9.

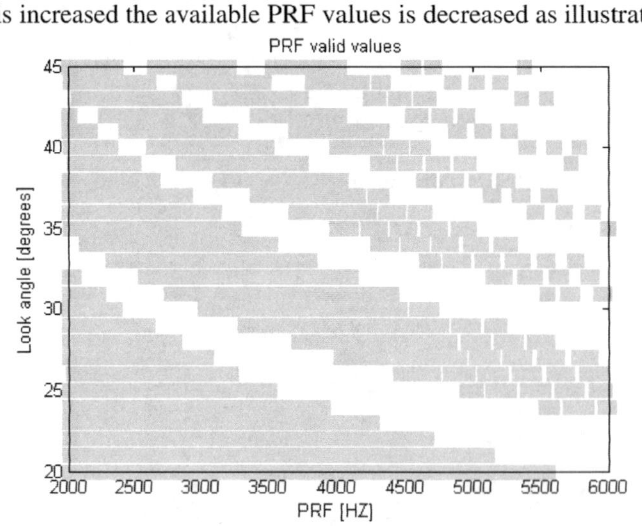

Figure 9. Noninterference PRF for 30 km swath width

Lower frequency bands need less power when scaling resolution with wavelength but requires large antenna sizes as illustrated in Figure 10 for 30 km swath.

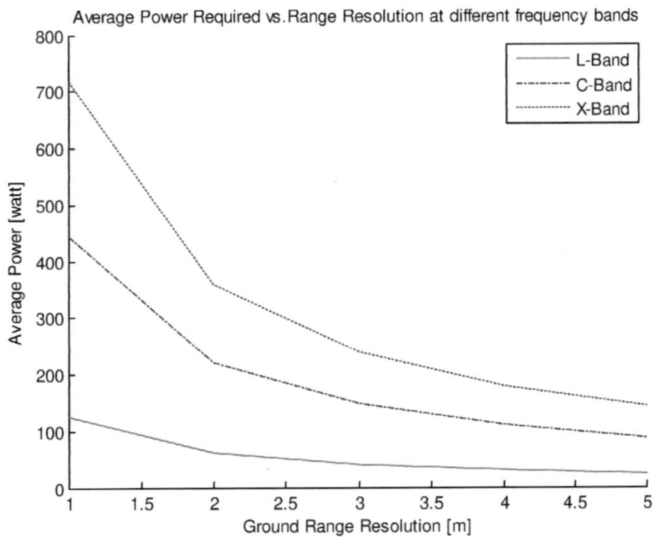

Figure 10. Transmission power at L, C and X bands

6. Results

The resulting values of the required system parameters are listed in Table 4 at the different frequency bands.

Table 4. SAR design configurations results

Parameter	L-Band	C-Band	X-Band	
Frequency f_0 [GHz]	1.3	5.3	9.65	
Altitude h [km]	510			
Look angle [°]	[20:45]			
Min antenna Area [m^2]	[5.14:23.8]	[1.26:5.71]	[0.69:3.14]	
Azimuth resolution δx [m]	3	2.5	2.5	30 km Swath width
Range resolution δR_g [m]	3			
Antenna length [m]	6	5	5	
Antenna width [m]	3.8	1.2	0.7	
Average power [w]	42	148	239	

7. Conclusion

In the SAR system design process, the operating frequency of the SAR system affects mainly the antenna dimensions and the average transmitted power with the same required mission parameters.

For more compact antenna size and dimensions the X-band is preferred, but it requires higher power transmission which requires heavier power supply system. On the other hand, for power transmission reduction and hence lighter power supply system the L-band is preferred, but the antenna dimensions are larger which requires more size. So, as a trade between the required antenna dimensions and power transmission for the same mission requirements the C-band is recommended as an operating frequency band.

References

[1] Esposito C, Natale A, Palmese G, Berardino P, Lanari R and Perna S 2020 On the Capabilities of the Italian Airborne FMCW AXIS InSAR System Remote Sens. 12

[2] Koo V, Chan Y K, Gobi V, Chua M Y, Hun L, Lim C-S, Thum C, Lim T S, Ahmad Z, Mahmood K, Shahid M H, Ang C, Tan W, Tan P, Yee K, Cheaw W, Boey H, Choo A and Sew B 2012 A new unmanned aerial vehicle synthetic aperture radar for environmental monitoring Prog. Electromagn. Res. 122 245–68

[3] Philipp M B and Levick S R 2020 Exploring the Potential of C-Band SAR in Contributing to Burn Severity Mapping in Tropical Savanna Remote Sens. 12

[4] Mengen, D., Montzka, C., Jagdhuber, T., Fluhrer, A., Brogi, C., Baum, S., Schüttemeyer, D., Bayat, B., Bogena, H., Coccia, A. and Masalias, G., 2021. The SARSense campaign: air-and space-borne C-and L-band SAR for the analysis of soil and plant parameters in agriculture. *Remote Sensing*, *13*(4), p.825.

[5] Refice A, Zingaro M, D'Addabbo A and Chini M 2020 Integrating C- and L-Band SAR Imagery for Detailed Flood Monitoring of Remote Vegetated Areas Water 12

[6] El-Darymli K, Moloney C, Gill E, McGuire P and Power D 2014 Design and implementation of a low-power synthetic aperture radar 2014 IEEE Geoscience and Remote Sensing Symposium pp 1089–92

[7] Wang W-Q, Cai J and Peng Q 2009 Conceptual design of near-space synthetic aperture radar for high-resolution and wide-swath imaging Aerosp. Sci. Technol. 13 340–7

[8] Harger R 1970 Synthetic aperture radar systems : theory and design

[9] Graziano M D, Renga A, Grasso M and Moccia A 2020 PRF Selection in Formation-Flying SAR: Experimental Verification on Sentinel-1 Monostatic Repeat-Pass Data Remote Sens. 12

[10] Douidar I, Safy M and Saleh A 2017 Parametric evaluation of PRF availability for a space borne SAR 2017 International Conference on Control, Automation and Diagnosis, ICCAD 2017

[11] k. Li F and Johnson W T K 1983 Ambiguities in Spacebornene Synthetic Aperture Radar Systems IEEE Trans. Aerosp. Electron. Syst. AES-19 389–97

[12] Wang W 2013 Large-Area Remote Sensing in High-Altitude High-Speed Platform Using MIMO SAR IEEE J. Sel. Top. Appl. Earth Obs. Remote Sens. 6 2146–58

[13] Kang M, Won Y, Lim B and Kim K 2018 Efficient Synthesis of Antenna Pattern Using Improved PSO for Spaceborne SAR Performance and Imaging in Presence of Element Failure IEEE Sens. J. 18 6576–87

[14] Edwards M, Madsen D, Stringham C, Margulis A, Wicks B and Long D G 2008 MICRO ASAR : A SMALL , ROBUST LFM-CW SAR FOR OPERATION ON UAVS AND SMALL AIRCRAFT Brigham Young University , Provo , UT 84602 514–7

[15] Werninghaus R and Buckreuss S 2010 The TerraSAR-X Mission and System Design IEEE Trans. Geosci. Remote Sens. 48 606–14

[16] Bickel D L, Brock B C and Allen C T 1993 Spaceborne SAR Study (New Mexico:Sandia National Laboratories)

[17] Mondéjar A G 2009 Feasibility Study on SAR Systems on Small Satellites (Barcelona)

IOP Conf. Series: Materials Science and Engineering 1172 (2021) 012014 doi:10.1088/1757-899X/1172/1/012014

Updating land use and land cover classes of Blue Nile basin using landsat-8 images

A S Shobary[1], A Elsharkawy[1], H E M El-Hanafy[1] and O M Moussa[1]

[1]Department of Civil Engineering, Military Technical College, Cairo, Egypt.

E-mail: a.s.shobary@mtc.edu.eg

Abstract. Land Use and Land Cover are considered as a main input layer for the estimation of surface runoff along with the soil maps, rainfall data, DEM. The main purpose of this research paper is to update the Land Cover and Land Use of the Blue Nile basin that can be used for further processing to estimate the surface runoff of the Blue Nile basin. Recent satellite images of Landsat 8, dated May and June, 2020, were used with less than 10% clouds covering the full scene of the study area. Six land cover classes were targeted namely; water, urban, barren land, forest, grass and agricultural crops. Supervised (maximum likelihood) classification method was used and accuracy assessment with the help of google earth as ground truth was performed. With the aid of 30-m DEM, the Blue Nile basin was divided into fourteen sub-watersheds. Classification scheme was applied to each of the fourteen sub-watersheds and the classification results were introduced as a percentage of the total area of each sub-watershed. Moreover, confusion matrices were formed for the sub-watersheds.

1. Introduction

Image classification is the method used to generate thematic maps from imagery. For example, the themes may vary from layers like surface water, soil and vegetation in the general characterization of a rural region, to specific sorts of vegetation, soil and pureness or depth of water for a more comprehensive description.

The process of possession descriptive labels, which classes various surface materials and conditions, is the prospective and final result. The data have been probably translated into a form that has information meaning by virtue of the labelling process [1].

Land Use and Land Cover are considered as a main input layer for the estimation of surface runoff along with the soil maps, rainfall data, DEM. The main purpose of this research paper is to introduce a recent update to the Land Cover and Land Use of the Blue Nile basin that can be used for further processing to estimate the surface runoff of the Blue Nile basin.

Abbay basin, area of study, is located in northwest of Ethiopia. Its average annual runoff was estimated about 54.8 BCM [2]. Lake Tana is considered the main head of Abbay basin. With the help of 30-m resolution DEM, Geographic Information System(GIS) was used to delineate Abbay basin and divide it into fourteen sub watersheds.

Landsat-8 images were used in this study as they are freely available. They were downloaded from Earth Explorer Web browser and georeferenced to the Universal Transverse Mercator map projection (World Geodetic System, WGS 84 datum, UTM Zone 37 N). These images were captured in the dry season with less than 10% cloud cover.

Content from this work may be used under the terms of the Creative Commons Attribution 3.0 licence. Any further distribution of this work must maintain attribution to the author(s) and the title of the work, journal citation and DOI.
Published under licence by IOP Publishing Ltd

Most classes of land cover, identified by remote sensing, are organized by classification scheme [3]. In this study, six land cover classes were targeted and plotted in the scheme as water, urban, barren land, forest, grass and agriculture crops.

Supervised (maximum likelihood) classification was used to make the required thematic map. Training samples were drawn on the images and ground truth were featured with the help of google earth. This operation was repeated for the whole thirteen sub watersheds.

In this study, Environment for Visualizing Images (ENVI) was used to create confusion matrices for the whole sub watersheds and estimate measures of classification accuracy: user's accuracy, producer's accuracy and overall accuracy.

Objective of this study is to produce updated thematic maps of land use and land cover of the Blue Nile Basin that can be recently used for estimating surface runoff.

2. Area of Study

The Nile River is considered the longest one in the world with total length of 6670 km. The river drains a catchment with area of 3.4 million km2. There are two main tributaries feeding The Nile, the White Nile and Blue Nile. The White Nile originates from The Equatorial Lakes plateau in Uganda and meets the Blue Nile, which rises from the Ethiopian Highlands, at Khartoum. The Nile River has another main tributary, Atbara River, originating from the Ethiopian Plateau to the north of Khartoum where it joins the main Nile [4].

The Blue Nile Basin and its main tributaries drain an estimated area of 324,000 km^2, about 250,000 km^2 in the Ethiopian Plateau. The Blue Nile Basin is divided into two parts; the part in Ethiopia called Abbay Basin or Ethiopia-Blue Nile, situated in the northwestern district of Ethiopia. The Blue Nile Basin exists between $34^0\ 0'\ 25''$ E and $39^0\ 0'\ 49''$ E longitude, and $7^0\ 0'\ 40''$ N and $12^0\ 0'\ 51''$ N latitude. There are other four basins have common boundaries with Blue Nile Basin: Tekeze basin in the north, the Awash basin in the east and south east, the Omo-Gibe basin in the south, and the Baro-Akobo basin in the south west. The main head of the Blue Nile (Abbay) river, Lake Tana, the greatest freshwater lake in the country is situated in the north part of the basin[5].

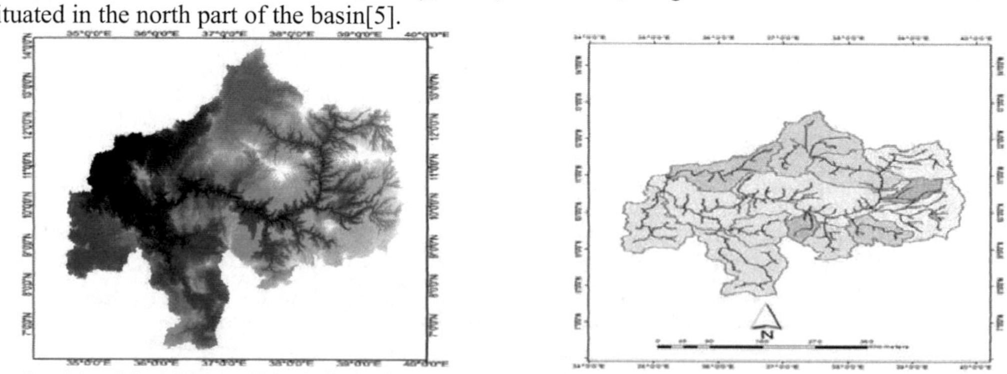

a) Mosaicked DEM of Blue Nile basin b) Watersheds forming the Blue Nile basin

Figure 1 Location of study area

Figure 1 shows transformation of DEM of study area (remote sensing data) into 14 sub watersheds by Arcgis. Area of sub watersheds, forming the study area, is illustrated in the following table.

Table 1. Area of sub watersheds.

Watershed	Beshelo	Tana	Jemma	Weleka	Mugger	Guder	South Gojam
Area (km^2)	12085	15935	15248	4448	6907	6594	27635
Watershed	Anger	Wenbera	Didessa	Dabus	Beles	Fincha	North Gojam
Area (km^2)	7475	16859	17855	13716	12327	3308	12045

South of lake Tana, a small spring which is considered the main source of the Blue Nile drains at height 2900 m and approximately 100 km length. The Abbay basin has a main portion in the irrigation and hydropower potential of Ethiopia with 815,581 ha and 78,820 GWH/yr respectively. Awlachew et.al estimated an average annual runoff of the basin about 54.8 BCM [2].

The Abbay Basin consist of 14 sub watersheds as shown previously in Figure 1. There are two unlike features form the topography of the Abbay basin; in the center and the eastern part, amorphous mountains areas form the highlands and the lowlands exist in the western part of the basin. The Ethiopian Plateau rises at elevations ranges from 1500 m up to 4000 m above mean sea level with average slope more than 25% in the eastern part, highlands, whereas the lowlands in the west prostrates 1000 m to 500 m above mean sea level with average slope less than 7%. Latitude, altitude, slope and exposure are factors that affect the climate of Abbay basin. The duration from June to September form the rainy season [6].

The large amount of rainfall is concentrated in the Ethiopian highlands ranging from 1500 to 2200 mm with peaks in August, on the other hand, the rainfall in the lowlands is less than 1500 mm. The climate of the basin differs from one region to another. The western part possesses the highest temperature with maximum and minimum range 280C-380C and 150C-200C respectively. In the eastern and central basin's part, lower temperature is monitored. The maximum temperature extents 120C – 200C whereas the minimum being -10C to 80C. There is an interconnection between temperature and potential evapotranspiration. In the lowlands, high temperature, high potential evapotranspiration ranging between 1800 mm and 2232 mm per year, lower potential evapotranspiration is observed between 1200 and 1800 mm per year in the eastern and southern parts. The highlands, lowest temperature, has the least potential evapotranspiration below 1200 mm per year [4].

Most of the study area is covered with volcanic materials, which are greatly common in the north central highlands in Ethiopia. Referring to uninterrupted events of storms through the rainy season, tough erosion is occurred, so the composition of these volcanic materials has a main significance for hydrologists and engineers. Huffnagle(1961) characterized two major types of siol: the black soils and the red to reddish browny clayey loams [7].

In depressions of the study area, the black soil exists. This type of soil is resulted from dolerites decay and is trended to fast crack and dry out. The red soil is greatly permeable. In the Blue Nile basin, there are various types of soil which differ from fine granular to medium or coarse. The fine granular has slow and moderate permeability but the medium and coarse has very slow one [8].

3. Methodology

3.1. Satellite image data sets

There are many factors for the choice of the most convenient satellite image data sources such as; images cost and availability of funding, available long time image series to analyze the change dynamics of the land cover and the area can be covered by suitable image data. According to the above criteria, Landsat 8 was found to be the most appropriate satellite to be used in this study. On February 11, 2013, Landsat 8 or Landsat Data Continuity mission, LDCM, was launched from Vandenberg Air Force Base, California, carried the Atlas-V vehicle. The Operational Land Imager, OLI, and the Thermal Infrared Sensor, TIRS, are the two sensors that installed on the satellite payload. OLI instrument develops pictures at nine wavelengths operating in the domain of 0.43-2.29 μm. Images that possessed from OLI instrument is at 15 m resolution so, they have a main role in studying many fields such: water quality control and cirrus clouds studying. TIRS acquires imagery at two infrared wavelengths with resolution of 100 m. The features of the image data are termed in Table 2.

Table 2. Landsat OLI and TIRS sensor system features [9].

sensor	Band number	Band name	Wavelength(μm)	Resolution
OLI	1	Coastal	0.43-0.45	30 m

OLI	2	Blue	0.45-.051	30 m
OLI	3	Green	0.53-0.59	30 m
OLI	4	Red	0.63-0.67	30 m
OLI	5	NIR	0.85-0.88	30 m
OLI	6	SWIR 1	1.57-1.65	30 m
OLI	7	SWIR 2	2.11-2.29	30 m
OLI	8	Pan	0.50-.68	15 m
OLI	9	Cirrus	1.36-1.38	30 m
TIRS	10	TIRS 1	10.60-11.19	100 m
TIRS	11	TIRS 2	11.50-12.51	100 m

All products of Satellite 8, including images used in this study, were corrected and georeferenced to the Universal Transverse Mercator map projection (World Geodetic System, WGS 84 datum, UTM Zone 37 N). These images are freely obtainable and can be downloaded from the United States Geological Survey, USGS, Earth Explorer web browser. To avoid cloud cover, these imags were acquired in the dry season. Less than 10% clouds cover the full scene of the study area as the images acquired in May and June,2020 (dry season).

3.2. Image pre-processing

Land sat images used in this study were of level 1 (on-demand imagery). A single multi-band image obtained by stacking the reflective bands with attached image file format (band 2, 3, 4, 5). After stacking layers of images, it was essential to mosaic these images to form the full scene of the study area as show in **Figure 2**. To simplify land cover analysis processing, the study area (Abbay basin) was divided into 14 sub-basins by importing shape file (vector) from **ArcGIS10.3**. Land cover classification was applied to each sub-basin separately.

Figure 2 Land sat images of Blue Nile basin

3.3. Classification scheme

Adequate attention must be paid to classification scheme used in such planning project that includes remotely sensed data. It is important for such classification scheme to be exclusive and in detailed [10].USGS, European Coordination of Information on the Environment (CORINE), and Africover used remote sensing data for developing classification systems. However, land use and land cover have no typical classification [11]. USGS Level I classification scheme is the basic and fundamental for organizing most types of land cover which had been identified by remote sensing [3]. For classification, in this study, six land cover classes were plotted in the scheme as water, urban, barren land, forest, grass and agricultural crops. Table 3 describes in detail the land cover classes.

Table 3. Land cover classes established and their description after [9].

Land cover class	description

IOP Conf. Series: Materials Science and Engineering 1172 (2021) 012014 doi:10.1088/1757-899X/1172/1/012014

Water	Including lakes, ponds, rivers and other open water areas.
Urban	Residential, commercial and industrial buildings are components of urban category.
Barren land	Sand, rocks and any land surface free of vegetation.
Forest	Trees, bushes, shrubs and any woody vegetation.
Grass	Areas covered by herbaceous vegetation.
Agricultural crops	Farms and bare crop fields.

3.4. *Image classification*
Land cover maps are produced from combination of characteristics of remotely sensed imageries and particular classes of land cover [12]. Supervised classification method is enormously used for classification of most land cover monitoring [13]. In supervised (maximum likelihood) classification, training samples should be featured as they are essential for assigning brightness values for computer algorithm symbolizing land cover class in each band [14]. In order to apply this method, ground truth should be well known and mind has a set of appropriate categories. Ground truth is characterized with the help of Google Earth. Training samples were drawn on images. Regions of Interest (ROIs) for training samples were selected graphically in irregular polygon shape. They were carefully chosen to be proportion to classes on the output map. Accordingly, maximum likelihood was performed using these ROIs to produce the required thematic map. This operation was performed repeatedly through 13 sub-basins. Because of shortage of spatial coherency in vast majority of classified images, clumping classes was performed to make them smooth. Afterwards, these images were classified to vector to make each class such a single layer to be easily exported to the ArcGIS for preparation the map.

3.5. *Accuracy assessment*
It is obligatory to provide accuracy assessment in remote sensing technique [15]. The produced classification quality is significant so, information should be provided about it. Error matrix is considered one of the most popular means that represent accuracy assessment [10, 14, 16-18]. It can be defined as a square array of numbers put in the form of rows and columns represent number of training samples that can be either pixels, groups of pixels or polygons set out to definite class relative to the real class (ground truth samples). Overall accuracy, Kappa statistics, user's accuracy and producer's accuracy are classification measures that are commonly used [9]. It is important to estimate the percentage of correct classified pixels and incorrect ones per land cover class on the produced thematic maps, based on Landsat images, prior to authorization the classification results in the final form. In this study, to perform accuracy assessment, other random ROIs were selected on the same Landsat images. They were used as reference data. Environment for Visualizing Image (**ENVI 4.5**) was used to create the error matrix and to derive measures of classification accuracy.

Overall accuracy is an essential classification accuracy measure defined as follows:

$$\text{overall accuracy} = \frac{\text{the sum of major diagonal of matrix (total correct pixels)}}{\text{total number of pixels in the matrix}} \quad (1)$$

Besides overall accuracy, user's accuracy (commission errors) and producer's accuracy (omission errors) are also derived to be used in estimating accuracy.

$$producer's\ accuracy = \frac{the\ number\ of\ correct\ pixels\ in\ a\ class}{\text{total number of column pixels}} \quad (2)$$

where, total column pixels represent the number of total pixels in the ground truth samples (reference data).

$$user's\ accuracy = \frac{\text{total number of pixels of a class}}{\text{total number of row pixels}} \quad (3)$$

Where, total row pixels represent pixels classified in that class.

The kappa statistic is another important technique used for classification accuracy assessment [19]. It is calculated by the following equation:

$$K = \frac{\sum_{i=1}^{r} xii - \frac{1}{N}\sum_{i=1}^{r}(x_{i+} * x_{+i})}{\mathrm{N} - \frac{1}{N}\sum_{i=1}^{r}(x_{i+} * x_{+i})} \quad (4)$$

Where, (r) represents the number of matrix rows, (x_{ii}) represents notices number in row i and column i, (x_{i+}) represents peripheral totals of row i, (x_{+i}) represents peripheral totals of column i and (N) represents total number of notices.

4. Results and discussion

The Abbay Basin was divided into fourteen sub watersheds using 30-m DEM. Satellite images of landsat-8, covering the whole scene of study area, were used. The aimed six land cover classes were water, urban, barren land, forest, grass and agriculture crops. Supervised (maximum likelihood) classification was performed. The resulted land use/ land cover maps are shown in the following figure.

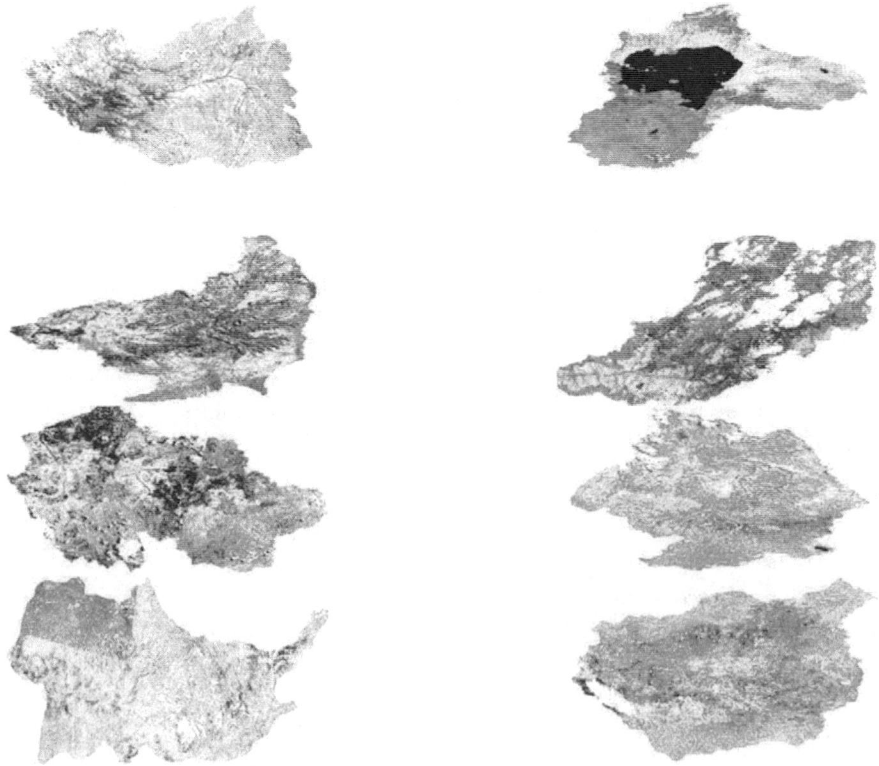

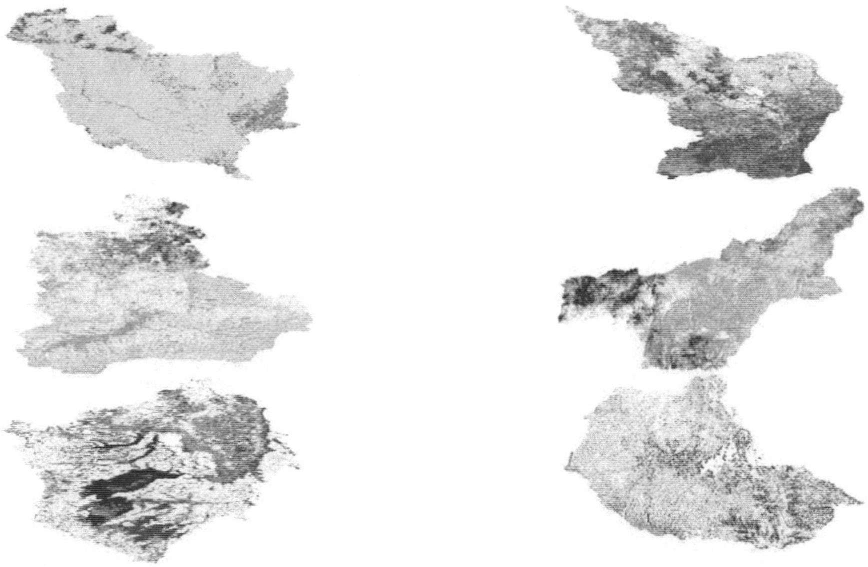

Figure 3 Final resulted land use and land cover maps of whole sub watersheds.

It was found that Weleka and Anger were extracted from images contain clouds cover bigger area compared to clouds in the other images. This makes percentage of urban class in these sub watersheds is large. The following figure shows land sat images of Weleka and Anger sub watersheds.

(a) Land sat image of Weleka sub watershed (b) Land sat image of Anger sub watershed

Figure 4 land sat images.

In order to compare easily among the percentages of land cover/ land use of the whole 14 sub watersheds of the Blue Nile basin, please see Table 4.

Table 4 percentage of each class for the whole sub watersheds.

feature / watershed	Water	Urban	Barren	Forest	Grass	Crops
Beshelo	15.66%	50.89%	3.33%	1.16%	7.71%	21.2%
Tana	19.1%	4.25%	29.35%	15.6%	13%	18.55%
Jemma	10.92%	41.91%	12.76%	1.3%	6.3%	26.72%
Weleka	20.67%	32.65%	28.96%	2.32%	5.29%	10.11
Mugger	22.93%	54.5%	4.66%	2.43%	10.32%	5.16%
Guder	1.6%	42.21%	9%	2.22%	37.46%	7.52%
S- Gojam	3.51%	10.88%	60.35%	5.51%	14%	5.75%
Anger	4.33%	26.82%	13.54%	8.1%	40%	7.22%
Wenbera	0.42%	—	86.4%	7.9%	5.28%	—
Didessa	23.5%	18.295	24.31%	17.65%	16.24%	—
Dabus	4%	28.1%	37.1%	18.6%	12.2%	—
Beles	11.9%	7.15%	40.8%	16.65%	21.5%	2%
Fincha	6.63%	12.58%	41.8%	18.44%	18.23%	2.33%
N-Gojam	9.83%	21.13%	37.18%	4.06%	14.66%	13.14%

Figure 5 display the estimated accuracy measures of the watersheds.

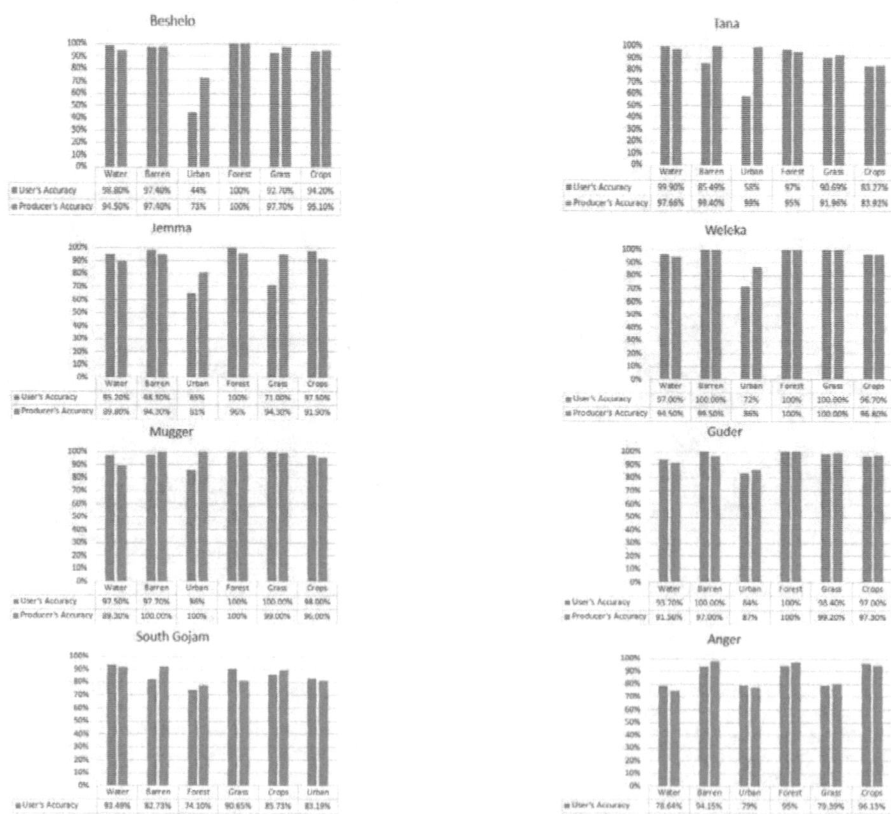

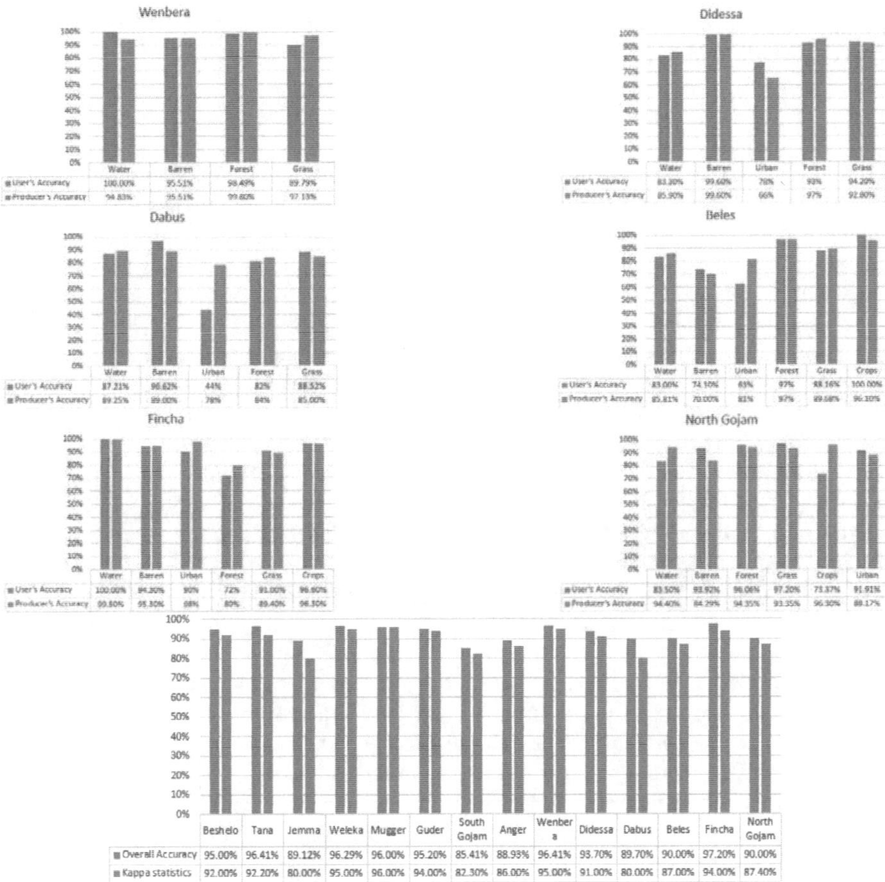

Figure 5 Estimated accuracy measures of the watersheds.

From the previous figure, it was found that the user's and producer's accuracies for all classes, except urban class, are reasonable as they range from 70% to 100%. The user's accuracy for urban class was low in some sub watersheds such Beshelo (44%), Tana (57.9%) and Dabus (44%). Such low accuracies occurred because these sub watersheds were extracted from more than one image at different times unlike the other sub watersheds which were extracted from one image.

5. Conclusions

From this research, we can conclude that:

- Landsat-8 images are the promising choice of satellite image data source because of their low cost and availability of funding.
- As landsat-8 images are available long time image, they can be used for analyzing the change dynamics and updating of land use and land cover.
- Geographic Information System (GIS) is effective in delineating basins as it is the most convenient tool to integrate with remote sensing for mapping.
- The commercial software Environment for Visualizing Images (ENVI) is suitable for image classification process as it can create error matrix and derive measures of classification accuracy.

- Accuracy measures for all sub watersheds are found acceptable except user's accuracy for Beshelo, Tana and Dabus because the images covered these sub watersheds were multi temporal images.

References

[1] Sadek M, Li X, Mostafa E, Freeshah M, Kamal A, Sidi Almouctar M A, Zhao F and Mustafa E K 2020 Low-cost solutions for assessment of flash flood impacts using Sentinel-1/2 data fusion and hydrologic/hydraulic modeling: Wadi El-Natrun Region, Egypt *Advances in Civil Engineering*, Vol. **2020**.
[2] Awulachew S B, Yilma A D, Loulseged M, Loiskandl W, Ayana M and Alamirew T 2007 *Water resources and irrigation development in Ethiopia*, Vol. **123** Iwmi.
[3] Jensen J R 2005 Introductory digital image processinga remote sensing perspective Tech rep.
[4] Yilma A D and Awulachew S B 2009 Characterization and Atlas of the Blue Nile Basin and its Sub basins.
[5] Shahin M M 1985 *Hydrology of the Nile basin, Elsevier*.
[6] Hurst H E and Phillips P 1938 *The hydrology of the Lake Plateau and Bahr el Jebel*, Government Press.
[7] Huffnagel H et al 1961 Agriculture in Ethiopia *Agriculture in Ethiopia*.
[8] Ashour A S, Moussa O M, Al-Tohame F, and El-Hanafy H E 2014 Determination of Grand Ethiopian Renaissance Reservoir (GERR) Life Span Using Remote Sensing Techniques *The International Conference on Civil and Architecture Engineering* Vol. **10** Military Technical College pp. 1–28.
[9] Wondrade N, Dick Ø B, and Tveite H 2014 GIS based mapping of land cover changes utilizingmulti-temporal remotely sensed image data in Lake Hawassa Watershed, Ethiopia *Environmental monitoring and assessment* Vol. **186** No. 3 pp. 1765–1780.
[10] Congalton R G A 1991 review of assessing the accuracy of classifications of remotely sensed data *Remote sensing of environment* Vol. **37** No. 1 pp. 35–46.
[11] Anderson J R 1976 A land use and land cover classification system for use with remote sensor data US Government Printing Office Vol. **964**.
[12] Aplin P 2004 Remote sensing: land cover *Progress in Physical Geography* Vol. **28** No. 2 pp. 283–293.
[13] Rogan J and Chen D 2004 Remote sensing technology for mapping and monitoring land-cover and land-use change *Progress in planning* Vol. **61** No. 4 pp. 301–325.
[14] Lo C and Yeung A 2002 Concepts and techniques of geographic information systems.
[15] Okeke F and Karnieli A 2006 Methods for fuzzy classification and accuracy assessment of historical aerial photographs for vegetation change analyses. Part I: Algorithm development *International journal of remote sensing* Vol. **27** No. 1 pp. 153–176.
[16] Congalton R G, Green K et al 2009 Assessing the accuracy of remotely sensed data.
[17] Foody G M 2002 Status of land cover classification accuracy assessment *Remote sensing of environment* Vol. **80** No. 1 pp. 185–201.
[18] Lillesand T and Kiefer R W 2000 Remote sensing and image interpretation.
[19] Congalton R G 2001 Accuracy assessment and validation of remotely sensed and other spatial information *International Journal of Wildland Fire* Vol. **10** No. 4 pp. 321–328.

IOP Conf. Series: Materials Science and Engineering 1172 (2021) 012015 doi:10.1088/1757-899X/1172/1/012015

Assessment of top of atmosphere reflectance generation algorithm for world view-2 high resolution images

M Allam[1], T Mahmoud[1], A Elsharkawy[2] and B Sheta[1]

[1]Electrical Engineering, Military Technical College, Cairo, Egypt.
[2]Civil Engineering, Military Technical College, Cairo, Egypt.

E-mail: eng.ma.mo.allam@gmail.com

Abstract. Recently, there are many developments in remote sensing techniques using satellite imagery as a result of the variety of remotely sensed image sources with different spatial and spectral resolutions and from different sensors. Optical satellite products are affected by an error of payload, atmosphere (scattering and absorption), and variations relative to positions of Sun, Earth, and satellite during capturing data. Top of Atmosphere (TOA) correction is the process of removing the effects of variations relative to sensor error and positions of Sun, Earth, and satellite. In this paper, maximum benefits from high spectral and spatial resolution images are demonstrated and analyzed or compared with images from different. TOA correction algorithm, which was implemented using Python environment, is applied to high resolution images from WorldView-2 (WV-2) satellite. TOA reflectance is considered the first step in any algorithm dedicated to the change detection process. Reflectance conversion is also performed on the same images using Orfeo ToolBox (OTB), the open-source software. The performance and efficiency of the proposed algorithm are compared with that of the Orfeo ToolBox (OTB) TOA reflectance output. The achieved results show that the proposed algorithm, which is automatically performed, is faster and provides significant results for WV-2 images, and can be adapted to be applied on different optical satellite sensors.

1. Introduction

Remote sensing satellite images offer important information about earth resources and the environment used in different applications image classification, fusion, and change detection[1, 2]. The popularization of satellite images is due to availability online for visual inspection or download free for further analysis. Continuous development of optical satellite production provides a variety of satellite image products with different spectral and spatial resolutions. High spectral resolution satellite images provide dense information about different features in the image. Moreover, the high spatial resolution provides more details of earth observations. However, sometimes we have to combine data from different sensors due to restrictions of cloud coverage and low revisit time. A seamless combining of images from different sensors is not an easy or straightforward process. The combination of data from different sensors raises some worries about data compatibility. Radiometric differences at top of the atmosphere level could be one of these worries [3]. while traveling of solar spectrum electromagnetic radiation through atmosphere to satellite sensors atmosphere gases and aerosols causes absorption and scattering of radiations[4]. Atmospheric correction of impacts on optical images is a very important pre-processing step for further analysis or comparison with different sensors [5]. Achieving accurate surface

Content from this work may be used under the terms of the Creative Commons Attribution 3.0 licence. Any further distribution of this work must maintain attribution to the author(s) and the title of the work, journal citation and DOI.
Published under licence by IOP Publishing Ltd

reflectance becomes a mandatory step in Effective processing. TOA corrections process working on minimizing the effects of atmospheric on images and generate TOA reflectance products. TOA corrections account for variations due to acquisition time, atmospheric conditions, and positions of the Sun, Earth, and satellite. TOA correction of satellite images is very important for precise analysis and remote sensing applications. WV-2 satellite products represent a high spectral resolution image with multiple bands. The variety of spectral bands enable using of different applications such as image analysis and classify land cover according to the spectral signature of each feature[6]. WV-2 sensor provides products consists of eight multispectral bands which are useful in spectral analysis, monitoring, land-use land cover mapping, disasters, defense and intelligence, exploration, and visualization[7]. TOA reflectance conversion applied to three WV-2 images for different cities Miami, Rio De Janeiro and Brisbane captured at different times with different environments using the proposed algorithm. Orfeo Toolbox (OTB) is an open-source library for remote sensing applications and image processing initiated by the French space agency National Centre for Space Studies (CNES) [8]. The library originally targets specific images captured by satellites Pleiades and Cosmo-Skymed but also support other sensors. OTB is a rich source of variety applications available for image calibration, enhancement, ortho-rectification or pansharpening, classification, SAR processing, and much more[9]. OpticalCalibration application performs TOA calibration and supports specific sensors.

The next sections of this paper are structured as follows: Section 2 describes the area of study. Section 3 demonstrates the proposed methodology. The experimental results and discussion are illustrated in Section 4. Section 5 provides some concluding remarks, and ideas for future work in this area are also discussed.

2. Area of Study

2.1. WV-2 satellite images

WorldView-2 satellite is DigitalGlobe's third satellite in orbit launched in October. WV-2 provides a 0.46m resolution panchromatic image, and a 1.84m resolution eight-band multispectral which awards advantages compared with other DigitalGlobe's satellites as shown in figure 1[10].

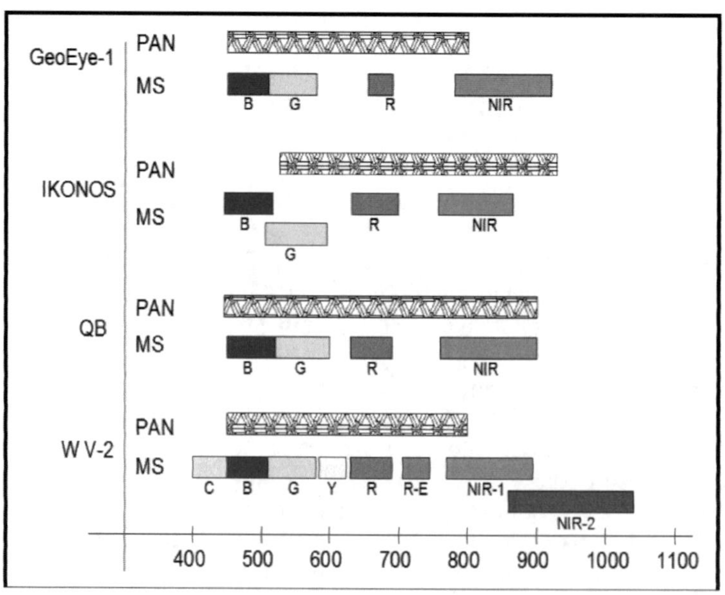

Figure 1. DigitalGlobe's satellites with different spectral bands [10].

2.2. WV-2 multispectral bands specifications

WV-2 satellite products provide eight multispectral bands six in the visible range and two in the near-infrared range figure 2 All bands have different sensitivity to features on the ground. This combination of bands improves remote sensing applications. The Coastal blue band is absorbed by healthy plants; least absorbed by water and significantly affected by atmospheric scattering. Thus, this given information makes it useful for vegetative analysis and identification, bathymetric studies, and improves atmospheric correction techniques. Yellow band measures the "yellow-ness" of particular vegetation and very useful for feature classification. The red-edge band is very useful in detecting plants. The two infrared bands partially overlapped but the second infrared band was characterized by less affection by atmospheric impact[7].

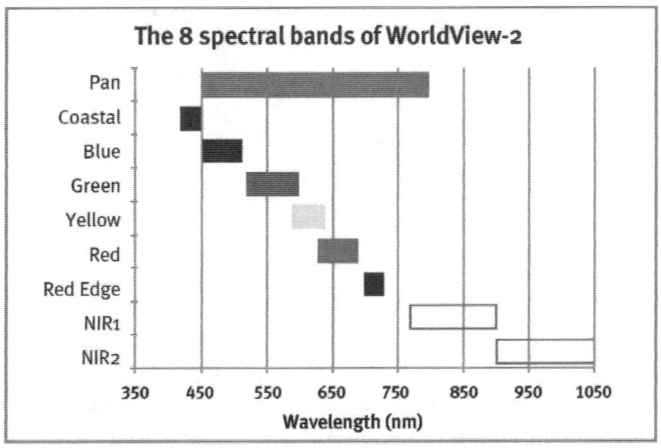

Figure 2. WV-2 image spectral bands[7].

2.3. Input data

The input data are about three WorldView-2 (WV-2) satellite images which were captured at different times of cities Miami, Rio De Janeiro, and Brisbane as shown in figure 3(a), (b), and (c), respectively. Data are downloaded from the DigitalGlobe website and captured in different environments in table 1 [10].

(a) (b) (c)

Figure 3. WV-2 satellite images: (a) Miami, (b) Rio De Janeiro, and (c) Brisbane. [10]

Table 1. Data parameters.

	Miami	Rio De Janeiro	Brisbane
Projection	UTM	UTM	UTM
Datum	WGS-84 zone 17N	WGS-84 zone 17N	WGS-84 zone 17N
Upper left corner	Lat 25° 47` 37.6841`` N Lon 80° 14` 1.9210`` W	Lat 22 ° 51` 19.6499`` S Lon 43° 13` 53.6890`` W	Lat 27 ° 26` 9.4311`` S Lon 153° 0` 31.3479 `` E
Lower left corner	Lat 25° 46` 14.6881`` N Lon 80° 11` 35.5308`` W	Lat 22 ° 63` 31.1101`` S Lon 43° 11` 28.3343 `` W	Lat 27 ° 28` 22.3549`` S Lon 153° 3` 0.4217 `` E
Size	8x(2048x1266)	8x(2048x2048)	8x(2048x2048)
Time of capturing	2010-11-12 16:08:11.809487	2011-01-25 13:11:53.815364	2011-07-13 00:19:45.597708
Sun elevation angle (deg)	44.5	63.3	35.8

3. Methodology

This research work concentrates on converting the WV-2 raw image Digital Numbers (DNs) to TOA reflectance. This conversion is performed by removing atmospheric effects from satellite images and account for variations relative to positions of Sun, Earth, and satellite. figure 4 illustrates the procedures that are utilized in the experiment. Python environment is utilized to implement the proposed TOA conversion algorithm to transform DNs to TOA reflectance which is performed in two main steps: conversion to TOA spectral radiance step followed by conversion to TOA spectral reflectance.

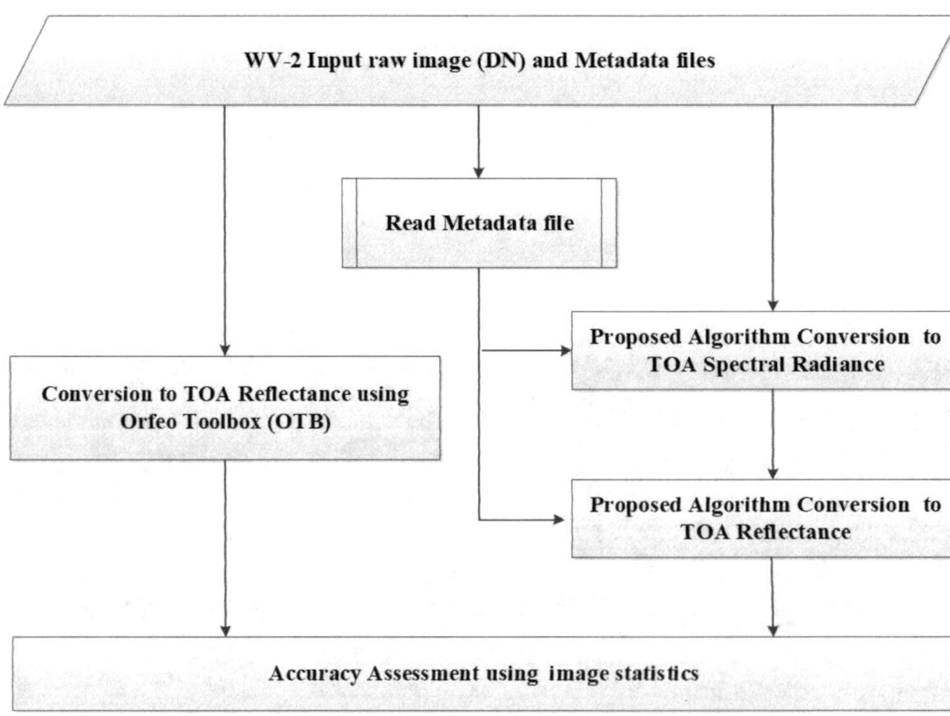

Figure 4. Research methodology.

3.1. Conversion to TOA spectral radiance

IOP Conf. Series: Materials Science and Engineering 1172 (2021) 012015 doi:10.1088/1757-899X/1172/1/012015

WV-2 images delivered to customers radiometrically corrected pixels. Pixel values represent the amount of spectral radiance acquired by the sensor converted to DN value [12]. Consequently, WV-2 image pixel values are particular for each sensor and could not directly compare to images from a different sensor. TOA spectral radiance of WV-2 represents the spectral radiance entering the satellite sensor. Each band of WV-2 is converted to TOA spectral radiance according to equation (1) [11].

$$\mathbf{L}_{\lambda_{\text{pixel,band}}} = \frac{\mathbf{k}_{\text{band}} \cdot \mathbf{q}_{\text{pixel,band}}}{\Delta\lambda_{\text{band}}} \quad (1)$$

Where $\mathbf{L}_{\lambda_{\text{pixel,band}}}$ is the TOA spectral radiance pixels $[\mathrm{W} - \mathrm{m}^{-2} - \mathrm{sr}^{-1} - \mu\mathrm{m}^{-1}]$ (watts/unit source area* unit solid angle* unit wavelength), $\mathbf{k}_{\text{band}}$ is the absolute radiometric calibration factor $[\mathrm{W} - \mathrm{m}^{-2} - \mathrm{sr}^{-1} - \mathrm{count}^{-1}]$ (watts/unit source area* unit solid angle* Digital Numbers counts) for each band, $\mathbf{q}_{\text{pixel,band}}$ are WV-2 image pixels, and $\Delta\lambda_{\text{band}}$ is the effective bandwidth $[\mu\mathrm{m}]$. Both $\mathbf{k}_{\text{band}}$ and $\Delta\lambda_{\text{band}}$ can be found in the metadata file with the extension (*.IMD) under names absCalFactor and effectiveBandwidth, respectively.

3.2. Conversion to TOA spectral reflectance

Obtained TOA spectral radiance varies with Earth-Sun distance, solar zenith angle, topography, bi-directional reflectance distribution function (varies depending on the illumination and observation geometry), and effect of the atmosphere (absorption and scattering). So, the obtained TOA spectral radiance must be converted into reflectance before applying spectral analysis techniques or comparing it with other sensor data. For each product, some factors must be calculated: the distance from the Sun to Earth during capturing the image, the Julian Day of capturing time, and the solar zenith angle during capturing [12].

$$\mathrm{A} = \mathrm{int}\left[\frac{\mathrm{year}}{100}\right] \quad (2)$$

$$\mathrm{B} = 2 - \mathrm{A} + \mathrm{int}\left[\frac{\mathrm{A}}{4}\right] \quad (3)$$

$$\mathrm{JD} = \mathrm{int}[365.25 * (\mathrm{year} + 4716)] + \mathrm{int}[30.6001 * (\mathrm{month} + 1)] + \mathrm{day} + \frac{\mathrm{UT}}{24.0} + \mathrm{B} - 1524.5 \quad (4)$$

$$\mathrm{D} = \mathrm{JD} - 2451545.0 \quad (5)$$

$$\mathrm{g} = 357.529 + 0.98560028 * \mathrm{D} \quad (6)$$

$$\mathrm{d}_{\mathrm{ES}} = 1.00014 - 0.01671 * \cos(\mathrm{g}) - 0.00014 * \cos(2\mathrm{g}) \quad (7)$$

The from earth to sun d_{ES} is calculated according to equation (7), in Astronomical Units (AU). The average solar zenith angle is calculated for the whole scene at the time of acquisition according to equation (8).

$$\theta_s = 90.0 - \mathrm{sunEl} \quad (8)$$

Where the acquisition time and sunEl are also included in the image metadata file with the extension (*.IMD). Finally, the process of converting the radiance to reflectance is performed according to equation (9).

$$\boldsymbol{\rho}_{\lambda_{\text{pixel,band}}} = \frac{\mathbf{L}_{\text{pixel,band}} * d_{ES}^{2} * \pi}{Esun_{\lambda\text{band}} * \cos\theta_s} \qquad (9)$$

Where $\boldsymbol{\rho}_{\lambda_{\text{pixel,band}}}$ is the TOA reflectance, $\mathbf{L}_{\text{pixel,band}}$ is the calculated TOA radiance, d_{ES} is the calculated Earth-Sun distance, θ_s is the calculated average solar zenith angle and $Esun_{\lambda\text{band}}$ is the WV-2 solar spectral irradiance.

The study uses two different programs to apply TOA reflectance on three data sets shown in figure 3.

- Conversion using open-source Orfeo Toolbox (OTB)
- Conversion using implemented python program

3.3. Orfeo ToolBox (OTB)

Orfeo Toolbox (OTB) is an open-source library/software for any remote sensing user, developer, or scientists. OTB has various functions for image processing such as Optical calibration application which permits converting pixel values from Digital Numbers (DN) to TOA reflectance figure 5 Optical calibration must be defined by satellite image and metadata files (IMD, TIL, XML, and RPC) to calculate TOA reflectance automatically [9]figure 6.

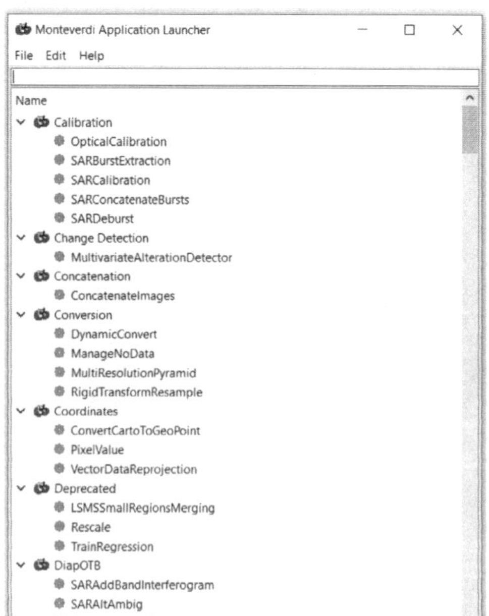

Figure 5. Orfeo ToolBox (OTB) has different applications: calibration, change detection, concatenation, conversion, Segmentation, etc…

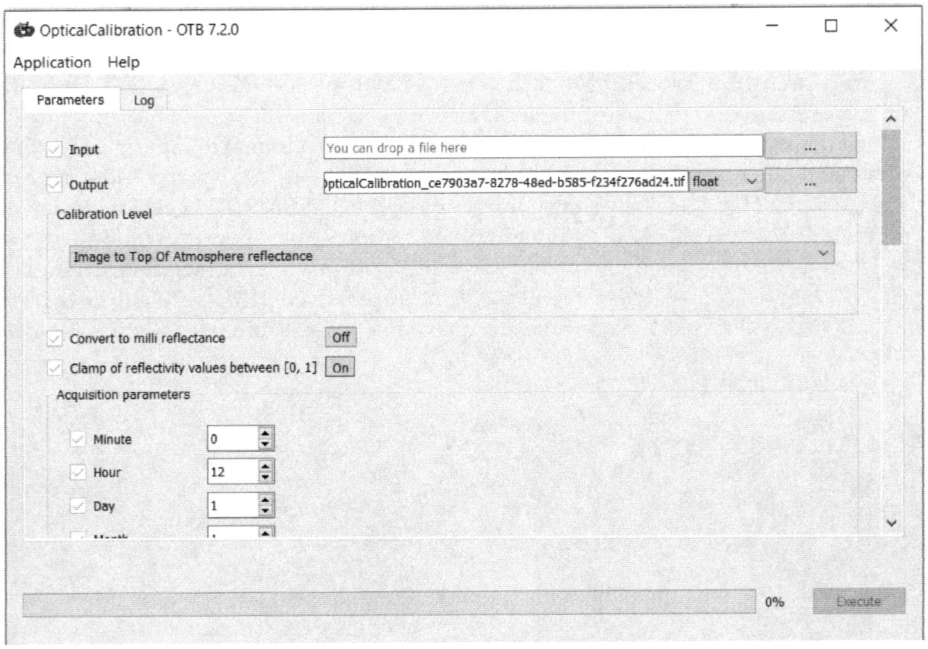

Figure 6. Optical calibration application.

3.4. Proposed algorithm

TOA reflectance conversion algorithm is implemented using high-level and general-purpose programming language Python. The implemented graphical user interface (GUI) of the proposed algorithm is designed to read the image file and attached metadata as inputs and convert the input image DN values to the TOA reflectance values in the range [0, 1] figure 7.

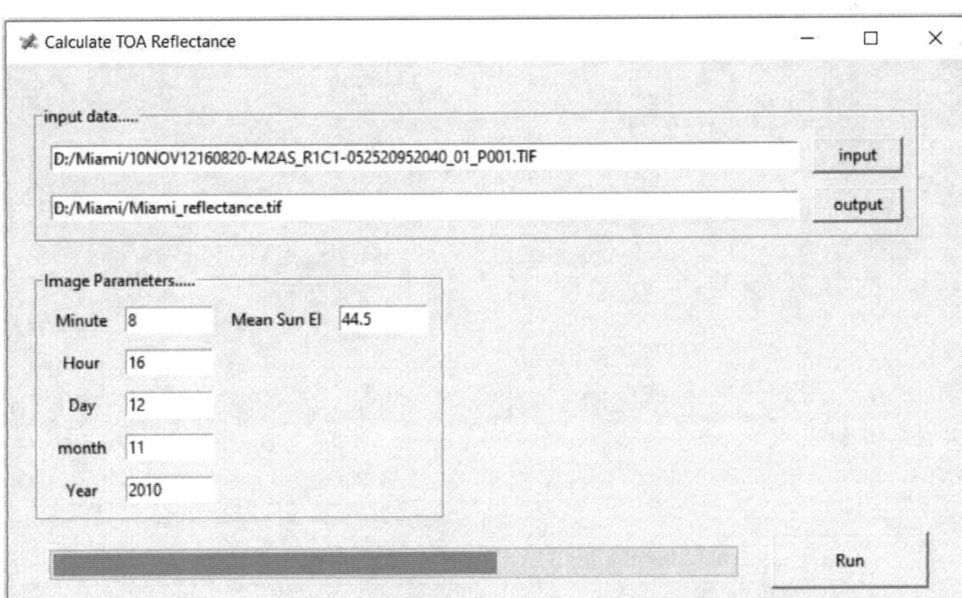

Figure 7. Proposed algorithm GUI.

4. Results and Discussion

A histogram curve is a type of histogram that represents the distribution of pixel values as a graphical representation by plotting the frequency of each value. The histogram for a specific image enables the viewer to judge the entire distribution of the image. The histogram curve is defined by the horizontal axis (X-axis) that represents the range of minimum (darkest values) to maximum (lightest values) pixel values, DN sampled in the statistics, and vertical axis (Y-axis) of a histogram that represents the frequency of each value [13]. The histograms of output reflectance for Red, Green, and Blue bands are calculated for three WV-2 satellite images in figure 3 As samples to analyze the results, figure 8, figure 9, and figure 10 show the histogram curves of the output reflectance from the proposed algorithm and OTB for selected sample bands Red, Green, and Blue respectively. The resulted histogram curves indicate that the X-axis value and Y-axis value are approximately the same distribution.

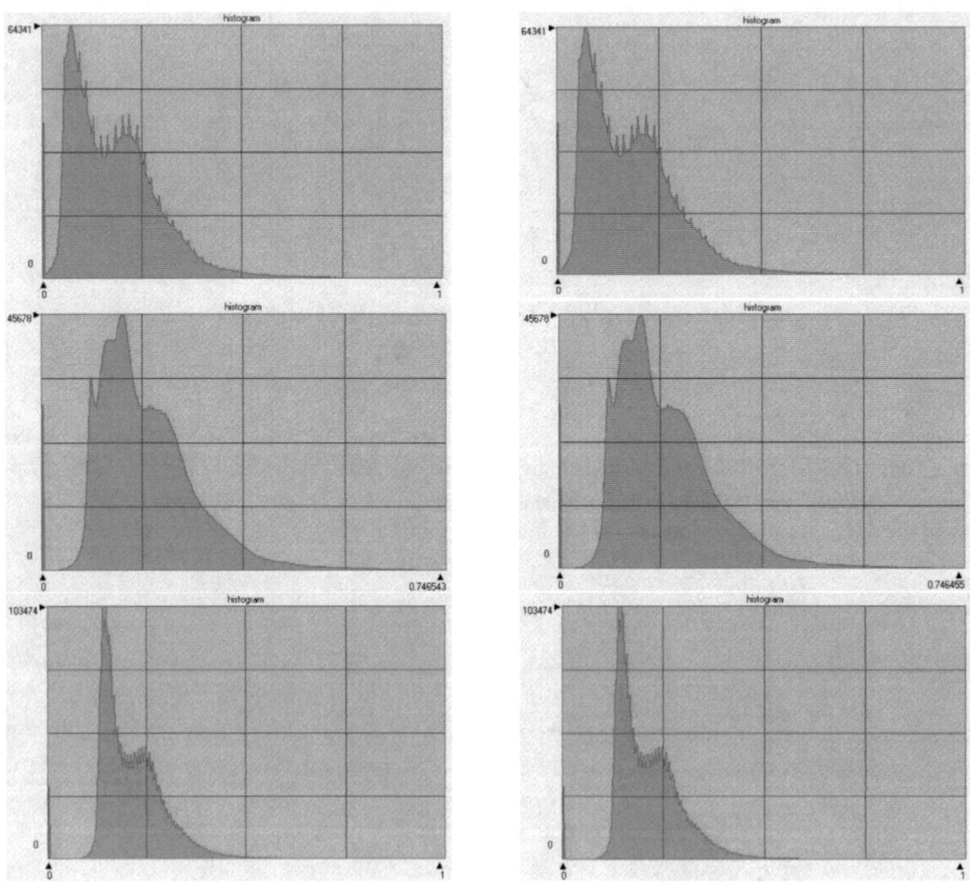

(a) (b)

Figure 8. Histogram of TOA reflectance applied on figure 3(a): (a) using OTB, (b) using the proposed algorithm.

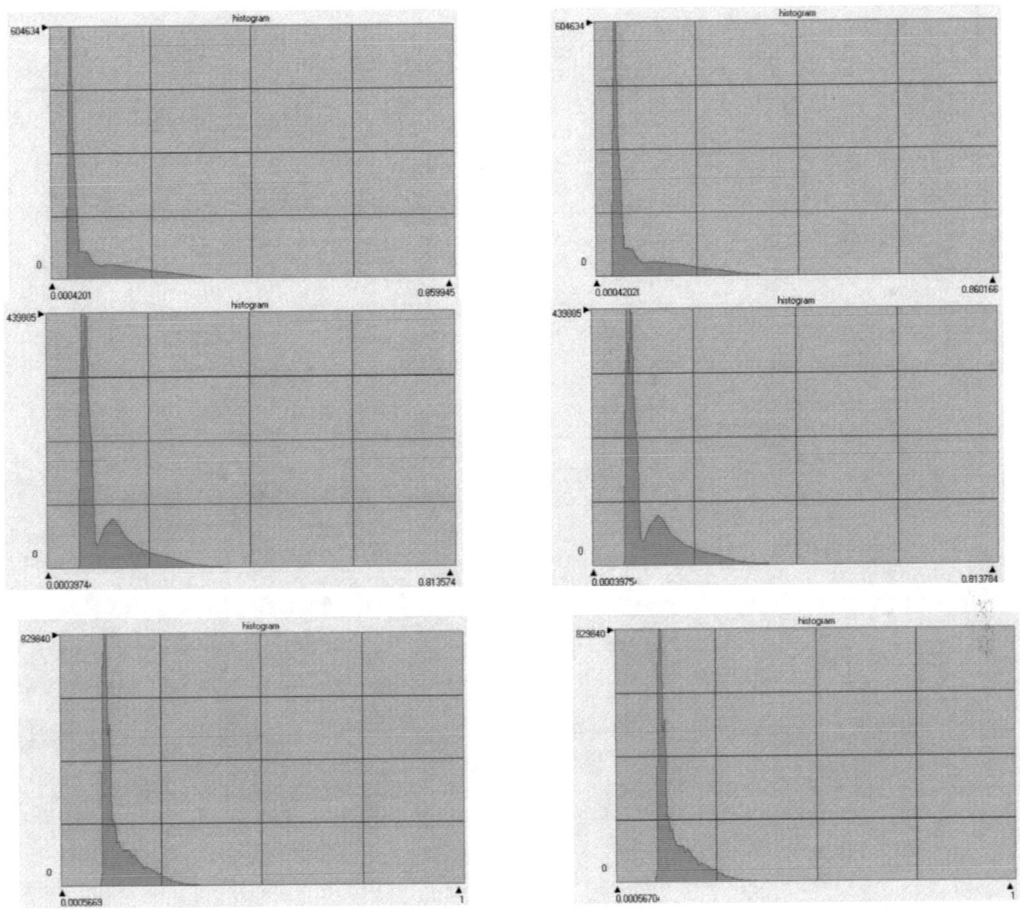

(a) (b)

Figure 9. Histogram of TOA reflectance applied on figure 3(b): (a) using OTB, (b) using the proposed algorithm.

4.1. Statistical assessment

The statistical properties of an image provide useful information, such as the total, mean and standard deviation of the pixel values. The statistical properties are calculated for the output reflectance from both software for each image data according to the minimum value seen in the column (Min), maximum value seen in the column (Max), middle value in the column (Mean), an average of the values in the column (Median), most commonly observed value in a set of data (Mode) and standard deviation (STD) calculated according to equations (10), (11), and (12) [13]. A careful study of statistical properties of generated TOA reflectance of WV-2 satellite images in figure 3, from OTB and proposed algorithm to measure the similarity, shows that they are approximately had the same statistical properties as shown in table 2.

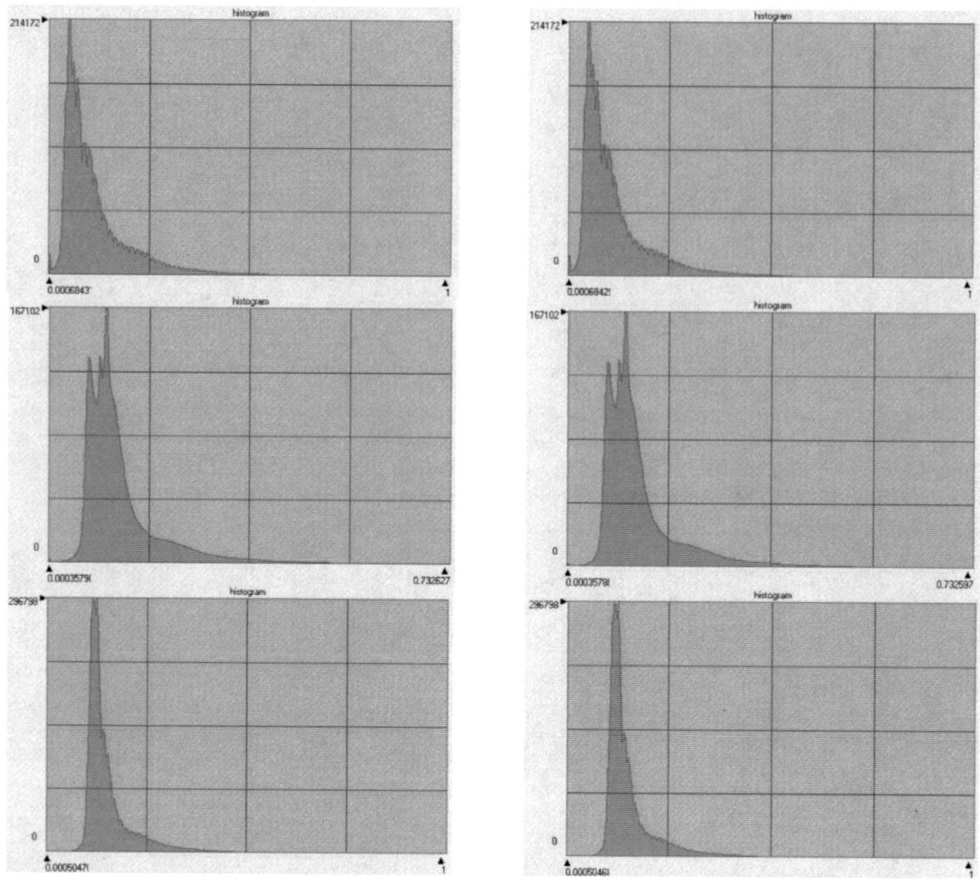

(a) (b)

Figure 10. Histogram of TOA reflectance applied on figure 3(c): (a) using OTB, (b) using the proposed algorithm.

$$\text{Mean} = \frac{\sum p_{i,j}}{N} \tag{10}$$

$$\text{Median}(P) = \begin{cases} \dfrac{X\left[\frac{N}{2}\right] + X\left[\frac{N+1}{2}\right]}{2} & \text{if N is even} \\ X\left[\dfrac{N+1}{2}\right] & \text{if N is odd} \end{cases} \tag{11}$$

$$\text{STD} = \sqrt{\frac{\sum (p_{i,j} - \text{Mean})^2}{N}} \tag{12}$$

Where $p_{i,j}$ is the pixel value of row and column, P is the ordered list of pixel values and N is the total number of pixels.

Table 2. Statistical properties

	TOA reflectance of figure 3(a)		TOA reflectance of figure 3(b)		TOA reflectance of figure 3(c)	
	OTB	Proposed algorithm	OTB	Proposed algorithm	OTB	Proposed algorithm
Min	0	0	0.00038086	0.00038095	0.00062039	0.000623037
Max	1	1	0.77961	0.77981	1	1
Mean	0.227	0.227	0.168	0.168	0.173	0.173
Median	0.21484	0.21484	0.14953	0.14957	0.15287	0.15287
Mod	0.17969	0.17969	0.1404	0.14044	0.14506	0.14506
STD	0.063	0.063	0.039	0.039	0.052	0.052

Moreover, the processing time for any application is an important factor that must be accounted for. Therefore, the processing time for each method is calculated for each data calculated according to table 3 The achieved results indicate that the proposed algorithm implemented using Python environment is faster than that of OTB.

Table 3. Processing time.

	Image size	OTB (sec.)	The proposed algorithm (sec.)
TOA reflectance of figure 3(a)	8x(1266x2048)	1.069	0.977
TOA reflectance of figure 3(b)	8x(2048x2048)	1.465	1.329
TOA reflectance of figure 3(c)	8x(2048x2048)	2.204	1.82

5. Conclusion and Future work

In this study, TOA atmospheric correction of WV-2 images is assessed by using the proposed algorithm and OTB for study areas over cities Miami, Rio de Janeiro, and Brisbane captured at different times. The output TOA reflectance of both the proposed algorithm and OTB are then compared using image statistics to measure the accuracy of the proposed algorithm output. The statistical results show that the TOA reflectance of the proposed algorithm and OTB are approximately having the same statistical properties. The TOA reflectance conversion time of the proposed algorithm is faster than OTB by 8.6%, 9.3%, and 17.4% for WV-2 satellite images in figure 3(a), (b), and (c), respectively. The overall results provide accurate, decent capacities, fast and automated algorithm for Conversion of satellite image DN values to TOA reflectance. The proposed algorithm already supports different satellite sensors and flexible to support more satellite sensors since OTB only supports specific satellite sensors. Top of Atmosphere (TOA) correction is a very important step to remove effects results from variations relative to sensor error and positions of Sun, Earth, and satellite. TOA correction pre-process is required to either analyze or compare imagery from different sensors. In future work, the proposed algorithm can be adapted to support different satellite sensors or used as the first step in another algorithm for remote sensing image classification or change detection.

References

[1] Mani V 2020 *International Journal of Sensors and Sensor Networks* **8** 1-10
[2] Dhanaraj K and Angadi D P 2020 *GeoJournal* 1-27
[3] Sterckx S and Wolters E 2019 *Remote Sensing* **11** 1-17
[4] Doxani G, Vermote E, Roger J-C, Gascon F, Adriaensen S, Frantz D, Hagolle O, Hollstein A, Kirches G and Li F 2018 *Remote Sensing* **10** 1-18
[5] Franch B, Vermote E F, Roger J-C, Murphy E, Becker-Reshef I, Justice C, Claverie M, Nagol J, Csiszar I and Meyer D 2017 *Remote Sens (Basel)* **9** 1-14
[6] Maglione P, Parente C and Vallario A 2014 *European Journal of Remote Sensing* **47** 685-99
[7] Coffer M M, Schaeffer B A, Zimmerman R C, Hill V, Li J, Islam K A and Whitman P 2020 *Remote Sensing of Environment* **250** 112036
[8] Mclnerney D and Kempeneers P 2014 *Open source geospatial tools: applications in earth observations* (Delhi Book Store: Springer International Publishing)
[9] Lee K, Kim K, Lee S-G and Kim Y 2020 *ISPRS International Journal of Geo-Information* **9** 1-16
[10] Elsharkawy A, Elhabiby M and El-Sheimy N 2012 In: *Int. Arch. Photogramm. Remote Sens. Spatial Inf. Sci.*, pp 191-5
[11] Updike T and Comp C 2010 *Technical Note* 1-17
[12] Elhabiby M, Elsharkawy A and El-Sheimy N 2012 *International Refereed Journal of* Engineering *and Science (IRJES)* **1** 01-10
[13] Nelson S A and Khorram S 2018 *Image Processing and Data Analysis with ERDAS IMAGINE®* (Boca Raton: CRC Press)

IOP Conf. Series: Materials Science and Engineering 1172 (2021) 012016 doi:10.1088/1757-899X/1172/1/012016

Hybrid fusion using Gram Schmidt and Curvelet transforms for satellite images

M Hammad[1], T Ghoniemy[1], T Mahmoud[1] and A Amein[2]

[1]Electrical Engineering, Military Technical College, Cairo, Egypt.
[2]Electrical Engineering, Senior IEEE, FISCS, October 6 University, Cairo, Egypt.

E-mail: Radok2003@hotmail.com

Abstract. Optical satellites generally provide high-resolution panchromatic but low-resolution multispectral images which provide structural details of features and spectral information respectively. Nowadays, fusion of the two types of resolutions, to have complementary information, becomes increasingly essential for many applications such as microscopic, astronomical and satellite imagery. In this paper, a novel hybrid pixel-level image fusion method is proposed for benefiting from both panchromatic (PAN) and multispectral (MUL) images. The proposed method integrates Gram Schmidt (GS) and curvelet transforms (CVT), by the aid of local energy and maximum fusion rules, for reducing individual method limitations and achieving both better spectral consistency and spatial details preservation. After a pre-processing stage, orthonormal bases are obtained for low spatial resolution images by using GS transform. Then, high-resolution and low-resolution images are fused using CVT by the aid of histogram matching. Finally, the fused image is obtained by applying both curvelet and GS inverse transforms. The performance of the proposed method is evaluated using publicly available Pleiades benchmark-datasets. Consequently, the spectral and spatial qualities of the fused images are assessed subjectively as well as objectively using different quality metrics. Moreover, the proposed method is compared with state-of-the-art fusion techniques and results show the robustness of the proposed method that has the best result in spatial and spectral evaluation metrics such as, Quality with No Reference (QNR), Peak Signal to Noise Ratio (PSNR), Standard Deviation (SD), Entropy (ENT) and Spectral Correlation Coefficient (SCC) metrics.

1. Introduction

Remote sensing satellite images have different types of resolutions due to the technical and physical constrains of satellites. Optical satellites have high spatial/low spectral resolution images over wide bandwidth and low spatial/high spectral resolution images over narrow bandwidth [1]. Therefore, there is a need to integrate the desired information from various images into a single more informative image. This is performed in order to have a clear understanding of these images for better using in different applications such as urban mapping, topographic map updating, mineral exploitation, land use/land cover mapping and natural hazards and disasters [1], and providing efficient decision-making. Accordingly, image fusion meets these needs. It achieves credibility by preservation of all relevant information, elimination of irrelevant information and noise. Moreover, fusion process should succeed in minimizing artifacts, shift and rotational invariance, stability, consistency and reducing time complexity [2]. Fused image is expressed mathematically as follows:

$$FI = F(I_1, I_2, \dots, I_N) = c_1 I_1 + c_2 I_2 + \dots + c_N I_N \quad (1)$$

Content from this work may be used under the terms of the Creative Commons Attribution 3.0 licence. Any further distribution of this work must maintain attribution to the author(s) and the title of the work, journal citation and DOI.
Published under licence by IOP Publishing Ltd

Where FI is the fused image, $I_1, I_2, \ldots, I_N$ are N input images, F represents the fusion rule, and $c's$ are constants such that $\sum_{n=1}^{N} c_n = 1$.

Nowadays, a huge number of image fusion techniques have been proliferated that are classified in several distinct ways such as acquisition mode, level of fusion, fusion domain that can be either spatial or transform, type of resolutions which the fusion process emphasizes, and type of images which are utilized in the fusion process [3-6]. On the basis of a level at which fusion is carried out, fusion methods are classified into three categories namely pixel level, feature level, and decision level [3] as shown in figure 1.

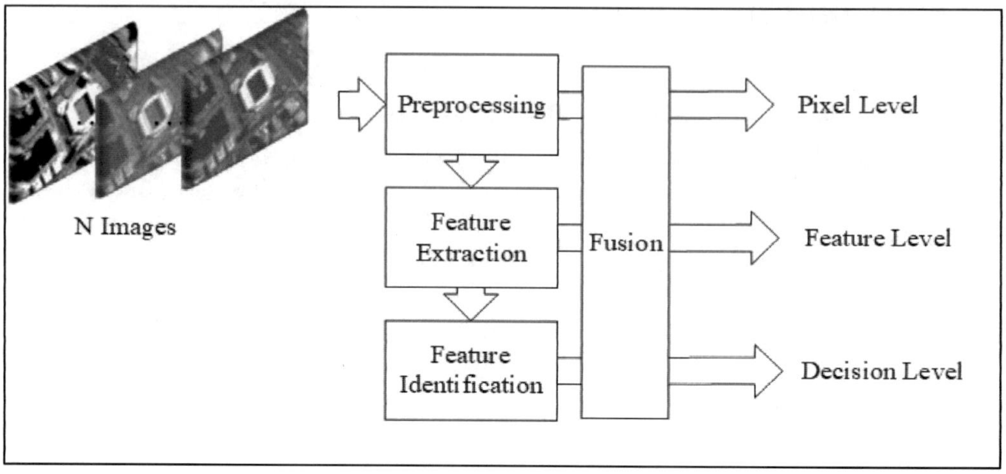

Figure 1. Fusion levels.

The reset of this paper is structured as follows. Section 2 discusses the related work. Section 3 overviews the methodology. Section 4 illustrates the data description and experimental setup. Section 5 gives the experimental results and discussion. Section 6 the conclusion is given. In Section 7, the future work is proposed.

2. Related works

Pixel level classification is relevant to this paper that is categorized into [3, 7]: (1) component substitution based (CS), (2) multiscale decomposition (MSD), (3) hybrid methods, and (4) model-based methods. CS-based methods project the low spatial resolution image into other space in which the spatial and spectral components are separated, and then the spatial component is substituted by high spatial component from high spatial resolution image. CS-based methods have high spatial fidelity, but high spectral distortion. Intensity-Hue-Saturation (IHS), principal component analyses (PCA), Brovey transform (BT), and Gram-Schmidt (GS) belong to this category. MSD-based methods try to overcome the spectral deformation problem of CS-based category and achieves better signal to noise ratio at the cost of complexity. It differs from CS-based category by the spatial details extraction method. To achieve the main target of image fusion, CS-based methods and MSD-based methods are combined to synergy the advantages of these categories and relieving individual method limitations, namely, hybrid category [7]. Typical examples of hybrid category are IHS–BT fusion, wavelet-IHS fusion, wavelet-PCA fusion, modified BT-wavelet fusion, curvelet-PCA fusion and Laplacian Filtering and Multiple Regression, etc. In [8], SAR and multispectral images are fused using Spare Representation (SR) and non-subsampled Shearlet transform (NSST) that preserve the edge and detailed information, as well as sharpness and contrast. In [9], Xiaole Ma presents new fusion method based on SR and guided filtering, the proposed approach is evaluated over a different kind of images, it preserve detail information and has a good visual analysis however, this method is inefficient if the low frequency in the image is greater than its high frequency. In [10], SAR and optical images fusion using IHS and wavelet transform Integration is introduced that made the image more sharpener and interpretable. In [11] Multispectral and panchromatic image are fused using IHS and CVT with different fusion rules that ensure the efficiency of the integration method. In [12], a combination of PCA and CVT is proposed that preserves spectral and improves spatial information. CVT is also combined with adaptive neuro-fuzzy inference system in [13] for improving the fusion process.

Motivated by the advantages of hybrid category, a new integration of GS-based and CVT-based methods is proposed as a fusion method in this paper, which uses matching criteria with a predefined threshold and local energy as a saliency measure. The proposed fusion algorithm is analyzed and compared visually and statistically with numerous CS-based and MSD-based methods over ten panchromatic (PAN) and multispectral (MUL) Pleiades satellites samples images. Result reveals that the proposed method significantly enhances the spectral as well as the spatial information.

3. Methodology

The proposed fusion method combined GS-based and CVT-based methods to overcome the limitation of individual method and have high spatial details while preserving the spectral information as much as possible as will be illustrated in the following sections. Table 1 gives the definition of the used symbols.

Table 1. Definition of the used symbols.

symbol	definition	symbol	definition
b	Number of low spatial resolution image band	$EHR_{x,y}$	Local energy around center coefficient of high resolution image
GS	Gram Schmidt component $1 \leq GS \leq$ b+1	$ELR_{x,y}$	Local energy around center coefficient of low resolution image
HR_{sim}	Simulated high spatial resolution image	T	threshold
LR	Low spatial resolution image	W_{max}	Max weight
$C_{x,y}$	Curvelet coefficient	W_{min}	Min weight
$CHR_{x,y}$	Curvelet coefficient of high spatial resolution image	SD	Average standard deviation of image
$CLR_{x,y}$	Curvelet coefficient of low spatial resolution image	$Mean$	Average mean of image
j	Scale decomposition	D_λ	Spatial distortion
$\emptyset[x,y]$	Curvelet function	D_S	Spectral distortion
θ	Orientation of curvelet	L	The dynamic range of the image being analyzed
x,y	Pixel location in image	R	Referenced image
$k1,k2$	Spatial location of curvelets	FI	Fused image
$I[x,y]$	Input image	μ_{FI}	Local mean of image FI
m	Number of rows in image	μ_R	Local mean of image R
n	Number of column in image	$p(g)$	The probability of occurrence of gth gray level
$\widehat{LR}$	Resampled low spatial resolution image	RF	Row frequency
$[X,Y]$	Dimension of used local window	CF	Column frequency
M	Matching function		

3.1. GS-based method

Since the published patent in [14], GS-based fusion method becomes one of the most interesting fusion methods [1, 15, 16]. First a simulated low-spatial resolution from the high spatial resolution image is computed, then GS-transform is applied to the simulated result and the low spatial resolution image bands as in equations (2- 5) [1], the simulated result is employed as a first component and the remaining components are computed orthogonal to it for solving the radiometric accuracy shortage in PCA-based fusion method, so it is described as a generalized PCA-based method [3]. Then the high spatial resolution image band replaces the first Gram Schmidt band. Finally, an inverse Gram Schmidt transform is applied to create the fused image. Its advantages are high preserving spectral information with high spatial details and deals with not limited numbers of images.

$$GS_1 = HR_{sim} \tag{2}$$

$$GS_2 = \widehat{LR}_1 - \frac{\langle HR_{sim}|\widehat{LR}_1\rangle}{\langle HR_{sim}|HR_{sim}\rangle} \times HR_{sim} \tag{3}$$

$$GS_3 = \widehat{LR}_2 - \frac{\langle HR_{sim}|\widehat{LR}_2\rangle}{\langle HR_{sim}|HR_{sim}\rangle} \times HR_{sim} - \frac{\langle GS_2|\widehat{LR}_2\rangle}{\langle GS_2|GS_2\rangle} \times GS_2 \tag{4}$$

$$\vdots$$

$$GS_{b+1} = \widehat{LR}_b - \frac{\langle HR_{sim}|\widehat{LR}_b\rangle}{\langle HR_{sim}|HR_{sim}\rangle} \times HR_{sim} - \sum_{k=1}^{b} \frac{\langle GS_k|\widehat{LR}_k\rangle}{\langle GS_k|GS_k\rangle} \times GS_k \tag{5}$$

3.2. CVT-based method

Curvelet transform is a multi-scale representation of image at different angle, and is a higher dimensional generalization of the wavelet transform that overcomes its limitation. CVT had been preceded through many development stages beginning from the derivation from ridge-wave theory that is complex until the second generation namely fast discrete curvelet transform algorithm that can be implemented in two ways via unequally spaced fast Fourier transform or wrapping function. In this work, a fast discrete curvelet transform algorithm is used via wrapping method which is faster and widely used [11, 12, 17]. These curvelets are arranged in various orientation and scale from (course to fine), this arrangement ensures that the entire image plane is covered to avoid signal losses as shown in figure 2.

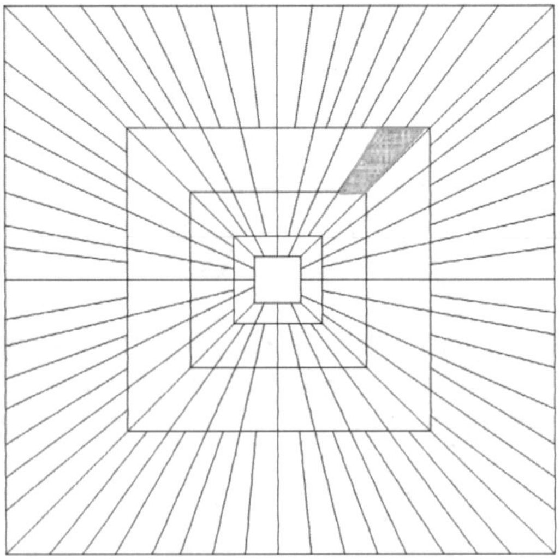

Figure 2. Basic digital tiling of curvelet transform frequency spectrum [18].

In general CVT can be expressed in time domain as in the following equation:

$$C(j,\theta,k1,k2) = \sum_{\substack{0\le x\le m\\ 0\le y\le n}} I\,[x,y]\,.\,\emptyset_{j,\theta,k1,k2}[x,y] \tag{6}$$

In figure 2, an example of 5 level curvelet digital tiling and the shaded wedged at scale 4 and at orientation 5, the center square gives approximation curvelet coefficient and the other scales give detailed curvelet coefficient.

3.3. The proposed GS-CVT method

The main contribution of this paper is to develop fusion algorithm capable of preserving the spectral as well as the spatial information. This work integrates the advantage of two fusion methods (CVT and GS) with suitable fusion rule as illustrated in the following steps, and in figure 3:

1- Resampling the low spatial resolution image with one of the interpolation methods to have the same size as high spatial resolution image. Bicubic interpolation method is used in this paper.
2- Simulated low-spatial resolution from the high spatial resolution image is computed using gaussian filter, the blurred output of the filter is used as the first component in the next step.
3- GS-transform is applied to the simulated result from step 2 and the orthogonal low spatial resolution image bands as in equations (2- 5).
4- Histogram matching is applied on the high spatial resolution image in order to have the same mean and variance of the low spatial resolution image bands GS-components.
5- Curvelet transform is applied on both the low spatial resolution image bands GS-components and high spatial resolution output from histogram matching in step 4, and the decomposition coefficients are obtained.
6- Decomposition coefficients output from curvelet transform for both images are divided to three layers, low frequency layer namely coarse-scale layer, detail scale layer, and fine scale layer.
7- The selection of fusion rule is a crucial factor for success of image fusion algorithm, this paper applies different fusion rules for each layer to rich the information of the final fused image as follows:
   a- Low frequency layer fusion rule: the energy distribution of image is reflected by its low frequency, so local energy based method is used. First matching function M is computed over the low and high spatial resolution images bands for each coefficient $[x,y]$over local neighborhood $[X,Y]$ in j^{th} level of decomposition

$$M^j_{x,y} = \frac{2\sum_X \sum_Y CHR^j_{x+X,y+Y} CLR^j_{x+X,y+Y}}{EHR^j_{x,y} ELR^j_{x,y}} \tag{7}$$

The local energy at center coefficient is:

$$E^j_{x,y} = \sum_{x-X}^{x+X}\sum_{y-Y}^{y+Y} C^2_{x,y} \tag{8}$$

If the matching function value at a certain coefficient exceeds a predefined threshold T, then the coefficients of maximum local energy function are chosen, else if the matching function value is below that threshold, a weighted addition of coefficients is taken to generate fused coefficients. The shifted sigmoid function is used for the weights to ensure that the weights fall in range from [0, 1] according to the matching values. The following equation describes fusion rule for low frequency coefficients:

$$W_{max} = 2 \times \left(\frac{1}{(1+e^{-M}) - 0.5}\right) \tag{9}$$

$$W_{min} = 1 - W_{max} \tag{10}$$

$$C^j_{x,y} = \begin{cases} CLR^j_{x,y}, & M^j_{x,y} > T, ELR^j_{x,y} > EHR^j_{x,y} \\ CHR^j_{x,y}, & M^j_{x,y} > T, ELR^j_{x,y} < EHR^j_{x,y} \\ W_{max} \times CLR^j_{x,y} + W_{min} \times CHR^j_{x,y}, & M^j_{x,y} < T, ELR^j_{x,y} > EHR^j_{x,y} \\ W_{max} \times CHR^j_{x,y} + W_{min} \times CLR^j_{x,y} & M^j_{x,y} < T, ELR^j_{x,y} < EHR^j_{x,y} \end{cases} \tag{11}$$

b- Detail scale layer, and fine scale layer coefficients fusion rule: decision map is used in order to obtain the detailed coefficients of images as it uses the average window based strategy to be more robust to mis-registration and choose the maximum strategy to get the fused coefficients map.

8- Apply inverse curvelet transform to reconstruct GS-components.

9- Apply inverse GS-transform in order to obtain the final fused image.

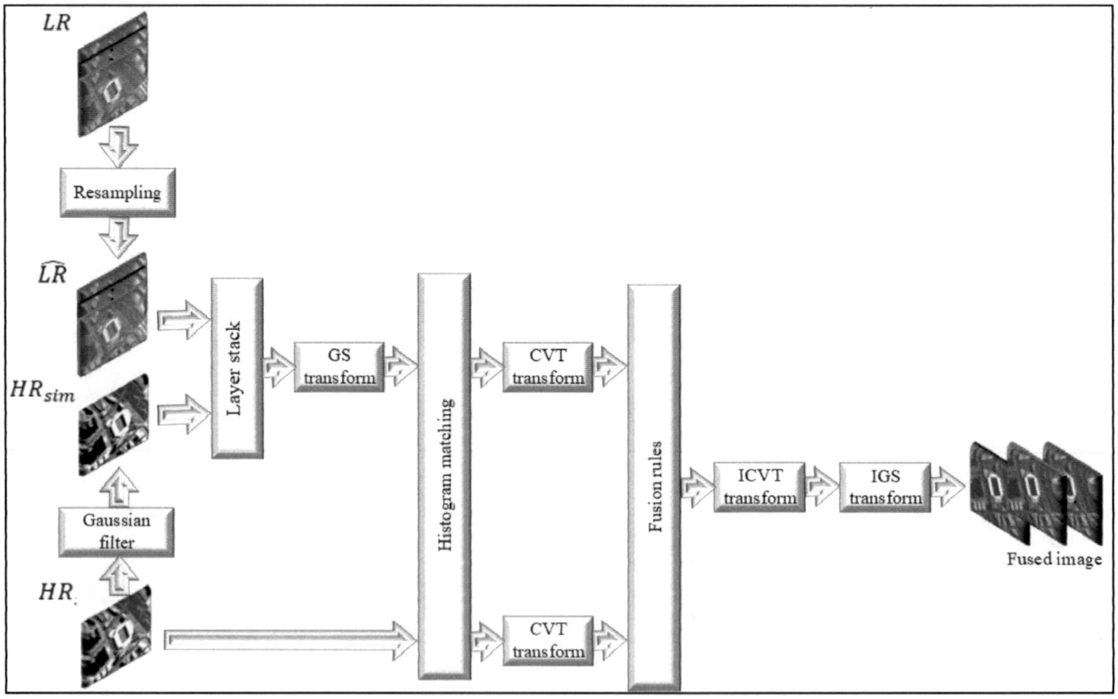

Figure 3. The proposed GS-CVT method flow chart.

4. Experimental setup

4.1. Dataset description and preprocessing

The proposed method is evaluated over ten pairs of Pleiades dataset as shown in figure 4, which is downloaded from [19] with the specifications shown in table 2. Dataset images size are 512×512 pixel for panchromatic images and 128×128 pixel for multispectral images as shown in figure 4. All the images are pre-processed such that they are registered then multispectral images are resampled using bicubic interpolation method by using ERDAS/ MATLAB software.

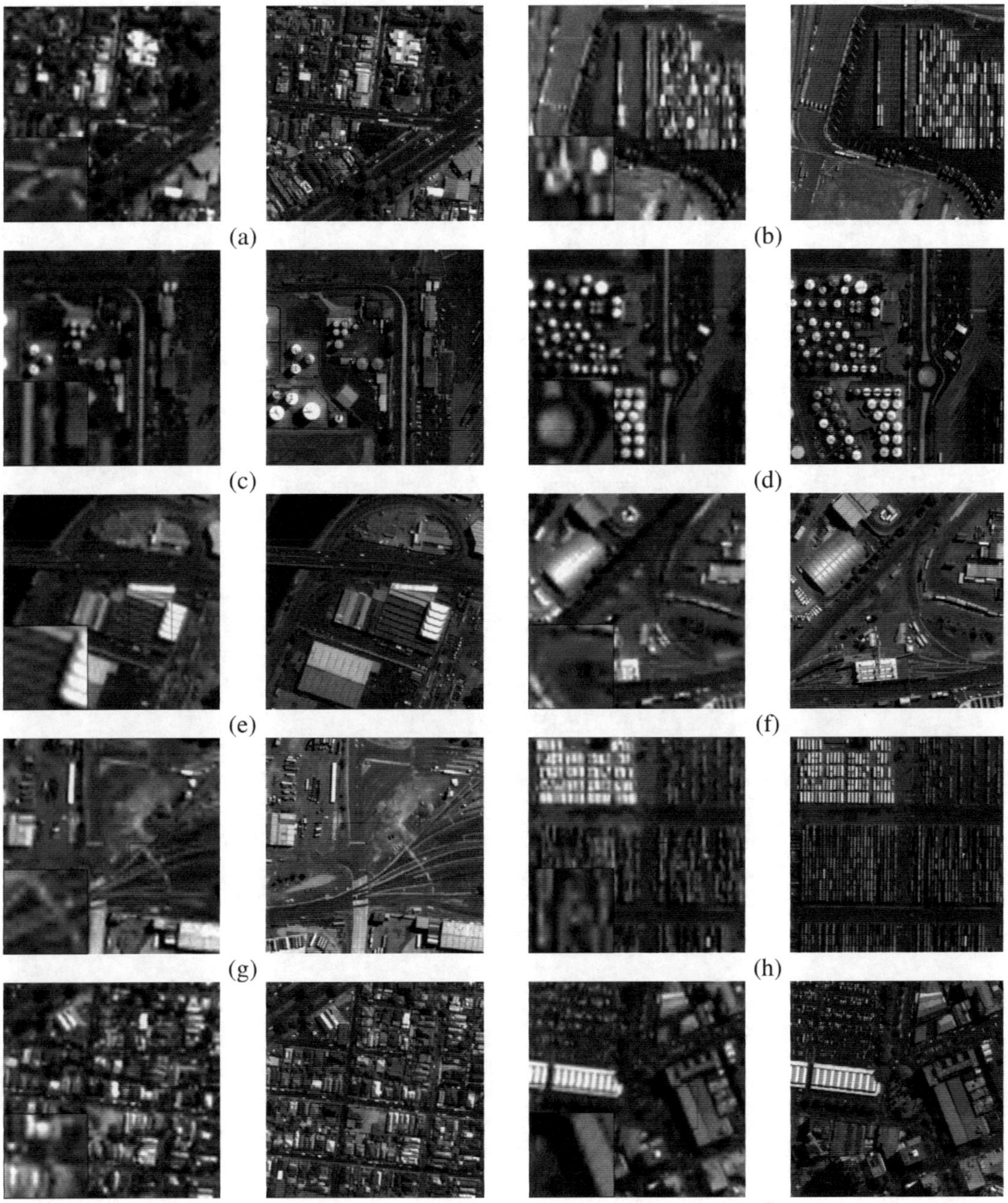

Figure 4. Ten pairs of panchromatic and multispectral Pleiades satellite dataset.

Table 2. Satellite images specification.

Band	Resolution (meter)	Projection	Datum	Date

Panchromatic **Four multispectral bands (Red, Green, Blue & NIR)**	0.5 2	UTM, Lat. / Long.	WGS 84	04-05-2012

4.2. Metrics
The visual analysis of the image does not represent the real quality analysis metric of images as it depends on human eyes, which are sensing different from an interpreter to another. In this paper the analysis is extended by using spatial and spectral metrics as illustrated in the following sections [3].

4.2.1. Quality with no reference (QNR) index. It calculates the distortion of image quality as a relation between the complement of spatial and spectral distortion multiplication D_λ and D_s respectively. Its value from zero to one, the higher value is the better.

4.2.2. Peak Signal to Noise Ratio (PSNR). It is the ratio between the intensity levels in the referenced source image and the related pixels in the fused image. The higher value of PSNR means higher quality.

$$PSNR = 10\log_{10}\frac{l^2}{\left(\frac{\sum_{i=1}^{m}\sum_{j=1}^{n}[FI(i,j) - R(i,j)]^2}{m \times n}\right)} \tag{12}$$

4.2.3. Standard Deviation (SD). It is a statistical measure that computes the diversity of the image as a contrast value that indicates the information availability in the image. The higher value is the better.

$$SD = \sqrt{\frac{\sum_{i=1}^{m}\sum_{j=1}^{n}(FI(i,j) - \mu_{FI})^2}{m \times n}} \tag{13}$$

4.2.4. Correlation Coefficient (CC). It measures statistically the strength and direction of the linear relationship between the referenced source image and the fused image. It indicates the spectral integrity if the reference image that is used is the multispectral image while it indicates the spatial integrity if the reference image that is used is high spatial image and then CC called spectral correlation coefficient (SCC). The ideal value is (+1) in range from [-1: 1].

$$CC= \frac{\sum_{i=1}^{m}\sum_{j=1}^{n}(FI(i,j)-\mu_{FI})(R(i,j)-\mu_R)}{\sqrt{\sum_{i=1}^{m}\sum_{j=1}^{n}((FI(i,j)-\mu_{FI})^2((R(i,j)-\mu_R))^2)}} \tag{14}$$

4.2.5. Entropy and Cross Entropy (ENT). They are statistical measures that indicate the information availability in the image. The higher value is the better.

$$ENT = -\sum_{g=1}^{L-1} p(g)\log_2 p(g) \tag{15}$$

4.2.6. Spatial Frequency (SF). It computes the spatial information level in both row and column directions. The higher value is the better.

$$RF = \sqrt{\frac{1}{m \times n}\sum_{i=1}^{m}\sum_{j=2}^{n}(FI(i,j) - R(i,j-1))^2} \tag{16}$$

$$CF = \sqrt{\frac{1}{m \times n}\sum_{i=2}^{m}\sum_{j=1}^{n}(FI(i,j) - R(i-1,j))^2} \quad (17)$$

$$SF = \sqrt{RF^2 + CF^2} \quad (18)$$

4.2.7. The blur effect. This metric is sensitive to focal and motion blur and also to edges, also it is sensitive to the presence of noise [20].

5. Results and evaluations

5.1. Subjective evaluation

As shown in figure 5, the fused images are obtained by seven different state-of-the-art fusion techniques either CS-based method (Brovey, Ehler, GS, HIS and PCA) or MRA-based method (DWT and CVT) are different in their spatial and spectral resolution according to the used fusion methods. In CS-based method generally, the main problem is the spectral deformation, while in MRA is the loss of spatial details. The proposed GS-CVT fusion method enhances the spectral information as well as enriches the spatial details. It is noticed that the fused images using the proposed technique have the best result visually in comparison with the other competing techniques.

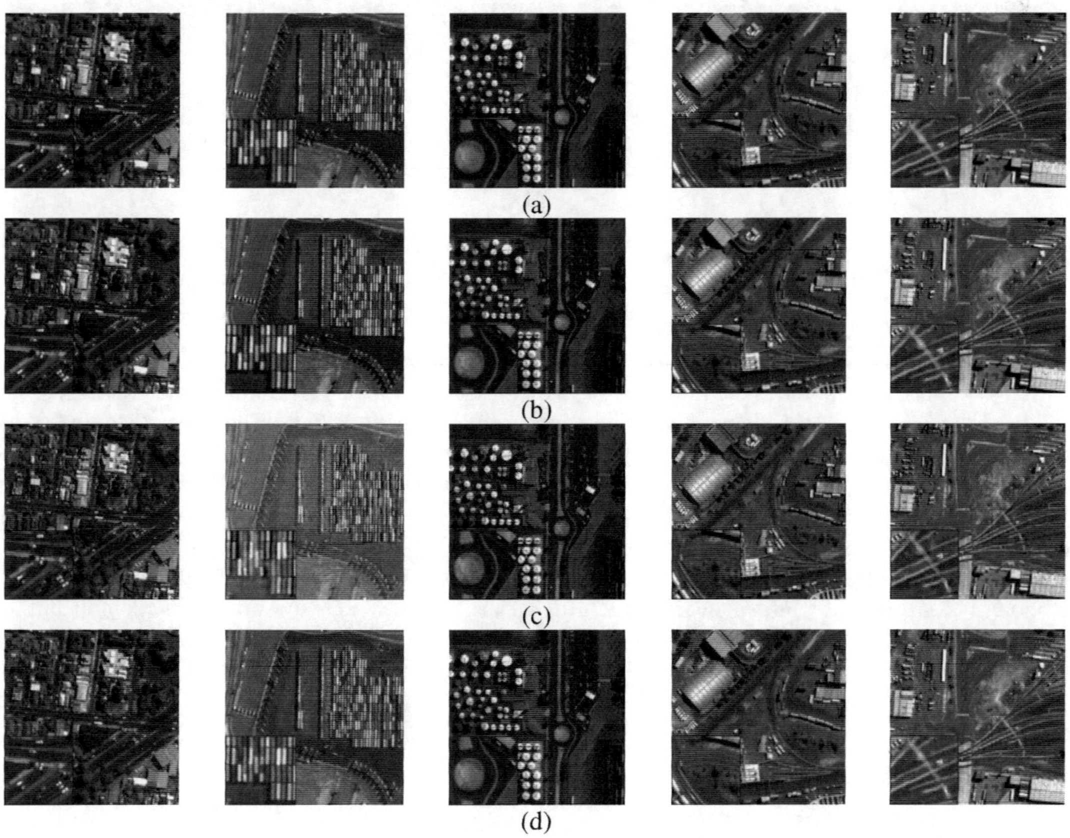

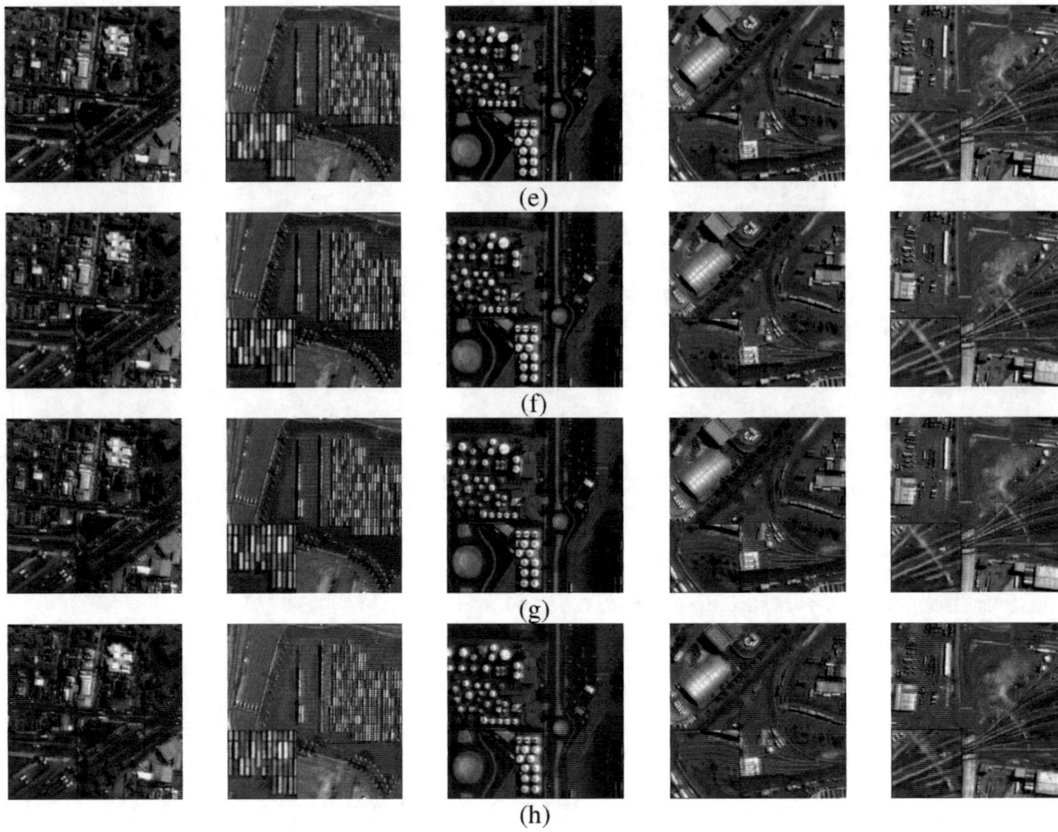

Figure 5. Example of applying different fusion techniques on Pleiades satellite dataset; (a) Brovey, (b) Ehler, (c) GS, (d) IHS, (e) PCA, (f) DWT, (g) CVT, (h) GS-CVT.

5.2. Objective evaluation

Table 3 shows the average values of objective evaluation metrics over the Pleiades benchmark-dataset, the proposed fusion method has the highest value of QNR that indicates minimum distortion; DWT and CVT have lower values and higher distortion. The proposed method has the highest value of PSNR indicates the high quality of the fused image followed by GS- based method then Brovey, the lowest value is DWT-based method as shown. CS-based methods have the lowest values of SD while the proposed method has the highest value that indicates its high contrast and image sharpener. For CC-metrics the best value is Ehler-based method followed by MRA-based method, then CS-based methods but for SCC the best value for the proposed method followed by MRA-based method, the CS-based methods. The measure of information availability ENT indicates that the proposed method has the highest value followed by MRA-based methods, the CS-based methods. Ehler-based method has the highest SF value followed by the proposed method, and the lowest values for the CS-based methods. The best blur values are assigned to MRA-based method followed by the proposed method, then CS-based method.

Table 3. Objective evaluation of fusion results.

Fusion method	blur	CC	SCC	Dλ	DS	QNR	Entropy	SF	PSNR	SD
PCA	0.343	0.976	0.777	0.063	0.237	0.716	1.646	6.248	40.166	228
IHS	0.344	0.978	0.767	0.073	0.242	0.704	0.649	6.103	40.43	228
Brovey	0.344	0.974	0.762	**0.053**	0.237	0.723	1. 659	5.913	40.584	229
Ehler	0.337	**0.99**	0.761	0.119	0.311	0.61	1.022	**11**	34.979	312
GS	0.334	0.944	0.731	0.086	0.242	0.694	1.73	6.943	40.716	246
DWT	0.320	0.985	0.769	0.16	0.313	0.581	1.917	8.695	34.894	305
CVT	**0.314**	0.983	0.788	0.164	0.295	0.594	1.858	9.855	35.981	289
GS-CVT	0.327	0.936	**0.804**	0.069	**0.219**	**0.728**	**2**	10	**41.992**	**329**

5.3. Competitive approaches

In this paper, the proposed GS-CVT-based method not only compared by the state-of-the-art methods but also compared with a newly published method that combines PCA and CVT methods [12]. The proposed method uses GS transform which is more generic than PCA in the hybrid fusion to overcome the radiometric accuracy shortage. Also the benefits from local fusion strategy that are proposed to make the fused image more robust than global fusion strategies. The comparison is done subjectively and objectively. The subjective evaluation indicates that PCA-CVT-based method is blurrier than this paper proposed method and also the spectral information are more preserved by the proposed method so the fused image looked more interpretable as shown in figure 6. It is worth mentioning that PCA-CVT method in [12] is implement by MATLAB where there is no open source code for it, also the used dataset for [12] is not uploaded, so there is no unified dataset.

(a)

(b)

Figure 6. PCA-CVT-based method and GS-CVT-based method over sample Pleiades dataset.

It is clear from figure 6 that the GS-CVT-based method has sharpener image and good coloured compared to the other method. Table 4, represents different average values of objective evaluation metrics that reflect that PCA-CVT-based method can be objectively better than GS-CVT-based method in some metrics such as QNR, SCC , and Entropy while GS-CVT-based method is better in $CC, SD, SF, and\ Blur\ metrics$. From ranking table 3 and table 4 using the values of individual methods and their integration arranged such that best values take small values and the worst values take the large values in order as shown in figure 7, the proposed GS-CVT integration is the best that can improve the performance of the fusion process as it has the lowest value.

Table 4. Objective evaluation of fusion results by PCA-CVT and the proposed method.

Fusion method	blur	CC	SCC	QNR	Entropy	SF	PSNR	SD
GS-CVT	**0.327**	**0.936**	0.804	0.728	2	**10**	41.992	**329**
PCA-CVT	0.441	0.876	**0.962**	**0.864**	**2.068**	7	**43.937**	323

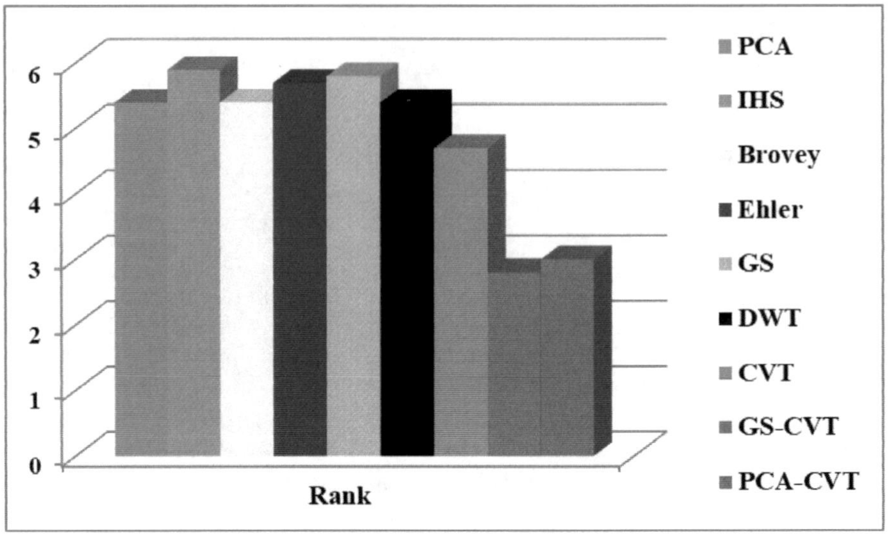

Figure 7. Ranking fusion methods in order

6. Conclusion

In this work, a proposed method based on GS and CVT methods is used in merging high spatial resolution panchromatic image and low spectral resolution multispectral image from remote sensing satellite Pleiades dataset. The fusion is done using a novel fusion rule that depends on predefined matching criteria over local window. The proposed method is compared with not only state-of-the-art methods but also a newly published work that combines PCA and CVT. The superiority of the proposed method is validated subjectively and objectively using $QNR, PSNR, CC, SCC, SD, SF, and\ Blur\ metrics$. The proposed method presents an efficient solution for fusion process of Pleiades satellite dataset, and provides high quality fused images spatially and spectrally.

7. Future work

Fusion process is a sophisticated operation that has different parameters as it can be done in many domains (spatial, frequency), and can be used in many applications using numerous types of images, etc. so traditional fusion methods are not the optimum fusion methods. It leads to using new fusion methods to get the optimum quality fused image. In order to estimate the optimum fused image we will take in consideration in future work to use the cost function in order to optimize the weights by using iterative methods based on statistical parameters to be more generalized in the fusion process.

Reference

[1] Pohl C and Van Genderen J 2016 *Remote sensing image fusion: a practical guide*: CRC Press

[2] Poobalasubramanian M, V S, Karthick S, Haribaabu V, Santhanam R and Samuel J 2020 A Review of Multi-resolution Analysis (MRA) and Multi-geometric Analysis (MGA) Tools Used in the Fusion of Remote Sensing Images *Circuits, Systems, and Signal Processing* **39**

[3] Kulkarni S and Rege P 2020 Pixel Level Fusion Techniques for SAR and Optical Images: A Review *Information Fusion* **59**
[4] Belgiu M and Stein A 2019 Spatiotemporal image fusion in remote sensing *Remote Sensing* **11** 818
[5] Ghamisi P, Rasti B, Yokoya N, Wang Q, Hofle B, Bruzzone L, Bovolo F, Chi M, Anders K and Gloaguen R 2018 Multisource and Multitemporal Data Fusion in Remote Sensing *arXiv preprint arXiv:1812.08287*
[6] Bhataria K C and Shah B K 2018 A Review of Image Fusion Techniques. In: *2018 Second International Conference on Computing Methodologies and Communication (ICCMC),* pp 114-23
[7] Imani M 2020 A collaborative representation-based approximation method for remote sensing image fusion *International Journal of Remote Sensing* **41** 974-95
[8] Liu K and Li Y 2019 SAR and multispectral image fusion algorithm based on sparse representation and NSST. In: *AIP Conference Proceedings*: AIP Publishing LLC p 020059
[9] Ma X, Hu S, Liu S, Fang J and Xu S 2019 Remote sensing image fusion based on sparse representation and guided filtering *Electronics* **8** 303-20
[10] Zhang W and Xu M 2019 Translate SAR data into optical image using IHS and wavelet transform integrated fusion *Journal of the Indian Society of Remote Sensing* **47** 125-37
[11] Dong L, Yang Q, Wu H, Xiao H and Xu M 2015 High quality multi-spectral and panchromatic image fusion technologies based on Curvelet transform *Neurocomputing* **159**
[12] Chen C, He X, Guo B, Zhao X and Chu Y 2020 A pixel-level fusion method for multi-source optical remote sensing image combining the principal component analysis and curvelet transform *Earth Science Informatics* 1-9
[13] Devulapalli S and Krishnan R 2019 Synthesized pansharpening using curvelet transform and adaptive neuro-fuzzy inference system *Journal of Applied Remote Sensing* **13** 034519
[14] Laben C A and Brower B V 2000 Process for enhancing the spatial resolution of multispectral imagery using pan-sharpening. Google Patents)
[15] Zhang D-D, Xie F and Zhang L 2018 Preprocessing and fusion analysis of GF-2 satellite Remote-sensed spatial data. In: *2018 International Conference on Information Systems and Computer Aided Education (ICISCAE)*: IEEE pp 24-9
[16] Yilmaz V, Serifoglu Yilmaz C, Güngör O and Shan J 2020 A genetic algorithm solution to the gram-schmidt image fusion *International Journal of Remote Sensing* **41** 1458-85
[17] Chen C, He X, Chu Y and Zhao X 2020 A new remote sensing image fusion method combining principal component analysis and curvelet transform *MS&E* **780** 032054
[18] Candès E, Demanet L, Donoho D and Ying L 2006 Fast Discrete Curvelet Transforms *SIAM Journal on Multiscale Modeling and Simulation* **5**
[19] Harrisgeospatial.com. 2021. L3Harris Geospatial | Remote Sensing Technologies | Geospatial Data and Imagery.
[20] Crété-Roffet F, Dolmière T, Ladret P and Nicolas M 2007 The Blur Effect: Perception and Estimation with a New No-Reference Perceptual Blur Metric *Human Vision and Electronic Imaging* **12**

IOP Conf. Series: Materials Science and Engineering 1172 (2021) 012017 doi:10.1088/1757-899X/1172/1/012017

Output power boosting of a photovoltaic panel based on various back pipe structures: a computational study

A Bayoumi[2], G M Abdo[2], and A A Emara[1,2]

[1] Mechanical Engineering Department, Faculty of Engineering, Mataria- Helwan University in Cairo, Egypt.
[2] Mechanical Engineering Department, Faculty of Engineering and Technology, Badr University in Cairo (BUC), Cairo, Egypt.

E-mail: gamal.abdo@buc.edu.eg

Abstract. The efficiency of photovoltaic panels (PV) drops due to the rise in temperature, which leads to a decrease in the PV output power. A PV/Thermal system is therefore used as a solution to increase the output power from the PV panels. Comsol Multiphysics software program, a simulator-based Finite Element Method (FEM) tool, and Matlab program simulations are used to perform this electro-thermal model. The simulation process for different back pipes follows a serpentine, and a new shape of pipes called (square shapes) which are attached under the PV module. These shapes were specifically chosen for higher conversion efficiency and increase the heat transfer of the system. Additionally, a comparison between electrical parameters and heat transfer characteristics of water and CuO/water nanofluid in a PV cooling system has been studied. The new shape leads to improve the photovoltaic (PV) module parameters, such as short circuit current ISC, open-circuit voltage VOC, and maximum power Pmax for a new shape more than the serpentine shape. These parameters are calculated under Air Mass 1.5 G (AM1.5G) with 1000 W/m2 of irradiance which is considered the average irradiance at the MENA (The Middle East and North Africa) region throughout the year. The results exhibit the PV module's total enhancement by using the new shape with CuO nano-fluid about 24.7 %.

1. Introduction

The PV module is one of the most significant sources of renewable energy in countries that have high-temperature weather. This is due to its cheapness, availability, and eco-friendly behavior. On the contrary, the efficiency of PV modules dropped which leads to the output power reduction in this region. Therefore, the goal is to achieve a high photovoltaic conversion efficiency by reducing the PV module temperature in general,[1].

Reviewing the Previous Works, offered different ways to decrease the PV module temperature. To cope with the rather high module temperature, different passive in addition to the active cooling techniques which have been verified in previous works are investigated [2] and [3]. As for active cooling, the cooling fluid is one of the main factors that affects the system performance [4]. Air, water, and nano-fluids have been recorded as a cooling fluid to decrease the PV module temperature [5],[6] [7][8][8].

Teo et al. investigated a PV module cooling system with an air duct in parallel pipes attached under the modules, where ever the PV modules were cooled down by forced convection. The performance enhanced by about 4-5% [9].

Yang et al. improved a new material called Functional gradient material (FGM) attached as PV layers under the module, copper pipes underneath the layers that were attached by thermal paste under

Content from this work may be used under the terms of the Creative Commons Attribution 3.0 licence. Any further distribution of this work must maintain attribution to the author(s) and the title of the work, journal citation and DOI.
Published under licence by IOP Publishing Ltd

the backside of a PV/T system. Water as a cooling fluid is pumped with different flow rates through the copper pipes to reduce the PV module temperature, therefore, increasing PV/T system electrical power by about 2%. This work also tested the thermal efficiency of the closed system design and the results showed that the electrical efficiency increased to 71% of the output power of the system. Herein, a simulation model for a PV/T open system is introduced. The attached water pipe is heated by the effect of the temperature increase of the PV modules due to the Sun irradiance. The results exhibited an advancement in the total efficiency of the system by raising the temperature of water output [10].

Afroza Nahar et al. enhanced the PV module output performance by a 3D numerical method (Comsol Multiphysics package). The new shape (pancake shape) is attached to the backside of the PV module to increase PV output power by using the water as a cooling fluid. The new channel was made from two different materials (copper & aluminum) under various inlet velocities (0.0009 to 0.05 m/s) and compared with the system without cooling. The simulation results showed an increase in the overall efficiency of about 14.6% for aluminum and 16.3% for copper pipes [11].

Nooshin Karami and Masoud Rahimi introduced a novel method to decrease the temperature of the PV cells experimentally using different shapes of pipes (Straight and Helical) connected in the backside of PV cells. K-type thermocouples were placed in the flow path of nano-fluids (Boehmite AlOOH_xH2O) with a mixture concentration of 0.1 wt% to show the temperature variation when the fluids moved inside the pipes. The temperatures of the cell are reduced by 3°C in each temperature sensor with a growing in electrical power over the entire day. Whereas the straight shape showed an increase of 20.57% while the helical shape increased by 37.67 %, [12].

This work aims to enhance the output power of the PV panels in the MENA region [13]. This study shows an innovative simulation model for a PV panel with attached different backside pipe structures (Serpentine and square). The shapes of pipes by the Comsol Multiphysics package are illustrated. The suggested simulation model is capable of simulating the water and CuO nano-fluid thermal behavior and expressing its effect on the PV module performance.

2. Model Description

A computational-based model using the Comsol Multiphysics package for a Photovoltaic/Thermal hybrid system (PV/T) is executed as given in Figure ***1***. It is divided into a closed technique laminar flow and sequence heat transfer physics package, which is being connected to show the convective heat exchange of the fluid inside the copper pipes from the PV panel to cool it by finite Element Method (FEM). In this studied model, a mesh independence test of 0.00542 m and a maximum size of 0.212 m has been utilized [14].

The simulated system includes a mono-crystalline PV panel with a total area of 1.5013 m^2 and the maximum output power of 280 Watt as shown in Figure ***1a***. Various shapes of pipes bonded on the copper plate as the same area of PV panel with 0.001 m thickness. The copper plate is attached to the backside surface of the PV panel to allow the fluid to move through the pipes under the PV panel. All suggested shapes of pipes are made from copper with dimensions 0.005 m and 0.006 m for both inner and outer diameters. The copper pipes are chosen because of their high thermal conductivity and wide use [15]. The volumetric flow rate inside the pipe 10 L/m to speed up the enhancement performance. ***Figure 1***-b exhibits the structure of the first backside water pipe, serpentine shape, with a length of 16.1 m long and a total area of 0.631 m^2. Figure ***1***- c shows the second type, square shape, with an area of 1.193 m^2 and length 31.78 m long. The nano-fluid is a mixture of water and CuO nanoparticles. In the present study, the concentration of nano-particles by about 0.01 wt% is used. The goal for selecting CuO nano-particles is for higher thermal properties [16].

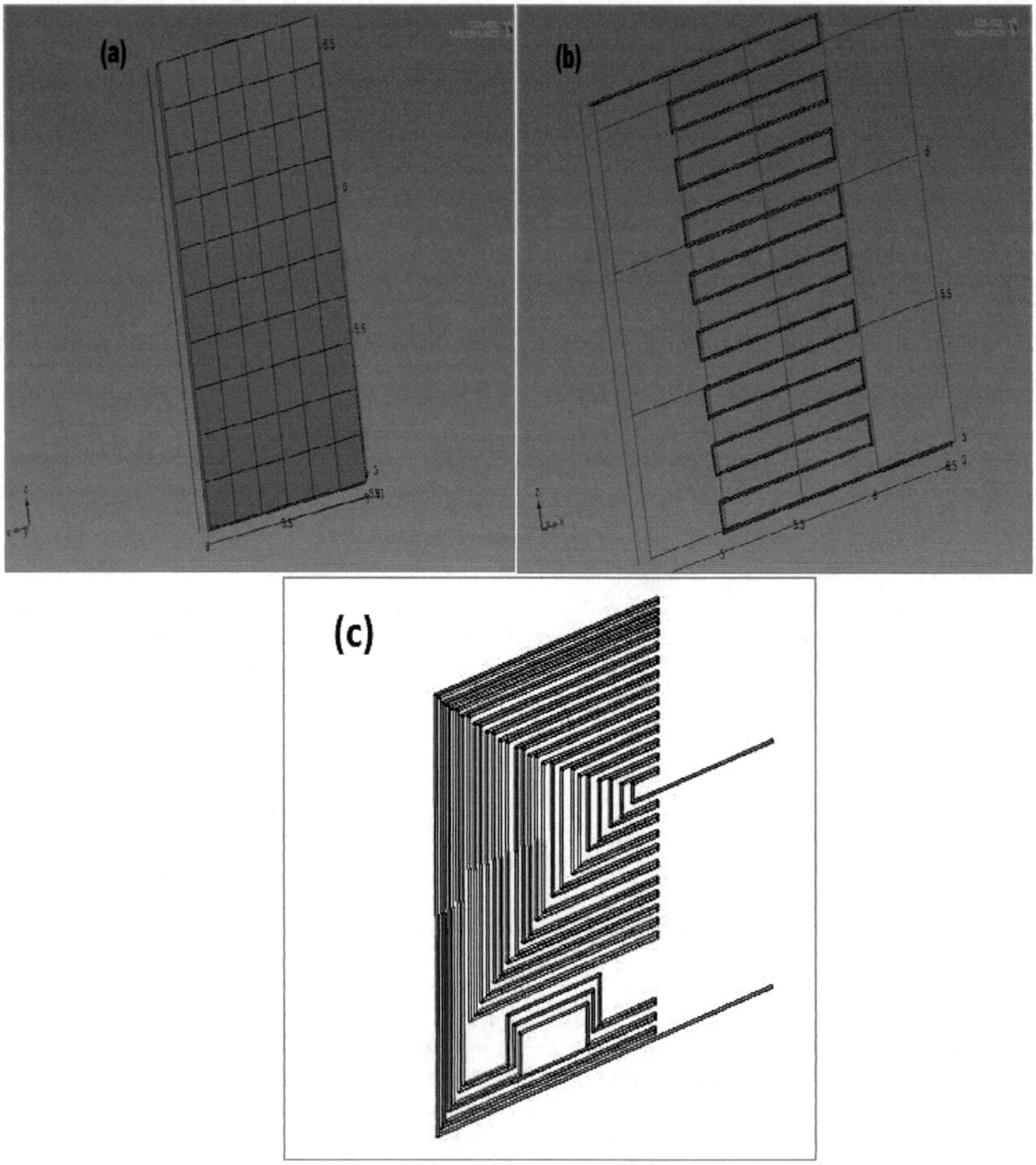

Figure 1: (a) Comsol software program model for a PV panel, (b) Serpentine back pipes, and(c) Square back pipes.

3. Methodology

The modeling in this work is divided into two parts. Firstly, the thermal modeling to decrease the PV module temperature. Secondly, the electrical modeling to show the enhancement in $\mathbf{I_{sc}}$, $\mathbf{V_{op}}$, and the output power of the system.

In the thermal modeling, the sun irradiance under (AM1.5G) standard is increased the silicon PV panel temperature which is the body (PV panel) passes through its heat to the copper plate and pipes (external and internal) by conduction. At last, the cooling liquid (water or nano-fluid) is heated by convection. The thermal parameters are calculated for a closed thermodynamic technique [17]. The overall heat transfer coefficient is given by equations (1, 2, and 3) and the rate of heat transfer is calculated by equation (4) for an open thermodynamic system [18].

$$\frac{1}{U\,A} = \frac{t_{th}}{K_{Si}\,A} + \frac{t_{th}}{K_{cu}\,A} + \frac{ln(\frac{r_{in}}{r_{out}})^2}{2\,\pi\,K_{Pipe}\,L} + \frac{1}{h_f\,a} + \frac{1}{h_{air}\,a} \quad (1)$$

$$A = l \times w \quad (2)$$

$$a = 2\pi\, r\, L \quad (3)$$

$$Q^{\cdot} = A\,U\,\Delta\,T \quad (4)$$

The electrical parameters are calculated after the thermal parameters to show the influence of the improvement in the output power on the PV performance. PV module efficiency $\boldsymbol{\eta_{PV}}$ is calculated by dividing the output electrical power (at maximum output power) by the total input power under the AM1.5G standard. The short current based on FEM is calculated by using equation (5) [19]:

$$I_{sc} = I_{PV} - I_{01}\left(eq\frac{V}{K_B T} - 1\right) - I_{02}\left(eq\frac{V}{K_B T} - 1\right) \quad (5)$$

The open-circuit voltage is given by equation [19](6) as follow:

$$V_{oc} = n\,\frac{KT}{q}\ln[\frac{I_{PV}}{I_{01}} + 1] \quad (6)$$

The solar cell efficiency can be represented as a variation in the solar output electrical power, the module surface area, and the solar irradiance (see equation (7))[19].

$$\eta_{PV} = \frac{Pout}{Pin} = \frac{V_{oc} I_{sc} FF}{q_{rad}\,A} \quad (7)$$

The tested nanofluid is considered a mixture between the water and the CuO nanoparticles. In the present study, the CuO oxide of 0.01% mass concentration is used. The purpose for selecting Copper Oxide is that openly known thermal properties and easy dispersion [20]. The nano-material properties have a density = 6400 kg/m^3, size = 40 nm, thermal conductivity = 40 W/m.K, the specific heat 892 J/kg.K and the shape type of particles is a spherical shape. The nano-fluid density is determined from Pak and Cho's equation (8) [21].

$$\rho_{nf} = \rho_f(1\text{-}\emptyset) + \rho_p.\,\emptyset \quad (8)$$

The specific heat of the nano-fluid is deduced from Xuan and Roetzel's equation (9).

$$(C_P\rho)_{nf} = (C_P\rho)_f(1\text{-}\emptyset) + (C_P\rho)_p\emptyset \quad (9)$$

The nano-fluid thermal conductivity is calculated from Maxwell's equation (10) [22].

$$k_{nf} = k_f\frac{(k_p + 2k_f - 2\emptyset\,k_f - (k_p)}{((k_p + 2k_f + \emptyset\,k_f + (k_p)} \quad (10)$$

At the results, the mixture of nano-fluid properties are density= 1201.7 kg/m^3, thermal conductivity= 0.7431 W/m.K and specific heat= 5202.11 J/kg.K.

4. Results and Discussion

This section demonstrates the results and the descriptive analysis included in the present work. The calculation of the thermal parameters of backside temperatures is performed to improve the performance of the PV system. Figure 2 shows the comparison of the module temperature (with (square and serpentine shapes) without cooling (STC) at different shapes attached in the backside of the PV module by using water along with daytime. This is to study the influence of the fluid used to minimize the temperature of the PV/T performance by using water and to show the best shape that can be performed by nano-fluid by comparing with (STC) Standard Temperature Condition without cooling the PV panel temperature reduced by water flew inside the backside pipes attached under the module. The temperature of the PV module backside without cooling reached 53.9°C while the PV module temperature with cooling by using serpentine shape decreases by 5.1°C and the same PV module temperature with square shape reduces by 15.2°C. Surly, the square shape enhanced the PV module performance by about 17.7%. The system improved due to a rise in the total area of pipes of square shape.

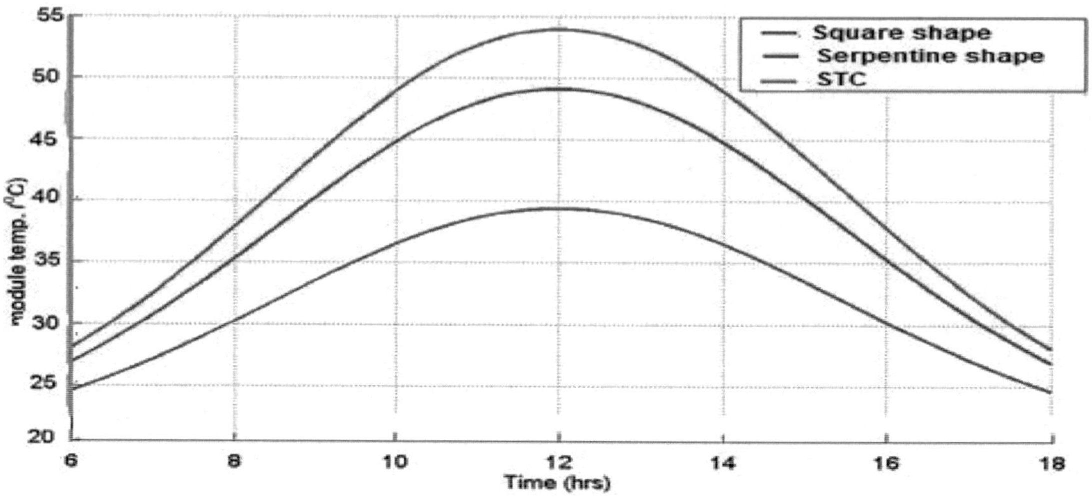

Figure 2. A comparison of the module temperature (with (square and serpentine shapes)/without cooling (STC)) at different shapes attached in the backside of the PV module by using water along with the daytime.

Figure 3 presents the **I-V** and **P-V** curves to compare among different shapes of pipes by water as a cooling fluid, concerning the standard curve. The **I-V** curves in Figure 3-a show only the variation across the voltage range from 0 to 15**V** and the current from 8 to 9.5**A** to observe the fluctuations in the short circuit current, as the open-circuit voltage is almost maintained constant for all the simulated shapes concerning standard curve (not shown) [12]. By comparing with STC module performance, the Figure shows an increase in module output power by using serpentine shape 0.2**A** while with square shape the system output power raised about 0.64 **A**. That increase happened due to the reduction of PV module temperature by using a square shape.

The P-V curves in Figure 3-b showed the power variation across the maximum power point from 26 to 29**V**. The short circuit current at STC condition without any cooling reached 8.29**A** and output power of 219.26**W** while $\mathbf{I_{sc}}$ of serpentine shape enhanced to a maximum of about 8.49**A** at the output power of 223**W**. The new shape with a larger area than the serpentine one has $\mathbf{I_{sc}}$ = 8.95**A**, and $\mathbf{P_{out}}$ = 236**W** by about a 5.7 % increase compared with the serpentine shape. The results show the best performance was the square shapes of pipes to enhance the PV output power.

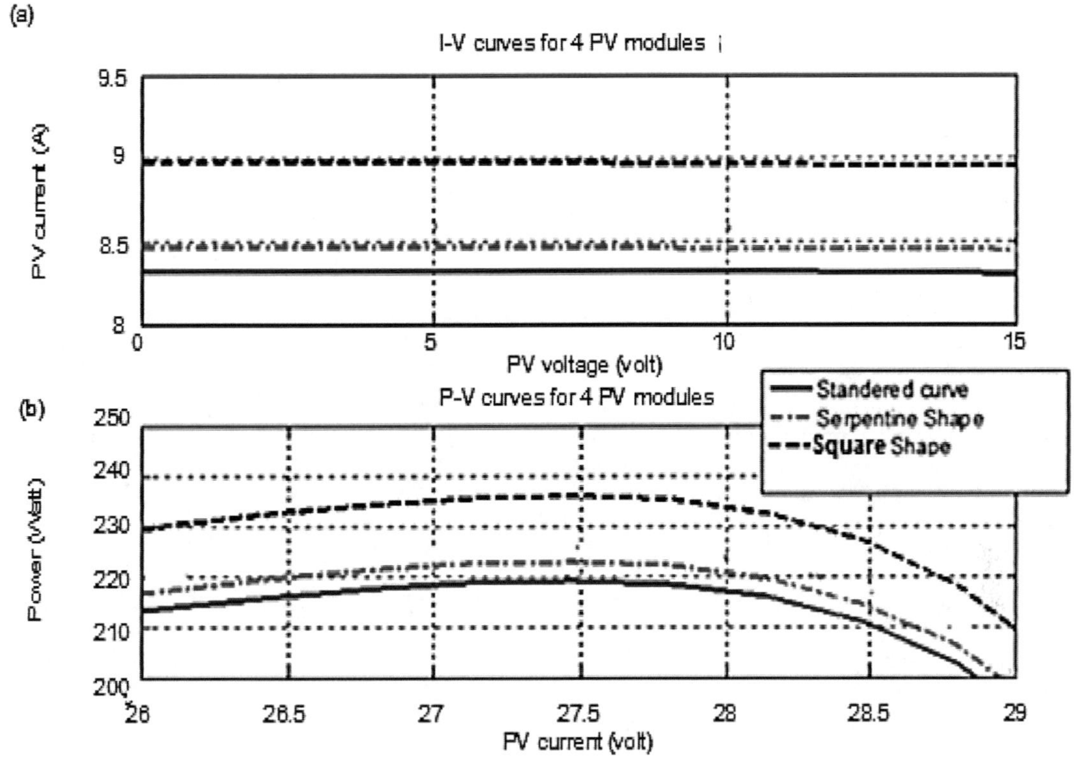

Figure 3 Simulation results for I-V and P-V characteristic curves without cooling besides with cooling by using water with different shapes attached in the backside of PV modules.

Figure 4-a shows the comparison between water and nano-fluid (CuO) cooling technique through the best shape of pipes (square shape). By comparing the results with water that introduced the I_{sc} of serpentine shape with nano-fluids observed to 9.35**A** and Pout = 246**W**. All graphs of I-V and P-V characteristics curves are plotted using the Matlab program [15]. The results showed an enhancement in the PV panel by using nano-fluid with a new shape of about 5% in the output power compared with water as a cooling fluid as shown in Figure 4-b.

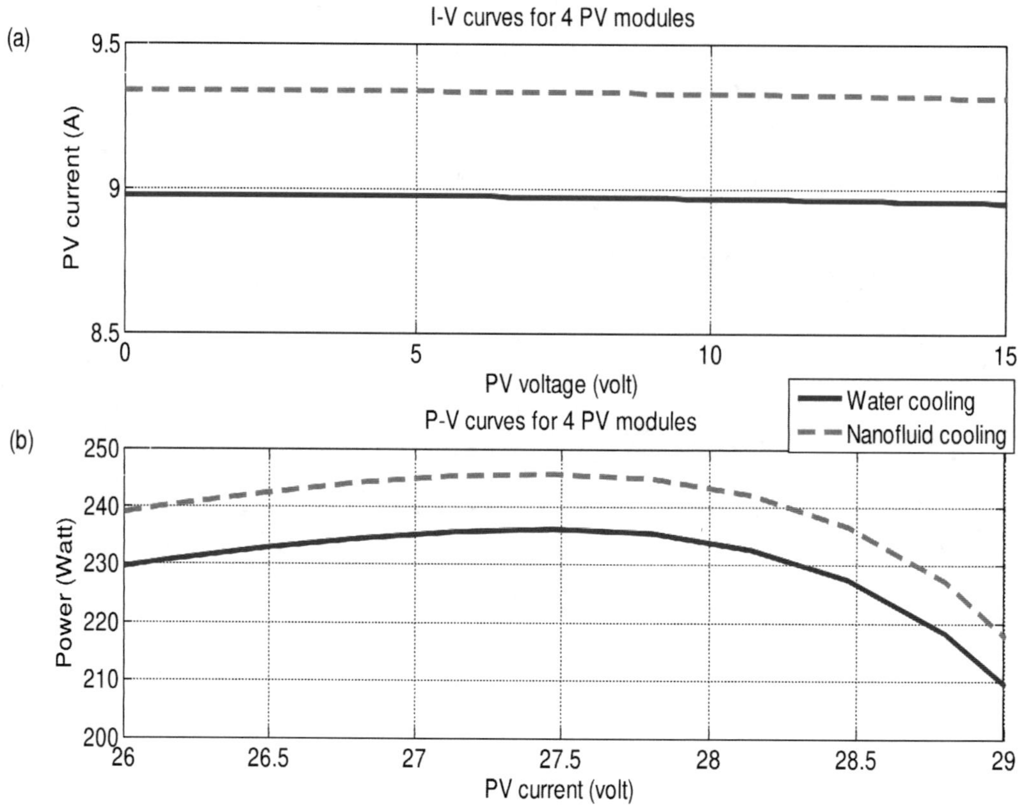

Figure 4 A comparison between water and nano-fluids (CuO) by using a square shape attached to the backside of the PV panel.

5. Conclusion

This study demonstrates the effect of an active cooling system with water and CuO nano-fluid as a cooling fluid. Two different shapes of pipes (serpentine and square) were presented with the insertion of a structured targeting maximum thermal heat transfer and output power of the **PV/T** system. The main conclusions are summarized as follow:

- By using water as a cooling technique, the results showed that the module temperature with serpentine and square shapes decreased about 8.4% and 17.7% compared with STC conditions.
- This increase reflects on the maximum output power which reaches 223 Watt for serpentine shape and 236 Watt for a square shape.
- After choosing the square shape (the best cooling one), the CuO nanofluid is the solution to investigate the system output power.
- By comparing the output power of square shape by using CuO nanofluid with water as a cooling fluid, the results exhibit an improvement in overall system performance by about 24.7%.
- The results in this paper showed a fair agreement with what is published in the literature.

Nomenclature

Symbol	Description	Units

U	Overall heat transfer coefficient	W/m^2*K
A	Plate surface area and PV area	m^2
t_h	The thickness of the plate	m
A	Pipe surface area	m^2
K_{si}	Silicon thermal conductivity	W/m^2*K
K_{cu}	Copper thermal conductivity	W/m^2*K
K_{Pipe}	Pipe thermal conductivity	W/m^2*K
h_f	fluid thermal conductivity	W/m^2*K
h_{air}	Air thermal conductivity	W/m^2*K
r_{in}	Pipe inner radius	m
r_{out}	Pipe outer radius	m
L	Plate length	m
W	Plate width	m
l	Total pipe length	m
$\dot{Q}$	The rate of heat transfer	W
I_{sc}	Short circuit current	A
I_{PV}	PV current	A
I_{01}	Dark saturation Current	A
V	Voltage	Volt
K_B	Boltzmann constant	$1.38 \times 10^{-23} m^2 . kg \times s^{-2} \times K^{-1}$
I_{02}	Reactive current	A
V_{oc}	Open circuit voltage	V
q	The electron charge	1.6×10^{-19} C
η_{PV}	PV efficiency	%
Pout	The maximum power output	W
Pin	The input power	W
FF	Fill factor	%.
ρ_{nf}	Density of nano-fluid	Kg/m^3
ρ_f	The density of cooling fluid	Kg/m^3
ρ_P	Density of nano-particles	Kg/m^3
$\emptyset$	Volume concentration	%
C_{pnf}	Specific heat of nano-fluid	J/kg.K
C_{pf}	Specific heat of cooling-fluid	J/kg.K
C_{Pp}	Specific heat of nanoparticles	J/kg.K
k_{nf}	Thermal conductivity of nano-fluid	W/m.K
k_f	Thermal conductivity of cooling-fluid	W/m.K
k_p	Thermal conductivity of nanoparticles	W/m.K
q_{rad}	Sun irradiance at MENA region	1000 W/m^2

Reference

[1] Bakker M, Zondag H A, Elswijk M J, Strootman K J, and Jong M J M 2005 *Performance and costs of a roof-sized PV/thermal array combined with a ground-coupled heat pump* (Solar Energy, vol 78) pp 331-339

[2] Hosseini R, Hosseini N, and Khorasanizadeh H 2011 *An experimental study of combining a photovoltaic system with a heating system* (World Renewable Energy Conference, Sweden) pp 2993-3000
[3] Kerzmann T and Schaefer L 2012 *System simulation of a linear concentrating photovoltaic system with an active cooling system* (Journal of Renewable Energy vol 41) pp 254-261
[4] Kunnemeyer R, Anderso T N, Duke M, and Carson J K 2014 *Performance of a V-trough photovoltaic/thermal concentrator* (Solar Energy Journal, Science Direct, vol 12), pp 19-27
[5] Hussain F, Othman M Y H, Sopian K, Yatim B, Ruslan H, and Othman H 2013 *Design development and performance evaluation of photovoltaic/thermal (PV/T) airbase solar collector*, (Renewable and Sustainable Energy Reviews, El Sevier Journal) , pp 431-441
[6] El-Seesy E and Khalil T 2014 An *Experimental Study of a Composite Photovoltaic/Thermal Collector with a Tracking Concentrator System*, (World Applied Sciences Journal vol 31) pp 988-992
[7] Nguyen G, Gauthier C, and Galanis N 2017 *Heat transfer enhancement using* Al_2O_3*– water nanofluid for an electronic liquid cooling system* (Applied Thermal Engineering vol 26) pp 1273-1501
[8] Hussein A K, Walunj A, and Kolsi L 2016 *Applications of nanotechnology to enhance the performance of the direct absorption solar collectors*, (Journal of Thermal Engineering, vol 2 No. 1) pp 529-540
[9] Teo H G, Lee P S, and Hawlader M N A 2012 *An active cooling system for photovoltaic modules* (Applied Energy vol 90) pp 309-315
[10] Yang D J, Yuan Z F, Lee, P H, Yin H M 2012 *Simulation and experimental validation of heat transfer in a novel hybrid solar panel* (International Journal of Heat and Mass Transfer vol. 55) pp 1076-1082
[11] Nahar A, Hasanuzzaman M, and Rahim N 2017 *A 3D comprehensive numerical investigation of different operating parameters on the performance of a PVT system with pancake collector* (Journal of Solar Engineering ASME vol 55) pp 245-266
[12] Karami N and Rahimi M 2014 *Heat transfer enhancement in a PV cell using Boehmite nanofluid* (Energy Conversion and Management vol 86) pp 275-285
[13] Hossam El din A, Ahmed C F G, and Alin H 2014 *Effect of Ambient Temperature on The Performance of Different Types of PV Cells at Different Locations in Egypt* (Sixteenth International Middle East Power Systems Conference MEPCON'14 Egypt) pp 331-339
[14] Comsol Version 4.2 http://www.comsol.com
[15] Benabderrahmane A, Benazza A, and Hussein A K 2020 *Heat transfer enhancement analysis of tube receiver for parabolic trough solar collector with central corrugated insert* (Journal of Heat Transfer-Transactions of the ASME vol 142) pp 062001-1-062001-8
[16] Hwang Y, Park H S, Lee J K, and Jung W H 2006 *Thermal conductivity and lubrication characteristics of nano-fluids* (Applied Physics Elsevier Journal, vol 6) pp 67–71
[17] Li D, Li Z, Zheng Y, Liu C, Hussein A K, and Liu X 2016 *Thermal Performance of a PCM-filled double-glazing unit with different thermophysical parameters of PCM* (Solar Energy vol 133) pp 207–220
[18] Liu C, Wu Y, Li D, Ma T, Hussein A K, and Zhou Y 2018 *Investigation of thermal and optical performance of a phase change material filled double-glazing unit* (Journal of Building Physics vol 42 No. 2) pp 99-119

[19] Bayoumi A, Abdelaziz M A, Abdelhameed M M 2013 *Modeling and Simulation of Tracking Photovoltaic/Thermal Hybrid System* (The Ninth International Conference on Computer Engineering and Systems ICCES IEEE Egypt) pp 127-132

[20] Hussein A K 2015 *Applications of nanotechnology in renewable energies-A comprehensive overview and understanding* (Renewable and Sustainable Energy Reviews vol 42) pp 460-476

[21] Pak B C and Cho Y 1998 *Hydrodynamic and heat transfer study of dispersed fluids with submicron metallic oxide particle* (Journal of Heat Transfer vol 11) pp 151–170

[22] Hussein A K 2016 *Applications of nanotechnology to improve the performance of solar collectors – Recent advances and overview* (Renewable and Sustainable Energy Reviews vol 62) pp 767-792

IOP Conf. Series: Materials Science and Engineering 1172 (2021) 012018 doi:10.1088/1757-899X/1172/1/012018

Numerical investigation of four photovoltaic/thermal integrated structures from energetic point of view

M Abd El-Hamid[1, 2], S Moustafa[1, 2] and G Wei[1]

[1] Key Laboratory of Power Station Energy Transfer Conversion and System of Ministry of Education, School of Energy, Power and Mechanical Engineering, North China Electric Power University, Beijing 102206, China
[2] Mechanical Power Engineering Department, Faculty of Engineering – Mattaria, Helwan University, Cairo 11718, Egypt

E-mail: eng.mohamed.abdelhamed90@gmail.com

Abstract. Photovoltaic/Thermal (PV/T) module is considered to be one of the most recent technologies which offers harness and production of both electric and thermal energy. In the current study an energetic analysis is carried out to compare between four different configurations which are: the single pass single glazed (PV/T-I), the single pass double glazed with air gap (PV/T-II), the single pass double glazed with argon gap (PV/T-III) and the double pass double glazed (PV/T-IV) hybrid photovoltaic/Thermal air collector systems. A 3 dimensional numerical model is built up and validated with both the numerical and experimental results coming from the literature. The numerical simulations have been accomplished to investigate the energetic performance with a detailed thermal and electrical study taking in account an inlet coolant temperature the same as the ambient temperature for a coolant (air) mass flow rate equals to 0.025 kg/s of a typical day in August from 9:00 to 17:00 under the ambient conditions of Beijing, China. The results show that the single pass single glazed configuration has the highest electrical efficiency, whereas the double pass configuration has the greatest thermal and energy efficiencies among the proposed configurations. The average daily energy efficiencies are 53.14%, 75.92%, 77.63% and 82.19% for the (PV/T-I), (PV/T-II), (PV/T-III) and (PV/T-IV) configurations, respectively.

1. Introduction

The lack of energy, along with the environmental challenges associated with conventional energy production using fossil fuels, as well as the rapid growth of the world population, offer a clear motivation for a much extended sustainable energy supply represented in the renewable energy resources [1]. Solar energy is considered to be as the most relevant renewable energy source for the whole world. Solar energy can be captured by two different means separately: the solar photovoltaic system (PV) which transforms the solar energy to an electric power and the solar thermal structure that transforms the solar energy in to a thermal energy. As a result, the photovoltaic thermal hybrid structure (PV/T) is a method in which the PV not only provides electricity but also functions as a thermal absorber by collecting heat from the PV system, resulting in a higher electrical gain.

In the past 50 years, a wide variety of the PV/T investigations have been performed since the early 1970s using water or air as a coolant starting from Wolf [2], Kern and Russell [3] in the early 1970s who use water or air as cooling fluids. Many studies have been implemented in recent decades to boost

Content from this work may be used under the terms of the Creative Commons Attribution 3.0 licence. Any further distribution of this work must maintain attribution to the author(s) and the title of the work, journal citation and DOI.
Published under licence by IOP Publishing Ltd

the overall energetic efficiency of the PV/T integrated configurations, however, the design and the layers arrangement of a PV/T integrated collector need further improvement to boost the overall efficiency. It was found from the literature review that the coolant passage underneath the proposed PV design achieved the highest overall efficiency among the evaluated different designs of the PV/T structures as shown by Zondag et al. [4]. Using the glazed PV/T configuration for the purpose of water-heating can enhance the thermal efficiency till 30% compared to the unglazed systems as dedicated by Tripanagnostopoulos et al. [5] and observed that, the electrical efficiency was reduced by about 16%.

Joshi et al. [6] gave a thermal performance comparison between two different types of the hybrid PV/T air collector systems, which were the PV module with a glass to tedlar form and the other was glass to glass form under the weather climates of India. In order to improve the thermal and electrical measurement models, a novel PV/T air collector was tested by Sarhaddi et al. [7] with some heat loss coefficient corrections. The study of the uni pass hybrid PV/T and the dual hybrid PV/T configurations with a metal plate and a glass cover at the top was done by Slimani et al. [8]. The findings revealed a significantly higher thermal efficiency, but a significantly lower electrical efficiency of the second configuration. The average energetic efficiency and the annual energy output were found to be the highest for the PV/T dual-pass air collector as concluded by Slimani et al. [9] who investigated a comparative analysis between different four structures: the photovoltaic traditional module, the PV/T conventional structure, the integrated PV/T dual-pass structure and the PV/T double glazing structure.

Latterly, Sainthiya et al. [10] have studied the effect of cooling the upper surface of the hybrid PV/T collector by water numerically and experimentally under the Indian weather conditions. Recently, an electrical and thermal performances of PV/T air configuration which had one-pass and dual-flow air passage have been proposed experimentally by Choi et al. [11], and it was found that the overall energetic efficiency increased with rising the coolant flow rate, a slight enhancement in the electrical efficiency and significant increase in the thermal efficiency were induced.

According to the previous literature review, it will be an effective study to concentrate on the influence of adding an upper glass layer for single pass configurations with a static air or argon above the PV structure and use a double pass channel for the cooling fluid of PV/T configurations, that may enhance the overall energetic performance of the hybrid PV/T configuration. Thus, a thermal and electrical performance investigation performed in the current study to compare between different four configurations, which are: the single pass single glazed (PV/T-I), the single pass double glazed with air gap (PV/T-II), the single pass double glazed with argon gap (PV/T-III) and the double pass double glazed (PV/T-IV) hybrid photovoltaic/Thermal air collector systems.

2. System design and description

In the current study four PV/T structures are numerically analyzed and compared. The proposed hybrid PV/T configurations are using glazed PV module with different positions of cooling channels as illustrated in Fig. 1. The PV module is of mono-crystalline type with 50 watt maximum power rating as released by the manufacturer [12]. The first configuration is the single pass single glazed module (PV/T-I) which is composed of: the PV and an air passage beneath the PV module with an aluminum absorber plate attached under the air duct. The tedlar layer below the PV cells and the top glass cover both act as protective layers. The aluminum absorber plate is to lower the temperature of the module through the convective and radiative heat transfer [9].

The second configuration is the single pass double glazed with air gap module (PV/T-II) which is the same as the first configuration, but with adding confined air gap above the PV module and enclosing this gap with an upper protective glass layer. The third configuration is the single pass double glazed with an argon gap module (PV/T-III) with the same layers as the second configuration but with the confined gap filled with argon instead of air. Whereas, the fourth configuration is the double pass with double glazed module (PV/T-IV) consisting of top glass cover with dual air passages; one above the PV module and the other is underneath it attached to an aluminum absorber plate as seen in Fig. 1.

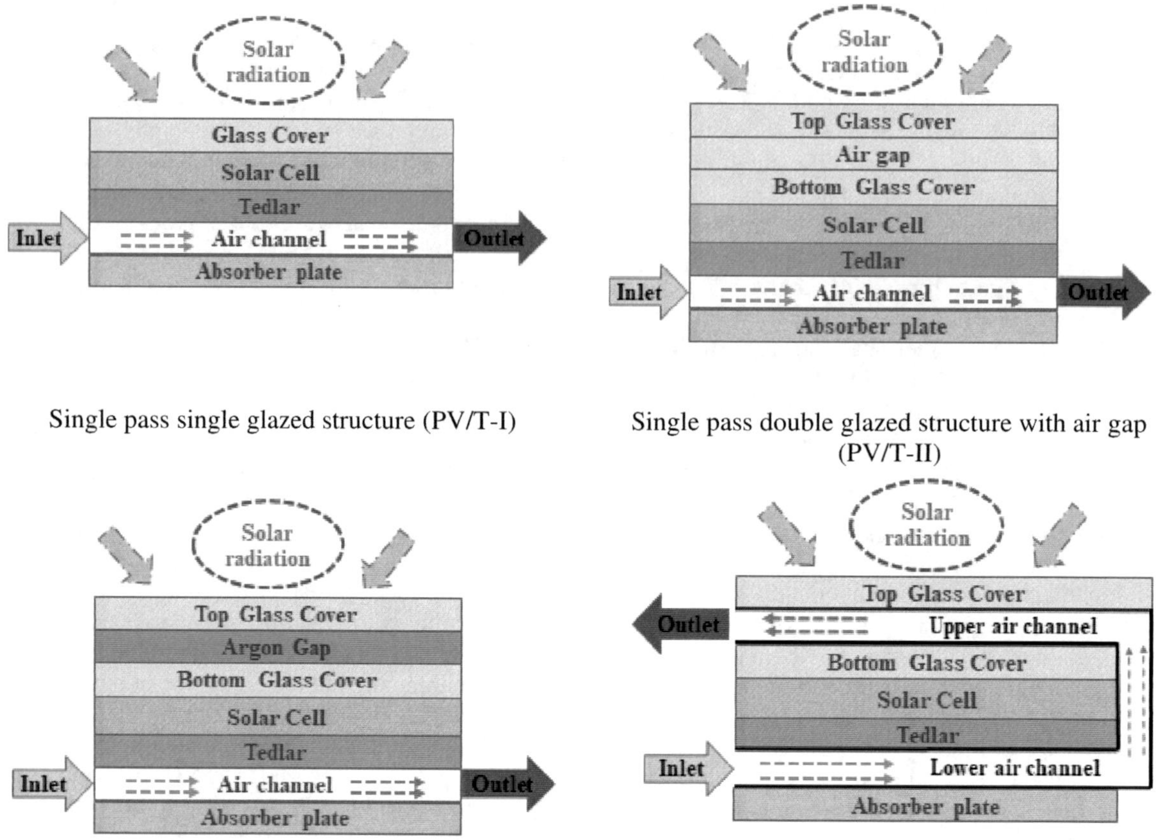

Single pass single glazed structure (PV/T-I)

Single pass double glazed structure with air gap (PV/T-II)

Single pass double glazed structure with argon gap (PV/T-III)

double pass double glazed structure (PV/T-IV)

Figure 1. Schematic diagram of the suggested hybrid PV/T air collector configurations.

3. Numerical analysis

The governing equations, the boundary conditions, the proposed model accuracy including the mesh independence test and the ambient conditions for the current study are presented in this section.

3.1. Governing equations and performance evaluation parameters

In the current study, a three-dimensional numerical model is established and simulated by using ANSYS FLUENT 19.1. The governing equations can be derived as:

Continuity equation:

$$\nabla.(\rho\vec{V}) = 0 \tag{1}$$

Momentum equation:

$$\vec{V}.\nabla\,(\rho\vec{V}) = \nabla P + \nabla.(\mu\nabla\,\vec{V}) \tag{2}$$

Energy equation for the cooling fluid:

$$\vec{V}.\nabla(\rho C_p T_f) = \nabla.(k_f \nabla T_f) + \nabla q_{rad} \tag{3}$$

Energy equation for the PV/T module layers:

$$\nabla.(k_f \nabla T_f) + \nabla q_{rad} = 0 \tag{4}$$

where $\vec{V}$, P, μ, ρ, C_p, k_f and T_f are the symbols for the velocity vector, pressure, dynamic viscosity, density, specific heat at constant pressure, thermal conductivity and temperature of the cooling fluid, respectively. q_{rad} refers to the radiative energy net loss through a control volume and named as the radiative term which equals to the absorbed incident radiation on a control volume subtracted from the emitted energy from that volume.

In order to investigate the system performance, the thermal efficiency, electrical efficiency and overall energetic efficiency is evaluated according to Eqs. (5), (8) and (10), respectively.

$$\eta_{th} = \frac{E_{th}}{E_s} \tag{5}$$

where E_s is the total energy coming from solar radiation. E_{th} is the absorbed thermal energy by the coolant, which is air in the current work. E_s and E_{th} are calculated as Eqs. (6) and (7), respectively.

$$E_s = G\, A_s \tag{6}$$

where G is the radiation intensity per unit area and A_s is the net absorbed surface area of PV module.

$$E_{th} = \dot{m}\, C_p\, (T_{out} - T_{in}) \tag{7}$$

where $\dot{m}$, C_p, T_{out} and T_{in} are the mass flow rate, specific heat capacity, outlet temperature and inlet temperature of the coolant (air) , respectively.

$$\eta_{ele} = \frac{E_{ele}}{E_s} = \eta_{ref}([1 - \delta_{sc}(T_{sc} - T_{ref})] \tag{8}$$

$$E_{ele} = \eta_{sc}\, A_{sc}\, G_{sc}\, [1 - \delta_{sc}(T_{sc} - T_{ref})] \tag{9}$$

where E_{ele} is the produced electric power from the PV module, η_{sc} refers to the PV cell efficiency at standard temperature, A_{sc} is the PV cell surface area, G_{sc} is the portion of incoming solar radiation that reaches to the PV cell by taking in account both the solar cell absorptivity, the above layers transmissivity and the PV module packing factor, δ_{sc} refers to the temperature coefficient of the mono crystalline silicon cell which can be taken as 0.45%, T_{sc} and T_{ref} are the solar cell temperature and the reference temperature, respectively.

The overall energetic efficiency (η_o) for the hybrid system can be deduced by adding both electrical and thermal efficiencies as follows:

$$\eta_o = \eta_{ele} + \eta_{th} \tag{10}$$

3.2. Boundary conditions

A thermal coupled boundary condition for solid-solid interface in PV/T systems is employed as:

$$(k_o \nabla T_o) = (k_{ó} \nabla T_{ó}) \quad (11)$$

where k refers to the thermal conductivity, subscripts o and $ó$ refer to the proposed layer and the upper layer or the layer beneath it. For the solid-fluid interface, the no slip condition is applied as:

$$u=v=w=0, \qquad T_{air} = T_{duct} \quad (12)$$

The boundary conditions at both the inlet and outlet sections of the air channels are as follows:

$$w=w_{in}, \qquad T=T_{in}, \quad (13)$$

$$P_o = 0\ (gage) \quad (14)$$

Eq. (15) represents the calculation method of the hydraulic diameter and the turbulence intensity at both inlet and outlet sections of the air duct, which are activated on using the K-epsilon turbulence model used in the current study [13] :

$$D_h = \frac{2\,(L_{duct} \times B_{duct})}{(L_{duct} + B_{duct})}, \qquad z = 0.16\,Re^{-1/8} \quad (15)$$

where D_h refers to the hydraulic diameter, L_{duct} refers to the duct length, B_{duct} is the duct width, z refers to the turbulence intensity and Re is Reynolds number. By treating each layer in PV/T hybrid system as a control volume, the energy is gained by conduction from the above layer and by radiation via the reached portion of solar radiation to this layer, which is calculated according to the absorptivity of this layer and the transmittance of the above layer to the incoming solar radiation according to thermal and optical properties stated in Table 1 and 2.

Table 1. Dimensions and thermal properties of the used materials.

Layer	Dimensions (mm)	Density (kg/m^3)	Thermal conductivity (W/m K)	Specific heat (J/kg K)
Top and bottom Glass covers	1280×320×3.2	2770	2	500
Solar cell	1280×320×0.3	2330	148	677
Tedlar	1280×320×0.5	1200	0.15	1250
Air duct (Aluminum)	1280×300×20	2719	202.4	871
Aluminum plate	1280×300×2	2719	202.4	871

Table 2. Optical properties of the used materials [14].

Layer	Reflectivity	Absorptivity	Transmissivity	Emissivity
Glass	0.04	0.04	0.92	0.85
Solar cell	0.08	0.90	0.02	-
Aluminum	-	-	-	0.90

3.3. Mesh independence study

In order to check the grid independence test, ten mesh systems are done with different cell numbers for a PV/T module with the same dimensions and material properties as the PV/T module that was investigated by Joshi et al. [6]. The grid independence test is conducted by applying 600 W/m2 radiation heat flux, 300 K ambient temperature, 300 K cooling fluid inlet temperature and 1/m/s wind velocity. It can be seen that the solution represented in both solar cell and cooling fluid outlet temperatures becomes independent on grid size after 2,116,851 elements as illustrated in Fig. 2.

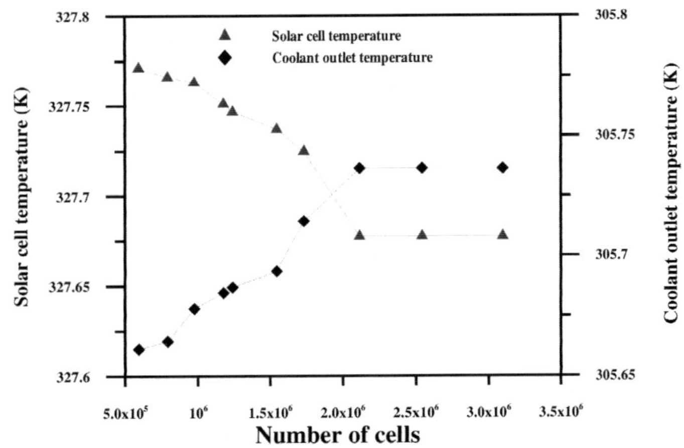

Figure 2. Mesh independence test for solar cell and coolant outlet temperatures.

3.4. Model validation

The present numerical work has been validated with both numerical and experimental works that has been carried out by Joshi et al. [6] and numerical results investigated by Slimani et al. [9] and Sarhaddi et al. [7] at an identical ambient and operating conditions. The validation is carried out to verify the accuracy of the current numerical model. The numerical data for the solar cell temperature of the present numerical model achieves a maximum errors of 6%, 7.5%, 16.5% and 17.5% ,but the maximum errors attained through the present work for the coolant outlet temperature are 5.8%, 1.8% , 7.4% and 1.7% with Joshi et al. [6] experimental study, Slimani et al.[9], Sarhaddi et al. [7] and Joshi et al. [6] predicted results, respectively, as displayed in Fig. 3 (a) and (b).

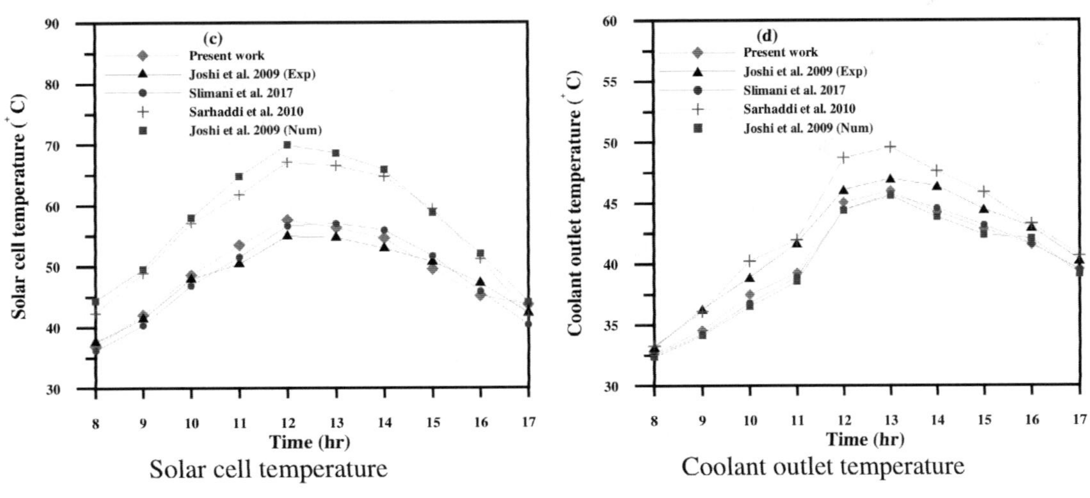

Figure 3. Comparison of the present numerical work with the previous work hourly.

IOP Conf. Series: Materials Science and Engineering 1172 (2021) 012018 doi:10.1088/1757-899X/1172/1/012018

3.5. Weather conditions
Figure 4 gives the typical weather conditions in August of Beijing, China. These data is used in the following numerical simulations.

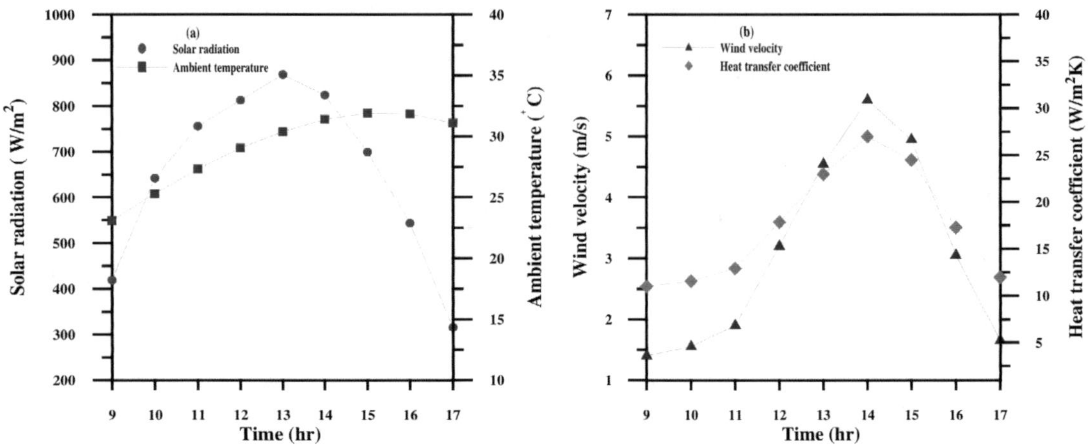

Solar radiation and ambient temperature | Wind velocity and heat transfer coefficient

Figure 4. Variations of the solar radiation and the ambient temperature for a typical day in August.

4. Results and discussion
In this section the temperature gradient of the layers and the energetic performance of the four proposed configurations are numerically investigated with an air mass flow rate of 0.025 kg/s and the temperature of the incoming coolant is the same as the ambient temperature for a certain day in Beijing, China, in August.

4.1. Temperature distribution
Figure 5 represents the temperature change with time of the suggested four configurations. It is depicted that the highest temperatures occur under the highest solar radiation intensity, which has a major effect rather than wind speed and ambient temperature for all layers, except for the temperature of air outlet and the top glass layer. The temperatures of air outlet and the top glass layer are apparently affected by the wind velocity rather than the incoming solar radiation. The PV cell layer has the highest temperature among all layers for all configurations and the daily peak values are 345.46, 344.55, 324.14 and 323.32 K for the (PV/T-III), (PV/T-II), (PV/T-I) and (PV/T-IV) configurations, respectively. However, the highest daily outlet air temperatures are 312.97, 313, 312.8 and 308.42 K for the (PV/T-IV), (PV/T-III), (PV/T-II) and (PV/T-I) configurations, respectively.

IOP Conf. Series: Materials Science and Engineering 1172 (2021) 012018 doi:10.1088/1757-899X/1172/1/012018

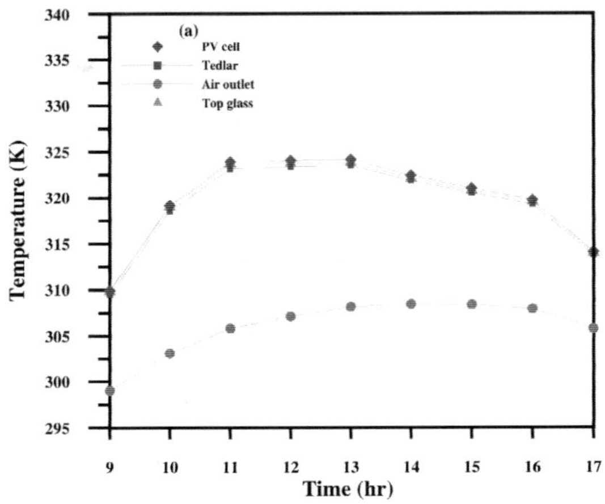

(a) PV/T-I

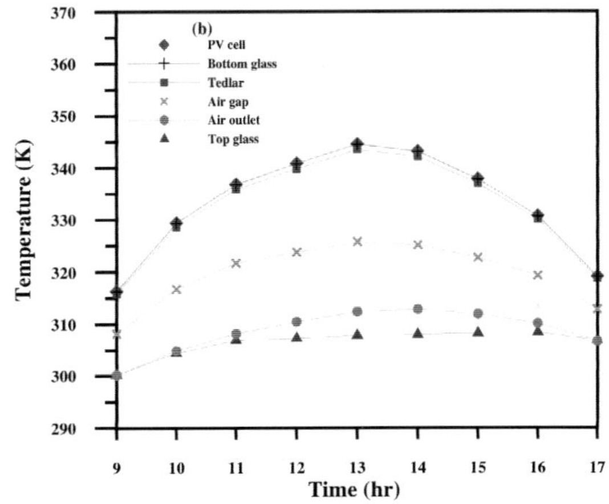

(b) PV/T-II

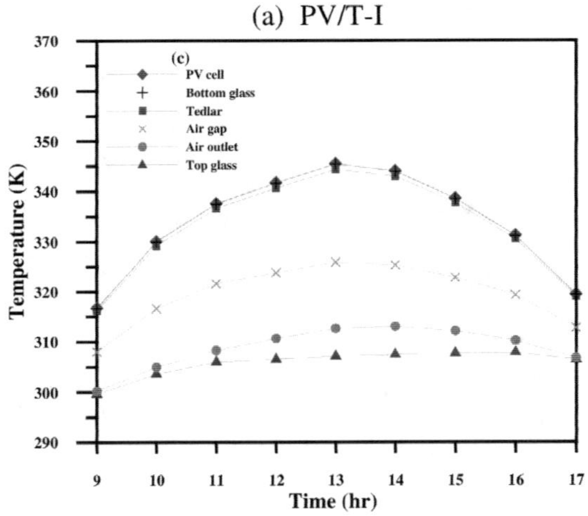

(c) PV/T-III

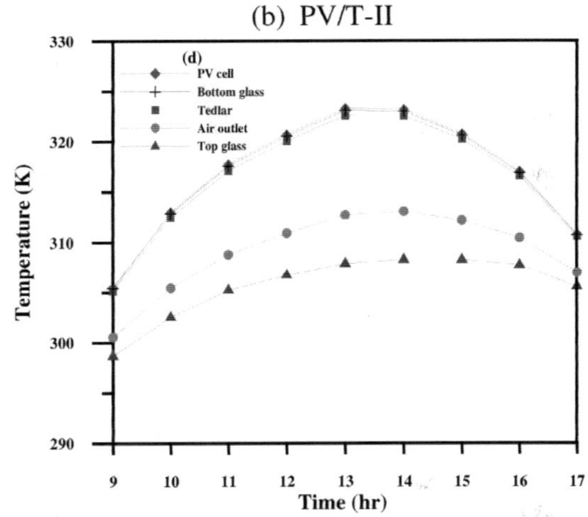

(d) PV/T-IV

Figure 5. Temperature variation manners at different positions with time of the four PV/T configurations.

Figure 6 displays the temperature distribution contours along the surface of the PV cell layer. It is plain from these contours that, for all designs, the natural convection at the edges of the PV surface is not effective near the inlet portion of the air channel. However, the natural convection effect at the edges of the PV surface begins to increase gradually when the cooling fluid begins to remove the heat from the PV cell layer as it goes far from the inlet section of the air passage. At mid parts, the temperature of the PV surface is significantly higher than that at the edges of the PV surface, as the natural convection effect at the edges of PV surface is stronger than the forced convection effect of the cooling fluid which is air in the present study. The temperature of the PV surface ranges from 318 to 326 K, 326 to 352 K, 326 to 354 K and 316 to 325 K for the (PV/T-I), (PV/T-II), (PV/T-III) and (PV/T-IV) configurations, respectively.

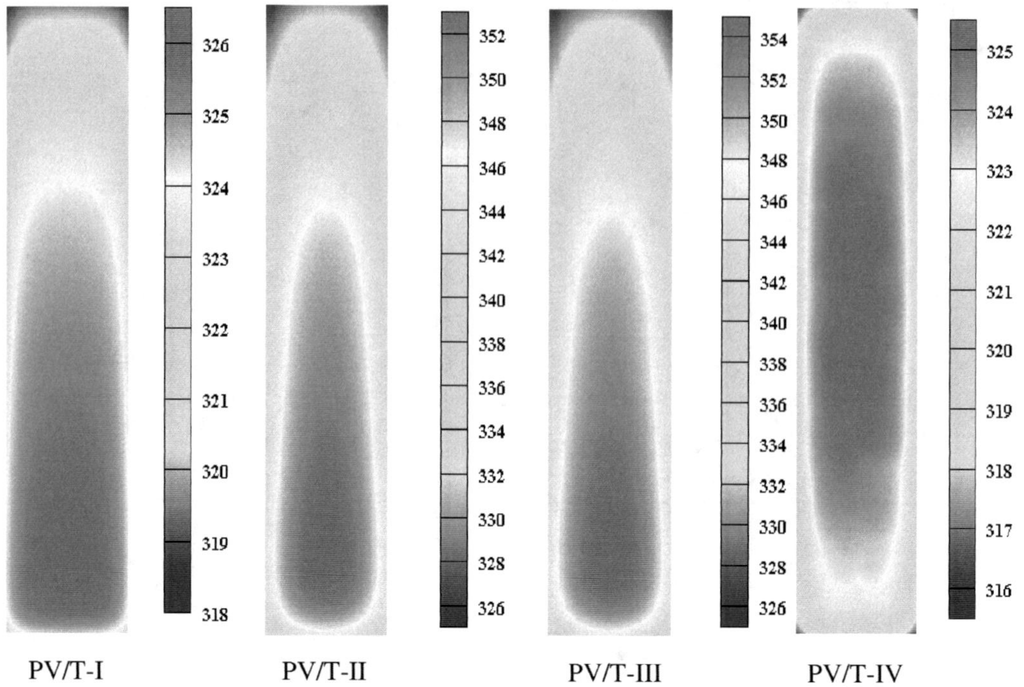

Figure 6. Contours of PV cell surface temperature for all configurations.

4.2. Energetic performance

Figure 7 displays the hourly changes of the thermal power and the electrical power for all the suggested PV/T structures. It is apparent from Fig. 7 (a) that the thermal power marginally increases with increasing the incident solar radiation, as the coolant at the air duct outlet section induces higher temperatures. The highest thermal power values for all configurations are at 13:00 o'clock corresponding to the highest solar radiation, except for the (PV/T-I) configuration the highest thermal power happens at 11:00 o'clock, due to minimizing the layers of this configuration causes the major effect of the wind velocity rather than the solar radiation intensity. The highest thermal power is 240.47 W for the (PV/T-IV) configuration, followed by 238.43, 231.19 and 140.98 W for the (PV/T-III), (PV/T-II) and (PV/T-I) configurations, respectively.

It is shown from Fig. 7 (b) that the highest daily electrical power is reached at 13:00 o'clock corresponding to the maximum daily solar radiation for all the suggested structures. The highest electrical power is 41.65 W for the (PV/T-I) configuration, followed by 35.73, 30.03 and 29.8 W for the (PV/T-IV), (PV/T-II) and (PV/T-III) configurations, respectively.

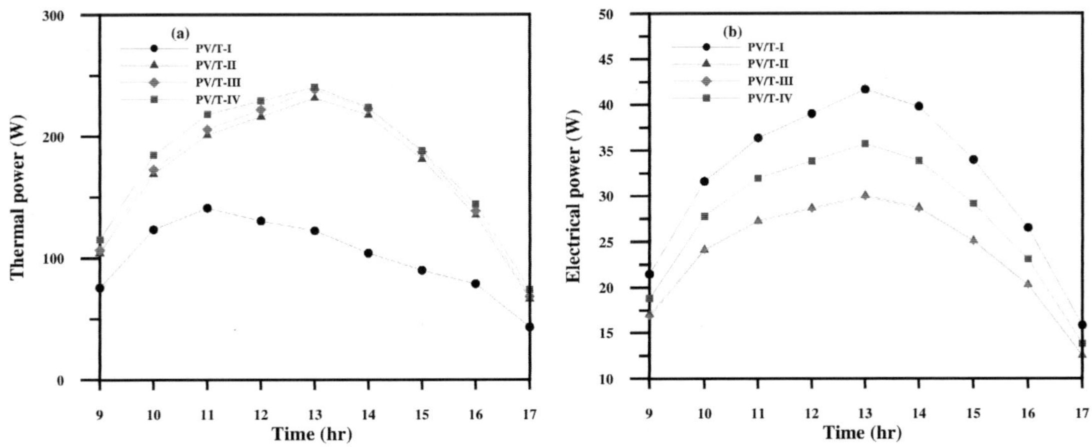

(a) Thermal power (b) Electrical power

Figure 7. Variations of (a) thermal power and (b) electrical power with time.

Figure 8 presents the electrical, thermal and overall energetic efficiencies change with time for all configurations. It is shown in Fig. 8 (a) that the electrical efficiency dramatically decreases with increasing the solar radiation intensity that consequently increase the PV cell temperature. Else, the electrical efficiency reaches its minimum value at 13:00 o'clock corresponding to the highest PV cell temperature for all configurations. It is illustrated in Fig. 8 (a) that the (PV/T-I) configuration offers the highest daily electrical efficiency with a values ranging from 12.49 to 13.33% followed by the (PV/T-IV), (PV/T-II) and (PV/T-III) configurations, with values variation from 10.89 to 12.05%, 9 to 10.57% and 8.95 to 10.42%, respectively.

It can be deduced from Fig. 8 (b) that for the (PV/T-I) and (PV/T-IV) structures the highest daily thermal efficiency is not corresponding to the peak solar radiation time confirming that the wind velocity effect appears significantly more than the solar radiation intensity with different percentages, however for the (PV/T-II) and (PV/T-III) structures the highest daily thermal efficiency is corresponding to the highest daily solar radiation, meaning that, the solar radiation has a greater effect more than the influence of the wind velocity on both modules. The highest daily thermal efficiency is 75.05% for the (PV/T-IV) configuration, followed by 71.48, 69.31 and 50% for the (PV/T-III), (PV/T-II), (PV/T-I) configurations, respectively.

It is obvious from Fig. 8 that the thermal efficiency has the major effect on the behavior of the overall energetic efficiency with a slight contribution by the electrical efficiency for all the designed structures. The overall energetic efficiency values range from 72.86 to 86.4%, 66.7 to 80.44%, 65.1 to 78.68% and 45.47 to 62.83% for the (PV/T-IV), (PV/T-III), (PV/T-II) and (PV/T-I) structures, respectively, as presented in Fig. 8 (c).

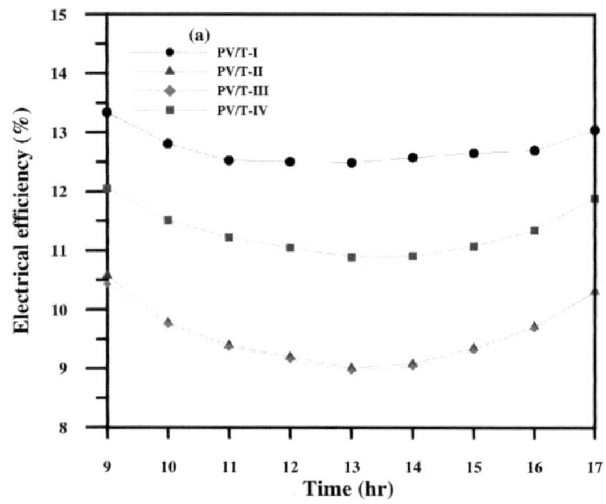

(a) Electrical efficiency

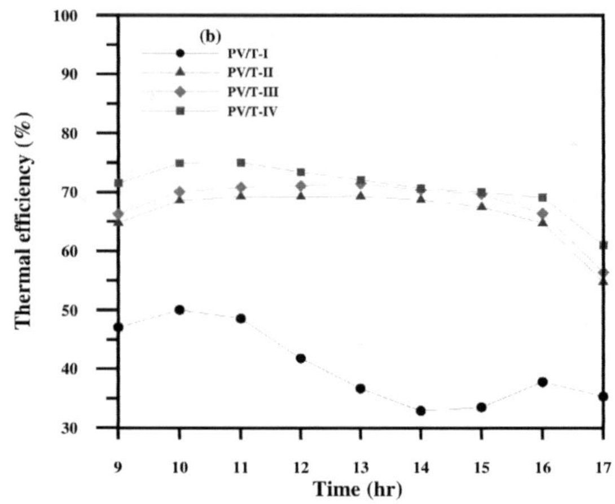

(b) Thermal efficiency

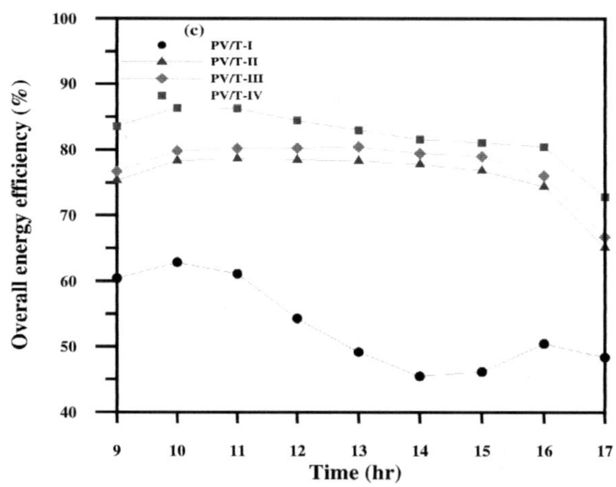

(c) Overall energetic efficiency

Figure 8. Variations of (a) electric efficiency, (b) thermal efficiency and (c) overall energetic efficiency with time for all suggested configurations.

Figure 9 investigates the daily average electrical, thermal and overall energetic efficiencies for each configuration. It is evident from this figure that the contribution of the thermal efficiency is somewhat superior to the electrical efficiency, due to the performance evaluation using the energetic concept accounts for the quantity of energy only as stated by the thermodynamics first law. The percent of share for the average daily thermal efficiency in the average daily overall energy efficiency is 86.2, 87.69, 87.35 and 76% for the (PV/T-IV), (PV/T-III), (PV/T-II) and (PV/T-I) configurations, respectively.

Figure 10 demonstrates the thermal and electrical energy outputs for each configuration. It is clearly shown that the highest average daily energy output is for the (PV/T-IV) configuration with a value of 4856.7 Wh/m^2 followed by the (PV/T-III), (PV/T-II) and (PV/T-I) structures with values of 4614.27, 4519.6 and 3111.32 Wh/m^2, respectively.

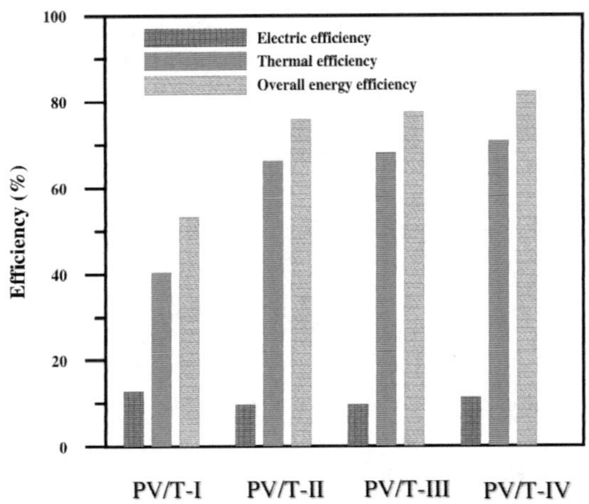

Figure 9. Daily average electrical, thermal and overall energy efficiencies for each configuration.

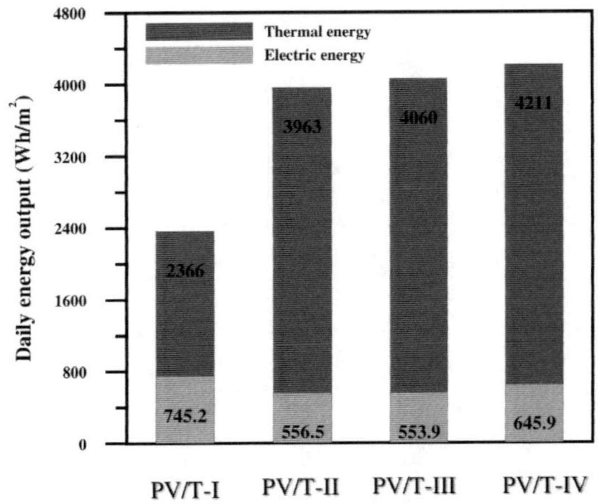

Figure 10. Daily thermal and electrical energy output for each configuration.

5. Conclusions

To assess the PV/T system energetic performance, four PV/T configurations have been investigated and studied numerically in the current work by considering a typical day of ambient conditions of Beijing, China in August. The following points can be concluded:

- The thermal and energetic efficiencies for the double pass configuration are the highest among the suggested structures.
- The single pass single glazed structure offers the highest electrical efficiency, in contrast the lowest thermal efficiency is achieved among the proposed structures.
- The overall energetic efficiency increases by adding a glass cover at the top surface of the PV module and marginally increases with a double passage of the coolant.
- Using argon instead of air as a gap between the PV module dramatically increases the thermal efficiency and the output, but both the electrical efficiency and the output are silghtly decreased, consequently raising the daily overall energy efficiency and the output.
- The double pass configuration exhibits the highest daily overall energetic efficiency by 82.19% followed by the double glazed with an argon gap, double glazed with air gap and single glazed structures with values of 77.63, 75.92 and 53.14%, respectively.

Acknowledgement

This research was financially supported by the National Natural Science Foundation of China (No. 51776066 and No.51806064).

References

[1] Grossman G and Zaltash A 2001 Absim—modular simulation of advanced absorption systems Int. J. Refrig. **24** 531-543.

[2] Wolf M 1976 Performance analyses of combined heating and photovoltaic power systems for residences *Energy Convers.* **16** 79-90.

[3] Kern J and Russell M 1978 Combined photovoltaic and thermal hybrid collector systems *13th IEEE Photovoltaic Spec. Conf.* 1153-1157.

[4] Zondag H, De Vries D, Van Helden W, Van Zolingen R, and Van Steenhoven A 2003 The yield of different combined pv-thermal collector designs *J. Sol. Energy* **74** 253-269.

[5] Tripanagnostopoulos Y, Nousia T, Souliotis M and Yianoulis P 2002 Hybrid photovoltaic/thermal

solar systems *J. Sol. Energy* **72** 217-234.
[6] Joshi A, Tiwari A, Tiwari G, Dincer I and Reddy B 2009 Performance evaluation of a hybrid photovoltaic thermal (pv/t)(glass-to-glass) system *Int. J. Therm. Sci.* **48** 154-164.
[7] Sarhaddi F, Farahat S, Ajam H, Behzadmehr A and Adeli M 2010 An improved thermal and electrical model for a solar photovoltaic thermal (pv/t) air collector *Appl. Energy* **87** 2328-2339.
[8] Slimani M, Amirat M and Bahria S 2015 Analysis of thermal and electrical performance of a solar pv/t air collector: energetic study for two configurations *3rd Int. Conf. Control Eng. & Info. Technol.* 1-6.
[9] Slimani M, Amirat M, Kurucz I, Bahria S, Hamidat A and Chaouch W 2017 A detailed thermal-electrical model of three photovoltaic/thermal (pv/t) hybrid air collectors and photovoltaic (pv) module: comparative study under algiers climatic conditions *Energy Convers. Manag.* **133** 458-476.
[10] Sainthiya H and Beniwal N 2019 Efficiency enhancement of photovoltaic/thermal (pv/t) module using front surface cooling technique in winter and summer seasons: an experimental investigation *J. Energy Resour. Technol. Trans. ASME* **141** 091201.
[11] Choi H and Choi K 2020 Performance evaluation of pv/t air collector having a single-pass double-flow air channel and non-uniform cross-section transverse rib *Energies* **13** 2203.
[12] Siemens solar industries 2009 Siemens solar module SP50 <www.siemenssolar.com>.
[13] Mohammadi B and Pironneau O 1993 Analysis of the k-epsilon turbulence model.
[14] Zhou J, Yi Q, Wang Y and Ye Z 2015 Temperature distribution of photovoltaic module based on finite element simulation *J. Sol. Energy* **111** 97-103.

Comparison of dust and high-temperature effects on mono and poly photovoltaic panels

E Elshazly[1], A Abd El-Rehim[1] and I El-Mahallawi[1]

[1]Centre for Renewable Energy Faculty of Engineering, British University in Egypt (BUE),El-Sherouk 11837, Cairo, Egypt

E-mail: iemahallawi@bue.edu.eg

Abstract. Solar Photovoltaic Panels are considered as one of the most powerful alternative renewable and sustainable energy sources. However, a major challenge is the effect of dust accumulation on the photovoltaic panels in natural outdoor environment as it reduces the transmissivity of the light on the surface of the solar panels. For many small communities, the decision of implementing mono or polycrystalline PVs should consider economic aspects. This study is a case study that is held at The British University in Egypt at El Sherouk city to study the effect of different parameters such as dust accumulation, water cooling and coating on the performance of both mono- and poly-crystalline panels at El-Sherouk City. The effects of high temperature and dust accumulation on different solar panels placed in natural outdoor conditions at El- Sherouk City are studied and the electrical performance of the solar panels is represented by measuring several characteristic parameters of dusty and cooled PV panels compared to cleaned and non-cooled panels. The effect of the tilt angle on the accumulation of dust on the surface of the solar panels is, also, studied. The mono-crystalline solar panels are installed at tilt angles 0°, 15°, 30°, 45°, and 60° for one month without cleaning, by any method. The results shows that the power reduction percentage is 17%,20%,25%,27% and 30% for tilt angles 60°,45°,30°,15° and 0°; respectively. Tilt angles 15° and 30° show to be optimal for the installation of the PV solar system, as they produce the highest amount of output power. It is found from the study that the accumulation of dust on the surface of different types of solar panels can reduce the efficiency by 30%. While the high temperature can reduce the efficiency by up to 10 %.

1. Introduction

Integrating renewable energy has become a key role in the growth and development of societies. Advanced and emerging societies are working towards increasing the renewable energy input from different sources such as solar energy, wind, hydro, etc.. as the movement towards renewable energies as an alternative to fossil fuels has become an important energy sustainability requirement. Considering the vast potential of renewable energy resources in the Mediterranean region, the development of efficient technologies for using renewable energy represents a key issue for sustainable growth in the region. In line with its 2030 development vision, the Government of Egypt has encouraged small investors and local communities to integrate solar energy in their residential facilities, as well as large investors to contribute in building power stations [1-4].

Content from this work may be used under the terms of the Creative Commons Attribution 3.0 licence. Any further distribution of this work must maintain attribution to the author(s) and the title of the work, journal citation and DOI.
Published under licence by IOP Publishing Ltd

Figures 1 and 2 show the great development in solar PV capacity during the period from 2010 to 2019. Where, Figure 1 shows the world's capacity from solar Photovoltaics while, Figure 2 shows the recent massive development in Egypt's Solar PV capacity especially during the recent 5 years to achieve Egypt's vision 2030 which aims to achieve a diversified, competitive, and balanced economy within the framework of sustainable development as detailed in the Integrated Sustainable Energy Strategy in 2035, released by the Ministry of Electricity and Renewable Energy in 2015 [5].

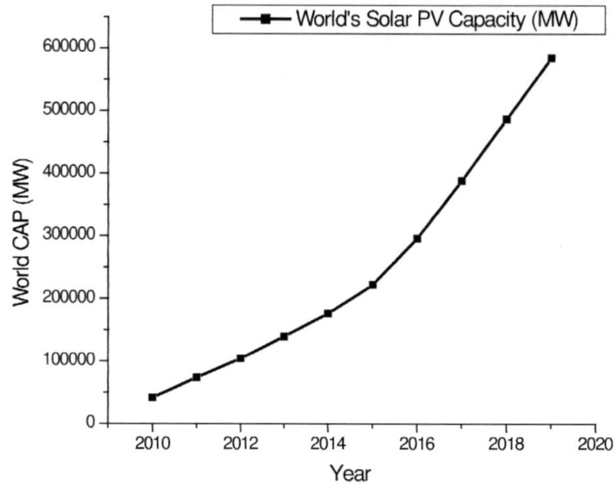

Figure 1: The World's Solar PV Capacity [5]

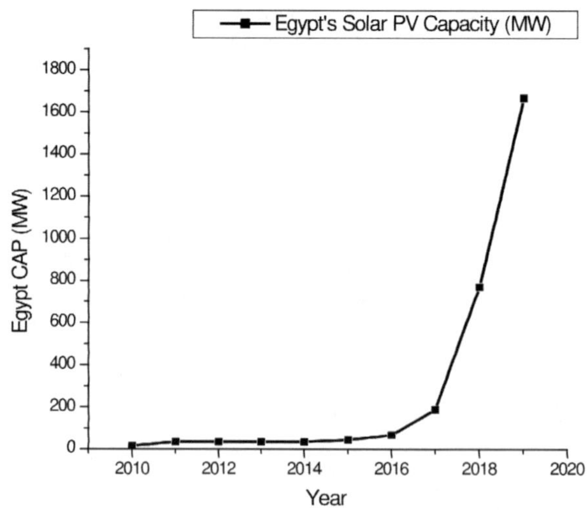

Figure 2: Egypt's Solar PV Capacity [5]

However, a major challenge associated with PV technologies in Egypt as well as other regional environments is its degraded performance with environmental conditions such as temperature, wind, dust, clouds, and shade, encouraging research on mitigation methods [6]. Examples include the concept of photovoltaic/thermal (PV/T) systems. The PV/T system can improve the cell's production, and electrical efficiency along with taking advantage of stored heat for other applications using air, water, nanomaterials, and phase-changing materials with nanofluids to cool the PV cell [7].

Among these weather condition factors that negatively affect the performance of PV cells is the accumulation of dust and pollutants on the cell surface, which reduces the transmissivity of the light on the surface of the solar panels and acts as a barrier between PV and solar irradiation. The impact of Dust on PV efficiency is one of the most important problems facing PV utilization in dusty countries. Additionally, the amount of dust suspended in the air, which reduces PV efficiency in desert areas causes efficiency degradation reaching 8.41% from the maximum power in dusty PV modules compared to cleaned ones [8].

The impact of dust accumulation on the PV panel surface has been studied with excessive concentration because of its great importance, particularly in the solar countries. Dust accumulation in desert countries can cause a cell electrical efficiency degradation up to 40% [9]. Studying the impact of dust accumulation on PV systems in Egypt has indicated up to 17.4%/month degradation in PV efficiency for the PV systems installed at 45° angle to the south. The effect of the PV tilt angle on accumulated dust was studied experimentally using 100 glass samples. The results revealed that dust deposition decreases with 70% for tilt angle 90° compared to tilt angle 0° for the same conditions [10]. On the other hand, the degradation in PV productivity due to dust accumulation was reported to reach 31–35% in Qatar, which is characterized by its high relative humidity and it's attached to the desert [11]. The intensity of the dust increases as the sun's surface warms and wind speed increases. Dust follows the density of the type of land that passes through it, In addition to, humidity, wind deviation, and air altitudes. These factors are changeable with different locations consequently, dust accumulation effects can differ from one place to another [12].

Most of the photovoltaic installations in desert areas suffer from a loss in efficiency due to the accumulation of dust, airborne dirt, sand particles from sandstorms, and high temperature in the summer season. So, to sustain a steady performance of PV panels, regular cleaning is required. However, the traditional methods of manual cleaning are energy, labor, and time-consuming processes. Another disadvantage of manual cleaning is the introduction of cracks and scratches on the PV surface. Researchers and engineers around the globe have been developing new cleaning methods, namely electrostatic, mechanical, and coating methods. Cost-wise, the electrostatic and mechanical methods are more expensive than most coating methods. The development of low-cost coatings is, therefore, an attractive goal for many researchers. The adhesion of soiling on PV arrays results in a significant loss in their energy harvesting efficiency and can be drastically reduced by applying a hydrophobic or hydrophilic coating to the cover glass surface. Even if soiling occurs, the effort needed for cleaning is reduced, reducing costs significantly. However, the effect of sand erosion can cause the coating to deteriorate.

In this context, the British University in Egypt (BUE), which is located in a newly built area, works on solving the problems of its natural environment and community and has been working during the last year on evaluating and analyzing data showing the impact of dust and temperature rise on the performance of three different types of photovoltaic solar panels, mounted at BUE premises at El-Sherouk area. El Shorouk essentially has a hot desert climate, generally extremely dry all over the capital in addition to rarity of rain, extreme heat during summer months is also a general climate feature of El-Shorouk although daytime temperatures are more milder during autumn and winter.

2. Experimental Setup

The experimental work includs three different phases. The first phase aims at studying the effect of dust and cooling. Figure 3 shows the Mono and poly-crystalline panels, where they are held on 30° tilted holders and directed to the South direction. The experiment is held on two different solar panels; monocrystalline with 280 Wp and polycrystalline with 270 Wp. The tested panels are of dimensions 1.6 m height * 1m width and total area 1.6 m^2. The system is installed in a natural outdoor area.

Figure 3: Tested solar panels held on 30° tilted holders facing South direction

A PV solar analyzer is used to measure solar radiation, the temperature of solar panels through wireless sensors, and develop the I-V characterization curve. The measured parameters are the short circuit current, the maximum current, the open voltage, the output power, and the weather parameters, solar radiation, and the panel's surface temperature. Table 1 shows the specifications for PROVA 1011 solar analyzer.

Table 1: PROVA 1011 Specifications

Dimension:	257(L) x 155(W) x 57(H) mm
Battery Type:	Rechargeable Lithium Battery (3400mAh)
Solar Detector	Remote Solar Detector
Temperature Sensor	Thermometer (connected wireless to the solar Analyzer)
Data Transfer	USB cable

The second phase included the study of the tilt anlgle effect in the range of 0° to 60° (by changing the facing angle by manual methods) on dust accumulation and the efficiency degradation for dusted panels. The measurements are taken for one month (May). While the third phase includs seven- days measurements of the same panels after applying a simple formula coating layer on the panels with average radiation of 950 W/m^2. The third part of experiments was performed to study the effect of accumulated dust and cooling effect on mono and poly crystalline solar panels where measurements were performed for three consecutive days in July, each with different conditions as shown in Table 2:

Table 2: Test conditions

Day (average Irr. = 950 W/m^2)	***Tested conditions***
Day 1	Panels with accumulated dust for one month
Day 2	Cleaned panels
Day 3	Cleaned and cooled panels

3. Results and Discussion

The radiation and temperature were measured for one year to calculate the average solar energy kWh/m^2/day over the months of the year in Egypt and is shown in Figure 4. The data were measured using the BUE weather station (WS-GP1) and data was logged at 5-minute intervals. The Figure shows the monthly incident solar radiation at a module inclination angle of 30° in the BUE location at Sherouk City Cairo, Egypt. Which is located at the northeast of Cairo and characterized by its desert hot climate with latitude 30.144212 and longitude 31.639718. The Figure shows high solar irradiance for the specified location, especially in the summer months. Also, the figure displays an example for a typical daily average solar radiation for this location for the month of March which ranges from 295 W/m^2 to 652 W/m^2.

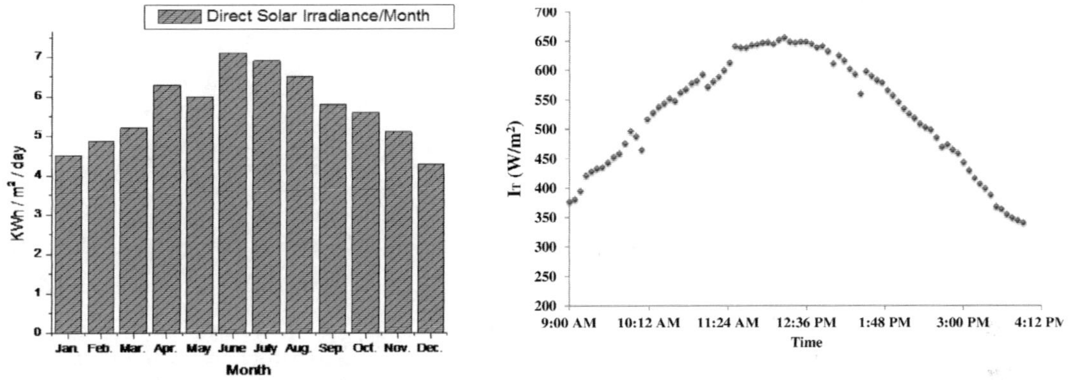

Figure 4. a) The average solar energy kWh/m^2/day over the months of the year in Egypt, b) Typical daily solar radiation measurement for El-Shorouk city, Egypt.

3.1 Effect of tilt angle on dust accumulation

A test was performed to figure out the optimal tilt angle of the PV modules to be installed in the specified location. Figure 5 shows that the efficiency of angles 15° and 30° are the highest compared to the output power of the other angles. Therefore, the optimal tilt angles to be used in the installation of the solar panels are preferred to be in the range of 15° to 30°, followed by angle 45°. The results of angle 0° are approximately near to the results of angle 45°. Therefore, it can be also used in specific conditions if there are no options in choosing the installation angle. Angel 60° has the lowest power output because of the high tilt angle and the lower radiation reaching the solar panel. Therefore, angle 60° is not preferred to be used.

The results presented in Figure 5 show that the panels with tilt angle 30° produce higher power than those with tilt angle 15°, at the start of the experiments, which indicates that the solar panels with a 30° tilt angle produce the highest power in clean conditions. Though the tilt angle of 30° remains the highest in power output, the difference between the inclination angles tends to decrease towards the end of the one-month experiment upon becoming dusty. The Figure shows that the difference between the output power values decreases till it reaches a minimal after 30 days with dust accumulation on the solar panels. The maximum value for difference occurs in the first week in favor of the clean panels. It can be concluded from the curve that the output power is highly influenced by the tilt angle for the clean panels, while the effect of the tilt angle tends to decrease for the dusted conditions for both inclination angles of 15° and 30°.

Figure 5.a PV panels at different tilt angles

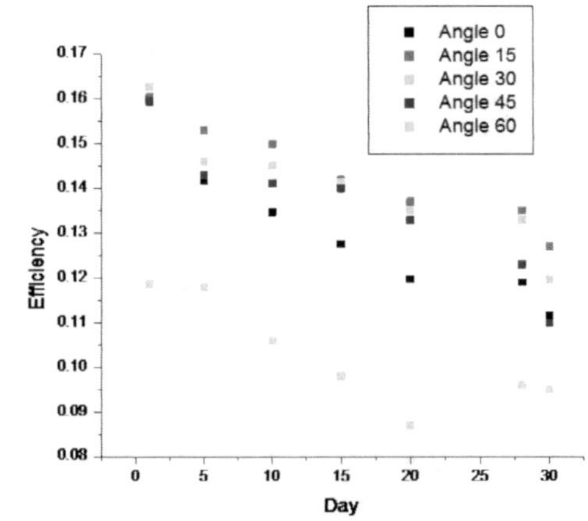

Figure 5.b The output power for different tilt angles for Mono/290 Wp

Figure 6 shows the average reduction in the PV out power over a month for the given tilt angles. The average reduction is about 14 % in the first two weeks for the tilt angle of 30°, which is reduced to about 8 % for the next two weeks of the month. However these values increaded for the lower tilt angles and reduced for the higher tilt angles. This implies that, the best cleaning period for the panels is two weeks of dust accumulation to avoid the effective reduction in the output power.

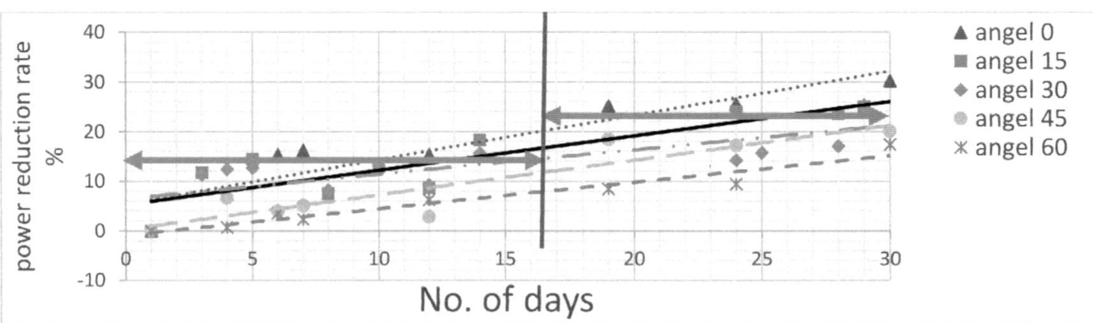

Figure 6. Power reduction rate for the Mono/290 Wp panel at different tilt angles

3.2 Combined effect of dust and temperature rise

Figures 7 and 8 show the performance of mono and poly-crystalline solar panels with the accumulation of dust and the cooling effect. First is the irradiance effect, which can be seen in the 2 graphs, as the efficiency of the panels increase from 09:00 Am reaching its maximum at 12:30 PM, which is approximately the time at which there is maximum irradiance. Efficiencies of the panels then start to decrease till the end of the testing interval as the irradiance also starts decreasing with time. Second, is the dust effect which is shown by the efficiency increase by 20 ~ 30% after it was cleaned on the second day of testing.

Third, is the effect of water cooling, where the panels were cooled every 2 hours. The curves for all types show higher efficiency by approximately7 % than the cleaned panels. The temperature of the panels at noon is usually around 58°C on a normal day without cooling compared to an average of 44°C obtained directly after cooling. The solar panels' efficiency was calculated by the following equation:

$$\eta = \frac{V_{OC} * I_{SC} * FF}{P_{in}} \qquad (1)$$

Where V_{oc} is the open-circuit voltage; I_{sc} is the short-circuit current; FF is the fill factor and η is the efficiency.

As shown from Figures 7 & 8, the most important factor that affected the solar panels' performance was the dust accumulation on the surfaces. Thus, another test was performed to study the performance degradation due to dust accumulation by placing initially cleaned panels in the same place and keeping them without removing dust for four weeks.

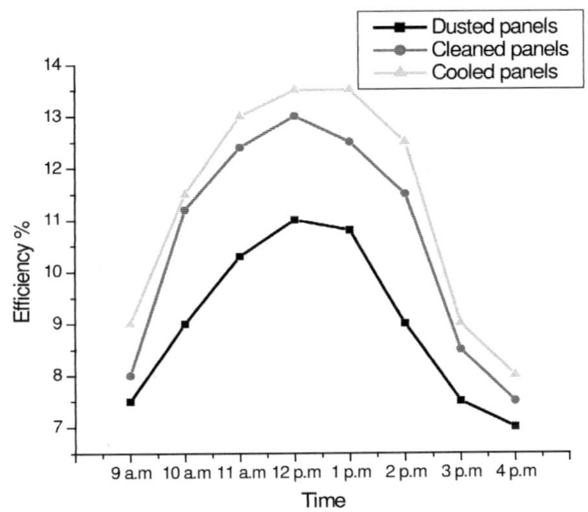

Figure 7: Monocrystalline-280Wp solar panel's efficiency measurements for three different conditions

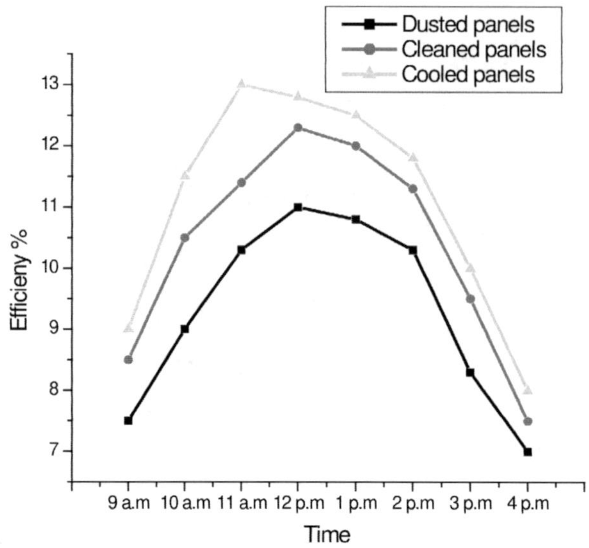

Figure 8: Polycrystalline-270 Wp solar panel's efficiency measurements for three different conditions

Figure 9 shows the electrical efficiency degradation during the examined period under approximately the same temperature and solar radiation conditions for the two different solar panels. The Figure shows that linear degradation starts at a maximum value of 10 % with clean conditions and the degradation keeps increasing up to the four testing weeks without cleaning, which reflects the big influence of dust accumulation on the solar panels.

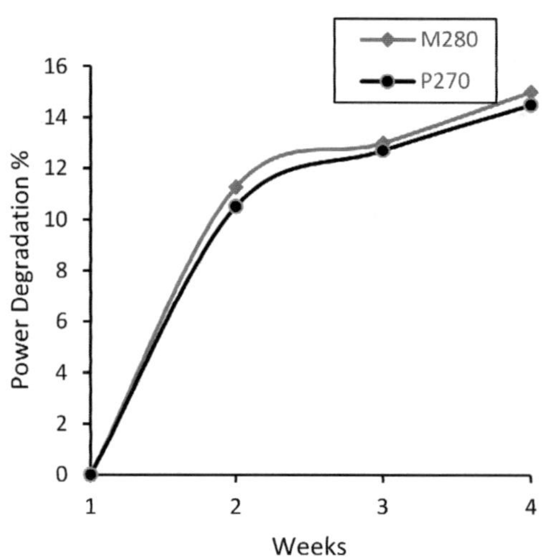

Figure 9: The electrical efficiency degradation during the examined period for poly- and monocrystalline Si modules

Finally, the effect of heat on solar PV efficiency is illustrated in Figure 10, where the temperature was measured for M/280 and P/270 solar panels at the same solar radiation (900 W/m^2) for different panels' temperatures. The curves show that the increase in temperature by 10° results in a degradation of approximately 7 % in the panel's efficiency

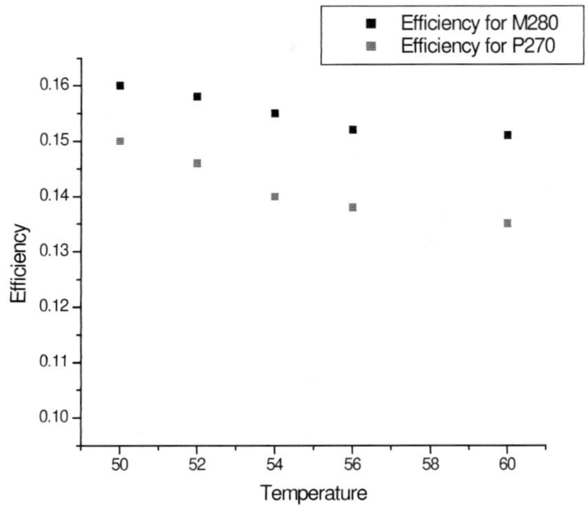

Figure 10: Effect of Temperature on PV efficiency

3.3 Effect of applying self- cleaning material

The previous results emphasize the significance of adopting self-cleaning technologies for PV systems installed in Egypt. Cleaning solar panels is one of the most important and economically urgent problems for the Mediterranean countries. Currently, most solar panel installations rely on one or a combination of these cleaning methods; natural methods (rain and wind), mechanical methods (manual cleaning, or scrubbing), and passive self-cleaning technologies. The cleaning method is summarized as natural means, mechanical means, self-cleaning nano-film, and electrostatic means [13]. However, there is an urgent need to develop a coating material that will mitigate PV cells' degradation in efficiency with soiling and increased temperature during service.

A simple formula for a self-cleaning coating was investigated at the Center of Renewable Energy Lab (CRE). The prepared material was made-up of Glycerin and Al_2O_3 nanoparticles. The nanoparticle concentration was 5wt% of the base carrier and was prepared by applying magnetic stirring for one hour. The design of the coating material was based on nano-materials properties that have been observed in other applications (e.g., enhanced absorbance of nanoparticles observed in selective solar applications, or good heat storage capacity observed in polymeric based materials used for phase change materials. Figure 11 shows the P/270 solar panels; one without coating and the other after applying the prepared nano-coating, to compare between measured efficiencies under the same weather conditions.

Figure 11: Solar Panel after applying the prepared Nano-coating

The results which were obtained after applying the prepared coating for one week and repeating the measurements at noon each day are shown in figures 12 and 13. The reported results show an improvement in the solar panel's obtained output power by about 1%. The results at the beginning of the experiments show a decrease in the panel's temperature of approximately 5°C and a 1 % increase in electrical efficiency. Figure 13 (showing the change in electrical efficiency through the seven testing days) confirms that the electrical efficiency of the coated panels increased during all the examined periods above the uncoated panels, even though the temperature records showed lower temperatures for the uncleaned conditions.

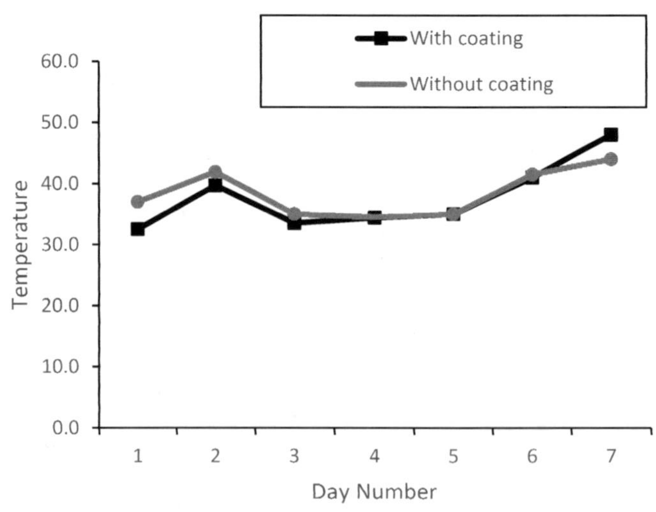

Figure 12: The temperature improvement for Mono/290 Wp

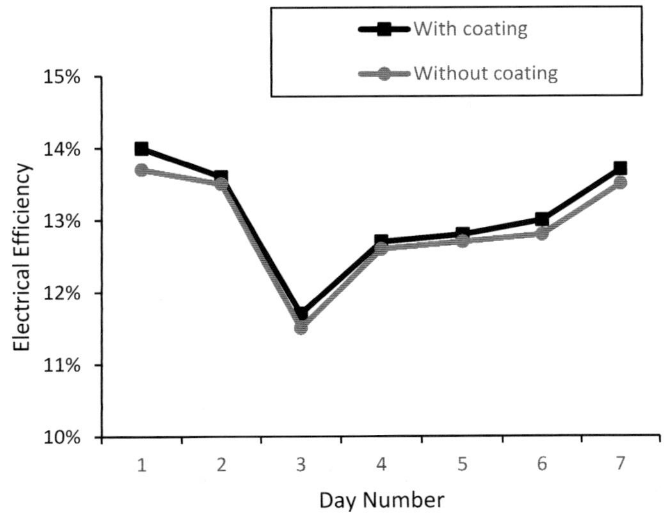

Figure 13: The output power improvement for Mono/290 Wp

The fact that all methods based on coatings have problems with environmental stability requires need for automatic sensing and cleaning of solar panels to keep them free of contamination all the time. This could be realized using technology, which would sense contamination and start cleaning immediately, without human intervention. The new emerging technologies that manage the transition of the ultrasonic cleaning process from a bath environment directly to the object being cleaned—solar panel—are poised to have great success in satisfying growing solar power generation market requirements [14]. The problems associated with manual cleaning result from mechanical impact like scratching by additional wiping, removal of stubborn deposits, the impact of hail, etc. And unfortunately, the better the self-cleaning performance (better super-hydrophobicity), the worse the environmental stability. Better hydrophobicity means a higher aspect ratio of nanostructures, which are more prone to damage by mechanical impact. Chemical resistance of hydrophobic coatings, especially in prolonged contact with water and acids in the rain is also problematic. Photocatalytic coatings contain nanoparticles, which could be released into the environment [14].

4. Conclusions

The effect of irradiance, high temperature, and dust accumulation has been studied on three different solar panels. The reduction in efficiency due to high temperature can be up to 5% while, it reaches 20% due to dust accumulation in all studied types. The degradation in electrical efficiency has been also studied and shows that solar panels may reach half their maximum values after only one month without cleaning in Egypt's weather conditions. The optimum tilt angle has been shown to be 30° for Egypt's conditions. The effect of the tilt angle tends to decrease for the dusted conditions. The application of self-cleaning coating has increased the efficiency by about 1%.

Acknowledgment

The authors would like to express their deepest appreciation to the Centre of Renewable Energy in the British University in Egypt for their technical assistance and support in this project.

References

[1] Al- Maamary et al. 2017 *Ren. and sus. energy* The impact of oil price fluctuations on common renewable energies in GCC countries **75** 989–1007.

[2] Darwish Z et al. 2013 *Inter. j. of energy and envir.* Impact of some environmental variables with dust on solar photovoltaic performance: review and research status **7** 152–159.
[3] Niazy, M.M., Energy management system Comparative study between Egypt and Germany.
[4] Nassar, I.A., Hossam, K. and Abdella, M.M., 2019. Economic and environmental benefits of increasing the renewable energy sources in the power system. *Energy Reports*, *5*, pp.1082-1088.
[5] Dhabi, A., 2020. IRENA,“. *Renewable energy statistics.*
[6] Chaichan M T and Kazem H A 2016 *Inter. J. of scien. & eng. research* Experimental analysis of solar intensity on photovoltaic in hot and humid weather conditions **7** 91–96.
[7] Ibrahim A et al. 2018 *Energy* Comparison study of indoor/outdoor experiments of a photovoltaic thermal PV/T system containing SiC nanofluid as a coolant **151** 33–44.
[8] Dida M et al. 2020 *Ren. and sus. energy* Output power loss of crystalline silicon photovoltaic modules due to dust accumulation in Saharan environment **124** 109–787.
[9] Dorobantu L et al. 2017 *Ren. energy power quality* The effect of surface impurities on photovoltaic panels **1** 622–626.
[10] Elminir H K et al. 2006 *Energy conv. and manag.* Effect of dust on the transparent cover of solar collectors **47** 3192–3203.
[11] Essalaimeh S et al. 2013 *Energy conv. and manag.* Electrical production for domestic and industrial applications using hybrid PV-wind system **65** 736–743.
[12] Guo B et al. 2015 *SGRE* Effect of dust and weather conditions on photovoltaic performance in Doha, Qatar 1–6.
[13] He G et al. 2011 *Procedia eng.* Review of Self-Cleaning Method for Solar Cell Array **16** 640 – 645.
[14] Kobrin B 2018 *n-Tech research publication* self-cleaning technologies for solar panels.

IOP Conf. Series: Materials Science and Engineering 1172 (2021) 012020 doi:10.1088/1757-899X/1172/1/012020

Effective electromagnetic interference shielding using polyurethane / graphite composite foam materials

H Oraby[1], H R Tantawy[1], I Naeem[1], M Magdy[2], A A Ezz[3] and M Senna[4]

[1]Department of Chemical Engineering, Military Technical College, Cairo, Egypt.
[2]Department of Radar, Military Technical College, Cairo, Egypt.
[3]El-Nasr Company for Intermediate Chemicals, Industrial Area, Giza, Egypt.
[4]Radiation Chemistry Department, National Center for Radiation Research and Technology, P.O.Box 29, Nasr City, Cairo, Egypt.

E-mail: hussein.mohamed4544@yahoo.com

Abstract. Conversely, to metallic based electromagnetic interference (EMI) shielding materials; composite polymer-based foams are normally lightweight, cheaper than metals, and less sensitive to types of environmental degradation. In this work polyurethane - Graphite composite (PU-G) foam materials were prepared with different filler concentrations. Different characterization tools such as Fourier Transform Infra-Red (FTIR) and scanning electron microscope (SEM), were used to identify the structural and topological construction of the prepared composites. Further mechanical properties for the prepared samples were studied to elucidate the opportunity of utilizing these composites in applied applications, specifically for electromagnetic interference (EMI) shielding efficiency (SE) for aerospace applications. Also, in order to adjust this research in the area of aerospace EMI SE, the evaluations were executed in the X-band at (8-12) GHz. The obtained data indicated that the moreover in filler concentration enhanced the compressive strength and compressive modulus of the prepared samples. Moreover EMI SE reached -44 dB with 30 wt % graphite concentration. Finally polyurethane – graphite composite foam material can be taken into consideration a gratifying material to be utilised in EMI SE.

Keywords: electromagnetic interference; shielding; polyurethane foam; mechanical properties.

1. Introduction

The exuberant electromagnetic (EM) pollution caused by the general usage of high-frequency electronic equipment has impacted society as a whole, owing to its direct danger to human health and communication stability [1]. A system is assumed to be electromagnetically dependable with its surroundings of marginal electromagnetic interference (EMI) but, it interferes with further systems

Content from this work may be used under the terms of the Creative Commons Attribution 3.0 licence. Any further distribution of this work must maintain attribution to the author(s) and the title of the work, journal citation and DOI.
Published under licence by IOP Publishing Ltd

and is not influenced by releases from additional systems [2, 3]. Decent shielding tools must also resist all received and emitted EMI [2].

Various shielding strategies were designed not only to reduce the danger of EMI but also to enhance the life time and routine of the system [4]. However, typical metal shields, particularly for aerospace applications, impose significant weight penalties [5, 6]. Due to their small weight, flimsy, low price, and processability, polymer foam composite materials (PFCM) are further desirable materials [7, 8]. PFCM have fairly large conductivities (σ) that can be easily monitored through its chemical composition [9].

Several kinds of carbon resources can be used as a conducting filler for PFCM such as graphite, carbon black, carbon fibers, carbon nanotubes, and graphene that have excellent electrical efficiency, mostly combined with polymers to protect or obstruct EM waves because of their superior properties over typical metals, such as mechanical stability, lightweight, corrosion resistance, etc… [10].

Graphite has many advantages other than carbon materials such as (1) Lower cost, (2) Doesn't need additional chemical treating, (3) Simplicity of combining with polyurethane. These composites could be turned into foams with the following advantages [11], (1) greater versatility and (2) greater EMW absorption [12].

The present study details the preparation of polyurethane – Graphite composite (PU-G) foam materials with various filler concentrations. Different characterization tools such as Fourier transform infra-red (FTIR) and scanning electron microscope (SEM), was used to identify the chemical and topological composition of the prepared composites. Further mechanical properties for the prepared samples were tested to elucidate the probability of utilizing these composites in applied applications, particularly for aerospace uses. Also, to adjust this work in the area of aerospace EMI SE, the evaluations were executed in the X-band at (8-12) GHz.

2. Theory of shielding

Proportion of remaining power to imposing power, generally equal to the overall electromagnetic (EM) waves emitted (T) by the shielding separated by the incident (i) pulse, is shielding effectiveness (SE). Electrical (E), magnetic (H) electromagnetic force (P) can be determined from any of the energy intensities:[13, 14].

$$SE = -20\log [E_T / E_i] = -20\log [H_T / H_i] = -10\log [P_T / P_i] \quad (1)$$

The EMI reduction is dependent on the wavelength, the space between the protective material and the origin of the interference, the width of the protective material and shielded content [2]. There were 3 main pathways of EM/material exchange causes SE, portion of the incident waves was redirected from the front surface of protective material, portion is taken in the shield, and portion was redirected from the rear surface, Figure 1 illustrates the mechanism of shielding efficiency. The overall efficiency of the shielding substance (SE_{Total}) is thus proportional to the sum of the absorption losses (SE_A), the reflection losses (SE_R), and the correction factor (M) for several reflections, which can be important for thin shields [15].

$$SE_{Total} = SE_A + SE_R + M \quad (2)$$

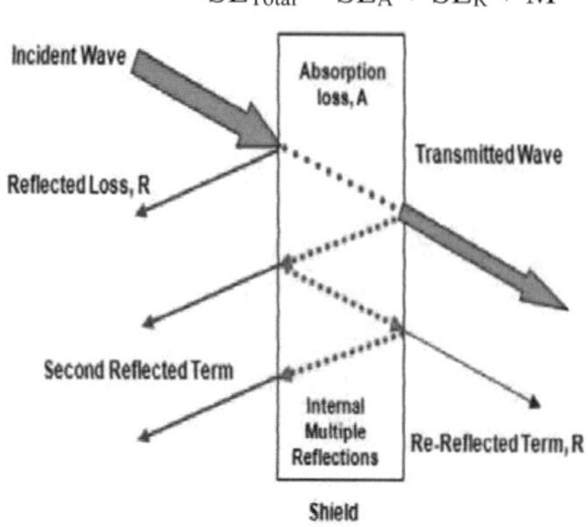

Figure 1. Mechanism of shielding efficiency

Generally, the reflection factor M was neglected at absorption loss value greater than -10 dB [13]. For simpler equations, electrical fields and plane waves can be ignored. The lack of absorption or decay takes place when currents caused in the intermediate create ohmic losses then substrate warming [16]. Reflection loss, was belonged to the relative difference between the incident wave and the protective material surface impedance. Besides, several correction factors for reflection may be negative or positive (it is usually -ve in practice) and become negligible when SE_A is > 6 dB. Usually, it is simply important by low frequencies (i.e. lower than 20 kHz) [17].

SE could be measured in two port analysis in terms of diffusion factors that detect the way voltages pass in a transmission line.

In a waveguide, as the incident EMW at a point k move to other point h, a radiated energy undergoes SE (transmission, absorption, and reflection). These wave scattering values were expressed as S_{hk}. E.g., S_{12} is the energy originating from point 2 at point 1. Thus, the S parameters S_{11} and S_{21} express the portion of redirected energy and the portion of transmitted energy, respectively [2]. The transmittance (T_{coef}), reflectance (R_{coef}), and absorbance (A_{coef}) coefficients can be expressed as follows [18, 19]:

$$T_{coef} = [E_T / E_i]^2 = S_{12}^2 = S_{21}^2 \quad (3)$$

$$R_{coef} = [E_R / E_i]^2 = S_{11}^2 = S_{22}^2 \quad (4)$$

$$A_{coef} = 1 - T_{coef} - R_{coef} \quad (5)$$

The terms A_{coef} was given here regarding the power of the EM energy. Even If the impact of various redirections among shield's two boundaries was insignificant, [20]. Positive absorbance could be well-defined as follows:

$$A_{eff} = (1 - T_{coef} - R_{coef}) / (1 - R_{coef}) \quad (6)$$

SE_A and SE_R could be defined with respect to reflectance and transmittance, respectively, by the application of power equilibrium findings [20].

$$SE_R = 10\log (1 - R_{coef}) \quad (7)$$

$$SE_A = 10\log (1 - A_{coef}) = 10\log [T_{coef} / (1 - R_{coef})] \quad (8)$$

$$SE_T = 10\log (1 - R_{coef}) + 10\log [T_{coef} / (1 - R_{coef})] \quad (9)$$

3. Experimental

3.1. Materials

For rigid PU foam formulation, the polyol mixture (with a hydroxyl number of 343.5 mg KOH g) was purchased from Alcupol R-4520 (Repsol Quimica). Methylene diphenyl diisocyanate (MDI) (Suprasec ® 5005, dark brown liquid, viscosity 220 cP at 25 ° C) was purchased from Huntsman. Graphite TPC powder (M.W: 12.01 g/mol) was obtained from CHNO, Haryana (India). All reagents were used as purchased without more treatment.

3.2. Preparation of (PU - G) composite foams

In the first step, graphite was dispersed in a polyol matrix at 5, 15 and 30 weight percent by intense sonication with an ultrasonic homogenizer for 5 min at 50 percent amplitude to achieve an identical diffusion. Polyol is carefully chosen for graphite dispersion since it is less viscous. The particulate mixture was blended with MDI in 250 ml paper cups at a ratio of 1:1.1 (polyol / MDI) in next step. For post-curing and potential research, the samples were allowed to stay at room temperature for 24 h. The prepared samples with different graphite concentration will denoted here after as follows PU, PU / G 5%, PU / G 15% and PU / G 30%.

3.3. Characterization of (PU / G) composites foam

3.3.1. FTIR spectroscopic analysis. Using a JASCO 4100 FTIR spectrometer over the 400-4100 cm^{-1} spectrum at a 5 cm^{-1} resolution, IR analysis was conducted.

3.3.2. Scanning Electron Microscopy (SEM). The topology was studied by ZEISS EVO MO-10 Germany scanning electron microscope. Samples were prepared for the SEM images of the composite foam cross-section by fracturing the related composite into liquid nitrogen.

3.3.3. Compression test. The compressive characteristics of neat PU and PU composite samples were measured with respect to ASTM D 1621-94. For the compression research, test samples with a sectional area of 25.7 cm^2 and a length of 2.53 cm were tested. For each 25.4 mm of specimen thickness, the load was applied to the sample with a crosshead movement of 2.6 mm/min. For aforementioned procedure were repeated three times for each sample and the average was approved as the final result. The compressive strength was determined automatically.

3.4. EMI SE measurements

The efficacy of the SE was calculated by ZVA-24 vector network analyzer in 8 to 12 GHz frequency. Composite foams were sliced into the aluminum waveguide sample holder into a rectangle (2.02 cm X 1.02) cm. For each foam composite, the final recorded results were the combination of three samples tested.

4. Results and Discussion

4.1 FTIR spectroscopic Analysis

The change in chemical composition structure was detected from FTIR spectra. Figure 2 displays the FT/IR spectrum of PU and PU / G 30 %. At (3427 and 3520) cm^{-1}, the pure PU foam sample showed its characteristic bands, which are due to the stretching pulsation of hydrogen bonding and (N-H) groups. At 2925 and 2865 cm^{-1}, the absorption peaks related to the symmetric and non symmetric stretching pulsation of CH_2 were detected. The bands at 1723, 1227, and 1070 cm^{-1} were belonged to carbonyl (C = O) stretch pulsation, aromatic (C-O) stretch pulsation, and (C-O-C) non symmetric stretch pulsation. At (1535 and 1225) cm^{-1}, there is a mixture of (N-H) distortion and (C-N) stretching pulsations [21]. By comparing pure PU and PU / G 30 % FTIR spectra, we can notice that the relative position of distinctive composite bands were similar to that of tidy PU foam. This suggests that the use of graphite as a filler has not changed the segmented PU foam structure, which ensures that graphite does not interact with the PU molecules [22-24].

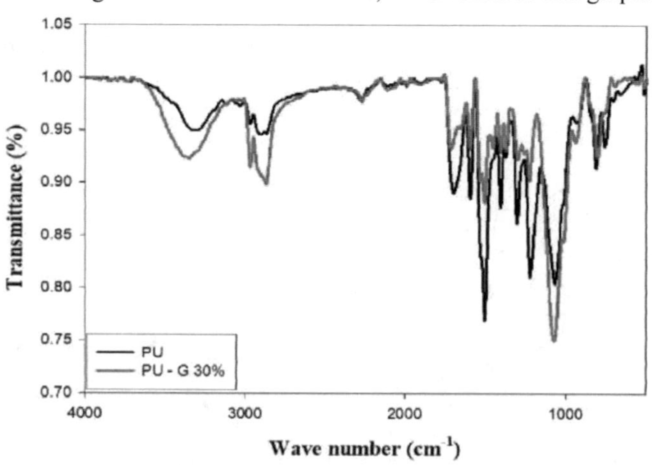

Figure 2. FTIR spectrum for PU and PU / G 30 %

4.2. Morphology of the prepared samples

The microstructure of PU, PU / G 5%, PU / G 15% and PU / G 30% foams is seen in Figure 3 (a, b, c and d). The cell was circular in the PU foam, with diameter of 315 μm. The foam cell becomes thin and irregular with the incorporation of G. The cell diameter of PU / G 5%, PU / G 15% and PU / G 30% composites, for instance, is (191, 120, and 85) μm respectively. Extra pores can be produced during the foaming phase in composite foam because of nucleation influence of G, causes reduced cell diameter [25]. Also, the great graphite/polyol mixture viscosity facilitates the joining of cells in the foaming phase, contributing to the irregular morphology of cells in composite foams. Organic materials and sonication have been usually used to prepare these foams to keep identical spreading and decrease the viscosity of the PU / G composites.

Extreme usage of organic materials, however, will cause environmental problems and more expenses, while, plentiful usage of sonication allows it more economical to manufacture large-scale nanocomposites [25].

The dispersion of graphite in polyurethane foam is shown in figure 3.d. In the polyurethane matrix, the graphite dispersion is fairly uniform but in a strongly aggregated state. The wide G agglomerates could be fragmented by the heavy shear strength produced by sonicator. Slightly aggregated G state has aided to create electrical conductive pathways [26]. Figure 3.d shows that the nucleation continues at high G content at a high rate that allows the cells to converge with each other and creates channels between cells to create an open-cell foam structure that harms the mechanical properties, and this will be addressed in the next section.

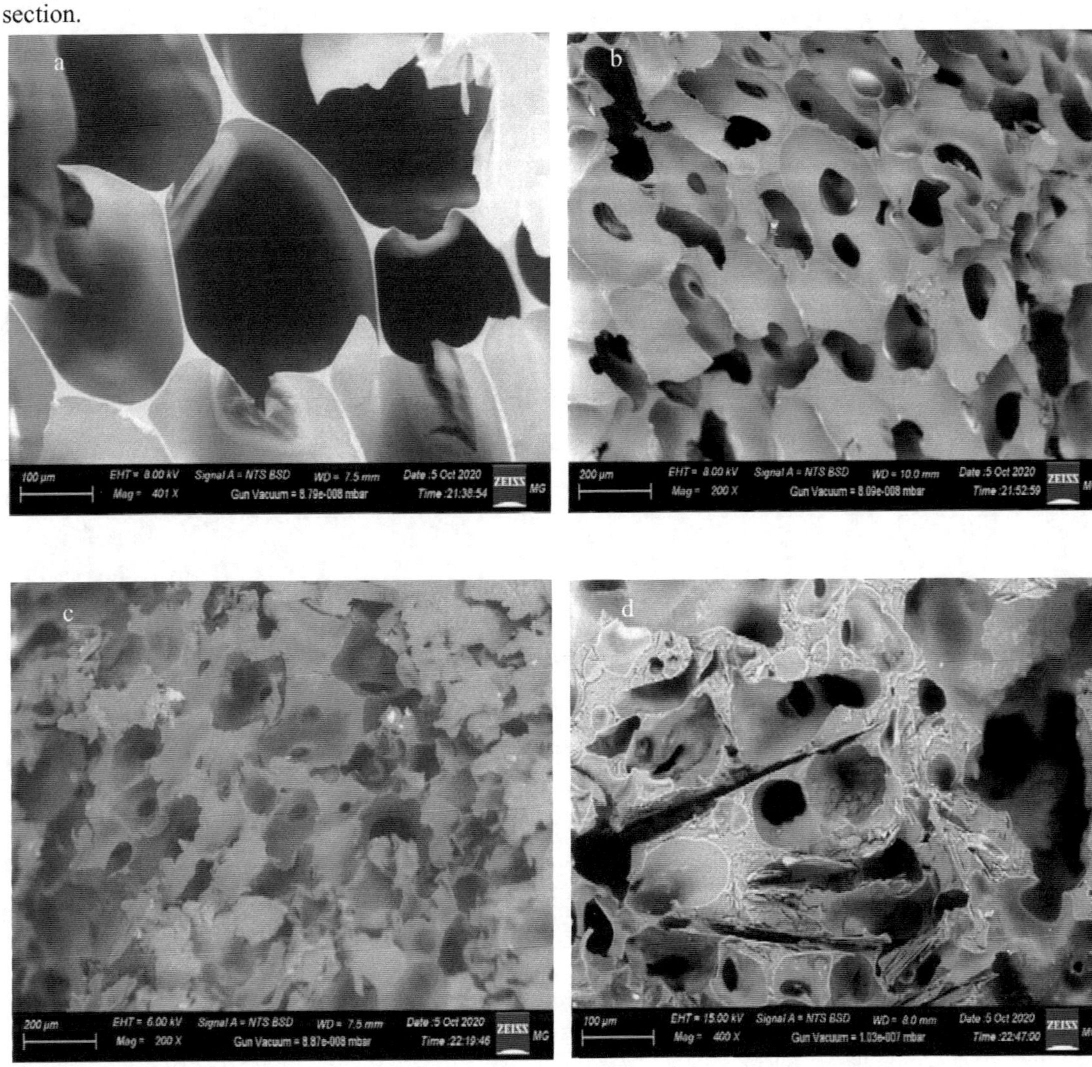

Figure 3. SEM images for a) PU, b) PU / G 5 %, c) PU / G 15 % and d) PU / G 30 %.

4.3. Mechanical compression test

Figure 4 demonstrates stress-strain curves for PU and PU-graphite composite foams. Three deformation phases were seen; original linear action, linear plateau area, and finally, densification. To estimate foam compressive modulus, the slope of the first region is used and intersection point is used to measure the compressive strength between the initial linear region and the plateau region [22]. For PU / G 5% and PU / G 15%, together compressive modulus and strength were increased, as we note that the modulus raised by 320 % and 505 % respectively and the strength also raised by 125 % and 255 %, but at 30 % concentration, the compressive module increased marginally with a rise of 3.7 % and a decrease in compressive strength of -29.7 %. We can notice that the slope of the linear plateau region is higher for all PU-G composite foams under careful observation and it is improved by increasing graphite wt percentage [22]. we can notice that the profile of the plateau region depends on structures of cells [27].

The deformation of cells happens as a merging of cells folding and collapsing in the event of a linear plateau. Therefore, the post-yield action of the PU - graphite composite foam can be enhanced until some concentration due to dispersion of graphite particles. However at higher concentrations PU-G 30% the nucleation is very strong and cell merger occurs and the foam structure becomes open cell structure with a higher percentage, thereby decreasing the mechanical properties, which confirmed by the SEM results. Table 1 summarize the compression test results.

Table 1. Compression test results

PU/G	**Compressive modulus (MPa)**			**Compressive strength (MPa)**		
	Results	**Average**	**Gain %**	**Results**	**Average**	**Gain%**
Neat	0.54 0.53 0.55	0.54	-	5.21 5.67 5.99	5.62	-
PU / G 5%	2.32 2.28 2.26	2.28	320 %	12.64 13.01 12.22	12.62	125 %
PU / G 15%	3.29 3.31 3.25	3.28	505 %	20.47 19.96 19.35	19.92	255 %
PU / G 30%	0.58 0.55 0.57	0.56	3.7 %	3.99 4.01 3.87	3.95	-29.7 %

4.4. Electromagnetic interference shielding effectiveness

Electromagnetic waves could be inhibited by using good shield material which could attenuates the imitating energy [28]. The shielding effectiveness for polyurethane composites with different graphite contents in the X-band frequency range (8-12) GHz is presented in Figure 5. In PU composites, SE rises with graphite percentage enrichment. A small amount of graphite containing composite PU-G 5% displayed a SE of 17 dB followed by 36 dB at PU-G 15%, and a limit of 44 dB was observed at the X-band for PU-G 30%.

As is obvious from the increasing conductivity, G particulates come to be more nearby with raising G content [28, 29]. The link of G pathways with very near gaps was formed as extra G particulates were dispersed in PU matrix which was confirmed by SEM results. Compared to PU-rich environments, these particles redirected or absorbed more radiation, causing SE value rises with higher concentration of graphite.

In this work, we find that shielding routine increases with growing frequency for samples with distinct G concentrations (5, 15, and 30) wt percent. Graphite-prepared PU composites have demonstrated relatively greater SE results rather than these synthesized for different fillers via compression molding method of melt manufacturing [30, 31]. Figure (6) shows the variance of reflection loss (RL) as a function of frequency. With the inclusion of graphite material, the highest RL value is 5.5 dB for PU / G 5% at (8-12) GHz, decreases and is found to be 3 dB for PU / G 15% and 1 dB for PU / G 30%. With frequency, it is randomly rising, and the variation for greater compositions is further disorderly. Structural properties, such as the filler scattering, through the relation of EM energy with G, may be responsible for such frequency dependency. It is obvious that PU composites with a high SE value output a smaller RL value [28].

The transmission loss variance (TL) as a function of frequency is shown in Figure 7. The minimum TL value for PU - graphite composite foam is 10 dB at PU / G 5% at (8-12) GHz, with the inclusion of graphite material, it rises to 30 dB for PU / G 15% and 40 dB for PU / G 30% is observed. We note that TL increases along the frequency spectrum, and because of the enhanced conductivity and good G scattering in PU, this variance can be called systematic [28].

Table 2 illustrates a comparison between this work and other work in literature using graphite in different polymeric matrices.

Table 2. Comparison between this work and results in the literature

No. #	Filler	Matrix	Filler	Thickness (mm)	Frequency range(GHz)	SE (dB)	ref.
1	Graphite	SEBS	20 wt%	5	8-12	-15	[32]
2	Graphite	PE	30 wt%	3	8-12	-33	[33]
3	Graphite	PE	7.5 vol%	2.5	8-12	-51.6	[34]
4	Graphite	Epoxy	2 wt%	5	8-18	-11	[35]
5	Graphite	PU	15 wt%	-	8-12	-21.7	[36]
6	Graphite	PU–Si rubber	5 wt%	-	8-12	-25	[37]
7	Graphite	PU	30 wt%	5	8-12	-44	This work
8	Graphite	PU	15 wt%	5	8-12	-35	This work

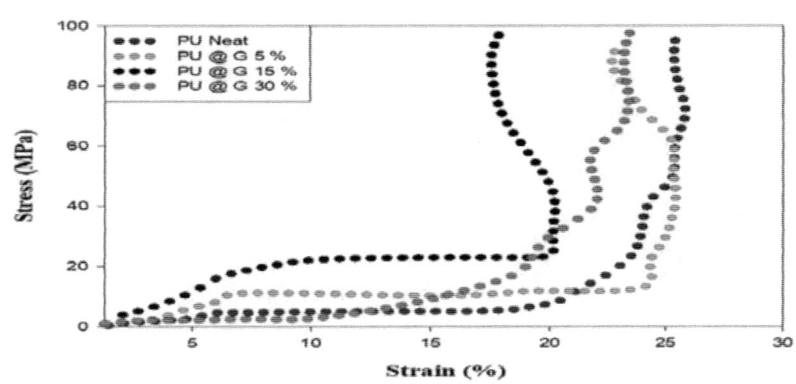

Figure 4. Compression Stress-Strain curve for a) PU, b) PU / G 5 %, c) PU / G 15 % and d) PU / G 30 %.

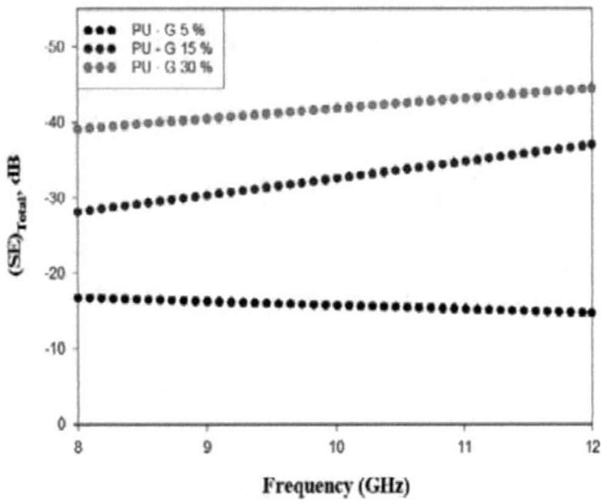

Figure 5. Total shielding efficiency (SE) as a function of graphite concentration for PU / G composite foam.

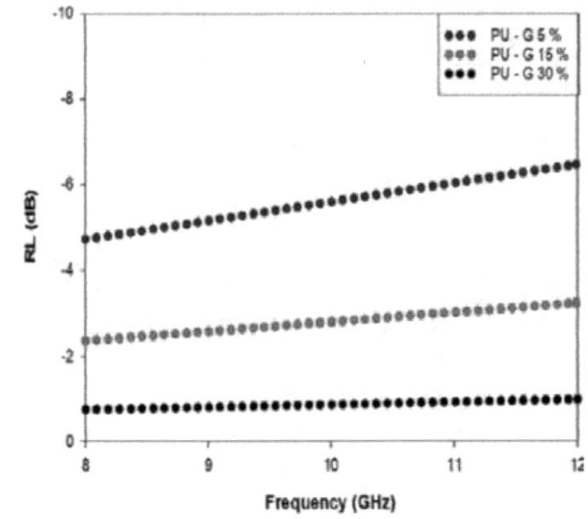

Figure 6. Reflection Loss (RL) as a function of graphite concentration for PU / G composite foam.

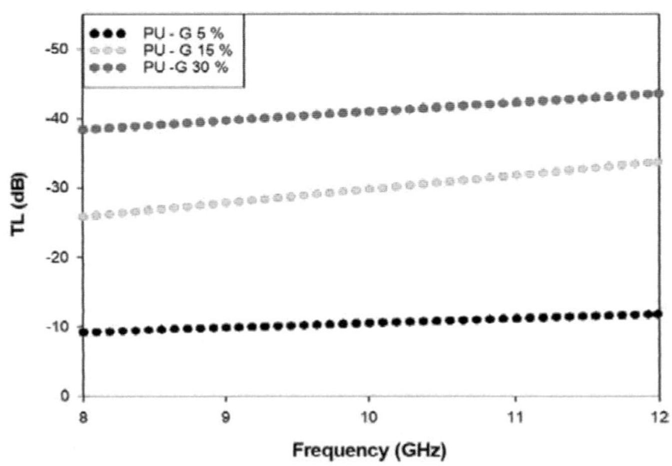

Figure 7. Transmission Loss (TL) as a function of graphite concentration for PU / G composite foam.

5. Conclusion

PU - G composite foams with great SE values were successfully prepared. SEM micrographs showed that that cell size decreases as filler concentration increases and at larger G concentration, foam construction begins to be diminished which has a negative influence on the compressive properties. Also, the mechanical properties (compressive modulus and strength) for the prepared composite foams enhanced by loading with G but, at increased G percentage the compressive characteristics weakens as SEM indicated. Also, FT/IR showed the typical bands for the prepared samples and illustrated that graphite didn't affect the PU segments structure. The prepared foam samples showed brilliant results in SE of 44 dB with PU / G 30% G concentration. Finally, the previous results introduce an optimization between compressive characteristics and SE for PU / G samples with different concentrations. It illustrated that the optimum G % in PU foam was 15 wt % which involves the great total SE -35 dB besides an acceptable compressive properties that could be considered as a promising material in EMI SE for aerospace applications.

References

[1] Y. Li, B. Shen, D. Yi, L. Zhang, W. Zhai, X. Wei, *et al.*, 2017 "The influence of gradient and sandwich configurations on the electromagnetic interference shielding performance of multilayered thermoplastic polyurethane/graphene composite foams," *Composites Science and Technology,* vol. **138**, pp. 209-16.

[2] H. R. Tantawy, D. E. Aston, J. R. Smith, and J. L. Young, 2013 "Comparison of electromagnetic shielding with polyaniline nanopowders produced in solvent-limited conditions," *ACS applied materials & interfaces,* vol. **5**, pp. 4648-58.

[3] C. R. Paul, 2006 "*Introduction to electromagnetic compatibility"* vol. **184**: John Wiley & Sons.

[4] L. Li and D. Chung, 1991 "Electrically conducting powder filled polyimidesiloxane," *Composites,* vol. **22**, pp. 211-18.

[5] P. Liu and G.-F. Chen, 2014 "*Porous materials: processing and applications*" Elsevier.

[6] H. Nakajima, 2019 "Fabrication, Mechanical and Physical Properties, and Its Application of Lotus-Type Porous Metals," *MATERIALS TRANSACTIONS,* vol. **60**, pp. 2481-89.

[7] M. M. A. Nikje, M. A. F. Nejad, K. Shabani, and M. Haghshenas, 2013 "Preparation of magnetic polyurethane rigid foam nanocomposites," *Colloid and Polymer Science,* vol. **291**, pp. 903-9.

[8] S. Petlitckaia and A. Poulesquen, 2019 "Design of lightweight metakaolin based geopolymer foamed with hydrogen peroxide," *Ceramics International,* vol. **45**, pp. 1322-30.

[9] H. Zhang, G. Zhang, Q. Gao, M. Zong, M. Wang, and J. Qin, 2020 "Electrically electromagnetic interference shielding microcellular composite foams with 3D hierarchical graphene-carbon nanotube hybrids," *Composites Part A: Applied Science and Manufacturing,* vol. **130**, p. 105773.

[10] G. Cao, 2004 "*Nanostructures and nanomaterials: synthesis, properties and applications*" World scientific.

[11] G. Viola and W. Schmeal, 1994 "Isocyanate trimerization kinetics and heat transfer in structural reaction injection molding," *Polymer Engineering & Science,* vol. **34**, pp. 1173-86.

[12] J.-M. Thomassin, M. Trifkovic, W. Alkarmo, C. Detrembleur, C. Jérôme, and C. Macosko, 2014 "Poly (methyl methacrylate)/graphene oxide nanocomposites by a precipitation polymerization process and their dielectric and rheological characterization," *Macromolecules,* vol. **47**, pp. 2149-55.

[13] X. C. Tong, 2016 "*Advanced materials and design for electromagnetic interference shielding*" CRC press.

[14] H. S. Nalwa, 1997 "*Handbook of organic conductive molecules and polymers*" Wiley.

[15] L.-F. Chen, C. Ong, C. Neo, V. Varadan, and V. K. Varadan, 2004 "*Microwave electronics: measurement and materials characterization*" John Wiley & Sons.

[16] D. Weston, 2001 "*Electromagnetic Compatibility: Principles and Applications, Revised and Expanded*" CRC Press.

[17] J. Li, G. Zhang, Z. Ma, X. Fan, X. Fan, J. Qin, *et al.*, 2016 "Morphologies and electromagnetic interference shielding performances of microcellular epoxy/multi-wall carbon nanotube nanocomposite foams," *Composites Science and Technology,* vol. **129**, pp. 70-78.

[18] M. Verma, P. Verma, S. Dhawan, and V. Choudhary, 2015 "Tailored graphene based polyurethane composites for efficient electrostatic dissipation and electromagnetic interference shielding applications," *RSC advances,* vol. **5**, pp. 97349-58.

[19] G. Wang, G. Zhao, S. Wang, L. Zhang, and C. B. Park, 2018 "Injection-molded microcellular PLA/graphite nanocomposites with dramatically enhanced mechanical and electrical properties for ultra-efficient EMI shielding applications," *Journal of Materials Chemistry C,* vol. **6**, pp. 6847-59.

[20] B. Reddy, 2011 "*Advances in Nanocomposites: Synthesis, Characterization and Industrial Applications*" BoD–Books on Demand.

[21] J. N. Gavgani, H. Adelnia, D. Zaarei, and M. M. Gudarzi, 2016 "Lightweight flexible polyurethane/reduced ultralarge graphene oxide composite foams for electromagnetic interference shielding," *RSC advances,* vol. **6**, pp. 27517-27.

[22] M. Saha, M. E. Kabir, and S. Jeelani, 2008 "Enhancement in thermal and mechanical properties of polyurethane foam infused with nanoparticles," *Materials Science and Engineering: A,* vol. **479**, pp. 213-22.

[23] R. Seymour, G. Estes, and S. L. Cooper, 1970 "Infrared studies of segmented polyurethan elastomers. I. Hydrogen bonding," *Macromolecules,* vol. **3**, pp. 579-83.

[24] H. S. Lee, Y. K. Wang, W. J. MacKnight, and S. L. Hsu, 1988 "Spectroscopic analysis of phase-separation kinetics in model polyurethanes," *Macromolecules,* vol. **21**, pp. 270-73.

[25] T. Zhai, D. Li, G. Fei, and H. Xia, 2015 "Piezoresistive and compression resistance relaxation behavior of water blown carbon nanotube/polyurethane composite foam," *Composites Part A: Applied Science and Manufacturing,* vol. **72**, pp. 108-14.

[26] J. Aguilar, J. Bautista-Quijano, and F. Avilés, 2010 "Influence of carbon nanotube clustering on the electrical conductivity of polymer composite films," *Express Polym Lett,* vol. **4**, pp. 292-99.

[27] W. Wang, X. Liao, Y. He, J. Li, Q. Jiang, and G. Li, 2020 "Thermoplastic polyurethane/polytetrafluoroethylene composite foams with enhanced mechanical properties and anti-shrinkage capability fabricated with supercritical carbon dioxide," *The Journal of Supercritical Fluids,* p. 104861.

[28] V. Sachdev, K. Patel, S. Bhattacharya, and R. Tandon, 2011 "Electromagnetic interference shielding of graphite/acrylonitrile butadiene styrene composites," *Journal of Applied Polymer Science,* vol. **120**, pp. 1100-05.

[29] C. Y. Huang and J. F. Pai, 1997 "Studies on processing parameters and thermal stability of ENCF/ABS composites for EMI shielding," *Journal of applied polymer science,* vol. **63**, pp. 115-123.

[30] C.-Y. Huang and T.-W. Chiou, 1998 "The effect of reprocessing on the EMI shielding effectiveness of conductive fibre reinforced ABS composites," *European polymer journal,* vol. **34**, pp. 37-43.

[31] C.-Y. Huang and C.-C. Wu, 2000 "The EMI shielding effectiveness of PC/ABS/nickel-coated-carbon-fibre composites," *European polymer journal,* vol. **36**, pp. 2729-37.

[32] S. Kuester, C. Merlini, G. M. Barra, J. C. Ferreira Jr, A. Lucas, A. C. de Souza*, et al.*, 2016 "Processing and characterization of conductive composites based on poly (styrene-b-ethylene-ran-butylene-b-styrene)(SEBS) and carbon additives: A comparative study of expanded graphite and carbon black," *Composites Part B: Engineering,* vol. **84**, pp. 236-47.

[33] V. Panwar and R. Mehra, 2008 "Analysis of electrical, dielectric, and electromagnetic interference shielding behavior of graphite filled high density polyethylene composites," *Polymer Engineering & Science,* vol. **48**, pp. 2178-87.

[34] X. Jiang, D.-X. Yan, Y. Bao, H. Pang, X. Ji, and Z.-M. Li, 2015 "Facile, green and affordable strategy for structuring natural graphite/polymer composite with efficient electromagnetic interference shielding," *RSC advances,* vol. **5**, pp. 22587-92.

[35] G. De Bellis, A. Tamburrano, A. Dinescu, M. L. Santarelli, and M. S. Sarto, 2011 "Electromagnetic properties of composites containing graphite nanoplatelets at radio frequency," *Carbon,* vol. **49**, pp. 4291-300.

[36] J. M. Kim, Y. Lee, M. G. Jang, C. Han, and W. N. Kim, 2017 "Electrical conductivity and EMI shielding effectiveness of polyurethane foam–conductive filler composites," *Journal of Applied Polymer Science,* vol. **134**.
[37] J. Jeddi and A. A. Katbab, 2018 "The electrical conductivity and EMI shielding properties of polyurethane foam/*silicone* rubber/carbon black/nanographite hybrid composites," *Polymer Composites,* vol. **39**, pp. 3452-60.

IOP Conf. Series: Materials Science and Engineering 1172 (2021) 012021 doi:10.1088/1757-899X/1172/1/012021

Parametric optimization of non-prismatic micro-plates to reduce stiffening and curling initiated during fabrication

B Bakeer[1], A Elsabbagh[1] and M Hedaya[1]

[1]Design and Production Engineering Department, Ain Shams University, Cairo, 11517 Egypt

Email: bahi.bakeer@eng.asu.edu.eg

Abstract. Microelectromechanical devices such as accelerometers, gyroscopes, pressure sensors, and radiofrequency (RF) switches are widely used in aerospace applications. Reduction of stiffening and curling initiated during fabrication of these devices is one of the challenging issues in MEMS design. Reducing response time is also favorable in some applications such as RF MEMS switches. This paper aims at reducing stiffening, curling, and increasing the natural frequency for three well-known designs of micro-plates with fixed-fixed supports. To achieve these objectives, a parametric size optimization is carried out. For comparison purposes, same volume is set as a constraint for all three designs. Compared to conventional rectangular micro-plate, a reduction of 34% in stiffening in design 2, and 44% in curling in design 3. Design 1 showed the maximum fundamental natural frequency. Thus, it is predicted to have the lowest switching time. Moreover, design 2 showed the maximum critical buckling temperature, extending the operation range of the device. The effect of changing micro-plate material is also studied in this paper.

1. Introduction

Microelectromechanical systems (MEMS) are used in wide variety of aerospace applications such as inertial navigation systems, structural health monitoring, and communication systems. Accelerometers, gyroscopes, pressure sensors , and radio-frequency (RF) MEMS switches are examples of devices used in these systems[1,2]. These devices have superior accuracy, light weight, low power consumption and lower cost than the macro-scale counterparts. The main research challenges facing the use of MEMS devices are related to their reliability and other performance parameters such as switching time in RF-MEMS switches. One of the main problems affecting the reliability of MEMS devices is the existence of residual stress due to fabrication processes [3]. The residual stresses cause two main problems 1) stress stiffening due the biaxial stress in axially restrained beams/plates, and 2) curling due to the stress gradient across the thickness. Both stress stiffening and curling affect the performance parameters of the MEMS devices.

Current technologies allow for different configurations for MEMS microplates. MEMS devices are fabricated either by surface micromachining or by bulk micromachining. In surface micromachining, the device is built on the surface of the substrate. Physical vapor deposition (PVD), chemical vapor deposition (CVD), and sputtering are examples of processes used extensively in surface

Content from this work may be used under the terms of the Creative Commons Attribution 3.0 licence. Any further distribution of this work must maintain attribution to the author(s) and the title of the work, journal citation and DOI.
Published under licence by IOP Publishing Ltd

micromachining. On the other hand, the bulk micromachining processes done by etching the substrate at selected positions.

An example of MEMS device fabricated from microplates is RF MEMS switches. RF MEMS switches are extensively used in satellite systems, communication system networks, instrumentation, wide band spectral analysis and radar systems. These switches are used for control of radio frequency signals and signal routing [4]. Many switches are fabricated using surface micromachining processes by deposition and etching of different layers till the required design is implemented. Deposition techniques result in stress gradients in the switch resulting in stiffening and curling. Different deposition conditions of the top and bottom layers of the beam/plate are the reason behind this stress gradient [5]. Stiffening and curling in RF MEMS switches negatively affect their performance by increasing the needed voltage for actuation.

A lot of research was done to MEMS structures in order to improve their performance. Stress gradient causes considerable curling for devices in the form of cantilever beams. In addition, the performance of the device is affected by the non- symmetric orientation of the cantilever beam [6]. Fixed-Fixed beams [7] is less vulnerable to the stress gradient, but it usually experiences stress stiffening due to the residual in-plane stress. Joglekar and Pawaskar [8] optimized the micro-beam width profile in order to extend the static and dynamic travel range of cantilever and fixed-fixed beams. Shekhar et al.[9] studied previously known geometries of RF MEMS switches in order to enhance their dynamic characteristics. Peroulis et al. studied the effects of residual stress in RF MEMS switches and proposed methods to alleviate them [10]. Mahameed and Rebeiz developed a stable electrostatic RF MEMS capacitive switch based on a thermal buckle-beam design [11]. Gupta et al. reduced the out-of-plane warpage in micromachined beams using corrugation [12]. Nieminen et al. designed a temperature-stable RF MEMS capacitor. The design is based on a square micro-plate structure with slits at specific positions [6]. Demirel et al. developed RF MEMS shunt switches for special radio frequencies and studied their structure and fabrication [13,14]. Philippine et al. used topology optimization to minimize the stiffening and curling on capacitive RF MEMS switches [15]. Finally, Bansal et al. introduced reinforcement for reduced buckling in MEMS switches [16].

In this paper, three designs of fixed-fixed micro-plates reported in literature [9,17] are investigated. The objective is to minimize stress stiffening and curling arising from residual stresses. Two of the designs under investigation have stepped variation in thickness while the third one is characterized by gradual variation in thickness according to a parabolic function. Non-linear Finite Element plate model considering the midplane stretching and stress gradient is developed in order to capture the effects of the biaxial residual stresses. A parametric optimization to the ratio between the width of the plate at the mid-length to the width at the supports is done for all three designs. Stiffening, curling, and natural frequencies are computed. Constant volume is set as a constraint to all designs for fair comparison purposes. The effect of changing the material is also studied. The total stiffness versus the operating temperature is simulated in order to determine the critical buckling temperature.

The rest of this paper is organized as follows. Section 2 describes the three designs of the micro-plate. Section 3 represents the details of the developed Finite element model. The results are discussed in section 4. Conclusions and future work are stated in section 5.

2. Micro-Plate Design

2.1. Micro-Plate Configuration

The three designs under consideration with their main dimensions are shown in Figure 1. In the first and second designs, the width of the middle third of the micro-plate is different from the width at the support according to a step function. The third design has a parabolic width profile.

The overall length L, and thickness of all three designs are 200, 1 μm, respectively. The width at mid-length and at the anchors are denoted as W_1 and W_2, respectively. It is worth mentioning that for the first and third designs, W_1 is not necessarily larger than W_2, as will be shown in the next sections.

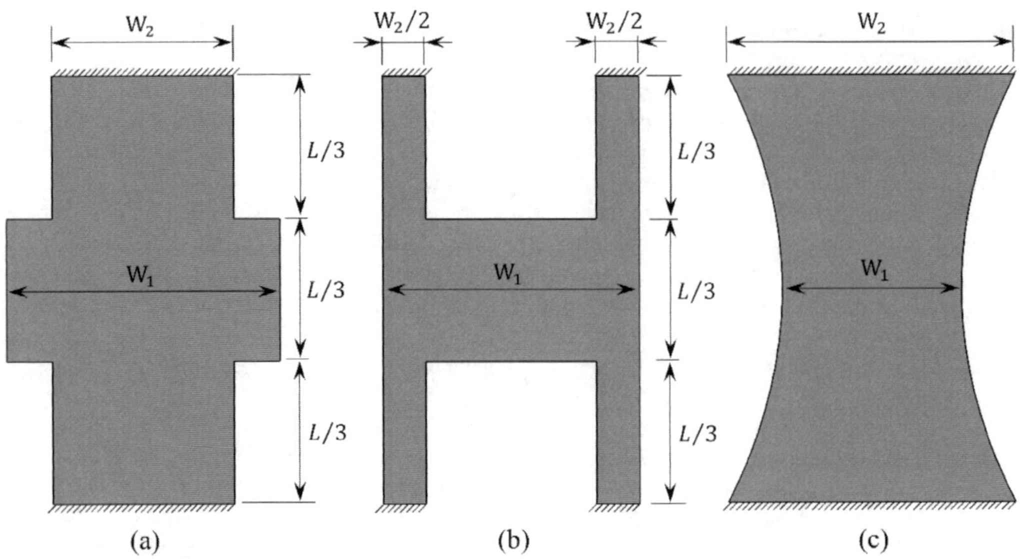

Figure 1. Schematic drawing of the three designs of the micro-plates.

2.2. Parametric optimization

To determine the effect of the widths W_1 and W_2 on mechanical performance of the micro-plates, a parametric optimization is carried out. Stiffening ratio, curling due to residual stresses, and the fundamental natural frequency are computed at different values of ratio a which is equals to W_1/W_2 . The ratio a ranges from 0.5 to 2 for designs 1 and 3, and ranges from 1 to 2 for design 2. The used materials and their main properties are shown in Table 1 [18,19]. For comparison purposes, the volume of the three designs is kept constant and equals to 20,000 μm^3, whatever the value of a.

Table 1. The used materials and their main properties.

	Young modulus E $10^9[N/m^2]$	**Poisson's ratio ν**	**Thermal expansion coefficient α 10^{-6}[1/K]**
Aluminum	77	0.3	23.1
Nickel	219	0.31	13.4
Gold	70	0.44	14.2

3. Finite Element Model of The Micro-Plate

The evaluation of the mechanical performance of the micro-plates is carried out by developing a 3D non-linear Finite Element plate model considering the midplane stretching and stress gradient developed using COMSOL Multiphysics.

Stiffening, curling, and natural frequencies are computed. Constant volume is set as a constraint to all designs for comparison purposes. The critical buckling temperature is determined by computing the stiffness versus operating temperature. All simulations are done with aluminum as a micro-plate material. Nickel and Gold are used only in stiffness versus temperature simulation. The details of each simulation are elaborated in the next subsections.

3.1. Stiffening and Curling

The micro-plate domain is meshed into triangular prism elements. The maximum and minimum element sizes are 4 and 0.04 μm, respectively. The actuation area is the lower surface of the micro-plate

A_{ac}=20,000 μm^2, taking into consideration the constant volume and thickness. The effective spring constant of the micro-plate is thus

$$k = \frac{p\,A_{ac}}{w_c} \tag{1}$$

where p is the force per unit area, and w_c is the deflection of the center point.
Stiffening is the ratio between the increase in stiffness due to the midplane stretching $\boldsymbol{k_2}$ to the stiffness without axial force $\boldsymbol{k_1}$. To calculate $\boldsymbol{k_1}$, a small force per unit area of 10 $\mathbf{N/m^2}$ is applied to the actuation area such that the non-linear effects are negligible. To calculate $\boldsymbol{k_2}$, the total stiffness is computed after applying the biaxial mean residual stress $\boldsymbol{\sigma_o}$. Hence, the stiffening ratio $\boldsymbol{k_2/k_1}$ can be calculated. The geometric non-linearity is included in the FE model to capture the coupling between the deflection and the midplane stretching. The total stiffness versus the operating temperature is simulated. This simulation is important to determine the critical buckling temperature for different designs and materials.

For simplicity, the substrate is not included explicitly in the model. The coefficient of thermal expansion CTE for the micro-plate is defined as the difference between CTE of the micro-plate material and alumina substrate (6.8 ppm/°C).
Curling is calculated when the micro-plate is subjected to the residual stresses $\boldsymbol{\sigma_{res}}$ without applying the actuation force such that

$$\boldsymbol{\sigma_{res}} = \begin{bmatrix} \boldsymbol{\sigma_o + \Gamma z} & \mathbf{0} & \mathbf{0} \\ \mathbf{0} & \boldsymbol{\sigma_o + \Gamma z} & \mathbf{0} \\ \mathbf{0} & \mathbf{0} & \mathbf{0} \end{bmatrix} \tag{2}$$

where $\mathbf{\Gamma}$ is the stress gradient across the thickness.
The values of $\boldsymbol{\sigma_o}$ and $\mathbf{\Gamma}$ are chosen similar to [18] to be 60 **MPa** and 10 $\mathbf{MPa/\mu m}$, respectively. Curling is calculated as the difference between maximum and minimum values of deflections of the whole micro-plate.

3.2. Natural Frequencies

The fundamental natural frequency f_o is computed considering the residual stress σ_{res} and geometrical non-linearity. The importance of this simulation is the relation between natural frequency and the switching time which is a key performance parameter in some MEMS devices such as RF MEMS switch. Equation (3) shows a simplified formula of the switching time [5]

$$t = 0.585\frac{V_p}{f_o\,V_s} \tag{3}$$

where V_p and V_s are the pull-in and the applied voltages. This equation shows that the switching time is inversely proportional to the fundamental natural frequency f_o.

4. Results and Discussion

Figure 2 shows the stiffnesses k_1, k_2, and the total stiffness k_t versus the width parameter a for designs 1 (D1), 2 (D2), and 3 (D3). It is obvious that increasing a decreases all these stiffness components. This is attributed to the fact that the stiffness is more affected by the width at the anchor W_2 more than in the middle W_1. On the other hand, the stiffening ratio k_2/k_1 differs from design to another as shown in Figure 3. Figure 3 shows the stiffening ratio and curling versus the width ratio parameter a for all three designs. For the first and third designs (D1 and D3), the stiffening increases with the increase of a. The third design has a slight change. On the contrary, the stiffening ratio decreases with the increase of a for the second design (D2).

For design 1 and design 2, the minimum value of curling is found at a = 1 and a = 1.25, respectively. On the other hand, curling of design 3 increases with the increase of a. The second design has a negligible variation of curling with respect to a.

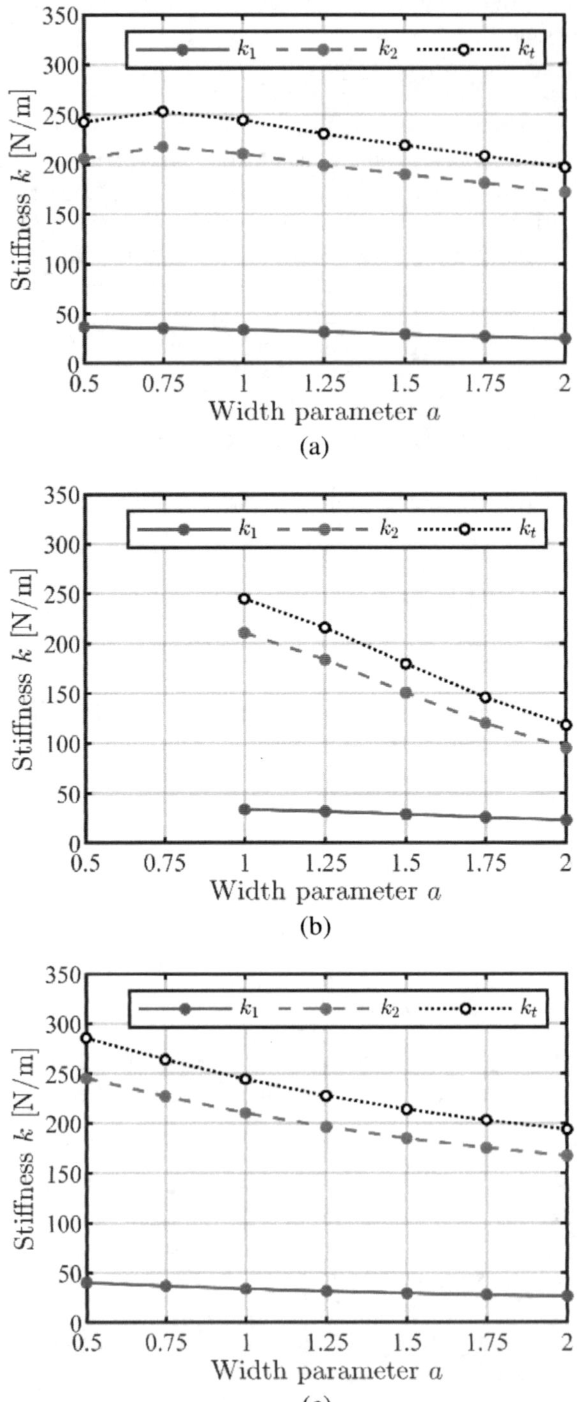

Figure 2. Stiffnesses k_1, k_2, and k_t versus the width parameter a for (a) Design1, (b) Design 2, and (c) Design 3.

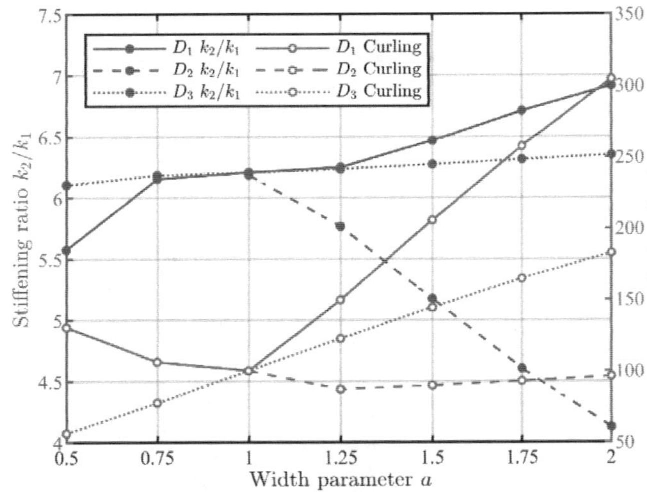

Figure 3. Stiffening and curling for the three designs versus the width parameter a.

The variation of the total stiffness with the operating temperature for the first design at a=0.5 and different types of materials is shown in Figure 4. Nickel (Ni) shows more stiffness than aluminum (Al) but both have close values of slope. Gold (Au) is less sensitive to temperature variation than Al and Ni. The variation of stiffness with temperature for all designs and materials are shown in Table 2.

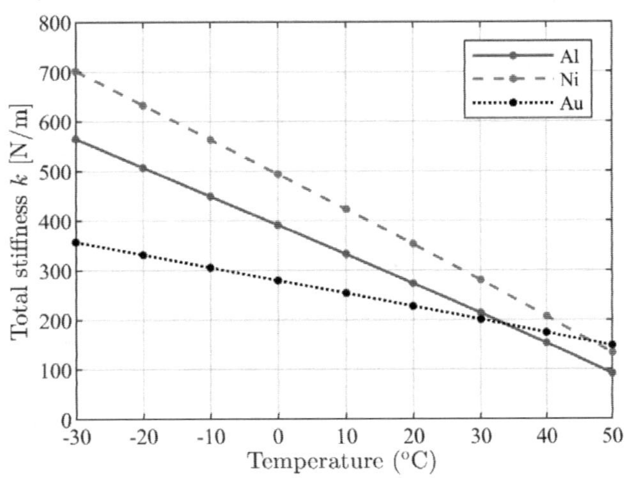

Figure 4. Total stiffness versus operating temperature for the first design at a=0.5 for different materials.

Table 2. Stiffness variation with temperature for all designs using different materials.

Stiffness variation with temperature $[(N/m)/°C]$								
Design 1			Design 2			Design 3		
Al	Ni	Au	Al	Ni	Au	Al	Ni	Au
-5.9	-7.1	-2.6	-2.4	-3.2	-1.1	-6.9	-8.5	-3.1

The second design (D2) shows the lowest variation. Thus, it has the highest critical buckling temperature. The critical buckling temperature is the temperature at which the bending stiffness of the plate vanishes. It is preferred to increase the critical buckling temperature to extend the operation temperature range of the MEMS device. Figure 5 shows the critical buckling temperature for all designs and materials at a=0.5 for D1 and D3, and at a=2 for D2.

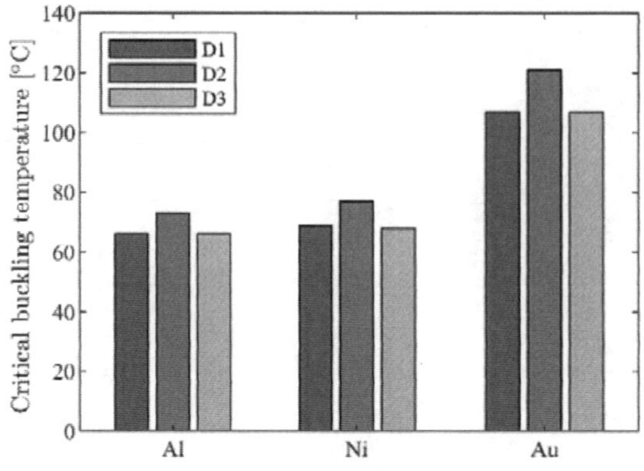

Figure 5. Critical buckling temperature for all designs and materials.

Figure 6 shows the fundamental natural frequency f_o versus the width ratio parameter a. Maximum f_o is found at minimum a for all designs. Design 1 shows higher f_o than design 3 at a less than 1. For higher values of a, the fundamental frequencies of design 3 are higher. The values of the fundamental natural frequency of design 2 coincides with that of design 1.

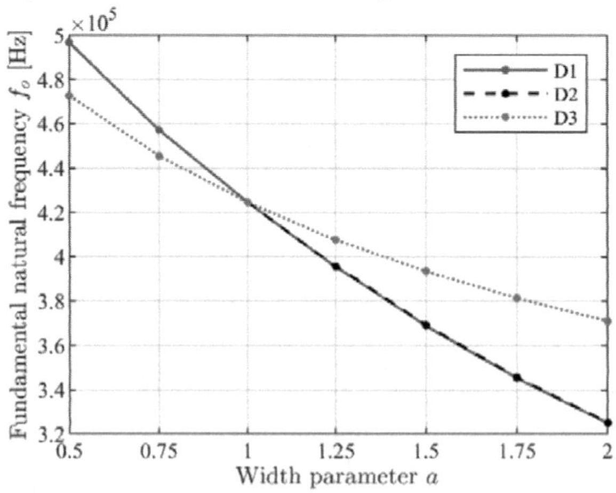

Figure 6. Fundamental natural frequency f_o for all designs and aluminum material versus width parameter a.

5. Conclusions

This work investigates the reliability issues related to residual stresses initiated during fabrication of MEMS devices in the form of micro-plates. The paper focuses on three well-known micro-plate designs found in literature. A parametric optimization for stiffening, curling, and fundamental natural frequency is carried out. The study variable a is the ratio between the width at the mid-length and the width at the supports. The first design has the highest natural frequency compared to other designs at the lowest a. Thus, this design can be used in applications that require high actuation speeds such as RF MEMS switches. The second design D2 has the lowest stiffening compared with other designs at the highest a. Thus, D2 has more stable operation with temperature variation. Moreover, it has a wider range of operating temperatures due to its higher critical buckling temperature. The third design D3 has lower curling than the first and second designs at a=0.5. According to the specifications of the device and the fabrication process, one can select the required design. As a future work, experimental verification is needed to evaluate the performance of the proposed configurations. Moreover, topology optimization of a fixed-fixed plate for specific applications is also proposed.

References

[1] Osiander R, Darrin M A G and Champion J L 2006 MEMS and Microstructures in A erospace Applications (Boca Raton: CRC Press Taylor & Francis Group)
[2] Kraft M and White N M 2013 MEMS for Automotive and Aerospace Applications (Cambridge: Woodhead Publishing Limited)
[3] Saleem M M and Nawaz H 2019 A Systematic Review of Reliability Issues in RF-MEMS Switches *Micro Nanosyst.* **11** 11–33
[4] Rebeiz G M, Patel C D, Han S K, Ko C H and Ho K 2013 The search for a reliable MEMS switch? *IEEE Microw. Mag.* **14** 57–67
[5] Rebeiz G M 2003 *RF MEMS, Theory, Design, and Technology* (NewYork: Wiley)
[6] Nieminen H, Ermolov V, Silanto S, Nybergh K and Ryhänen T 2004 Design of a temperature-stable RF MEM capacitor *J. Microelectromechanical Syst.* **13** 705–14
[7] Goldsmith C L, Yao Z, Eshelman S and Denniston D 1998 Performance of Low-Loss RF MEMS Capacitive Switches *IEEE Microw. Guid. Wave Lett.* **8** 269–71
[8] Joglekar M M and Pawaskar D N 2012 Shape optimization of electrostatically actuated microbeams for extending static and dynamic operating ranges *Struct. Multidiscip. Optim.* **46** 871–90
[9] Shekhar S, Vinoy K J and Ananthasuresh G K 2017 Surface-Micromachined Capacitive RF Switches with Low Actuation Voltage and Steady Contact *J. Microelectromechanical Syst.* **26** 643–52
[10] Peroulis D, Pacheco S P, Sarabandi K and Katehi L P B 2001 Alleviating the Adverse Effects of Residual Stress in RF MEMS Switches *31st European Microwave Conference* (London, England)
[11] Mahameed R and Rebeiz G M 2010 A high-power temperature-stable electrostatic RF MEMS capacitive switch based on a thermal buckle-beam design *J. Microelectromechanical Syst.* **19** 816–26
[12] Gupta A, Barron L, Brainin M and Lee J B 2014 Reduction of out-of-plane warpage in surface micromachined beams using corrugation *J. Micromechanics Microengineering* **24** 065023
[13] Demirel K, Yazgan E, Demir S and Akin T 2016 A New Temperature-Tolerant RF MEMS Switch Structure Design and Fabrication for Ka-Band Applications *J. Microelectromechanical Syst.* **25** 60–8
[14] Demirel K, Yazgan E, Demir Ş and Akın T 2017 A folded leg Ka-band RF MEMS shunt switch with amorphous silicon (a-Si) sacrificial layer *Microsyst. Technol.* **23** 1191–200
[15] Philippine M A, Sigmund O, Rebeiz G M and Kenny T W 2013 Topology Optimization of Stressed Capacitive RF MEMS Switches *J. Microelectromechanical Syst.* **22** 206–15
[16] Bansal D, Bajpai A, Kumar P, Kaur M, Kumar A, Chandran A and Rangra K 2017 Low voltage driven RF MEMS capacitive switch using reinforcement for reduced buckling *J. Micromechanics Microengineering* **27** 024001
[17] Goldsmith C, Forehand D, Scarbrough D, Johnston I, Sampath S, Datta A, Peng Z, Palego C and Hwang J C M 2009 Performance of molybdenum as a mechanical membrane for RF MEMS switches *IEEE MTT-S International Microwave Symposium Digest* pp 1229–32
[18] Reines I, Pillans B and Rebeiz G M 2011 Thin-film aluminum RF MEMS switched capacitors with stress tolerance and temperature stability *J. Microelectromechanical Syst.* **20** 193–203
[19] Multiphysics C 2017 Comsol Multiphysics Library

On the effect of post-processing techniques of the additively manufactured aluminum alloy parts

A H Maamoun[1]

[1] Technical Research Center, Cairo, Egypt

E-mail: maamouna@mcmaster.ca

Abstract. Additive manufacturing (AM) or 3D printing of metals promises a significant impact on the upcoming industrial revolution "Industry 4.0" as it is considered one of its main pillars. However, some challenges related to high initial cost and the fabricated part's quality still existed. Selective laser melting (SLM) is one of the effective techniques used for additive manufacturing to fabricate metal products. This paper presents the impact of different post-processing treatments on the microstructure and surface quality of the AlSi10Mg parts fabricated using SLM. This work illustrates the analytical view of the results obtained from two studies in a previous work by the author. A process map is presented for thermal post-processing treatment to customize the required quality and material characteristics of the AlSi10Mg parts. In addition, the shot peening results in a significant improvement for both surface roughness and hardness is illustrated. This work is a part of developing the manufacturing process of additively manufactured lightweight parts in some critical applications specifically for the metallic mirrors used in high power laser systems or wide view space telescopes.

1. Introduction

The upcoming industrial revolution known as Industry 4.0 paves a new road for advanced manufacturing. Industry 4.0 is powered by the current development in some fields, such as additive manufacturing (also known as 3D printing), autonomous robots, predictive maintenance and analytics, software integration, and cybersecurity [1]. Additive manufacturing of metals offers a variety of solutions to reduce current design and manufacturing limitations and therefore will improve part performance and customization [2].

Selective laser melting (SLM) is one of the commonly used techniques for additive manufacturing of metals, using a laser beam to melt layer-by-layer according to the generated slices from the 3D CAD model [3]. This technique can provide near-net shape objects compared to subtractive manufacturing (CNC machining), and thus will have a significant impact on workpiece development. SLM can also be used to produce efficient tools for forming and die casting technologies.

Additive manufacturing of aluminum (Al) alloys presents the production of efficient, flexibly designed, and lightweight parts using the SLM technique [4]. However, some challenges to widen SLM applications, such as obtaining consistent material properties, improving the quality of fabricated parts, and reducing their production cost, are active research issues [5]. SLM of these alloys promises a significant performance enhancement of the lightweight critical components used in different space and aerospace applications such as metallic optics and optomechanical components used for high energy laser systems [6]. However, the surface and inside defects of the additively manufactured parts present a challenge to product quality requirements.

Content from this work may be used under the terms of the Creative Commons Attribution 3.0 licence. Any further distribution of this work must maintain attribution to the author(s) and the title of the work, journal citation and DOI.
Published under licence by IOP Publishing Ltd

Literature studies evaluated the SLM process parameters' impact to fabricate high quality parts. Galy et al. addressed most defects obtained from the aluminum (Al) alloy parts fabricated by SLM [5]. The selection of the SLM process parameters, such as energy density, power of laser beam, speed of laser scanning, the thickness of powder layer could affect the quality of the as-built parts [7]. Various studies in the literature presented a process optimization to improve the mechanical properties of the as-built products. Read et al. studied the impact of the SLM process parameters on the mechanical properties of AlSi10Mg parts [8]. Maamoun et al. presented a comprehensive study that illustrated the impact of the process parameters on different performance characteristics such as mechanical properties, microstructure, surface roughness, relative density, and dimensional accuracy [9,10]. Their study developed a process map that investigated the optimum process parameter range to reach the required quality of parts produced. However, higher requirements of some performance characteristics for specific AlSi10Mg products are still needed. Consequently, the post-processing of SLM produced Al alloy parts becomes an essential step to improve the microstructure characteristics and to reduce the as-built defects.

According to ASTM F3301−18, post-processing for the additively manufactured parts is a treatment that could be applied after the AM process to achieve the desired material properties or surface finish to satisfy the engineering requirements. Post-processing techniques for additively manufactured parts can be divided into surface quality improvement and elimination of the internal microstructure defects. Shot peening (SP), laser shock peening (LSP), polishing, or chemical treatment processes can be implemented to improve the surface roughness of the final product [11–16]. Thermal post-processing and friction stir processing could be applied to control the microstructure and mechanical characteristics of the fabricated parts [17,18].

The current paper presented the impact of different post-processing techniques on both surface and microstructure characteristics of the SLMed AlSi10Mg parts. A guideline for selecting the optimum post-processing technique along with the desired quality is also discussed.

2. Materials and methods

A gas atomized EOS AlSi10Mg powder was used. The powder was characterized according to ASTM F3049-14 [19]. An energy X-ray dispersive Spectroscopy (EDS) test was performed to investigate the chemical composition of the selected powder. Powder particle shape was detected using a Scanning Electron Microscope (SEM). A laser diffraction test was used to estimate the powder size distribution (PSD) by dispersing the powder in water.

The machine used for SLM to fabricate the AlSi10Mg parts is EOSINT M290 where the laser power reaches up to 400 W. The processing parameters are presented in table 1.

Table 1. The parameters used for the SLM process to fabricate AlSi10Mg samples.

Power of Laser (W)	Laser Scan speed (mm/s)	Distance of hatch spacing (mm)	Angle of Layer orientation (°)	powder Layer thickness (μm)	Strategy of scan
370	1300	0.19	67	30	Stripes

The impact of thermal post-processing was investigated through three heat treatment processes by applying different heat treatment conditions. The annealing process was conducted at 200 and 300°C. Solution heat treatment (SHT) was applied to the as-built samples at 530°C that is held for 1 and 5 hours. T6 heat treatment was conducted on the solution heat treated samples by water quenching followed by artificial aging at 160°C for 11 hours. Each heat treatment process parameter was performed on three samples.

Shot peening was conducted on the top surface of the AlSi10Mg samples using a 19 mm diameter single nozzle of a Lance machine. Almen intensity was detected for the AlSi10Mg sample surface which was bombarded by glass beads. A high-intensity SP 20.3 N was applied using glass beads standard of Gp50. The shot peened sample surface was covered using coverage a 200% factor to increase the area of the peened surface. A 152 mm distance was adjusted from the nozzle to the sample surface using an impact angle of 90-degree.

The surface integrity of the SP surfaces was investigated using an Alicona optical microscope. The surface roughness and waviness of the selected areas were investigated.

A 10 × 10 mm measured area was selected on the top of each sample using a lens of 10X magnification. The microstructure characterization was performed along the building direction cross-section using wire-cut samples. The procedures of polishing and etching were selected as recommended by Maamoun et al. [17].

Image analysis of samples' microstructure was tested by Nikon Microscope and SEM. A Clemex micro-hardness instrument was used to investigate the hardness of the samples before and after applying post-processing. The micro-hardness values were obtained at 200 gf load and were calculated from the average of 10 measurements conducted on each sample.

An in-situ tracking of phase change was conducted using XRD along with variable temperatures of SHT conditions. A DHS 900 domed hot stage was used to perform the XRD experiment. The sample was heated inside the domed stage under vacuum to temperatures of 30, 100, 200, 300, 400, and 530°C.

3. Results and discussion

The characteristics of the gas atomized AlSi10Mg powder were obtained according to ASTM F3049-14 as shown in Figure.1. The particle size distribution shows that the size of powder particles ranges from 2 to 80 μm as illustrated in Figure.1 (a). Spherical particle shape was determined for different particles' sizes using SEM as presented in Figure.1 (b). The spherical shape of the AlSi10Mg powder could result in a higher relative density and improve the flowability of powder compared to the irregular powder shape [20]. The chemical composition of the fresh powder used was detected using EDS analysis, the weight % of elements is listed in table 2.

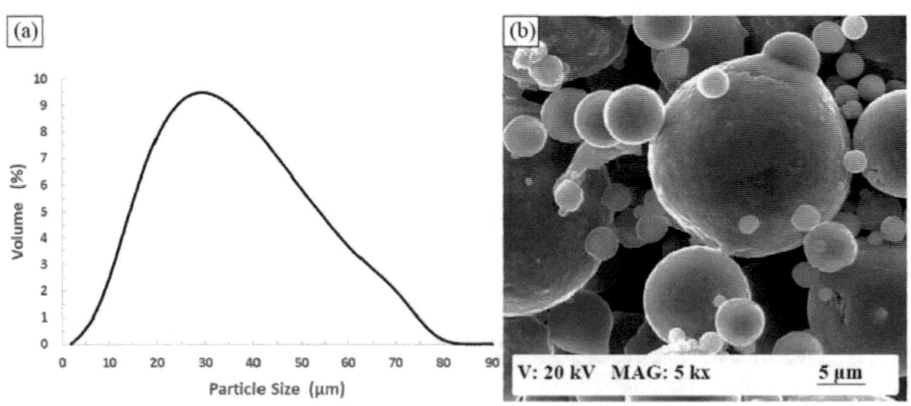

Figure 1. AlSi10Mg powder characteristics; a) particle size distribution, b) Powder morphology.

Table 2. The chemical analysis (wt. %) obtained from the AlSi10Mg powder using EDS.

Si	Mg	Fe	Cu	Sn	Pb	Zn	Al
10.6	0.36	0.4	0.09	0.06	0.06	0.05	Balance

Figure. 2 shows the images obtained by the optical microscope for the as-built AlSi10Mg etched samples along the building direction (Z-direction), the parallel plane to the layers (X-Y plane).

These images illustrate the melt pool shape obtained after the material solidified. Figure. 2 (a) shows that the melt pool shape is affected by the Gaussian shape of the laser beam through the formation of the semi-elliptical melt pool shape.

The laser scan pattern is clearly appeared in Figure.2 (b) where the intersection angle between the melt pool shapes represents the orientation angle between the building layers. Figure. 2 confirms the microstructure inhomogeneity of the additively manufactured AlSi10Mg samples.

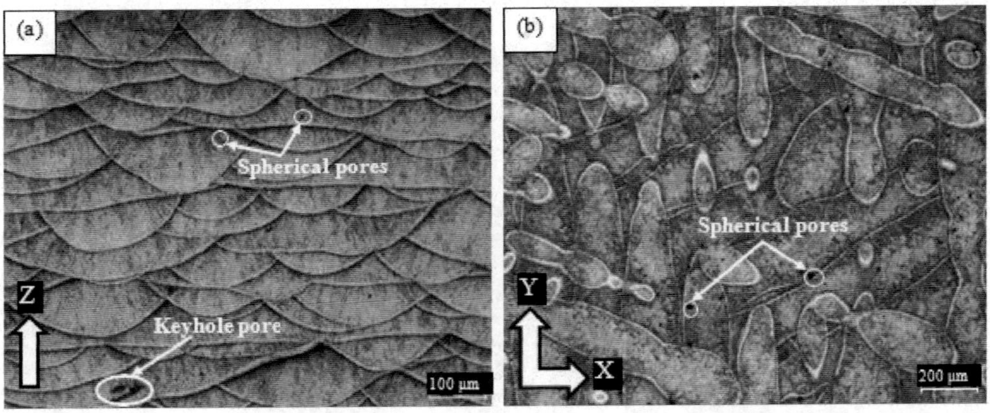

Figure 2. The as-built AlSi10Mg microstructure images: a) along Z-direction (building direction), b) through the x-y plane (the parallel plane to deposited layers) [17].

The observations of the defects above the upper surface of the additively manufactured sample are demonstrated in Figure.3 (a). These main defects are represented in the effect of balling phenomena Figure.3 (b), unmelted powder particles Figure3. (c), the effect of laser scan tracks Figure.3 (d), and the open pores on the sample surface Figure.3 (e).

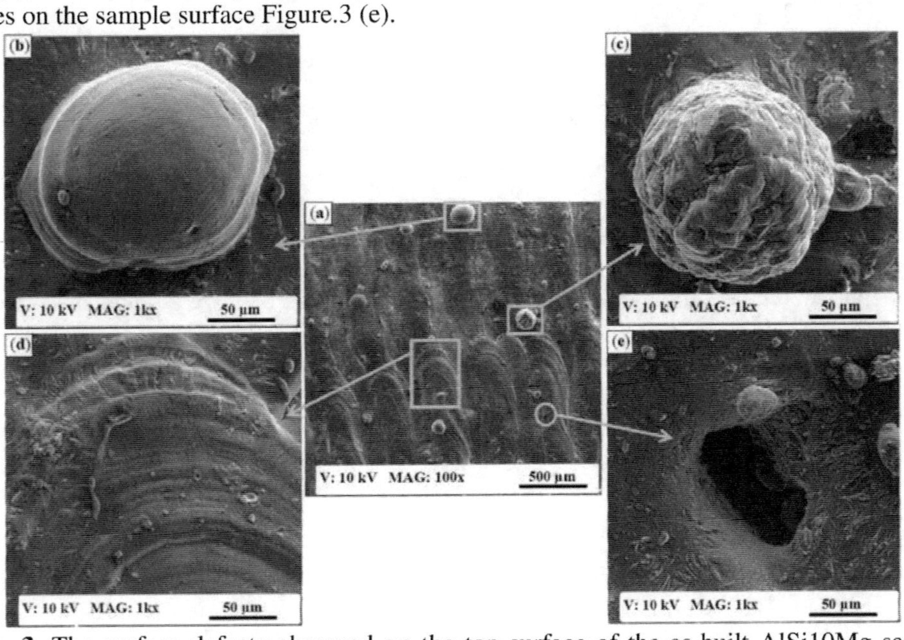

Figure 3. The surface defects observed on the top surface of the as-built AlSi10Mg samples; a) balling, b) partially or unmelted powder particle, c) laser scanning track effect, d) open pore on the sample surface [11].

The SLM process parameters used are the optimized process parameters as presented in the most recent studies [9,10]. However, inhomogeneity of the microstructure and surface defect still exists. The post-processing of the as-built parts becomes necessary to achieve the desired part quality for some critical applications.

Thermal post-processing affects the microstructure of the as-built parts related to the applied parameters as shown in Figure.4. The developed map shows that the microstructure homogeneity can be obtained using SHT or T6 HT. The microhardness gradually decreased after annealing up to 50% after SHT due to the decomposition of the Si network structure characterizes the as-built microstructure. However, the microhardness increased after the T6 treatment reaching almost the microhardness values obtained from the additively manufactured parts.

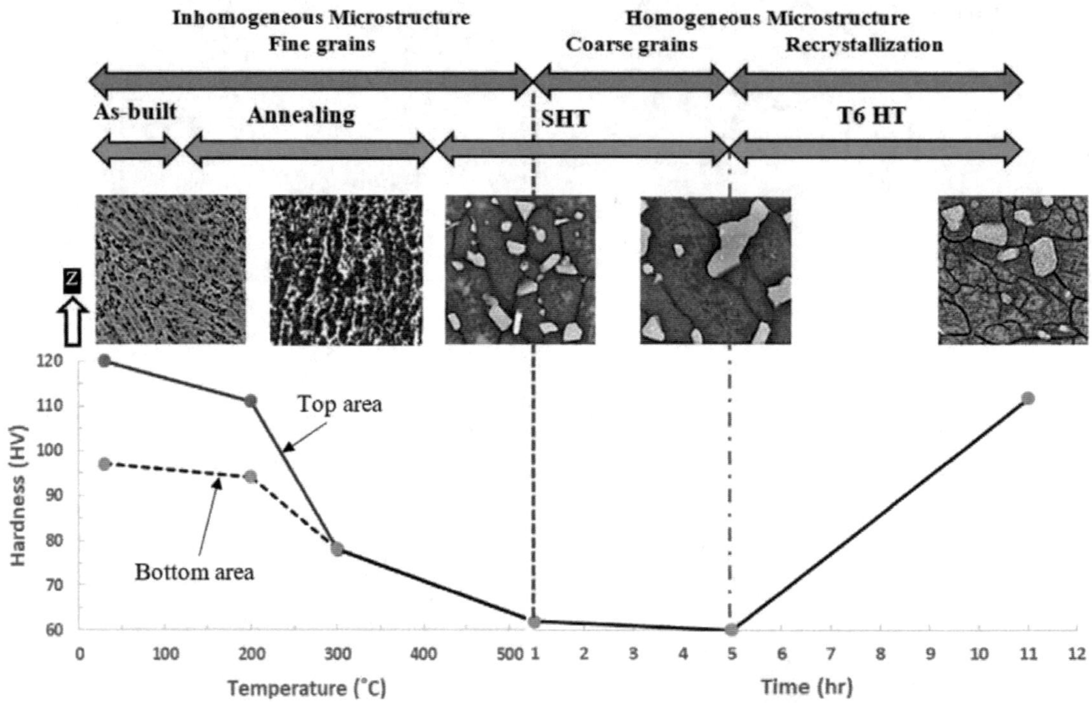

Figure 4. Microhardness map illustrating the impact of heat treatment post-processing on the microstructure along the building direction cross-section of the AlSi10Mg fabricated samples [17].

Figure.5 shows the XRD analysis of the in-situ phase change tracking under variable SHT conditions. The results illustrate that no phase change was obtained under the applied temperatures. The obtained results validate the Al and Si grain size change that was significantly obvious in the process map presented in Figure.4. In addition, the peak broadening of Al and Si increase along with the increase of temperature confirms the gradual decrease of microhardness values illustrated in Figure.4 [21].

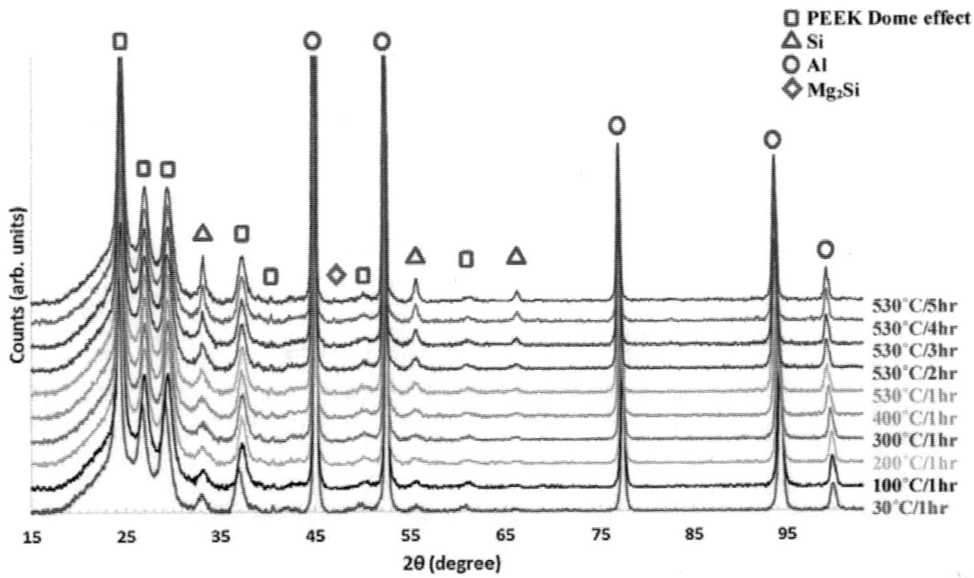

Figure 5. In-situ XRD phase pattern of the AlSi10Mg sample through variable SHT temperatures.

The shot peening significantly improves the as-built surface characteristics as illustrated in Figure.6. Thus was obtained due to the uniform pattern of the Gp165 glass beads applied at 200% surface covering factor using the optimized Almen intensity of 22.9A.

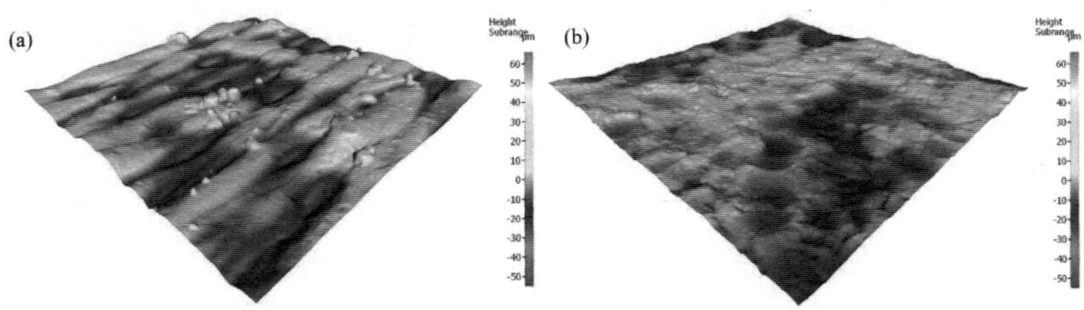

Figure 6. The surface texture of the AlSi10Mg sample; a) additively manufactured, b) shot peened.

Surface hardening and highly compressive stresses were investigated. SP generated severe plastic deformation on the sample surface which results in high compressive stresses. Figure.7 illustrates the surface hardening of the shot-peened area which extends into a depth of 500 μm from the sample surface.

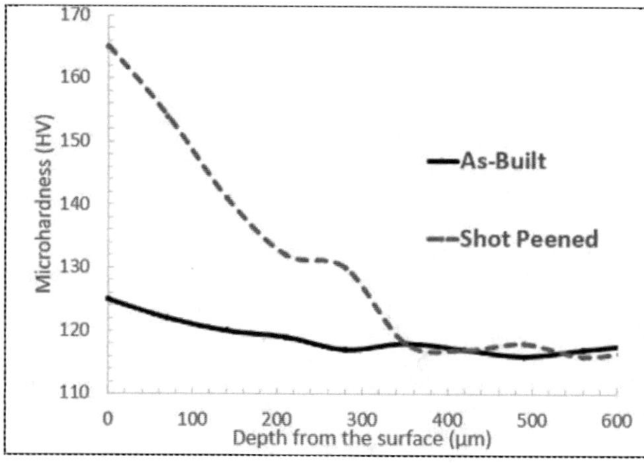

Figure 7. The microhardness profile in depth from the surface of the AlSi10Mg sample before and after SP.

Figure 8 shows that SP significantly results in high stresses. The high compressive residual stresses generated could lead to enhance mechanical properties and fatigue strength.

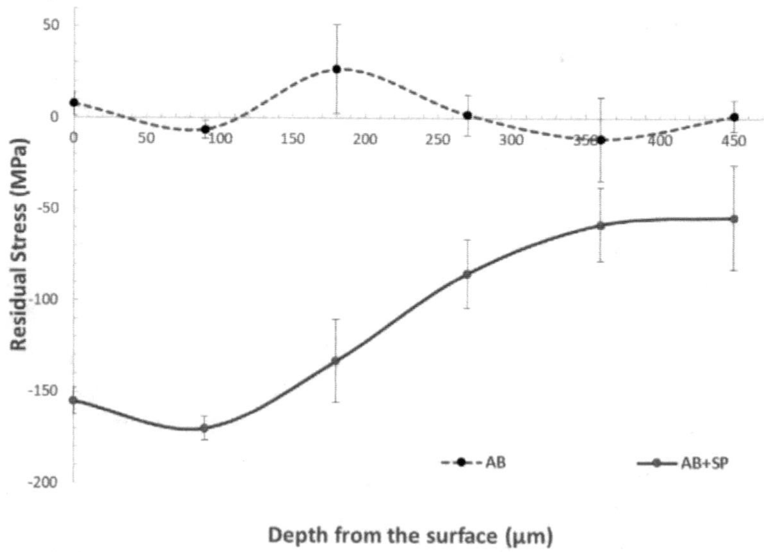

Figure 8. The residual stress profile along the cross-section of the shot peened and as-built AlSi10Mg samples [11].

4. Summary and conclusion

The current study shows the impact of different post-processing treatments for the additively manufactured AlSi10Mg parts fabricated by SLM. It was found that post-processing treatment is considered an essential step to solve the issues of microstructure inhomogeneity and to improve the as-built surface quality by reducing the defects obtained. AlSi10Mg powder was characterized according to ASTM F3049-14; then the parts were fabricated according to the optimized SLM parameters obtained by the authors from another study. The results can be summarized as follows:

- Microstructure inhomogeneity of the additively manufactured parts was illustrated through different areas and orientations. An anisotropic material structure could be obtained due to this microstructure inhomogeneity.
- The defects obtained on the as-built part surface were characterized.

- Heat treatment of solution heat treatment at 530 °C and T6 treatment eliminates the microstructure inhomogeneity obtained from the as-built AlSi10Mg parts.
- SHT also increases the ductility and improves the microstructure homogeneity. The microhardness was reduced by 50% compared to the values measured from the SLMed conditions. However, T6 HT resulted in a higher rate of microhardness with a homogenous and refined microstructure.
- A microhardness map was developed through different heat treatment conditions to assist in satisfying the design requirements of the desired part characteristics. The trend obtained from this map was validated using the in-situ XRD tracking of the phase change along SHT.
- SP eliminates the defects on the top of the peened surface of as-built parts and improves the shot-peened surface roughness.
- After SP, surface hardening was achieved along with generating high in-depth compressive stress up to 500 μm from the sample surface. Thus could improve the mechanical properties and of the processed parts.

Acknowledgments
The author would like to appreciate the support presented by Prof. Mohamed Elbestawi and Prof. Stephen Veldhuis.

References

[1] Schwab K 2017 *The fourth industrial revolution*
[2] Srivatsan T S and Sudarshan T S 2015 *Additive manufacturing: innovations, advances, and applications*
[3] Gibson I, Rosen D and Stucker B 2015 Development of additive manufacturing technology *Additive manufacturing technologies* (Springer) pp 19–42
[4] DebRoy T, Wei H L, Zuback J S, Mukherjee T, Elmer J W, Milewski J O, Beese A M, Wilson-Heid A, De A and Zhang W 2018 Additive manufacturing of metallic components – Process, structure and properties *Prog. Mater. Sci.* **92** 112–224
[5] Galy C, Le Guen E, Lacoste E and Arvieu C 2018 Main defects observed in aluminum alloy parts produced by SLM: From causes to consequences *Addit. Manuf.* **22** 165–75
[6] Maamoun A 2019 Selective laser melting and post-processing for lightweight metallic optical components
[7] Aboulkhair N T, Simonelli M, Parry L, Ashcroft I, Tuck C and Hague R 2019 3D printing of Aluminium alloys: Additive Manufacturing of Aluminium alloys using selective laser melting *Prog. Mater. Sci.*
[8] Read N, Wang W, Essa K and Attallah M M 2015 Selective laser melting of AlSi10Mg alloy: Process optimisation and mechanical properties development *Mater. Des.* **65** 417–24
[9] Maamoun A, Xue Y, Elbestawi M, Veldhuis S, Maamoun A H, Xue Y F, Elbestawi M A and Veldhuis S C 2018 Effect of Selective Laser Melting Process Parameters on the Quality of Al Alloy Parts: Powder Characterization, Density, Surface Roughness, and Dimensional Accuracy *Materials (Basel).* **11** 2343
[10] Maamoun A H, Xue Y F, Elbestawi M A and Veldhuis S C 2018 The effect of selective laser melting process parameters on the microstructure and mechanical properties of Al6061 and AlSi10Mg alloys *Materials (Basel).* **12**
[11] Maamoun A, Elbestawi M, Veldhuis S, Maamoun A H, Elbestawi M A and Veldhuis S C 2018 Influence of Shot Peening on AlSi10Mg Parts Fabricated by Additive Manufacturing *J. Manuf. Mater. Process.* **2** 40
[12] Ahmed H. Maamoun, Mohamed A. Elbestawi, Stephen C. Veldhuis 2018 Post-processing of the additively manufactured AlSi10Mg parts produced by Selective Laser Melting *7th Int. Conf. Virtual Mach. Process Technol. (VMPT 2018)* **May**
[13] Damon J, Dietrich S, Vollert F, Gibmeier J and Schulze V 2018 Process dependent porosity and

the influence of shot peening on porosity morphology regarding selective laser melted AlSi10Mg parts *Addit. Manuf.* **20** 77–89

[14] Zupanc U and Grum J 2011 Surface integrity of shot peened aluminium alloy 7075-T651 *Stroj. Vestnik/Journal Mech. Eng.* **57** 379–84

[15] Hetram L S and Om H 2015 Shot Peening Effects on Material Properties: A Review *Int. J.* **1** 480–4

[16] Hassanin H, Elshaer A, Benhadj-Djilali R, Modica F and Fassi I 2018 Surface Finish Improvement of Additive Manufactured Metal Parts *Micro Precis. Manuf.* 145–64

[17] Maamoun A H, Elbestawi M, Dosbaeva G K and Veldhuis S C 2018 Thermal Post-processing of AlSi10Mg parts produced by Selective Laser Melting using recycled powder *Addit. Manuf.* **21** 234–47

[18] Maamoun A H, Veldhuis S C and Elbestawi M 2019 Friction stir processing of AlSi10Mg parts produced by selective laser melting *J. Mater. Process. Technol.* **263** 308–20

[19] F3049 – 14 A 2014 Standard Guide for Characterizing Properties of Metal Powders Used for Additive Manufacturing Processes *F3049 - 14* 1–3

[20] Tan J H, Wong W L E and Dalgarno K W 2017 An overview of powder granulometry on feedstock and part performance in the selective laser melting process *Addit. Manuf.*

[21] Langford J I and Wilson A J C 1978 Scherrer after sixty years: A survey and some new results in the determination of crystallite size *J. Appl. Crystallogr.* **11** 102–13

IOP Conf. Series: Materials Science and Engineering 1172 (2021) 012023 doi:10.1088/1757-899X/1172/1/012023

Factors affecting Failure mode and load of rubber polyester composite joint

B H Mousa[1], N A Azab[1], G Bassioni[2] and M H Abdellatif[1]

[1]Mechanical Design and Production Engineering Department, Faculty of Engineering, Ain Shams University, Cairo, Egypt
[2]Chemistry Division, Faculty of Engineering, Ain Shams University, Cairo, Egypt

E-mail: bassant.hany@eng.asu.edu.eg

Abstract. The joining of rubber polyester composite is affected by design and geometry parameters. These parameters affect the failure load of the composite joint and the failure mode. Catastrophic failure modes should be avoided, and the optimum parameters are selected to achieve a progressive failure that could be observed. Geometry parameters such as the edge distance and the center distance between holes are tested under tensile loading to get the optimum values. Design parameters such as tightening torque of bolt and washer size are also tested under tensile loading, and their effect on the failure load is obtained and studied. The failure modes are investigated under the digital microscope in each case. It is found that the failure load increases by increasing the edge distance, the tightening torque, and decreasing the center distance between holes. The optimum washer size is varied according to the used tightening torque. The optimum obtained failure mode in tests is a mixed - mode of bearing and net tension modes.

1. Introduction

Compo site materials, especially the reinforced polymer composites, have wide use nowadays in many applications [1,2]. Fiber-reinforced composites are used in yachts, aviation, and automotive industries[3]. Glass fiber reinforced polyester is a commonly used composite, and fillers are added to reduce the cost of the final product[4, 5]. One of these fillers is waste tire rubber particles which improve the specific impact strength of the composite and reduce the cost of the final product[6]. The rubber particles volume fraction is selected to be 10% of the total volume of the composite as recommended by another researcher[7]. In the above-mentioned applications, the composites are joined with metals or any other materials.

Mainly, there are two types of joining materials which are adhesive and bolted joints. The bolted joints have many advantages compared to the adhesively bonded joints. Bolted joints can easily be assembled and disassembled for maintenance, easily replaced, and repaired[3]. Bolted joints have a disadvantage as the joint between the joined parts are the weakest part of the structure due to the fact of drilling the holes needed for joining.

On the other hand, many factors are affecting the strength of this joint and its failure load. The failure load is the maximum load the joint can withstand before failure. These factors could be categorized into four main groups, which are material, fastener, design, and geometry factors[8]. The material parameter includes the fiber type, form, and laminates sequence. The fastener parameter includes the fastener type

Content from this work may be used under the terms of the Creative Commons Attribution 3.0 licence. Any further distribution of this work must maintain attribution to the author(s) and the title of the work, journal citation and DOI.
Published under licence by IOP Publishing Ltd

and size. The design parameter includes the loading type and direction. Geometry parameter includes the distance of the hole from the end, the distance from the edge, the hole size, and part dimensions [8].

The width of the specimen (W) and hole size belong to geometry factors that can affect the strength of the joint. Width to hole diameter (W/D) is a ratio used to easily study these two factors and their effect on the strength of the joint. Okutan et al. proved that the increase of this ratio led to the increase of the bearing strength. Most studies and researches used W/D ranged from 2 to 5[8, 9]. The suitable value that is recommended to use is more than 3, as for smaller ratios, the joint becomes weak[8]. After a certain ratio of W/D, depending on the material, the mode of failure changed from tension to bearing mode, which is a desirable mode as it is a sign of failure before a catastrophic behavior is observed [8].

The distance of the hole center to the edge (E) is also studied as a ratio to hole diameter. Esmaeil et al. concluded that by increasing this ratio, the bearing strength increases [10]. Edge distance to hole diameter ratio (E/D) affects the strength of the joint and its failure mode [11]. There are different failure modes which are catastrophic modes such as net tension, shear out, and cleavage modes. The desirable mode is the bearing mode which is a progressive failure mode. The tested specimens are examined to consider failure mode. Up to a certain point, the increase in E/D ratio increases the strength of the material, then the value stabilizes, and the mode of failure changed to bearing mode. The studied ratio ranges from 1 to 5. At smaller values, the failure mode is shearing out mode, which is a catastrophic, not desirable mode. The suitable recommended ratio is beyond 2[8].

The effect of clearance between bolt and hole on the joint strength is also investigated where two different types of clearances are compared to each other. Esendemir et al. proved that the bigger clearance has a lower bearing strength of the bolted joint [12].

Bolt tightening torque (T) and washer size are two design factors that affect the bearing strength of the joint. It is observed that by increasing the tightening torque, the bearing strength increases, and by increasing the washer's outer diameter, the bearing strength increases up to a limit that may differ from one material to another and then decreases[13].

In applications, more than one hole is needed in a joint. The effect of the hole-to-hole distance (M) on the bearing strength of the joint becomes a point of interest. This geometrical parameter is studied as a ratio between hole-to-hole distance and hole diameter (M/D) [14, 15]. The bearing strength of the joint increases by increasing this ratio, and the studied range is from 2 to 5[14, 15].

This study aims to study the design and geometry factors affecting the failure load of the novel composite joint. This work is a collective study of different factors that affect the joint. The optimum factor is obtained in one test and set as a constant for other tests to obtain a final joint of the composite of optimum loading conditions and grantee desirable failure mode.

The specimens are drilled and the effect of different factors on the failure load of composites are studied. The width of the specimen and the hole diameter are kept constant in all tests. The studied factors are as follows; the distance between the edge and the hole center (E), the tightening torque of the bolt, the washer size, and the distance between two parallel holes (M) on drilling more than one hole.

2. Experimental work

2.1 Materials

A glass fiber reinforced polyester with rubber particles (of volume fraction 10%) composite is prepared in laminates as specified in detail in a previous study[6].

IOP Conf. Series: Materials Science and Engineering 1172 (2021) 012023 doi:10.1088/1757-899X/1172/1/012023

2.2 *Set up preparation*

Specimens -of 120 mm length, 25 mm width, and 3.5 mm thickness- are prepared then drilled according to the needed dimensions specified in each test. A tensile test attachment is manufactured according to ASTM D 953 – 02 [16] to hold the specimen during the test, as shown in Figure 1a. A drilling jig is also manufactured to ensure accurate dimensions of holes in each specimen, as shown in Figure 1b.

The specimen dimensions nomenclature is shown in Figure 1c. The hole diameter is equal to 5 mm. The width to hole diameter ratio is kept constant in all performed tests and equals 5.

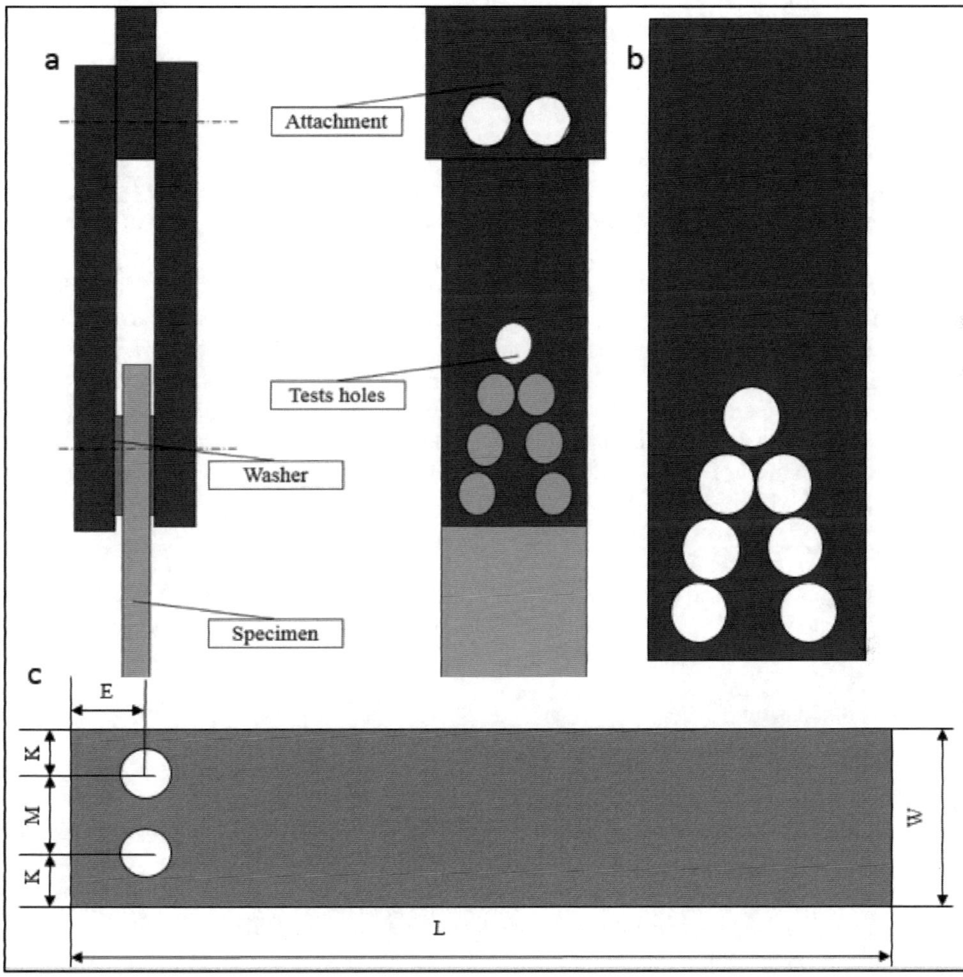

Figure 1. Test set up; a) Test attachment, b) Drilling jig, and c) Specimen dimensions nomenclature

2.3 *Tests procedure*

Tensile tests are performed on a LLOYD universal testing machine-30 tons, and a load-displacement curve is obtained in each case. The crosshead speed is 0.5 mm/min, in agreement with another research[17]. Four groups of tests are performed to study four different factors that affect the failure load of composites. Specimens are tested in each case, and a load-extension curve is obtained. The four factors are as follows;

2.3.1 *The ratio between the edge distance and the hole diameter (E/D).* This test is performed to get the suitable distance of the hole from the edge to avoid undesired failure modes. The studied ratios are 1, 3, and 5. A pin of outer diameter 5 mm is used. The failure modes are then investigated according to ASTM D 5961/D 5961M – 01 [18]. The optimum distance which gives the maximum failure load and the desired failure mode is chosen and set for the next test.

2.3.2 *The tightening torque (T).* Different tightening torques are tested to study their effect on the failure load of the composite. The studied torques are 0, 2, 4, and 6 N.m. A bolt of outer diameter 5 mm is used. A constant washer size of an outer diameter of 25 mm is used in this test. E/D is equal to 5. The optimum tightening torque, which gives maximum failure load depending on washer size, is selected and set for the next test.

2.3.3 *The washer size (S).* Four washer sizes of different outer diameters 10, 15, 20, and 25 mm respectively are used in this test to study the effect of washer size on the failure load of the composite. The tightening torque is 6 N.m, and the E/D is equal to 5. A bolt of outer diameter 5 mm is used.

2.3.4 *The ratio between the center distance of holes and the hole diameter (M/D).* The center distance between holes is restricted by the width, the diameter of holes, and the outer diameter of the used washer. The selected dimensions are E/D = 5, tightening torque = 2 N.m and the outer diameter of the washer is 15 mm. These values are chosen to fulfill design considerations.
This test is performed to check the optimum center distance between two holes to use a pin and bolt together to restrict the rotation of the specimen and avoid the effect of clearance in the bolt. The center distances of two parallel holes are varied depending on the width of the specimen, where 25 =2K+M. The tested M/D ratios are 2.2, 2.6, and 3.

2.4 Specimens examination
A digital microscope is then used to examine the tested specimens to evaluate the specimen condition and define its failure mode.

3. Results and discussion
The results and discussion of the four performed tests are as follows;

3.1 Effect of edge distance/hole diameter (E/D)
The average failure load of tested specimens is calculated in each case and the results are plotted as shown in Figure 2a. The results show that by increasing E/D, the failure load increases up to a limit then the load stabilizes. The decision is not clear. Therefore, the average displacement at failure in each case is calculated and plotted as shown in Figure 2b. This figure shows that the displacement at the highest tested E/D is the highest which means that the specimen has more displacement before failure, and this could be the effect of bearing mode. Therefore, E/D equals 5 is selected to be the optimum in this case.

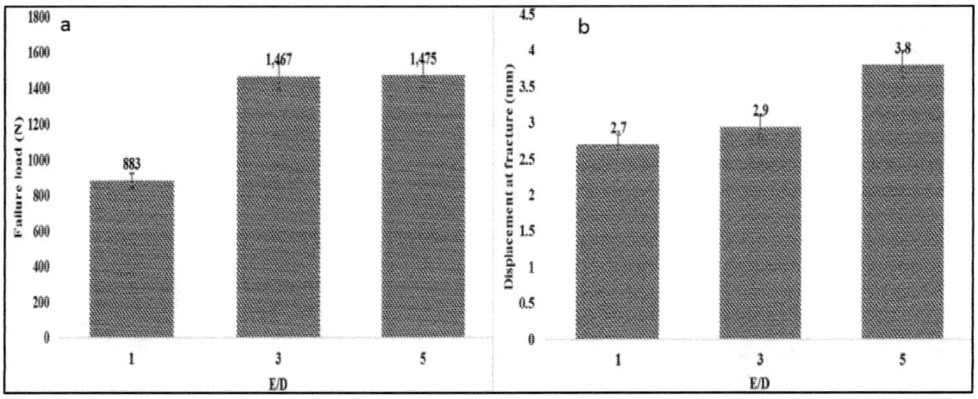

Figure 2. a) Failure load at different E/D and b) Displacement at failure at different E/D

The tested specimens are investigated under the microscope, and the mode of failure in each case is detected. Figure 3a shows that the failure mode at E/D = 1 is cleavage mode, Figure 3b shows that the failure mode at E/D = 3 is a mixed-mode of cleavage-net tension failure. Both failures are sudden failures that are undesirable modes. Figure 3c shows that the failure mode at E/D = 5 is a mixed-mode of bearing-net tension failure, which is a desirable mode as it does not result in a catastrophic failure. This figure confirms the results of failure loads and displacements at failure.

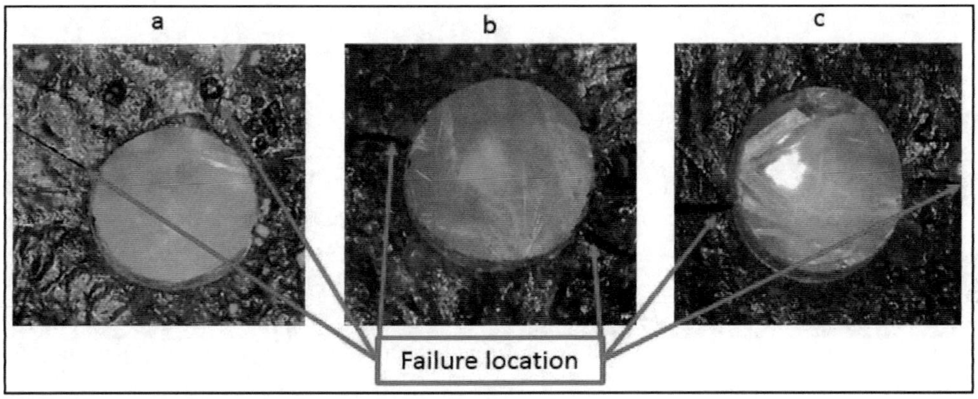

Figure 3. Fractography of specimens at different E/D ratios; a) E/D = 1, b) E/D = 3, and c) E/D = 5

The decision here is that the optimum value for E/D is equal to 5 which corresponds to E = 25 in the performed tests. The corresponding failure load at E/D = 5 is approximately 1474 N. This result is set as constant for the next tests.

3.2 Effect of the tightening torque (T)

The average failure load is calculated in each case and plotted as shown in Figure 4. The figure shows that by increasing the tightening torque, the failure load increases. This is due to increasing the axial force. The axial force increases the contact pressure, which increases the frictional force between the jointed members. The frictional force prevents slipping of the members, which prevents the bearing of the bolt on the edges of the hole. Therefore, the failure load of the joint is increased. This is agreed with other researchers[19].

The figure also shows that the failure load at T = 0 is approximately 1450 N which is slightly lower than the optimum value (1474 N) that is obtained from the previous test. This may be due to the clearance found in the case of using a bolt. The tight fit obtained in the case of using a pin increases the failure load compared to the case of a bolt with clearance. The contact area between the pin and joint increases so, the stress on the joint decreases, and the joint load-carrying capacity increases in agreement with another researcher[20].

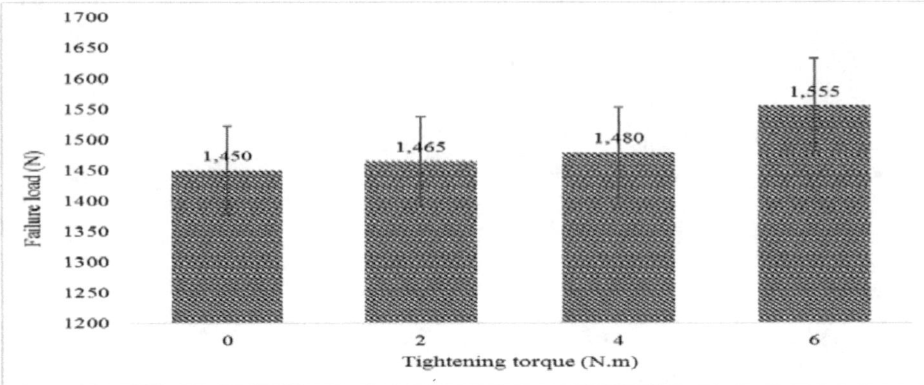

Figure 4. Failure load at different tightening torques

The tested specimens are investigated under a digital microscope. The figure shows that the failure mode is unchanged by varying the tightening torque. The observed failure mode is a mixed-mode of bearing-net tension, as shown in Figure 5.

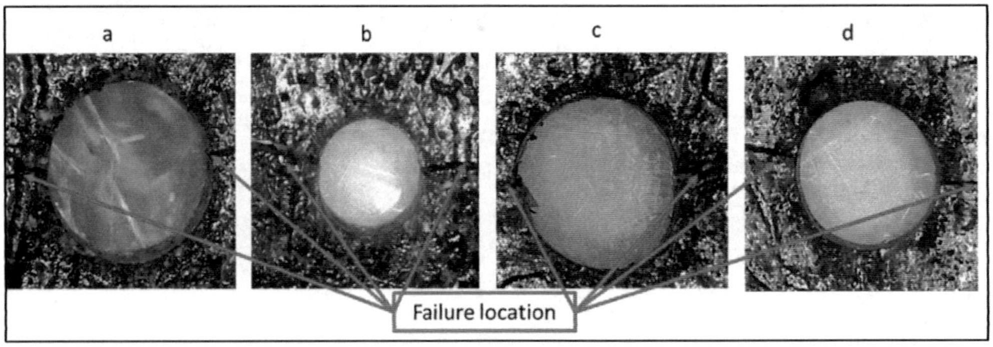

Figure 5. Fractography of specimens at different tightening torques; a) T = 0, b) T = 2, c) T = 4, and d) T = 6 N.m

The optimum tightening torque from this test is 6 N.m as it increases the failure load of the joint. The corresponding failure load is approximately 1555 N.

3.3 Effect of the washer size (S)

The effect of washer size on the failure load is studied, and the results are plotted as shown in Figure 6. The available washer sizes are categorized into two groups; big washers of outer diameters 20 mm and 25 mm and small washers of outer diameters 10 mm and 15 mm.

It is noticed that for each group of washers, the decrease of washer's size, leads to an increase in failure load. This is due to the larger pressure obtained for a smaller contact area having the same axial load resulting from the use of the same tightening torque. The tightening torque exerted an axial force, and the pressure is distributed on the washer's contact area, as illustrated in Figure 7

Small-sized washers have lower failure loads than big-sized washers while they have lower contact area and higher pressure. The high pressure that exceeds the material's strength leads to microcracks under the washer after applying the tightening torque.

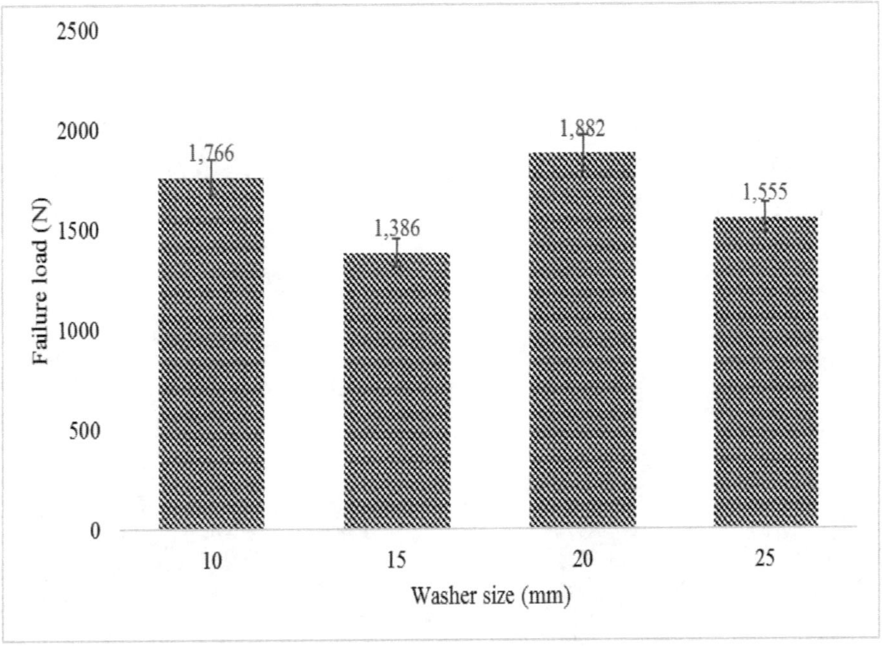

Figure 6. Failure load at different washer sizes

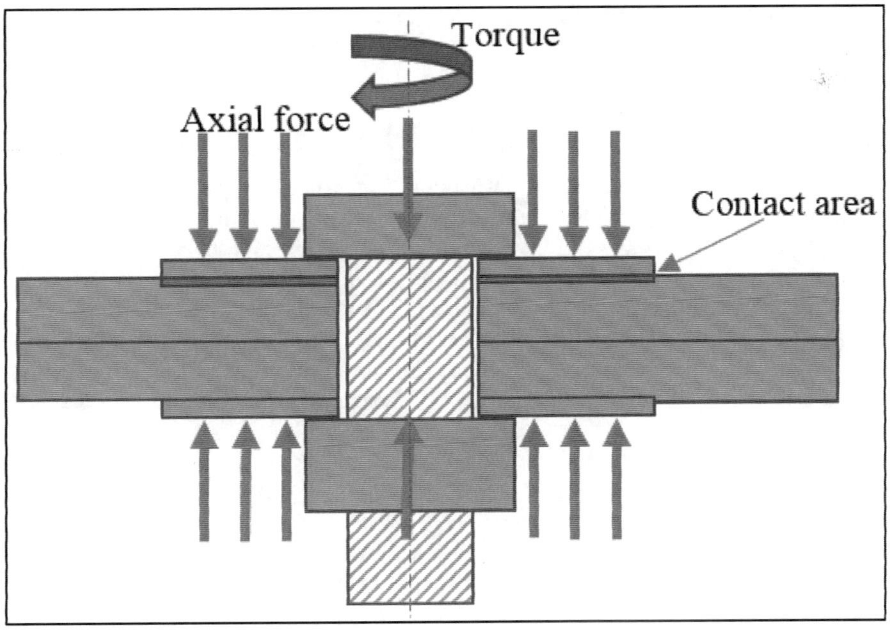

Figure 7. Illustration for the axial force exerted on the joint

The tested specimens are investigated under the microscope, as shown in Figure 8. Figure 8a and Figure 8b are corresponding to big washer sizes of Do = 20 and 25 mm respectively, the mode of failure is a mixed-mode of net tension and bearing. Figure 8c and Figure 8d are corresponding to small washer sizes of Do = 10 and 15 mm, respectively. Many cracks appeared under the contact area of the washer, and these cracks propagate during tensile loading and resulting in lower bearing strength of the joint. These figures comply with the results discussed in Figure 6.

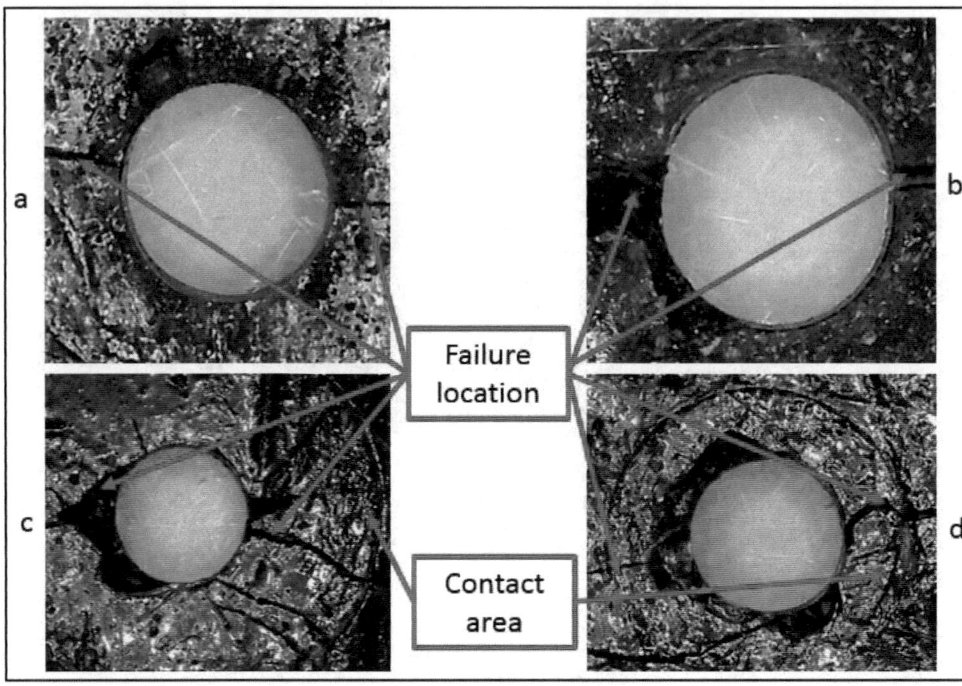

Figure 8. Fractography of specimens at different washer's outer diameter; a) Do = 25, b) Do = 20, c) Do = 15, and d) Do = 10 mm

The optimum result from this test is, using a washer of outer diameter 20 mm with a tightening torque of 6 N.m. It is also concluded that the optimum washer size depends on the used tightening torque.

3.4 Effect of center distance of holes to the hole diameter (M/D)
The effect of different M/D ratios on the failure load of the specimen is tested and the results are plotted as shown in Figure 9. The figure shows that by increasing the center distance, the failure load decreases, and this is because the lateral distance from the edges (K) decreases. When the distance K decreases, the probability of the crack propagation increases, and the separation of the joint is observed at lower loads.

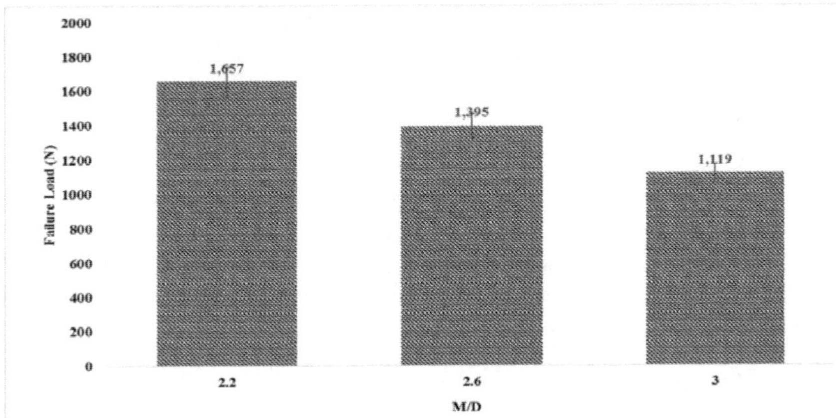

Figure 9. Failure load at different M/D

The tested specimens are investigated under the microscope as shown in Figure 10. The figure shows that changing M/D does not affect the failure mode. Mixed - mode of net tension and bearing is observed in all studied cases.

It is also observed that by using a washer of outer diameter equals to 15 mm with tightening torque equals to 2 N.m., there are not any cracks under the contact area of the washer. This proves that the washer size depends mainly on the axial force results from the tightening torque.

Figure 10. Fractography of specimens at different M/D ratios; a) M/D = 2.2, b) M/D = 2.6, and c) M/D = 3

The optimum result obtained in this test is that the minimum possible center distance -depending on the width and the outer diameter of the washer- gives higher failure loads compared to the bigger center distances.

4. Conclusion

Increasing the edge distance results in increasing the failure load and obtaining a mixed-mode of bearing and net-tension. Increasing the tightening torque increases the failure load as the axial load on the joint members' increases. The failure load increases by decreasing the washer size due to increasing the pressure on the joint. The small-sized washers have a bad effect on the joint as increasing the pressure too much can exceed the compressive force resulting in microcracks under the washer. The failure load increases with decreasing the center distance and increasing the lateral distance from the edges.

References

[1] Elenien K F A, Abdel-Wahab A, ElGamsy R and Abdellatif M H 2018 Assessment of the properties of PP composite with addition of recycled tire rubber *Ain Shams Engineering Journal* **9** 3271-6

[2] Mousa, B.H. and Latif, M.H.A., 2019, September. Mechanical behaviour of rubber hybrid composites. In *IOP Conference Series: Materials Science and Engineering* **Vol. 610, No. 1**, p. 012064.

[3] Galińska A and Galiński C 2020 Mechanical Joining of Fibre Reinforced Polymer Composites to Metals—A Review. *Part II: Riveting, Clinching, Non-Adhesive Form-Locked Joints, Pin and Loop Joining Polymers* **12** 1681

[4] Baskaran R, Sarojadevi M and Vijayakumar C T 2011 Mechanical and thermal properties of unsaturated polyester/calcium carbonate nanocomposites *Journal of reinforced plastics and composites* **30** pp1549-56

[5] Wahono S, Irwan A, Syafri E and Asrofi M 2018 Preparation and characterization of ramie cellulose nanofibers/CaCO3 unsaturated polyester resin composites *ARPN Journal of Engineering and Applied Sciences* **13** pp 746-51

[6] Mousa, B.H., Azab, N.A., Bassioni, G. and Abdellatif, M.H., 2020. Assessment of the Damage Resulting from Drilling Holes in Waste Tire Rubber Polyester Composite Laminates. *Waste and Biomass Valorization*, pp.1-12.

[7] Elenien, K.A., Azab, N.A., Bassioni, G. and Abdellatif, M.H., 2020, November. The effect of tire rubber particles on the mechanical and physical properties of polyester. In *IOP Conference Series: Materials Science and Engineering* **Vol. 973, No. 1,** p. 012019.

[8] Okutan B, Aslan Z and Karakuzu R 2001 A study of the effects of various geometric parameters on the failure strength of pin-loaded woven-glass-fiber reinforced epoxy laminate *Composites science and technology* **61** 1491-7

[9] Kumar M, Saini J and Bhunia H 2020 Investigations on the strength of mechanical joints prepared from carbon fiber laminates with addition of carbon nanotubes *Journal of Mechanical Science and Technology* 1-12

[10] Ghanbari E, Sayman O, Ozen M and Arman Y 2012 Failure load of composite single-lap bolting joint under traction force and bending moment Australian *Journal of Basic and Applied Sciences* **6** 683-92

[11] Awad A, Abdel-Ghany A W, Abd El-Wahab A A, El-Gamasy R and Abdellatif M H 2019 The influence of adding marble and granite dust on the mechanical and physical properties of PP composites *Journal of Thermal Analysis and Calorimetry* 1-9

[12] Esendemir Ü and Öndürücü A 2011 Comparison of bolted joints with two different clearance types

[13] Khashaba U, Sallam H, Al-Shorbagy A and Seif M 2006 Effect of washer size and tightening

torque on the performance of bolted joints in composite structures *Composite Structures* **73** 310-7

[14] Karakuzu R, Taylak N, İçten B M and Aktaş M 2008 Effects of geometric parameters on failure behavior in laminated composite plates with two parallel pin-loaded holes *Composite structures* **85** 1-9

[15] Dehghanpour S, Hosseini Safari K, Barati F and Attar M 2020 Investigation on Failure Loads and Failure Modes for Two Parallel Pin-Loaded Holes Made from Unidirectional *Glass-Epoxy Nanoclay Laminates ADMT Journal* **13** 119-27

[16] ASTM, D., 953-02. Standard method of test for bearing strength of plastics.

[17] Tong L 2000 Bearing failure of composite bolted joints with non-uniform bolt-to-washer clearance Composites *Part A: Applied science and manufacturing* **31** 609-15

[18] ASTM D 2005 5961/D 5961 M-05 Standard test method for bearing response of polymer matrix composite laminates *ASTM International, West Conshohocken, PA*

[19] Mínguez J M and Vogwell J 2006 Effect of torque tightening on the fatigue strength of bolted joints *Engineering Failure Analysis* **13** 1410-21

[20] Rosales-Iriarte F, Fellows N and Durodola J 2011 Experimental evaluation of the effect of clamping force and hole clearance on carbon composites subjected to bearing versus bypass loading *Composite Structures* **93** 1096-102

IOP Conf. Series: Materials Science and Engineering 1172 (2021) 012024 doi:10.1088/1757-899X/1172/1/012024

Water recycling system based on adsorption by activated carbon synthesised from c. verum for space exploration; an estimated design

M N Ettish[1], O Abuzalat[1], G S El-Sayyad[2] and M A Elsayed[1]

[1] Chemical Engineering Department, Military Technical College, Kobry Elkobbah, Cairo, Egypt.
[2] Drug Microbiology Laboratory, Drug Radiation Research Department, National Center for Radiation Research and Technology (NCRRT), Egyptian Atomic Energy Authority, Cairo,Egypt.

E-mail: mnasr4810@gmail.com

Abstract. Water recycling is a crucial component of space flights. In this study, c.verum, a low-cost agricultural by-product abundant in Egypt, which was not utilized before for the preparation of porous carbons, and its ability for recycling water in space stations was estimated. The prepared samples show high porosity and surface area by physical activation. The influences of the pyrolysis temperature and activation hold-up time on the activated carbon's porosity were studied. The BET surface area and the total pore volume of the prepared carbon were used as the criteria for selecting the optimum preparation parameters. The optimum temperature for pyrolysis was found to be at a temperature of 900°C, hold-up time of two-hour, a nitrogen flow rate of 150 cm3/min, and a heating rate of 10^{0}C/min. However, the optimum activation conditions were at a temperature of 900^{0}C, a CO2 flow rate of 150 cm3/min, a heating rate of 20^{0}C/min, and a hold-up time of 120 min. Equilibrium data is used for fitting to Freundlich, Langmuir, and Temkin isotherms models. The result revealed that the Langmuir model was the finest match for the equilibrium data, with an extreme monolayer adsorption capability of 12.37 mg/g at 25^{0}C. The maximum monolayer adsorption capacity decreased with increasing temperature confirmed the exothermic character of the adsorption interaction.

1. Introduction

Water is as crucial for existence in outer space as on Earth. Around 75% of the cost of established human life on the international space-station (ISS) is related to safe drinking water availability. A space traveler on the ISS uses approximately one gallon of water per day, and approximately estimated cost of $83,000 per gallon to lift to space, the costs will hastily be subjected to be increased in the future. Due to the high water recycling cost in space, NASA has developed and implemented research to design systems to recycle water at a low cost. Water recycling preserves costs down on resupply and makes space missions more Earth-independent and self-sufficient. However, more extended activities will demand much from the waste management system than from storage [1].

Activated carbon (AC) is broadly used for water effluent treatment. The activated carbon's utilization becomes restricted by its elevated cost due to using non-renewable and moderately-expensive origin. So, this has pointed to a growing concern in the construction of low-cost activated carbon. Additionally, the production method is a moderately-easy because of the great-reactivity of the performed biomass,

Content from this work may be used under the terms of the Creative Commons Attribution 3.0 licence. Any further distribution of this work must maintain attribution to the author(s) and the title of the work, journal citation and DOI.
Published under licence by IOP Publishing Ltd

reducing the waste-disposal values and the harmful environmental influences. The manufacture of activated carbons from agricultural waste has gained much attention lately, such as almond shell, waste apricot, bean pod, cherry stone, date palm seed, rice husk, corn cob, coconut, and Olive stone.

The beginning substance's nature is an essential function in determining the formed activated carbon's quality, features, and characteristics [2].

Activated carbon synthesis from cinnamon sticks by-product has not been estimated before in the literature. As a result, looking at the viability of using a new by-product is the purpose of this study. Cinnamon sticks, as a candidate precursor to be used in the manufacturing of activated carbon by physical method. A study of its capability of removing Chlorpyrifos (CPS) - an organo-phosphorus organic compound.

2. Experimental section

2.1. Materials

Cinnamon sticks were purchased locally, as it is abundant in Egypt. N_2 and CO_2 gases were obtained from El Nasr Company. Sigma Aldrich delivered the acetonitrile. KZ Company supplied the chlorpyrifos.

2.2. Characterization and production of cinnamon sticks-based activated carbon (CSAC)

2.2.1. Pyrolysis process. The obtained raw-materials (Cinnamon sticks) were washed with purified water to discharge the contaminants after dissolving the unwanted substances, dehydrated at 110°C for 24 h, sieved after grounding to achieve particle size below 2 mm diameter. The dehydrated cinnamon granules were pyrolysed (tube furnace) at various temperatures (600-1000^0 C), with a heating flow of 10^{0}C/min for 120 min in the closeness of a nitrogen flow (150 $cm^3.min^{-1}$).

2.2.2. Activation progression. The char was activated at 900 ^{0}C with different holding time intervals, 30-150 min, using a CO_2 flow rate was 150 $cm^3.min^{-1}$, and heat rate was 20 $^0C.min^{-1}$.

2.3. Porosity & surface analysis

Adsorption isotherms were accomplished from nitrogen adsorption at -196^{0}C over the processed activated carbon utilizing an automatic adsorption apparatus (Nova 3000e series, USA). Brunauer–Emmett–Teller (BET) equation was applied for determining the relative pressure (P/P^0) within 0.05 – 0.2. The obtained results from the isotherms were applied to define the S_{BET}. The Horvath–Kawazoe (HK) method was applied to determine the micropore volume. The pore size division and the mesopore volume were defined using the Barrett–Joyner–Halenda (BJH) design.

2.4. Ultimate and Proximate analysis

The cinnamon, char, and prepared CSAC samples' proximate analysis were conducted using (TGA; Perkin Elmer TGA 7). The proximate analysis was studied according to. For CHN-O elemental analysis, about 30–50 mg of a sample was analyzed, and the data was recorded using a Perkin-Elmer Series II CHNS/O Analyser model 2400.

2.5. FTIR analysis

The samples' FTIR spectra were registered between 4000 and 450 cm^{-1} in a mobile IR- portable FT-IR (Bruker Optics). This measurement was used for the starting material (cinnamon stick), char, and CSAC samples. Firstly, before the analysis, samples were dried out in an oven at 50–60°C (at atmospheric pressure) for 3–5 hours for complete water removal.

2.6. SEM analysis

Surface morphology was surveyed utilizing a scanning electron microscope (SEM) ZEISS, EVO-MA10.

2.7. Adsorption equilibrium studies

In a set of 100 ml conical Erlenmeyer flasks, by adding a 25 ml solution of different initial concentrations (20-50 mg/L) CPS in acetonitrile adsorption isotherms were performed. An orbital shaker (SI-300R) continually stirred the mixture [3]. Similar masses of activated carbon (0.1 g) of mesh size (18*30) were attached. Samples were regularly-obtained every 10 min, and the concentration of CPS was defined by applying a double beam UV-Vis Spectrophotometer (Shimadzu, Japan). A wavelength of about 289 nm (λ_{max}) was adopted for all absorbance examinations for all specimens. For measurement of the removal capability (%) R, Eq*(1)* was adopted:

$$\%R = \frac{C_0 - C_t}{C_t} \times 100 \tag{1}$$

Where C_o and C_t (ppm) imply the initial concentration and concentration at time interval *t*.
The highest adsorption potential at equilibrium, q_e (mg. g^{-1}), was calculated applying Eq. (2):

$$q_e = \frac{(C_0 - C_e)v}{w} \tag{2}$$

Where C_e (mg. L^{-1}) signifies the concentration after the equilibrium of the CPS solution, w denotes the adsorbent mass (g), and v signifies CPS solution volume (L).

3. Results and discussion

3.1. Ultimate and Proximate analysis

The ultimate and proximate examinations of the raw-material, char, and CSAC specimens are seen in Table 1. The results revealed that the raw-material gives extended carbon content, high volatile material, and low-ash content, which is an excellent candidate for producing activated carbon. The liberation of rising volatile material by carbonization and activation were the targets, ending in chars and initiated carbon, a carbon element. The humidity percent in the chars and CSAC did not show any unusual model with different pyrolysis and activation because all these processes were done over 100°C. Consequently, their dehydration impacts would most reasonably be related to the various specimens.

Table 1. Properties of raw material, char, and CSAC samples.

Sample	Ultimate (wt.% dry-ash-free basis)				Proximate (wt.%)			
	C	H	N	O	Moisture (%)	Volatiles (%)	Carbon (%)	Ash (%)
Raw material (cinnamon)	46.35	18.65	0.32	34.68	7	70	21.6	1.40
Char	68.09	1.08	0.45	30.38	7.5	28	62.48	1.49
(CSAC)	80.05	0.92	0.58	18.45	8	11	79.48	1.52

3.2. FTIR analysis.

Surface groups created by heteroatoms attached to the carbon atoms provide additional chemical characteristics. Infrared spectroscopy presents knowledge on the chemical composition and the same functional groups of the carbon element. Figure 1 shows FTIR spectra for different samples showing the difference between the raw cinnamon, char, and the CSAC. The prominent specimen peak located at 3570– 3000 cm^{-1} represents the (–OH) group and may be accompanying to the water adsorbed. The different peaks cannot solely be relevant to certain specific groups. It is appropriate that the 1460 –1490 cm^{-1} peak agrees to the (–CH–) group, the 1603 cm^{-1} peak describes the aromatic group, and the (C- O-C) group should be located between 1000 and 1300 cm^{-1} (many peaks look in this interval). For the char and CSAC samples, the challenging preparation conditions of temperature treatment in inert and oxidizing environments, respectively, end in a large variety of the peak strengths of all the related definite absorption bands, commonly with peak ever-changing or peak merging. This is because of converting the organic form of cinnamon into a nearly-absolute carbon construction by eliminating any lasting of oxide and hydrogen groups at these higher temperatures.

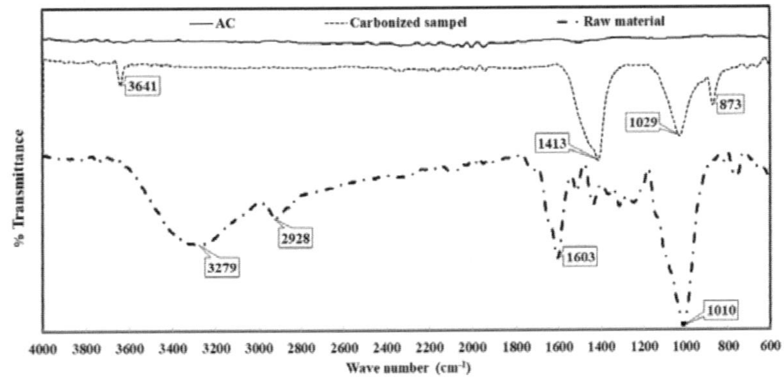

Figure 1. FTIR Spectra for the raw material, char, and CSAC samples

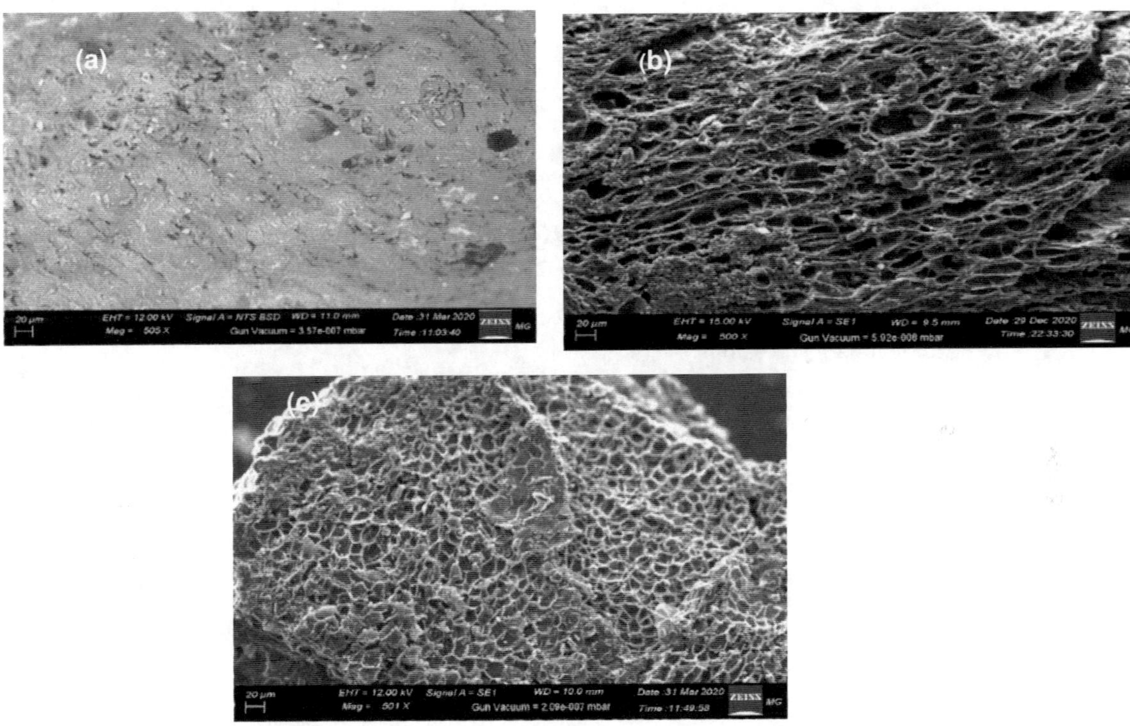

Figure 2. SEM images of (a) raw material (b) carbonized sample (c) CSAC.

3.3. SEM study for the preparation of CSAC.

Figure 2 a, b, and c illustrate the sample's SEM picture before and after the process of pyrolysis and activation. Porous solid is formed through these two processes. In general, during pyrolysis and cinnamon activation, the largest of the non-carbon components, hydrogen, and oxygen are primarily-extracted in volatile mode by the pyrolytic breakdown of the initial components. The innocent atoms of primary carbon are classified into fixed crystallographic structures recognized as elementary graphite crystallites. The crystallite's shape is different so that free spaces (pores) endure among them.

3.4. Yield, pore size, and surface area study

3.4.1. Impact of pyrolysis. The CSAC's surface area and micropore volume were the primary indicators for determining pyrolysis's most appropriate conditions. The yield was further brought into concern when choosing these excellent states. In this research, a dual-stage tuning was applied. The initial step or the primary optimization method was to choose a suitable condition for the pyrolysis temperature. When the pyrolysis temperature parameter had been optimized, its benefit would be applied for following activation tests in which one of the other parameters to be optimized was different (hold up time) whilst maintaining the balance fixed. The Impact of the temperature of pyrolysis on the surface area, carbon yield, and porosity of the char samples are given away in Figure 3 and Table 2.

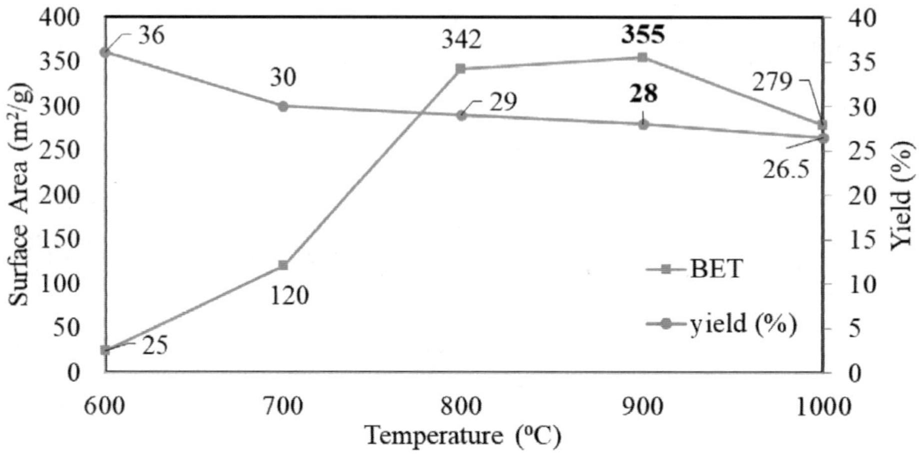

Figure 3. Effect of pyrolysis temperature on the BET surface area ($m^2.g^{-1}$) and the yield of carbonised samples

Table 2. Surface area ($cm^3.g^{-1}$) and porosity ($m^2.g^{-1}$) volume distribution of carbonized samples.

Sample	S_{BET} (m^2/g)	Pore volume (cm^3/g)				Mean pore diameter (nm)
		Total pore volume	Micropore volume	Meso pore volume	Macro pore volume	
600 °C	25	0.029	0.008	0.021	0	4.83
700 °C	120	0.087	0.037	0.045	0.005	3.45
800 °C	342	0.196	0.137	0.047	0.012	2.58
900 °C	355	0.198	0.144	0.046	0.008	2.23
1000 °C	279	0.173	0.129	0.043	0.001	2.48

Figure 4 demonstrates the isotherms of the different carbons prepared under unlike pyrolysis temperatures. Nitrogen adsorption at 77 K increased with increases in pyrolysis temperature from 600 to 900 ^{0}C. This is because the pore volume and surface area (BET) have increased from 0.029 to 0.198 cm^3/g and 25 to 355 m^2/g. However, any increase in the pyrolysis temperature (1000 ^{0}C) leads to a reduction in the nitrogen (N_2) adsorption due to decreased surface area (S_{BET}) and total pore volume to 279 m^2/g 0.173 cm^3/g, respectively.

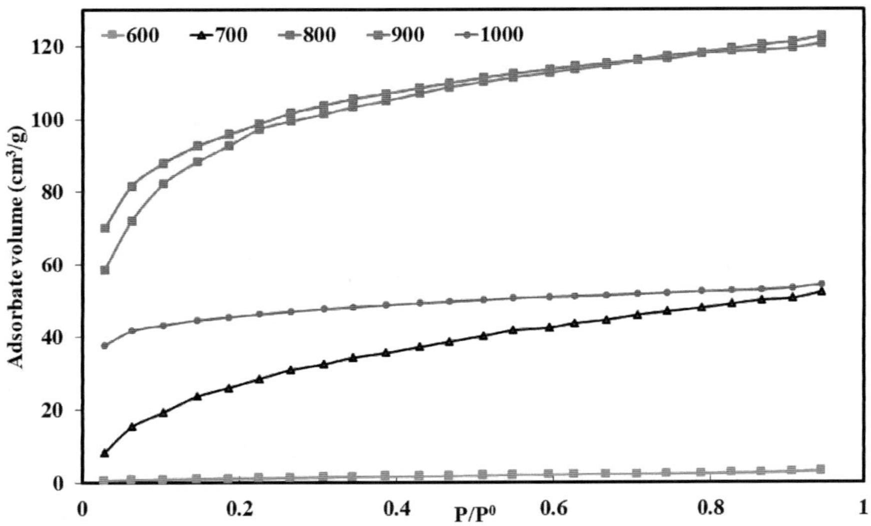

Figure 4. Isotherms of char samples prepared under different pyrolysis conditions [600-700-800-900-1000 denotes pyrolysis temperature (°C)].

3.4.2. Process of Physical Activation. At this step of the work, the pyrolyzed sample at 900°C, which has a profusely surface area, was chosen for seeking the Impact of the activation progression. Time of activation affects the surface area, and porosity, as presented in Table 3. the BET surface area's optimum values and total porosity were at holding time for 2h.

Table 3. The surface area by BET method and pore volume distribution for CSAC samples.

Sample	S_{BET} (m^2/g)	Pore volume (cm^3/g)				Mean pore diameter (nm)
		Total pore volume	Micropore volume	Meso pore volume	Macro pore volume	
30 min	640	0.32	0.254	0.062	0.004	2.31
45 min	652	0.36	0.257	0.1	0.003	2.55
60 min	696	0.37	0.278	0.054	0.038	2.46
90 min	842	0.455	0.33	0.125	-	2.51
120 min	**894**	0.49	0.35	0.14	-	2.67
150 min	741	0.38	0.29	0.087	0.003	2.48

Figure 5 demonstrates the isotherms at 77 K by N_2 adsorption for CSAC created by physical activation using CO_2. All isotherms of all activated carbons series belong to a union of type (I) and (IV) isotherm, rendering to IUPAC. Moreover, the nitrogen uptake significantly-improves at a much more crucial relative pressure. The adsorption of nitrogen increased with increases of activation hold-up time from 30 to 120 min due to increased porosity and surface area. However, protracted activation of the sample at a temperature of 900°C and a staying at this temperature for 150 min leads to diminishing porosity and surface area and, consequently, the nitrogen adsorption (Table 3).

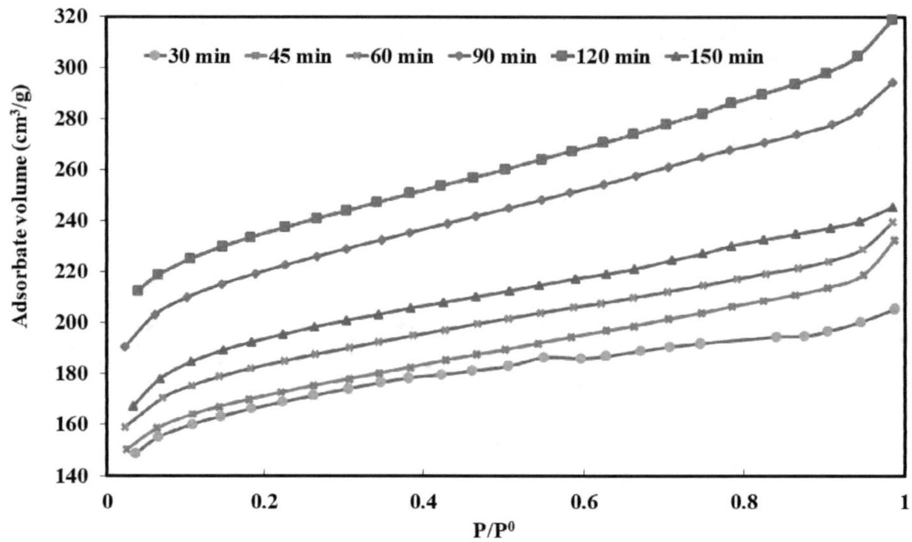

Figure 5. Adsorption isotherms of CSAC samples at 900⁰ C under a different holding time [30-45-60-90-120-150 min].

3.5. Batch adsorption studies

Chlorpyrifos (CPS) is used in most agrochemicals in urban areas as an organophosphorus compound. CPS is extant in various forms, such as condensed emulsions, wet-table powders, granules, dust, controlled-release polymers, etc.[3].

3.5.1. Impact of CPS's initial concentration. In this part of the work, the optimized prepared sample was chosen for studying CPS adsorption. This sample displays the uppermost surface area, 894 m^2/g, and a whole pore volume of 0.49 cm^3/g. The adsorption of CPS by CSAC was studied at different initial CPS concentrations (20,30,40, and 50ppm). The experimentations were conducted out at a fixed adsorbent(CSAC) dose (0.1 g), at room temperature (25 ±1 ℃), pH (7.0), and 100 rpm. Figure 6 demonstrates the result for the outcome of CPS's initial concentration adsorption onto CSAC. According to the sorting of adsorption isotherms, the adsorption of CPS by the CSAC is of class L (Langmuir-type)[4]. The maximum value of sorption capacity was detected at 50 ppm. The adsorption capacity improved increasingly as the concentration increased because of the vital role of initial concentrations on the adsorption of CPS on AC[5]. Figure 7 demonstrates the result of the CPS initial concentration effect on the elimination efficiency by CSAC samples. The initial CPS concentration-effect hinge on the direct relationship between the CPS concentration and the existing accessible sites on the exterior of an adsorbent [6]. Frequently, the rate of CPS replacement declines with an improvement in initial dye concentration, which maybe because of the fullness of adsorption places on the adsorbent exterior [7]. At low concentration, there will be unfilled active places on the adsorbent cover, and when the first CPS concentration develops, the active places needed for adsorption of the CPS molecules will close [8]. Still, the development in the original CPS concentration will improve the adsorbent's filling quantity on the surface of the CSAC, which may be due to the great driving power for mass at a tremendous primary CPS concentration [9].

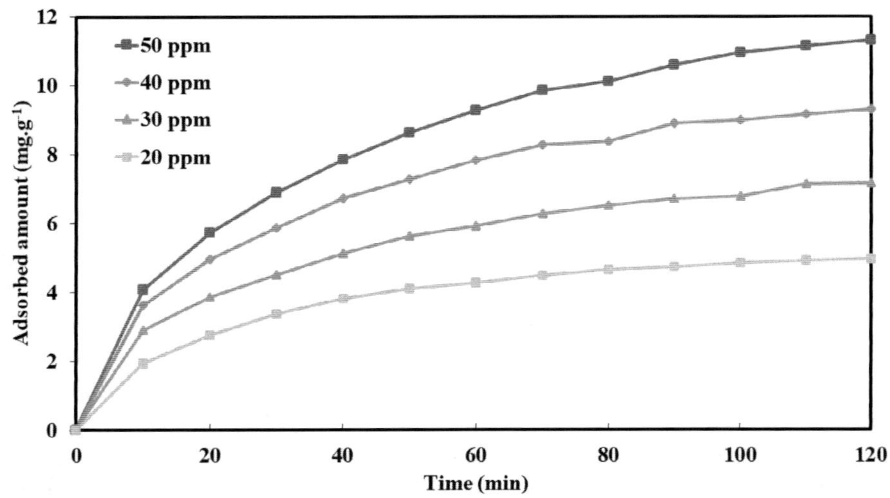

Figure 6. Initial adsorbate concentration effect on the extreme adsorption capacity of CPS using CSAC.

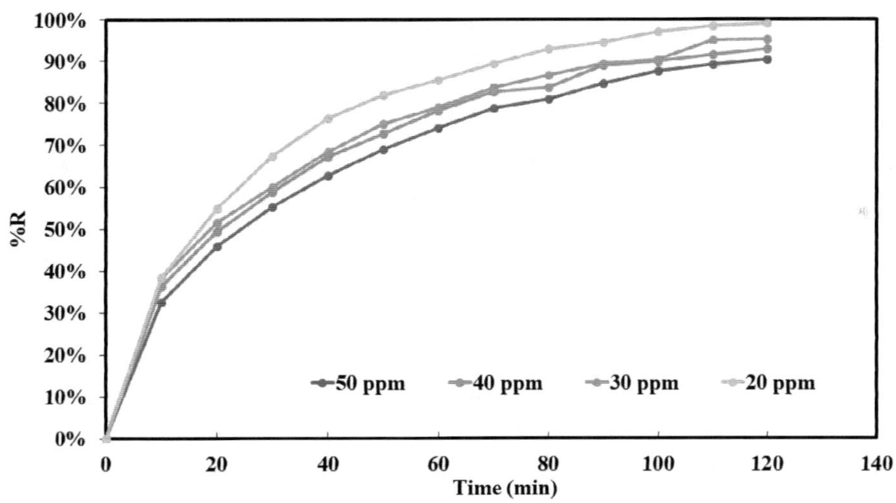

Figure 7. Effect of initial adsorbate concentration on the % removal of CPS using cinnamon sticks-based AC.

3.5.3. Adsorption Temperature consequence on the elimination of CPS. As demonstrated in figure 8, the adsorption of CPS onto CSAC samples decreases with temperature increase, indicating an exothermic process. This decrease of removal at high temperature possibly is due to the adsorption forces' failing between the surface of CSAC and the CPS species' active sites and between adjacent CPS molecules on the adsorbed phase[10].

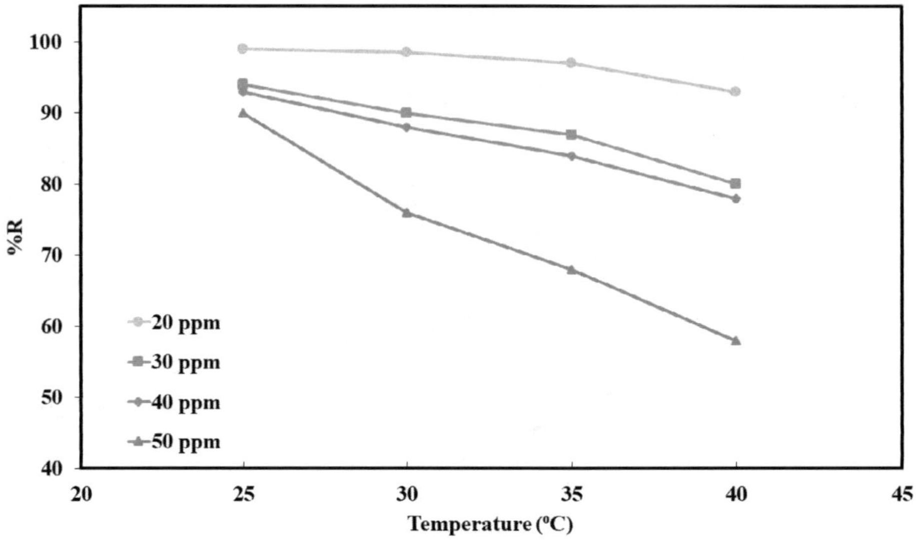

Figure 8. Effect of temperature on the removal of CPS with different initial concentrations.

3.6. Adsorption equilibrium isotherms

Three isotherms were verified for their capability to define the investigational consequences, namely the Langmuir, Freundlich, and the Temkin isotherm. Eq (3) expresses the Langmuir model.

$$\frac{C_e}{q_e} = \frac{1}{q_{max} \cdot K_L} + \frac{C_e}{q_{max}} \quad (3)$$

Where q_e(mg/g) and C_e(mg.L^{-1}) are the CPS adsorbed amount per unit mass of CSAC and CPS concentration at steadiness, respectively. q_{max} is the maximum quantity of the CPS per gram CSAC to form an entirely monolayer perfusion on the surface-bound at high C_e, and the constant K_L is referred to as the binding sites' affinity (L.mg^{-1}). The outline of specific adsorption (C_e/q_e) against the steadiness concentration (C_e) (figure 9) illustrates that the adsorption conforms to the Langmuir model. The Langmuir constants q_{max} and K_L were determined from the plot's slope and intercept and are existing in Table 4. The separation factor R_L is considered one of the critical characteristics of the Langmuir isotherm that is donated by Eq (4).

$$R_L = \frac{1}{1 + K_L C_0} \quad (4)$$

Where; C_0 is the uppermost initial CPS concentration (mg/L), and K_L (L/mg) is Langmuir constant. The R_L value is a sign of the form of the isotherm to be whichever unfavorable ($R_L > 1$), linear ($R_L = 1$), irreversible where ($R_L = 0$), or favorable ($0 < R_L < 1$). The values of R_L between 0 and 1 specify favourable adsorption. R_L's value in the current search was found to be 0.0154, demonstrating that the adsorption of CPS on CSAC is favourable.

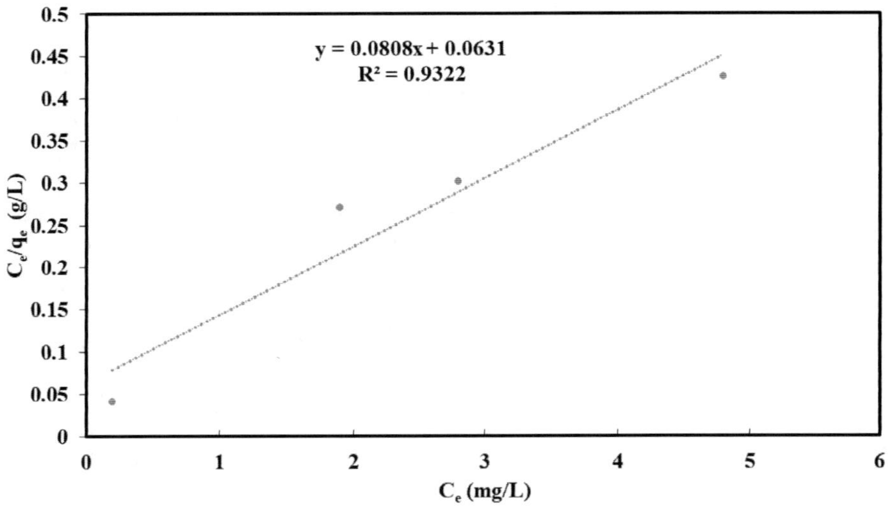

Figure 9. Langmuir adsorption isotherm of CPS on CSAC.

Eq (5) is used to describe a heterogeneous system (Freundlich isotherm) [5].

$$log\, q_e = log\, K_F + \left(\frac{1}{n}\right) log\, C_e \quad (5)$$

Where K_F and n are, Freundlich coefficients with K_F (mg/g (L/mg)$^{1/n}$) are the sorbent's adsorption capacity, and n indicating how is adsorption is esteemed. The extent of the exponent 1/n gives favourable adsorption. Values of n more than 1 represent favourable adsorption. Values of n and K_F are computed from the plot's slope and intercept (figure 10) and recorded in Table 4.

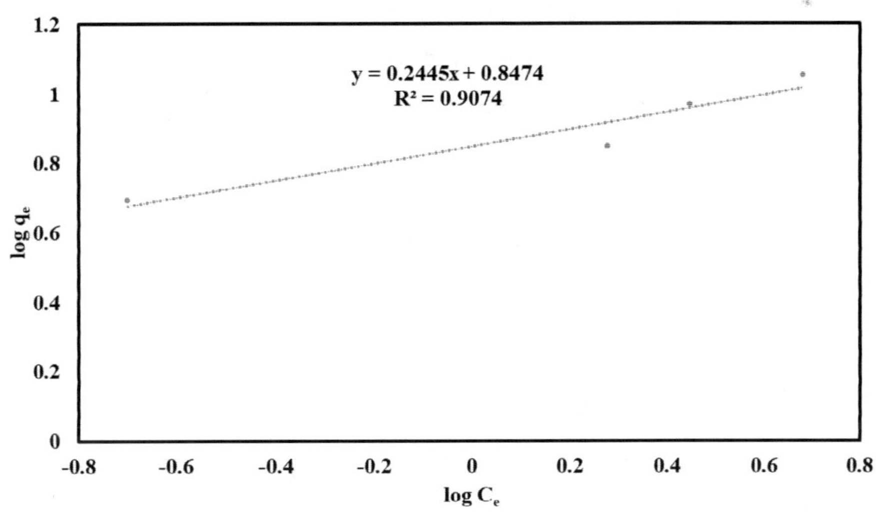

Figure 10. Freundlich adsorption isotherm of CPS on CSAC.

The Temkin isotherm is expanded for the determination of physical or chemical sorption characteristics. It has generally been applied in the linear form as described in Eq (6).

$$q_e = \frac{RT}{b}\ln K_T + \frac{RT}{b}\ln c_e \quad (6)$$

Where b is a constant link to the sorption heat (J/mol), K_T is the Temkin isotherm constant (L/g), R the gas constant (8.314 $J.mol^{-1}K^{-1}$), and T the absolute temperature (K). Therefore, by plotting q_e versus ln C_e (figure 11), the constants K_T and b can be determined. The constants K_T and b are listed in Table 4. If the heat of the sorption value is less than 1.0 kcal/mol, then physical adsorption occurs. Besides, its value of 20-50 kcal/mol, then chemical adsorption is occurring. If the sorption value is in-between (1 - 20 kcal/mol), then both chemical and physical adsorption is entangled in the adsorption. As it can be seen in Table 4, the Langmuir model's isotherm matches fairly well with the investigational data (correlation coefficient $R^2 > 0.93$). The capacity of monolayer CPS sorption rendering to this model was 12.37 $mg.g^{-1}$. This is owed to the homogeneous distribution of active sites onto the CSAC surface since the Langmuir equation suppose that the surface is similar or homogenous.

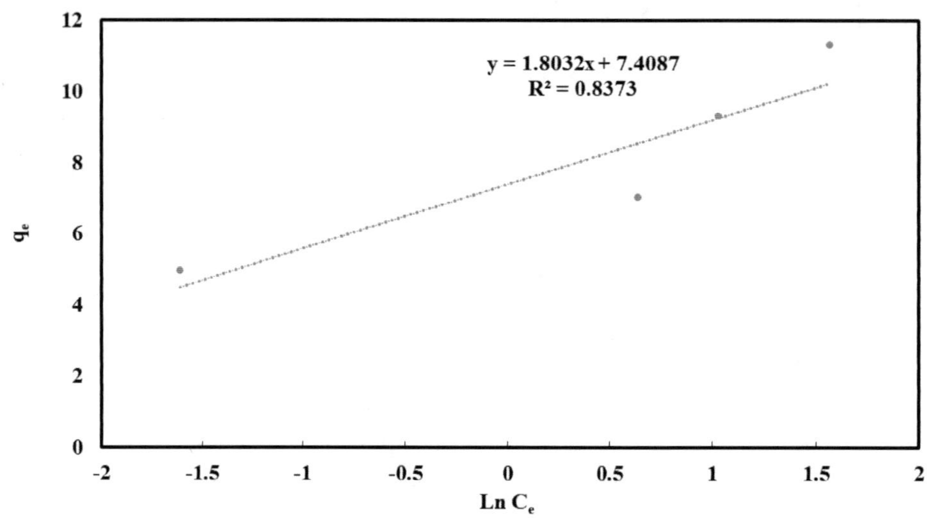

Figure 11. Temkin adsorption isotherm of CPS on CSAC

Table 4. Isotherm parameters for removal of CPS by CSAC

Isotherm	Parameters
Langmuir	
q_{max}(mg/g)	12.37
K_L(L/mg)	1.28
R_L	0.0154
R^2	0.9322
Freundlich	
K_F	7.04
n	4.09
R^2	0.9074
Temkin	
K_T	1.413
b(kJ/mol)	1.373
R^2	0.8373

3.7. Adsorption kinetics

Kinetic adsorption results for different CPS concentrations were examined using a pseudo-first-order (PFO) model and a pseudo-second-order (PSO) model. For the PFO kinetic model, k_1 values were obtained from the linear log plots' slopes (q_e-q_t) against t (figure 12). The values obtained for the correlation coefficient were comparatively small. The values of experimental q_e were not compatible with the estimated values obtained from the linear plots, as perceived in Table 5. This demonstrates that the sorption of CPS to CSAC did not conform to the PFO equation.

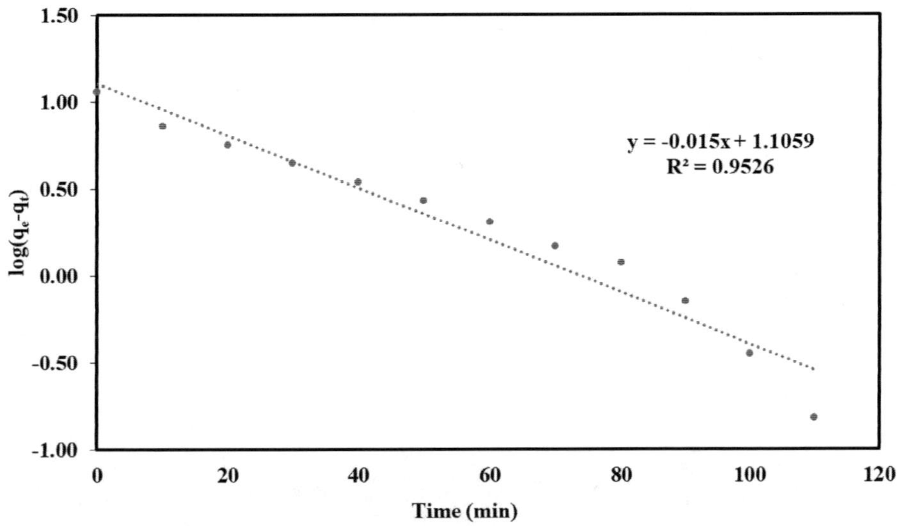

Figure 12. Pseudo-first-order kinetics for CPS adsorption onto CSAC at 25 °C.

But for the (PSO) pseudo-second-order kinetics, the linear plot of t/q_t against t, as seen in figure 13, indicates a strong relationship between the investigational and the measured q_e values (Table 5). Besides, the correlation coefficients for the PSO model were larger than 0.99 for the adsorption of CPS, indicating the applicability of this model to explain the CPS sorption mechanism on the CSAC samples.

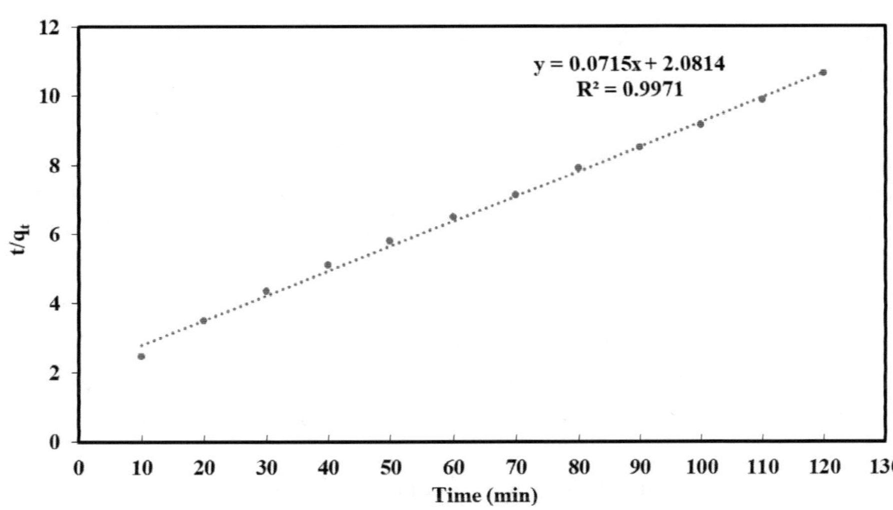

Figure 13. Pseudo-second-order (PSO) kinetics for adsorption of CPS onto CSAC at 25 °C.

Table 5. Adsorption kinetic parameters

Kinetic model	parameters	value
Experimental	q_t	11.3
Pseudo first order (PFO)	q_e	12.76
	k_1	0.0345
	R^2	0.9526
Pseudo-second-order (PSO)	q_e	13.98
	k_2	0.0025
	R^2	0.9971

3.8. Design of water recycling system based on the batch sorption process

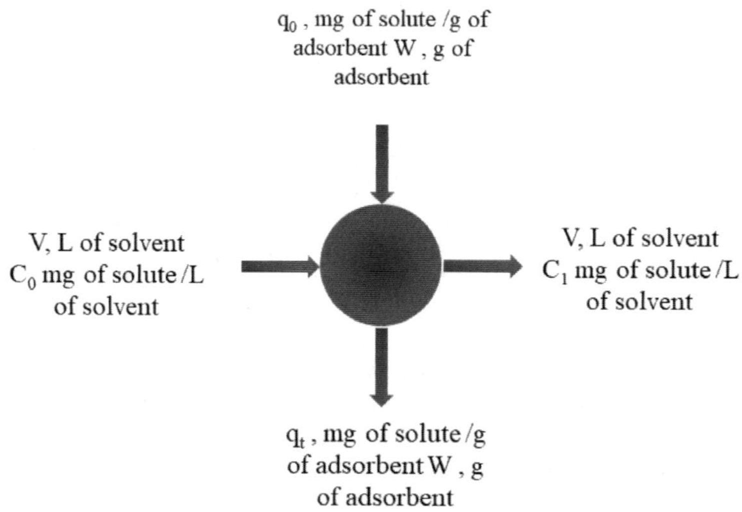

Figure 14. A single-stage water recycling batch adsorber

Figure 14 depicts a representation diagram of a single-step process (batch sorption). The effluent holds V (L) of wastewater and a CPS initial concentration C_0 is reduced to C_1 in the adsorption process. In the management stage W (g) CSAC is added to the solution, and the adsorbate concentration on the CSAC deviations from $q_0 = 0$ (at the beginning) to q_1. Eq. 7 shows the mass balance for the adsorbate in a single-stage water recycling system.

$$V(C_0-C_1) = W\,(q_0- q_1) \tag{7}$$

Under equilibrium conditions, $C_1 \rightarrow C_e$ and $q_1 \rightarrow q_e$

$$VC_0+Wq_0 = VC_e + Wq_e \tag{8}$$

For the adsorption of CPS on CSAC. The Langmuir isotherm best reflects experimental results. As a result, the equation works better when q_1 is replaced in the repositioned form of Eq.(*8*), giving adsorbent material/solution shares for this certain system,

$$\frac{W}{V} = \frac{(C_0 - C_1)}{q_e} = \frac{C_0 - C_e}{q_{max}K_L C_e/(1 + K_L C_e)} \quad (9)$$

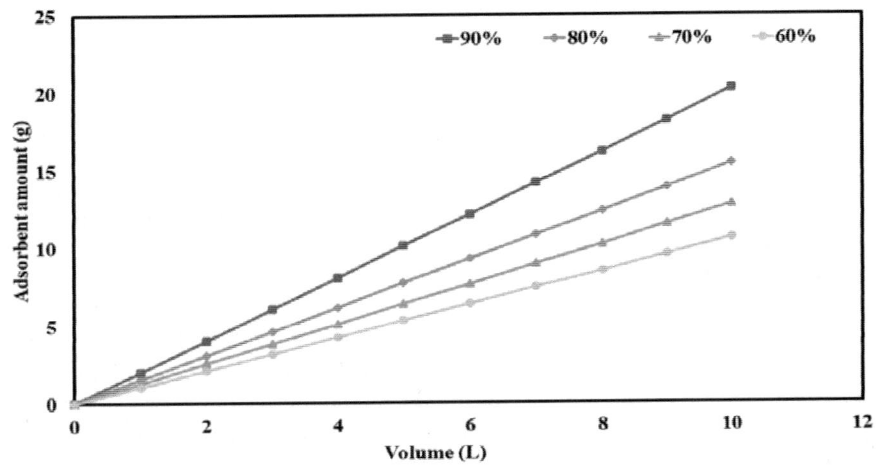

Figure 15. Adsorbent mass (W) against the volume of effluent (V) treated for different percentages of CPS removal

Figure 15 shows a sequence of plots (60, 70, 80, and 90% CPS removal at different solution volumes, i.e., 1, 2, 3, 4, 5, 6, 7, 8, 9, and 10 L) derived from Eq. (9) for the adsorption of CPS on CSAC at the primary concentration of 20 ppm. The amount of CSAC required for the 90% removal of CPS solution of concentration 20 mg/L was 2.01, 4.02, 6.07, and 8.09 g for CPS solution volumes of 1, 2, 3, and 4 L, respectively.

4. Conclusions

This study revealed that water recycling systems used in space flights could depend on the usage of the prepared CSAC. Besides, CSAC is cost-effective and considered an encouraging adsorbent for wastes such as CPS discharge across a broad array of concentrations. The surface area was moderately-high (894 m^2/g) with a high pore volume (0.49 cm^3/g) and was observed to be micro-porous (V_{mic}= 0.35 cm^3/g). Equilibrium data were related to Langmuir, Freundlich, and Temkin isotherms. The data were adequately-represented by the Langmuir isotherm form, with the highest monolayer adsorption potential of 12.37 mg/g at 25°C. The maximum monolayer adsorption capacity reduced with rising temperature proved the exothermic character of the adsorption cooperation. The adsorption kinetics was decided to closely-reflect the pseudo-second-order kinetic design. The adsorption completion of the CSAC was similar to the popular charcoal and some other adsorbents described in related investigations.

References

[1] J. W. Fisher, S. Pisharody, M. J. Moran, K. Wignarajah, Y. Shi, and S.-G. Chang, 2000, "Reactive carbon from life support wastes for incinerator flue gas cleanup," *SAE Technical Paper*0148-7191,.

[2] A. SE, C. Gimba, A. Uzairu, and Y. Dallatu, 2013"Preparation and characterisation of activated carbon from palm kernel shell by chemical activation," *Research Journal of Chemical Sciences* ISSN, vol. **2231**, p. 606X.

[3] M. M. Jacob, M. Ponnuchamy, A. Kapoor, and P. J. J. o. E. C. E. Sivaraman, 2020"Bagasse based biochar for the adsorptive removal of chlorpyrifos from contaminated water" *Journal of Environmental Chemical Engineering*, p. 103904.

[4] R. C. Bansal and M. Goyal, 2005, Activated carbon adsorption. CRC press.

[5] I. Tan, A. Ahmad, and B. Hameed, 2009,"Adsorption isotherms, kinetics, thermodynamics and desorption studies of 2, 4, 6-trichlorophenol on oil palm empty fruit bunch-based activated carbon," *Journal of hazardous materials*, vol. **164**, no. 2-3, pp. 473-482.

[6] L. Wang, Zhanga, J., Wang, A, 2008"Removal of methylene blue from aqueous solution using chitosan-g-poly (acrylic acid)/montmorillonite superadsorbent nanocomposite. *Colloids Surf.," A: Physicochem. Eng Aspects* **322**, 47-53.

[7] L. J. Yu, Shukla, S.S., Dorris, K.L., Shukla, A. and Margrave, J.L., 2003, "Adsorption of Chromium from Aqueous Solutions by Maple Sawdust," *J. Hazard. Mater.,* B**100**, 3-63.

[8] A. J. K. Algidsawi, 2011" A Study of Ability of Adsorption of Some Dyes on Activated Carbon From Date' S Stones," *Australian Journal of Basic and Applied Sciences*, **5**(11): 1397-1403.

[9] J. R. S. Annadurai G, Lee D. J. , Ruey-Shin Juang , Duu-Jong Lee., 2002, "Use of cellulose-based wastes for adsorption of dyes from aqueous solutions," *Journal of Hazardous Materials* B**92** , 263–274.

[10] S. Vigneshwaran, J. Preethi, and S. J. I. j. o. b. m. Meenakshi, 2019, "Removal of chlorpyrifos, an insecticide using metal free heterogeneous graphitic carbon nitride (g-C3N4) incorporated chitosan as catalyst: Photocatalytic and adsorption studies," *International Journal of Biological Macromolecules* vol. **132**, pp. 289-299.

An investigation of plasma sprayed nickel-based and pure aluminum coatings on austenitic stainless steel AISI 304

K. Hamed[1], N. El-Mahallawy[2] and M.O.A. Mokhtar[3]

[1] M.Sc. Student, Design and Production Engineering Department, Faculty of Engineering, Ain Shams University, Egypt
[2] Professor Dr., Design and Production Engineering Department, Faculty of Engineering, Ain Shams University, Egypt
[3] Professor Dr., Department of Mechanical Design and Production, Faculty of Engineering, Cairo University, Egypt

E-mail: 13084@eng.asu.edu.eg

Abstract Atmospheric Plasma Spray (APS) is one of the most leading industrial techniques for protective coating, by improving the performance of parts in the thermal barrier, and wear resistance. Ni-Al alloys are very effective players in the field of design of protective coatings. Accordingly, mixed Al, Ni/Al, and Ni5Al powders were applied on 304stainless steel substrate to develop plasma sprayed coatings. The effect of different compositions on microstructure, microhardness, and porosity was measured. The microstructures of the as-deposited films were characterized utilizing X-ray diffraction (XRD), scanning electron microscopy (SEM), and microhardness measurements. The results showed the formation of two intermetallic compounds, namely: NiAl and Ni3Al. The existence of NiAl is inevitable in all samples, despite the amount of Ni-based alloys in mixtures, or even the atomic percentage of nickel, where the appearance of Ni3Al depends only on increasing the amount of the Ni-based alloy to 50 % percent in mixtures. As regards the steel substrate, the microhardness of the interdiffusion zone of the substrate has been significantly enhanced. Results have shown that the microhardness of the different tested coatings is increased directly with the increment of Ni-based percentage in the coating mixture. The average porosity of the plasma sprayed coatings has proven to be within the normal range.

1. Introduction

Thermal spraying serves a wide range of industrial applications through different coating materials. Metallic, ceramic, and some polymeric materials in the form of powder, ceramic rod, or wire can be used to develop a coating. The diversity in materials and techniques is the main advantage of thermal spraying technology. Besides, the process has the ability of coating production without extremely heating the substrate [1], [2]. Applications of protection have always been a great concern to research society when it comes to mechanical parts that are subjected to structural failures during operation. This can be widely seen in gas turbine systems. Thus, boosting the efficiency can be maintained through the design of a thermal barrier coating (TBC). The TBC structure is based on two layers regardless of the substrate underneath. The layer just above the substrate is the bond coat while the final layer is named the topcoat. One of the most leading industrial techniques for depositing protective coatings is Atmospheric plasma spraying (APS); APS enhances part's performance in wear, corrosion, and can deposit layers as a thermal barrier, which have an impact on parts' life, value, and machinery downtime.

Content from this work may be used under the terms of the Creative Commons Attribution 3.0 licence. Any further distribution of this work must maintain attribution to the author(s) and the title of the work, journal citation and DOI.
Published under licence by IOP Publishing Ltd

During the spraying process, powders are heated to be molten droplets and then accelerated by the heat source, whereupon the molten droplets impact the substrate and rapidly solidify to form a coating layer. Porosity and cracks are the two critical features that must be tightly controlled. They are formed during the process of solidification of impacting particles. Porosity level depends on the process parameters; speed, spray distance, and particle size [3]–[5]. The porosity of thermal spray coatings is usually between 5 – 15% by volume. However, it could be controlled in APS to reach approximately 7%. The major disadvantage of plasma spraying technology is the line-of-sight. Accordingly, the substrate surface must be visible to the area of motion of the APS torch [6]–[9].

Austenitic stainless steel (304) has been chosen as a substrate due to its good corrosion resistance and excellent creep rupture strength; it has been extensively utilized in the manufacturing of many components of conventional power plants [10]–[13]. However, its low corrosion resistance in high-temperature environments makes it essential to develop a protective layer.

Using Aluminum-based coatings in different industries, e.g., aerospace and automotive is attributable to its high strength-to-weight ratio, high electrical and thermal conductivity, adequate corrosion resistance, and affordable metal. These properties made Aluminum-based coating is an optimum choice for steel protection against corrosion with suitable cost reduction. Despite all these advantages, low hardness could be the main limitation for Aluminum to be widely used [14], [15].

On the other hand, according to Kadir [16], nickel alloys are good candidates for bond coat layers in high-temperature applications due to their high melting temperature, good mechanical properties, and oxidation resistance. Ni-Al-based alloys, in particular, have high oxidation resistance, high hardness, and low density [17], [18]. As stated by many studies, the most common stable intermetallic compounds are NiAl and Ni3Al; their excellent mechanical strength, desirable wear-resistant, and sufficiently high melting temperature (1640 ̊C and 1390 ̊C, respectively), provide sustainable performance at elevated temperature. [19]–[27].

Ni/Al[28] and Ni5Al [3], [15] powders are extensively configured with APS [29]–[31], with adequate processing parameters and appropriate measures that exhibit adequate adherence to the substrate and required thickness [17]. They exothermically react and provide nickel aluminides phases, which is mainly used as an effective bond coat for thermal barrier coating [32]. For instance, a study of Ni5Al coating on three types of superalloys in [33] stated that the coating provided good protection in an aggressive environment, especially for Superfer 800.

Ni3Al [34]–[37] and NiAl [6], [10], [31], [38] are potential intermetallic alloys that have an adequate yield strength that proportionally increases with temperature, low density, and high melting point. Enough Aluminum in their compositions could form, in oxidizing environments, thin films of Alumina (Al2O3), which in turn results in excellent oxidation resistance, compact and protective coating.

In practice, the coating is not affected only by a single service condition, but many conditions appear simultaneously [32], [39]–[41]. The coating must sufficiently resist various types of wear and corrosion. Therefore, it has come to mind to use Ni-Al powders in the Aluminum coating matrix, which in turn, could be a suitable additive to pure Aluminum to enhance coating hardness [42]–[44].

The present work is an investigation on stainless steel coating performance, morphology, and microstructure adopting mixtures of Ni/Al or Ni5Al with pure Aluminum as coating materials prepared by atmospheric plasma spraying technique deposited on a stainless-steel substrate.

2. Experimental procedures

Austenitic Cr-Ni stainless steel (AISI 304) has been selected as a Substrate due to its practical application in cases where high wear and corrosion resistance are of prime importance. Atmospheric Plasma Spray (APS) has been applied for surface coating with preassigned powders. Different samples of powders of Ni/Al or Ni5Al have been mixed with pure Aluminum powder at different percentages. Table 1 shows compositions of powder mixtures of tested samples. To maintain uniformity during the deposition, all the powders were thoroughly mixed via ball milling for 6 hours at 200 rpm.

Table 1 Composition of different sprayed coating powders

Sample	Compositions (wt. %)		
	Ni/Al (80/20)	Ni 5Al (95/5)	Pure Al
1A	10	-	90
3A	30	-	70
5A	50	-	50
1B	-	10	90
3B	-	30	70
5B	-	50	50

The substrate samples were flat-strips 30 x 100 x 2.5 mm in size with 2.5 mm sheet thickness. The first step before coating, surface cleaning, and preparation was precisely held to maintain good coat quality. The surfaces of samples were degreased with TCE (Trichloroethylene C2HCL3) for removing surface impurities, then roughened using a blasting procedure with Corundum 500 and blow pressure 5 bar. This surface preparation procedure helped to activate the surface and offered an increased surface area for full adhesion of the sprayed particles. The main spraying parameters employed are shown in Table 2.

Table 2 APS process parameters

Parameter	Value
Gun	F4 – MB
Gas	Ar /H_2
Main gas [Ar]	≈ 46 SLPM 1.3 bar
Secondary gas [$\mathbf{H_2}$]	≈ 8.1 SLPM 1.5 bar
Current	600 A 64 V
Spray distance	150 mm

After coating the specimens by the prepared mixtures using the APS technique, the samples were subjected to metallurgical, morphological, mechanical (micro-hardness), and surface texture (Porosity and roughness) investigation.

Standard procedures of specimen preparation for metallographic examination were employed according to the following sequence; the metallographic cross-section samples for coating structure evaluation were cut from the coated specimens using a wire cutting machine. Then, the specimens were mounted in conventional Bakelite hot mounting material and pre-ground on the water as a –lubricant. Silicon Carbide abrasive, grade "120" was used as abrasive media to remove about 0.5 mm of the material. Standard grinding and polishing operations were carefully applied for metallographic observations.

The phase composition of composite coating was determined using an X-ray Diffractometer (XRD) (PANanalytical, X'Pert PRO) having Kα radiation with generator settings of 30 kV and 40 mA. Diffraction data were collected over a 2θ range of 5° - 100°, with a step width of 0.02°. The morphologies of as-sprayed coatings were examined using a Secondary Electron Microscope model FEI (Inspect S 50-Netherlands) with Energy Dispersive X-Ray Spectroscopy EDS attachment (Bruker AXS-Flash Detector 410-M, Germany). The microhardness of the as-sprayed coating was measured on polished surfaces using a standard (HV-1000) micro-hardness tester at a load of 300 g and a dwell period of 10 s. For each specimen, hardness tests were repeated ten times and the average value was recorded. The porosity percentage in the coatings was measured by using ImageJ analyzer software applying the thresholding procedure. Surface roughness was measured using (TR 110) tester with five readings for each sample at least while recording the average values.

3. Result and Discussion

3.1 *Powder microstructural analysis*

Figure 2a, b shows the SEM micrographs of the mixtures of Ni/Al (80/20) and Ni5Al (95/5) with pure Aluminum, respectively. All powders have uneven morphologies, elongated straight, twisted, or spheroidal shapes, and irregular particle sizes. The particle size of pure Aluminum powder is 22 to 84 μm, while particle size of as-received Nickel-based powders Ni/Al (80/20) and Ni5Al (95/5) are 45 to 88 μm, and 50 to 110 μm, respectively. This diversity of Aluminum particle size affects directly the composition of coating and the formation of intermetallic compounds [7].

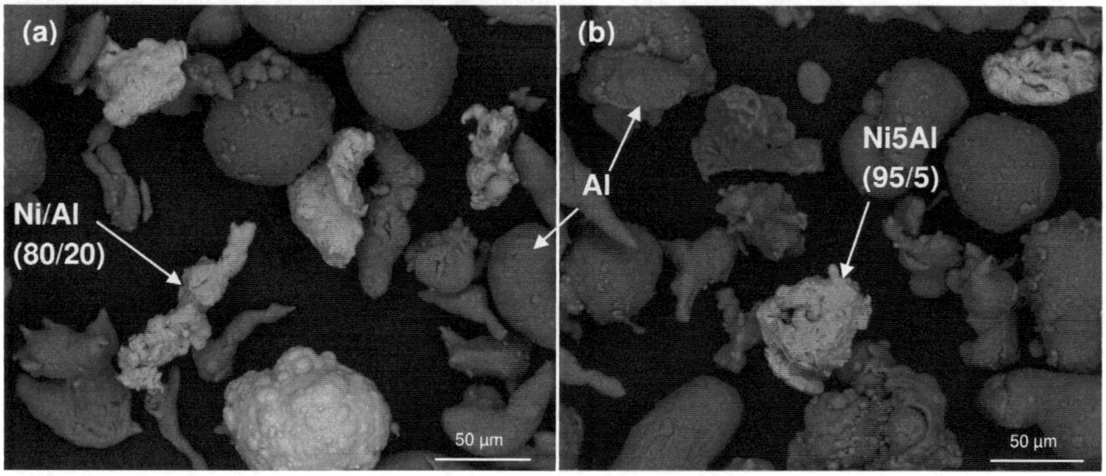

Figure 1SEM micrographs of (Metco™54-NS-1) Al powders with (a) Metco™404-NS, (b) Metco™450-NS

3.2 *XRD analysis of coatings*

Surveying previous studies have highlighted some important points to be considered before coating analysis. According to [7], the particle size of alumni affects directly the composition of coatings. NiAl formation is preferred with the aluminum size distribution 75 -100 m, whereas Ni3Al formation is preferred with small particles in the mixtures. Hussain in [37] stated that the ordered phases NiAl and Ni3Al appear at 50 and 75 atomic percent of Ni respectively based on the equilibrium phases diagram of Al-Ni.

Figure 3. shows the X-Ray Diffraction patterns of the coatings as deposited on stainless steel substrates. By comparing all the previously stated studies with the XRD analysis shown in Figure 2, two intermetallic compounds NiAl and Ni3Al appeared (most common phases in the phase equilibrium diagram of Ni-Al). The predominant phase was NiAl β-phase [23], which is found in all samples, even in the samples with the lowest Ni percentage (samples: 1A and 1B), despite the above-mentioned fact that states NiAl formation in the range of atomic percentage of Ni which is 50, sssssssssssin aligning with the same research criteria of applying Ni5Al and Ni/Al powders by APS, the XRD patterns of Rizaee et al. [20] and Q. Jia et al. [45], respectively, revealed the formation of NiAl. Nevertheless, [20], [45] have stated the absence of Ni3Al ϒ-phase, which appears in samples 5A and 5B only. In these particular samples, the Ni-based powders are reached 50% of the coating.

Pure Al and Ni appeared in all patterns, whereas Al could be considered the coating matrix, while the intensities of Ni increased directly with the increment of Ni-based alloys in the mixture. XRD patterns shown in Fig. 2.a. indicate the absence of Al2O3 for samples with Ni-5Alpowders as indicated by Rizaee [20] in his research of Ni5Al by APS, whereas samples with Ni-Al powder in Fig. 2.b. shows the same result of Al2O3 absence, except in sample 5A at 2θ~66.8°, similar results were also illustrated by Kubatik research [17] with Ni10Al and Ni40Al by APS. It has been observed that NiO appears with the increase of the Ni-based percentage in samples 3B, 5A, and 5B. This mixture of metal and metal oxide could be harder than the metal coating itself but to a certain limit as stated in [2], [32].

3.3 Cross-sectional characterization of coatings

Figure 3 shows cross-sectional microstructures of all samples that exhibit some typical APS features similar to other studies [17], [31], [32], [46]. The coatings contain lamellar or layered splat structures (as a consequence of molten particles' impact on the substrate), entrapped unmelted particles, micro-pores, and oxides inclusions. Abu-Warda et al. [10], developed an Al2O3-30(Ni20Al) layer on 304 stainless steel with High-Velocity-Oxy-Fuel (HVOF) technique; the coating morphology was characterized by an irregular surface with quite similar morphology to our deposited layers with APS technique.

Aluminum can be considered as the coating matrix for all samples, the existence of lamellar light gray splats can be overseen. It increased gradually by increasing the content of the Ni-based compositions that reached its highest percentage in samples (5A and 5B). Oxides (Al2O3 and NiO) are generally seen as dark, elongated phases that appear as strings in the coating cross-section, parallel to the substrate. Neither micro-cracks nor large pores exist in all samples. No evidence of micro-cracks to be observed, but almost all samples show a small number of micro-pores. Generally, increasing the percentage of Nickel-based alloys enhances the uniformity of coating. Coatings of samples 1B, 3B, and 5B, Ni-based alloys were denser and bulkier, while they tended to be thinner and more flattened. All deposited APS layers had an average thickness of 350 ± 19 μm.

3.4 *Physical and mechanical properties of coatings*

Porosity and surface roughness of coating are connected, and both could be affected by different parameters, characteristics of powder, coating compositions, spraying distance, gun, or even substrate temperature [7]. Figure 4 shows that, with 15 cm spraying distance and different compositions of Ni-Al with pure Al, all samples had a normal range of porosity [47], although sample 5A has the maximum porosity percentage (7.443 %), It is still within the normal range from 40 to <1% [2]. Other literature has stated a similar porosity range. Javadi et al. [25] stated that the optimum spray distance for Ni-Al powders is 11 cm which resulted in 4.5% porosity of coating volume, Wang et al. [40] reported a low porosity percentage resulted for NiAl10. Nonetheless, Kubatík et al. [17] produced more porous phases of NiAl10 and NiAl40 with 22 % and 17 %, respectively.

Moreover, Figure 4 displays surface roughness which was approximately 6.242 ± 0.689 μm for all samples, where they are quite similar to the commercial alloy coatings [7].

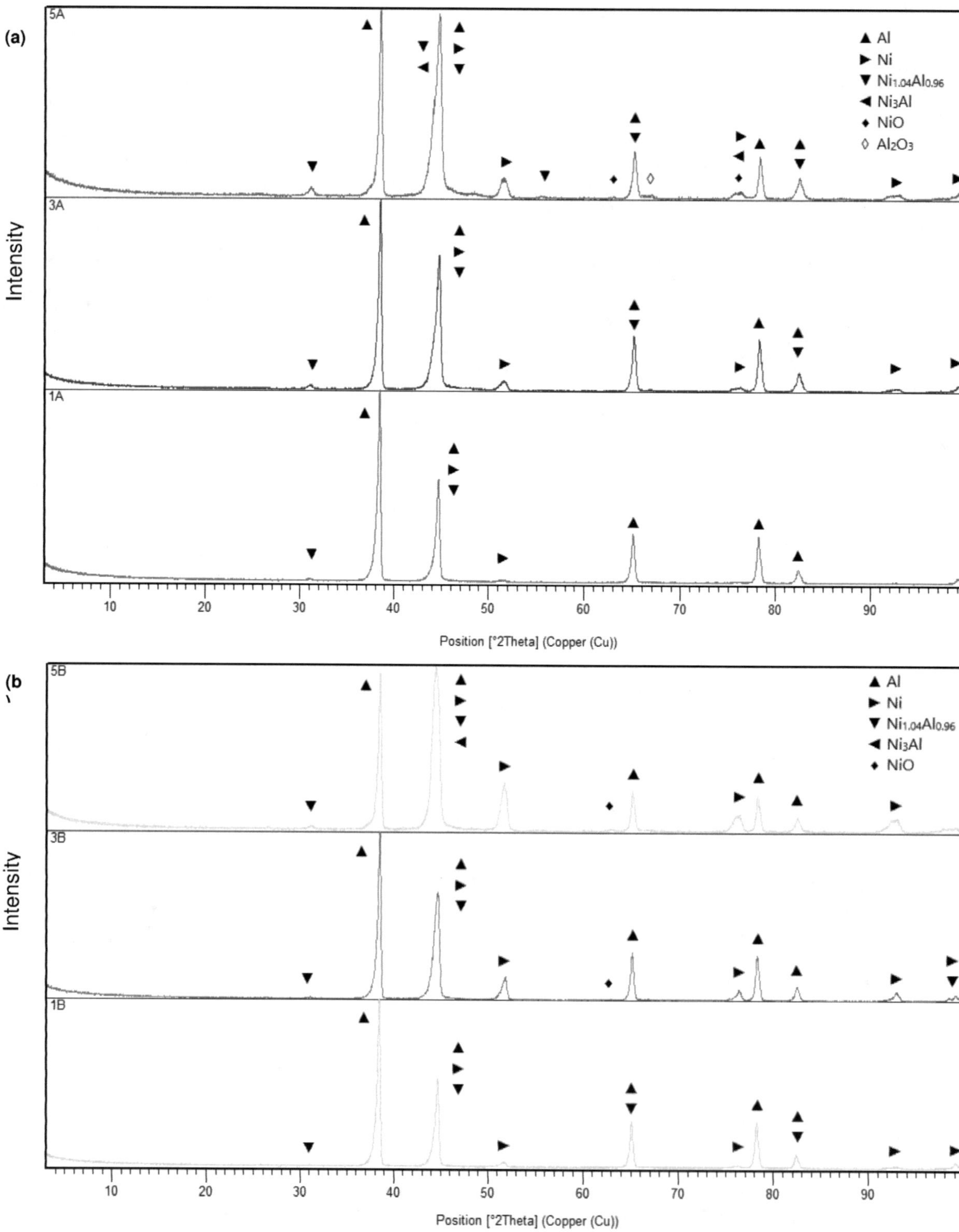

Figure 2 XRD patterns for coatings, (a) 1A, 3A, 5A; (b) 1B, 3B, 5B

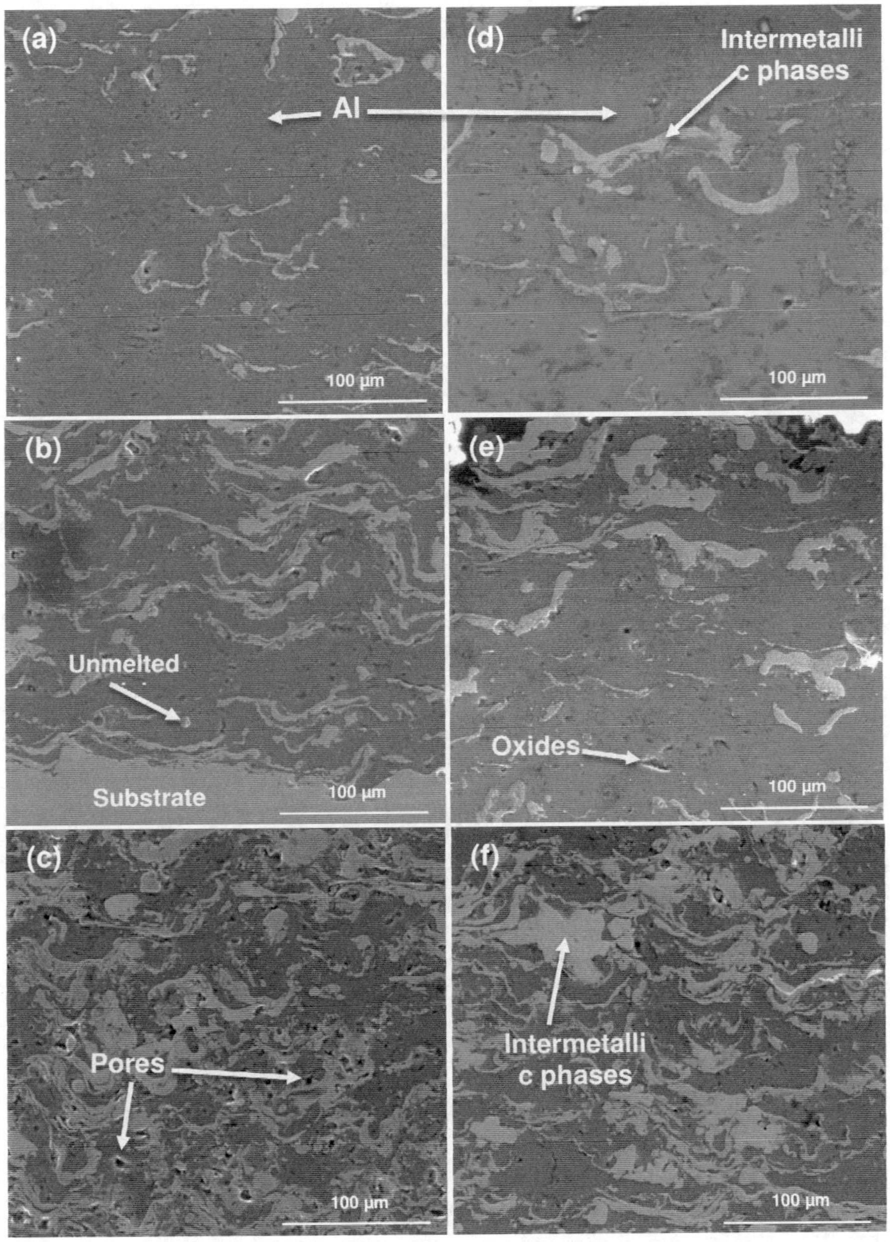

Figure 3 SEM micrographs of as-sprayed samples; (a) 1A, (b) 3A, (c) 5A, (d) 1B, (e) 3B and (f) 5B

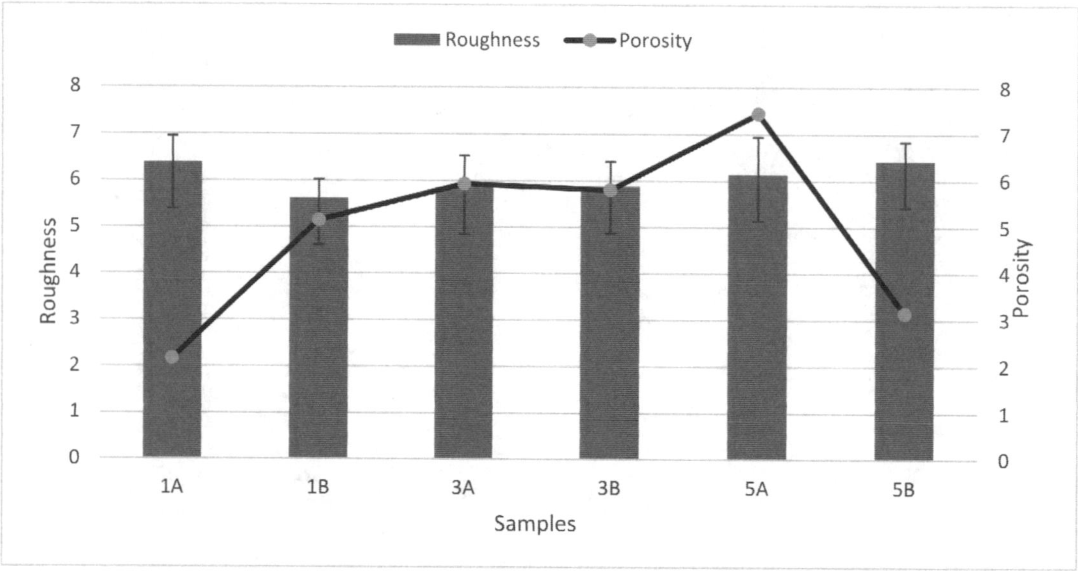

Figure 4 Coating properties; Porosity and Surface Roughness

Figure 5,a shows micrograph of microstructure of APS coating and substrate with indenter indentation. The cross-sectional zone of the substrate near the deposited layer has a noticeable color change, it could be attributed to the heat developed through the APS process which affects directly the physical characteristics. Moreover, microhardness values of the substrate in this interdiffusion zone (50 µm alongside coating) increase from 125 to 163 ± 8 HV as shown in Table 3, which is considered as proof positive for good interaction between substrate and coatings layers. Microhardness measurements of all samples are listed in Table 3 Both values of substrate and coating layers are included in Figure 5,b.Despite the increment in microhardness of this zone of interdiffusion, all samples exhibit different behavior, which is affected directly by the percentage of nickel [48]. Microhardness values of samples 1A and 1B are slightly higher than commercially pure Aluminum coating [49]. Therefore, the microhardness of both mixtures with the amount of Ni-based alloys below 30% is not a recommended type of as-sprayed coating for an application that requires adequate mechanical performance. By increasing the amount of Nickel-based alloys above 30 %, microhardness values reached an average of 164.5 HV with a standard deviation of 10.31 HV [21], [50]. It could be observed that the microhardness of the six different coatings was enhanced obviously by increasing the amount of Ni-based alloys in mixtures above 30%.

Table 3 Microhardness measurements of coatings and substrate

Sample No.	Interdiffusion Area (Substrate)	Coating layer
1A	155	74
1B	165	81
3A	158	149
3B	163	165
5A	167	167
5B	170	178

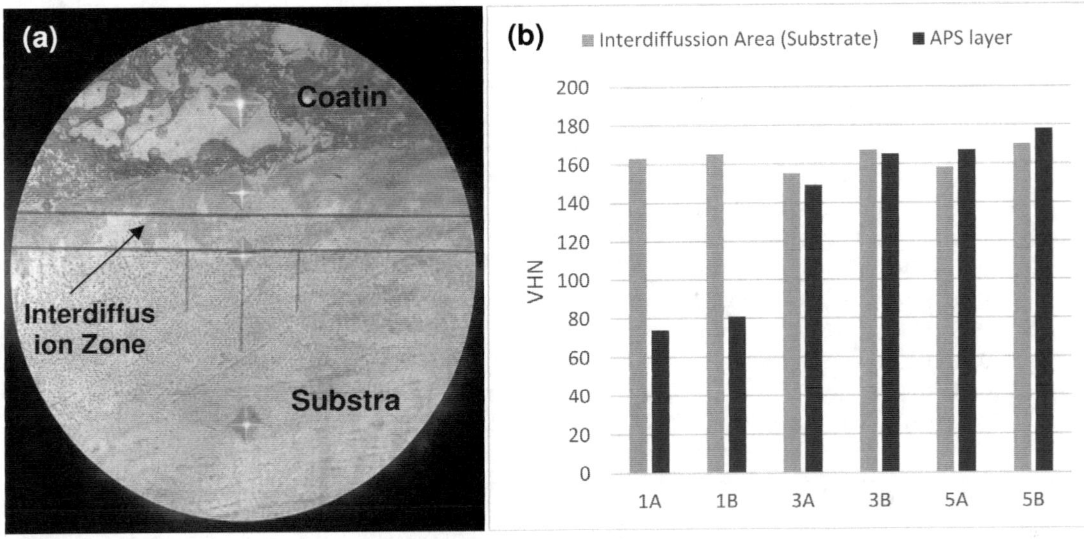

Figure 5(a) Details of the microstructures of both APS layer and substrate with indenter imprint after measurement. (b) Microhardness values for substrate's interdiffusion area and APS layers.

4. Conclusion:

This work has investigated the coating of stainless steel 304 by APS using Al, Ni/Al, and Ni5Al. Several observations were recorded as follows: Deposition of Aluminum powder separately mixed with Ni-Al and Ni 5Al results in a dense coat with no cracks or other major defects. The sample of Ni/Al and pure Al(5A) with 50% each has resulted in the highest porosity percentage which is still in the normal range stated in previous literature. Microhardness of the substrate (interdiffusion zone, 50 μm alongside coating) extensively increase compared to the average value of 304 SST microhardness. This is due to heat developed in the APS process. The measured microhardness of proposed coatings is affected directly by the percentage of Ni-based alloys in mixtures. Sample with Ni-based alloys in mixtures less than 30% have shown poor results. Formation of intermetallic compounds such as NiAl is inevitable in all samples, despite the amount of Ni-based alloys in mixtures, or even the atomic percentage of Ni; besides the existence of Ni3Al depends only on increasing the percentage of the Ni-based alloy to 50 % percent in the mixtures.

References

[1] Fauchais, P.L., Heberlein, J.V.R. and Boulos, M.I. (2014). Thermal Spray Fundamentals. Boston, MA: Springer US.

[2] Tucker, R.C. ed., (2013). Thermal Spray Technology. ASM International.

[3] Sampath, S., Jiang, X.Y., Matejicek, J., Prchlik, L., Kulkarni, A. and Vaidya, A. (2004). Role of thermal spray processing method on the microstructure, residual stress and properties of coatings: an integrated study for Ni–5 wt.%Al bond coats. Materials Science and Engineering: A, **364(1-2)**, pp.216–231.

[4] Guo, R.Q., Zhang, C., Yang, Y., Peng, Y. and Liu, L. (2012). Corrosion and wear resistance of a Fe-based amorphous coating in underground environment. Intermetallics, **30**, pp.94–99.

[5] Gupta, M., Markocsan, N., Li, X.-H. and Peng, R.L. (2017). Improving the lifetime of suspension plasma sprayed thermal barrier coatings. Surface and Coatings Technology, **332**, pp.550–559.

[6] Wang, Y., Yang, Y. and Yan, M.F. (2007). Microstructures, hardness and erosion behavior of thermal sprayed and heat treated NiAl coatings with different ceria. Wear, **263(1-6)**, pp.371–

378.
[7] Xanthopoulou, G., Marinou, A., Vekinis, G., Lekatou, A. and Vardavoulias, M. (2014). Ni-Al and NiO-Al Composite Coatings by Combustion-Assisted Flame Spraying. Coatings, **4(2)**, pp.231–252.
[8] Lukauskaitė, R., Černašėjus, O., Škamat, J., Zabulionis, D., Stonys, R., Kalpokaitė-Dičkuvienė, R. and Antonovič, V. (2017). The effect of Al Mg substrate preparation on the adhesion strength of plasma sprayed Ni Al coatings. Surface and Coatings Technology, **316**, pp.93–103.
[9] Ang, A.S.M. and Berndt, C.C. (2014). A review of testing methods for thermal spray coatings. International Materials Reviews, **59(4)**, pp.179–223.
[10] Abu-warda, N., López, A.J., López, M.D. and Utrilla, M.V. (2019). High temperature corrosion and wear behavior of HVOF-sprayed coating of Al2O3-NiAl on AISI 304 stainless steel. Surface and Coatings Technology, **359**, pp.35–46.
[11] Sun, C., Hui, R., Qu, W. and Yick, S. (2009). Progress in corrosion resistant materials for supercritical water reactors. Corrosion Science, **51(11)**, pp.2508–2523.
[12] Li, Y., Wang, S., Sun, P., Xu, D., Ren, M., Guo, Y. and Lin, G. (2017). Early oxidation mechanism of austenitic stainless steel TP347H in supercritical water. Corrosion Science, **128**, pp.241–252.
[13] Rodriguez, D. and Chidambaram, D. (2015). Oxidation of stainless steel 316 and Nitronic 50 in supercritical and ultrasupercritical water. Applied Surface Science, **347**, pp.10–16.
[14] Subramanian, C. (2010). Wear properties of aluminium-based alloys. Surface Engineering of Light Alloys, pp.40–57.
[15] Khandanjou, Sh., Ghoranneviss, M. and Saviz, Sh. (2017). The detailed analysis of the spray time effects of the aluminium coating using self-generated atmospheric plasma spray system on the microstructure and corrosion behaviour. Results in Physics, **7**, pp.1440–1445.
[16] Kadir, NF., Manap, A., Satgunam, M. and Afandi, N.M. (2018). Review on Nickel Aluminide based Bond Coat Properties and Oxidation Performance for Thermal Barrier Coating (TBC) Application. International Journal of Engineering & Technology, **7(4.35)**, p.624.
[17] Kubatík, T.F., Lukáč, F., Stoulil, J., Ctibor, P., Průša, F. and Stehlíková, K. (2017). Preparation and properties of plasma sprayed NiAl10 and NiAl40 coatings on AZ91 substrate. Surface and Coatings Technology, **319**, pp.145–154.
[18] Orban, R.L., Lucaci, M., Rosso, M. and Grande, M.A. (2007). NiAl Oxidation and Corrosion Resistant Coatings Obtained by Thermal Spraying. Advanced Materials Research, **23**, pp.273–276.
[19] Properties and Selection: Nonferrous Alloys and Special-Purpose Materials. (1990). Asm International.
[20] Rezaee Hajideh, M., Farahani, M., Pakravan, M. and Shahmirzalo, A. (2019). Corrosion resistance and hydrophilic properties of plasma sprayed Ni+5%Al coatings. Heliyon, **5(6)**, p.e01920.
[21] Wang, X., Feng, X., Lu, C., Yi, G., Jia, J. and Li, H. (2018). Mechanical and tribological properties of plasma sprayed NiAl composite coatings with addition of nanostructured TiO2/Bi2O3. Surface and Coatings Technology, **349**, pp.157–165.
[22] Tao, X.P., Zhang, S., Zhang, C.H., Wu, C.L., Chen, J. and Abdullah, A.O. (2018). Effect of Fe and Ni contents on microstructure and wear resistance of aluminum bronze coatings on 316 stainless steel by laser cladding. Surface and Coatings Technology, **342**, pp.76–84.
[23] Fan, X., Zhu, L. and Huang, W. (2017). Investigation of NiAl intermetallic compound as bond coat for thermal barrier coatings on Mg alloy. Journal of Alloys and Compounds, **729**, pp.617–626.
[24] Han, Y., Yang, K., Jing, P., Xue, B. and Ma, W. (2018). Mechanical and tribological properties of NiAl/muscovite composites. Journal of Alloys and Compounds, **741**, pp.765–774.
[25] Javadi, M.M., Edris, H. and Salehi, M. (2011). Plasma Sprayed NiAl Intermetallic Coating Produced with Mechanically Alloyed Powder. Journal of Materials Science & Technology,

27(9), pp.816–820.

[26] Bochenek, K. and Basista, M. (2015). Advances in processing of NiAl intermetallic alloys and composites for high temperature aerospace applications. Progress in Aerospace Sciences, **79**, pp.136–146.

[27] Li, B., Gao, Y., Han, M., Guo, H., Jia, J., Wang, W. and Deng, H. (2017). Tribological properties of NiAl matrix composite coatings synthesized by plasma spraying method. Journal of Materials Research, **32(9)**, pp.1674–1681.

[28] Wu, Q., Li, S., Ma, Y. and Gong, S. (2013). Study on behavior of NiAl coating with different Ni/Al ratios. Vacuum, **93**, pp.37–44.

[29] Li, B., Jia, J., Gao, Y., Han, M. and Wang, W. (2017). Microstructural and tribological characterization of NiAl matrix self-lubricating composite coatings by atmospheric plasma spraying. Tribology International, **109**, pp.563–570.

[30] Poliarus, O., Morgiel, J., Umanskyi, O., Pomorska, M., Bobrowski, P., Szczerba, M.J. and Kostenko, O. (2019). Microstructure and wear of thermal sprayed composite NiAl-based coatings. Archives of Civil and Mechanical Engineering, **19(4)**, pp.1095–1103.

[31] Fan, X., Huang, W., Zhou, X. and Zou, B. (2020). Preparation and characterization of NiAl–TiC–TiB2 intermetallic matrix composite coatings by atmospheric plasma spraying of SHS powders. Ceramics International, **46(8)**, pp.10512–10520.

[32] Shi, P., Wan, S., Yi, G., Sun, H., Yu, Y., Xie, E., Wang, Q., Shen, S.Z. and Alam, N. (2020). TiO2–ZnO/Ni–5wt.%Al composite coatings on GH4169 superalloys by atmospheric plasma spray techniques and theirs elevated-temperature tribological behavior. Ceramics International, **46(9)**, pp.13527–13538.

[33] Mahesh, R.A., Jayaganthan, R. and Prakash, S. (2008). A study on hot corrosion behaviour of Ni–5Al coatings on Ni- and Fe-based superalloys in an aggressive environment at 900°C. Journal of Alloys and Compounds, **460(1-2)**, pp.220–231.

[34] La, P., Bai, M., Xue, Q. and Liu, W. (1999). A study of Ni3Al coating on carbon steel surface via the SHS casting route. Surface and Coatings Technology, **113(1-2)**, pp.44–51.

[35] Marinou, A., Xanthopoulou, G., Vekinis, G., Lekatou, A. and Vardavoulias, M. (2015). Synthesis and heat treatment of sprayed high-temperature NiAl–Ni3Al coatings by in-flight combustion synthesis (CAFSY). International Journal of Self-Propagating High-Temperature Synthesis, **24(4)**, pp.192–202.

[36] Mishra, S.B., Chandra, K., Prakash, S. and Venkataraman, B. (2005). Characterisation and erosion behaviour of a plasma sprayed Ni3Al coating on a Fe-based superalloy. Materials Letters, **59(28)**, pp.3694–3698.

[37] A. Hussain, M. A. Choudhry, and S. S. Hayat. (2009). Effects of ordering on the thermal properties of an Ni3Al intermetallic alloy system: A molecular dynamics approach. Chinese J. Phys., **vol. 47, no. 3**, pp. 344–354.

[38] M. I. K, S. S. Havaldar, and A. Hiriyannaiah. (2021). Materials Today : Proceedings Composition optimization for NiAl + Al 2 O 3 + CeO composite coating on bearing steel by air plasma spray. Mater. Today Proc, pp. 2–7.

[39] Bhosale, D.G. and Rathod, W.S. (2020). Investigation on wear behaviour of SS 316L, atmospheric plasma and high velocity oxy-fuel sprayed WC-Cr3C2-Ni coatings for fracturing tools. Surface and Coatings Technology, **390**, p.125679.

[40] Wang, C., Gao, P., Liu, T., Li, L., Wang, E. and Wang, Q. (2019). Microstructure and wear performance of Ni-10 wt.%Al coatings plasma sprayed on Ni-based superalloys with a sound field. Surface and Coatings Technology, **370**, pp.157–162.

[41] Esmaeili, Z., Loghman-Estarki, M.R., Ramezani, M., Naderi, M. and Mohammad Sharifi, E. (2020). Toward hardening of NiCrAlY alloy by spark plasma sintering of NiCrAlY-nanoSi3N4-graphite nanocomposite. Journal of Alloys and Compounds, **847**, p.155802.

[42] Ma, Y., Li, W., Guo, M., Yang, Y., Cui, Y., Sun, W., Dong, Y. and Yan, D. (2021). TiC-TiSi2-Al2O3 composite coatings prepared by spray drying, heat treatment and plasma

spraying. Journal of Alloys and Compounds, **857,** p.158221.

[43] Beyhaghi, M., Kiani-Rashid, A.-R., Kashefi, M., Khaki, J.V. and Jonsson, S. (2015). Effect of powder reactivity on fabrication and properties of NiAl/Al2O3 composite coated on cast iron using spark plasma sintering. Applied Surface Science, **344**, pp.1–8.

[44] Yu, Y., Zhou, J., Ren, S., Wang, L., Xin, B. and Cao, S. (2016). Tribological properties of laser cladding NiAl intermetallic compound coatings at elevated temperatures. Tribology International, **104**, pp.321–327.

[45] Jia, Q., Li, D., Li, S., Zhang, Z. and Zhang, N. (2018). High-Temperature Oxidation Resistance of NiAl Intermetallic Formed In Situ by Thermal Spraying. Coatings, **8(8)**, p.292.

[46] Rukhande, S.W. and Rathod, W.S. (2020). An isothermal oxidation behaviour of atmospheric plasma and high-velocity oxy-fuel sprayed nickel-based coating. Ceramics International, **46(11)**, pp.18498–18506.

[47] Pawlowski, L. (2008). The Science and Engineering of Thermal Spray Coatings.

[48] Yamanoglu, R., Karakulak, E., Zeren, M. and Koç, F.G. (2013). Effect of nickel on microstructure and wear behaviour of pure aluminium against steel and alumina counterfaces. International Journal of Cast Metals Research, **26(5)**, pp.289–295.

[49] Kubatík, T.F., Ctibor, P., Mušálek, R. and Janata, M. (2017). Mechanical properties of plasma-sprayed layers of aluminium and aluminium alloy on AZ 91. Materiali in tehnologije, **51(2)**, pp.323–327.

[50] Gautam, R.K.S., Rao, U.S. and Tyagi, R. (2019). High temperature tribological properties of Ni-based self-lubricating coatings deposited by atmospheric plasma spray. Surface and Coatings Technology, **372**, pp.390–398.

Studying the distribution of hardness values in friction drilled 7075 Al-alloy sheets at different conditions

N Hamzawy [1], M Khedr [1], T S Mahmoud [1], I El-Mahallawi [2] and T A Khalifa [1]
[1] Mechanical Engineering department, Faculty of Engineering at Shoubra, Benha University, Cairo 11629, Egypt.
[2] Department of Metallurgical Engineering, Faculty of Engineering, Cairo University, Cairo 12613, Egypt.

E-mail: Mahmoud.khedr.mk@gmail.com

Abstract. The thickness of sheet metal parts can be locally increased by friction drilling technology via forming of a hole with a bush by a special drilling tool. Here, a 7075 Al-alloy was drilled by friction using tool cone angles with values of 40, 45 and 50^0 under different feed rates (100, 200 and 315 mm/min) and rotational speeds (1000, 1250 and 1600 rpm). The present study investigates the hardness distribution in the thermally-formed bush and in the heat-affected zone around the bush. It was found that the hardness of the bush was slightly increased with increasing of the tool cone angle and reduction of the tool rotational speed. However, the hardness of the thermally-induced bush showed values lower than the parent metal. The hardness near the drilling surface was approximately 65±10 HV, while it recorded hardness values of 75±10 HV at 5 mm away from the drilling surface. In addition, the microstructure of the friction drilled specimens showed a very fine structure in the drilling zone due to crushing of the original structure during the friction drilling process.

1. Introduction

EN AW 7075 (Al-Zn-Mg-Cu) high strength alloy is used in many industrial applications such as aerospace components, bike parts, and building construction elements which require excellent mechanical performance. However, this alloy has a poor weldability, which limits the choices of joining methods available for 7075 Al-alloys assembly [1].

Thermal drilling is a non-traditional hole-making method used in manufacturing of thin sheet metals. It creates holes with extra height more than the base metal without generation of chips via a non-traditional tool-drill. Hence, it is a suitable technology to produce a clean hole with high efficiency and good surface quality [2, 3].

The presence of a bushing (the extra height) in thin sheets can be further threaded; therefore, temporarily joining of the thermally drilled specimens can be performed using studs or bolts [4-6]. The feed rate and spindle speeds of the tool are the most vital factors during the friction drilling process. The thermal drilling parameters directly affect the axial thrust force and torque values during the drilling process [7, 8].

Friction drilling begins when the rotary tool contacts with the work piece. Under the influence of the rotating tool, friction is induced which increases the temperature of the sheet material in contact with the drilling tool. The temperature reaches 750 °C and 380 °C during thermal drilling of austenitic stainless steels and Al-alloys respectively [9,10]. The induced temperature softens the material in front

Content from this work may be used under the terms of the Creative Commons Attribution 3.0 licence. Any further distribution of this work must maintain attribution to the author(s) and the title of the work, journal citation and DOI.
Published under licence by IOP Publishing Ltd

of the tool drill. The tool has a conical geometry that gradually expands the diameter of the hole and lengthens the molded bushing. The geometry of the tool affects the shape and size of the bushing [5].

On one hand, 7XXX Al-alloy series are heat-treated via T6 treatment; it produces spherodized precipitates of eutectic Si, in addition to existence of Cu-, Fe- and Mn- intermetallic compound particles which enhance the hardness, strength and ductility of these alloy-series [11-13]. On the other hand, the friction drilling of 7075 Al-alloy is expected to create thermal conditions that may affect the heat treatment influence.

The aim of the present study is to investigate the distribution of hardness values in friction drilled 7075 Al-alloy sheets at different working conditions via the manipulation of the rotational speeds, feed rates and tool cone angles.

2. Experimental setup

2.1 Materials and Methods

The chemical composition of the investigated 7075-T6 Al-alloy is 5.6 wt.% Zn, 2.3 wt.% Mg, 1.4 wt.% Cu,0.5 wt.% Fe, 0.4 wt.% Si, 03 wt.% Mn and 89.5 wt.% Al. The thickness of the drilled specimens is 4 mm. It worth mentioning that, the melting temperature of the 7075 Al-alloy is approximately 635 °C. A vertical milling machine was used to perform the friction drilling processes of the specimens.

2.2. Drilling tool

The details of the friction drilling tool were discussed in our previous work [10]. The friction angle (is also defined as the tool cone angle) and the length of the conical region are referred as β and h_n respectively. In the present study, three different friction angles of 40°, 45° and 50° are considered. The drilling tool has a diameter d equals 8 mm and a cylindrical length h_l equals to 16 mm.

2.3. Drilling conditions

Table 1. Sample codes corresponding to the different drilling parameters values.

Sample no.	*R, rpm*	*F, mm/min*	*β, o*	*Sample no.*	*R, rpm*	*F, mm/min*	*β, o*	*Sample no.*	*R, rpm*	*F, mm/min*	*B, o*
S1	1000	100	40	**S10**	1250	100	40	**S19**	1600	100	40
S2	1000	100	45	**S11**	1250	100	45	**S20**	1600	100	45
S3	1000	100	50	**S12**	1250	100	50	**S21**	1600	100	50
S4	1000	200	40	**S13**	1250	200	40	**S22**	1600	200	40
S5	1000	200	45	**S14**	1250	200	45	**S23**	1600	200	45
S6	1000	200	50	**S15**	1250	200	50	**S24**	1600	200	50
S7	1000	315	40	**S16**	1250	315	40	**S25**	1600	315	40
S8	1000	315	45	**S17**	1250	315	45	**S26**	1600	315	45
S9	1000	315	50	**S18**	1250	315	50	**S27**	1600	315	50

The investigated friction drilling parameters imply the tool rotational speed R, feed rate F and tool cone angle β. The present investigation considers three values of each drilling parameter: R equals 1000, 1250 and 1600 rpm, F equals 100, 200 and 315 mm/min, and β equals 40°, 45° and 50°. The experiments were carried out based on three factors with three levels of full factorial experimental design. The developed design matrix produced 27 experiments as shown in table 1.

2.4 Vickers micro-hardness measurements.

Vickers micro-hardness test has been performed with an applied load of 100 gf at a holding time of 15 s according to the ASM standard E384 [14]. Hardness values were measured by using PHASE- II Vickers hardness tester. The average of three readings was recorded. Hardness was measured in numerous points in the thermally-formed bush (drilling zone referred as DZ) and in the heat-affected zone (HAZ) around the bush. Figure 1 displays a schematic drawing showing the HAZ, DZ, and the edges of the friction drilled hole.

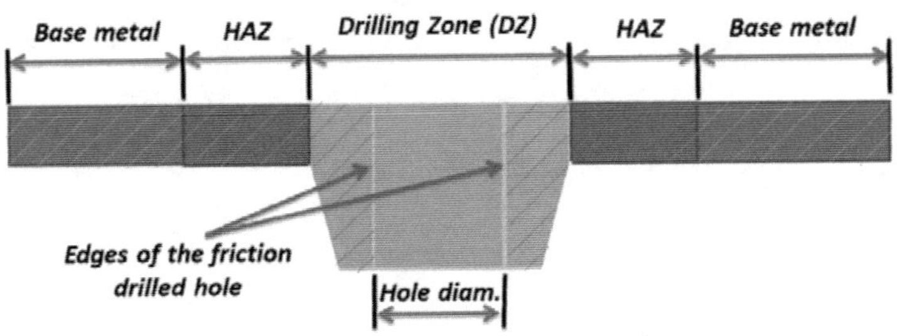

Figure 1. Schematic drawing showing a cross section after friction drilling processing at where base metal, HAZ and DZ are identified.

2.5. Microstructure analysis.

Samples for metallography were mounted, ground, polished and then etched for 5-15 s with Keller agent according to the ASTM standard E407 [15]. The microstructure was recorded with KERN optical microscope.

3. Results and discussion

3.1. Hardness results

Figure 2 shows the results of hardness values near the edge of the friction drilled bush and through a distance of 6 mm away from it with a step equals 1 mm in the 27 specimens. It is clear that, both of the DZ and the HAZ have hardness values lower than the base metal. Moreover, the hardness values increase with moving away from the edge of the induced bush.

Specimen S8 shows a maximum value of hardness (77.2 HV) near the edge of the friction drilled hole. While S21 shows a minimum value of hardness (57.3 HV) near the edge of the hole. It leads us to conclude that the hardness of the bush is a maximum when R equals 1000 rpm, F equals 315 mm/min and β equals 45°. However, the hardness of the bush is a minimum when R equals 1600 rpm, F equals 100 mm/min and β equals 50°.

On the other hand, the maximum temperature during friction drilling of S8 and S21 recorded 292 ^{0}C and 376 ^{0}C respectively [10]. It seems that the higher the induced temperature during the friction drilling process, the lower the hardness of the induced bush of the 7075 Al-alloy.

Figure 3 shows the evolution of hardness values at near the edges of the friction drilled holes at different rotational speeds, cone angles and feed rates. Clearly, when F equals 100 mm/min, the micro-hardness values increase with the increase of the cone angle from 40° to 45° respectively, while it decreased with the increase of the cone angle from 45^0 to 50^0 at 1000 rpm and 1600 rpm (figure 3.a). Moreover, when F equals 200 mm/min, the micro-hardness increased with the increase of the tool cone angles from 40° to 45°, then, it decreased with the increase of the tool cone angles from 45° to 50° (figure 3.b).

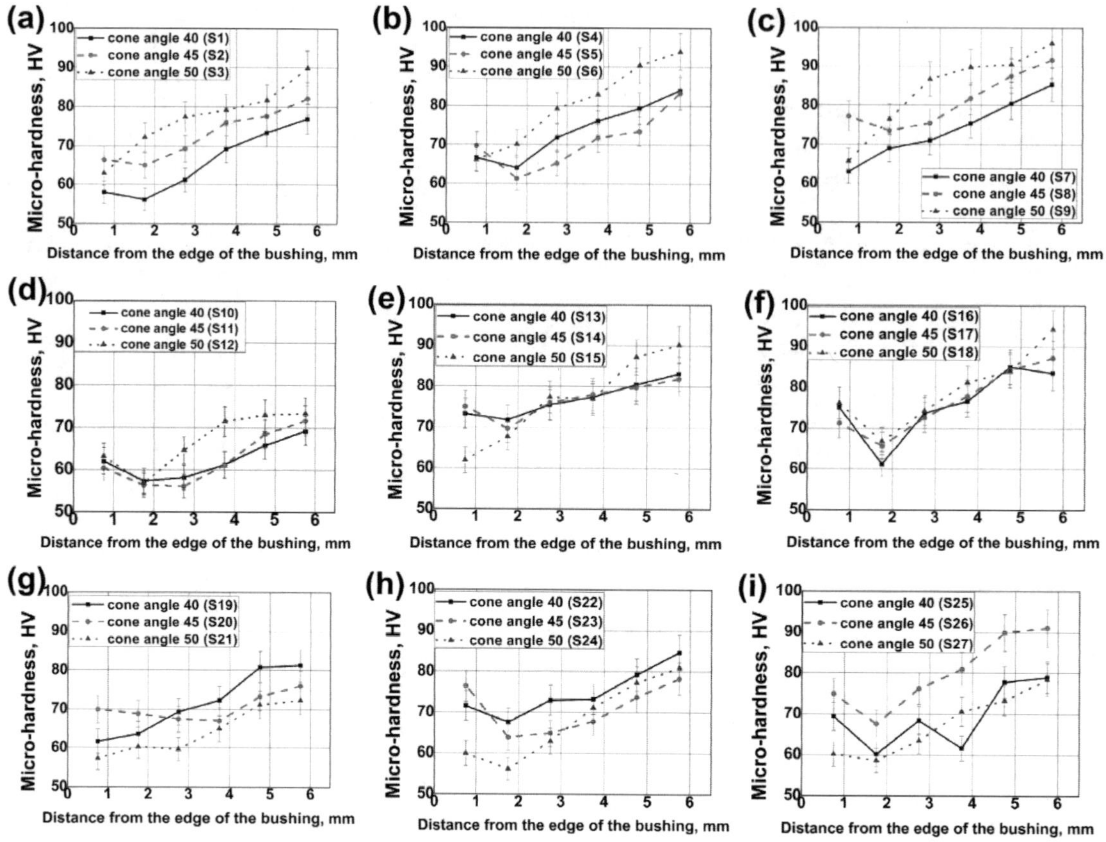

Figure 2. Micro-hardness values at different distances from the edge of the friction drilled bush under different friction drilling conditions: a) *R*=1000 rpm, *F*=100 mm/min, b) *R*=1000 rpm, *F*=200 mm/min, c) *R*=1000 rpm, *F*=315 mm/min, d) *R*=1250 rpm, *F*=100 mm/min, e) *R*=1250 rpm, *F*=200 mm/min, f) *R*=1250 rpm, *F*=315 mm/min, g) *R*=1600 rpm, *F*=100 mm/min, h) *R*=1600 rpm, *F*=200 mm/min, i) *R*=1600 rpm, *F*=315 mm/min,

Furthermore, at *F* equals 315 mm/min, the micro-hardness values increased with the increase of the tool cone angle from 40° to 45°, while it decreased with the increase of the cone angle from 45° to 50° at 1000 rpm and 1600 rpm. However, the micro-hardness values decreased with the increase of the the cone angle from 40° to 45°, while it increased with the increase of the cone angle from 45 1250 rpm (figure 3.c). The highest value of the micro-hardness recorded at tool cone angle of 45°.

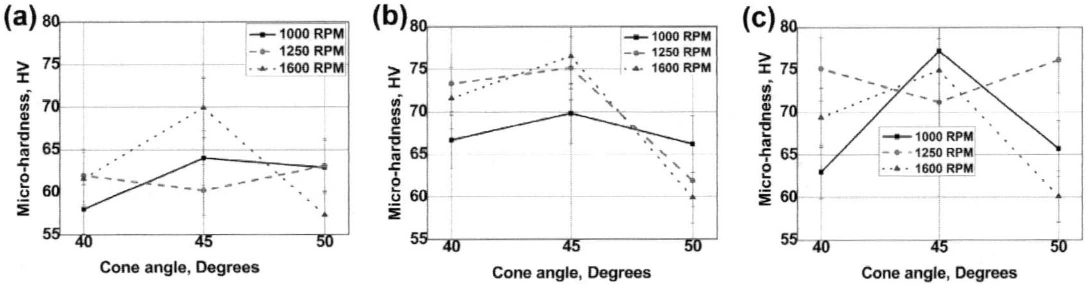

Figure 3. Hardness values at near the edges of the thermally induced bush under different tool rotational speeds and cone angles at different feed rates: a) F=100 mm/min, b) F=200 mm/min and c) F=315 mm/min.

3.2. Microstructure analysis

Figure 4 shows the microstructure at the edge of the induced bush and the as received condition before etching. The graphs show existence of intermetallic compound particles which mainly consists of Cu, Fe and Mn as reported in the literature [11]. It is clear the intermetallic particles are homogeneously distributed in the microstructure of the as-received condition, the average size of the particles is 8±5 µm. However, the microstructure of the induced bush shows refining of the intermetallic particles with average size of 1±0.5 µm. It leads us to conclude that the dissolution and/or fragmentation of the intermetallic particles at the drilling zone decreased the hardness of the formed bush and the HAZ in comparison with the base metal, as discussed above.

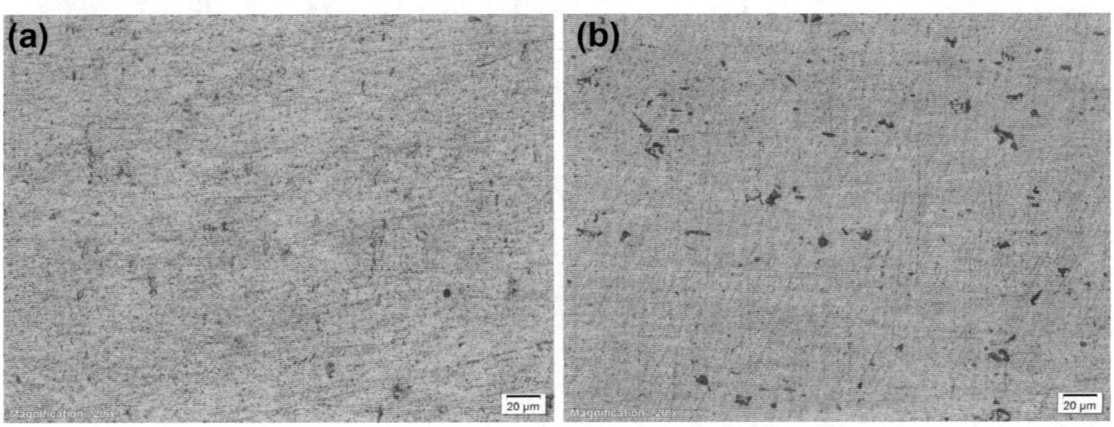

Figure 4. The microstructure of 7075 Al-alloy before etching at: (a) the induced bush, (b) the as-received condition.

Figure 5 displays the microstructures of the DZ, HAZ and the base metal after etching. It is apparent that the base metal has a grain size of approximately 50±15 µm. The grain size at the HAZ is 25±5 µm, while the grain size near the DZ is approximately 15±2 µm. The refinement of the grain size near the DZ is attributed to the dynamic recrystallization which occurs during the friction drilling process.

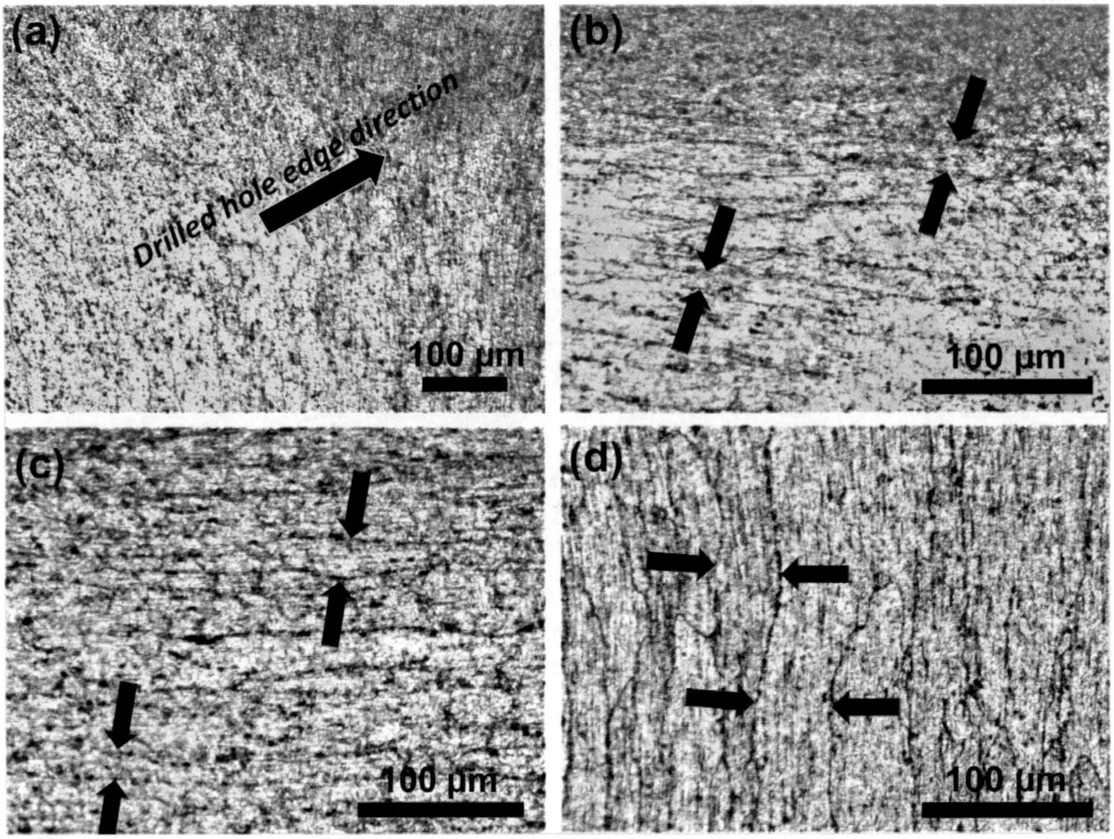

Figure 5. The microstructure of 7075 Al-Alloy after etching at: a), b) near the edge of the friction drilled specimen S21 at different magnifications, c) HAZ and d) base metal.

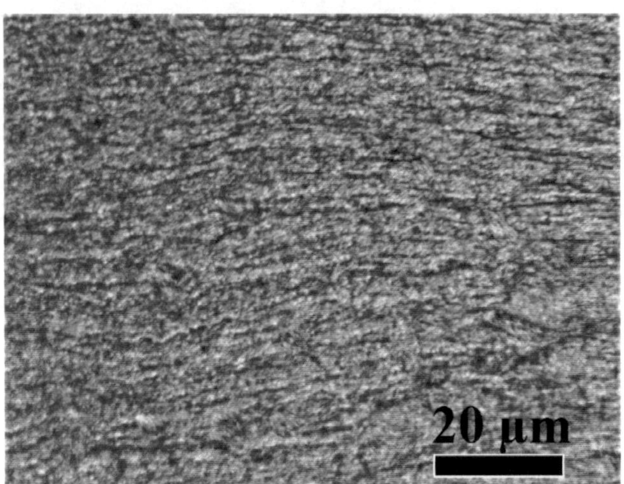

Figure 6. The microstructure at the edge of the induced bush of S21specimen.

Figure 6 exhibits the microstructure at the edge of the induced bush of S21. The grain size is approximately 2±0.5 μm. In spite of the grains refinement at the DZ and the HAZ, the hardness values of these zones are lower than the as-received condition. It seems that the existence of the intermetallic particles (figure 4.b) is the main reason behind the high values of hardness of the base metal.
The current results highlight the importance of the existing intermetallic particles in 7075 Al-alloy. The intermetallic particles are refined and/or crushed after friction drilling and consequently the friction drilled bush and the surrounding zone have hardness values lower than the base metal. On the other hand, we recommend performance of heat-treatments of 7075 Al-alloy after friction drilling processes in order to raise its hardness.

4. Conclusions
In the present study, the hardness values of thermal induced bushes of 7075 Al-alloy were investigated under different rotational speeds, feed rates and tool cone angles.

- The hardness value at the drilling zone was a maximum (77.2 HV) at R equals 1000 rpm, F equals 315 mm/min and β equals 45°. However, the hardness of the bush is a minimum (57.3 HV) when R equals 1600 rpm, F equals 100 mm/min and β equals 50°.
- The hardness values at the drilling zone and the heat affected zone are lower than that of the base metal.
- The dissolution/fragmentation of the intermetallic particles as well as material recrystallization associated with the friction drilling process are responsible for the reduction of the hardness values of the induced bush in spite of the refinement of the grain sizes at the drilling zone.

References:

[1] Peter I and Rosso M 2017 Sci. Bullet. Valahia Univ. Mater. Mech. 15 7.
[2] Özek C and Demir Z 2013 TEM J 2(1) 170.
[3] Yu K , Welo T and wang J 2018 Proc. Manuf. 26 663.
[4] Kaya M, Aktas A, Beylergil B and Akyildiz H 2014 Trans. Canad. Soc. Mech. Eng. 38(3) 319.
[5] McQueen H and Jonas J 1975 Treatise on Mater. Sci. Tech. 6 393.
[6] Shalamov P V and Kazantseva Ju V 2017 Proc Eng. 206 985.
[7] Miller S F, Li R, Wang H and Shih A J 2006 *J. Manuf. Sci. Eng*. 128(3) 802.
[8] Cavaliere P, Nobile R, Panella F W and Squillace A 2006, Int. J. Mach. Tools Manuf. 46, 588.
[9] Lee S M, Chow H M, Huang F Y and Yan B H 2009 Int. J. Mach. Tools Manuf. 49 81.
[10] Chen G, Chen Q, Wang B and Du Z 2015 Met. Mater. Int. 21(5) 897. doi: 10.1007/s12540-015-5139-6.
[11] Chen K, Zhang Z, Liu W, Song L and Huang L 2003 Cent. South. Univ. Tech. 34(2) 114.
[12] ping H, Kang C, Song L and Hong L 2005 Chinese J nonfer. Met. 15(2) 727.
[13] Hamzawy N, Khedr M, Mahmoud T S, ElMahallawi I, Khalifa T 2020 Min. Metal. Mater. Soc. (TMS) 471. doi: 10.1007/978-3-030-36408-3_67.
[14] ASTM Standard E384, Standard Test Method for Knoop and Vickers Hardness of Materials.
[15] ASTM Standard E407, Standard Practice for Micro-etching Metals and Alloys.

Low thermal emissivity coating based on aluminium / acrylic composite coatings

E E Ateia[1], M A Metwaly[1], H R Tantawy[2]

[1] Physics Department, Faculty of Science, Cairo University, Giza, Egypt
[2] Chemical Engineering Department, Military Technical College (M.T.C)

E-mail: cap.mahmoud2011@gmail.com

Abstract. Low emissivity coatings were synthesized by using fine flake Aluminium (Al) powder as a filler within acrylic resin to from the desired composite coatings. In the current work, ball milling was applied to prepare Al with different shapes and sizes. The technical parameters of prepared Al powders with respect to the initial raw Al are characterized by scanning electron microscope (SEM), X-Ray diffraction (XRD), and Energy Dispersive X-Ray Analysis (EDX). Moreover, parameters that affect the emissivity of the coating was investigated; such as coating thickness, particle size, spin coating, ball milling time and the content of coated Al powder. The thermal signature is highly affected by the variation in Al content (5%, 20%, 25%, 35%, 40% and 60%) at different temperatures (50°C, 70°C, 90°C). The results indicate that the perfect percentage for the filler (Al) in the matrix within the range (35 wt. - 40 wt. %) and Al fine flake powder particle, which gives the lowest infrared emissivity of 0.385 μm and 0.412 μm for (3-5) μm, (8-12) μm, respectively.

1. Introduction

Two main goals for the art of military camouflage. Firstly, concealing a possible target such as building, aircraft, tank, weapon emplacement. Secondly, simulating a false target or decoy to divert attention away from real targets [1]. Previously, camouflage nets or mats were used to conceal the target by covering it and were only intended to prevent visual target detection by providing an image that was dissimilar to that of the surrounding terrain. This Simple visual camouflage formed of a covering surface that was painted on and otherwise designed to look like a normal product or terrain surface while concealing the target underneath it [2, 3]. Minimizing external heat absorption and reducing intensity of infrared radiation emitted by military targets can be achieved by using Low-emissivity materials [4]. Low infrared emissivity coatings have attracted a lot of interest because of their military and civilian purposes, such as infrared camouflage, energy saving, and vehicles protection from infrared detection [5, 6]. The infrared Emissivity should be as low as possible in order to achieve excellent IR camouflage efficiency [4]. With the increasing development of infrared detection techniques and infrared guided weapons low infrared emissivity have been achieved by using different types of materials such as composite ceramic[7-9], bio-electret composites [10, 11], and metals such as silver and aluminum [12-15], AlSiN and Ag-alloy multilayer films [16]. In general, metallic pigments, polymer binder, organic solvent and additives are the mainly composition for preparing low infrared emissivity coatings [17-19]. While visual camouflage is still important, the production of other types of military sensing and visibility devices has complicated the requirement for concealment. Older types of camouflage can be ineffective against radar surveillance and interrogation systems which use electromagnetic wavelengths in the infrared and ultraviolet bands [20]. Infrared sensors sense heat (infrared) radiation generated by

Content from this work may be used under the terms of the Creative Commons Attribution 3.0 licence. Any further distribution of this work must maintain attribution to the author(s) and the title of the work, journal citation and DOI.
Published under licence by IOP Publishing Ltd

exhaust system and hot engine of the machines, this heat arises from unprotected and running military tank or truck. when the target concealment is the aim, understanding the terrain characteristics (e.g. snow, farmland, marsh, desert, woods etc.) is important if the risk of target detection is to be minimized [21]. The camouflage should reduce or remove contrast between the target and the surrounding terrain, eliminate transmission of energy (e.g. far infrared) emanating from the target, and reflect, scatter, or absorb incoming target illuminating energy beams (e.g. , laser coherent, sunlight, radiation, radar, etc.) in a manner that mimics the return or signature of the surrounding terrain [22].

2. Materials and Method

2.1 Materials.

Al powder (fine powder, purity 99.5%, 35μm) was purchased from Alpha Chemika (India). Solvents such as toluene (>99%) and ethanol (≥99.5%) were purchased from Al-Ahram Chemicals (Egypt). Commercial acrylic coat was purchased KAPCI coatings company (Egypt), it consists mainly from clear coat (Kapci 3100 2k MS) and hardener (kapci MS 651) with optimum mixing ratio (2:1). The resultant acrylic coat delivers high gloss, excellent finish, weathering resistance and offers high durability.

2.2 Preparation of controlled particle size Al Powder.

The high energy ball milling (Fritsch GmbH, Milling and Sizing, Industriestrasse8, 55743 Idar-Oberstein. Germany) was utilized for preparing of flake Al. The ratio between initial Al fine powder and milling balls during the preparation period was fixed at 1:15. Typically, 15g of Al powder with 225 gram balls were milled for 30 hours at steady speed of 350 rpm in a wet medium of 70 ml of toluene. The crucial rule of toluene was to prevent oxidation and agglomeration of powder. Moreover, milling was carried out interrupted cycle involving 15 min on and 30 min off for an effective time of 15 and 30 hours. The samples were taken at different times (0, 15, 30 hours) to inspect the impact of milling time on Al particle size and shape. Details of milling process were summarized in Table (1).

2.3 Preparation of Al/ acrylic composite coatings.

teel substrates (10 cm×10 cm, thickness 1.0 mm) were used to test the variations in the emissivity after the application of various coatings as shown in Fig. (1). All the substrates were pretreated with 10% dilute H_2SO_4 then distilled water at ambient temperature before applying the desired coating. Moreover, fixed amounts of polymer (clear coat, hardener) and Al powder were mixed together under continuous ultra-sonication for 10 min in order to prepare the Al/ acrylic composite coatings suspensions. The ratio between clear coat and hardener about (2:1) according to kapci coatings company regulations. Then the obtained coatings suspensions mixtures was painted onto the steel substrates via a spin coater technique (KW-4A, U.S.A.). Spin coater was adjusted at two speeds, low speed (speed 1) up to 2500 rpm and high speed (speed 2) up to 8000 rpm. Finally, the coatings were solidified completely after curing for 24 h at room temperature and kept for further characterization.

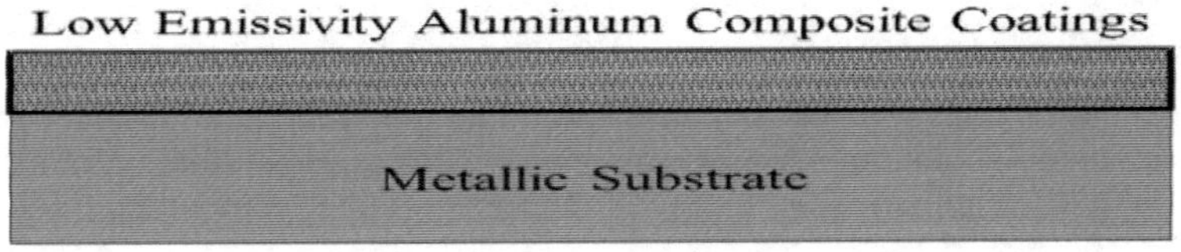

Figure 1. The schematic diagram of the low emissivity coating.

Table 1. Milling conditions and details of planetary ball mill

No	Parameter	Value
1	Rotation speed of vial (rpm)	350
2	Capacity of vial (ml)	500
3	Vial material	Zirconium oxide
4	Ball material	Zirconium oxide
5	Diameter of ball(mm)	10
6	Ball to powder weight ratio	15:1
7	Times of milling (h)	0-30
8	Types of milling	Wet
9	Process control agent	Toluene
10	Initial powder mass (g)	15

3. Characterizations

Morphological characterization and particle size analysis of the Al samples were carried out by scanning electron microscope (SEM, MIRA 3 XMU, TESCAN). The powder was suspended in ethanol. The suspension was ultrasonicated for 20 minutes and waiting until it dries. The dried samples were then mounted on to a SEM sample stub. X- Ray diffractometer (XRD, ADX 2500) with Cu Kα radiation source was utilized to examine the crystallinity and investigate the impact of milling process on crystallinity of Al. The Elcometer 456 (England) coating thickness gauge scan probe was used for drying film thickness. ET-10 (Surface Optics Corporation 410-0045 ET10, China) used to measure infrared emissivity values within spectral regions 3 -5 and 8 -12 μm. The thermal signature was captured using a fluke thermography (Model: Fluke Ti25 thermal imager, accuracy: ± 2 °C, Measurement temp range: -20 °C to +350 °C, Germany)

4. Results and Discussion

4.1 Morphological analysis.

The impact of milling time (0-30h) on the shape and size of the ball milled synthetized Al powders was investigated with SEM as shown Fig. (2). The SEM images represent the variation of the grain sizes with ball milling time. Typical initial Al powder with size around 35μm lateral length flakes as shown in Fig. (3-a). During milling initial Al powder loss its morphology and transformed to irregular-size particle. The observed particles size for 15h and 30h milling were 3μm and ~300nm, respectively as shown in Fig. (3-d) and (3-c). It is clear that the size of the samples is significantly affected by milling time. The amount highly flake Al, decreases with increasing the milling time. It is obvious that flake size decreases to the nano flake size (~300nm) with increasing milling time till 30h. Table (2) represents the variation of average grain size with ball milling time for Al powder.

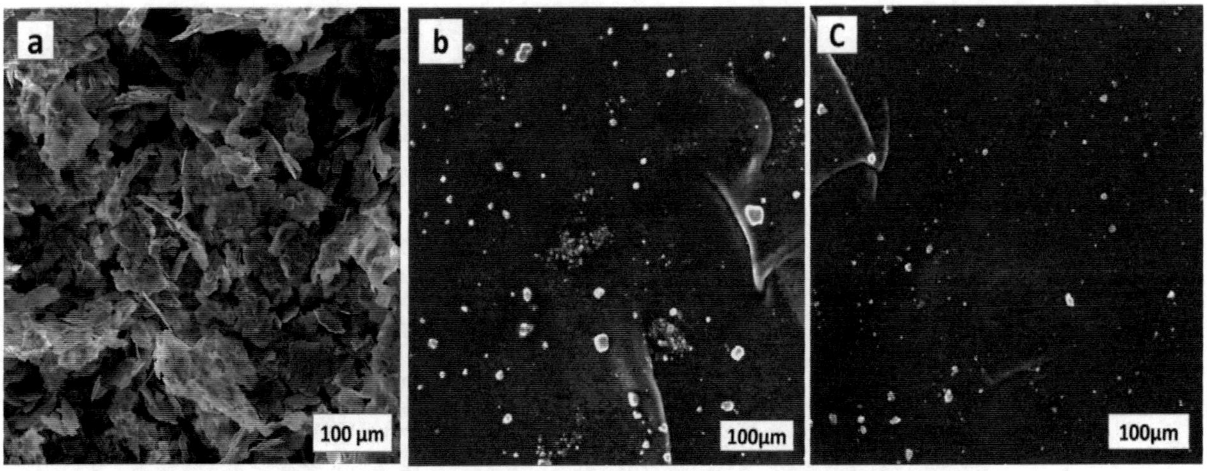

Fig 2. SEM images Al powder samples after milling 0h (a), 15h (b), and 30h (c)

Table 2. Variation of average grain size with ball milling time for Al powder

Ball milling time (h)	Grain size (μm) from SEM
0	35
15	3
30	0.3

4.2 *Energy Dispersive X-ray (EDX) analysis.*

EDX analysis was carried out to examine the degree of oxidation of the obtained samples. EDX analysis of the prepared samples for Al powder is shown in Fig. (3). It elucidates the impact of milling time on the oxidation degree of the samples. Inset of figures shows the relative abundance, between Aluminum and oxygen as a function of the milling time. It is obviously recorded that the oxygen content increases with increasing milling duration. The formation of Aluminum oxides (Al_2O_3) during milling is expected. It is found that by increasing milling time from 0h to 30h. The oxygen content increases from 11% to 39%. In this case, the highest increase in oxygen content is obtained at milling time 30h.

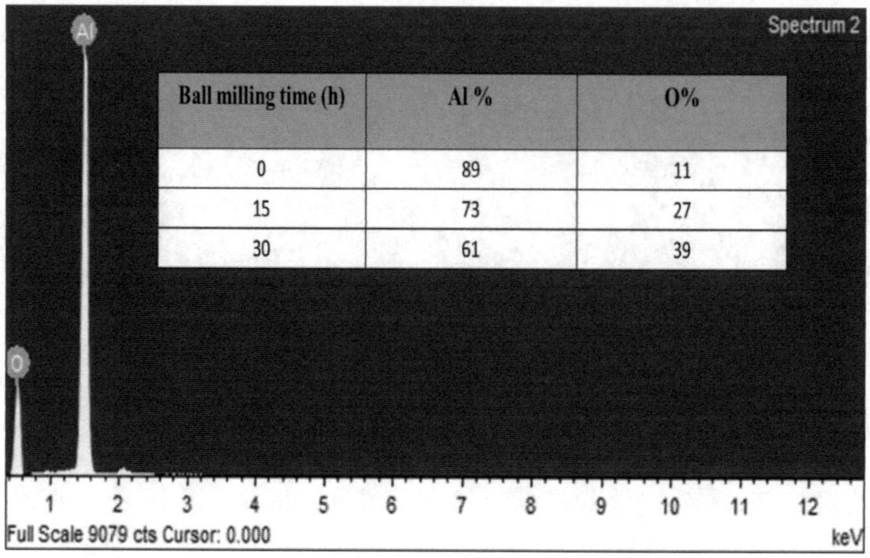

Ball milling time (h)	Al %	0%
0	89	11
15	73	27
30	61	39

Figure 3. EDX of Al powder samples after milling 0h, 15h, and 30h

4.3 *X-ray analysis.*

Figure. (4) Shows the XRD patterns of the ball milled synthesized Al flake samples compared to the starting raw of Al. The sharp diffraction peaks are detected at 2θ values of 38.5o, 44.7o, 65.1o and 78.2o corresponding to (111), (200), (220) and (311) planes respectively. The obtained structure fully matches with diffraction file (JCPDS card No. 99-0005) [23, 24]. Additionally, a diffraction peak appears at 30.4o, 33.6o and 55. 9o corresponding to aluminum oxide (Al_2O_3). The obtained data shows that the increase of the milling time increases the percentage of the aluminum oxide (Al_2O_3) which is clearly detected. The appearance of aluminum oxide (Al_2O_3) starts after 15h of milling. The high increase in the percentage of the aluminum oxide (Al_2O_3) with increasing milling time from 15h to 30h. The obtained data is in a good agreement with EDX analysis.

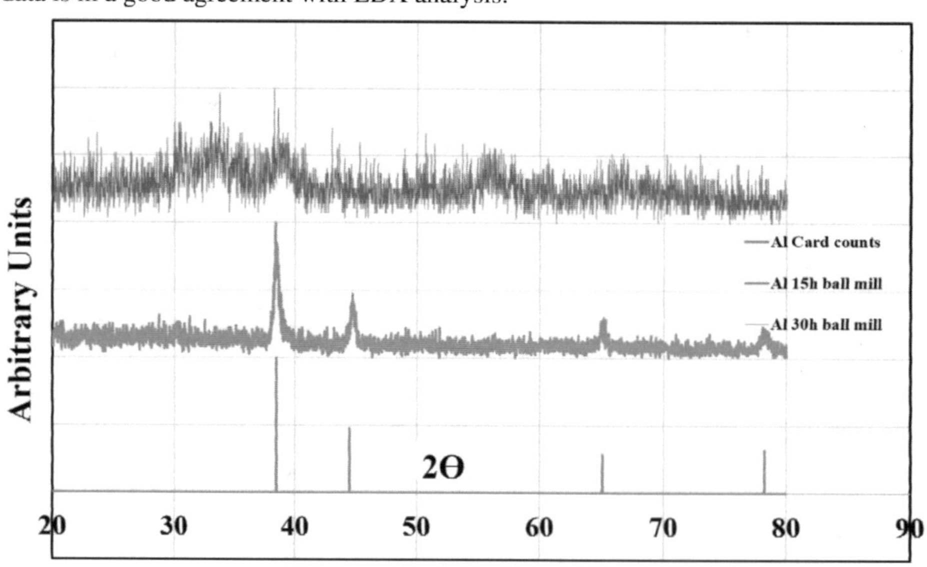

Figure. 4. X-ray diffraction patterns of synthesized Al samples with different milling times

4.4 Factors affecting emissivity.

4.4.1 Coating efficiency. Figure.(5) illustrates the relation between thickness (μm), emissivity (3-5, 8-12μm) and spin coating operational conditions (rpm) where the percentage between the Al powder and the polymer is (1: 9) g. As shown from the figure, the thickness of the coating slightly decreases from 13μm to 10μm with increasing speed (rpm). The infrared emissivity very slightly increases with decreasing thickness and could see that the infrared emissivity nearly at constant value. Therefore, depending on the visual inspection for samples instead of depending on the infrared emissivity at the wavelength of (3-5 μm, 8-12 μm). Figure (6: a-d) shows the dependence of the visual inspection of the samples on the spin coating operational condition (rpm). The difference between the images of the spin coater conditions (rpm) of speed1 and speed 2 is clearly observed. Fixing speed1 at 2, speed2 at 1 and the visual inspection of sample (1) is ok as shown at Fig 6.a. fixing speed1 at 8, speed2 at 1and the visual inspection of sample (4) is ok as shown at Fig 6.b. but sample (1) is more smoothing than sample (4). Fixing speed1 at 2, speed2 at 3 and the visual inspection of sample (5) is not ok as shown at fig 6.c. fixing speed1 at 8, speed2 at 3 and the visual inspection of sample (8) is not ok as shown at Fig 6.d. it was found that by increasing spin coater conditions (rpm) to a high speed where fixing speed2 at 3, the visual inspection of the samples shows that the samples are not smooth, becomes rougher so that the visual inspection of samples (5-8) are not ok as shown at table 3 and the infrared emissivity of this samples slightly increase as the samples becomes rougher. Therefore, the optimum spin coater conditions for commercial resin (clear coat & hardener)/10% Al coatings will be fixed with speed1 (low speed) at 2, time 18 sec, 0.8*1000 rpm and Fixing speed 2 (high speed) at 1, time 20 sec, 1.54*1000 rpm.

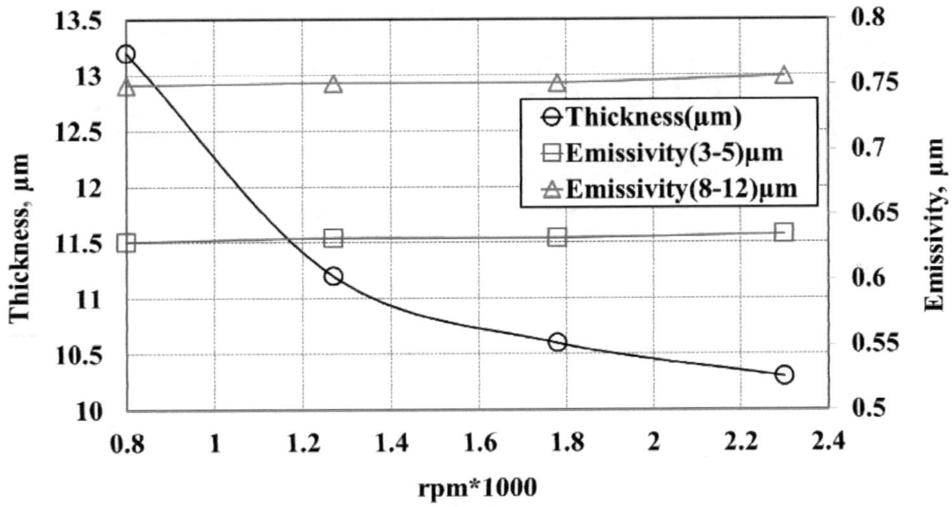

Figure 5. Emissivity at the wavelength of (3-5 μm, 8-12 μm.) and thickness (μm) dependence on spin coating operational condition (rpm).

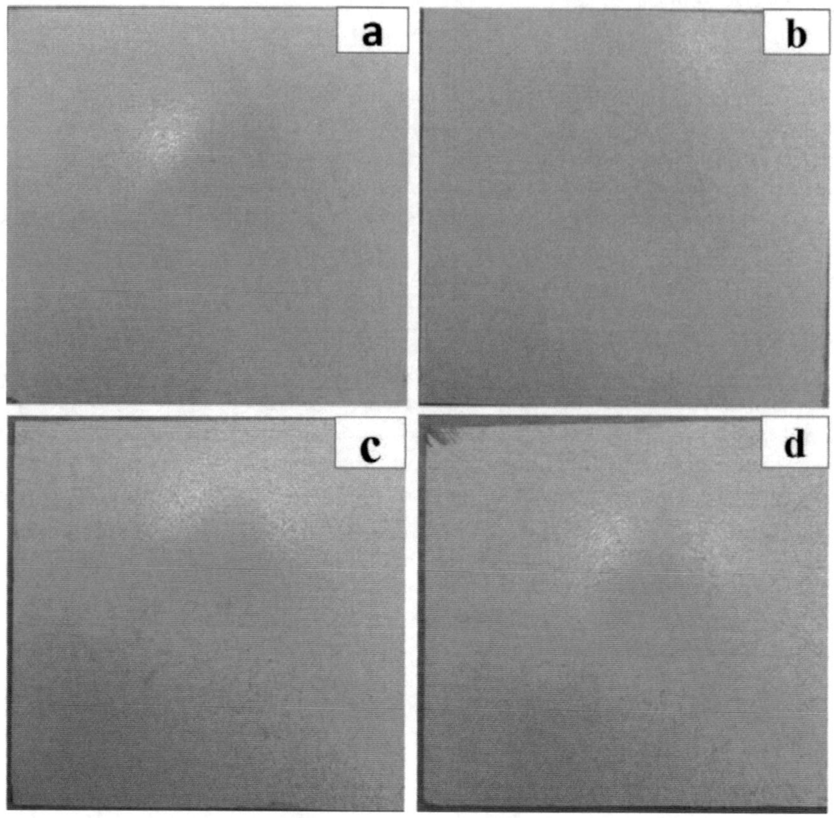

Figure (6: a-d). Shows the visual inspection of samples dependence on spin coating operational condition (rpm) where samples as 10*10 cm

Table 3. The complete set of spin coating operational condition (rpm).

Sample ID	speed 1	Rpm*1000 of speed 1	Speed 2	Rpm*1000 of speed 2	Thickness (μm)	Emissivity (3-5)μm	Emissivity (8-12)μm	Visual inspection
1	2	0.8	1	1.54	13.2	0.629	0.749	ok
2	4	1.27	1	1.54	11.2	0.632	0.751	ok
3	6	1.78	1	1.54	10.6	0.632	0.751	ok
4	8	2.3	1	1.54	10.3	0.635	0.756	ok
5	2	0.84	3	3.01	9.7	----	-----	not ok
6	4	1.3	3	3.01	9.2	-----	----	not ok
7	6	1.75	3	3.01	8.6	----	----	not ok
8	8	2.32	3	3.01	8.2	-----	------	not ok

4.4.2 The effect of thickness on the emissivity of commercial resin/Al composite coatings. In general, the yield of thickness with any composite coatings is nearly fixed and therefore different thickness can't be obtained. To solve this problem, one layer can be coated on the substrate, and then after a few minutes another layer can be coated, etc. waiting for long time can lead to separating the coats. Due to the previous action, the effect of coating thickness can be examined. Figure. (7) Shows the thickness dependence of infrared emissivity of commercial resin /10% Al coatings where fixing speed1 (low speed) at 2, time 18 sec, 0.8*1000 rpm and Fixing speed 2 (high speed) at 1, time 20 sec, 1.54*1000 rpm. As shown from the figure, when the thickness increases from 13μm to 39μm, the infrared emissivity very slightly decreases from 0.629μm to 0.623μm and from 0.74μm to 0.743μm at two regions (3-5, 8-12) μm. respectively. Therefore, the emissivity profile becomes nearly steady as the thickness reaches 40 μm. With the successive increase of the thickness from 39μm to 115μm, the infrared emissivity increases from 0.623μm to 0.641μm and from 0.743μm to 0.774μm at two regions 3-5 μm and 8-12 μm, respectively. Finally, the optimum thickness for commercial resin (clear coat& hardener) /10% Al coatings is found to be from 13 μm to 39μm.

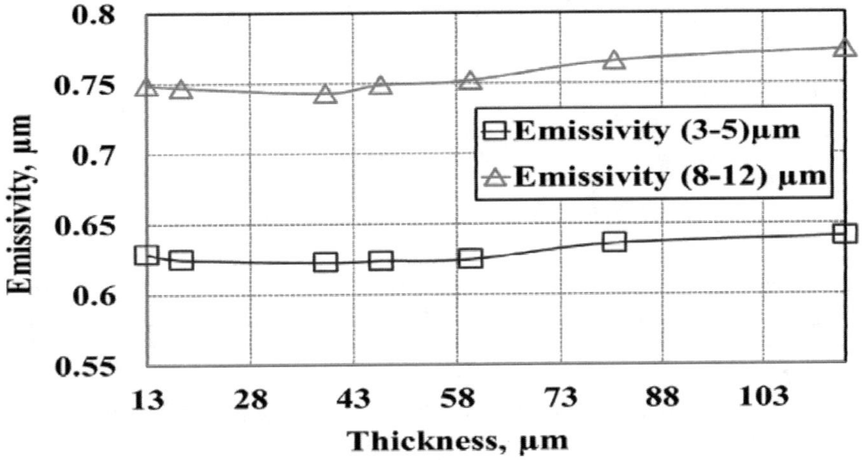

Fig 7. Relation between thickness (μm) and infrared emissivity radiation in two regions (3-5, 8-12) μm

4.4.3 The effect of Al powder content on the emissivity of commercial resin /Al composite coatings. Figure. (8) Shows the relationship between the infrared emissivity of commercial resin/ Al coating and concentration of Al powder. As shown from the figure the infrared emissivity (at band 3-5μm and 8-12μm) of the samples has the same behavior and can be divided into three regions. In the first region, the percentage of Al powder in the mixture is low (5 to 20 wt. %.), hence there are inter distance between the coated Al particles and increase the porosity between particles makes them low reflective to the incident radiation. Therefore, the infrared emissivity decreases with little percentage from values of 0.634μm to 0.589μm and from values of 0.784μm to 0.660μm at two regions 3-5 and 8-12 μm, respectively.

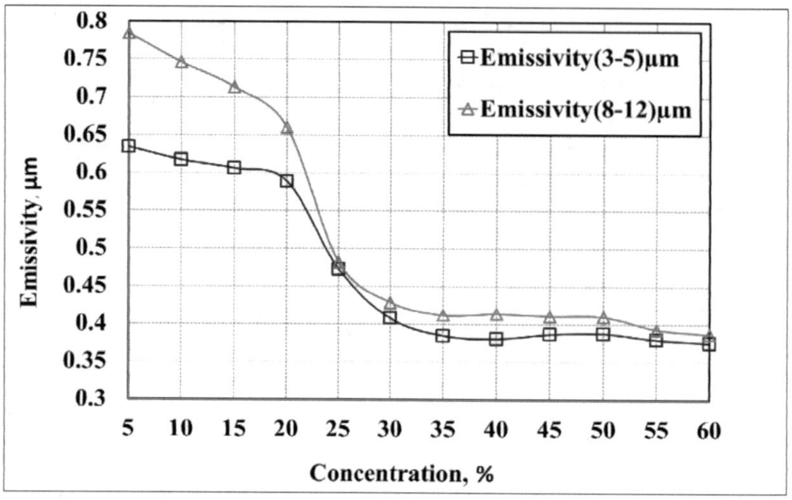

Fig 8. The relation between Al content and infrared emissivity radiation at two regions (3-5, 8-12 µm)

In the second region, the infrared emissivity sharply decreases with increasing the Al content from 20 wt. % to 35 wt. %. Infrared emissivity measurement shows that the emissivity of the coatings sharply decreased from values of 0.589µm to 0.385µm and from values of 0.660µm to 0.412µm at two regions 3-5 µm and 8-12 µm, respectively. High Al content causes the closer packing of the coated Al particles and reduces the porosity between particles making them highly reflective leading the infrared emissivity to decrease sharply. In the third region, the infrared emissivity slightly decreases with increasing Al concentrations from 35 wt. % to 60 wt. %. Infrared emissivity of the coatings slightly decreased from values of 0.385µm to 0.376µm and from values of 0.412µm to 0.387µm at two regions (3-5, 8-12) µm. respectively. Coating was saturated with the filler (Al) and it slightly decreases in emissivity. The perfect percentage for the filler (Al) in the matrix within the range (35 wt. % to 40 wt. %). At high coated Al concentration (35 wt. % Al), the agglomeration of coated Al powder causes non-uniform dispersion, which causes a noticeable increase in surface roughness, which causes an increase in the infrared emissivity. Both theoretically and experimentally, the relationship between emissivity and surface roughness has been investigated [8, 25, 26]. It can be easily represented from equation according to Kirchhoff's law and energy conservation [27].

$$\varepsilon = 1 - \rho_r = 1 - \rho_p \, exp\left(\frac{-4\pi\delta}{\lambda}\right)^2 \quad (1)$$

Where ρ_r and ρ_p are the reflectance for a rough surface and a polished surface, respectively, δ is surface roughness, ε emissivity and λ is wavelength. Eq. (1) shows that rough-surfaced coatings have a larger infrared emissivity. Where the content of coated Al powder rises 35 wt. %, the negative effect of increased coating surface roughness offsets the emissivity reduction. As a result, the infrared emissivity continues to decrease, but at a slower rate. Although it is not possible to minimize coating emissivity by raising coated Al powder content to extremely high levels, a moderate amount of 35 wt. % is beneficial for the coating's low infrared emissivity and good physical properties [27].

In this section, we fixed the thickness of the coating in the range 13µm to 39µm and fixing speed1 (low speed) at 2, time 18 sec, 0.8*1000 rpm, fixing speed 2 (high speed) at 1, time 20 sec, 1.54*1000 rpm. Thermal signature management takes the form of varying a material's emissivity or adjusting a surface's temperature to effect the quantity of energy radiated from its surface. Figure (9: a-c) illustrates that the thermal signature with various Al content (5%, 20%, 25% 35%, 40%, 60%) at different temperatures (50°C, 70°C, 90°C). The difference between the thermal signatures of Al content is clearly detected in which different temperature ranges are appointed different colors. Table 4 shows the apparent temperature with different Al content.

Table 4. Shows the apparent temperature of different Al content

Al content	T (50 °c)		T (70°c)		T (90°c)	
	°F	°C	°F	°C	°F	°C
5%	107.7	42.0	149.7	65.3	184.8	84.8
20%	104.9	40.5	142.6	61.4	174.9	79.3
25%	101.6	38.6	128.9	53.8	152.4	66.8
35%	100.5	38.0	125.8	52.1	147.4	64.1
40%	99.7	37.6	123.4	50.7	144.1	62.2
60%	99.6	37.5	122.6	50.3	143.4	61.8

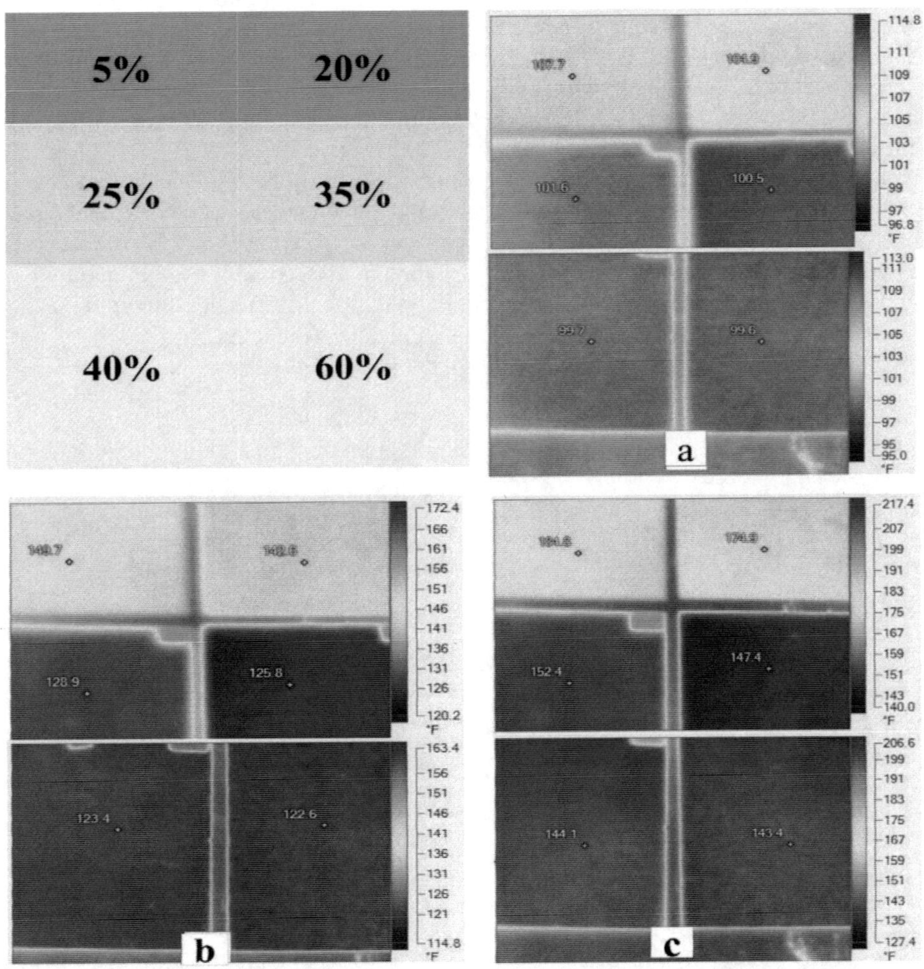

Fig (9: a-c). Shows the thermal signature with various Al content (5%, 20%, 25%, 35%, 40%, and 60%) at different temperature degree (50°C, 70°C, 90°C) for samples 10*10 cm.

As shown from the figure, it confirms the concept of the three regions in fig. (8). At fig. (9: a) with temperature 50c°, increasing Al content from 5 wt.% to 20 wt.% found that the temperature decrease with little percentage from 107.7 F° to 104.9 F° where there is difference in temperature equal (2.8 F°) which is identical with the first region. With successive increasing content from 20 wt.% to 35 wt.%

found that the temperature sharply decreases from 104.9 F° to 100.5 F° where there is difference in temperature equal (4.4 F°) which is nearly double the value than the previous value (2.8 F°) and typical with the secondly region. With increasing content from 40 wt. %. to 60 wt.%. Found that the temperature slightly decreases from 99.7 F° to 99.6 F° where there is difference in temperature equal (0.1 F°) which is identical with the thirdly region. Finally, the perfect content percentage for the filler (Al) in the matrix with the concentration range from (35 wt. % to 40 wt. %).

4.4.4 Emissivity profile of Nano shape Al powder Figure. (10) Shows the relation between ball milling time, particle size and the infrared emissivity of the coating in two regions band 3-5μm & 8-12μm for Al powder at concentration 30%. The spin coater conditions for Al coatings will be fixed with speed1 (low speed) at 2, time 18 sec, 0.8*1000 rpm and fixing speed 2 (high speed) at 1, time 20 sec, 1.54*1000 rpm. From Figure it can be seen that when milling time increases from 0h to 15h, the average particle sizes decrease from 35μm to 3μm, the infrared emissivity sharply increases from 0.408μm to 0.921μm at region 3-5 μm and sharply increases from 0.429 μm to 0.934μm at region 8-12 μm. By increasing the milling time up to 30h, average particle size is reduced to 300 nm. Therefore, the infrared emissivity slightly increases from 0.921 μm to 0.929μm at region 3-5 and slightly increases from 0.934μm to 0.95μm at region 8-12 μm.

It was found that as the Al particles size reaches the Nano level, the number of particles multiplies dramatically. However, the reunion of Al particles, on the other hand, is seen because the particles are active on the surface of small particles, which reduces successful pigment particle contact and increases porosity. Additionally, higher emissivity may be attributed to high oxidation of Al due to the presence of aluminum oxide (Al_2O_3) with high percent reaches about 39% at corresponding milling time 30h.

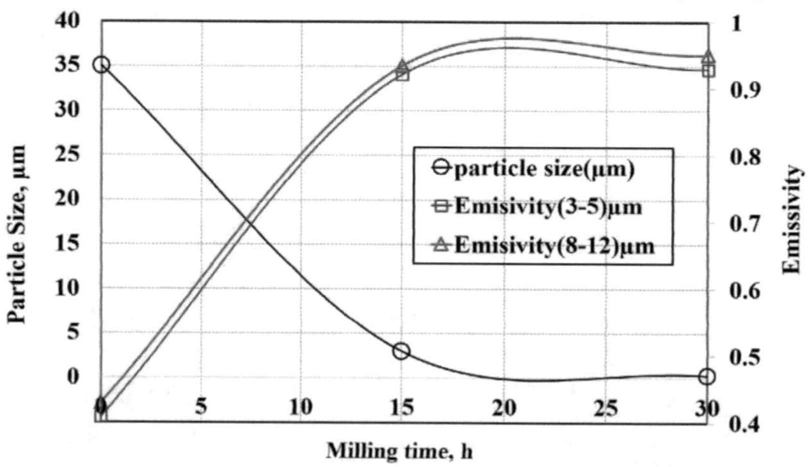

Fig 10. Relation between ball milling time and particle size, infrared emissivity radiation at two regions (3-5, 8-12) μm for Al powder

5. Conclusion

Ball milling approach was utilized to control shape and size of Al powders, which were examined to evaluate their infrared emissivity profiles within the band 3-5 and8-12μm. It was found that particle size has a significant impact on the emissivity of the coatings. Almost Nano Al particles are unfavorable for the formation of low emissivity coatings. On the other hand, the highly flake Al pigments are more suitable for the low infrared emissivity composite coatings. As coat thickness was optimized, it was concluded that the optimum performance achieved 13 μm to 39μm. Furthermore, the optimum Al content to achieve the maximum performance (lowest emissivity) was found to be 35 - 40 wt. %. Moreover, the micro size Al flakes achieves higher performance than the sub micro (Nano size). The formation of Aluminum oxides (Al_2O_3) as results show; due to increase of milling time has a negative impact on emissivity.

References

[1] R. Liu, C. Ji, J. Mock, J. Chin, T. Cui, and D. Smith, 2009 "Broadband ground-plane cloak," *Science,* vol. **323**, pp. 366-69.

[2] L. McDougall, 2012 "*Complete Tracker: Tracks, Signs, and Habits of North American Wildlife*: " Rowman & Littlefield.

[3] C. Mikunda, 2004 "*Brand lands, hot spots & cool spaces: Welcome to the third place and the total marketing experience*: " Kogan Page Publishers.

[4] L. Yuan, X. Weng, and L. Deng, 2013 "Influence of binder viscosity on the control of infrared emissivity in low emissivity coating," *Infrared Physics & Technology,* vol. **56**, pp. 25-29.

[5] P. K. Biswas, A. De, N. Pramanik, P. Chakraborty, K. Ortner, V. Hock, *et al.*, 2003 "Effects of tin on IR reflectivity, thermal emissivity, Hall mobility and plasma wavelength of sol–gel indium tin oxide films on glass," *Materials letters,* vol. **57**, pp. 2326-32.

[6] F.-Y. Zhang, Y.-M. Zhou, Y. Cao, and J. Chen, 2007 "Preparation and characterization of KGM/CdS nanocomposite film with low infrared emissivity," *Materials letters,* vol. **61**, pp. 4811-14.

[7] A. Licciulli, A. Maffezzoli, D. Diso, M. Mazzer, G. Torsello, and S. Tundo, 2004 "Porous garnet coatings tailoring the emissivity of thermostructural materials " *Journal of sol-gel science and technology,* vol. **32**, pp. 247-51.

[8] B. Rousseau, M. Chabin, P. Echegut, A. Sin, F. Weiss, and P. Odier, 2001 "High emissivity of a rough Pr 2 NiO 4 coating," *Applied Physics Letters,* vol. **79**, pp. 3633-35.

[9] G. Leftheriotis and P. Yianoulis, 1999 "Characterisation and stability of low-emittance multiple coatings for glazing applications," *Solar energy materials and solar cells,* vol. **58**, pp. 185-97.

[10] Y. Shan, Y. Zhou, Y. Cao, Q. Xu, H. Ju, and Z. Wu, 2004 "Preparation and infrared emissivity study of collagen-g-PMMA/In2O3 nanocomposite," *Materials letters,* vol. **58**, pp. 1655-60.

[11] L. Fagundes, T. Sousa, A. Sousa, V. Silva, and E. Sousa, 2006 "SBA-15-collagen hybrid material for drug delivery applications," *Journal of non-crystalline solids,* vol. **352**, pp. 3496-01.

[12] G. Wu and D. Yu, 2013 "Preparation and characterization of a new low infrared-emissivity coating based on modified aluminum," *Progress in Organic Coatings,* vol. **76**, pp. 107-12.

[13] C. Hu, G. Xu, X. Shen, C. Shao, and X. Yan, 2010 "The epoxy-siloxane/Al composite coatings with low infrared emissivity for high temperature applications," *Applied surface science,* vol. **256**, pp. 3459-63.

[14] K.-S. Chou and Y.-C. Lu, 2007 "The application of nanosized silver colloids in far infrared low-emissive coating," *Thin Solid Films,* vol. **515**, pp. 7217-21.

[15] H. Chu, Z. Zhang, Y. Liu, and J. Leng, 2016. "Silver particles modified carbon nanotube paper/glassfiber reinforced polymer composite material for high temperature infrared stealth camouflage," *Carbon,* vol. **98**, pp. 557-66.

[16] Y. Xin, H. Zhou, X. Ni, Y. Pan, X. Zhang, J. Zheng, *et al.*, 2015 "The optical properties of low infrared transmittance WO 3− x nanocrystal thin films prepared by DC magnetron sputtering under different oxygen ratios," *RSC Advances,* vol. **5**, pp. 57757-63.

[17] T. Hallberg, T. Heikkila, H. Karlsson, P. S. Salonen, C. Nilsson, and A. Janis, 2005 "*Development of low-emissive camouflage paint*: Sensor Technology", Swedish Defence Research Agency.

[18] L. E. Nunes Almeida, A. Martins, S. R. Gomes, S. Aramaki, R. O. Pinto, and F. Cunha, 2013 "Infrared Camouflage of the Warfighter's Equipment through Enhanced Coating Technology," in *49th AIAA/ASME/SAE/ASEE Joint PropulsionConference*, ed, p. 4101.

[19] R. F. Supcoe, 1982 "Blue-gray low infrared emitting coating," ed: Google Patents.

[20] W. Mazurczyk, S. Wendzel, S. Zander, A. Houmansadr, and K. Szczypiorski, 2016 "*Information hiding in communication networks: fundamentals, mechanisms, applications, and countermeasures*": John Wiley & Sons.

[21] Kim, T., Bae, J.Y., Lee, N. and Cho, H.H., 2019. Hierarchical metamaterials for multispectral camouflage of infrared and microwaves. *Advanced Functional Materials*, ***29*(10)**, p.1807319.

[22] J. R. White, 2012 "Aircraft infrared principles, signatures, threats, and countermeasures," NAVAL AIR WARFARE CENTER WEAPONS DIV CHINA LAKE CA.
[23] B. W. Krueger, C. S. Dandeneau, E. M. Nelson, S. T. Dunham, F. S. Ohuchi, and M. A. Olmstead, 2016 "Variation of Band Gap and Lattice Parameters of β−(AlxGa1− x) 2O3 Powder Produced by Solution Combustion Synthesis," *Journal of the American Ceramic Society,* vol. **99**, pp. 2467-73.
[24] C. Wu, H. Tan, W. Huang, W. Li, K. N. Dinh, C. Yan, *et al.*, 2020 "A new scalable preparation of metal nanosheets: potential applications for aqueous Zn ion batteries anode," Advanced *Functional Materials,* vol. **30**, p. 2003187.
[25] F. Ghmari, T. Ghbara, M. Laroche, R. Carminati, and J. J. Greffet, 2004 "Influence of microroughness on emissivity," *Journal of Applied Physics,* vol. **96**, pp. 2656-64.
[26] J. O. Sacadura, 1972 "Influence de la rugosité sur le rayonnement thermique émis par les surfaces opaques: Essai de modèle," *International Journal of Heat and Mass Transfer,* vol. **15**, pp. 1451-65.
[27] G. Wu and D. Yu, 2012 "Preparation of a novel infrared low-emissive coating from the Cu powder modified by the polyethylene wax," *Infrared Physics & Technology,* vol. **55**, pp. 26-31.

IOP Conf. Series: Materials Science and Engineering 1172 (2021) 012028 doi:10.1088/1757-899X/1172/1/012028

Enhancement of carbon fiber/epoxy composite electrical, optical and thermal properties by using different types of nano-additives.

L Bassiouny[1], T Samir[2], S Abdallah[2], H Ashour[1] and A Anwar[3]

1 Faculty of Engineering, Misr for Science and Technology, Cairo, Egypt.
2 Faculty of Engineering, Benha University, Shobra, Cairo, Egypt.
3 Space Technology Center (STC), Cairo, Egypt.

E-mail: msc.a.anwar@gmail.com.

Abstract. Environmental space threats are becoming more critical as they affect the optical, thermal, and electrical properties of the reinforced fiber polymeric-based materials in spacecraft. Three different Nano-particles Alumina (Al_2O_3), Multiwall Carbon Nanotubes (MWCNT), and Reduced Graphene Oxide (RGO) were added to the epoxy matrix and then reinforced by bidirectional carbon fiber plain to form carbon fiber/epoxy by hand lay-up using autoclave curing technique to make three different reinforced materials. In this paper, the electrical, optical, and thermal properties of the carbon fiber/Epoxy Nanocomposite were studied.

Fourier transform infrared (FTIR) was performed to evaluate the structural changes in the newly synthesized materials. The optical, thermal, and electrical properties were tested by UV-visible Spectroscopy, Photo-acoustic spectroscopy (PA), and Keithley 2635A respectively. The results showed an enhancement in the electrical, optical, and thermal properties of the epoxy matrix after the addition of Nano-particles.

The optical test showed that the neat epoxy and epoxy/Nano-particles absorption spectra were in the infrared range. The thermal test indicated that the three thermal parameters diffusivity, effusivity, and conductivity showed the best enhancement after the addition of MWCNTs. The electrical test pointed out that after the addition of Nano-particles, neat epoxy changed from an insulating material to a semi-conductive material.

1. Introduction

Many environmental threats cause the degradation of materials used on the external panels of spacecraft [1]. For this reason, it is important to keep up-to-date and improve the existing materials and establish new materials with enhanced properties for proper use in space applications [2]. Space radiation causes the spacecraft material properties degradation and this degradation affects the spacecraft orbit lifetime. There are two kinds of space radiation, charged particles radiation and electromagnetic radiation [3].

Epoxy resin is the most widely used matrix in polymer-based composites. The reason behind this, the epoxy matrix contains good chemical, electrical, thermal, and mechanical properties [4]. However, the typical thermal conductivity of epoxy ranged between (0.17–0.21 W/mK) and often exhibits a brittle nature [5]. Therefore Nano-reinforcements such as Nano-ceramic, carbon nanotubes, and graphene structures are usually employed to enhance the deficiencies in neat epoxies [3], [4].

In 2016, a study by Tetjana Tomaskova et. al tested the effect of aluminium oxide nanoparticles (Al2O3) on the mechanical and physical properties of epoxy resin. Various weight percentages of 2, 4,

Content from this work may be used under the terms of the Creative Commons Attribution 3.0 licence. Any further distribution of this work must maintain attribution to the author(s) and the title of the work, journal citation and DOI.
Published under licence by IOP Publishing Ltd

6, 8, and 10 wt. % were added to epoxy resin. The steady-state method, Dielectric strength tests, and tensile tests were applied to study the physical and mechanical properties of epoxy/Al2O3 Nano-composite. The results revealed an enhancement in the thermal and electrical properties of epoxy resin after the addition of Al2O3 nanoparticles. The highest increase in thermal conductivity and dielectric strength was achieved by 8 wt. % of nanoparticles [6].

In 2017, Ganiu B. et. al studied the effect of reduced graphene oxide (RGO) on the thermal and mechanical properties of the epoxy matrix. Different thermal and mechanical behaviors of the epoxy Nano-composites were investigated. The result revealed an outstanding increase in the thermal conductivity when a small amount of RGO was added. An enhancement of 40% compared with that of the unmodified epoxy polymer was obtained by adding a value of 0.06 wt. % of RGO. This value represents one of the highest increases in the thermal conductivity per wt. % when adding RGO ever reported [7].

In 2019, a study was performed by Tagreed M. Al-Saadi et.al to show the use of multi-walled carbon nanotubes (MWCNTs) for the strengthening of epoxy resin with different MWCNTs wt. % (0.0, 0.02, 0.04, and 0.06) to fabricate the epoxy/MWCNTs nanocomposites. This research observed the effect of MWCNTs on the mechanical, electrical, and thermal properties of the epoxy matrix. Hardness, electrical properties such as dielectric constant, dielectric loss factor, dielectric strength, electrical conductivity, and thermal properties such as thermal conductivity were studied. The results revealed an enhancement in hardness, thermal and electrical conductivity, and break down strength with increasing MWCNTs wt. %. However, the dielectric loss factor and dielectric constant are decreased when the concentration of MWCNTs increased [8].

In this work, aluminium oxide (Al_2O_3), multi-wall carbon nanotubes (MWCNTs), and reduced graphene oxide (RGO) were dispersed in an epoxy matrix to study their effect on the physical property of epoxy resin. These three Nano-particles were chosen because they showed an enhancement in the physical and mechanical properties of the epoxy matrix [6] [8].

2. Experimental Work

2.1. Material

Epoxy Nanocomposites samples were made using biresin® two parts matrix; part A CR82 (resin) and part B CR80-6 (hardener). Three different nanoparticle materials were added to the epoxy matrix are listed in Table 1.

Table 1 summarize different types and characteristics of nanoparticle used

Nanoparticles	Characteristics
Aluminum oxide (Al ε by Sky Spring Nanomaterials Co.	White powder of Gamma- Al ε diameter range (0.4 to 1.5) Nano-meter.
Multi-walled carbon nanotubes (MWCNT). It was purchased from the Egyptian Petroleum Research Institute.	MWCNT's were formed by a high-yield catalytic process based on chemical vapor deposition (CVD) with an outer mean diameter of (8-10 nm) and inner mean diameter (4nm) and length from (5-10 μm). The purity of neat MWCNTs was greater than 90%.
Reduced graphene oxide (RGO) Three different types of reduced graphene oxide used in this study were (G24N, G33C, and G270). They were purchased from the City of Scientific	**Sample G24N** is an N-doped reduced graphene oxide with 3D structure. Its atomic composition is 83.3% C, 13.9% O, and 2.8% N. It is synthesized by thermal dissociation of polyethylene tetra-phthalate (PET) waste bottles with urea at 800°C for 5 hr. **Sample G33C**. It was prepared as sample 9C as described in reference [9], [10] but not in the same batch. Its atomic composition is 90.49% C, 2.131% H, and 7.379% O. **Sample G270**. It is an N-doped multi-layered graphene Nano-sheet. Its atomic composition is 89% C, 7.2% O and 3.8% N. The graphene

Research and Technological Applications in Alexandria	Nano-sheets are well exfoliated. The doped N has dominated pyloric conformation followed by pyrdinic and lastly graphitic. It is synthesized by hydrothermal treatment of glucose solution under mild synthesis conditions.

2.2. *Specimen Preparation*

2.2.1. Sonication process. Epoxy resin (50g) was mixed with the three different types of nanoparticles (Al_2O_3, RGO, and MWCNTs) each type was added at a specified weight percent (1%, 0.3%, and 0.5% respectively) in epoxy resin after stirring for (5 min) at room temperature. The nanoparticles were well dispersed in epoxy resin by using "SONICS VCX750" sonicator for a constant time (1.5hr.) with (9kHz) frequency and (750W) power.

2.2.2. Production of nanocomposites sheet. Firstly, about (27) wt. % of hardener was added to the mixture of (50g) epoxy and nanoparticles. Then using a well-waxed glass sheet with a double-faced tape border; the epoxy/nanoparticle mixture was rolled into a uniform layer. A carbon fiber fabric layer of (0.4mm) thickness was then rolled on the epoxy/ nanoparticle layer. Followed by a second layer of the epoxy/nanoparticle mixture. This process was repeated three times then a layer of foam was added to absorb the excessive resin used in the hand layup process. The whole glass was wrapped in a vacuum bag and sealed. Finally, the whole sheet was vacuumed with (-1) bar at (45℃) for (24hr.). The final sheet obtained contained four layers of carbon fiber fabric with (2.4mm) thickness.

2.3. *Experimental methods*

2.3.1. Fourier-Transform Infrared Spectroscopy (FTIR). The FTIR spectrometer "JASCO 4100" was used to characterize the neat epoxy and epoxy/nanoparticles. In this technique, two parameters such as the chemical bonds and molecular compositions can be obtained for each specimen. The FTIR showed a number of peaks for the tested specimens that are hard to be recognized. Therefore, All the FTIR analysis was accomplished on the basic constituents of each composite type. The wavelength obtained was in the range of 400 to 4000 cm-1 with an average resolution of 4 cm^{-1}. The peak profile was sharpened by background scans to improve the resolution and remove any spectrum noise. At a room temperature of 23±3°C and relative humidity of 50±10 % according to the ASTM D-3039-06 [11], the investigated materials are prepared.

2.3.2. Optical Test (Spectrophotometer (JASCO 670)). A double beam Spectrophotometer (V-670) was used to obtain the optical behavior of the samples by measuring the amount of electromagnetic radiation transferred through a sample as a function of wavelength. The wavelength ranged from 200-2700nm. All samples were extracted from the fabricated sheet in a square of dimension (10 mm x10 mm).

2.3.3. Thermal Test (Photoacoustic spectroscopy (PA)). Thermophysical properties (Thermal diffusivity, Thermal effusivity, and Thermal conductivity) of each sample were determined by a laser photoacoustic spectroscopy (Stabilite 2017 model 2550) as Figure 1. Moreover, the output beam was subjected to a sample dimension (10x10) mm that was attentively mounted inside PA cell.

2.3.4. Thermal diffusivity (α). It is a thermos-physical parameter that determines how adequately photons transfer heat through the sample. Also, it indicates the rate of heat distribution inside the material. This rate depends mainly on the thermal conductivity (k) of the material and the energy stored inside the material [12];

$$\alpha = {}^{k}/_{\rho c} \tag{1}$$

Where: (c) is the specific heat capacity and (ρ) is the density of the sample.

Thermal diffusivity is experimentally determined by two commonly used methods [13]; the first method is known as the transient heat-flow method [14] and the second method is known as the periodic heat-flow method [15]. The PA method used in this research is related to the second method. Moreover, to determine the thermal diffusivity (α), the PA signal is obtained for various chopping frequencies.

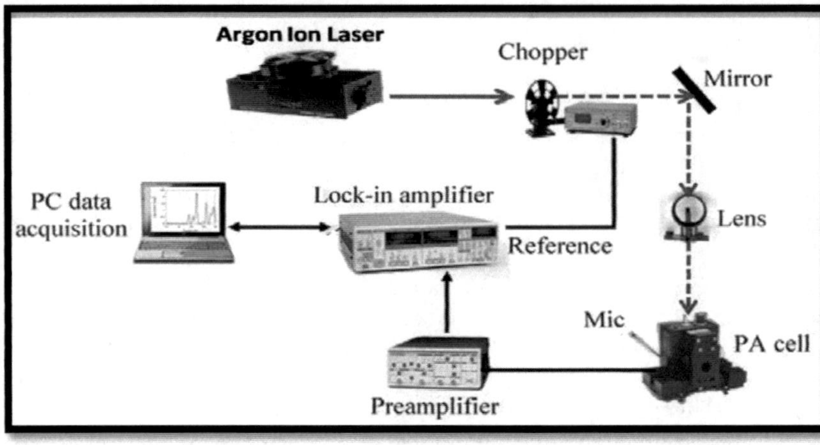

Figure 1 Photoacoustic spectroscopy

2.3.5. Characteristic frequency (fc). It is defined as the critical frequency at which the sample passes from the thermally thin region to the thermally thick region [16]. The thermal diffusivity is computed from the characteristic frequency. The PA signal varies with the thickness and the optical absorption of the sample. The characteristic frequency is obtained by plotting (ln PA) amplitude versus (ln f) for neat epoxy and epoxy/nanoparticles. As an example, Epoxy/G33C Nano-composite is drawn in Figure 2.

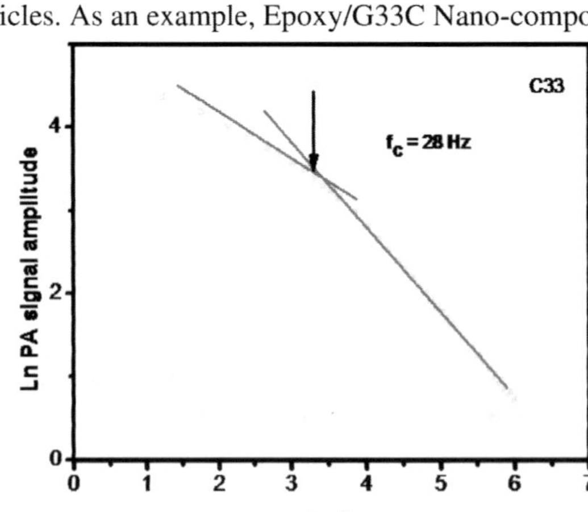

Figure 2 Characteristic frequency (f_c) of Epoxy/ G33C nanocomposites

The marked change in the slope is where the characteristic frequency (f_c) crossover takes place at which the sample changes from being thermally thin to thermally thick. The thermal diffusivity (α) was then estimated by using this equation [17].

$$\alpha = f_c l^2 \tag{2}$$

Where: (f_c) is the characteristic frequency and (l) is the sample thickness. The values of thermal diffusivity (α) are calculated and written in Table 2 for the neat epoxy and epoxy Nano-composites samples. From Table 2, the diffusivity (α) value for epoxy/carbon fiber (neat epoxy) is ($2.4x10^{-7}$) m^2s^{-1} which is much greater than pure epoxy of ($0.5x10^{-7}$) m^2s^{-1} [18]. Also, the highest thermal diffusivity is obtained by epoxy/MWCNTs of ($3.1x10^{-7}$) m^2s^{-1}.

2.3.6. Thermal effusivity (e). It is an important thermal parameter that indicates the surface heating of the substances. The amplitude of the PA signal "q" in case of thermally thick and optically opaque samples is given by:

$$\mathbf{q} = \mathbf{B}/_{\mathbf{ef}} \tag{3}$$

$$\mathbf{B} = \frac{\mathbf{I_o}}{\mathbf{2}} \frac{\gamma \mathbf{P_o} \alpha_{\mathbf{g}}^{1/2}}{\mathbf{2\pi l_g T_o}} \tag{4}$$

Where: (I_o) is the incident light intensity, (γ) is the ratio of specific heats, (P_0) is the ambient pressure, (α_g) is the gas thermal diffusivity, (l_g) is the length of the gas column, (T_o) is the ambient temperature.

The thermal effusivity is obtained by Equation 5 of the sample and the modulation frequency (f) [19].

$$\mathbf{e} = \sqrt{\mathbf{k\rho c}} \tag{5}$$

The different values of the thermal effusivity of neat epoxy and epoxy/Nano-particles are obtained by linear fitting the relation between PA amplitude (q) and (1/f) as shown in Figure 3.

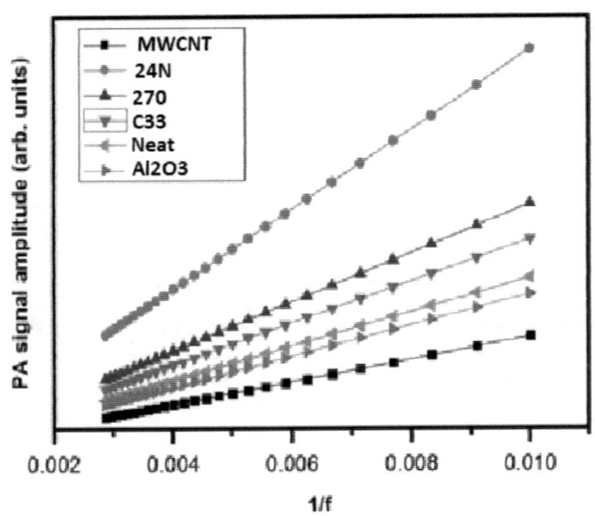

Figure 3 PA spectra for epoxy and epoxy/nanoparticles

2.3.7. Electrical Test. The samples' electrical resistances were measured by (Keithley 2635A) System Source Meter. Each sample was cut into a small sheet of (10x10) mm as per ASTM D-7264-15. The samples were subjected to a current of (1mA) and a voltage of (20V). The clamps were connected to positive and negative electrical current. The samples were placed between the clamps that contain a subjected circular area of (12.57mm^2). The electrical conductivity was calculated by Equation 6

$$\sigma = \frac{\mathbf{L}}{\mathbf{R*A}} \tag{6}$$

Where (σ) is the electrical conductivity in S/cm, (L) is the thickness in cm, (R) is the resistance in ohms, and (A) is the subjected area in cm2.

2.4. *Results and Discussion.*

2.4.1. Fourier Transform Infra-Red (FTIR) results. The FTIR analysis was used to investigate the chemical compositions and the functional groups obtained after the addition of nanoparticles to neat epoxy. The infra-red radiation causes stretching and bending of different bonds. These Bending bonds require high energies such as the part of the fingerprint region (1000-500) cm^{-1}. The other functional groups are characterized by their stretching frequencies. For example, C=O shows a peak at 1700 cm^{-1}

but for a C-H sp3 displays a peak at 3000 cm^{-1} [20]. The FTIR results depend on two parameters the wavenumber and the absorption intensity variation (abs).

From Figure 4 and Figure 5, FTIR results revealed that after the addition of nanoparticles, there was no change in the epoxy fingerprint region which means that the chemical structure of the epoxy remained the same. Moreover, in the epoxy matrix, there was no loss of functional groups, even after the addition of nanoparticles. This proves that the reactive sites of the nanocomposites are preserved for future manipulation. In addition, new bonds of (N-H) and (NH_2) appeared for epoxy/G24N and epoxy/ G270 nanocomposites as a consequence of using nitrogen doping to reduce the oxygen inside the graphene. Finally, it can be stated that there was a physical change not a chemical change after the addition of nanoparticles.

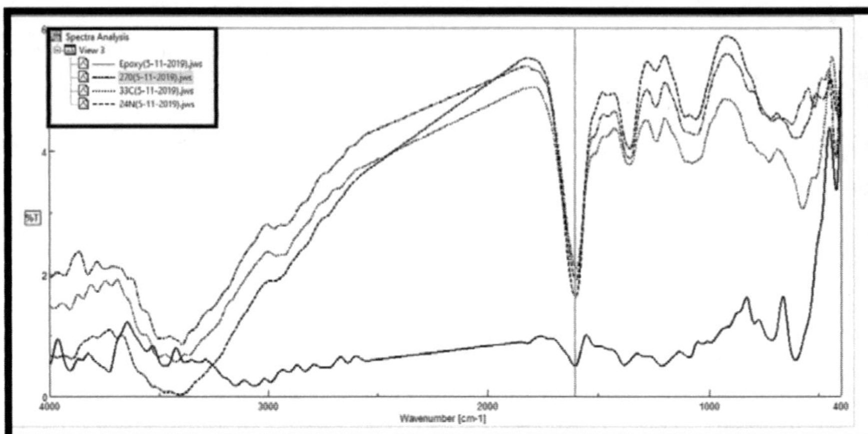

Figure 4 FTIR Results neat epoxy vs epoxy/RGO

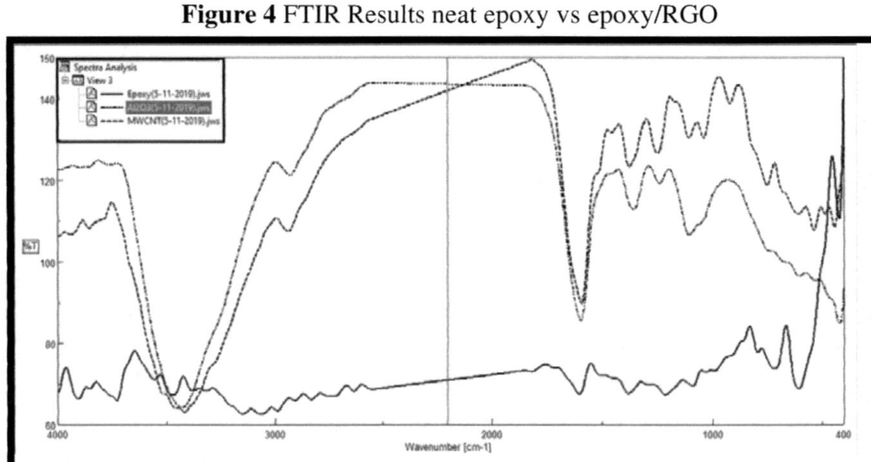

Figure 5 FTIR Results neat epoxy vs epoxy/MWCNT and epoxy/Al_2O_3

2.4.2. Optical test results. The optical absorption spectra of neat epoxy and epoxy/nanoparticles were obtained by regular UV-Vis. The range of wavelength was from (200-2700) nm to show the different types of light absorbed by each sample such as Ultraviolet (200-400) nm, visible light (400-700) nm, near-infrared (700-1100) nm, Infrared light (1100-2700) nm. As shown in Figure 6, it is easily observed that the epoxy matrix and all nanocomposite samples absorption ranges were in the infrared wavelength range. This is because of the opaque nature of the epoxy matrix. The absorption range for neat epoxy, epoxy/Al_2O_3, and epoxy/G270 Nano-composites laid at an absorption peak rang of (1800-2700) nm. This is because of the presence of oxygen and nitrogen that causes polarity charges on the sample surface which facilitate the movement of the infrared light [23], [24]. But epoxy/G33C and epoxy/MWCNTs showed a higher wavelength peak at (2200) nm because the charges on the surface are equivalent. The highest wavelength peak was obtained by G24N where it was observed at (2350) nm. This could be due to the small amount of doped Nitrogen used to reduce the graphene oxide as mention in **Table 1**.

Table 2 FTIR peaks indicating bonds for all samples

Bond	Wavelength (cm^{-1}) [21], [22]	Samples Epoxy	MWCNT	Al2O3	G33C	G24N	G270
Finger print	~<1000	421.37 - 715.461	442.583 - 921.807	419.442 - 603.61	410.763 - 727.996	419.442 - 718.354	424.263 - 622.895
sp³ C-O	1025-1200	-	1040.41	1106.94	1077.05	1061.62	1060.66
sp² C-O	1200	1215.9	1253.5	1241.93	1240.97	1242.9	1242.9
Sp³ C-H	2850-3000	-	2937.06	2929.34	-	-	-
Sp² C-H	3000-3100	3151.11	-	-	-	-	-
C=C-H	1607- 1510	1605.45	1594.84	1598.7	1598.7	-	-
CH2-CH2	1450-1470	-	-1457.92	-	-	-	-
CH2-CH3	1360-1390	-	1378.85	1359.57	1359.57	1357.64	1357.64
NH2	1590-1650	-	-	-	-	1601.59	1599.66
N-H	3350-3500	-	-	-	-	3406.64	3394.1
O-H	3200-3900	3728.69	3417.24 - 3938.89	3428.81 - 3895.5	3423.99 - 3907.07	-	3824.15

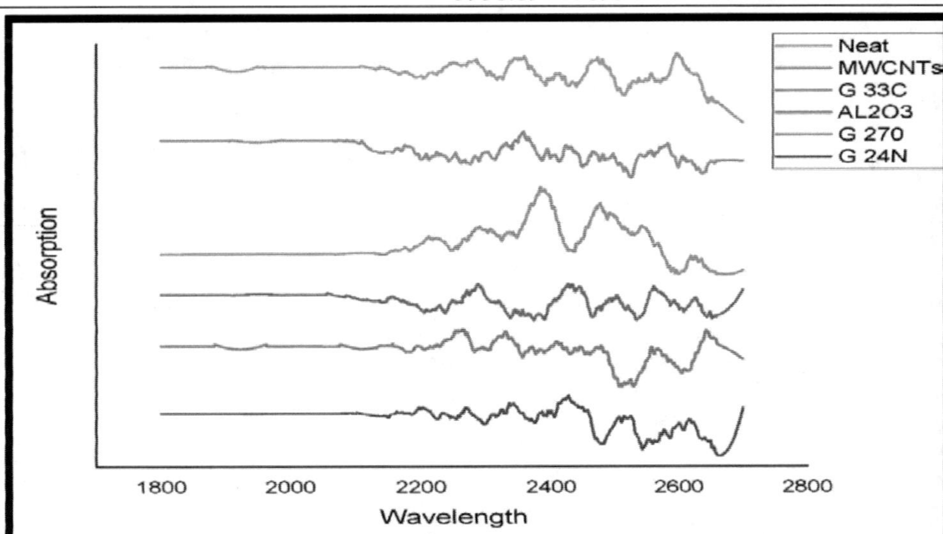

Figure 6 optical test for neat epoxy and epoxy nanocomposite

The absorption spectra shown in Figure 6 varies as a function of the additive type in the wavelength discussed above. It is clear that the neat epoxy absorption value is greater than the remaining nanocomposite samples. In addition, it is noticed that the epoxy/MWCNT nanocomposite has the greatest value among nanocomposites. While the epoxy/G33C has the highest value among the different RGO samples, which indicate about 50% of the neat epoxy absorption value. The absorption values of the remaining RGO samples (epoxy/G270, epoxy/G24) were decreased to approximately (25%, 10%)

of the neat epoxy, respectively. Moreover, the epoxy/AL2O3 nanocomposite shows a low absorption value around 40% of the neat epoxy. Thus, it may reveal that the absorption value varies from the value of neat epoxy regarding the additive chemical structure.

2.5. *Thermal test results*

The thermal properties of the neat epoxy and epoxy nanocomposites have been investigated by Photo Acoustic (PA) technique, which is of great importance for space application [25]. The sample's dimension (10x10) mm was mounted carefully inside the PA cell. Based on the depth profile analysis, the PA signal amplitude was obtained at various chopping frequencies (f) for each sample. The values of thermal properties are summarized in Table 3.

Table 3 Thermal properties of Neat epoxy, Epoxy/nanoparticles

Samples	Thermal diffusivity (α) ($10^{-7}m^2/s$)	STD	Thermal effusivity (e) ($Ws^{1/2}m^{-2}K^{-1}$)	STD	Thermal Conductivity (k) (W/m.k)	STD
Neat	2.4	±0.12	1080.8	±54	0.52	±0.03
Al2O3	2.85	±0.14	950.60	±48	0.50	±0.025
MWCNT	3.1	±0.15	1141.89	±57	0.63	±0.03
G24N	1.1	±0.06	581.35	±29	0.19	±0.01
G270	1.28	±0.06	700.37	±35	0.25	±0.012
G 33C	1.3	±0.06	793	±40	0.28	±0.014

As presented in Table 3, the thermal effusivity of neat epoxy is increased by the addition of MWCNTs nanoparticles. However, it decreased by Al2O3 was added and reduced graphene oxide nanoparticles (G270, G33C, and G24N)

2.5.1. *Thermal conductivity (k).*
It is a thermal parameter that can indicate the ability of the material to conduct heat. It is obtained by Equation 7 that relates to the thermal effusivity and diffusivity [17].

$$k = e\sqrt{\alpha} \tag{7}$$

Thermal conductivity was calculated by using the determined values from (e), (α) and listed in Table 3. Thermal conductivity of CF/epoxy (neat epoxy) was increased to 0.52 W m-1 K-1 compared to the thermal conductivity of the epoxy matrix of 0.12 Wm-1K-1. Moreover, as expected [8], the highest increase in thermal conductivity was obtained by epoxy/MWCNTs nanocomposite by 83% compared to neat epoxy. This is because of the presence of the volume of the new phase with high thermal conductivity, which increases the heat flow [26]. However, for reduced graphene oxide (G33C, G270, G24N) nanoparticles the thermal conductivity decreased by (54%, 48% 37%) compared with neat epoxy. As known in epoxy/RGO nanocomposite, the thermal conductivity of a material depends on the interfacial resistance between the RGO graphene and epoxy matrix. When this interfacial resistance is low, the epoxy matrix creates an interfacial layer with nanoparticles. This layer reduces the heat flow inside the epoxy matrix causing the reduction of thermal conductivity [27], [28]. Moreover, the thermal conductivity decreased for epoxy/Al2O3, which could be a result of small agglomeration that reduces the heat flow.

2.6. *Electrical test results*

Electrical resistivity is a measurement of the resistance of a given size of a specific material to electrical conductivity. Electrical resistivity for each sample was measured by (Keithley 2635A) system source meter. Table 4 shows the electrical conductivity calculated by

$$\sigma = \frac{L}{R * A} \tag{8}$$

Where (σ) is the electrical conductivity S/cm, (L) is the thickness in cm, (R) is the resistivity in ohms, and (A) subjected area = 0.125664 cm^2.

Table 4 Electrical conductivity of Neat epoxy, Epoxy/nanoparticles

Sample	Resistivity Ω	Electrical conductivity S/cm
Neat	0.2076×10^{9}	8.69×10^{-9}
MWCNT	7433	2.57×10^{-4}
Al2O3	633	3×10^{-3}
G-270	669	3.09×10^{-3}
G-24N	1047	1.84×10^{-3}
G-33C	957	2.08×10^{-3}

The electric conductivity of the epoxy matrix increased after nanoparticles were added and changed from an insulating material into a semi-conductive material [29]. The greatest increase in electrical conductivity was obtained by epoxy/G-270 nanocomposites relative to neat epoxy. The reason behind this improvement is the reduction of the graphene oxide because as the oxygen decreases the electrical conductivity increases [30]. In addition, as mentioned in FTIR results, new bond such as (N-H) was formed at different peaks at 3394.1 cm^{-1} and 3429.78 cm^{-1}. This new bond increases the flexibility of the epoxy matrix and facilitates the movement of electrons inside the epoxy matrix.

3. Conclusion

In this article, the electrical, optical, and thermal properties of carbon fiber/epoxy composite were enhanced by using various types of Nano-additives. The effect of adding nanoparticles in carbon fiber/epoxy (neat epoxy) composite has been investigated to prove the ability of the materials to sustain the space working conditions.

The optical test showed that the neat epoxy and epoxy/nanoparticles absorption spectra were in the infrared wavelength range of (1100-2700) nm. The highest absorption peak was obtained by Epoxy/G24N at 2350 nm.

The thermal test indicated that the three thermos-physical parameters; diffusivity, effusivity, and thermal conductivity were enhanced after the addition of MWCNTs. Moreover, thermal conductivity of carbon fiber/epoxy (neat epoxy) was enhanced to 0.52 $Wm^{-1}K^{-1}$ in compared with 0.12 $Wm^{-1}K^{-1}$ of neat epoxy. However, adding of Reduce Graphene Oxide (RGO) decreased the thermal conductivity of neat epoxy.

The electrical test revealed that after the addition of nanoparticles neat epoxy changed from an insulating material to a semi-conductive material. The highest increase in electrical conductivity was achieved by Epoxy/G270.

This means that adding nanoparticles increases the physical and electrical properties of carbon fiber/epoxy. In addition, the improvement was more precise for its optical and electrical properties than its thermal property where more research in this field should be taken into consideration.

References

[1] Dever J, Banks B, de Groh K and Miller S 2005 *Handbook of Environmental Degradation of Materials* (Ohio, NASA) 465-501.

[2] Jiacheng W, Thuc V and Inam F 2015 *Royal Society of Chemistry (RSC Adv.)* **5**, 73510-24.

[3] Sharon K. Miller S and Banks B 2010 *MRS* **35**-1 20-24.

[4] Zakaria M and Thirmizir M 2017 *Composites Part B: Engineering* **119**, 57-66.

[5] Permal A, Devarajan M, Hung L, Zahner T, Lacey D, Ibrahim K 2016" *Journal of Materials Science* **51** 7415–7426

[6] Bláha S, Tomášková T, Trnka P and Shlykevich A 2016 *18th European Conference on Power Electronics and Applications* (Karlsruhe) (IEEE) DOI: 10.1109/EPE.2016.7695425.

[7] Olowojoba G and Taylor A 2017 *Journal of Materials Science,* **52-12** 7323–44.

[8] Al-Saadi T, Aleabi S, Alabodi E and Abdul-Jabbar H 2018 *Journal of Physics* (Baghdad) (IOP Publishing*)* DOI: 10.1088/1742-6596/1003/1/012102.

[9] El Essawy N and Hamad H 2017 *Ecotoxicology and* Environmental *Safety* **145** 57-68.

[10] El Essawy N, Konsowa A, Elnouby M and Farag H 2017 *J Air Waste Manag Assoc* **67-3** 358–370.

[11] American Society for Testing and Materials, 2006, ASTM D3039 USA,.

[12] Almond D P and Patel P M 1996 *Photothermal science and techniques* (London) (Chapman and Hall).

[13] Zhao D, Qian X, Gu X and Jajja S 2016 *Journal of Applied Physics* **138-4** 1067-76.

[14] N. Yüksel 2016 *Insulation Materials in Context of Sustainability* http://dx.doi.org/10.5772/64157

[15] Nagao Y, Yamada T, Yoshida A and Kagata K 2015 *International Journal of Thermophysics* **36-4** DOI: 10.1007/s10765-014-1830-y

[16] Sigrist M 2017 Journal *of the Physical Society of Japan* 602-605,.

[17] López J and Mandelis A 2001 *Journal of Applied Physics* **90-5** 2273-81.

[18] Terpiłowski J, Woroniak J and Romanowska J 2014 *archives of thermodynamics* **35-3** 117–128.

[19] Raykar V and Singh A. 2011 *Journal of Thermodynamics* **2011** DOI:10.1155/2011/464368.

[20] Basiouny M 2018 1240 IR University of Toledo OH,.

[21] Nandiyanto A, Oktiani R, Ragadhita R 2019 *Indonesian Journal of Science & Technology* **4-1** 97-118.

[22] Munajad A, Subroto C and Suwarno 2018 *Energies* **11** 264-276.

[23] Kibria G and Bhattacharyya B 2017 *Microfabrication and Precision Engineering* 99-142.

[24] Wypych G 2014 *handbook of solvents-Safer Solvent and Processes 2nd edition (*ChemTec Publishing) **2** 635-785.

[25] Rodriguez A, Snapp C, Pessin M, Bauer P and Stevenson C *Engineering Innovations (NASA)* 182-199.

[26] Anwar A, Essam A, Assem M and Abdallah S 2020 *International Journal of Scientific and Engineering Research* **10-12** 470 - 477.

[27] Aradhana R, Mohanty S and Nayak S, 2018 *Polymer* **141** 109-123.

[28] Kulkarni H, Tambe P and Joshi G 2017 *Composite Interfaces* (Taylor and francis Group).

[29] Johari G 1993 *Chemistry and Technology of Epoxy Resins* (Dordrech-Springer).

[30] Jung I, Dikin D, Piner R and Ruoff R, 2008 *NANO letters* **8-12** 4283-87.

IOP Conf. Series: Materials Science and Engineering 1172 (2021) 012029 doi:10.1088/1757-899X/1172/1/012029

Non-destructive aerial and terrestrial analysis of uranium isotopic activity by sensitive γ-spectrometer for El-Sella site

M Sabry[1], A Omar[1], M A Abdel-Rahman[1] and S A El-Mongy[2]

[1] Nuclear Engineering Department, Military Technical College, Kobry El-Koba, Cairo, Egypt
[2] Nuclear and Radiological Regularity Authority (ENRRA), Nasr City, Egypt

E-mail: mohy_sabry@mtc.edu.eg

Abstract. Any comprehensive radioactivity surveillance for a specific site should be carried out by a sensitive airborne system (e.g. carried on aircraft) together with terrestrial sampling and analysis. In the framework of ore, environmental and nuclear materials monitoring and investigation, the activity and isotopic compositions signature are considered crucial factors that must be considered, determined and evaluated. In this work and based on previous aerial monitoring, a very sensitive hyper pure germanium detector (Hp-Ge) of 50% efficiency with the Genie2000 software has been used for spectroscopic non-destructive assay (NDA) of radioactivity content for samples collected from the Egyptian eastern desert (El-Sella Site). The identification of specific signature isotopes through their characteristic gamma lines and a calculation of their specific activities, activity ratios of $^{235}U/^{238}U$ and $^{234}U/^{238}U$ and mathematically estimation of the natural enrichment percentage are very vital and nuclear forensic targets. The results of measurements, analysis and calculations show that the Site samples have high radioactivity, natural enrichment origin and high uranium concentration. The obtained results are given tabulated, depicted, discussed and compared with the recent nationally published work and the international levels and limits.

1. Introduction

Environmental sampling (ES) is considered a very important tool and one of the powerful technical verification measures presented for the International Atomic Energy Agency (IAEA) safeguards. Many safeguard verification measures aim to confirm the amount and type of nuclear material declared by a state, ES gives a strong means for identifying undeclared activities and nuclear material and helps the IAEA in drawing reliable conclusions [1]. The most common and ideal method for measurements is the non-destructive assay because of the importance of not losing the radioactive and nuclear material needed to the analysis itself. This technique is used by the nuclear industry, defense facilities, and safeguard inspectors to make rapid accurate measurements of sensitive nuclear materials [2].

The first analysis performed after receiving the material is the identification of the radioisotopes in the material, the risk assessment, uranium isotopic abundance, activity and enrichment. This kind of analysis performed by a high-resolution gamma spectrometer (HRGS); mostly through using high pure germanium (Hp-Ge) detector [3] which was used to determine the characteristics of nuclear and radioactive materials and identify its radioisotopes from its gamma lines[2]. Energy and efficiency calibration were performed at the detector to ensure accurate results. Additionally, from the uranium isotopes activity, the enrichment can be determined. It is well known that gamma rays emitted from the

Content from this work may be used under the terms of the Creative Commons Attribution 3.0 licence. Any further distribution of this work must maintain attribution to the author(s) and the title of the work, journal citation and DOI.
Published under licence by IOP Publishing Ltd

radioactive material have specific energies and intensities that differ from one material to another according to the radioisotopes that exist in the material and its activities[2]. ^{238}U can be identified from its daughter energy lines ^{234}Th (63.29 keV) and ^{234}Pa (1001.03 and 766.36 keV) and ^{235}U can be identified from its own gamma energy lines (143.7, 163.3, 185.7 and 205.3 keV) [4].

The enrichment of uranium in the environmental samples should be natural, 0.72%, except in locations near uranium enrichment facilities as the uranium separation technologies mostly use UF_6 gas. This leads to trace amounts of enriched uranium being released to the environment from those facilities[5]. The concentration of uranium varies according to the location where it is collected and the type of material that it is mixed with. For example, the uranium concentration approximately 4 parts of uranium per million when uranium mixed with granite which covers 60% of the earth crust [6]. High-grade uranium contains about 200,000 ppm (µg/g) of uranium, while low-grade uranium contains about 1000 ppm U. There are slightly uranium-rich locations in Egypt like Gattar mountain which contains uranium in the range from 1000 to 5000 ppm [7].

This work focuses on the analysis of high radioactive samples by performing a set of measurements to calculate specific activity, natural enrichment verification and uranium isotopic composition samples of the El-Sella site in the South Eastern desert.

2. Experimental Work

2.1. Sample collection and preparation

The sample was collected from the Egyptian eastern desert, the El-Sella area. The sample was prepared for measurements in the nuclear engineering department laboratory at the Military Technical College through a set of sequential steps. The sample was grounded, crushed, homogenized and sieved until passing through mesh number (200). Then the sample was dried by oven at 105 °C for 1 day and then was weighed to be 323 g to be the same as the reference material used for calibration. The sample was then sealed in a cylindrical plastic container for 28 days to attain secular equilibrium[8]. This step is very important to ensure that all the uranium daughters remain in the sample as the radon daughter Rn-222 is in a gaseous form so it may escape from the sample unless it is tightly sealed.

2.2. Gamma spectrometer setup

The non-destructive technique used in this work was based on using hyper pure germanium (Hp-Ge) detector that has a 50% efficiency and a resolution of 1.95 keV at 1332 keV of ^{60}Co source with its associated electronics such as the spectroscopic amplifier and high voltage supply. The detector high voltage (H.V.) is 3.3 kV is connected to a Canberra DSA-1000 multichannel analyzer (MCA). The Hp-Ge crystal was fully immersed in liquid nitrogen for cooling (–200 °C) to reduce noise and leakage current. The Hp-Ge detector was shielded by a 10 cm thick lead material to avoid background radioactivity. The system used for identifying the radioactive nuclides through their gamma energies and calculation of their activities by analyzing the spectrum obtained from the detector was the Genie-2000 software [4].

2.3. Gamma Spectrometer Energy Calibration

The energy calibration was performed by using standard IAEA sources (point sources) model RSS8 and MGS-1 which contain certain radioisotopes (^{133}Ba-^{137}Cs-^{60}Co-^{22}Na-^{152}Eu) which have known activities (Table 1).

Table 1. Radionuclides of standard sources used for energy calibration

Model	Radioisotope	Energy (keV)	$T_{1/2}$ (day)	I_γ	Activity (Bq)	Reference date
RSS-8	Ba-133	81	3942	0.342	37000	1/10/2014
		276		0.071		
		303		0.184		
		356		0.622		
		384		0.089		
	Cs-137	662	11023	0.852	37000	
	Co-60	1173	1923.55	0.999	37000	
		1333		1		
	Na-22	1275	949	0.995	37000	
MGS-1	Eu-152	121.78	4927.5	0.284	37000	
		344.3		0.266		
		1408.01		0.208		
		964.1		0.145		
		1112.07		0.136		
		778.9		0.13		

The reference sources were counted for 1800 seconds in the (Hp-Ge) detector to obtain proper results[9]. With known energy readings of each radioisotope, a relationship can be obtained between the channel number and the energy which can be shown as follows in equation 2.1:

$$Energy = 0.1355 + 0.1758 \times Channel\ number \tag{2.1}$$

2.4. Gamma Spectrometer Efficiency Calibration

Efficiency calibration is considered to be an essential factor to attain accurate and valid measurements. The standard sources used for efficiency calibration should have the same geometry and density and should be measured at the same conditions as the unknown samples. The efficiency calibration was performed by using standard IAEA sources; **S-14** (^{234}U, ^{235}U, ^{238}U) and **S-16** (Am-214, Cs-137, Co-60) for uranium samples which can be found in table 2.

Table 2. Radionuclides of standard sources used for efficiency calibration

Model	Radioisotope	Energy (keV)	$T_{1/2}$ (day)	I_γ	Activity (Bq)	Reference date
S-14	U-235	143.1	2.57E+11	0.105	588.1	06/09/2010
		163		0.047	588.1	
		185.7		0.53	588.1	
		205		0.047	588.1	
	U-238	63	1.63E+12	0.038	12780	
		1001		0.007	12780	
	U-234	53	8.97E+7	0.001	12520	
		120		0.00034	12520	
S-16	Am-241	59.5	158000	0.357	5518.1	01/04/2011
	Cs-137	661.7	10980	0.852	2140.8	
	Co-60	1173.2	1925	0.999	3490.9	
		1332.5		1	3490.9	

The standard sources were counted in the detector for 86400 sec (24 hr) [10]. The absolute efficiency can be calculated by using the following equations.

$$\varepsilon_{abs} = \frac{C}{A_s I_\gamma T} \tag{2.2}$$

Where ε_{abs} is the absolute efficiency, **C** is the net counts of photopeak for a certain isotope in the reference material, A_S is the known specific activity for each isotope in the reference material, $\mathbf{I_\gamma}$ is the branching ratio, and **T** is the counting time. The two IAEA standard sources were used to cover a range of low and high gamma energies; usually single efficiency calibration is carried out for all energies but for this work, the calibration was divided into two categories, low and high, according to the energy. The low energy efficiency calibration was used for energies from 59.5 to 120.9 keV whilst the high energy efficiency calibration was used for energies higher than 120.9 keV. Figure 1 shows the relationship between efficiency and energy for the low energies and can be applied in equation 2.3 to calculate the efficiency by the known energy. Moreover, since Figure 2 shows the relationship between efficiency and energy for the high energies, the results can be applied in equation 2.4. This method is used to reduce the error in the measurements as much as possible.

$$Eff = -0.1787 + 0.1934 \,.\, ln\, E \tag{2.3}$$

$$Eff = 0.0191 - \frac{7.425}{E} + \frac{3352.26}{E^2} + \frac{236871.9}{E^3} \tag{2.4}$$

Where Eff is the absolute efficiency of the detector and E is the energy for each radioisotope in (keV)

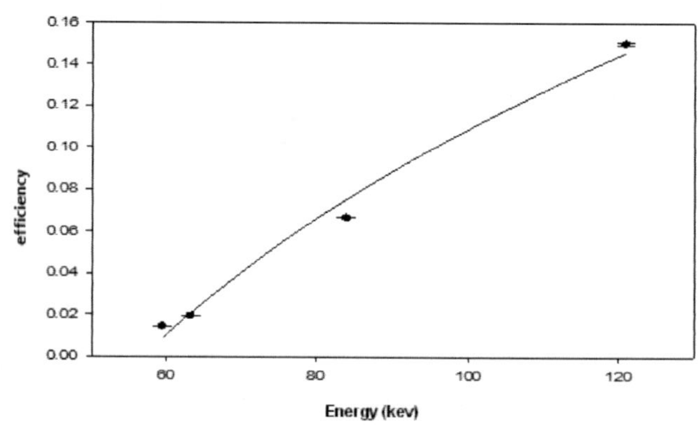

Figure 1. Efficiency calibration of the system for low energies

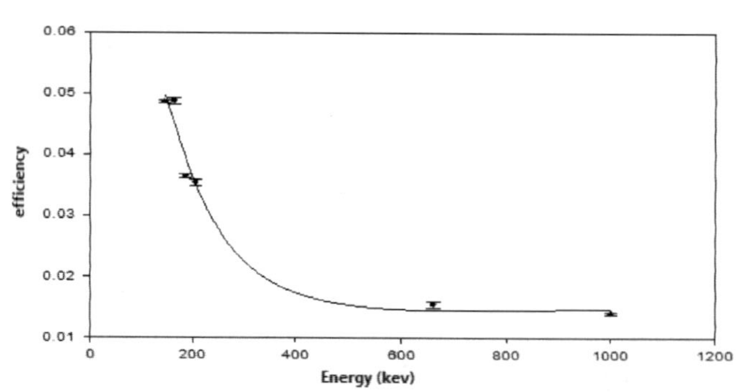

Figure 2. Efficiency calibration of the system for high energies

2.5. Spectrum analysis and activity calculation

Each uranium isotope emits gamma rays which have certain energies. These gamma rays when interacting with the detector are converted into signals and are processed to be shown as a spectrum.

Each peak in the spectrum indicates the gamma energy line and each energy line indicates the presence of a certain radioisotope. The sample was measured for 86400 sec. The photo-peaks which have high intensities were obvious in the spectrum which can be seen in Figure 3.

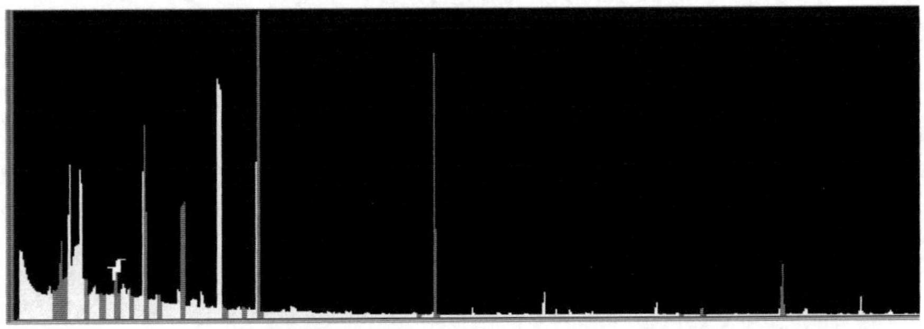

Figure 3. Typical spectrum for El-Sella sample using Genie2000 software

The activity can be calculated by estimating the counts from the radioactive sample spectrum which is counted three times to estimate the average and the standard deviation. Each line in the spectrum has a certain energy, efficiency and branching ratio. The uranium isotope ^{238}U itself is not emitting gamma lines so it can only be identified by the energy lines from its daughters, ^{234}Th (63.29 keV) and ^{234}Pa (1001.03 and 766.36 keV) which are considered to have the same activity of ^{238}U. The ^{235}U isotope can be identified by its energy lines (143.7, 163.3, 185.7 and 205.3 keV). On the other hand, ^{234}U can be identified from its two energy lines (53 and 120.9 keV). The sample activity can be calculated using the following equation:

$$A = \frac{C}{\varepsilon_{abs} \cdot I_{\gamma} T .m} \tag{2.5}$$

Where, **A** is the specific activity of the radioisotope (in Bq/kg). **C** is the net count of the photo-peak after subtraction of the background counts. $\boldsymbol{\varepsilon_{abs}}$ is the absolute efficiency of the detector, I_{γ} is the branching ratio, **m** is the mass of the sample and **T** is the time of counting (sec).

3. **Result and Discussion**

3.1. Specific Activity and Activity Ratio results

The results of specific activity obtained by using equation (2.5) for each isotope were summarized in Table 3. It was found that the average value of specific activity for the ^{235}U isotope was 1268.675±98 Bq/kg as the values of energy lines (143.7, 163.36, 185.72 and 205.23 keV) were 1167.4, 1353.3, 1201.2 and 1352.8 Bq/kg respectively. As for ^{238}U, the average specific activity was 27186.25±376 Bq/kg as the values of energy lines (63.29and 1001 keV) of daughters (^{234}Th and ^{234m}Pa) were 26919.77 and 27452.74 Bq/kg respectively. The average specific activity of ^{234}U was 22067.78±82 Bq/kg as its two energy lines (53.24 and 120.9 keV) specific activity were 22067.78 and 22125.81 Bq/kg respectively. The activity ratio of $^{235}U/^{238}U$ was also calculated and found to be in line with the natural reference value (0.046) [4]. The remarkable result was the $^{234}U/^{238}U$ activity ratio as its value was 0.81, smaller than unity <1, which means that there is disequilibrium between the parent and daughter in the decay series. The reason for this is that a disturbance occurred due to uranium migration probably by the groundwater interaction with rocks [11]. The result of specific activity and activity ratios were then compared with another work at the same location (El-Sela) and the results were in a comparable range. The previous work found that the specific activity for ^{238}U ranged from 500 to 80065 Bq/kg, for ^{235}U ranged from 10 to 3557 Bq/kg and the $^{234}U/^{238}U$ activity ratio ranged from 0.6 to 0.92 [12].

3.2. The Natural Enrichment Verification Results

The enrichment percentage of the sample can be estimated by the following atom ratio equation:

$$\%E = \left(\frac{N_{^{235}U}}{N_{^{234}U} + N_{^{235}U} + N_{^{238}U}} \right) \times 100 \qquad (3.1)$$

Where, $N_{^{235}U}$, $N_{^{234}U}$ and $N_{^{238}U}$ are the number of radionuclide of ^{235}U ^{234}U and ^{238}U respectively, which can be estimated from the decay constant and the activity by using the decay equations as follow:

$$N_{^{235}U} = \frac{A_{^{235}U}}{\lambda_{^{235}U}} = \frac{C}{\varepsilon . I_{\gamma} T . \lambda_{^{235}U}} \qquad N_{^{234}U} = \frac{A_{^{234}U}}{\lambda_{^{234}U}} \qquad N_{^{238}U} = \frac{A_{^{238}U}}{\lambda_{^{238}U}} \qquad (3.2)$$

So, the enrichment percentage equation (3.1) can be written as follow:

$$\%E = \left(\frac{\frac{A_{^{235}U}}{\lambda_{^{235}U}}}{\frac{A_{^{234}U}}{\lambda_{^{234}U}} + \frac{A_{^{235}U}}{\lambda_{^{235}U}} + \frac{A_{^{238}U}}{\lambda_{^{238}U}}} \right) \times 100 = \left(\frac{A_{^{235}U} . (\tau_{1/2})_{^{235}U}}{A_{^{234}U} . (\tau_{1/2})_{^{234}U} + A_{^{235}U} . (\tau_{1/2})_{^{235}U} + A_{^{238}U} . (\tau_{1/2})_{^{238}U}} \right) \times 100 \qquad (3.3)$$

Where, $\lambda_{^{234}U}$, $\lambda_{^{235}U}$, $\lambda_{^{238}U}$ are the decay constants and $(\tau_{1/2})_{^{234}U}$, $(\tau_{1/2})_{^{235}U}$ and $(\tau_{1/2})_{^{238}U}$ are the half-lives time of ^{234}U , ^{235}U and ^{238}U respectively.

The result of the sample natural enrichment was obtained from applying equation (3.3) by using the average specific activity of uranium isotopes and the half-life time of each isotope. The calculated results are shown in Table 3. The enrichment percentage of the verified sample is of natural origin (0.729% ±0.016) as expected for all the natural environmental samples.

Table 3. Activity results of uranium isotopes and natural enrichment percentage of the sample

Isotopes	Energy (keV)	$t_{1/2}$ (year)	S1 Activity (Bq/kg)	Av Activity ±SD	Calculated Enrichment %
U-235	143.7	7.038E+8	1167.4	1268.675±98	0.729±0.016
	163.36		1353.3		
	185.72		1201.2		
	205.23		1352.8		
Th-234	63.29	4.468E+9	26919.77	27186.25±376	-
Pa-234m	1001		27452.74		
U-234	53.24	2.455E+5	22067.78	22067.78±82	-
	120.9		22125.81		

3.3. Uranium Isotopic Mass and Total Uranium Concentration Calculations

The uranium mass of certain sample can be calculated by knowing the activity and the decay constant for each uranium isotope through applying the following equations:

$$N_{^{238}U} = n_{^{238}U} .N_A = \frac{m_{^{238}U}}{M_{^{238}U}} .N_A = \frac{A_{^{238}U}}{\lambda_{^{238}U}} \quad (3.4)$$

Then the mass of uranium isotopes can be expressed as follows

$$m_{^{238}U} = \frac{A_{^{238}U}}{\lambda_{^{238}U}} . \frac{M_{^{238}U}}{N_A} \quad , \quad m_{^{235}U} = \frac{A_{^{235}U}}{\lambda_{^{235}U}} . \frac{M_{^{235}U}}{N_A} \quad , \quad m_{^{234}U} = \frac{A_{^{234}U}}{\lambda_{^{234}U}} . \frac{M_{^{234}U}}{N_A} \quad (3.5)$$

Where, $m_{^{238}U}$, $m_{^{235}U}$ and $m_{^{234}U}$ are the uranium masses of ^{238}U, ^{235}U and ^{234}U respectively. $M_{^{238}U}$, $M_{^{235}U}$ and $M_{^{234}U}$ are the atomic weight of ^{238}U, ^{235}U and ^{234}U respectively. N_A is the Avogadro's number. The total uranium concentration in the sample can be calculated by using the ratio of the total masses of uranium isotopes and the total sample mass as finally calculated by the following Equation:

$$m_{U_t} = \frac{m_{^{238}U} + m_{^{235}U} + m_{^{234}U}}{m_{sample}} \quad (3.6)$$

The uranium isotopes masses of ^{234}U, ^{235}U and ^{238}U were calculated by using equation 3.5 which depends on the activity of the isotope in Bq, the decay constant in s^{-1}, molar mass in $g.mol^{-1}$ and Avogadro's number. The uranium isotopes masses results are found to be 0.0309 mg for ^{234}U, 5.112 mg for ^{235}U and 705.08 mg for ^{238}U and summarized in Table 4.

Table 4. The Uranium isotopes masses and total uranium concentration of the samples

Isotopes	Activity (Bq)	Decay constant $\lambda = \ln(2)/t_{1/2}$ (s^{-1})	Uranium isotopes masses (mg)	Uranium content in 323 g (mg)	Uranium concentration (ppm)
U-234	7127.64	8.94374 E-14	0.0309	710	2198.8
U-235	408.9	3.12161 E-17	5.112		
U-238	8781.07	4.9218 E-18	705.0835		

As shown from table 4. the total uranium concentration as calculated by using equation 3.6, which divides the summation masses of each uranium isotope over the mass of the sample, was found to be 710.2 mg in the sample weight of 323 g. Henceforth, the total uranium concentration was concluded to be 2198.8 ppm. The results compared with previous work and found to be relatively comparable with our results, where their calculated total uranium concentration ranged from 203 to 6147 ppm for this site (El-Sela) [12].

4. Conclusion

In this work, a non-destructive assay technique was applied based on a gamma spectrometer (Hp-Ge) detector, which was calibrated for energy and efficiency by a modified technique to reduce the errors in the results. The Hp-Ge detector was used to measure and analyze specific radioactive sample collected from the El-Sela area in the southeastern desert in Egypt. The sample was well precisely prepared for verification. The measured specific activity was then calculated and found to be very high; 1268.675±98 Bq/kg for ^{235}U, 27186.25±376 Bq/kg for ^{238}U and 22067.78±82 for ^{234}U. The activity ratio of ^{235}U/^{238}U was calculated and found to be a natural signature value (0.046). The ^{234}U/^{238}U activity ratio was less than unity indicating the occurrence of disequilibrium. The enrichment percentage of the sample was calculated from a derived equation and found to be of natural origin; 0.72%. The uranium isotopes masses were estimated from an equation deduced from the decay equation for ^{234}U, ^{235}U and ^{238}U and found to be 0.0309, 5.112 and 705 mg, respectively. The total uranium concentration in the sample was also calculated and found to have a high value of 2198.8 ppm (µg/g). The verified sample results were compared with previous work for the same Site.

References

[1] Axelsson A, Fischer D and Peńkin M V 2009 Use of data from environmental sampling for IAEA safeguards. Case study: Uranium with near-natural 235U abundance *J. Radioanal. Nucl. Chem* **282** pp. 725–729

[2] Tohamy M *et al* 2016 Passive Non-Destructive Assay based on gamma-ray spectrometry to verify UO2 samples in the form of powder and pellet *Ann. Nucl. Energy* **87** pp 186–191,

3] Keegan E *et al* 2014 Nuclear forensic analysis of an unknown uranium ore concentrate sample seized in a criminal investigation in Australia *Forensic Sci. Int* **240** pp 111–121

[4] Abdel-Rahman M A E, Shady A, and El-Mongy S A 2018 Analysis of ores and its purified constituents by γ -spectrometry with calculation of uranium isotopic atom, mass, and activity *Z. Anorg. Allg. Chem* **644** pp.477–482

[5] Belew W L 1993 Detection of uranium enrichment activities using environmental monitoring techniques *Department of Energy`s (DOE) international safeguard meeting* pp 2–4.

[6] IAEA 2009 Nuclear fuel cycle information system *Int. At. Energy Agency* pp. 1–82.

[7] Salman A Aet al 2019 Gebel Gattar prospect, an obvious model of intragranitic uranium mineralization *Second Arab Conference on the Peaceful Uses of Atomic Energy* pp 269–281

[8] Abdel-Rahman M A E, Sabry M, Khattab M R, El-Taher A and El-Mongy S A 2020 Radioactivity and risk assessment with uncertainty treatment for analysis of black sand minerals *Z. Anorg. Allg. Chem* **647** pp 1–9

[9] Shabaka A N, Omar A, El-Mongy Sayed A and Tawfic A F 2020 Analysis of natural radionuclides and 137Cs using HPGe spectrometer and radiological hazards assessment for Al-Nigella site, Egypt *Int. J. Environ. Anal. Chem* **100** pp 1–14

[10] Abdel-Rahman M A E and El-Mongy S A 2017 Analysis of radioactivity levels and hazard assessment of black sand samples from Rashid area - Egypt *Nucl. Eng. Technol* **49** pp 1752-1757.

[11] Ibrahim E M, 2017 "Effect of flash flood on uranium-series disequilibrium and groundwater residence time using gamma ray spectrometry," *Journal of Recent Scientific Research* **8** pp. 15126–15141.

[12] Gawad A E and Ibrahim E M 2016 Activity ratios as a tool for studying uranium mobility at El Sela shear zone, southeastern Desert, Egypt *J. Radioanal. Nucl. Chem* **308** (1) pp 129–142.

Investigation of foamcrete mechanical and physical properties

T Youssef [1], Y Elsayed [2], T A Elsayed [2] and S M Lawaty [2,3],

[1] Faculty of Engineering, l'Université Française d'Égypte, Cairo, Egypt
[2] Faculty of Engineering (Mataria), Helwan University, Cairo, Egypt
[3] Faculty of Engineering & Technology, Badr University in Cairo, Egypt

E-mail: Tarik.youssef@ufe.edu.eg

Abstract. Engineered Material Arresting Systems (EMAS) serve as a substitutional alternative/solution to airport runways when the Runway Safety Area (RSA) does not meet international Federal Aviation Administration (FAA) standards. The length of the runway can be shortened if an EMAS is installed on both ends of the runway. This paper provides experimental test results on foamcrete material used for such an important safety application. The objective of the paper is to present the required and measured properties of foamed concrete (density, compressive strength and water absorption) and results of an evaluation of a first phase of testing. In this study, a total of fourty eight mixes were conducted; yielding a range of densities, compressive strength(s) and water absorption characteristics that are: 554 to 1528 kg/m^3, 1.1 to 21 MPa and 7.4 to 28.3 %, respectively. It is demonstrated herein – though the Analysis of Means statistical method – that foam volume is predominantly the main factor affecting the observed output characteristics. This is followed by Sand/Filler and Filler/Cement that yield marginal effect compare to the former foam volume ingredient.

1. Introduction

Engineered Material Arresting Systems (EMAS) have been developed and enhanced as a solution to safely stop an over-running aircraft on runways; where obtaining the Federal Aviation Administration (FAA) standards of 1000 ft (304.8 m) length is not practical/applicable[1]. EMAS is typically composed of cellular cement-based blocks designed to stop an aircraft safely whilst over-running a runway at 70 knots (118 ft/sec) or less with minimal damage to the aircraft and equally minimal risk to passengers. It is mandatory that such system fully satisfies the safety requirements of “Part 139 Inspection” of the FAA Standards [2].

As the aircraft enters the EMAS, it makes a great transition from the paved ramp into the cellular cement arresting bed. Aircraft wheels then establish a tire to material interface; crushing the blocks as it moves through them (See **Figure *1*** and **Figure *2***). It is this contact that creates resistive drag force that quickly slows down and ultimately stops the aircraft. EMAS systems – in turn – are supposed to be customized to airport needs and limitations. They should also be tailored to accommodate specific aircraft types and a range of runway exit speed. As a result, arrester bed design(s) rely on complex computer analysis and modelling through incorporating many variables for each aircraft operating at the airport. The obtained models are continuously validated using the measured output of FAA testing.

Content from this work may be used under the terms of the Creative Commons Attribution 3.0 licence. Any further distribution of this work must maintain attribution to the author(s) and the title of the work, journal citation and DOI.
Published under licence by IOP Publishing Ltd

Figure 1. Coastal Carolina Regional Airport, ending with Arrested Bed Zone

Figure 2. Tire-to-Foamcrete Interface Crushing, ultimately stopping aircraft

On site, the arrester bed area is prepared for EMAS and individual blocks are set into place. They are adhered to the pavement and sealed to further maintain the integrity of the system, in a simple and straightforward manner. Many real-life incidents/accident, demonstrated the effectiveness of the EMAS system; one of which is when an Eastern Air Lines Boeing 737-700, with 37 people on board, landed on La Guardia's runway 22 in rain and reduced visibility, overran the end of the runway, was slowed by the runway arrestor bed (EMAS) and came to a stop about 60 meters/200 feet past the runway end and about 40 meters/130 feet to the right of the extended runway center line. One occupant received minor injuries, the aircraft received minor if any damage [3].

Preliminary radar data suggested that the aircraft was on a normal approach profile descending through 250 feet AGL at 135 knots over ground, after touch down was still doing 132 knots over ground about 1400 meters/4600 feet past the runway threshold with about 700 meters/2300 feet of runway remaining. Ground observers reported it was raining hard at the time of touch down, water was splashing everywhere from the aircraft (The Aviation Herald, 2016).

2. Study Objectives

Now that cellular concrete, a multi-functional key solution material [4] to [10], is gaining more popularity in the aviation safety domain and construction field, its characteristics need further investigation to exploit its capacity. Superior traits, such as low density, high thermal insulation and high fire resistance, are also attributed to foamed concrete; making it gain popularity in other fields such as the construction domain. Nonetheless, lack of knowledge remains regarding the prediction of the mechanical and physical properties of foamed concrete; necessitating further experimentation and modeling.

Many researchers have conducted investigations on foamed concrete to improve its mechanical properties and reported that the dry density and the pore structure have the most significant influence on the mechanical properties of foamed concrete [11], [12]. However, further research is yet necessary to: (i) increase the currently available knowledge, (ii) develop a mix-design approach to produce foamed concrete for different purposes and anticipated aircraft contingencies. In this study, the main objectives are to investigate the influence of different mix proportions of preformed foamed concrete on its physical and mechanical properties (particularly compressive strength, density and water absorption). The statistical approach used for this investigation is the graphical Analysis of Means (ANOM) approach that depicts significant differences among groups of information in a visual form. ANOM methodology compares the average of each group to the mean of the overall process to discover statistical differences of significance; thus, highlighting on the most influential input variable. Analysis of means is a systematic statistical procedure, active mostly in quality control.

This effort comprises a total of fourty eight mix proportions (experiments) with two different types of fillers (sand and lime powder), cement, water and protein-based foam. The resulting outputs, density, compressive strength and water absorption, are observed and discussed herein. This study is considered an initial phase in a more comprehensive study to produce such materials to serve the local Egyptian market whether in the aviation safety field or construction industry.

3. Experimental Program

3.1. materials

In this program, ordinary Portland cement (CEM I 42.5R) is used as a binder, where the chemical composition is displayed in Table 1 and mechanical-and-physical properties are displayed in Table 2. Natural sand of specific gravity (S.G = 2.59) - sieved to avoid particles larger than 1.18 mm - is used as filler. Lime powder (calcium carbonate, $CaCO3$) of (S.G = 2.59) serves as a partial and full replacement for natural sand. X-Ray Diffraction analysis (XRD), conducted for the latter, shows that it is predominantly formed of calcite (See Figure 3).

Table 1. Chemical composition of ordinary Portland cement.

Oxide	SiO_2	$A12O_3$	Fe_2O_3	CaO	MgO
%	19.29	4.52	3.59	62.08	1.80
Oxide	SO_3	K_2O	Na_2O	Cl	
%	3.61	0.29	0.45	0.09	

Table 2. Mechanical and physical properties of ordinary Port-land cement.

Compressive strength- 2 days	(MPa)	19.50
Compressive strength- 28 days	(MPa)	51.25
Setting time	(minutes)	123
Fineness (Blaine)	(cm^2/gm)	3732

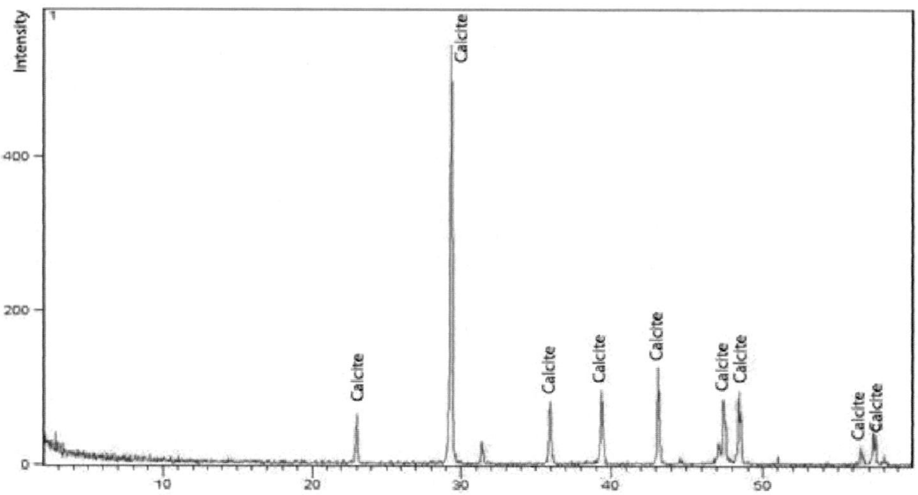

Figure 3. XRD test results of lime powder.

Foaming agent, LithoFoam SL 200-L, based on highly-active foam forming proteins, pre-foamed foam (at 80 kg/m3) was used. The foam is produced by blending foam agent, water and compressed air in a foam generator, as shown in Figure 4.

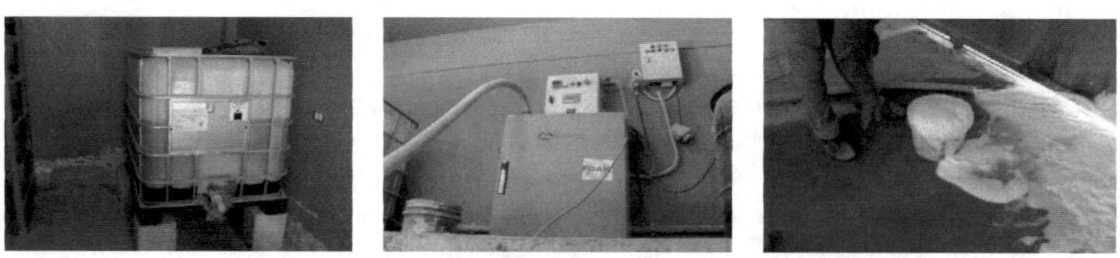

a. Foam agent tank. b. Generator set. c. Produced foam.

Figure 4. Foam Generator

3.2. Mix proportion

Forty-eight mix proportions were designed and conducted to investigate the influence of each ingredient on the foamed concrete physical/mechanical properties. These ingredients/factors are overall filler to cement ratio (F/C), sand-to-overall filler ratio (S/F) and foam volume per unit volume of concrete (V_f); noting that filler F, herein indicates both Sand and $CaCO_3$, that vary in percentage(s) depending on the mix (See Table3 below).

Table3 . Proportions of concrete mixes per one m^3

Ser.	Cement (kg)	Sand (kg)	Lime (kg)	Water (kg)	Foam (L)
S 73	798	80	319	389	479
S 75	641	64	257	312	641
S 77	526	52	211	256	736
S 79	424	42	169	206	763
S 81	345	35	138	169	760
S 83	343	34	137	167	892
S 85	598	119	478	379	478
S 87	506	102	404	321	675
S 89	455	91	363	289	848
S 91	343	69	274	218	822
S 93	269	54	215	171	791
S 95	254	51	203	161	878
S 109	736	147	221	348	441
S 111	642	129	192	303	642
S 113	701	140	211	331	981
S 115	462	92	139	218	831
S 117	347	69	104	164	763
S 119	336	67	101	159	876
S 121	572	229	343	357	458
S 123	240	96	144	150	320
S 125	447	178	268	279	834
S 127	342	136	206	213	820
S 129	336	134	201	209	984
S 131	273	109	164	171	949
S 145	754	226	150	356	452
S 147	636	191	127	300	636
S 149	569	170	114	269	797
S 151	537	161	107	254	967
S 153	426	127	85	201	936
S 155	351	105	70	166	915
S 157	597	358	239	369	478
S 159	411	247	164	254	548
S 161	395	236	158	244	737
S 163	335	201	133	207	803
S 165	297	178	119	184	873
S 167	263	158	105	163	912
S 181	694	277	69	308	417
S 183	562	224	56	249	562
S 185	567	227	57	252	794
S 187	438	175	44	194	787
S 189	371	149	37	165	817
S 191	316	126	32	141	821
S 193	547	438	109	282	438
S 195	435	348	87	225	580
S 197	367	294	73	190	687
S 199	310	248	62	160	745
S 201	312	250	62	161	914
S 203	262	211	53	136	910

3.3. Water-solid Ratio

Preformed foamed concrete is manufactured by adding preformed foam to cement mortar with a specific consistency (defined in terms of water-solid ratio). Optimal consistency is crucial; since using mortars at higher or lower consistency than the optimal, leads to foamed concrete with density ratio (defined as ratio of measured fresh density to design density) above unity. It is recommended - by the foam manufacturers - to use mortars with percent flow (measured by standard flow table [13]) in the range of 40 % to 50 % to obtain optimal consistency. Stiff mixes with low water-solids ratio causes bubbles to break. On the other hand, loose mixes with high water-solids ratio causes bubbles to merge and segregate [14]. Contrary to conventional concrete, water-cement ratio is not an influential factor on the compressive strength of foamed concrete [5], thus not considered in this study

3.4. Specimens Preparation

The process of foamed concrete manufacturing is described as follows: Portland cement and filler (sand and/or lime powder) were initially mixed in a horizontal mixer; water was added to the mixer; foam - at its final form was then added - to the homogeneous paste. Finally, full homogeneous foamed concrete was cast in 600 x 600 x 100 mm steel panels. The concrete panels were cured using wet burlap sheets for 28 days. Figure 5 shows the process of foamed concrete manufacturing.

a. Empty mixer. b. Solid materials. c. Adding water. d. Mixing paste. e. Adding foam. f. Final product.

Figure 5. Foamed concrete manufacturing process.

3.4 Description of Experimental Procedures

3.4.1. Density: Five cubes with dimensions 100 x 100 x 100 mm - (according to ASTM C513-89 R95) - were saw-cut from each concrete panel to be tested. Oven-dry mass (A), saturated sur-face-dry mass in air (B) as well as the immersed mass of saturated specimen in water (C) were recorded according to ASTM C642-97; from which the bulk density is calculated in Equation 1:

Equation 1:
$$\rho = (A / (B - C))\rho_w \quad (1)$$

where ρ is the density of foamed concrete and ρ_w is the density of water.

3.4.2. Compressive Strength: Four cubes of dimensions 100x100x100 mm were saw-cut from each concrete panel to be tested - (according to ASTM C513-89 R95) - to determine the 28-day compressive strength.

3.4.3. Water Absorption: Cubes with dimensions 100 x 100 x 100 mm were cut from each concrete panel to be tested. Oven dry mass (A), saturated surface-dry mass in air (B) and immersed mass of saturated specimen in water (C) were recorded according to ASTM C642-97 Equation 2.

Equation 2 $$\text{Water absorption (\%)} = (B-A)/A \times 100 \quad (2)$$

Porosity can be determined by the following Equation 3:

Equation 3 $$\text{Porosity (\%)} = (B-A)/(B-C) \times 100 \quad (3)$$

Where ρ = water density.

4. Results and Discussion

As per the experimental procedures explained above, **Table 4** below demonstrates the obtained results for density, compressive strength and water absorption of the tested 48-foamcrete mixes.

Table 4. Test Results

Ser.	Density (kg/m^3)		Compressive strength (MPa)	Water absorption (%)
	Fresh	Hardened		
S 73	1625	1487	21.00	13.42
S 75	1326	1249	11.28	12.38
S 77	1105	1020	7.50	16.94
S 79	902	754	2.52	24.16
S 81	747	631	1.68	26.35
S 83	753	666	1.77	21.01
S 85	1612	1528	17.45	19.41
S 87	1387	1309	11.30	9.57
S 89	1265	1131	6.26	18.97
S 91	969	859	3.54	27.2
S 93	773	698	1.94	24.99
S 95	738	593	1.62	23.21
S 109	1487	1415	18.32	12.03
S 111	1318	1257	12.89	13.32
S 113	1461	1377	6.86	18.15
S 115	977	880	2.95	16.22
S 117	745	647	2.27	31.34
S 119	734	609	1.56	26.3
S 121	1538	1451	11.05	10.62
S 123	655	596	6.46	14.05
S 125	1239	1133	4.45	25.88
S 127	963	826	2.65	27.34
S 129	959	814	2.18	21.33
S 131	793	660	1.65	26.18
S 145	1522	1374	14.46	7.4
S 147	1305	1179	10.35	16.62
S 149	1186	1049	6.47	17.11
S 151	1136	1001	4.79	19.86
S 153	914	791	2.41	21.26

Ser.	Density (kg/m^3)		Compressive strength (MPa)	Water absorption (%)
	Fresh	Hardened		
S 155	766	627	1.73	26.19
S 157	1600	1496	15.33	13.72
S 159	1120	1046	5.01	14.73
S 161	1092	953	3.8	18.25
S 163	940	845	3.19	19.47
S 165	849	721	1.84	23.35
S 167	762	614	1.38	26.38
S 181	1382	1233	12.16	13.63
S 183	1136	1030	5.34	14.18
S 185	1166	1029	4.78	16.7
S 187	913	785	2.65	19.72
S 189	787	692	1.32	21.11
S 191	680	554	1.09	25.45
S 193	1411	1305	9.76	13.44
S 195	1142	1038	4.19	15.58
S 197	980	876	2.49	13.96
S 199	841	735	1.53	24.14
S 201	858	710	1.12	21.24
S 203	734	617	1.12	28.31

Figure 6, below, demonstrates that the main effective parameter on density is foam volume (V_f) followed by Sand/Filler and finally Filler/Cement input parameters. The differences in mean responses of Filler/Cement, Sand/Filler and (V_f) are 16.2, 113.3 and 781.1, respectively. For this study, the optimal parameters for minimum density are Filler/Cement = 1.0, Sand/Filler = 0.2, and a (V_f) of 92%, by volume.

Foam volume dosage is the main influential factor on the density of foamed concrete. Filler/Cement has a little effect on density as resulted from analysis of means. It is inversely proportional to density which can be explained by the fact that specific gravity of filler is less than that of cement. This in turn results in lower density. Sand/Filler is significantly more effective than Filler/Cement. It is inverse proportional with density which can be explained by the fact that finer filler results in better distribution of air bubbles and less bubbles loss which results in positive effect on density.

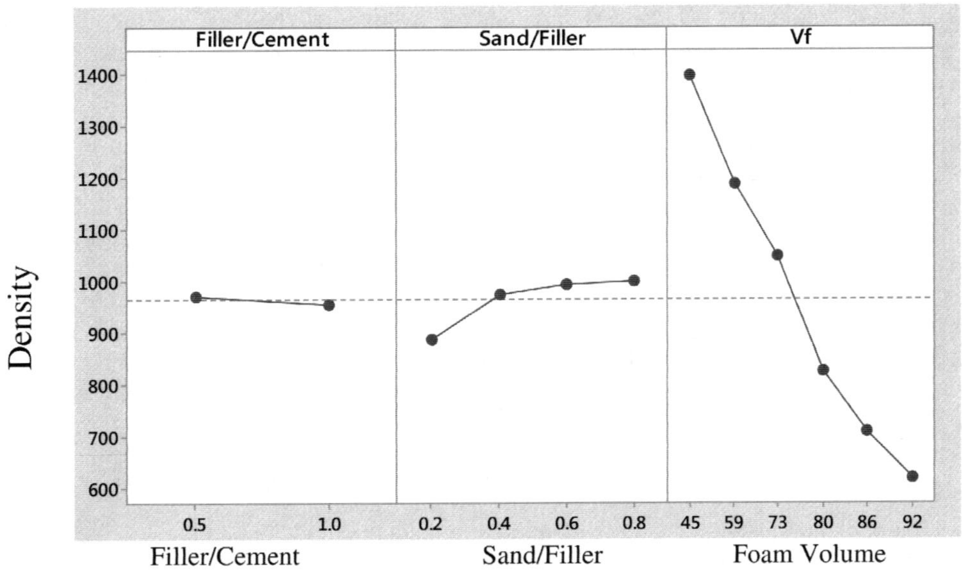

Figure 6. Factors affecting foamcrete density

Figure *7*, below, demonstrates that the main effective parameter on density is foam volume (V_f) then the Sand/Filler and Filler/Cement parameters. The differences in mean responses of Filler/Cement, Sand/Filler and (V_f) are 1.56, 3.61 and 12.98, respectively. For this study, the optimal parameters for maximum compressive strength are Filler/Cement = 0.5, Sand/Filler = 0.2, and a (V_f) of 45%, by volume. Foam volume dosage is the main influential factor on the compressive strength of foamed concrete. Filler/Cement has a marginal effect on compressive strength as shown by the analysis of means. It is inversely proportional to compressive strength whereas Sand/Filler is more effective than Filler/Cement.

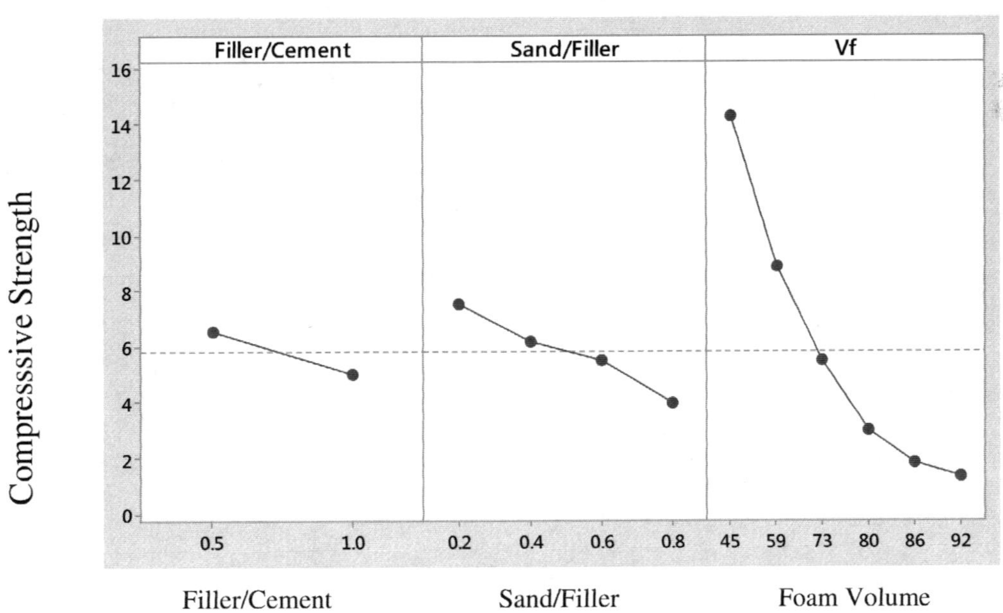

Figure 7. Factors affecting foamcrete compressive strength

The latter is also inversely proportional to compressive strength which can be explained by the fact that finer filler results in better distribution of air bubbles; resulting in positive effect on compressive strength. This is in addition to the micro filling effect. This explanation also meets the conclusions obtained in earlier research studies.

On the other hand, **Figure *8*** illustrates that the main effective parameter on water absorption is, once again, foam volume (V_f) then the Sand/Filler and Filler/Cement parameters. The differences in mean responses of Filler/Cement, Sand/Filler and (V_f) are 0.08, 1.58 and 13.57, respectively. For this study, the optimal parameters for minimum water absorption are Filler/Cement = 1.00, Sand/Filler = 0.80, and a (V_f) of 45%, by volume. Foam volume dosage is the main influential factor on the water absorption of foamed concrete. Filler/Cement has a little effect on water absorption as demonstrated through resulted analysis of means. Sand/Filler is more effective than Filler/Cement in this regards. The former is inversely proportional with water absorption which can be attributed to the increase in sand content, which leads to decrease in shrinkage and water demand; thus consecutive decrease in water absorption.

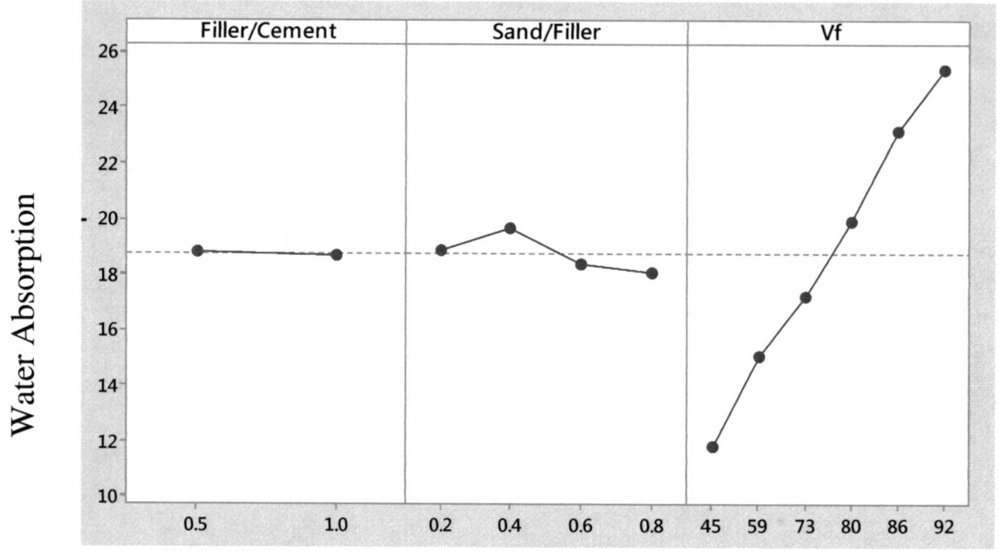

Figure 8. Factors affecting foamcrete water absorption

5. Recapitulation / Conclusion

In this study forty eight mixes were conducted; yielding a range of densities, compressive strength(s) and water absorption characteristics that are: 554 to 1528 kg/m^3, 1.1 to 21 MPa and 7.4 to 28.3 %, respectively.

Using the statistical Analysis of Means (ANOM) method allowed to deduce that foam volume (V_f) is predominantly the main factor affecting the observed output characteristics. This is followed by Sand/Filler and Filler/Cement that yield marginal effect compare to the former foam volume ingredient.

Further research is ongoing using different types of foam and filler, in a quest for better performance for the previously mentioned and other characteristics geared to aircraft tire arresting.

References

[1] Fredral Aviation Adminstration (FAA), 2013. *Code of Regulation, Airport Certification Manual Part 139*, s.1: US Government Publishing Office.

[2] Fedral Aviation Administration (FAA), 2012. *Advisory Circular No150/5220-22B*.s.1:US Government Publishing Office.

[3] The Aviation Herald 2016. *Accident; Eastern Air Lines B737, overran runway on landing long*, (New York),The Aviation Herald,

[4] Yanbin Fu et al. 2020. *Advances* in *Materials science and Engineering* "Foam Concrete: A state of the Art and State of the Practice Review" **volume 2002**, p. 25.

[5] Kui Wu et al., 2020. Construction *and Building Materials,* "A solution for squeezing deformation control in tunnels using foamed concrete: A review" **volume 257**, p. 15.

[6] Neville A M and Brooks J J, 1987. *Pearson Education Limited* "Concrete Technology" (Harlow, United Kingdom) pp. 357-358.

[7] Ma C and Chen B, 2011. *Construction and Building Materials,* "Experimental Study on The Preparation and Properties of A Novel Foamed Concrete Based on Magnesium Phosphate Cement" **volume 137**, (ELSEVIER) pp. 160-168.

[8] Xu Y, Jiang J, Xu J and Li Y, 2012. *Cement Concrete Re-search* "Mechanical Properties of Expanded Polystyrene Lightweight Aggregate Concrete and Brick" **volume 27**, (ELSEVIER) pp. 32-38.

[9] Hamad A J, 2014. *International Journal of Materials Science and Engineering* "Materials, Production, Properties and Application of Aerated Lightweight Concrete" **volume 2**, pp. 152-157.

[10] Aliabdo A A, Abd-Elmoaty A M and Hassan H H, 2014. *Alexandria Engineering Journal* "Utilization of Crushed Clay Brick in Cellular Concrete Production" **volume 52**, (ELSEVIER) pp. 119-130.

[11] Esmaily H and Nuranian H, 2012. *Construction and Building Materials* "Non-Autoclaved High Strength Cellular Con-crete From Alkali Activated Slag" **volume 25,** (ELSEVIER) pp. 200-206.

[12] Panesar D K, 2013. *Construction* and *Building Materials* "Cellular Concrete Properties and The Effect of Ssynthetic and Protein Foaming Agents" **volume 44**, (ELSEVIER) pp. 575-584.

[13] ASTM C230-03, 2005. Annual *Book of ASTM standards* "Standard Specification for Flow Table for Use in Tests of Hy-draulic Cement" **volume 4.01. International**.

[14] Nambiar E K and Ramamurthy K, 2006. *Cement Concrete Composites* "Influence of Filler Type on The Properties of Foam" **volume 28**, (ELSEVIER) pp. 475-480.

IOP Conf. Series: Materials Science and Engineering 1172 (2021) 012031 doi:10.1088/1757-899X/1172/1/012031

On the unsteady behaviours of the adiabatic endwall film cooling effectiveness

M Qenawy [1], Y Liu [2] and W Zhou [2]

[1] Faculty of Energy Engineering, Aswan University, Aswan 81528, Egypt.
[2] School of Mechanical Engineering, Shanghai Jiao Tong University, Shanghai 200240, China.

E-mail: m_qenawy@aswu.edu.eg.

Abstract. The film cooling technique is introduced in modern gas turbines to protect the blade from the high temperature of the incoming hot gases by forming a thin coolant blank over the blade surface. However, it is known as a jet in crossflow (JICF), where coolant and mainstream interact intensively and generate complex vortices leading to highly unsteady coolant coverage over the blades surface. In this study, a fast-response pressure-sensitive paint technique (fast-PSP) was used to measure the coolant unsteadiness with a high-resolution camera. The measurements were performed in a novel single-passage transonic wind tunnel to uncover the unsteady effectiveness of the endwall surface. Such effectiveness was dramatically influenced by the blowing ratios (*M*), showing attached flow at a low blowing ratio and lift-off at a high blowing ratio. The effectiveness was asymmetrically distributed due to the pressure gradients, jet compounding angle, and associated complex flows. The unsteady effectiveness was highly influenced by the energetic vortical structures, which interacted with the mainstream flow immediately behind the holes. It was featured by secondary structures (horseshoe, passage, and counter vortices) beside the JICF structures. Meanwhile, the unsteadiness was originated from the middle of the passage behind the holes. It is suggested to pay close attention to the locations of the holes for further optimization. This study could help the designers to understand the characteristics of unsteady effectiveness, promoting advanced cooling strategies for enhanced protection of future gas turbines.

Introduction

Gas turbine is a heavy-duty engine that is used for many engineering applications, such as land-based power plants and civil/military aircraft, making the improvement of its output capacity of global concern. Based on the thermodynamic science [1], increasing the turbine thermal efficiency and corresponding gained power could be achieved at high turbine inlet temperature. Such temperature exceeds the melting point of the blade materials, leading to engine failure with its duty operations, which challenge the overall performance. Film cooling is one of the advanced cooling techniques that is used recently to protect the turbine blades from the incoming gases by introducing a secondary coolant flow over the hot surface, such as endwall region. However, it is known as a jet in crossflow phenomena (JICF), where the coolant penetrates the mainstream in a cross manner as discussed by Kelso et al. [2] and Andreopoulos and Rodi [3]. Once the coolant ejects from the holes, highly coolant and mainstream interactions happen immediately at the hole exit, producing a complex three-dimensional (3-D) vortical structure as discussed by Fric and Roshko [4] and Mahesh [5]. The vortical structures enforce the coolant to fluctuate over the protected surface (unsteady). Consequently, it could be completely or even partially

Content from this work may be used under the terms of the Creative Commons Attribution 3.0 licence. Any further distribution of this work must maintain attribution to the author(s) and the title of the work, journal citation and DOI.
Published under licence by IOP Publishing Ltd

exposed to hot gases, leading to local overheating. Such phenomena could thermally stress the turbine blades, leading to turbine failure with heavy-duty operations. Therefore, it is significantly desirable to understand the coolant unsteadiness (unsteady behaviour) behind turbine discrete holes.

In the past decades, a significant research work has been done for understanding of the physics behind the endwall film cooling, which was featured by highly complex flows. Such complexity was introduced by Langston et al. [6,7], who explored the detailed endwall structures. They captured interested structures accompanying the incoming flow boundary layers, which formed horseshoe vortex near leading-edge. These vortices are extended and developed as counter vortex near suction side (SS) and passage vortex near the pressure side (PS). These vortices interacted with the coolant flow from the discrete holes; leading to coolant deterioration and non-uniform distributions. Friedrichs et al. [8] visualized the interacting coolant and secondary flows and explained the local features of the hard cooling regions. Consequently, numerous efforts enhanced the coolant coverage over the endwall surface by optimized configurations. Friedrichs et al. [9] introduced an endwall configuration different from that of Friedrichs et al. [8], which revealed an improvement in the coolant spreading over the uncooled regions. Meanwhile, the optimized endwall configurations, which were introduced by Thole and Knost [10,11], achieved the desired enhancement, especially at the PS. Similarly, Li et al. [12,13] uncovered the endwall features and proposed optimized configurations to cool-down the hard cooling region at the leading-edge. Recently, Shiau et al. [14] proposed full-scale endwall configurations to improve the whole endwall surface, considering the upstream purge effects. The recent comparison of Shiau et al. [15] showed important features allied with the upstream injection angles through different configurations.

Although these studies successfully investigated the coolant behaviour behind the endwall configurations, they focused on the mean effectiveness; leaving a few unsteady studies. Kadotani and Goldstein [16] and Jumper et al. [17] examined the mainstream turbulence (Tu=0.3-20.6%) unsteadiness, which was found to decrease the effectiveness at low injection rates and increase it at high injection rates. In contrast, Mehendale et al. [18] and Du et al. [19] studied the upstream unsteady wakes, which showed palpable reductions in the effectiveness. Teng et al. [20] measured the mean and fluctuating thermal profiles within the holes influenced by upstream wakes, which were found to produce turbulence in the boundary layer profiles. The early study of Fawcett et al. [21] showed interesting unsteady coolant forms behind the discrete holes. They investigated the engine-representative film cooling unsteadiness on the mid-span of the blade PS, although the incoming flow Mach number was negligible. Their investigation was done behind at a wide range of coolant conditions using high-speed photography of particle image velocimetry measurement techniques. Their results revealed unsteady coherent shearing forms (structures) in the mainstream flow, which was strongly dependent on the coolant velocity. However, the unsteady JICF and complex flow unsteadiness over the endwall surface at transonic conditions have not been well understood yet.

This paper uncovered the unsteady endwall effectiveness at the engine-like conditions; aiming to capture the JICF and complex flow unsteadiness. High-resolution fast-PSP measurements are conducted over an endwall configuration at transonic condition using novel wind tunnel [22] (single-passage design). During the experiment, the coolant gas (i.e., CO_2, coolant to mainstream density ratio of $DR = \rho_c/\rho_\infty \approx 1.53$) is fed internally from a large plenum chamber through an upstream slot with blowing ratio of $M = DR(U_c/U_\infty)$=0.3 and row of cylindrical holes (located 20D downstream) with blowing ratio of M=0.5-2.0. The measured effectiveness is analysed in terms of mean and unsteady quantities over the endwall surface. This paper is expected to deepen our understanding of the endwall unsteadiness.

1. Experiment setup and validation

1.1. High-speed test facility

The current measurements are performed at transonic conditions using novel wind tunnel, which is designed by the author [22] as a single-passage at turbomachinery lab of Shanghai Jiao Tong University for film cooling studies. Figure (1a) shows the proposed design of the wind tunnel, which follows the design of the high-duty inlet guide vane of Arts et al. [23] (i.e., Figure 1b). It consists of only single-passage that shaping the vane configuration through bounding the passage by two side walls. The design has two bleeds at the PS and SS to achieve the flow periodicity by removing the additional mainstream flows. This proposed strategy was firstly introduced by Blair [24] for simplified film cooling measurements. The design procedures of the proposed tunnel are achieved numerically before manufacturing via the periodic flow simulation and passage walls simulation. These procedures are performed on a high-duty inlet guide vane, which was tested by von Karman Institute [23] for isentropic Mach number of M_{is}=0.84. The proposed vane is shown in Figure (1b), which has a chord of 67.6 mm and pitch of 0.85 with axial flow inlet and 76° exit flow angle (see Arts et al. [23]). For consistency, the detailed procedures are found in Qenawy et al. [22].

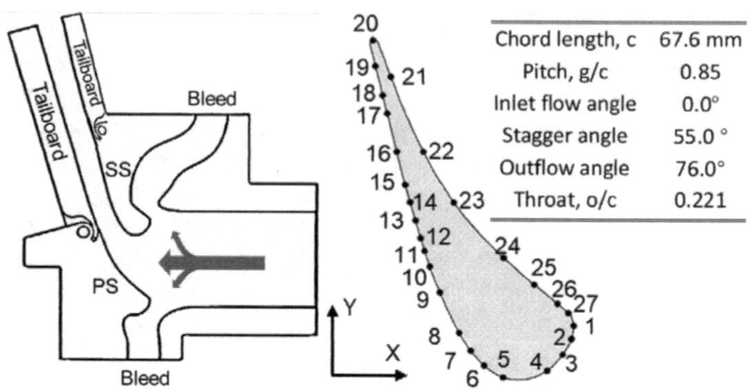

(a) Passage design (b) Passage shape

Figure 1. Novel single-passage transonic wind tunnel [22].

Figure (2a) shows the experiment setup of the current measurement. The mainstream flow is supplied into the wind tunnel by compressed air, being filtered to improve its quality. The desired flowrates, which are controlled using flow valve, are passed through big-chamber to reduce the acceleration and stabilize the flow before entering the passage. The air, then, is passed through three paths: two bleeds beside the passage. The majority of the air is flowed through the passage, while small amounts flow through the bleeds. The mainstream velocity is U_{∞}=51.9 m/s as measured by a five-hole probe upstream of the passage. The turbulence intensity was Tu=2%, while the boundary layer thickness was δ=3.1 mm as measured by (CTA-54T42, Denmark) hot wire anemometry. The suction bleed was observed by (SMC-PF3A706H, Japan) flowmeter, while the pressure bleed was observed by (SMC-PF3A703H, Japan) flowmeter, and both adjusted by flow control valves.

Figure (2b) shows the test model, which is installed underneath the passage with specific configurations. The model has a single-row of cylindrical holes with a hole diameter of D=1.0 mm (i.e., similar to engine conditions). The holes have 9D entry length, 3D spacing, and incline angle of 35° to the mainstream flow. Meanwhile, the model has a 1.5 mm slot width (located 20D upstream of the holes). The model interior surface is carefully refined with 2000 sandpaper to ensure surface quality.

During the experiment, the coolant is discharged from a high-pressure CO_2 tank through a long pipeline and water bath to maintain the ambient temperature before feeding into the slot and discrete holes. The slot flowrate is observed by (FMA-1610A-OMEGA) flowmeter and maintained fix by adjusting the flow control valve at a blowing ratio of M=0.3. Similarly, the hole flowrates are monitored by (FMA-1610A OMEGA) flowmeter and adjusted by flow control valve with M=0.5-2.0. The coolant is fed from a plenum chamber to the endwall. The plenum is located underneath the test model and sealed to prevent coolant leakage. with a 1-mm-thick rubber gasket.

1.2. Film cooling measurement

In film cooling, the coolant temperature (T_c) is relatively lower than the mainstream temperature (T_∞), while the film temperature depends on the mixing conditions, which contribute the wall temperature. With insulated walls, the wall temperature is considered as adiabatic wall temperature (T_{aw}). Consequently, the adiabatic effectiveness (η) measures the efficiency of the coolant distributions over the surface. It is a non-dimensional temperature scale, which is defined by Eq. (1) [1,25].

$$\eta = \frac{T_\infty - T_{aw}}{T_\infty - T_c} \quad (1)$$

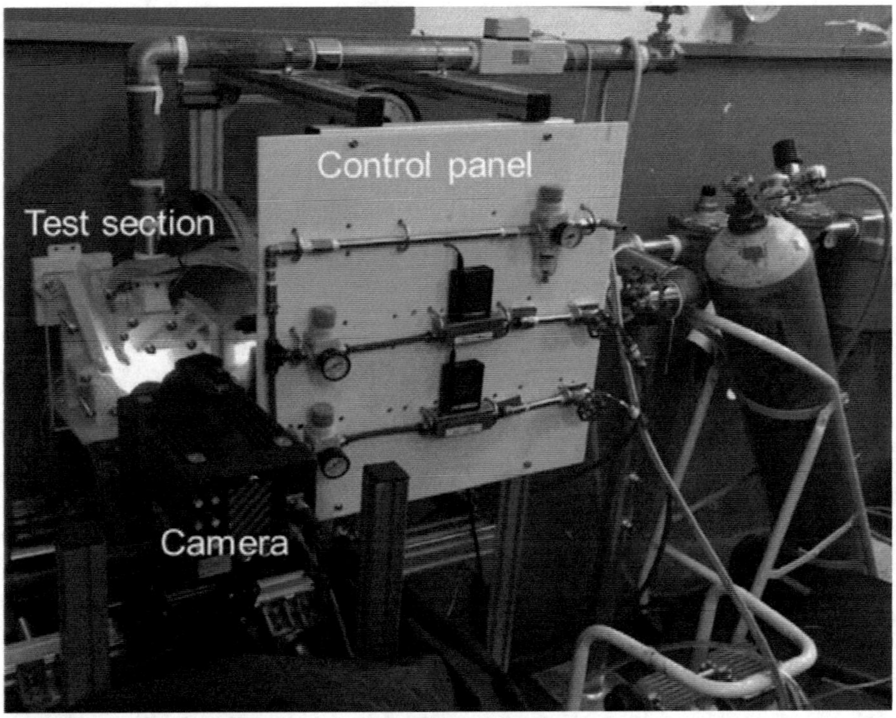

(a) Experimental setup

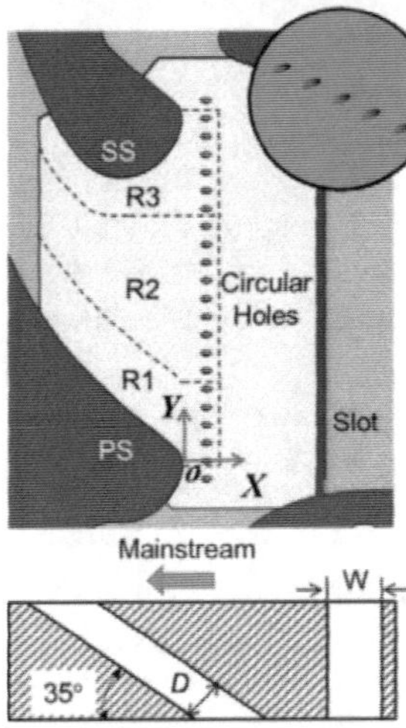

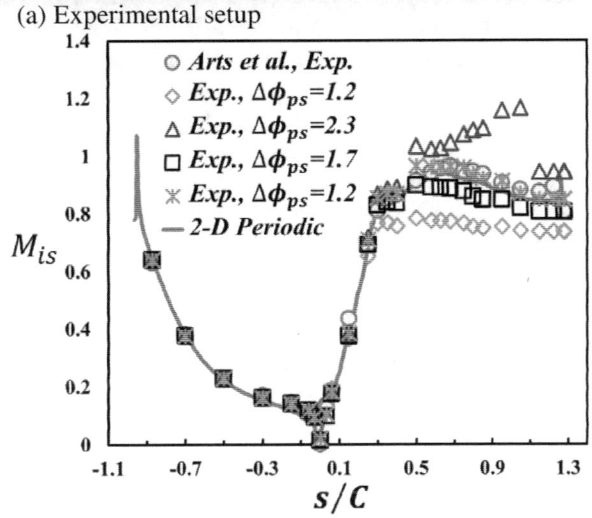

(b) Endwall configurations (c) Experimental validation

Figure 2. Current endwall measurement set up and validation.

In laboratories, heating the mainstream flow could increase the thermal losses through the walls, making the adiabatic conditions of hard lab achievement. Recently, the measurement is performed in isothermal conditions, using heat and mass transfer analogy. Theoretically, analogy of heat and mass transfer could be considered; if the concentration and thermal boundary layers are of the same order, which means unity of Lewis number (i.e., $Le = \alpha_s/D_s \approx 1.0$, where D_s is the concentration diffusion coefficient and α_s is the thermal diffusion). By making the coolant and mainstream flow different gases with the same temperatures (isothermal), the main focus will be the oxygen concentration. Consequently, the effectiveness (Eq. 1) can be expressed by Eq. (2) as reported by Charbonnier et al. [26]; where W_c/W_∞ is the coolant to mainstream molecular weight ratio and $P_{O2,\infty}/P_{O2,mix}$ is the partial oxygen pressure.

$$\eta = 1 - \frac{1}{\left[\left(\frac{P_{O2,\infty}}{P_{O2,mix}}\right)_{aw} - 1\right]\frac{W_c}{W_\infty} + 1} \quad (2)$$

Pressure sensitive paint (PSP), based on the mass transfer analogy ($Le \approx 1$), is quite a new kind of non-contact optical method for pressure measurement. It has been widely used to simulate the heat transfer process due to its ability to run in isothermal conditions, where heat conduction problems could be eliminated. Meanwhile, it is a reliable technique, which is based on the measurement of Wright et al. [27] and can verify the coolant effectiveness under adiabatic conditions. Besides, PSP can provide detailed coolant separations, lift-off, and reattachment phenomena. For example, Johnson et al. [28] captured a good agreement of PSP measurement in comparison with the open literature. Interestingly, our recent publications [29–34] showed an acceptable agreement of PSP immediately at the hole exit.

The present paint is a mixture of luminescent molecules, silicon dioxide particles, and polystyrene (binder). This study is performed using fast-response particle pressure-sensitive paint (fast-PSP) to resolve the dynamics of the fluctuating effectiveness. The silicon dioxide particles have a hollow and porous structure, leading to fast responding to the pressure fluctuations. It was developed and improved by Peng et al. [35]. It is prepared in-house with dynamic response of ~9.5 kHz (corresponding response time of ~106 μs) [35] on the shock tube. It is resolved by dichloromethane mixture, spread over the test surface by airbrushes, and kept in insulated/cleaned conditions for ~5 hours to dry.

In theory, PSP emits luminescence when exposed to an appropriate wavelength. However, the emitted intensity is influenced by the oxygen partial pressure, which is known as the oxygen quenching. Through it, especially via reduced emission in the presence of oxygen, excited molecules could be returned into their ground states. It could be caused by the quenching of some luminophores through the collision with the oxygen molecules. So, the intensity of the photo-luminescence is proportional inversely to the local oxygen concentration. Therefore, the emitted intensity decreases with increasing the oxygen partial pressures, which can be evaluated through the Stern–Volmer relationship [29,30], obtained by the calibration procedures. In contrast, the emitted intensity is captured by a CCD camera, which has a 650±25 nm filter to remove the background noise after exiting the paint by 385 nm UV light. During the experiment, the atmospheric air simulates the incoming hot gases in the real engine, while the oxygen-free foreign gas (e.g., CO2) is used as a coolant.

In this study, the endwall model is painted with black primer to reduce both transmission and reflection. Fast-PSP, which is used to measure the mean and unsteady effectiveness, is sprayed with an airbrush in an isolated condition and kept ~5 hours to dry. An ultraviolet LED (LM2X-DM, ISSI) as an excitation source was used with 385 nm wavelength. A complementary metal-oxide-semiconductor camera (PCO, HS4, high-speed) at a sampling rate of 10 Hz was used to record the photoluminescence light emission. A 650±25 nm wavelength band-pass filter is fixed ahead of the camera to filter the background noises.

1.3. Uncertainty analysis

In this study, the uncertainty arises from two main sources; measurement as well as device/instrument uncertainties. In the wind tunnel and test models, the uncertainty originally comes from the 3-D printer with a printing accuracy of ±1% and approximate resolution of ~50 μm based on manufacturing details. For measurement, the uncertainties in the instruments are taken from the supplier as:

- The uncertainty of the incoming flow pressure reading is ±2%, while temperature reading is ±2%.
- The uncertainty of the 5-probes pitot tube is ± 2%.
- The uncertainty of the SMC flowmeters is ±2%, while of OMEGA flowmeters is ±1%.
- The uncertainty of the pressure transducer is ± 1%.
- The uncertainty of the blowing ratios is ±2.6%, while of the measured M_{is} is ±2.8%.

For the PSP measurement, to minimize uncertainties, an interrogation window of 5 x 5 pixels and 50% overlapping are applied on the spatially-averaged procedures. The image resolution was ~0.06 mm/pixel (corresponding spatial resolution of ~0.1*D*). Based on the estimation of Johnson & Hu [36], the uncertainty of PSP measurement is 8% near the hole (high effectiveness regions) and 12% downstream (low effectiveness regions), which is probably related to mainstream and coolant conditions, PSP calibration procedures, and quality of the captured image. However, the accelerated mainstream (compressibility effects) increases the uncertainty of the measured effectiveness; resulting in 10% near the hole and 15% downstream.

1.4. Measurement validation

Figure (2c) shows the distributions of the isentropic Mach number (M_{is}) on the mid-span of the passage height in comparison with Arts et al. [23] and the periodic simulation. Notably, M_{is} was computed in Eq. (3) based on the local static pressure (*P*) of the distributed pressure taps along with the vane SS and PS (see Figure 1b), where P_0 is the stagnation pressure at the passage leading-edge and γ is the heat capacity ratio. By adjusting the tailboard angle, the results provide a good agreement with PS adjustable angle of $\Delta\Phi_{ps}$=2.0°, which denotes the validation of the current study.

$$M_{is} = \sqrt{\frac{2}{\gamma-1}\left(\left(\frac{P_0}{P}\right)^{\frac{\gamma-1}{\gamma}} - 1\right)} \qquad (3)$$

1.5. Time-average and statistical effectiveness

The measured instantaneous film cooling effectiveness (i.e., Eq. 2) is analysed into mean (i.e., $\bar{\eta}$) and fluctuating (i.e., $\acute{\eta}$) components (i.e., $\eta = \bar{\eta} \pm \acute{\eta}$). The mean effectiveness is calculated by averaging the captured instantaneous snapshots via Eq. (4).

$$\bar{\eta} = (1/t)\int_0^t \eta \, dt \qquad (4)$$

Meanwhile, the standard deviation (*SD*) is used to measure the variation of the data from the mean value. In statistical science, low *SD* means low fluctuations (close to the mean value), while high *SD* indicates high fluctuations (far from the mean value). As the measured effectiveness has large instantaneous data-sets, it would be useful to evaluate its fluctuations from the mean value. Therefore, *SD* is defined by Eq. (5); where *N* is the number of instantaneous snapshots.

$$SD = \sqrt{\frac{1}{N-1}\sum_{i=1}^{N}(\eta_i - \bar{\eta})^2} \qquad (5)$$

Figures (3-4) show the time-average and statistical distributions of the measured effectiveness over the endwall surface at *M*=0.5, 1.0, 1.5, and 2.0. For all figures, the coolant from the upstream slot exhibits almost the same footprints. It has accumulated through the passage's middle, flowing toward the SS, forming asymmetrical distributions. Such accelerated flow, as reported in Du and Li [37], could drive more coolant toward the passage vortex line, under the effects of pressure gradient [6,7]. Immediately, the coolant is widely spread behind the exit of the slot and develops narrower downstream before merge with the coolant flow from the holes, which agree with Chen et al. [38,39]; revealing the contribution of the upstream slot to the endwall coolant effectiveness. On the other hand, the corresponding *SD* of the measured effectiveness behind the holes is quite similar to the mean effectiveness. Noticeably, the *SD* signatures are influenced by both JICF and complex flows, where high effectiveness regions have low *SD* values; indicating inverse proportional to the mean distributions. However, the unsteady analysis is performed only behind the holes for a better understanding of the JICF and complex flow unsteadiness.

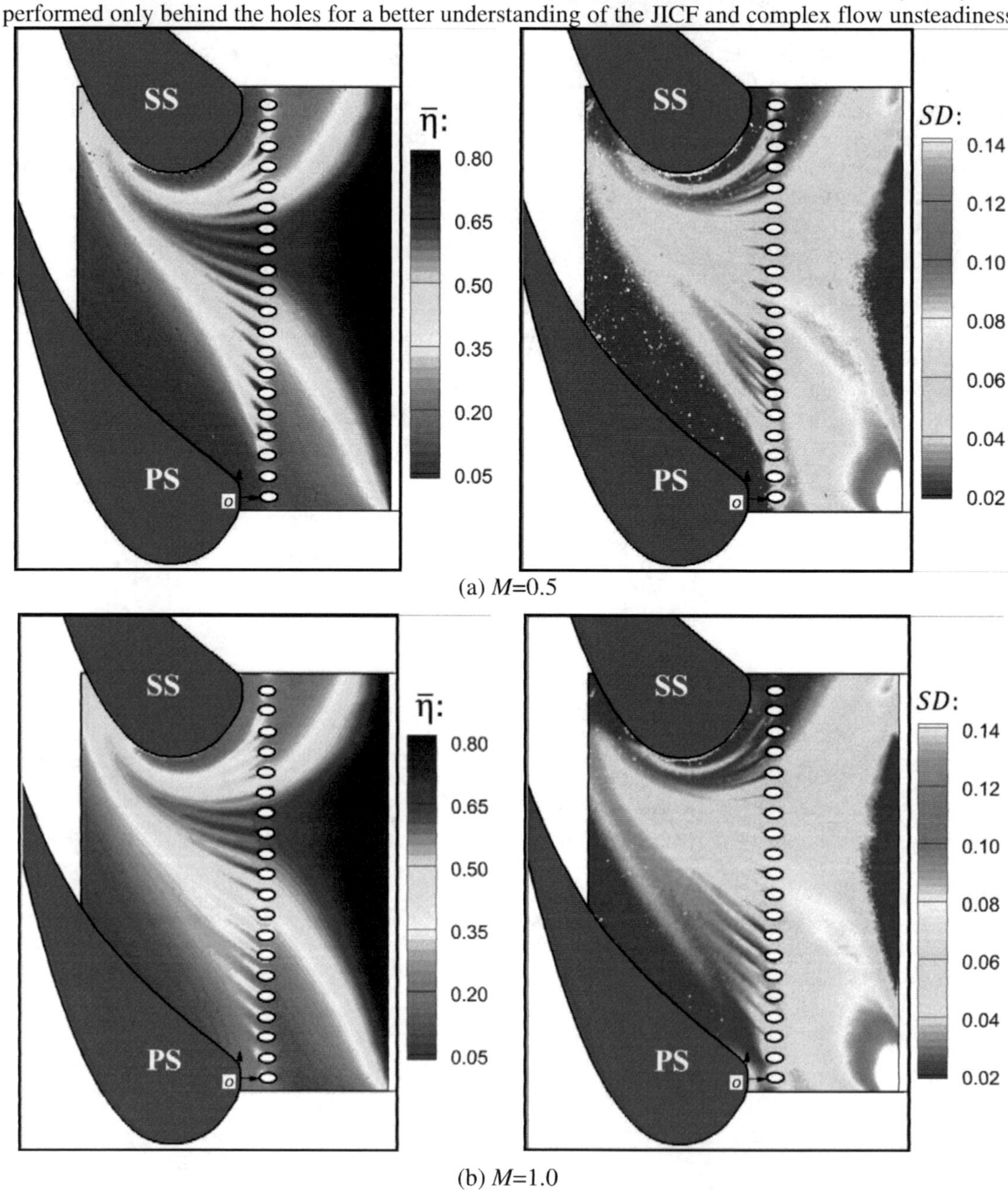

Figure 3. Endwall film cooling distribution at blowing ratio of M=0.5 and 1.0; (left) Time-average and (right) Standard deviation.

For M=0.5 (Figure 3a), the coolant asymmetrically spreads over the endwall, concentrating on the SS, where the passage vortex propagates [6,7]. The passage vortices at the PS force the coolant movement toward SS. Consequently, SS exhibits higher effectiveness due to the coolant merging from the holes and slot, while the PS, especially the downstream region is relatively uncooled. In contrast, at leading-edge, less coolant has observed, where the horseshoe vortices dominate the flow region, leading to blows coolant away, Friedrichs et al. [8,9]. Carefully examining the leading-edge at SS, rarely coolant is observed at the endwall junction indicating that the horseshoe vortex sweeps the coolant into the mainstream flow. Interestingly, no injection can be discharged through the PS holes (i.e., last two holes). Sundaram and Thole [40,41] reported a quite similar phenomenon, which was originated from low coolant momentums, leading to incapability of ejection through those holes. On other hand, the contour shows high *SD* values upstream the SS from the 2nd and 3rd holes. This *SD* signature probably originates from the horseshoe vortex (leading-edge) and associates with downstream counter vortex. Meanwhile, the contour shows high *SD* values near the PS from the 3rd and 4th holes; associated with the passage acceleration streamlines. Plausibly, the coolant accumulates along the passage vortex line and be highly unsteady downstream. In contrast, the middle region shows high *SD* values at the centre, which associates with the downstream counter-rotating vortex-pair (CRVP) and single-asymmetric main-vortex (SAMV) structures, as proposed by Li et al. [42]. Meanwhile, the lateral sides show low *SD* values; due to less coolant at the PS and due to the coolant accumulation at the SS.

For M=1.0 (Figure 3b), coolant is widely spread behind the holes across the endwall (the aforementioned converging coolant trace becomes much wider and more uniform). Due to enough coolant momentums, the coolant overcomes the mainstream flow, leading to obvious coolant distributions at the passage PS. The coolant partially lifts the endwall off in the near-holes region and diverts broadly downstream. On the other hand, the *SD* signatures are quite insignificant compared with M=0.5 due to the coolant partially lifts-off the surface. The unsteady horseshoe and corresponding counter vortices have weak *SD* signatures at the passage SS. Similarly, the passage acceleration line has weak *SD* signatures at the PS. Plausibly, this absence of the *SD* signatures is caused by insufficient coolant attachments to the endwall surface. In contrast, the passage middle shows apparent *SD* signatures associated with the coolant structures (i.e., CRVP and SAMV). Plausibly, the attached flow propagates along the passage acceleration lines; making the coolant highly unsteady downstream.

For M=1.5 (Figure 4a), the coolant footprints exhibit dissimilar events. The middle region, which characterizes highly accumulated flow, shows a slight reduction in the coolant effectiveness compared with M=1.0. The increased momentum of coolant jets can overcome the negative pressure within the passage. Obviously, the visible trace behind the holes shows coolant lift-off phenomena near PS. Interestingly, increasing coolant momentums behind the holes improves the effectiveness of the vane leading-edge. Although the coolant lifts the endwall surface, it interacts with the leading-edge vortices (i.e., horseshoe). The horseshoe vortex forces the coolant to flow downward and impinge onto the endwall leading-edge. On the other hand, the *SD* contour at M=1.5 exhibits almost similar behaviour as M=1.0 behind the holes. The coolant unsteadiness decreases at the PS and SS, while increases at the passage middle. Plausibly, the coolant is completely lift-off the endwall surface near the PS and SS with the unsteady accumulated flow at the passage middle. However, the passage leading-edge shows *SD* signatures at either PS or SS, which are caused by the unsteady reattached flow.

For M=2.0 (Figure 4b), the coolant has highest momentums. The coolant jets near the PS shows obvious separation from the endwall surface, observing from the effectiveness reductions in the contours. However, the hard to cool [11] region (i.e., near the leading-edge) has been pointedly enhanced by the coolant flow, indicating rebus coolant behaviours. On the other hand, the attached flow becomes more unsteady for M=2.0 compared with the aforementioned blowing ratios. The passage middle region shows highly unsteady *SD* signatures, which is probably caused by the unsteady accumulated flows. Meanwhile, the leading-edge shows clear signature of the horseshoe vortex from such a reattached flow at both PS and SS. In contrast, the *SD* signature at the PS extends little downstream, which could associate with the passage vortex interactions with the reattached flow.

1.6. Fluctuating Effectiveness

To fully understand the measured standard deviation, the fluctuating effectiveness (i.e., $\acute{\eta}$) through the interval time is presented in Figures (5-6). For consistency, only the fluctuating component at x/D=0.0 and 3.0 behind the holes is presented. Generally, the coolant and mainstream flows are in sparring; as indicated by blue and red colour maps over the endwall surface.

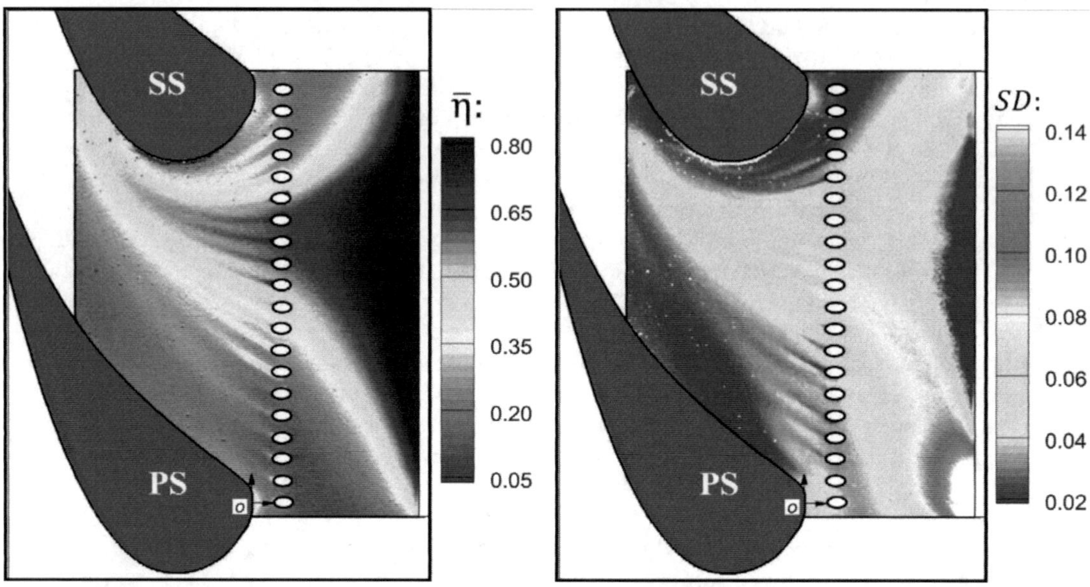

(a) M=1.5

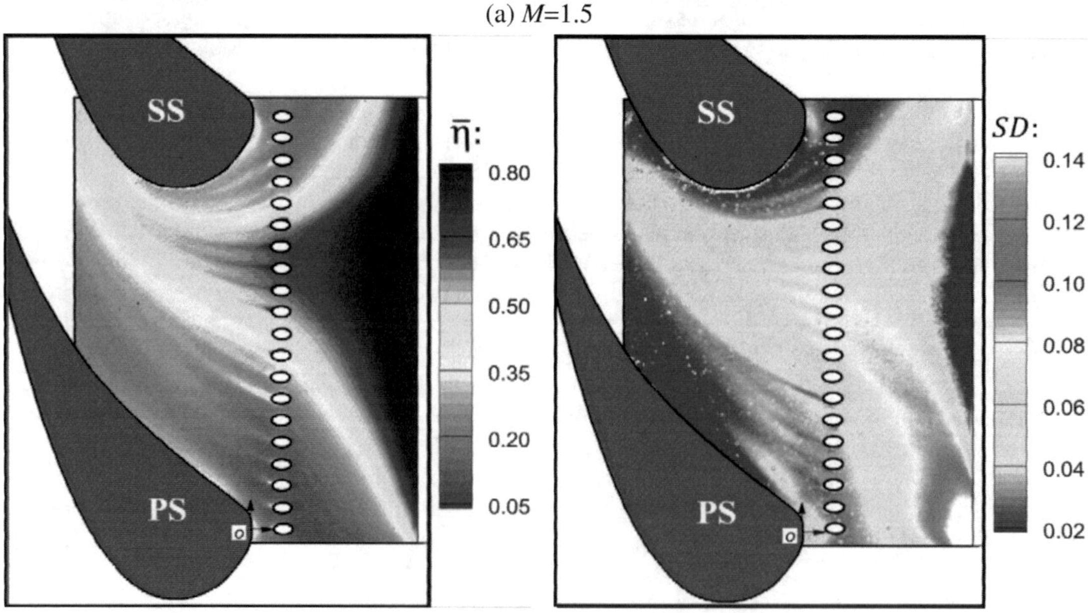

(b) M=2.0

Figure 4. Endwall film cooling distribution at blowing ratio of M=1.5 and 2.0; (left) Time-average and (right) Standard deviation.

Figure (5) shows the instantaneous fluctuating effectiveness downstream of the holes (i.e., x/D=0.0 and 3.0) at low blowing ratio (i.e., M=0.5 and 1.0). For M=0.5, the contours show obvious sparring between coolant and mainstream flows. The major fluctuations originate in the middle of the passage (i.e., close to the SS), which is consistent with the *SD* distributions (Figure 3a). Near the SS, there is a column of fluctuating effectiveness immediately at the hole exit (i.e., x/D=0.0) near $z/D\approx50$, which is caused by the passage horseshoe vortex. Meanwhile, this fluctuating column slightly moves toward the middle of the passage (i.e., $z/D\approx47$) at 3D downstream of the hole exit (i.e., x/D=3.0); leaving a blank column at $z/D\approx51$. Plausibly, the passage horseshoe vortex becomes a downstream counter vortex. In contrast, there is another column of fluctuating effectiveness immediately at the hole exit (i.e., x/D=0.0) near the PS (i.e., $z/D\approx11$), which associates with the passage acceleration line. This is observed from the inward orientation of the passage acceleration downstream (i.e., x/D=3.0), which originates at $z/D\approx17$. For M=1.0, the major fluctuations originate in the passage's middle near SS. This agrees with *SD* distributions (Figure 3b); revealing unsteady accumulation flow. Interestingly, no pure signature exists for either passage horseshoe/counter vortices at the SS or acceleration flow at the PS; revealing coolant lift-off. Generally, the cylindrical holes exhibit higher fluctuations in the middle of the passage.

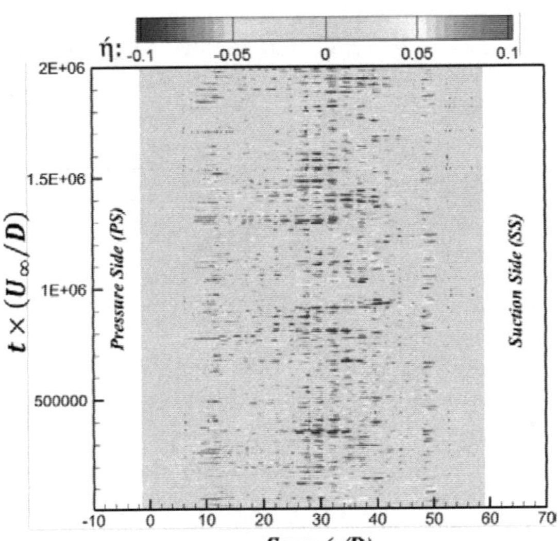

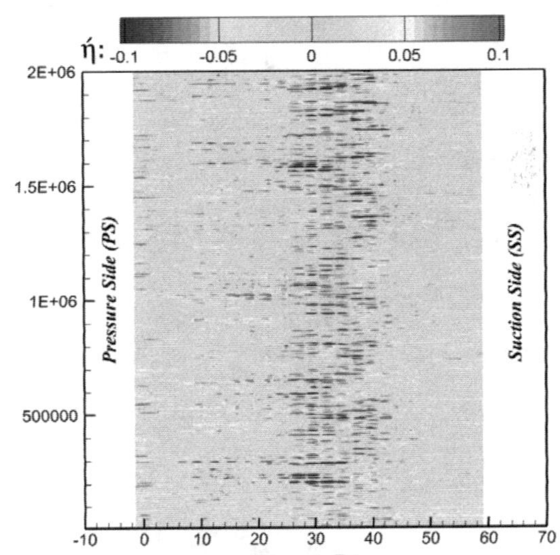

(a) Immediately at the hole exit x/D=0.0

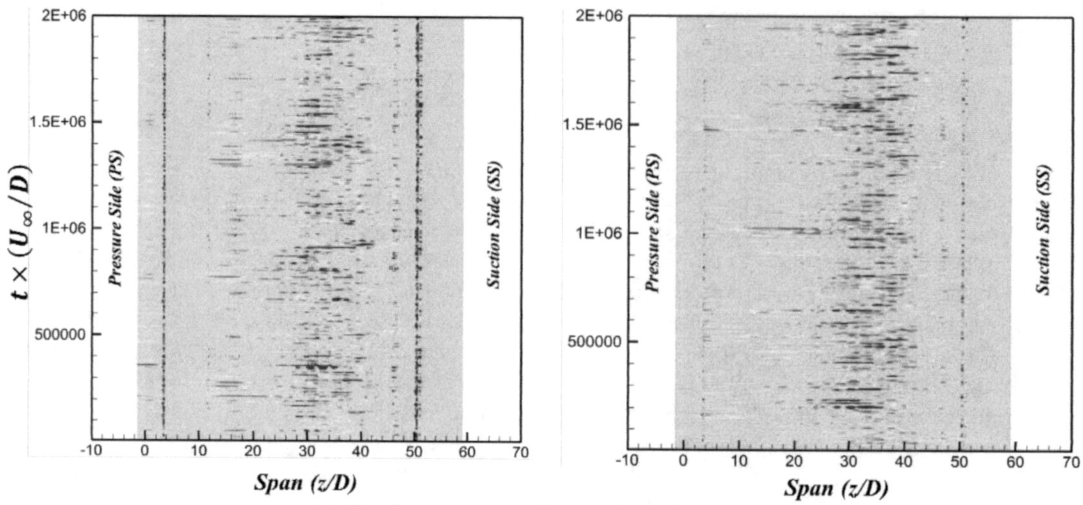

(b) $3D$ downstream of the hole exit (x/D=3.0)

Figure 5. Fluctuating effectiveness at (left) M=0.5 and (right) M=1.0.

Figure (6) shows the fluctuations of the measured effectiveness downstream of the discrete holes (i.e., x/D=0.0 and 3.0) at high blowing ratio (i.e., M=1.5 and 2.0). For M=1.5, the fluctuations still originate in the middle of the passage near the passage SS; revealing the unsteady coolant flow from the upstream slot. Interestingly, the passage horseshoe vortex is observed immediately at the hole exit (i.e., x/D=0.0) at $z/D\approx56$ and $z/D\approx0.0$ near the SS and PS, respectively. Plausibly, the reattached coolant flow near the leading-edge mixes with such horseshoe vortices; revealing unsteady effectiveness and explaining the SD signatures at the passage leading-edge in Figure (4a). Generally, the cylindrical holes still exhibit higher fluctuations near passage's middle compared with PS and SS. For M=2.0, the fluctuations increase at the middle, which reveal obvious unsteady coolant from the upstream slot. The reattached passage horseshoe vortex originates at x/D=0.0 near both SS ($z/D\approx58$) and PS ($z/D\approx0.0$). Meanwhile, the downstream behaviour (i.e., x/D=3.0) shows a weak signature near the PS, which originates from the reattached flow (see mean effectiveness distributions in Figure 4b).

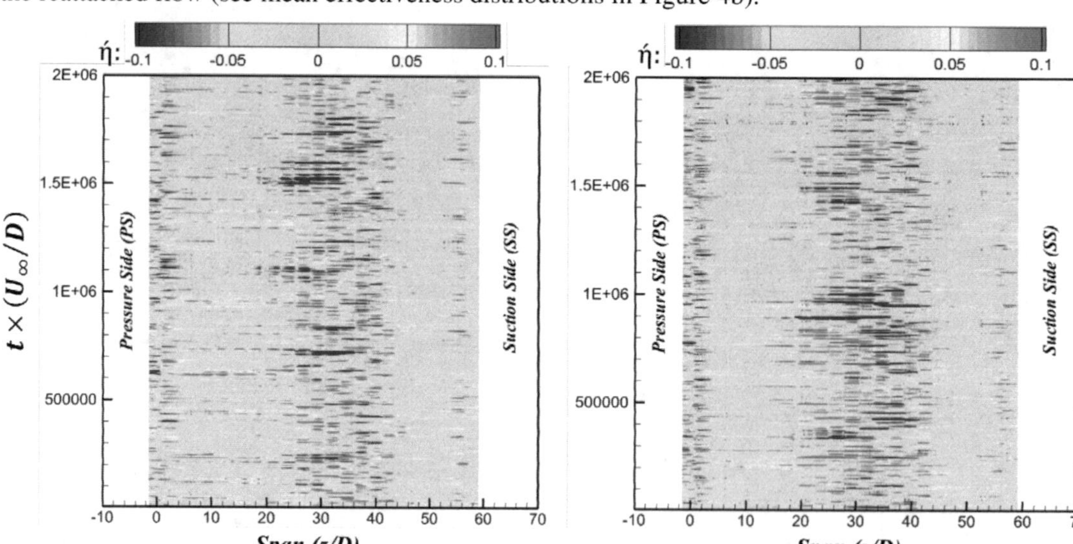

(a) Immediately at the hole exit (x/D=0.0)

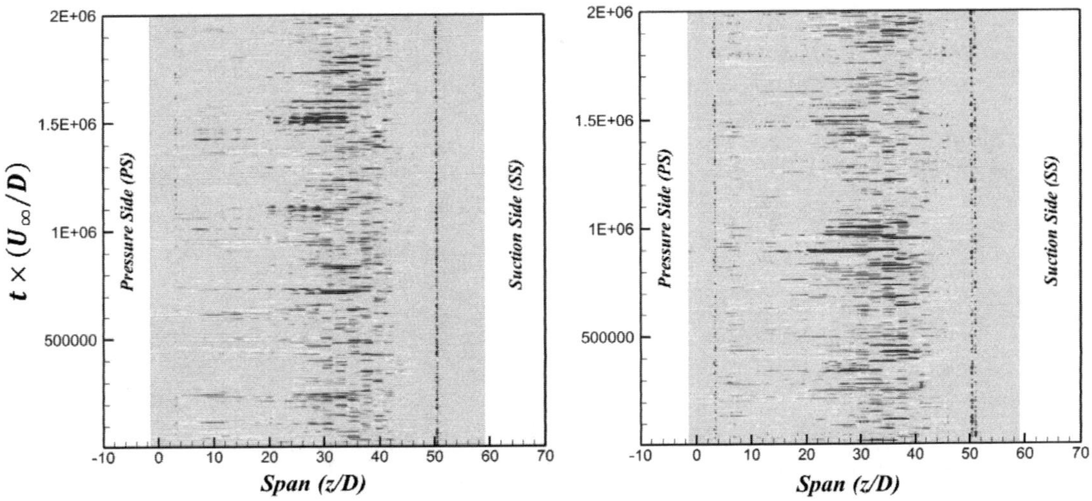

(b) 3D downstream of the hole exit (x/D=3.0)

Figure 6. Fluctuating effectiveness at (left) M=1.5 and (right) M=2.0.

2. Conclusion

In this contribution, unsteady endwall effectiveness behind the discrete holes was measured, considering the JICF and complex flow effects. The measurements were performed in a novel single-passage transonic wind-tunnel at M_{is}=0.84; showing acceptable agreements with literature's available data. High-resolution fast-PSP measurements were conducted to uncover the endwall unsteady effectiveness. Coolant gas (CO_2, $DR\approx1.53$) fed into the endwall model with slot and row of holes (M=0.3 for slot and M=0.5-2.0 for holes). Generally, endwall exhibited an asymmetrical effectiveness distribution, featuring higher effectiveness near the SS, following by middle region, and lower near PS. The passage vortices as well as flow structures at SS was responsible for cooling spreads, which was found to be stronger behind the holes, leading to decline endwall coverage. Due to pressure gradient, the coolant is enforced toward SS, resulting in higher cooling effectiveness. The unsteady analysis over the endwall surface revealed the contributions of large-scale structures to the unsteady effectiveness. In addition, the *SD* signatures were associated with the endwall passage, counter, and horseshoe vortices besides the CRVP and SAMV structures. The unsteadiness originated from the coolant accumulations in the middle of the passage along the coolant acceleration lines. While the current work focuses on the endwall unsteadiness, still additional investigation with various design configurations is required for increasing the coolant coverages with special considerations of local flow features.

Acknowledgments

The authors thankfully acknowledge the financial support of the current work from National Natural Science Foundation of China (51806138, 11725209).

Nomenclature

D	Hole diameter
DR	Density ratio
M	Blowing ratio
M_{is}	Isentropic Mach number
P	Static pressure
P_o	Total pressure
$\frac{P_{o2,\infty}}{P_{o2,mix}}$	Oxygen partial pressure

T_c	Coolant temperature
T_∞	Mainstream temperature
T_{aw}	Adiabatic wall temperature
W	Width
$\frac{W_c}{W_\infty}$	Molecular weight ratio
x, y, z	Cartesian coordinates
Greek symbols	
η	Instantaneous effectiveness
$\bar{\eta}$	Mean (time-average) effectiveness
$\acute{\eta}$	Fluctuating effectiveness
$\Delta\emptyset_{ps}$	PS deviation angle
Abbreviations	
CRVP	Counter-Rotating Vortex-Pairs
Fast-PSP	Fast Pressure-Sensitive Paint
JICF	Jet in Cross Flow
PS	Pressure Side
SD	Standard Deviation
SS	Suction Side
SAMV	Single-Asymmetric Main-Vortex

References

[1] Han J C and Rallabandi A P 2010 *Front. Heat Mass Transf.* **1** 1-21.
[2] Kelso R M, Lim T T and Perry A E 1996 *J. Fluid Mech.* **306** 111-144.
[3] Andreopoulos J and Rodi W 1984 *J. Fluid Mech.* **138** 93-127.
[4] Fric T F and Roshko A 1994 *J. Fluid Mech.* **279** 1-47.
[5] Mahesh K 2011 A*nnu. Rev.* Fluid *Mech.* **45** 379-407.
[6] Langston L S, Nice M L and Hooper R M 1977 *J. Turbomach.* **99** 21-28.
[7] Langston L S 1980 *J. Eng.* Gas *Turbines Power* **102** 866-874.
[8] Friedrichs S, Hodson H P and Dawes W N 1995 *Proc. ASME Turbo-Expo* Paper No. 95-GT-001.
[9] Friedrichs S, Hodson H P and Dawes W N 1999 *J. Turbomach.* **121** 772-780.
[10] Thole K A and Knost D G 2005 *Int. J. Heat Mass Transf.* **48** 5255-5269.
[11] Knost D G and Thole K A 2005 *J. Turbomach.* **127** 297-305.
[12] Li X, Ren J and Jiang H 2015 *Int. J. Heat Mass Transf.* **90** 1-14.
[13] Li X, Ren J and Jiang H 2016 *Int. J. Heat Mass Transf.* **102** 348-359.
[14] Shiau C C, Chen A F, Han J C, Azad S and Lee C P 2016 *J. Turbomach* **138** 051002.
[15] Shiau C C, Sahin I, Wang N, Han J C, Xu H and Fox M 2019 *J. Therm. Sci. Eng. Appl.* **11** 031012.
[16] Kadotani K and Goldstein R J 1979 *J. Eng. Gas Turbines Power* **101** 459-465.
[17] Jumper G W, Elrod W C and Rivir R B 1991 *J. Turbomach.* **113** 479-483.
[18] Menhendale A B, Han J C, Ou S and Lee C P 1994 *J. Turbomach.* **116** 730-737.
[19] Du H, Han J C and Ekkad S V 1998 *J. Turbomach.* **120** 808-817.
[20] Teng S, Sohn D K and Han J C 2000 *J. Turbomach.* **122** 340-347.
[21] Fawcett R J, Wheeler A P S He L and Taylor R 2012 *J. Turbomach.* **134** 021015.
[22] Qenawy M, Yuan L, Liu Y, Peng D, Wen X and Zhou W 2020 *J. Eng. Gas Turbines Power* **142** 071009.
[23] Arts T, Rouvroit M L D and Rutherford A W 1990 Von Karman Institute Report No. 174.
[24] Blair M F 1974 *Proc. ASME Turbo-Expo* Paper No. 74-GT-33.
[25] Bogard D G and Thole K A 2006 *J. Propuls. Power* **22** 249-270.
[26] Charbonnier D, Ott P, Jonsson M, Cottier F and Köbke T 2010 *Proc. ASME Turbo-Expo* Paper No. GT2009-60306.
[27] Wright L M, Gao Z, Varvel T A and Han J C 2009 *Proc. ASME Turbo-Expo* Paper No. HT2005-

72363.
[28] Johnson B, Tian W, Zhang K and Hu H 2014 *Int. J. Heat Mass Transf.* **76** 337-349.
[29] Khojasteh A R, Wang S F, Peng D, Yavuzkurt S and Liu Y 2017 *Int. J. Heat Mass Transf.* **110** 629-642.
[30] Cai T, Peng D, Yavuzkurt S and Liu Y Z 2018 *Int. J. Heat Mass Transf.* **120** 1325-1340.
[31] Zhou W, Peng D, Liu Y and Hu H 2019 *Exp. Therm. Fluid Sci.* **101** 16-26.
[32] Qenawy M, Chen H, Peng D, Liu Y and Zhou W 2020 *J. Turbomach.* **142** 041007.
[33] Zhou W, Qenawy M, Liu Y, Wen X and Peng D 2019 *Int. J. Heat Mass Transf.* **129** 569-579.
[34] Zhou W, Chen H, Liu Y, Wen X and Peng D 2018 *Phys. Fluids* **30** 127103.
[35] Peng D, Gu F, Li Y and Liu Y 2018 *Sensors Actuators, A Phys.* **279** 390-398.
[36] Johnson B and Hu H 2016 *J. Turbomach.* **138** 121004.
[37] Du K and Li J 2016 *Int. J. Heat Mass Transf.* **98** 768-777.
[38] Chen P, Alqefl M, Li X, Ren J, Jiang H and Simon T 2019 *Int. J. Therm. Sci.* **142** 233-246.
[39] Chen P, Li X, Ren J, Jiang H and Simon T 2019 *Int. J. Heat Mass Transf.* **137** 690-702.
[40] Sundaram N and Thole K A 2008 *J. Turbomach.* **130** 041013.
[41] Sundaram N and Thole K A 2009 *J. Turbomach.* **131** 041007.
[42] Li W, Li X, Ren J and Jiang H 2017 *Int. J. Therm. Sci.* **121** 410-423.

Enhancing the propulsion characteristics of rockets by adding the energetic Nitro-hydroxyl-terminated polybutadiene (NHTPB) in the propellant compositions

A Azazy[1], A Saleh[1], W F Aly[2], A Elbeih[2], A K Hussein[2], T Elshenawy[3], M Abdelhafiz[2], T Z Wafy[2], M G. Zaki[2] and H S Ahmed[1]

[1] Science and Technology Center of Excellence, Military Production, Cairo, Egypt
[2] Military Technical College, Kobry Elkobbah, Cairo, Egypt
[3] Technical Research Center, Cairo, Egypt.

E-mail: asaleh@stce-egypt.org

Abstract: Replacing the inert binder by an energetic one could increase the specific impulse of the propellants and enhance the propulsion characteristics of rockets. In this study, Nitro-hydroxyl-terminated polybutadiene (NHTPB) was prepared by a simple method. The prepared NHTPB in addition to HTPB binder were characterized. FTIR spectra of both HTBP and NHTPB was determined and compared. The thermal behavior of the prepared NHTPB was studied using DSC technique at heating rate 5 degree/min. A composite propellant based on AP/NHTPB was prepared and the specific impulse was measured for AP/NHTPB using two inch motor. It was concluded that the energetic nitro-hydroxyl-terminated polybutadiene has a clear max. exothermic peak at 203 ^{0}C with heat release of 323 J/g. By comparing the results, the prepared propellant AP/NHTPB has specific impulse higher than the traditional AP/HTPB propellant. NHTPB is a promising binder for the application of rocket propellants and needs more tests for its approval.

1. Introduction

The technology used for the production of composite solid rocket propellants (CSRP) is based on preparing a liquid laquar (binder system) where the solid powders, ammonium perchlorate (AP) and aluminum, are usually added into [1, 2]. The most usable traditional binder is hydroxyl terminated polybutadiene (HTPB) [1]. Other additives are usually used with the binder such as plasticizers and curing agents [3, 4]. HTPB has been used for several years as the main polymer system applied on explosives and CSRP [5-7]. The perfect thermal stability of HTPB and its ability to produce CSRP with good mechanical characteristics recommended its distribution in many countries [8-10]. BUT still the inert characteristics of HTPB is the disadvantage for its application [11, 12]. The new trend in the rocket propellants field is probably looking for highly energetic materials to enhance the energy output of the CSRP and also enhance the mechanical properties of the produced propellants [13-15]. Several energetic binders were prepared and studied to reach this goal such as glycidyl azide polymer (GAP) [16-18] and 3-nitratomethyl-3-methyloxetane (polyNIMMO) [18, 19]. Several publications discussed the performance of GAP in a propelling charges and other studied the thermal behavior of GAP in heterogenous compositions [20, 22]. The results proved that these binders have not good compatibility with the most usable oxidizer (AP) in comparison with HTPB binder. Also the cost of these advanced binders is high compared with the HTPB. On the other side, GAP has acceptable results in compositions based on ADN oxidizer [22].

Content from this work may be used under the terms of the Creative Commons Attribution 3.0 licence. Any further distribution of this work must maintain attribution to the author(s) and the title of the work, journal citation and DOI.
Published under licence by IOP Publishing Ltd

Nitro-hydroxyl-terminated polybutadiene (NHTPB) is a promising binder with energetic characteristics. It is predicted to have a higher performance than traditional used binder (hydroxyl terminated polybutadiene (HTPB)) [23, 24]. It has been published that thermal properties and propulsion characteristics of HTPB propellants are optimum and oblige the companies to use it in filling the rockets [25]. The energetic nitro-group of NHTPB is the main source of increasing the energy output of NHTPB in comparison with the inert HTPB. Few studies were carried out on the preparation of the energetic nitro-HTPB and reported that the characteristics of nitro-HTPB are suitable for its application as a binder for CSRP [26]. It was reported that nitro-HTPB has low viscosity and several isocyanates (aromatic or aliphatic) could be used as curing gent for it. The glass transition temperature of nitro-HTPB is higher than that of HTPB and also NHTPB is compatible with energetic plasticizers [27]. In this study, NHTPB was prepared using a simple method of production. The thermal behavior of NHTPB was checked using DSC technique. The FTIR spectra of both NHTPB and HTPB was obtained and compared with each other to check the presence of energetic group in NHTPB. A propellant based on NHTPB was produced and tested using two inch motor in comparison with CSRP based on the inert HTPB.

2. Results and discussion

2.1. Preparation of NHTPB

NHTPB has been prepared by the following steps; ethyl acetate was used to dissolve the HTPB sample. A mixture of water and ethylene glycol was used to dissolve sodium nitrite then this solution was added to the HTPB solution under continuous stirring. The temperature was decreased to 0 ^{0}C then iodine was poured to the mixture and left under continuous stirring for 4 days at 25 ^{0}C. The mixture was placed in separating funnel followed by separation of the organic layer and washing it with sodium thiosulphate then neutralized by sodium bicarbonate. The yield was washed by methanol to get rid of the impurities where a viscous liquid of NHTPB was obtained. Scheme 1 clarify the introducing of Nitro group to the polybutadiene chain. A photo of the separation step of NHTPB from the mixture is presented on Fig. 1.

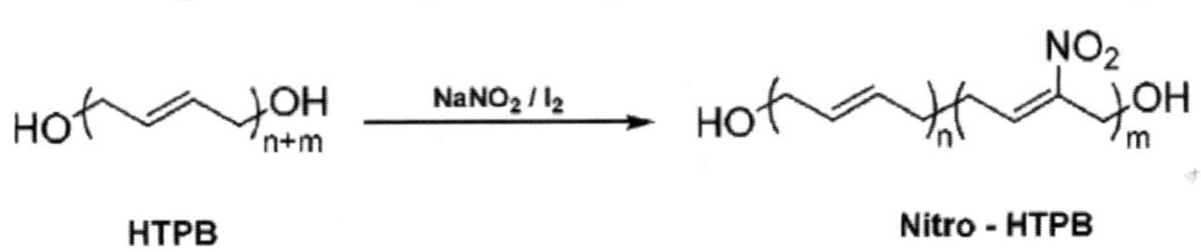

Scheme 1: NHTPB preparation reaction

Fig. 1 NHTPB separation step.

2.2. FTIR spectra

The spectra of both the prepared NHTPB and the raw material HTPB were determined using SHIMADZU 8400 FTIR spectrometer obtained from Japan. The range of the spectra were 400–4000 cm^{-1}, with 40 scans using the methodology of KBr pellet. Figure 2 includes The spectra of NHTPB in comparison with HTPB. It was observed that NHTPB has nearly the same stretching frequence as the original sample (HTPB) with only two more peaks appeared at 1521 cm^{-1} and 1336 cm^{-1}. These two peaks represent the NO_2 symmetric stretch and asymmetric stretch which did not appear in the spectrum of HTPB. These results confirm the the presence of nitro-group in the sample obtained after the preparation method.

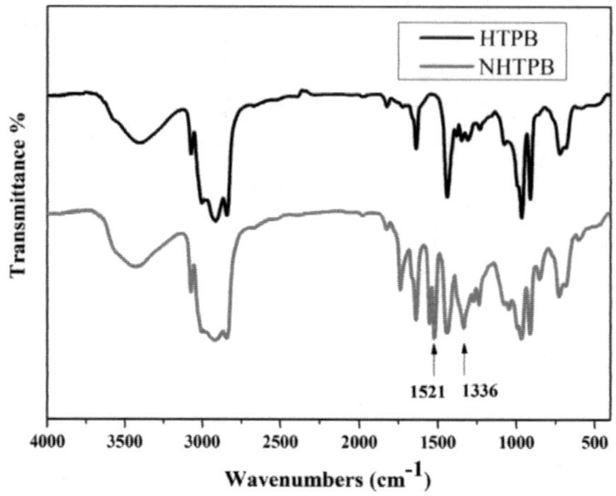

Fig. 2 The FTIR spectra of NHTPB in comparison with HTPB

2.3. *DSC measurement*

The measurement process is based on obtaining the heat flow at different temperatures for the studied sample. The prepared NHTPB sample was measured using DSC-Q200 of TA instruments. The sample weight was 2 mg placed on Al pan and the non-isothermal technique was used at heating rate of 5 ^{0}C/min.

A typical NHTPB thermogram is presented in Fig 3. The measurements started at 25 ^{0}C with heating rate 5 C/min. A decomposition peak of the sample started at onset decomposition temperature of 176 ^{0}C and the maximum decomposition peak was observed at 203 ^{0}C. The end peak of the decomposition process was at 246 ^{0}C. The heat release due to the exthothermic decomposition of NHTPB was 323 J/g. This exothermic peak represents the decomposition of the energetic NHTPB. This result confirms the high exothermicity of NHTPB and gives clear idea about the ability of NHTPB to energy out put to composition on its base.

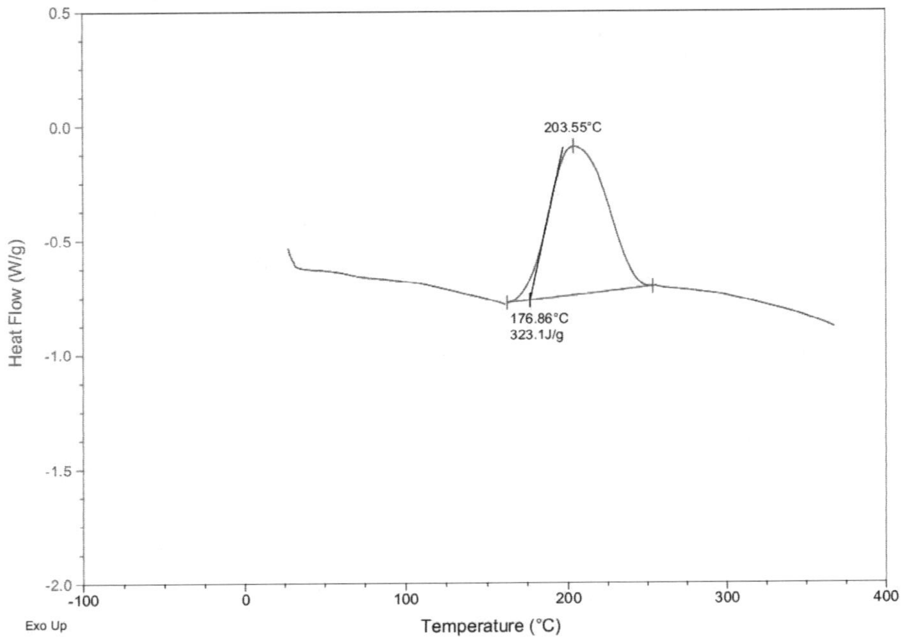

Fig. 3 typical NHTPB thermogram obtained by DSC

2.4. NHTPB composite propellant preparation

NHTPB-CSRP was prepared using 3 galloon mixer. The prepared NHTPB, dioctyl adipate (DOA) and MAPO were placed on the mixer and mixed for 10 min. at 60 ^{0}C, and then the solid phase (Ammonium perchlorate and Aluminum) was added in four portions, each portion takes 10 min. Finally the curing agent (HMDI) was added at 40 ^{0}C to the mixture during the stirring.

The prepared NHTPB-CSRP was casted and cured for two weeks. Two inch rocket motors were prepared to measure the ballistic performance of the prepared sample. The curing ratio of NHTPB-CSRP (NCO/OH) was 0.7.

2.5. Ballistic performance of NHTPB-CSRP

As mentioned in the preparation part, the prepared NHTPB-CSRP was casted and cured in selected steel cylinders. Two inch rocket motor was used to determine the ballistic characteristics (specific impulse, operating pressure and burning rate) of the prepared sample. The tests were performed using nozzle of 7 mm. In order to compare the results, a traditional HTPB-CSRP was tested under the same conditions used for NHTPB-CSRP. Typical charts of the pressure-time curve of both NHTPB-CSRP and HTPB-CSRP are shown in Fig. 4 and Fig. 5 respectively. Also the ballistic characteristics of both NHTPN-CSRP and traditional propellant are reported in table 1.

Table 1 The ballistic characteristics of the tested samples

Formulations	**Pressure (bar)**	**Burning rate (mm/sec)**	**Specific impulse I_{sp} (sec)**
HTPB-CSRP	109.2	9.86	243.6

NHTPB-CSRP	126.4	10.12	248.4

It is clear that the NHTPB-CSRP has higher operating pressure than the traditional propellant tested in this study. The burning rate of NHTPB-CSRP has slight increase in comparison with HTPB-CSRP while the specific impulse was significantly increased. These results confirm that the addition of NHTPB instead of HTPB caused a significant enhancement on the ballistic performance of the propellants which should lead to increase on the propulsion characteristics of the rockets.

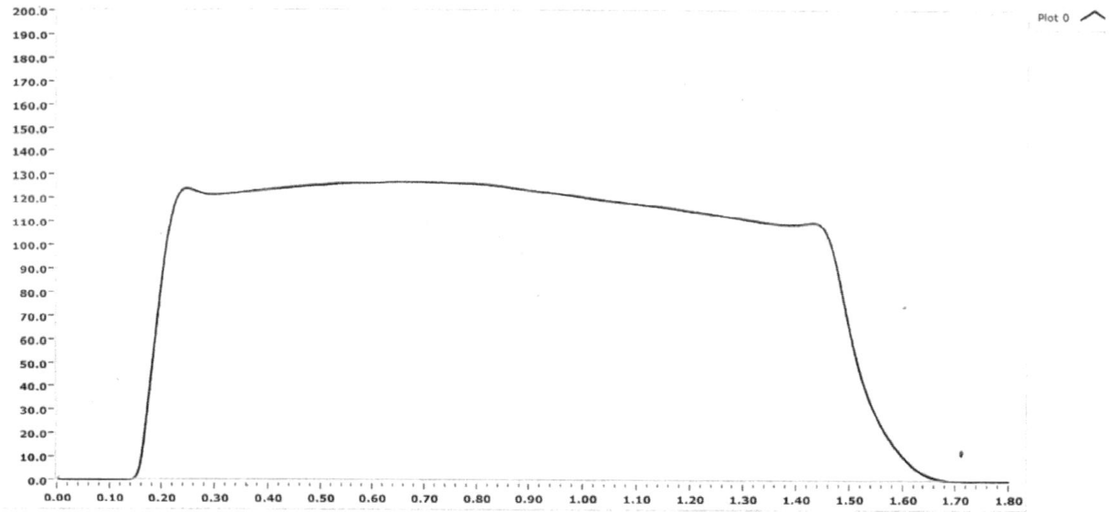

Fig. 4 Pressure-time curve of NHTPB-CSRP

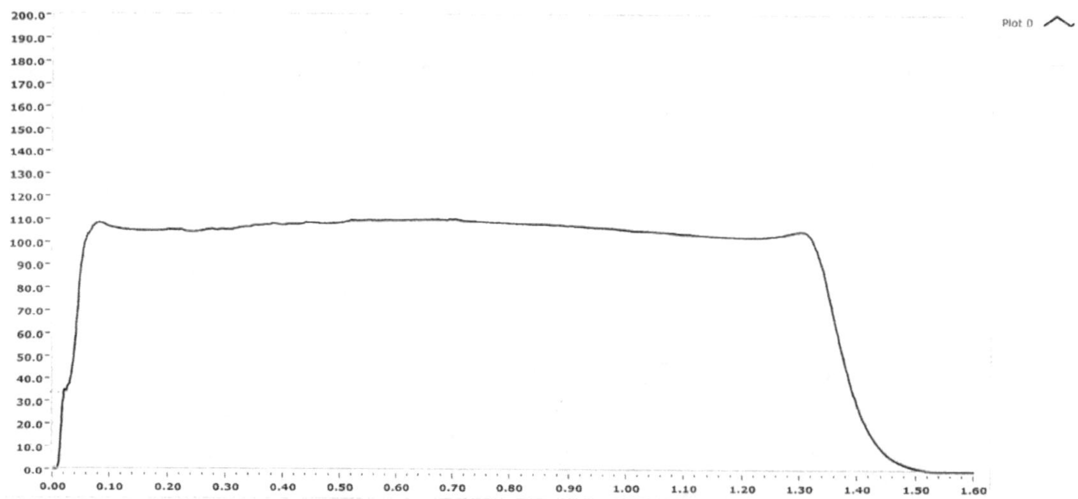

Fig. 5 Pressure-time curve of HTPB-CSRP

3. Conclusion

Nitro-HTPB was prepared successfully on our laboratories. The FTIR spectra confirmed the presence of nitro group on the building unites of HTPB after the nitration. On the other side, the high energetic characteristics of the prepared NHTPB was observed as a result of the exothermicity of the decomposition peak of NHTPB. The prepared CSRP based on NHTPN (NHTPB-CSRP) has higher performance than the traditional CSRP based on the inert HTPB; the operating pressure and the specific impulse was significantly increased, while a slight increase on the burning rate was observed in case of NHTPB-CSRP. NHTPB is a candidate to replace HTPB in the composite solid rocket propellants and it is recommended to be under more studies in the near future.

References

[1] Davenas A 1993, Solid Rocket Propulsion Technology, Pergamon Press, Oxford.
[2] Klapötke TM 2018 Journal of Engineering Science and Military Technologies **2**, 1-9.
[3] Tawfik S M, Saleh A, Elbeih A and Klapötke T M 2016, Zeitschrift fur Anorg. und Allg. Chemie, **642**, 1222–1229.
[4] Song M, Yang L, Yajin L, Guoping L and Yunjun L 2017 Central European Journal of Energetic Materials **14**, 708–725.
[5] Elbeih A, Wafy T Z and Elshenawy T 2017 Central European Journal of Energetic Materials **14**, 77–89.
[6] Elbeih A, Abd-Elghany M and Elshenawy T 2017, Acta Astronautica **132**, 124–130
[7] Chaturvedi S and Dave P N 2019, Arab. J. Chem. **12,** 2061-2068.
[8] Elbeih A, Abd-Elghany M and Klapötke T M 2017 Propellants, Explosives, Pyrotechnics **42**, 468-476.
[9] Abd-Elghany M, Klapötke TM, Elbeih A, Hassanein S and Elshenawy T 2017 Huozhayao Xuebao/Chinese Journal of Explosives and Propellants **40**, 24-32.
[10] Abd-Elghany M, Elbeih A and Hassanein S 2016 Central European Journal of Energetic Materials **13**, 349-356
[11] Abd-Elghany M, Elbeih A and Klapötke T M 2018 Journal of Analytical and Applied Pyrolysis **133**, 30–38.
[12] Larsson, A. and Wingborg, N. 2011 Advances in Spacecraft Technologies, 139-156.
[13] Abd-Elghany M, Klapötke T M and Elbeih A 2017 Journal of Analytical and Applied Pyrolysis **128**, 397-404.
[14] Gaur B, Lochab B, Choudhary V, and Varma IK. 2003 Journal of Macromolecular Science, Part C: Polymer Reviews **43(4)**, 505-545.
[15] Trache D, Maggi F, Palmucci I, DeLuca LT, Khimeche K, Fassina M and Colombo G. 2019 Arabian Journal of Chemistry **12(8)**, 3639-3651.
[16] Hussein A K, Zeman S and Elbeih A 2018, Thermochimica Acta **660** 110–123.
[17] Jensen TL, Unneberg E and Kristensen TE. 2017 Propellants, Explosives, Pyrotechnics, **42(4)**, 381-385.
[18] Leach C, Debenham D, Kelly J and Gillespie K. 1998 Propellants, Explosives, Pyrotechnics, 23(6), 313-316.
[19] Colclough ME, Desai H, Millar RW, Paul NC, Stewart MJ and Golding P. 1994 Polymers for Advanced Technologies, **5(9)**, 554-560.
[20] Boshra I K, Elbeih A and Mostafa H E 2019 Zeitschrift für anorganische und allgemeine Chemie, **645(5)**, 551-557.
[21] Hussein A K, Elbeih A and Zeman S 2018 RSC Advances **8** 17272–17278.
[22] Jones DE, Kwok QS, Vachon M, Badeen C, and Ridley W 2005 Propellants, Explosives, Pyrotechnics: An International Journal Dealing with Scientific and Technological Aspects of Energetic Materials, **30(2)**, 140-147.
[23] Colclough ME, Desai H, Millar RW, Paul NC, Stewart MJ, and Golding P 1994 Polymers for

Advanced Technologies, **5(9)**, 554-560.
[24] Ghayeni, H. R., Razeghi, R., and Olyaei, A. 2020 Polymer Bulletin, **77(9)**, 4993-5004.
[25] Dennis C, and Bojko B 2019 Fuel 254, 115646.
[26] Abusaidi H, Ghorbani M, and Ghaieni HR 2017 Propellants, Explosives, Pyrotechnics, **42(6)**, 671-675.
[27] Abusaidi H, and Ghaieni HR 2017 Journal of Thermal Analysis and Calorimetry, **127(3)**, 2301-2306.

Acoustic threat detection and direction finding system

A Abdallah[1] and H Tork[1]

[1] Naval Warning Engineering Department, Military Technical College, Cairo,Egypt

Email: adel.abdallah@mtc.edu.eg

Abstract. In this paper, an acoustic threat detection system is investigated as preliminary study of one of the anti-terrorism systems. The proposed system is a prototype of acoustic threats detection, classification, and direction finding system. For this purpose, an experiment is performed to test the capability of a proposed warning system to perform the task accurately. The direction of the sound source is determined by means of two microphones. The angle to the acoustic threat is calculated based on time difference of arrival (TDOA). The limitations of the system are discussed and some possible solutions are introduced to improve the performance of the system.

1. Introduction

Sound localization is the science concerned with determining the distance and direction of a sound source with the knowledge of the sound signals emitted from it [1]. Knowing these two parameters allows obtaining an accurate localization of a sound source which is important in many applications, especially at the present time, in the field of force protection and anti-terrorism. Technology that detects gunfire has existed for nearly two decades. For this purpose, many approaches have been proposed [2-4]. In many applications, the distance to the sound source is not as important as the direction information and this enables to design cost effective and less-complex systems. Some applications that ignore the distance and depend on the direction for locating the sound source like gunfire location [4], and intrusion detection systems [5]. As many other inventions all along the history, one of the first purposes of sound location was for military applications. A method called sound ranging was developed. Its principle is obtaining the coordinates of a hostile by the sound of its gun firing [6]. This technique allows the detection of the enemy even without direct vision. Other studies regarding sound localization are performed for closed or open environments, some of them introduced some techniques to minimize the errors [7], and echo or noise cancellation [8]. The common factors that control the system performance are either the number of microphones used in the array[9-13] or the technique of the signal processing after detecting the sound [14-16].The physical characteristics and the spectral content of the impulse noise from common weapons such as pistol 9 mm, military rifle (7.62 mm), and other weapons are discussed in [17]. The microphone arrays should be designed to make sure these frequencies are detected.

Optical or acoustic methods are the common techniques for acoustic threats detection such as sniper detection systems. Optical systems can be based on one or a combination of a laser system and a thermal camera. Laser systems used for sniper detection, based on cat's eye effect, can detect the sniper before firing but false alarm rate in these systems is still a limiting factor and needs further investigation [18, 19]. Thermal camera system can be used to detect the sniper before firing and just after the firing by

Content from this work may be used under the terms of the Creative Commons Attribution 3.0 licence. Any further distribution of this work must maintain attribution to the author(s) and the title of the work, journal citation and DOI.
Published under licence by IOP Publishing Ltd

detecting the muzzle flash[20]. Clear line of sight to the sniper's weapon is required to use the optical systems for sniper detection which is considered a limitation or a disadvantage of these systems. The acoustic systems used for sniper detection can't detect the sniper before shooting. So, using these systems alone can limit losses in troops and determine the direction of the threat to deal with the executor. The advantage of acoustic sniper detection systems is the unnecessity of the presence of a direct line of sight between the detector and the weapon if distributed sensor arrays are used. It is expected that best results can be obtained if the optical and acoustic systems are combined.
In this paper a simple technique to determine the direction of a sound source in real time by two microphones is proposed. The proposed warning system allows simultaneous acquisition, recording, processing and analysis of the incoming acoustic signals using two microphones for signal identification and determination of the direction of the acoustic source. Signal processing is performed using MATLAB.

2. Experimental setup

The proposed system is based on a single point sensor (SPS) which usually used to point a surveillance camera towards the gunfire. SPS has some disadvantages;

It generally requires a line-of-sight for accurate direction finding of the acoustic source.

It lacks the ability to correctly calculate range to the sound source instead it can just find the direction to the source. Even the correct direction can be misled by non-direct path propagation.

The advantage of using SPS is its speed of reaction. It can detect the acoustic threat within 1 second or less. For our purpose SPS is adopted. The system consists of two microphones, microcontroller, servo motor, LCD, cables, connectors and a computer for processing and display. At least two microphones are required to extract the time delay between the two measured signals and consequently to calculate the direction of arrival. The used microphones are omnidirectional, have frequency response range 50 Hz to 20 kHz, and sensitivity -46 ± 2.0 dB at 1 kHz.

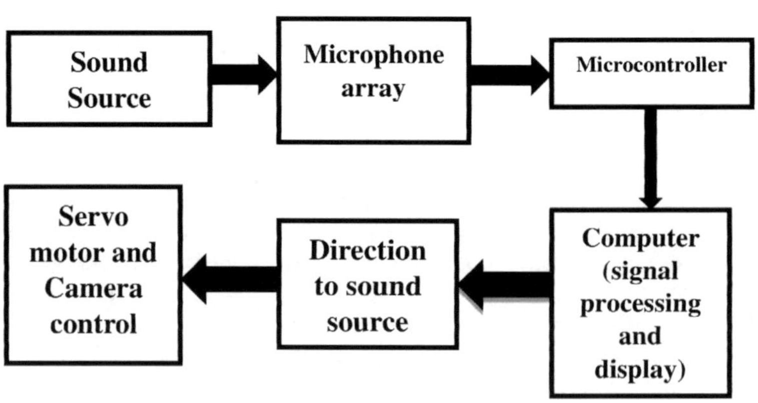

Figure 1. Block diagram of the proposed acoustic threat detection system

The Block diagram of the proposed acoustic threat detection system is shown in Figure 1. The sound source emits a signal that acquires some noise and reflections from the surrounding environment while traveling to the microphones. The output analog electrical signals from the two microphones are amplified and converted to digital signals using the microcontroller which is connected to the computer where the signals are acquired and processed in real time by MATLAB to determine the direction of the sound source. The signals are sampled at 44 kHz allowing analysis of the detected acoustic signals in the audible band. The angular information is sent back to the microcontroller that controls a servo motor and a camera. The initial experimental setup of the proposed acoustic threat detection and direction finding system is shown in Figure 2. This set up was modified many times during the study.

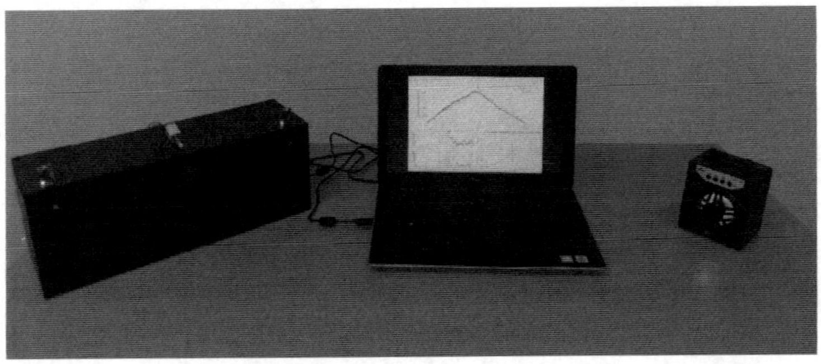

Figure 2. The initial experimental setup of the proposed acoustic threat detection and direction finding system.

3. Signal processing and direction finding

After detection of the incoming acoustic signals, cross-correlation between the recorded acoustic signals is used to calculate the bearing of the acoustic source. Two microphones with spacing (*L*) are used to receive the acoustic signal radiated from a source makes an angle (θ) with the normal to a reference line between the sound detectors as described in Figure 3. The spacing between the source and the sound detectors is much larger than (*L*). For linear arrays, the maximum frequency of operation is given by $c/2L$, c represents the velocity of sound, and L is the spacing between two elements of the array [21]. Taking into account the condition to avoid the effect of the grating lobes that the microphone spacing must be less than or equal to one-half of the minimum wavelength and is calculated from;

$$L \leq \frac{\lambda_{min}}{2} \ or \ L \leq \frac{c}{2f_{max}} \tag{1}$$

The two microphones are separated by distance L equal 10 cm and mounted on a box houses the electronic components. The signal radiated from the sound source received the two microphones with delay ΔT between them:

$$\Delta T = \frac{L \sin\theta}{c} \tag{2}$$

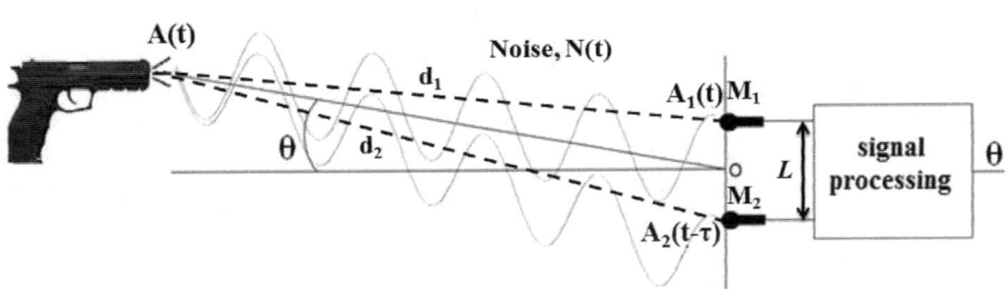

Figure 3. Description of the physical setup.

let $A(t)$ is the unknown acoustic signal emitted by the sound source, $A_1(t)$, and $A_2(t)$ are the observed acoustic signals recorded by the two microphones, $N(t)$ is the noise, and M_1, M_2 are the known positions of the microphones, $\tau_1=d_1/c$, and $\tau_2=d_2/c$ are the time of arrival from the sound source to microphone1, and microphone 2, respectively, and the time difference of arrival between the sound detectors is $\Delta T=\tau_1 - \tau_2$.The recorded signals can be expressed by;

$$A_1(t)= A(t-\tau_1)+N_1(t) \quad (3)$$
$$A_2(t)= A(t-\tau_2)+ N_2(t) \quad (4)$$

The signals $A_1(t)$ and $A_2(t)$ detected by the sound detectors are the same signals with delay ΔT:

$$A_2(t)= A_1(t-\Delta T) \quad (5)$$

The cross correlation $A_{12}(\tau)$ of the two signals;

$$A12(\tau)=\int_{-\infty}^{\infty} A1(t')A2(\tau - t')dt' \quad (6)$$

$$A_{12}=A_{11}(\tau - \Delta T) \quad (7)$$

The cross-correlation $A_{12}(\tau)$ of the two acoustic signals can be considered as the autocorrelation A_{11} of the detected signal from the microphone (M_1) but with delayed time $\tau = \Delta T$. The maximum of the autocorrelation of a signal is at $\tau = 0$, consequently, the maximum of $A_{12}(\tau)$ is at $\tau = \Delta T$. This maximum of the cross-correlation is be used to determine the direction of the sound source. Source bearing angle can be calculated from:

$$\theta = \arcsin(c\Delta T/L) \quad (8)$$

The following steps are followed for detecting and direction finding of the acoustic source:

- Studying the fundamental acoustic parameters to detect the acoustic threat such as acoustic signatures, the transmission loss of the signal as propagates through the channel, and the environmental noise.
- Signal from the source is detected and recorded (this simulates the acoustic signature of the acoustic source).
- Convert the signal from time domain to frequency domain, and identify the spectral components of the signal.
- Identify the received signal.
- If the signal is identified then calculate the cross-correlation of the detected signals, calculate delay time, and estimate the bearing of the sound source.

4. Experimental results and analysis

The system was integrated and tested in the lab. Serial communication is established between the microcontroller and the computer and the system is controlled by MATLAB code. The two signals were captured by the microphones and one of them is delayed version of the same signal. According to analysis and algorithms of cross correlation, the time delay and the angle of the source are calculated. The angle is displayed on the LCD, and the microcontroller is used to move the servo motor to the calculated direction of the sound source. For this purpose, three different audio signals represent different guns and rifles sounds are used to generate the sound and simulate the acoustic threats. In the paper, we denote these signals as gun 1, gun 2, and rifle. The spectral content of the received audio sound signals of the gun 1, and rifle in the form of a spectrogram is shown in Figure 4.

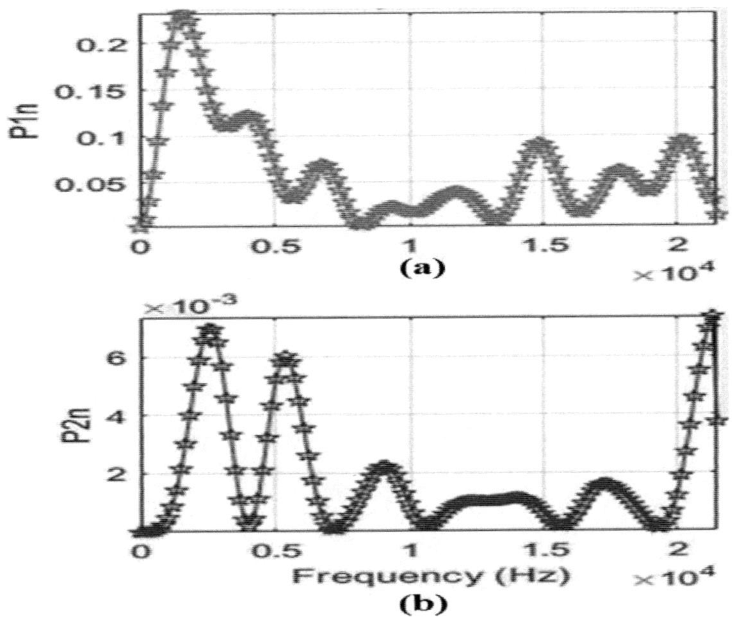

Figure 4. The spectral content of the received audio sound signals of the gun 1 (a), and rifle (b).

The cross correlation between output signal from two microphones for gun 1 as audio sound source is shown in Figure 5 (a). While the auto correlation of the same signal of single microphone is shown in Figure 5 (b) for comparison. It can be seen that at the lag, the peak of the correlation output occurs. This lag represents the time delay between the signals. The correlation sequence peaks when the lag is 0.3 ms which represent the time delay between the two signals. Delay between the received signals is calculated using cross correlation function to determine the sound source direction which is 15°.

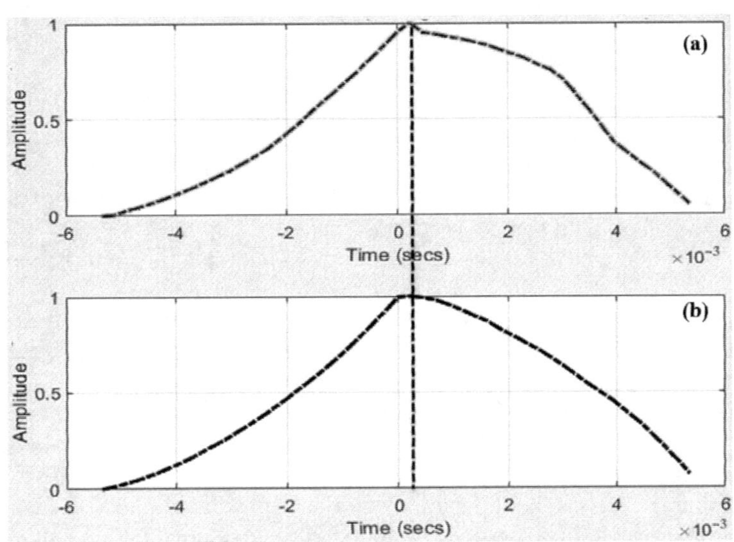

Figure 5. cross correlation between output signal from two microphones.

For signal identification, the three investigated signals are recorded as reference signals and the normalized cross correlation of the received sound signal and the recorded signals is calculated as shown in Figure 6. The matching level of the measured signals compared to the recorded reference signal of gun 1 is shown in Table 1. It is easy to set a certain matching level for signal identification.

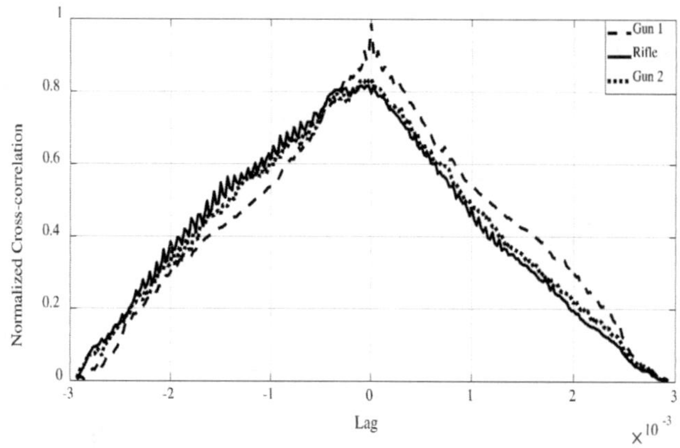

Figure 6. Normalized cross correlation of the three measured signals (gun 1, gun 2, and rifle) compared to the reference signal of gun 1.

Table 1. The matching level of the measured signals compared to the recorded reference signal of gun 1.

Measured signal	Matching level (%) (Reference signal is Gun 1)
Gun 1	99.2
Gun 2	83.56
Rifle	81.65

The accuracy of the signal identification is 100% but the accuracy of direction finding is about 60%. This can be attributed to using only two microphones in the experiment, and the room reverberation or the reflections from the room walls that form many sources of sound to the system. This causes inaccurate calculation of the sound source direction.

5. Conclusion

The proposed system is used to estimate the direction of acoustic threats using two microphones that can emulate the detection of the muzzle blast of a gunshot. In the experiment, the incoming signals were detected, recorded, identified and processed to determine the sound source direction based on TDOA. The proposed system showed a good performance. The accuracy of calculations can be enhanced by taking into account some considerations like using microphones with better sensitivity, number of the microphones in the array, optimizing the spacing between the microphones, and testing the system in open field to avoid the reflections of the signal from the flat surfaces or walls that causes inaccurate calculation of the sound source direction. The system can be developed to be mounted either at a fixed location or on a moving vehicle.

References

[1] Damarla T 2015 *Battlefield Acoustics* (USA: Springer)

[2] Scott J and Dragovic B 2005 *Proc. the 3rd Int. Conf. on Pervasive Computing* p 1
[3] Naz P Bouguereau J Lemer A and Ywanne F 2005 *Proc. SPIE Unattended Ground Sensor Technologies and Applications VII*
[4] Graves, J.R., 2012. *Audio gunshot detection and localization systems: history, basic design, and future possibilities.* University of Colorado at Denver.
[5] Zieger C Brutti A and Svaizer P 2009 *Proc. 6th IEEE International Conference on Advanced Video and Signal Based Surveillance* (Genova: Italy) p 314
[6] Ludeman L 1980 *Proc. Acoustics, Speech, and Signal Processing, IEEE International Conference (ICASSP)* p. 800
[7] Maganti H and Gatica-Perez D, 2006 *Proc. ICMI '06:the 8th international conference on Multimodal interfaces* p. 35
[8] Djendi M Gilloire A and Scalart P 2006 *Proc. IEEE Int. Conf. on Acoustics Speech and Signal Processing*
[9] McKinney E and DeBrunner V 1996 *Proc. IEEE Int. Conf. on Acoustics, Speech, and Signal Processing ICASSP-96* p 933
[10] Rubio J E *et al* 2007 *Proc. IEEE Int. Conf. on Acoustics, Speech and Signal Processing ICASSP* 2007 p 385
[11] Kurihara M *et al Proc. the 41st* SICE *Annual Conference (SICE 2002)* p 1100
[12] Song P Hao C Wu J and Yang C 2017 *Sensors and Actuators A: Physical* vol 267 376
[13] Kwok N M Buchholz J Fang G and Gal J 2005 *Proc. IEEE Int. Conf. on Industrial Technology* (Hong Kong: China) p 281
[14] Swartling M Grbic N and Claesson I 2006 *Proc. IEEE Int. Conf. on Acoustics, Speech and Signal Processing ICASSP*
[15] Argentieri S Danès P and Souères P 2015 *Computer Speech & Language,* vol 34 87
[16] Zhao Y Chen X and Wang B 2013 *Applied Acoustics* vol 74 1367
[17] Ylikoski M E *et al* 1995 *Scandinavian Audiology* vol 24 3
[18] Jiang C h *et al* 2019 *Proc. SPIE 14th National Conference on Laser Technology and Optoelectronics (LTO 2019)* vol 1117019
[19] Zheng Z 2013 *Proc. SPIE International Symposium on Photoelectronic Detection and Imaging 2013: Laser Sensing and Imaging and Applications* p 89051R
[20] Trzaskawka P Dulski R and Kastek M 2010 *Proc. SPIE Electro-Optical and Infrared Systems: Technology and Applications VII* vol 7834 (Toulouse: France) p 783414
[21] Damarla T 2015 *Acoustic arrays Battlefield Acoustics* (USA: Springer) chapter 6 pp 66-81

IOP Conf. Series: Materials Science and Engineering 1172 (2021) 012034 doi:10.1088/1757-899X/1172/1/012034

Tactics Overview for Implementing High-Performance Computing on Embedded Platforms

A Elshazly [1], A Elliethy [1] and M A Elshafey [1]

[1]Department of Computer Engineering and Artificial Intelligence, Military Technical College, Cairo, Egypt

E-mail: ahmedelshazly@mtc.edu.eg

Abstract. Future space missions will rely on novel high-performance computing to support advanced intelligent on-board algorithms with substantial workloads that mandates firm real-time and power constraints requirements. Consequently, these advanced algorithms require significantly faster processing beyond the conventional space-grade central processing unit capabilities. Moreover, they require careful selection of the target embedded platform from a diverse set of available architectures along with several implementation tactics to map the algorithms to the target architecture to fully unlock its capabilities. In this paper, we present a study of different architectures and embedded computing platforms for the satellite on-board computers. Moreover, we present a comprehensive overview of recent implementation tactics such as source code mapping and transformations. Additionally, we highlight some optimization techniques such as partitioning and co-designing using hardware accelerators. Finally, we discuss several implementation analysis methodologies to derive optimized code implementations. The top ranked YOLO-v3, as a deep learning based object detection algorithm, is selected as a case study model to be optimized using OpenVINO toolkit. The experimental results show an improvement ratios up to 73%, 41%, and 34% in terms of frames per second, CPU utilization, and cache memory, respectively. The study presented in this paper aims to guide the researchers in the field of high performance embedded computing in terms of different hardware architectures along with several implementation tactics.

1. Introduction

Recently, Cubesat satellites attracted attention on both commercial and academic levels. These satellites have restricted image processing capabilities due to their limited resources. Thus, they may restrain the exploitation of automated and intelligent image on-board processing tasks that can be considered as the future trends for satellite missions. Additionally, these tasks are crucial in some applications that require automated decision support or low latency conclusions [1, 2, 3, 4, 5].

One of the challenging automated task is oriented toward onboard deep learning based applications such as: telemetry compression [6, 7] and image processing [1, 2]. However, deep learning based models are characterized by a larger number of parameters with high degree of precision accuracy. Due to hardware and power limitations of the on-board computer (OBC) of the satellites, in general and Cubesats in particular, efficient optimization techniques should

Content from this work may be used under the terms of the Creative Commons Attribution 3.0 licence. Any further distribution of this work must maintain attribution to the author(s) and the title of the work, journal citation and DOI.
Published under licence by IOP Publishing Ltd

be applied on the target deep learning based approach taking into consideration the hardware specification of the current OBCs [8, 9, 10, 11, 12].

In this paper, we provide a survey of recommended tactics for efficient optimizations that can be utilized for performing complex computations on the OBC of the satellites. Our survey is followed by a case study of optimizing the pre-trained YOLO-v3 deep learning based model for object detection in order to be accommodated on the OBC of the satellite. In our case study, we provide an experimental comparison between the typical YOLO-v3 deep learning model and the optimized one, in terms of the processing time (frame per second). Moreover, the paper provides a survey about the OBCs that are commonly used in satellite systems.

The paper is organized as follows. Section 2 presents an overview of OBCs. In section 3, an overview of the deep learning model of the YOLO-v3 structure is demonstrated. General tactics for implementing efficient optimization are described in details in Section 4. In Section 5, we present our experimental results in which we apply the optimization tactics presented in the paper on the YOLO-v3 model [13] using the OpenVINO™ [14] toolkit. Conclusions and future work are suggested in section 6.

2. Satellites on-board computers

The satellite OBC is the heart subsystem of the satellite that connects, manages and controls other subsystems. It also handles payload and housekeeping data to accomplish the targeted mission of the satellite [15, 16, 17, 18, 19].

The main tasks of the OBC includes the span [20] of:

- Attitude Determination and Control (ADCS), such that the attitude of the satellite is measured by sensors. i.e. magnetic field and solar sensors, and corrected by magnetorquers.
- Platform monitoring and control (housekeeping functions) through periodically checking measured sensors values (voltage, temperature, attitude, and orbit) for safe operation and monitoring telemetry data sent to the ground station.
- Experimental measurement and analysis, in which operational data and experimental measurements of new sensors, actuators and control techniques are collected, processed and analysed.

The architecture of the OBC belongs mainly to one of the following three architectures showed in Fig. 1 and listed in Table. 1 associated with their advantages and disadvantages.

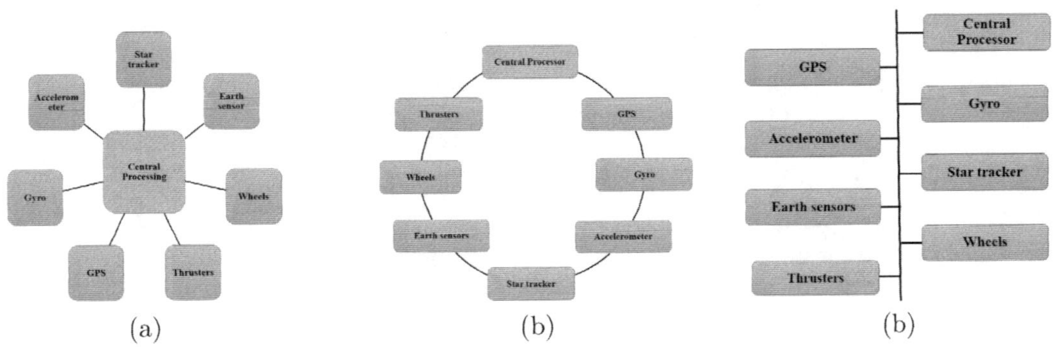

Figure 1. Different OBC architectures: (a) centralized architecture, (b) distributed architecture, and (c) bus architecture.

The hardware of the satellite OBC computing platforms [15, 16, 17, 18, 19, 20] has limited resources. Examples of such platforms are: (1) Microprocessor based, (2) Micro-controllers, (3)

Field Programmable Gate Array (FPGA), (4) Application Specific Integrated Circuit (ASIC) , and (5) System-on-Chip (SOC). The differences of these platforms are summarized below

- Microprocessor based platform. It focus on computation, requires peripheral ICs, supports fastest generic computation, and has typical power consumption higher than 10 Watt.
- Micro-controllers based platform. It focus on embedded systems, has memory integrated, supports integrated peripheral function: ADC, DAC, PWM, I2C, UART, etc, and has a typical power consumption less than 1 Watt.
- FPGA based platform. It is a Re-programmable logic, has IP-cores for specific and complete micro-controllers, provides fast and low power for specific functions, has higher power than micro-controllers.
- ASIC based platform. It is a complete hardware solution for specific application, supports IP-cores for specific function, and has fast and power efficiency.
- SOC based platform. It provides a complete hardware and generic application. It has the smallest complete systems.

Table 1. General OBC architectures presented in Fig. 1.

	Advantages	Disadvantages
Centralized	– Adequate for a small number of well-defined sub-systems with a direction interfaces to the central computer. – High reliability that is achieved by isolating sudden failure at any interface and preventing influence on others.	– Required changes in both hardware and software of the central node in order to connect to a new one. – Large wiring harnesses is needed for transmitting data to multiple recipients.
Distributed	– Small wiring harnesses with the capability of being distributed along the structure of the spacecraft . – Limited effect on the central node when adding new nodes.	– Reliability reduction is because each node is required to achieve data transmission to the next node.
Bus	– High processing capacity with high reliability can be achieved through the ability of adding multiple processing units.	– More complex testing procedures are relatively required for these systems.

3. YOLO-v3 deep learning model

Our case study in this paper is the optimization of the pre-trained YOLO-v3 deep learning based object detection model. It is a deep neural network that contains a set of 53 convolutional layers each followed by Batch Normalization (BN) layer and Leaky ReLU (LRELU) activation function. The YOLO-v3 deep network model, shown in Fig. 2, can recognize up to 80 different shapes in images and videos [21].

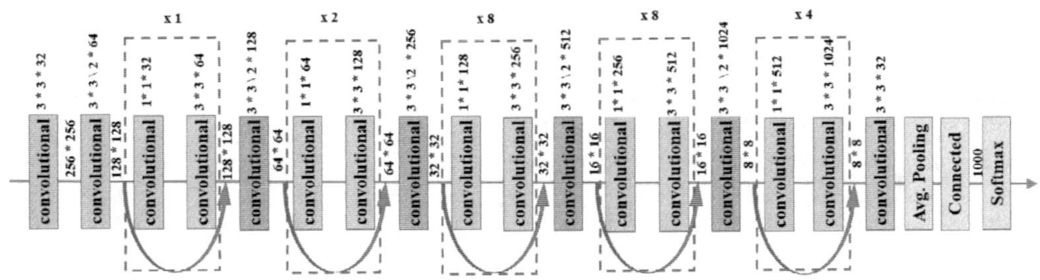

Figure 2. YOLO-v3 fully convolutional network.

The detection phase of YOLO-v3 model is achieved at three different scales by three strides of values 32, 16, and 8. At each scale, a count of 3 bounding boxes are detected by 3 anchors, different anchors for different scales, resulting a total number of 9 anchors is used. For such deep learning based detection model, detections at scales of 13 x 13, 26 x 26 and 52 x 52 are achieved for 416 x 416 input size [21].

4. Efficient optimization tactics for complex computations

Optimizations is used to mitigate the cost the expensive instructions, which results in reducing the execution time. It includes several components that can be either hardware or software. We will summarize them below.

4.1. Hardware accelerators

Hardware accelerators are used to offload the computations from a host processor to an accelerator element in order to enhance the performance and meet the nonfunctional requirements (such as strict execution time limit). There are many types of the H/W accelerators such as ASICs, GPUs and FPGAs. The programming models of these devices exhibit different levels of trade-offs between accuracy and the nonfunctional requirements.

GPU accelerators consists of lightweight cores and on chip memories are used to increase the degree of parallelism for a single program. Originally, GPUs are used for the acceleration of graphical applications but it has been evolved to support many intensive applications. Recently, GPUs are used in embedded systems as hardware accelerators, thanks to their increasing capabilities and capacities [22] [23].

The enhances introduced in CPU cores nowadays, allow them to further support parallelism which includes the support of multi-threading. Single Instruction Multiple Data units (SIMD) is one example, in which the CPU can concurrently operate on multiple data operands and perform the same operation. Another example is the Fused Multiple-add units (FMA) that are used to perform multiple-add or multiple-subtract operations [24].

To enhance the performance of a complex algorithm, it is common to develop the code on different architectures. The code in Fig. 3 shows a code that uses different directives to speak to different targets to measure the execution time for a PC or an embedded system using a Xilinx MicroBlaze [25, 26, 27, 28, 29]. To get the maximum performance from every computing element in the computing platform, it is required to split the code into different sections and map these sections of the code to distinct processing elements in the computing platform [30, 31], in order to meet the nonfunctional requirements. This process can be performed in two ways:

`#ifdef embedded_timing`
`XTime_GetTime(&timeStamp_206983496);`
`printf("TS-before 'gridsInit': %llu\n", timeStamp_206983496);`
`#endif`
	Timing for MicroBlaze

`#ifdef pc_timing`
`{ hTimer t ("t1");`
`#endif`
	Timing for PC

`#ifdef embedded_timing`
`XTime_GetTime(&timeStamp_206983496);`
`printf("TS-after 'gridsInit': %llu\n", timeStamp_206983496);`
`#endif`
	Timing for MicroBlaze

`#ifdef pc_timing`
`}`
`printf("TS-after 'gridsInit': %llu\n", hTimer::elapsed("t1"));`
`#endif`
	Timing for PC

Figure 3. Code developing for different targets.

static partitioning and dynamic partitioning. The static partitioning mapping is determined at compile time and remains unchanged at run-time. The static partitioning follows the following steps

(i) Mapping: in which the code is inspected and a static call graph is generated. The graph is G=(N, E), where nodes N represent the function calls while edges E represent the data flow and control that relates the two nodes connected by the edge E.
(ii) Mapping to the target heterogeneous architecture.
(iii) Finding a configuration in which the overall cost

$$\sum_{task} cost(PE_t, task), \tag{1}$$

is minimized. This minimization is required to meet a given performance goal.

Dynamic partitioning [32] on the other hand determines the mapping during run-time according to existing workload and device availability. We use dynamic partitioning because the input size is determined only during the run-time. The efficient processing element is determined based on an estimation model for a particular job.

4.2. Code transformations and optimizations

Developers need to perform the code optimizations in order to meet the nonfunctional requirements. Some basic code optimizations are internally done by the compilers [33] such as the strength reduction. There are many examples of strength reduction such as the replacement of multiplication and division by left and right shifts, respectively. Another example of the strength reduction is when we need to negate a value, as shown in Fig. 4.

`unsigned int p2 = * (unsigned int *) &a;`
`p2 = p2 ^ (1 << 31);`
`float neg_a = * (float *) &p2;`
	`B = 105 * A;`
	can be substituted by:
	`A1 = (A<<4)-A;`
	`B = (A1<<3)-A1;`

Negate a value.	The replacement of multiplication and division by left and right shifts.

Figure 4. Basic code optimizations.

However, compilers do not have all code transformation so some optimizations must be done manually. Example of such manual optimization is the Loops optimizations. Loops are considered as very important hot-spots as they have a very high impact on performance. There are many types of loop optimizations such as

- Loop splitting: It simply partitions a single loop into many other loops [34] which is very important when considering data parallelism, as shown in Fig. 5 (a).
- Loop Strip-mining: In this transformation we transform a loop into a nested loop[35], as shown in Fig. 5 (b).
- Loop Perforation: In this loop transformation, we only consider a part of the iterations to give approximate results but better performance [36] [37]. An example is shown in Fig. 5 (c).
- Loop Unrolling: To unroll a loop by a factor K is to repeat the body of the loop K times which is very important when considering the usage of SIMD units [38]. An example is shown in Fig. 5 (d).
- Loop UN-switching: In this transformation, the invariant conditions are moved outside the loop to avoid unnecessary processing[39]. An example is shown in Fig. 5 (e).

To perform the code optimization, first we need to perform source code analysis[40]. Source code analysis is used in understanding the structure of the code. In this phase, the procedures are being analyzed to determine which procedures invoke other procedures or which procedures allocate storage and understanding where the execution spends the bulk of its time. But before we speak about the source code analysis we need to understand the data dependencies first.

Data dependency is the relationship between the instructions. We must check for it in order identify the legal optimizations that can be done within the code[41]. There are four types of data dependencies. The first type is True-data dependency which happens when an operand is modified and then read in a later instruction. The second type is Anti-data dependency which happens when an operand is read and then modified in a later instruction. The third type is Output-data dependency which happens when an operand is modified by two instructions. The last type is Input-data dependency which happens when an operand is read by two instructions.

For a compiler to check for data dependency in a loop, the compiler first generates code to check for data dependency at run-time during the first iteration. If no dependencies are detected, then the rest of the iterations can run concurrently. Another optimization is called Loop Permutations, in which the order of execution is changed between two loops in a nested loop. This change reduces the dependency which enables us to use data parallelism as shown in Fig. 6.

Source code analysis can be classified into two types: Static and dynamic code analysis. Static code analysis aims at understanding the properties of the code without running the app. To perform the static analysis we must follow the following steps. First, we need to parse the code to generate a flow graph. Then, we need to identify the data dependencies to decide the legal optimizations that can be done within the code in order to exploit the characteristics of the target computing platform [42] [43]. In dynamic analysis, the goal is to uncover the metrics of performance during execution time. Specifically, the dynamic analysis tries to find the hot-spots which are the prime candidates for optimization. These spots can be found by profiling [44] and instrumenting the code by adding primitives to the code to inspect the code during the run-time. This can by done manually or by using the LARA language [45].

5. Experimental work

In this section, we provide an experimental comparison between the typical YOLO-v3 deep learning model and the optimized one using the optimization tactics presented in Section 4. We utilized the OpenVINO™ [14] toolkit to perform the optimization. First, we give an overview about the OpenVINO toolkit, then we demonstrate the performance metrics used for evaluating the optimization gain, and finally our results are figured out.

for(i = 0; i < N; i++)
A[i] = B[i] + C;
	for(i = 0; i < N/2; i++)
	A[i] = B[i] + C;
	for(i = N/2; i < N; i++)
	A[i] = B[i] + C;

Original loop	Splitted loop

(a) Loop splitting

for(i=0; i<N; i++) {
sum += A[i];
}
	for(is=0; is<N; is+=S) { // S is the size of the 1-
	dimension block
	for(i=is; i<min(N,is+S-1); i++) {
	sum += A[i];
	}
	}

Original loop	Loop after strip-mining

(b) Loop strip-mining

for(i=0; i<N; i++) {
sum += A[i];
}
	for(i=0; i<N; i+=k) {
	sum += A[i];
	}

Original loop	Loop after Perforation

(c) Loop perforation

for(i=0; i<4;i++) {
c[i] = a[i] + b[i];
}
	c[0] = a[0] + b[0];
	c[1] = a[1] + b[1];
	c[2] = a[2] + b[2];
	c[3] = a[3] + b[3];

Original loop	Loop after fully unrolled

(d) Loop unrolling

for(i=0;i<M;i++) {
if(op==1)
sum += A[i];
else
sum *= A[i];
}
	if(op==1)
	for(i=0;i<M;i++) {
	sum+=A[i];
	}
	else
	for(i=0;i<M;i++) {
	sum *= A[i];
	}

Original loop	Loop after Un-switching

(e) Loop un-switching

Figure 5. Different loop optimizations.

for (i=1; i < N; i++){
  for (j=1; j < N; j++){
    C[i][j] = C[i-1][j] + 1;
  }
}
	for (j=1; j < N; j++){
	  for (i=1; i < N; i++){
	    C[i][j] = C[i-1][j] + 1;
	  }
	}

D = (1,0) , Original loop	D=(0,1), Permutated loop

Figure 6. Loop Permutations

5.1. OpenVINO™ toolkit

OpenVINO toolkit is an Intel toolkit used to perform tasks such as automatic speech recognition and emulation of human vision using latest artificial neural networks [46, 47, 48]. The model

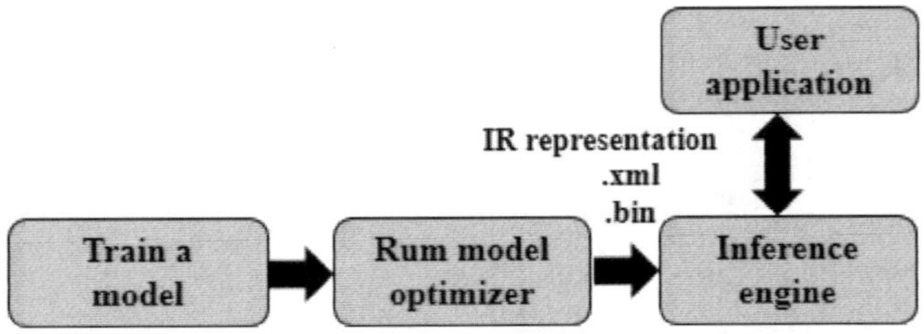

Figure 7. Overview of OpenVINO™ toolkit [14].

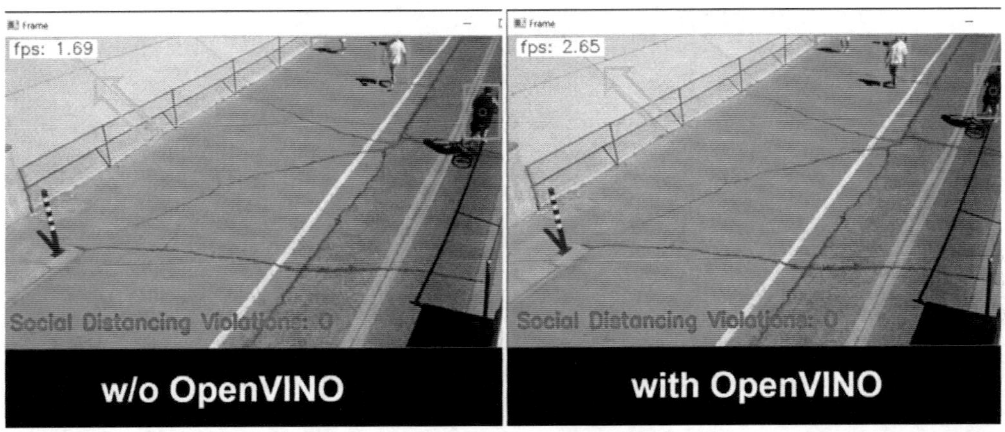

Figure 8. Visual result that shows two frames from the video used in our experiments.

optimizer which is the most important component in the OpenVINO™ toolkit converts the trained models to an Intermediate Representation (IR) (.xml and .bin) to be used in inference operations, as shown in Fig. 7. The Model Optimizer performs a group of code optimizations to produce simpler and faster models and it is applicable for models trained in popular frameworks.

5.2. Performance evaluation metrics

For evaluation the performance gain due to applying optimization techniques, we implement the following three metrics:

- Frames per second: The higher the better. It is the average count of frames accessed per second.
- CPU utilization: The lower the better. It measures the amount of CPU resource requirements for handling the task.
- Cache memory: The higher the better. It indicates the size of data cached, in repeated access cases, to avoid the delay in going back to access the storage memory.

5.3. Results

We have benchmarked the YOLO-v3 on CPU as a processing element without any optimizations and with the optimizations applied by the model optimizer from the OpenVINO toolkit. The application used here is to calculate the social distancing between persons using YOLO-v3. Fig. 8 shows two frames from the video used in our experiments.

Table. 2 represents the performance enhancement in frames per second (FPS), cache memory, and CPU utilization after implementing the optimization techniques using the OpenVINO

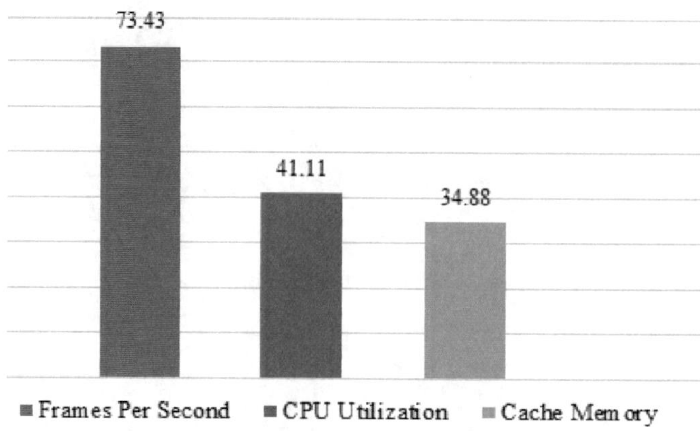

Figure 9. Enhancement ratios (%) due to optimization.

toolkit. The listed results, in table 2, show that the performance gain is about 73%, 41%,

Table 2. Performance enhancement after optimization.

	Without optimization	With optimization
Average FPS	1.28	2.22
Cache memory	911 MB	1.2 GB
CPU utilization	90%	53%

and 34% in terms of frames per second, CPU utilization, and cache memory, respectively, as shown in Fig. 9.

6. Conclusion

This paper demonstrates how we can improve the performance of complex and intensive algorithms, especially intelligent ones, using optimizations such as offloading the computations from the host processor to a co-processor and analysing the code to find the hot-spots and apply code optimizations to them. The paper showed experimentally that code optimization techniques can lead to a better performance in terms of frames per second, CPU utilization, and cache memory, that helps to implement the high-performance computing algorithms on embedded platforms with limited resources as in case of satellite on-board computers.

References

[1] Lim S and Xiaofang C 2014 *2014 IEEE International Conference on Aerospace Electronics and Remote Sensing Technology* (IEEE) pp 137–141
[2] Nzeugaing G N and Biermann E 2016 *Journal of Engineering, Design and Technology*
[3] Liu S and Yang J 2019 *Symmetry* **11** 1373
[4] Tsog N, Behnam M, Sjödin M and Bruhn F 2018 *2018 IEEE Aerospace Conference* (IEEE) pp 1–8
[5] Meß J G, Dannemann F and Greif F 2019 *European Workshop on On-Board Data Processing (OBDP2019)* (European Space Agency)
[6] Mahmoud T A, Shehab A F and Elshafey M A 2020 *Journal of Aerospace Information Systems* 1–8
[7] Shehab A F, Elshafey M A and Mahmoud T A 2020 *2020 IEEE Aerospace Conference* (IEEE) pp 1–11

[8] Huq R, Islam M and Siddique S 2018 *1st China Microsatellite Symposium*
[9] Ibrahim S K, Ahmed A, Zeidan M A E and Ziedan I E 2020 *Ain Shams Engineering Journal* **11** 45–56
[10] Li D, Wang M, Dong Z, Shen X and Shi L 2017 *Geo-spatial information science* **20** 134–140
[11] Xu X, Dong X and Xie Y 2020 *Remote Sensing* **12** 1216
[12] Jeon M J, Kim E and Lim S B 2014 *Aerospace Engineering and Technology* **13** 37–43
[13] Redmon J, Divvala S, Girshick R and Farhadi A 2016 *2016 IEEE Conference on Computer Vision and Pattern Recognition (CVPR)* pp 779–788
[14] Abhishek Nandy M B 2019 *Intel Deep Learning Frameworks: Applications of Open Vino and Other Intel Tools* (APress) ISBN 978-1484239810
[15] Tyugashev A and Sygurov Y M 2019 *Journal of Physics: Conference Series* **1368** 042032
[16] Ahmadi A, Kosari A and Malaek S 2018 *IEEE Aerospace and Electronic Systems Magazine* **33** 34–51
[17] Kim Y V 2020 *Satellite Control System: Part I-Architecture and Main Components* (IntechOpen)
[18] Tyugashev Andrey O and Sergei 2019 *International Symposium on Intelligent and Distributed Computing* (Springer) pp 404–413
[19] Tyugashev A and Sergei 2019 *Procedia Computer Science* **159** 1019–1026
[20] Schwenk K, Ulmer M and Peng T 2018 *High-Performance Computing in Geoscience and Remote Sensing VIII* vol 10792 (International Society for Optics and Photonics) p 1079205
[21] Redmon J and Farhadi A 2018 *arXiv preprint arXiv:1804.02767*
[22] Wu R, Zhang B and Hsu M 2009 *Proceedings of the Combined Workshops on UnConventional High Performance Computing Workshop plus Memory Access Workshop* UCHPC-MAW '09 (New York, NY, USA: Association for Computing Machinery) p 1–6 ISBN 9781605585574
[23] Reuther A, Michaleas P, Jones M, Gadepally V, Samsi S and Kepner J 2019 *2019 IEEE High Performance Extreme Computing Conference (HPEC)* pp 1–9
[24] Corp I 2016 *Intel® 64 and IA-32 architectures optimization reference manual* (Intel Corp) order number: 248966-033
[25] Cardoso J M and Weinhardt M 2016 *FPGAs for Software Programmers* vol 1 (Springer) ISBN 978-3-319-26408-0
[26] Pilato C, Mantovani P, Di Guglielmo G and Carloni L P 2016 *IEEE Transactions on Computer-Aided Design of Integrated Circuits and Systems* **36** 435–448
[27] Josipović L, Ghosal R and Ienne P 2018 *Proceedings of the 2018 ACM/SIGDA International Symposium on Field-Programmable Gate Arrays* pp 127–136
[28] Mantovani P, Di Guglielmo G and Carloni L P 2016 *2016 21st Asia and South Pacific Design Automation Conference (ASP-DAC)* (IEEE) pp 204–211
[29] Carloni L P, Cota E G, Guglielmo G D, Giri D, Kwon J, Mantovani P, Piccolboni L and Petracca M 2019 *Proceedings of the Workshop on Computer Architecture Education* pp 1–8
[30] Yan X H, He F Z and Chen Y L 2017 *Journal of Computer Science and Technology* **32** 340–355
[31] Hou N, Yan X and He F 2019 *Design Automation for Embedded Systems* **23** 57–77
[32] Stitt G, Lysecky R and Vahid F 2003 *Proceedings of the 40th Annual Design Automation Conference* DAC '03 (New York, NY, USA: Association for Computing Machinery) p 250–255 ISBN 1581136889
[33] Triantafyllis S, Vachharajani M, Vachharajani N and August D I 2003 *International Symposium on Code Generation and Optimization, 2003. CGO 2003.* pp 204–215
[34] Liu J, Wickerson J and Constantinides G A 2016 *2016 IEEE 24th Annual International Symposium on Field-Programmable Custom Computing Machines (FCCM)* pp 72–79
[35] Weiss M 1991 *Proceedings of the 5th International Conference on Supercomputing* ICS '91 (New York, NY, USA: Association for Computing Machinery) p 234–243 ISBN 0897914341
[36] Misailovic S, Sidiroglou S, Hoffmann H and Rinard M 2010 *2010 ACM/IEEE 32nd International Conference on Software Engineering* vol 1 pp 25–34
[37] Sidiroglou-Douskos S, Misailovic S, Hoffmann H and Rinard M 2011 *Proceedings of the 19th ACM SIGSOFT Symposium and the 13th European Conference on Foundations of Software Engineering* ESEC/FSE '11 (New York, NY, USA: Association for Computing Machinery) p 124–134 ISBN 9781450304436
[38] Murthy G S, Ravishankar M, Baskaran M M and Sadayappan P 2010 *2010 IEEE International Symposium on Parallel Distributed Processing (IPDPS)* pp 1–11
[39] Lokuciejewski P, Gedikli F and Marwedel P 2009 *Proceedings of Th 12th International Workshop on Software and Compilers for Embedded Systems* SCOPES '09 (New York, NY, USA: Association for Computing Machinery) p 11–20 ISBN 9781605586960
[40] Binkley D 2007 *Future of Software Engineering (FOSE '07)* pp 104–119
[41] Wu Y, Sun J, Liu Y and Dong J S 2013 *2013 28th IEEE/ACM International Conference on Automated Software Engineering (ASE)* pp 323–333
[42] Keith Cooper L T 2011 *Engineering: A Compiler 2nd Edition* (Morgan Kaufmann) ISBN 978-0120884780

[43] Louridas P 2006 *IEEE Software* **23** 58–61
[44] Dang K N, Meyer M, Okuyama Y and Abdallah A B 2017 *The Journal of Supercomputing* **73** 2705–2729
[45] Cardoso J a M, Carvalho T, Coutinho J G, Luk W, Nobre R, Diniz P and Petrov Z 2012 *Proceedings of the 11th Annual International Conference on Aspect-Oriented Software Development* AOSD '12 (New York, NY, USA: Association for Computing Machinery) p 179–190 ISBN 9781450310925
[46] Kustikova V, Vasiliev E, Khvatov A, Kumbrasiev P, Vikhrev I, Utkin K, Dudchenko A and Gladilov G 2019 *International Conference on Analysis of Images, Social Networks and Texts* (Springer) pp 11–23
[47] Mathew G, Ramachandran S S and Suchithra V 2019 *TENCON 2019-2019 IEEE Region 10 Conference (TENCON)* (IEEE) pp 1783–1788
[48] Andriyanov N 2020 *2020 Systems of Signal Synchronization, Generating and Processing in Telecommunications (SYNCHROINFO)* (IEEE) pp 1–5

IOP Conf. Series: Materials Science and Engineering 1172 (2021) 012035 doi:10.1088/1757-899X/1172/1/012035

Modelling and simulation of two axes gimbal fuzzy PI stabilization system in the presence of feedback sensors noise

A Fathi [1], A M Kamel[1], A O Elfarouk[1], Y Elhalwagy[1]

[1]Military Technical College, Cairo, Egypt.

E-mail: ahmedfathy111980@gmail.com

Abstract. The missile guidance in terminal phase is an optimization problem as the miss distance should be minimized. The optimization of missile miss distance is highly affected by the missile seeker performance. Design of gimbal control system suffers always from feedback sensors noise which leads to system instability. In this paper, a promising design of the fuzzy PID controller for a missile seeker gimbal is proposed considering the feedback sensor noise with practical gyro transfer function calculated based on real experimental measurements utilizing MATLB system identification toolbox. Also, the mathematical model of two stabilized axes gimbal. Stabilization is achieved considering the missile motion parameters such as rates, torques and coupling between yaw and roll channels. A Matlab simulation is carried out for evaluating the proposed system modelling and to test the robustness of the fuzzy PID controller in the presence of feedback sensor noise. A comparative analysis with PI based controller is conducted to evaluate the performance of the proposed controller which presents sufficient enhancement to the missile gimbal stability parameters.

1. Introduction

Latest years, stabilized platform-based seekers are widely used in ballistic missiles [1]. Usually, the seeker assembly is attached to the missile nose, the seeker coordinate axis is always aligned with a missile body axis. The seeker's field of view (FOV) is easily defined to be aligned with missile body axis. Furthermore, the hypersonic missile aerodynamic configuration with sharp nose. This configuration has narrow space for nose[2], which is not suitable with large-sized seekers to be fixed easily. However, when the seeker axis is not aligned to the missile body axis, FOV of these seeker lies in the direction of the body. The FOV of seekers are out of a missile axis, which always leads to target miss-tracking. The seeker detector's FOV is always constrained by minimum and maximum angles in missile's horizontal and vertical planes which presented in [3]. Derived from guidance laws, the maximum angle of look will not overdo the constraint angle's border by changing of guidance parameters [4], and a system identification for both electrical servo and gyro using PI controller, and fuzzy PID controller impact angle optimized in guidance law, which considers FOV angle constraint is presented in this paper. A sliding mode control guidance law considering constraints of FOV angle is represented [5]. By utilize a sliding mode theory, the guidance law and FOV angle-controller realizes impact as well designed. Control systems designed for gimbal stabilized platform to enhancement seeker stability in two planes, disturbance end results are produced undesirable influencing of system. Unbalance off mass, cable restraint and coupling between lateral and longitudinal planes are considered

Content from this work may be used under the terms of the Creative Commons Attribution 3.0 licence. Any further distribution of this work must maintain attribution to the author(s) and the title of the work, journal citation and DOI.
Published under licence by IOP Publishing Ltd

a main sources of the disturbance on system position, angular velocity [6]. Operational conditions and angular velocity are main parameters must be monitored. Before actually designing and producing system dynamics of the system are considered. Accordingly, two electrical light or more axes gimbal are required. LOS controller system systems with useful stability is very difficult, due to the coupling between azimuth and elevation planes [7]. Target tracking accuracy and system stabilization is highly affected by the missile dynamics, it's clear that system performance depends on system modelling accuracy [8]. System is rotated into two azimuth and elevation planes, IMU installed on inner gimbal to measure angular acceleration from two channels. Actually, these IMU is applied to develop accuracy of feedback angular velocities, and prevent disturbances due to seeker gimbals. Notably, the deviation produced is calculated utilizing installed target recognition camera or laser quadratic detector installed on inner gimbal axis [9]. PI controller is used to guarantee the system stabilization and improve its performance [10]. Fuzzy controller is presented in [11] for the design of two axes gimbal stabilization.

On the basis of the above, the most important aim of this article is structured as follow: through section 2, the modelling of two-DOF gimbal dynamic system is introduced. Section 3, the design of PI and direct fuzzy PID controllers is presented. Section 4, the simulation results is demonstrated.

2. Two DOF gimbal dynamic system modelling

The guidance system accuracy is critical for the performance of the overall system autonomy and accuracy. It's responsible for minimizing the deviation of missile actual trajectory from the pre-calculated reference trajectory, provide the proper actuation commands to execute the required maneuverer considering the missile time and frequency stability constraints [12].

Recently, proportional navigation guidance law is the most common used guidance law. The missile-target relative velocity measurement and LOS angular rates are essential to more accurate guidance law. The missile seeker is always devoted to angles measurement as described in Figure 1. The angles definitions related to missile-target during flight are θ velocity vector angle relative to inertial space, ϑ missile body x-axis angle with respect to inertial space, relative LOS angle to the inertial frame, ε error in angles of missile-target line and q_d angle of detector with respect to inertial frame.

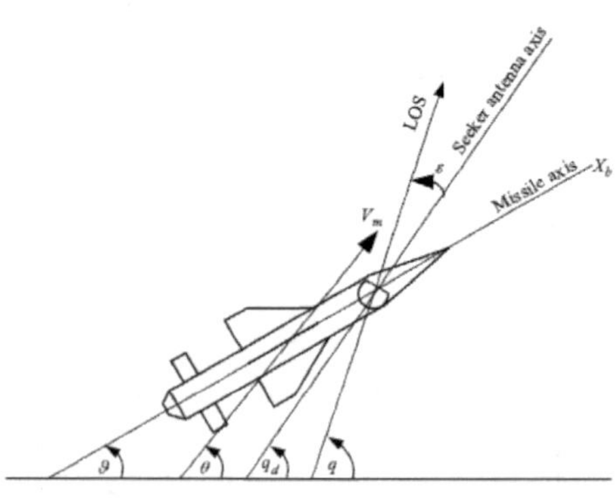

Figure 1. Definitions of angles

Essentially, current missiles which using seekers techniques in terminal phase commonly, categorized as dynamic gyro-based stabilization seekers technique, stabilized platform-seekers, semi-strap down seekers, strap down seekers and roll-pitch-seekers. This paper motivation is enhancement of stabilisation platform-based seeker controller using PI and fuzzy $-PID$ controllers further, a comparative investigation is carried-out for evaluating the two performances for time stability parameters such as, rising time $t_r(sec)$, settling time $t_s(sec)$ and overshoot $OV(\%)$.

Two-degree stabilized platform gimbal-based Laser Seeker, which gimbal is composed of two control loops.

The disturbance of the angular motion is captured and rejected using high-gain IMU sensor through the inner loop, the angular velocity stabilization provides the missile body motion stability and also,

IOP Conf. Series: Materials Science and Engineering 1172 (2021) 012035 doi:10.1088/1757-899X/1172/1/012035

acquire the missile LOS angular velocity output. The outer loop is dedicated for missile positional stabilization to ensure the target tracking as shown in Figure 2. Although, MEMS IMU based stabilized seeker provider slower response, lower load, and smaller gimbal angles when compared to stabilized platform-based seeker.

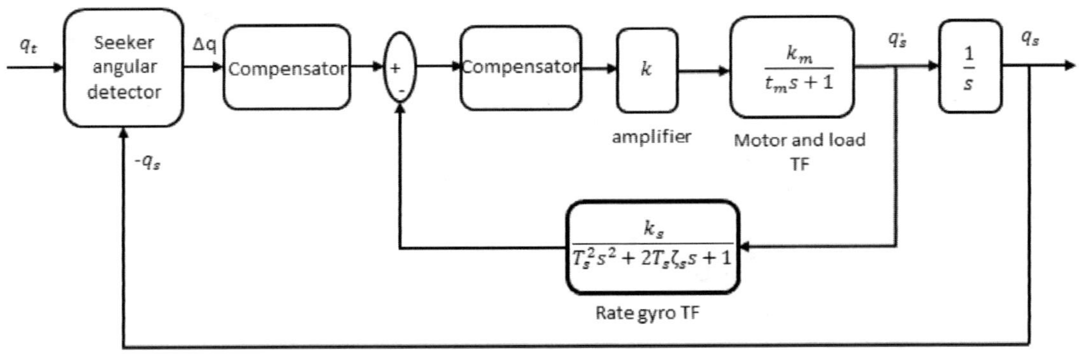

Figure 2. Block diagram of seeker stabilized platform controller

The practical gimbal servo and gyro transfer functions are essential for the modelling and implementation the two DOF gimbal system and applying different technique of controllers and compare such techniques with each other. The transfer function for MEMS gyro VN100 which investigated utilizing system identification tool box with fit estimation data 84.45% in MATLAB by tacking real 15000 measurements for different inputs and save it as experimental work. The rate gyro transfer function in equation 1 is derived from system identification experimental work can be described with second order transfer function as shown.

$$G_{Gyro}(s) = \frac{2.056e06}{s^2 + 2372s + 9.862e05} \quad (1)$$

Servo DC motor transfer function, which parameters are rated current consumed with load $I_e = 3.95\ \text{Ampere}$, rated voltage $U = 27$ Volt, inductance in motor $L = 0.5433 * 10^{-3}$H, motor internal resistance is $R = 0.9\ Ohm$, estimated friction coefficient for the motor $F_m = 0.8 * 10^{-4} N.M/rad/s$ and Voltage drop on Motor E

$$U(t) = RI(T) + L\frac{dI}{dt} + E \quad (2)$$

$$E = K_e w(t) \quad (3)$$

$$T = K_T\, I(t) \quad (4)$$

$$T = J\frac{d(w_t)}{dt} + F_m w(t) \quad (5)$$

Where E is inverse electrical potential in Motor, T is Torque, J Rotational Inertia, $J\frac{d(w_t)}{dt}$ Moment of Inertia and F_m Friction coefficient. By taking Laplace transformation of equations (2), (3), (4) and (5)

$$U(S) - E = (R + LS)I(S) \tag{6}$$

$$E(S) = K_e w(S) \tag{7}$$

Figure 3 present block diagram of servo system after Laplace transformation, which is fully proportional, to digital servo torque. The utilized HSR-2645CRH servo is a digital based servo, where the speed control is accomplished utilizing PWM technique. It's sufficiently used in wide range of applications such as missile terminal guidance, pedestal platform, camera and seeker stabilization devices and on robots.

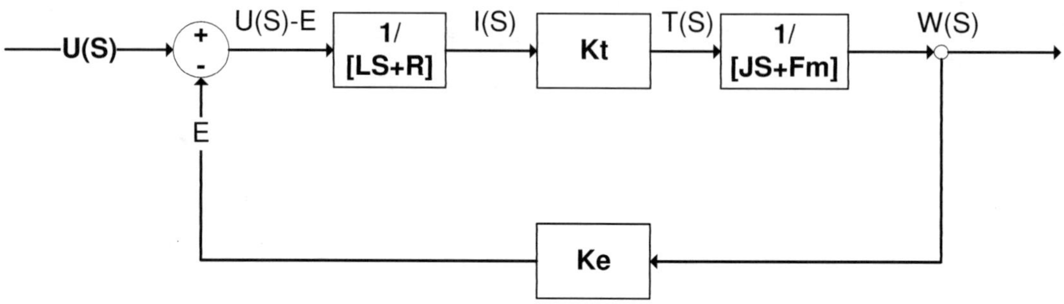

Figure 3. Servo model content transfer functions

From equation (4), (5)

$$T(s) = JSW(S) + F_m W(S) \tag{8}$$

$$T(s) = (JS + F_m)W(S) \tag{9}$$

$$G_d(S) = \frac{W(S)}{U(S)} = \frac{\frac{1}{LS+R} K_t \frac{1}{JS+F_m}}{1 + \frac{1}{LS+R} K_t \frac{1}{JS+F_m} K_e} \tag{10}$$

$$G_d(S) = \frac{K_t}{(LS + R)(JS + F_m) + K_t K_e} \tag{11}$$

$$G_d(S) = \frac{K_t}{LJS^2 + (R.J + L.F_m)S + (R.F_m + K_t K_e)} \tag{12}$$

$$G_d(S) = \frac{W(S)}{U(S)} = \frac{0.06402}{3.78e - 09 * S^2 + 5.762e - 06 * S + 0.013744} \tag{13}$$

As demonstrated in the previous section the effect of missile angular motion on seeker performance can be reduced by increasing the open-loop gain. Considering the seeker load, its limited driving bandwidth and due to the motion interference for missile dropped down in low range frequency, the low frequency gain is increased by adding a PI compensator to reduce the coupling effect of missile seeker system.

Following system parameters illustrates, how to implement this approach. The rate gyro transfer function includes the damping- coefficient and time-constant. Missile angular motion interference frequency is in range between 9.8*105 and 1.1*103. So, the PI compensator parameters (k_p andk_i) optimization under these design constraints is required, gain-margin range is 4 dB, a phase-margin is greater than 40, the stabilization loop optimization is to reject the angular motion disturbances located on missile body. The compensator $k_p$and k_i gains should be investigated for maximize loop gain at interference frequency two hertz. Problem of gimbal design is carried out as shown in Figure 4, the servo motor and gyro transfer functions presented into two equation number 1, 13.

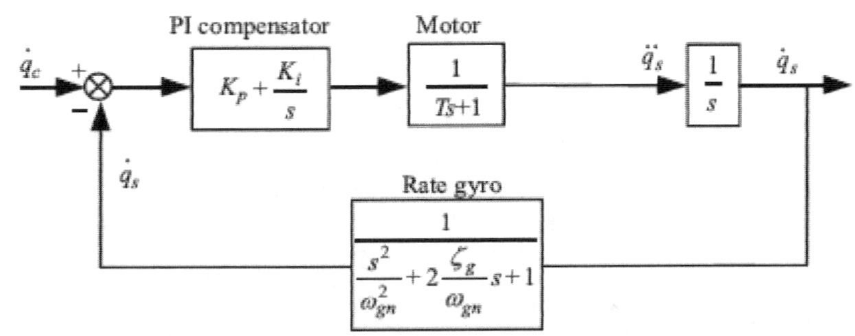

Figure 4. Block diagram of the stabilization loop with PI compensation

Two degree of freedom gimbal with a laser source system is assembled of two systems inner and outer gimbals. To navigate and control laser source systems in lateral and longitudinal axes, to controlling two different loops rate controller is developed. Complete scheme of gimbal system is shown in Figure 5.

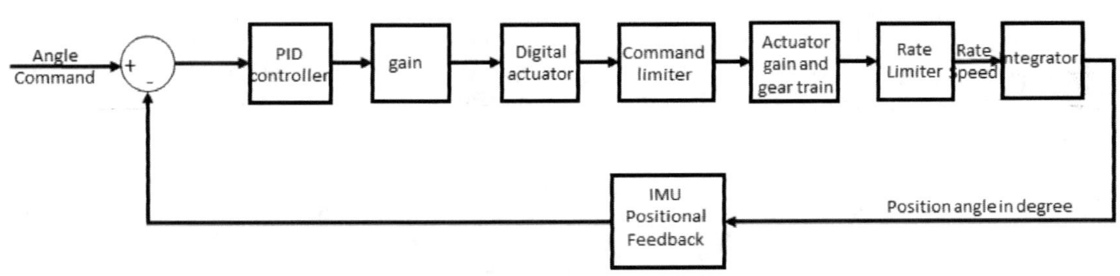

Figure 5. Two-DOF gimbal Block diagram.

3. Design of the PI controller and direct fuzzy PID controllers

The PI controller always shares a poor performance with two DOF plant due to system environmental varying conditions. Consequently, controller parameters have to be re-tuned to maintain good performance with inertia seeker stabilization platform systems proposed. Adaptive fuzzy logic approach is proposed to develop the seeker gimbal control system. Fuzzy controller contains four main stages as shown in figure 6 fuzzification, inference mechanism, rule base, de-fuzzification.

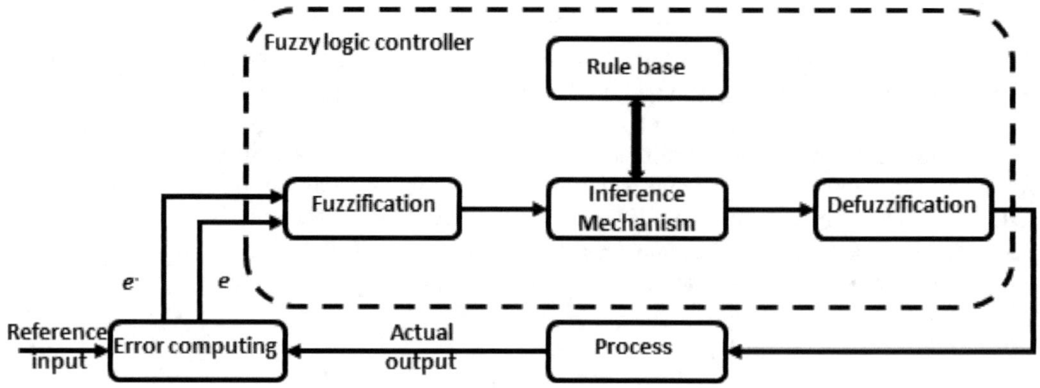

Figure 6. Fuzzy controller block diagram

The membership Fuzzy controller functions used, *PL, PS, ZR, NS, NL* denotes positive large, positive small, zero, negative small, and negative large. The PI controller gains highly affect the dynamic performance and stability. Centre of gravity method outputs had determined according to conditions in table 1.

Table1. Control response

Control Response	Rise Time	Over Shoot	Settling Time	S-S error
Kp	Decrease	Increase	Small change	Decrease
Ki	Decrease	Increase	Increase	Eliminate
Kd	Small change	Decrease	Decrease	Small change

When control parameter K_p gets so large, steady state is damaged and overshoot is too great, if K_i becomes too large, that affect the steady-state error by causing ripples on torque. The overshoot is reduced as K_d gain gets small, the response always suffers from overshoot and slower response (longer settling time). Then the above procedure is designed as an if-then rule statements indicated in Table 2.

The fuzzy-PID-controllers design parameters are classified into two groups, structural and tuning parameters which are calculated offline and online design respectively, the tuning parameters is the main difficult where it includes the membership function and scaling parameters. The difficulty for calculating the tuning parameters requires an experience where, it always calculated using trial and error to accommodate adaptive capability of system process considering the environmental disturbance and system uncertainties. When the compensator is being work and any of tuneable parameter is changed the adaptive is regarded, otherwise the fuzzy controller is working normally. The scaling factors always shares a global and critical effect on the control system performance as the missile angular rates w_{el} *and* w_{az} are considered the most dominant effecting system parameters.

Table 2. Rule bases

Δ_{K_P} Δ_{K_i} Δ_{K_d}	e˙					
		NL	NS	ZR	PS	PL
e	NL	PL	PS	PS	PS	ZR
		NL	NS	NS	NS	ZR
		NL	NS	NS	NS	ZR
	NS	PL	PS	PS	ZR	NS
		NL	NS	NS	ZR	PS
		NL	NS	NS	ZR	PS
	ZR	PS	PS	ZR	NS	NS
		NS	NS	ZR	PS	PS
		NS	NS	ZR	PS	PS
	PS	PS	ZR	NS	NS	NL
		NS	ZR	PS	PS	PL
		NS	ZR	PS	PS	PL
	PL	ZR	NS	NS	NS	NL
		ZR	PS	PS	PS	PL
		ZR	PS	PS	PS	PL

4. Simulation results

The comparative analysis of conventional PI and Fuzzy PID for digital DC servo controller system is accomplished utilizing various input angular rate commands into two planes azimuth and elevation$w_{el} = w_{az} = 10\ deg/sec.$ w_{az}, w_{el} are changed through interval [0 -12 deg/sec]. Figure 7 one case study the system response is analysed which clearly reflects the efficiency criteria of a fuzzy-PID compensator compared to PI controller. The results as summarized as shown in, tables 3,4 present comparison testes results utilizing overshoot (OV), rising time (t_r), settling time (t_s). While PI is utilized, angular rate creates an increase in settling time, large percentage in overshoot. While fuzzy-PID compensator met the variant in body rates of missile and improves steady state performance and by achieving lower overshoot with fast response as a compared to PI controller. Furthermore, the MATLAB system identification toolbox output for the rate gyro real data is presented in Figure 8 as the calculation of a practical transfer function for the rate gyro is highly affect the controller design, which leads to enhance the overall system performance.

Table 3. Comparison result for the elevation channel

w_{pj} (deg/sec)	PI controller			Fuzzy PID controller		
	$OV(\%)$	$t_s(sec)$	$t_r(sec)$	$OV(\%)$	$t_s(sec)$	$t_r(sec)$
2	10.8	0.195	0.08	3	0.078	0.099
4	19.2	0.184	0.063	5	0.085	0.092
6	28.3	0.175	0.053	7.3	0.099	0.073
8	35	0.265	0.033	6.2	0.156	0.053
10	49	0.276	0.02219	12.3	0.137	0.039

Table 4. Comparison result of horizontal channel

w_{pj} (deg/sec)	PID controller			Fuzzy PID controller		
	$OV(\%)$	$t_s(sec)$	$t_r(sec)$	$OV(\%)$	$t_s(sec)$	$t_r(sec)$
2	7.8	0.166	0.064	7	0.158	0.130
4	17.5	0.178	0.048	2.4	0.137	0.123
6	37	0.198	0.04	3.4	0.138	0.084
8	42	0.21	0.036	4	0.098	0.084
10	53	0.29	0.04	4,9	0.232	0.07

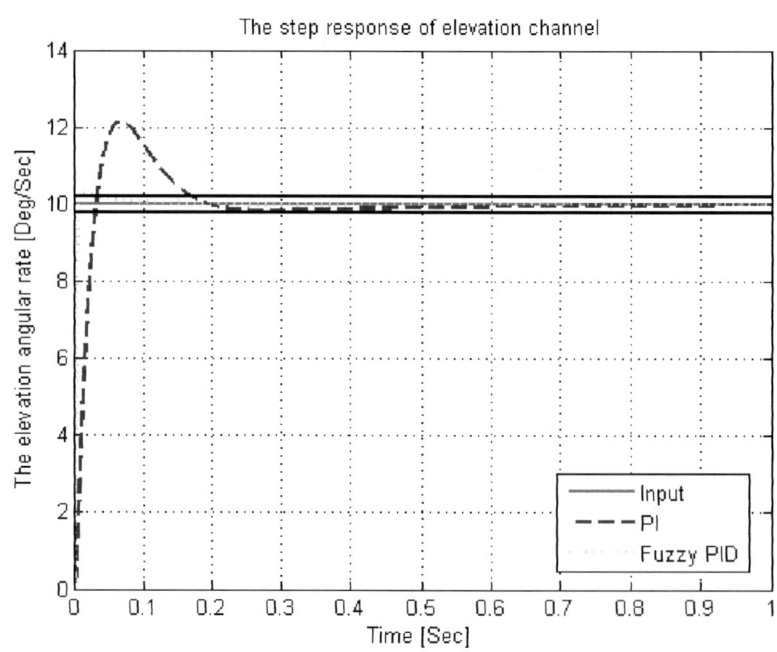

(a) Elevation plane

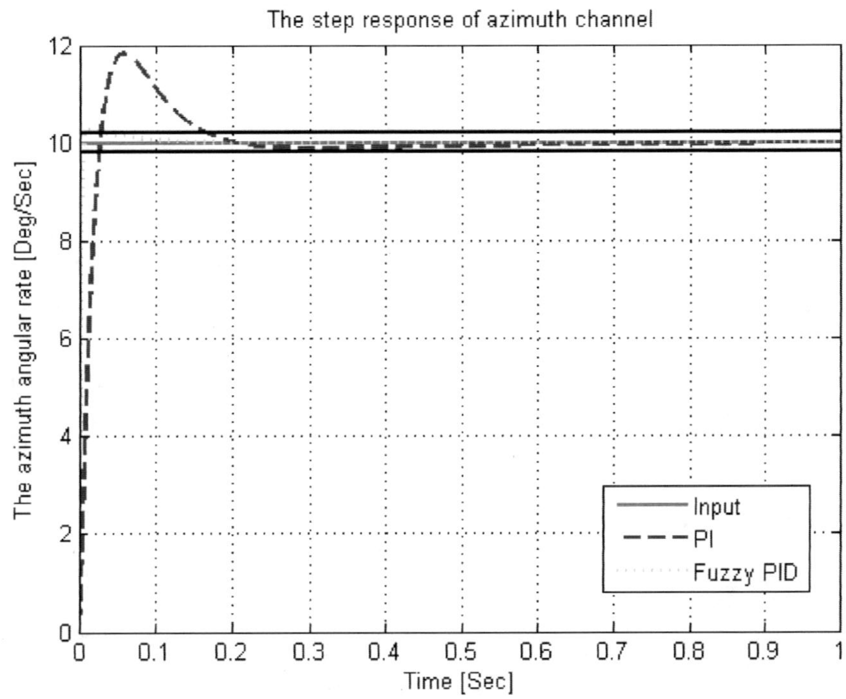

(b) Azimuth plane

Figure 7. Gimbal system response for $w_{EL} = w_{AZ} = 10\ deg/sec$

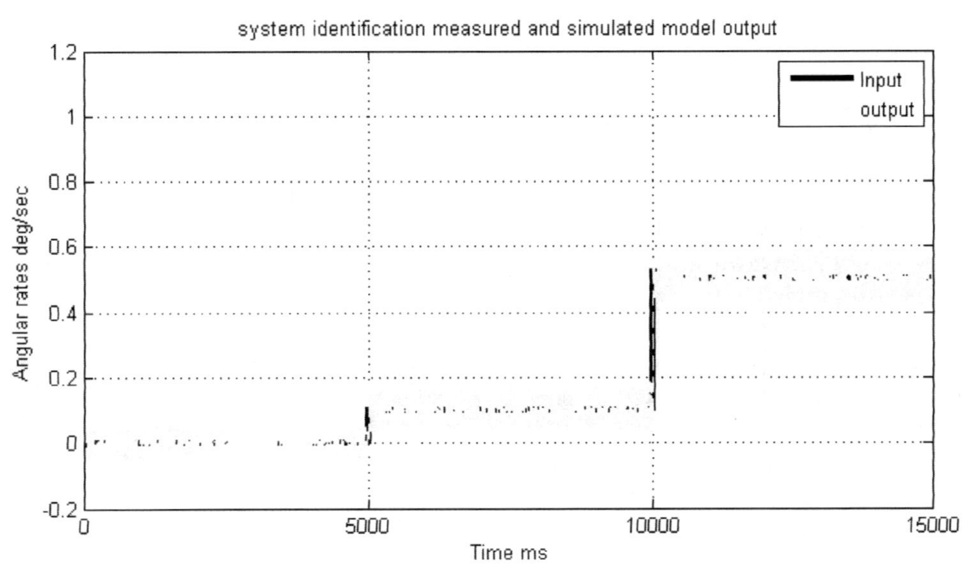

Figure 8. System identification

5. Conclusion

Two axis gimbal system stabilization is very essential for several applications in air borne and missile guidance. Accuracy in performance of these applications plays an essential task to design high performance gimbal control system. Then, computation of the control signal is applied to the two DOF gimbal system. Stabilization of a two-axis laser source gimbal system, a fuzzy PID controller is proposed and improving its performance by responses of gimbal system. Proposed controller parameters tuning method used to tune different accuracy variations performance, tuned fuzzy PID controller gives stable in convergence and accuracy. In a stabilization mode, imposed disturbance on gimbal system has to be removed by proposed controller to reserve the stability of gimbal system. Control signals in azimuth and elevation planes are within the drive power range and smooth. The feasibility of validity of a proposed controller was verified by experimental work by accomplishment an experimental work for capturing the response of gimbal control signal with different position and different angular positions.

References

[1] Fouad A, Taha H and Elhalwagy Y 2015 Automatic Aerial Target Recognition using A Robust SURF-MSER Feature-based Algorithm. In: *International Conference on Aerospace Sciences and Aviation Technology*: The Military Technical College) pp 1-14

[2] ARAFA I I, ELHALWAGY Y and FOUAD A 2015 Separable warhead mathematical model of Supersonic & Hypersonic Re-entry Vehicles. In: International Conference on Aerospace Sciences and Aviation Technology: The Military Technical College) pp 1-10

[3] Wen Q, Xia Q and Weixia S 2015 A parameter design strategy for seeker's field-of-view constraint in impact angle guidance *Proceedings of the Institution of Mechanical Engineers, Part G: Journal of Aerospace Engineering* **229** 2389-96

[4] Ran L, Qiuqiu W, Wangchun T and Yijie Z 2018 Adaptive weighting impact angle optimal guidance law considering seeker's FOV angle constraints Journal of Systems Engineering and Electronics 29 142-51

[5] Chen X and Wang J 2018 Nonsingular sliding-mode control for field-of-view constrained impact time guidance Journal of Guidance, Control, and Dynamics 41 1214-22

[6] Tsai N-C, Sue C-Y and Lin C-C 2008 Design and dynamics of an innovative micro gyroscope against coupling effects Microsystem technologies 14 295-306

[7] Chen X and Chen M 2013 Precise control of a magnetically suspended double-gimbal control moment gyroscope using differential geometry decoupling method Chinese Journal of Aeronautics 26 1017-28

[8] Hurák Z and Rezac M 2011 Image-based pointing and tracking for inertially stabilized airborne camera platform IEEE Transactions on Control Systems Technology 20 1146-59

[9] Abdo M, Vali A R, Toloei A and Arvan M R 2013 Research on the cross-coupling of a two axes gimbal system with dynamic unbalance International Journal of advanced robotic systems 10 357

[10] Kim S, Kim S and Kwak Y 2010 Robust control for a two-axis gimbaled sensor system with multivariable feedback systems IET control theory & applications 4 539-51

[11] Roshdy A A, Su C, Mokbel H F and Wang T 2012 Design a robust PI controller for line of sight stabilization system

[12] Kapeel E H, Abozied M A, Kamel A M and Hendy H 2020 Evaluation of Missile Terminal Guidance Using Software in Loop (SIL). In: 2020 12th International Conference on Electrical Engineering (ICEENG): IEEE) pp 389-95

IOP Conf. Series: Materials Science and Engineering 1172 (2021) 012036 doi:10.1088/1757-899X/1172/1/012036

Performance comparison among popular implementations of H.264 encoders

H Y El-Arsh, A S Elliethy, A M Abdelaziz and H A Aly
Dept. of Computer Engineering, Military Technical College, Cairo, Egypt, 11766

E-mail: hassan.yakout@ymail.com

Abstract. Remote sensing videos captured by Unmanned Aerial Vehicle (UAV) air-born high-resolution cameras require an efficient compression scheme that preserves the details of the visual contents of the videos while reducing the total size of the data to be managed in real-time. This paper presents a detailed comparison between different open-source implementations for the H.264 video compression scheme. While the high-resolution videos allow analysts to extract more descriptive interpretations and draw more conclusive results, the increase in the consequent data size consumes more storage, resulting in more channel bandwidth, more power, and encounters an extra delay in transmission time. An efficient implementation of video compression can alleviate these large data size effects. In this paper, we analyze and compare the JM-encoder, the X264, the FFmpeg, and Cisco's OpenH264 open-source implementations in terms of compression efficiency, video quality, and computational load. Moreover, we present the rate-distortion curves in terms of PSNR as a quality metric against the bit-rate for a combination of 20 videos with various resolutions and dynamic contents. Albeit H.64 is superseded by H.265, till now H.264 is used in more than 65% of video coding applications. For example, YouTube only allows H.264 for live streaming.

1. Introduction

Since the last decade, high-resolution videos are required in the field of surveillance and reconnaissance, especially in association with UAV air-born drones. This is due to the fact that increasing video resolution enables more meaningful analyses to obtain more rigorous results. Accordingly, raising video resolution for the sake of obtaining precise results will consume more storage, more channel bandwidth, and more delay in transmission time. Subsequently, efficient video codecs gain more attention in the field of remote sensing, especially for limited-resources implementations such as UAV.

H.264, the most ubiquitous video codec, was standardized as the *Advanced Video Codec* (AVC) through [1]. Although superseded by H.265 [2], H.264 is still the most widely used video codec [3]. More than 65% of security monitoring devices support H.264 (powered by the included hardware chips) [3]. Also, for software, statistics show that 67% of network videos utilize H.264 technologies [3]. Moreover, according to YouTube answer number 2853702, the only allowed video codec for YouTube live streaming is H.264.

The H.264 video codec performs the video decoding with low-complexity. However, the corresponding encoder is complex because the building blocks of the codec require intensive computations, especially the motion estimation part. Specifically, the motion estimation process

Content from this work may be used under the terms of the Creative Commons Attribution 3.0 licence. Any further distribution of this work must maintain attribution to the author(s) and the title of the work, journal citation and DOI.
Published under licence by IOP Publishing Ltd

consumes 60% to 80% of the total encoding resources [4] and this consumes processing and power resources. Hence, optimized hardware implementations are considered an excellent choice over software for implementing the H.264 encoder. Nevertheless, these implementations are not suitable for special applications where customized design of the codec is required, such as crypto-coding [5, 6] and steganography [7, 8, 9]. Crypto-coding is a class of techniques in which coding and encryption are done within the multimedia codec to reduce the time and complexity for both operations and providing extra selective encryption capabilities [10]. Steganography is the art and science of data hiding within an innocent-looking cover such as all multimedia types. However, the long time to market for the hardware implementations and their lack of flexibility when changes in design parameters are required, stand as barriers against the usage of the hardware implementations in these types of applications. Consequently, the software implementation of the codecs, especially the open-source, is a viable choice.

In this paper, we introduce a practical benchmarking for three open-source software implementations for H.264 video codec, which are: FFmpeg [11], X264 [12] and OpenH264 [13]. The benchmarking is conducted in terms of compression efficiency (measured by Rate-Distortion (RD) curves) and computational resources. Also, we use JM v19 H.264 encoder [14] as a reference implementation for H.264 encoder to evaluate the RD-curves of the former three codecs. Experimental results on 20 test video sequences with different resolutions and constructions show that OpenH264 is the only software encoder that achieves real-time performance. The reason for the real-time performance of OpenH264 implementation is the utilization of Single-Instruction-Multiple-Data (SIMD) extensions, which are available for most recent processors. These extensions allow parallel computations in vector representations to speed up the computations of the different operations of the codec.

This paper is organized as follows. Section 2 introduces a brief description of H.264 encoder. Section 3 explains the basic ideas for SIMD instruction sets and its dominant rule for amplifying software performance, especially video encoders. Section 4 discusses our experiments and explains their resulting outcomes. Section 5 highlights the conclusion of the paper.

2. Overview for H.264 encoder

In this section, we introduce a brief overview of H.264 encoder. For more technical details, readers can refer to [1, 15]. Fig.1 describes the general coding steps for H.264 video encoder [1, 15]. The motion estimation module (ME) estimates a motion vector (MV) for every macro-block (MB) in the current frame (I_n) by searching the best match for (MB) corresponding to minimum Sum of Absolute Differences (SAD) in a search window within the previously reconstructed frame (I_{n-1}), which is stored in the frame memory (FM) module. Then, the motion compensation module (MC) uses (MV) and (I_{n-1}) to reconstruct the predicted frame (P). Then, the prediction error (P_E) is calculated by subtracting (P) from (I_n). The (P_E) is then coded by transformation (T) then lossy-compressed by the quantization module (Q) to compose the residual data (R_D). R_D is fed to the inverse quantization module (Q^{-1}) followed by the inverse transform module (T^{-1}) to compose the compressed prediction error (P'_E), which is added to (P), then saved in FM. Additionally, R_D is also fed to the entropy coder (ENT) module to generate the final output bitstream (BS). Two types of entropy coding exist in H264: Context-adaptive binary arithmetic coding (CABAC) and Context-adaptive variable-length coding (CAVLC). Although CABAC is more complex than CAVLC, CABAC reduces the compressed stream size by more than 9% [15].

H.264 defines several groups of coding features, defined as *profiles*. Each profile defines the required decoder's features for a certain class of applications, which set limits to the encoding features. The most common profiles are the *Constrained-Baseline*, *Baseline*, *Main*, *Extended*, *High* and *Intra* profiles. The *Constrained-Baseline* profile is used by applications such as mobile videos and video conferencing running on limited computing resources platforms, in which the

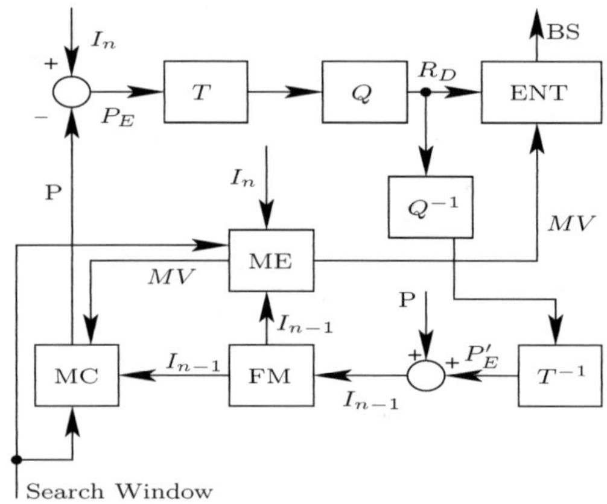

Figure 1. General description for H.264 encoder.

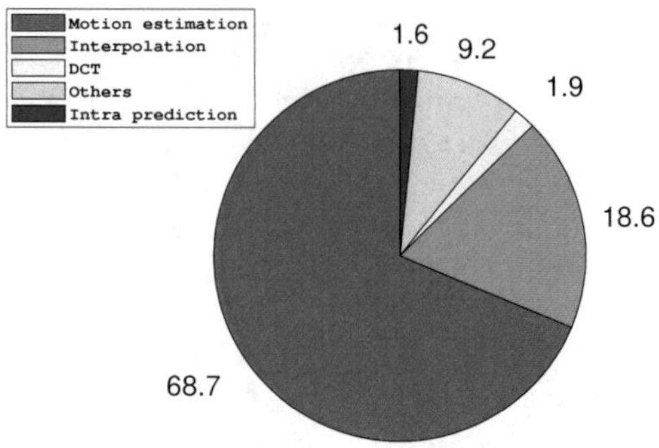

Figure 2. Profiling for H264 components [16].

error-resilience features are disabled. The *Baseline* profile has the same coding features as the *Constrained-Baseline* profile with additional error resilience features for applications that require robust data transmission. The *Main* profile is a superset for the *Baseline* profile, with additional complex features such as CABAC and B-frames [15], which make it suitable for SD TV broadcasting. The *Extended* profile is utilized for streaming applications, where switching frames features are enabled to allow extra error resilience performance. *High* profiles are used for HD TV broadcasting and HD video DVDs as they allow advanced coding features such as higher pixel precision and different chroma-subsampling schemes. *Intra* Profiles are subsets of *High* profiles without inter prediction (*i.e.* no P or B frames), used mainly for video editing applications. Other profiles exist for stereoscopic video coding.

H.264 also utilizes a term called *level* that defines the decoder performance, which depends on the amount of computational resources available for the decoder, and hence sets another limit for the encoder in accompaniment with profile number. Each *level number* specifies the maximum number of macro-blocks per frame, maximum decoding speed (macro-blocks/seconds) and the maximum video bit rate. For instance, level 1 limits are 99 macro-blocks/frame, 1485 macro-blocks/sec and video bitrate of 64 Kbits/s, while level 4 limits are 8192 macro-blocks/frame, 245760 macro-blocks/sec and video bitrate of 20 Mbits/s. Thus, for a video resolution of 1080p with 30 fps, a decoder with a minimum level number of 4 is required. This is due to that each frame with a resolution of 1080p contains 8100 macro-blocks and with frame rate of 30 fps, thus the minimum coding speed is 243000 macro-blocks/sec.

3. Video coding optimization with SIMD

In this section, we will explain how SIMD instruction sets can allow software-based encoder's implementations to achieve superior performance over traditional implementation techniques.

In recent years, CPUs' clock speed did not considerably increase, as it has reached some physical and power limitations. But as computational requirements still increasing over time, parallelization techniques are the only available solutions. By increasing the number of cores, the operating system can distribute its computational load among different cores and running programs can construct multiple threads to maximize usage efficiency. Another solution is by adding vector operation capabilities to each core, thus allowing the CPU to perform the same instructions, but for a vector of data.

SIMD instruction-sets extension is an example of vector processing akin to specialized Digital

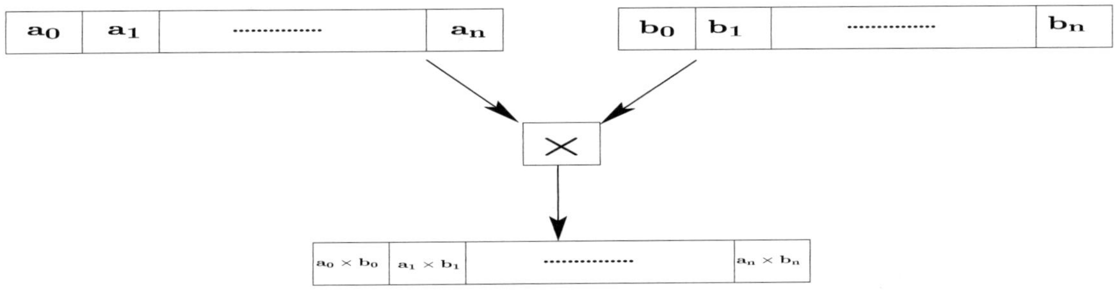

Figure 3. General description for SIMD operation (multiply).

Signal Processor (DSP). SIMD has become an essential technology for many modern CPU architectures. Intel MMX/SSE/AVX instructions for x86 architecture and NEON for ARM are all examples of SIMD instruction [17]. SIMD uses only one instruction to compute multiple data organized in a vector format, performing an operation in a single work cycle that was traditionally requires multiple working cycles, and hence can increase the performance for an application up to 11.5x [17]. Fig.3 illustrates the main idea of SIMD. Vector length for SIMD varies from 64-bits for MMX, 128-bits for SSE, 256-bits for AVX/AVX2 and 512-bits for the latest AVX-512 [18]. For ARM, NEON vector length is 128-bits [19]. Moreover, besides these notable performance enhancements, energy consumption is also reduced as it eliminates the number of instructions for fetch and decode processes. Many applications can be enhanced by SIMD instructions, such as machine learning, virtual reality and DSP applications especially video encoding, which is the main focus of this paper.

As described in fig.2, motion estimation algorithms consume 60% to 80% of the total encoding resources and both DCT and 1/4 pixel interpolation consume more than 20%. As SAD is the main atomic operation for the motion estimation process, optimizing SAD with "**_mm256_sad_epu8**" [20] for 32-bytes in a single cycle speeds up the calculations by 63x. This is due to calculating SAD for two 32-bytes arrays requires 32 subtraction operations and 31 addition operations. The convolution multiply and add (MullAdd) is the main atomic operation for both DCT and 1/4 pixel interpolation, utilizing "**_mm256_maddubs_epi16**" [20] for 16 signed 16-bit integers speeds up the calculations by 31x. Table 1 illustrates this comparison, where:

- *CPI* (Cycles Per Instruction): Measure of throughput by the number of CPU cycles required to perform single instruction, through which the part of CPU performing this instruction cannot execute another instruction [21].
- *Latency*: Number of CPU cycles from the start of the instruction until its result is available; usually considered only when this instruction is a part of a loop dependency chain [21].

Table 1. Performance evaluation for AVX2-intrinsic for Intel®Core i5-5200U in terms of CPI (Cycles Per Instruction) for throughput and latency [20].

Purpose	C/C++ Function Name	Assembly Instruction Name	CPI	Latency	Speed up factor
SAD for motion estimation	**_mm256_sad_epu8**	**vpsadbw**	1	5	63x
Convolution for DCT & interpolation	**_mm256_maddubs_epi16**	**vpmaddubsw**	1	5	31x

Table 2. Measured FPS for each encoder per each video.

Encoder \ Video file	Fig.4(a)	Fig.4(b)	Fig.4(c)	Fig.4(d)	Fig.4(e)	Fig.4(f)	Fig.4(g)	Fig.4(h)	Fig.4(i)
Openh264	44.12	62.856	73.509	102.041	108.828	155.801	152.127	183.199	245.976
X264	4.101	4.509	5.181	4.174	6.818	11.48	16.51	20.728	43.036
FFmpeg	4.114	4.518	4.958	3.84	6.166	10.068	15.137	20.128	30.595

4. Experimental Results

All experiments had been carried out with Intel®Core i5-5200U CPU @2.20GHz, except JM-encoder coding experiments were performed on higher performance PC, as JM-encoder is a single thread, implemented with pure C without any optimization, aiming to be a reference for compliance with the standard described in [1], in contrast to other implementations, especially OpenH264, with its optimized implementation utilizing SIMD instruction sets for CPUs such as SSE, AVX, MMX and NEON [13]. Our testbench environment is Ubuntu 18.04 x64 OS. We have utilized a combination of 20 HD, SD and low resolution online videos with various construction schemes from video datasets utilized in [22, 23]. Due to space limitations, we only demonstrate a sample of 9 videos representing different types of resolutions and construction schemes, as shown in table 3. All videos are coded with the *Main* profile with *level* 4.1 and CABAC enabled. It should be noted that although OpenH264 supports the *Main* profile within its "Layer Configuration File", and we have verified the output H264 stream from the encoder using both *FFmpeg* and *Mediainfo* utilities, the support for the *Main* profile is not included in the OpenH264 documentations.

Fig.4 illustrates the RD-curves with PSNR metric in dB for our selected 9 videos. It can be concluded from fig.4 that JM-encoder achieves higher RD performance than other codecs, as it allows Full-search technique as ME algorithm, in contrast with other codecs which implement more optimized ME algorithms such as 3-step search, 4-step-search [24] and diamond-search [25] algorithms. Moreover, fig.4 verifies that RD performance is approximately the same for FFmpeg, X264 and OpenH264.

For performance measurements, we have utilized the built-in Linux "*top*" utility for calculating CPU utilization for each coding operation per each video's bitrate and for OpenH264, we set the *UsageType* as camera video. Also, the Linux built-in "*taskset*" tool is used to dedicate a single logical CPU core for each coding operation. Table 2 illustrates the average frame per second (FPS) for each encoder per each video under the previously mentioned conditions. It can be concluded from table 2 that OpenH264 is the only encoder that achieves real-time constraints among other tested video encoders, as it achieves an encoding rate above 25 FPS for all selected 9 videos, as well as for all 20 video combination. This is due to its implementation utilizing SIMD extended instruction sets such as AVX/AVX2, MMX, SSE and NEON, discussed in section 3.

5. Conclusion and future work

Through this paper, we have introduced a detailed explanation for our practical video encoding experiments for the most common open-source implementations for H.264 video encoder. Our experiments show that OpenH264 encoder is the only software encoder that achieves real-time constraints with acceptable resources margins. Moreover, the reasons which led to these results are clearly explained. These practical results can be used as a base for future implementations for specially customized video usages, such as cryptocoding or steganography, utilized for streaming and broadcasting and other real-time-constrained applications.

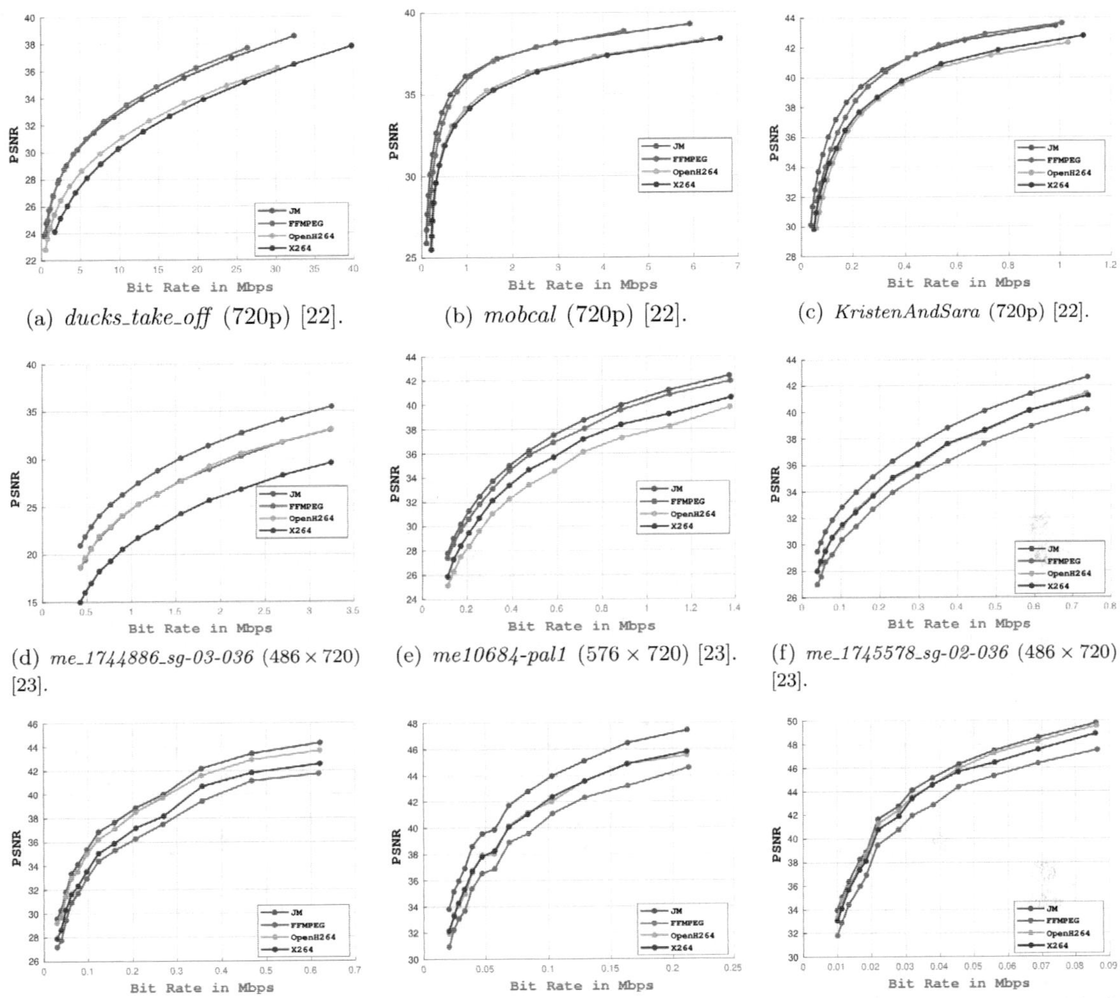

(a) *ducks_take_off* (720p) [22]. (b) *mobcal* (720p) [22]. (c) *KristenAndSara* (720p) [22].

(d) *me_1744886_sg-03-036* (486 × 720) [23]. (e) *me10684-pal1* (576 × 720) [23]. (f) *me_1745578_sg-02-036* (486 × 720) [23].

(g) *me_583335* (360 × 640) [23]. (h) *me_583380_muscle* (360 × 640) [23]. (i) *me_587716_birds* (360×640) [23].

Figure 4. RD curves for the 9 selected videos representing different dynamics and resolutions.

Table 3. Resolutions and dynamic contents for test video sequences in fig.4.

Resolution \ Dynamics	High dynamics with complex motions	Moderate dynamics	Low dynamics with simple motions
HD	Fig.4(a)	Fig.4(b)	Fig.4(c)
SD	Fig.4(d)	Fig.4(e)	Fig.4(f)
Low resolution	Fig.4(g)	Fig.4(h)	Fig.4(i)

References

[1] ITU-T 2013 *Advanced video coding for generic audiovisual services* E 38445 (www.itu.int)
[2] ITU-T 2016 *High efficiency video coding* E 41298 (www.itu.int)
[3] Jiao S, Luan L, Qu H and Zhang M 2019 *2019 IEEE 19th International Conference on Communication Technology (ICCT)* pp 510–513
[4] Chen Z, Zhou P and He Y 2002 *Joint Video Team (JVT) of ISO/IEC MPEG and ITU-T VCEG, 6th meeting: Awaji Island, JP* **JVT-F017rl**
[5] Neji N, Jridi M, Alfalou A and Masmoudi N 2013 *10th International Multi-Conferences on Systems, Signals Devices (SSD13)* pp 1–4
[6] Boyadjis B, Bergeron C, Pesquet-Popescu B and Dufaux F 2017 *IEEE Transactions on Circuits and Systems for Video Technology* **27** 892–906
[7] Aly H A 2011 *IEEE Transactions on Information Forensics and Security* **6** 14–18 ISSN 1556-6013
[8] Tew Y and Wong K 2014 *IEEE Transactions on Circuits and Systems for Video Technology* **24** 305–319 ISSN 1051-8215
[9] Kapotas S K and Skodras A N 2009 *Journal of Real-Time Image Processing* **4** 33–41 ISSN 1861-8219
[10] El-Arsh H Y and Mohasseb Y Z 2013 *MILCOM 2013 - IEEE Military Communications Conference* pp 1844–1849
[11] Tomar S 2006 *Linux J.* **2006** 10 ISSN 1075-3583
[12] Shamieh F and Wang X 2019 *IEEE Transactions on Multimedia* **21** 1893–1904
[13] Roux L and Gouaillard A 2020 *Principles, Systems and Applications of IP Telecommunications (IPTComm)* pp 1–8
[14] Jankar J R and Shah S K 2017 *International Conference on Intelligent Sustainable Systems (ICISS)* pp 250–255
[15] Richardson I E 2010 *The H.264 Advanced Video Compression Standard* 2nd ed (Wiley) ISBN 978-0470516928
[16] Shengfa Y, Zhenping C and Zhaowen Z 2006 *2006 International Conference on Communications, Circuits and Systems* vol 1 pp 126–129
[17] Fu S, Wu J and Hsu W 2015 Improving simd code generation in qemu *2015 Design, Automation Test in Europe Conference Exhibition (DATE)* pp 1233–1236
[18] Rucci E, Garcia Sanchez C, Botella Juan G, Giusti A D, Naiouf M and Prieto-Matias M 2019 *International Journal of Parallel Programming* **47** 296–316 ISSN 1573-7640
[19] Seo H, Liu Z, Großschädl J, Choi J and Kim H 2015 *Information Security and Cryptology - ICISC 2014* ed Lee J and Kim J (Cham: Springer International Publishing) pp 328–342 ISBN 978-3-319-15943-0
[20] Kusswurm D 2018 *Modern X86 Assembly Language Programming: Covers x86 64-bit, AVX, AVX2, and AVX-512* 2nd ed (Apress) ISBN 978-1484240625
[21] Liu H H 2009 *Software Performance and Scalability: A Quantitative Approach* (Wiley-Blackwell) ISBN 978-0-470-46253-9
[22] Gryzov G Y and Dvorkovich A V 2017 *2017 6th Mediterranean Conference on Embedded Computing (MECO)* pp 1–4
[23] Masanovic B, Bavcevic T and Prskalo I 2019 *Montenegrin Journal of Sports Science and Medicine* **8** 69
[24] Po L M and Ma W C 1996 *IEEE Transactions on Circuits and Systems for Video Technology* **6** 313–317 ISSN 1051-8215
[25] Zhu S and Ma K K 2000 *IEEE Transactions on Image Processing* **9** 287–290 ISSN 1057-7149

IOP Conf. Series: Materials Science and Engineering 1172 (2021) 012037 doi:10.1088/1757-899X/1172/1/012037

Modelling and transient response study of hydraulic servo actuator

E E Ibrahim[1], T Elnady[1] ,I Saleh[1] and S Hassan[1]

[1]Department of Mechanical Power & Energy, Military technical college, Cairo, Egypt

E-mail: ezzbos@gmail.com

Abstract. This paper presents a mathematical model of electro hydraulic servo system (EHS) using two-stage electrohydraulic servovalve (EHSV), by the aim of Matlab-simulink-simscape multibody to predict the flow characteristics (pressure/flow rate) with changing the parametric configuration of (EHSV), these parameters could be used to estimate the error at different operating conditions. It includes establishing mathematical model, controller model and validating the performance of the (EHSV) by experimental result of published work using the data of servovalve labelled B.31.210.12.1000.U2V (mechanical feedback) manufactured by PPT – Trstenik. The effect of changing servovalve orifice diameter (0.25, 0.30, and 0.35) mm, show that increasing the orifice diameter of the servovalve leads to increase the transient time but decrease the system overshoot. The pressure behaviour is plotted and shows that increasing the orifice diameter leads to increasing the pressure inside the system as to reach the relieve pressure.

1. Introduction

The electro hydraulic servo systems (EHS) are the best solution for many applications due to its power supply for large force, its accuracy and its response. There are many applications of (EHS) such as remote lifting, materials handling, machine tool drives, flight control, rolling mills, aircraft simulator and etc. [1]. Servo valves are used in controlling the hydraulic circuits where the force or position on the hydraulic cylinder is measured and feed back to a controller that varies the signal sent to the servo valve which allows very precise position control.

Modelling of two-stage electrohydraulic servo valve with spool position feedback was created in [2], and the model were verified experimentally using (B.31.210.12.1000.U2V) Moog servovalve. This model can predict the performance of the servo valve in a wide range of expected working conditions. A development of an improved servovalve model with both leakage and orifice flows were established in [3], it is verified experimentally.

Derivation, simulation, and implementation of a nonlinear tracking control law for a hydraulic servo system was made by [4], it provides a controller that achieve a position tracking which is simulated and experimentally tested to test the limits of its performance and the realistic effects of friction.
The flow through spool orifices in different working regimes were analytically described in [5] which calculate the internal leakage through the spool orifice.
Dynamic modelling of an electro hydraulic servo actuator with constant load was established with a controller model in [6], which assigned a control parameter on the experimental model by using fuzzy PID control controller.

Content from this work may be used under the terms of the Creative Commons Attribution 3.0 licence. Any further distribution of this work must maintain attribution to the author(s) and the title of the work, journal citation and DOI.
Published under licence by IOP Publishing Ltd

Motion control of an aircraft electro-hydraulic servo actuator was obtained in [7], it provided a mathematical model of an aircraft integrated electro-hydraulic servo-actuator and designed a classical PID controller by Zeigler-Nichols tunning method.
An online controller with least-mean-square algorithm is controlled to the force closed-loop system of the electro hydraulic servo system controlled by offline designed controller in [8].
Mathematical model is developed in [9], with the characteristics of pump, hydraulic cylinder, dynamic characteristic of the valve and position feedback. Adaptive Neuro Fuzzy is used to control servo valve with position feedback (-10 to 10 V) made by Rexroth Bosch to control a hydraulic actuator in [10].

2. Model description

The system **illustrated in figure (1), consists of hydraulic power unit (hydraulic pump), which delivers hydraulic to the control unit (hydraulic servo valve) which is adapted by the controller (PID), to deliver the required hydraulic to the hydraulic actuator, which track the required trajectory.**

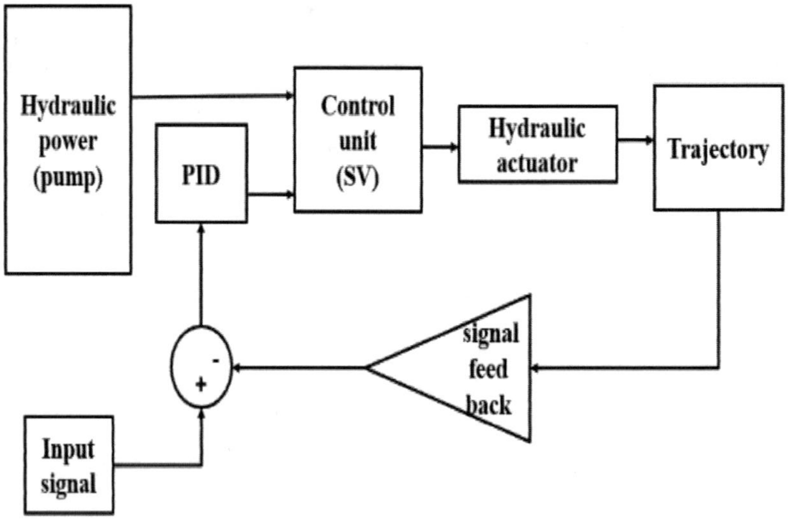

Figure 1. block diagram for the overall system

Figure (2) focus on the hydraulic circuit of the proposed system which is composed of a source of hydraulic (hydraulic pump) (1), connected with pressure relieve valve (2), that delivers fluid to the electro hydraulic servo valve (EHSV) (3) via a one way valve (4), (EHSV) controls the output hydraulic that delivers to the hydraulic actuator (5) to allow the hydraulic actuator tracking the required input position by the aim of PID controller (6), that take a feedback from position sensor (LVDT) (7) the figure illustrates the hydraulic circuit and the specification of this circuit is shown in table (1).

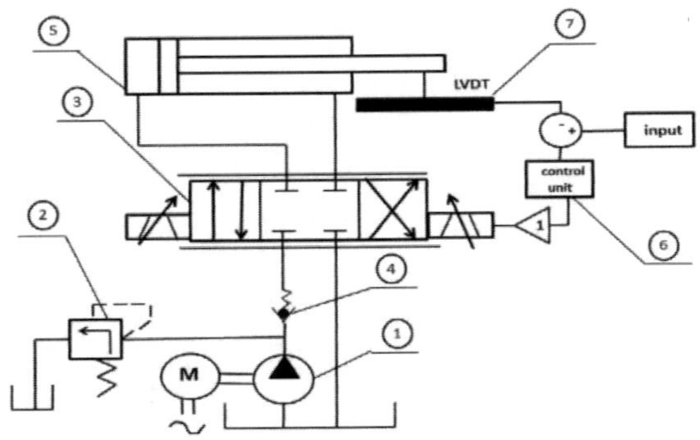

Figure 2. Electro hydraulic servo actuator circuit.

Figure (3-a) illustrates a typical servo valve used in this study where it consists of torque motor (1), that drives flapper (2), which makes a difference in pressure between servovalve nozzles (3), to force the sliding spool (4), to allow pressure port (5), to feed port (6) or (7).
The spool other side allows hydraulic to go back through the return port (8), while the two orifices (9) apply a pressure difference.

Figure (3-b) focuses on the servo valve torque motor which contains:

1	polar piece	4	flapper
2	magnet	5	control coils
3	flexible tube	6	Armature

While figure (3-c) illustrates the flapper-orifice area of the servo valve to focuses on the important of this area on the hydraulic behaviour, it also illustrates the valve input flow rate, flow rates before valve orifices (Q_1 and Q_2) and flow rates after valve orifices (Q_3 and Q_4).

Table 1. Parameters table.

Hydraulic element	symb	value	Hydraulic element	symb	value
Hydraulic cylinder	Ap (A)	1.2e-3 (m^2)	Pump	Q	12 (L/min)
	Ap (B)	1.2e-3 (m^2)		n	1800 (rpm)
	L	0.5 (m)		V_g	7.07e-7 (m3/rad)
	V_0	1e-5 (m^3)		η_v	92 %
Servovalve	As	16.76 (mm^2)	System	B	1.5e9 (Pa)
	D_n	0.28 (mm)		P_s	210 (bar)
	m	3.1 (Kg)		P_T	0 (bar)
	D_o	0.18 (mm)		ρ	871 (kg/m3)

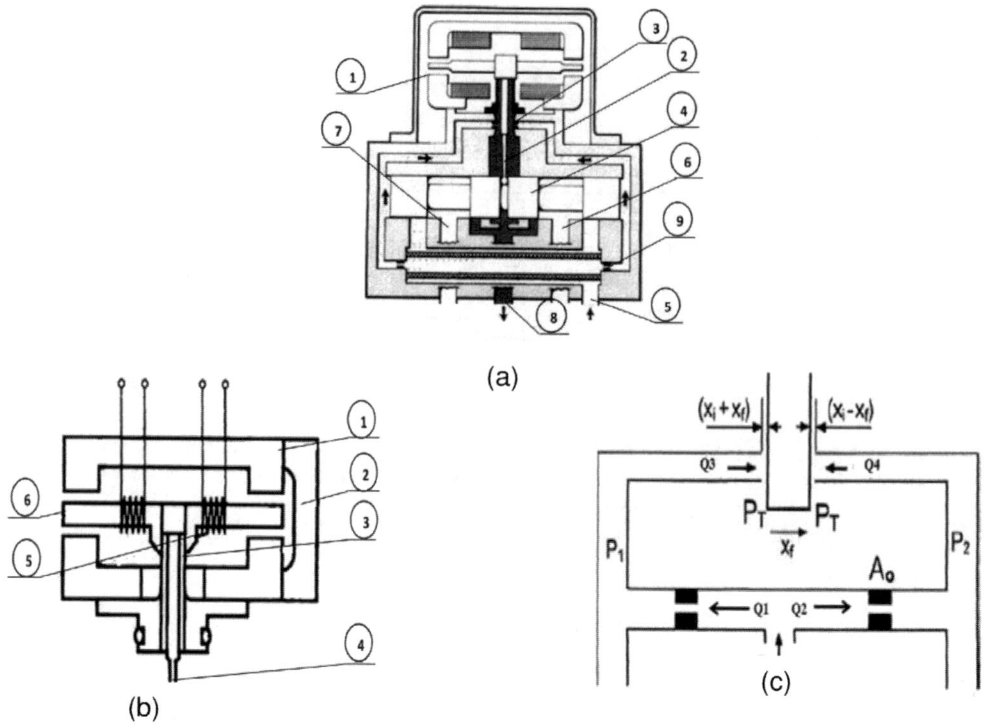

Figure. 3: Typical two stage electro hydraulic servo valve. (a) Valve cross section. (b) Valve torque motor. (c) Orifice and flapper section.

3. Mathematical model

The servo valve mathematical model is derived in 2009, by G. Rabie, [11].
The effective torque delivered by the electromagnetic motor is given by the equations (1) and (2).

$$T = K_i * i + K_\theta * \theta \qquad (1)$$

Where

$$K_i = \frac{N\,\lambda_p \mu_0\, AL}{2{X_0}^2}$$

$$K_\theta = \frac{{\lambda_p}^2\, \mu_0\, AL^2}{4{X_0}^3}$$

i input current. N. number of turns of the coil, λp... magneto-motive force of the permanent magnet, μ0... permeability of free space, A... area of air gap.

L.. armature length, X_0.. length of the air gap in the neutral position of the armature.
The armature equation of motion is

$$T = J\frac{d^2\theta}{dt^2} + f_v\frac{d\theta}{dt} + K_t\theta + T_l + T_p + T_F \qquad (2)$$

The flow rate affecting on the flapper and make a pressure difference between both sides of the spool is obtained by equations (3), (4) and (5).

$$Q_1 = C_d A_0 \sqrt{\frac{2}{\rho}}(P_s - P_1) \qquad Q_2 = C_d A_0 \sqrt{\frac{2}{\rho}}(P_s - P_2) \qquad (3)$$

$$Q_3 = C_d\pi\, d_f\left(X_i + X_f\right)\sqrt{\frac{2}{\rho}}(P_1 - P_3) \tag{4}$$

$$Q_4 = C_d\pi\, d_f\left(X_i + X_f\right)\sqrt{\frac{2}{\rho}}(P_2 - P_3) \tag{5}$$

The flapper chamber continuity equations had shown in equations (6), (7) and (8).

$$Q_1 - Q_3 + A_s\frac{dx}{dt} = \frac{V_0 - A_s x}{B}\frac{dP_1}{dt} \tag{6}$$

$$Q_2 - Q_4 - A_s\frac{dx}{dt} = \frac{V_0 + A_s x}{B}\frac{dP_2}{dt} \tag{7}$$

$$Q_3 + Q_4 = \frac{V_3}{B}\frac{dP_3}{dt} \tag{8}$$

The spool equation of motion is shown in (9)

$$A_s(P_2 - P_1) = m_s\frac{d^2x}{dt^2} + f_s\frac{dx}{dt} + F_j + F_s \tag{9}$$

Continuity equations through the hydraulic actuator:

$$Q_{AR} - a\frac{dy}{dt} - Q_i - Q_e - \frac{(V_R + ay)}{B}\times\frac{dp_R}{dt} = 0 \tag{10}$$

$$A\frac{dy}{dt} + Q_i - Q_{psB} - \frac{(V_{ps} - Ay)}{B}\times\frac{dp_R}{dt} = 0 \tag{11}$$

Actuator Equation of motion:

$$a.p_R - A.p_{ps} = m_c\frac{d^2y}{dt^2} + f_c\frac{dy}{dt} + F \tag{12}$$

These equations represent the mathematical model which are presented by the aim of Matlab-simulink, then using System Identification Toolbox and hence the transfer function is defined as

$$\frac{63.12}{S^2 + 16.44S + 63.09} \tag{13}$$

Then by trial and error with PID controller toolbox system tunned, the three coefficients (KP,KI and KD) are adopted to KP (20) KI (1) KD (0.01).

4. validation

The performance of (EHSV) is validated by experimental result of published work in [2] as seen in figure (4), with the same data in table (1) by applying ±10% of the rated current signal, with square periodic signal (frequency was 10 Hz).

The result shows that the simulation result have a good agreement with the experimental one.

The validated simulation of (EHSV) has been used to estimate the proposed servo actuator performance in this study.

5. Result and discussion

The controller is responsible for reducing error during running condition as it adopts the input signal to compensate the required position and also adopts the transient response of the actuator.

In this study attempts to predict the response of the actuator with varying the geometric configurations of the control unit (EHSV) is done by changing valve orifice diameter to simulate the effect of impurities such as burrs, metal particles, and particles produced during operation on servo valve operating condition, these types of impurities are considered in this study to discuss its effect on servovalve response after long time of operation and hence predict the servovalve life time.

Figure (5) shows the position tracking of three servo valves with three different orifice diameters (0.25, 0.30and 0.35) mm.

As it shown when using orifice diameter 0.25 mm it effects on the overshoot of the actuator to reach 0.35 m and also reach the steady state condition after 0.6 second.

By using orifice diameter 0.30 mm the actuator overshoot is reduced to 0.31 m and the settling time is increased to 0.65 second. By using orifice diameter 0.35 mm the actuator overshoot can be neglected and the settling time is increased to 0.8 second. This increasing in the settling time is due to increasing the pressure difference caused by the orifice to reach the relief valve pressure (200 bar) as shown in figure (8). This behaviour "by using step input" is obvious appeared and also by using pulse input, same overshoot and same delay is done as shown in figure (6).

Figure (7) and (8) show the pressure variation through the servovalve orifices (left and right side) while acting with input signal 0.3 m and it explains that by increasing the orifice diameter to 0.30 and 0.35 mm the pressure increases in the spool left side to the relief pressure which is not suitable to compensate the required force to push the spool of (EHSV) to deliver fluid to the actuator, while by using orifice diameter 0.25 the system is stable.

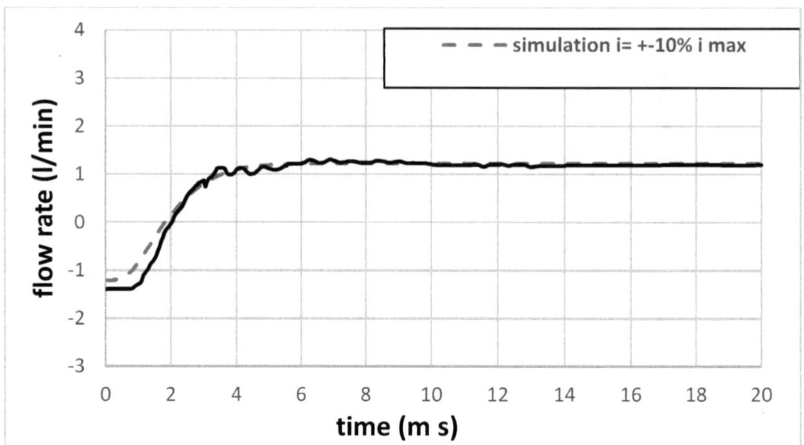

Figure. 4: Experimental and simulation servovalve flowrate behavior with input signal ±10% of rated current signal

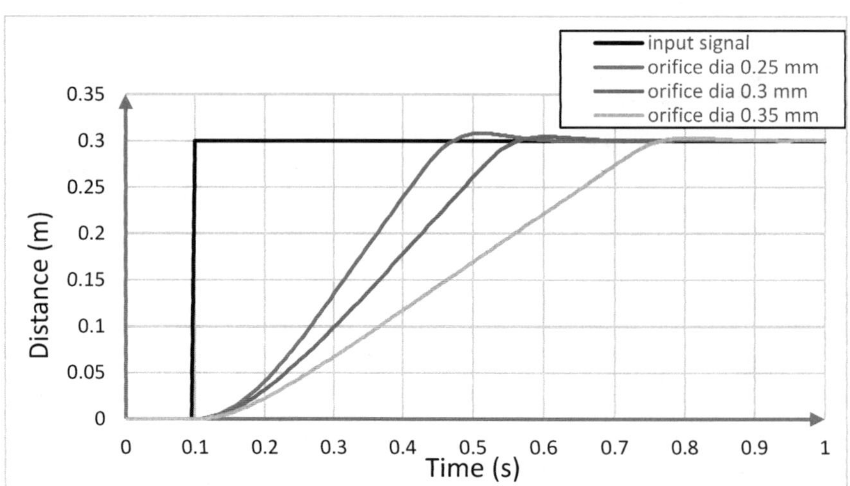

Figure 5: position tracking by step input 0.3 m with varying the orifice diameter (0.25, 0.30 and 0.35) mm.

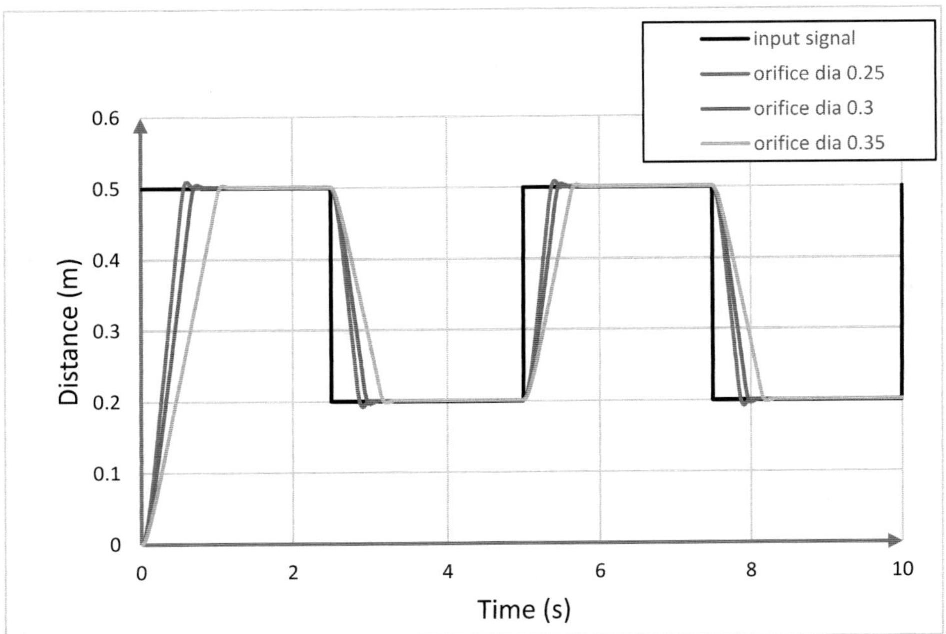

Figure 6: position tracking by square pulse input with varying the orifice diameter (0.25, 0.30 and 0.35) mm.

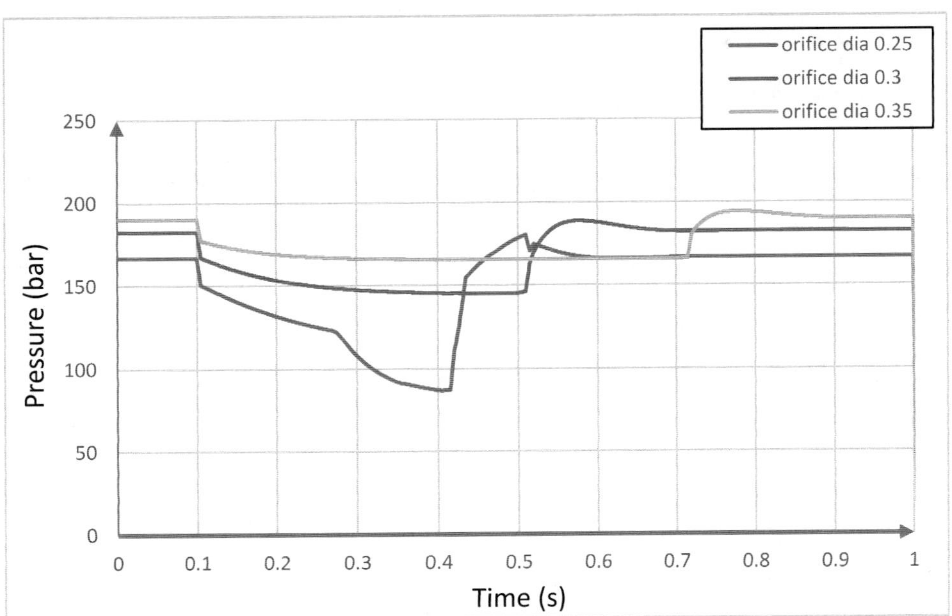

Figure 7: pressure through servo valve, input displacement (0.3 m) with varying the orifice diameter (0.25, 0.3, and 0.35) mm. left side.

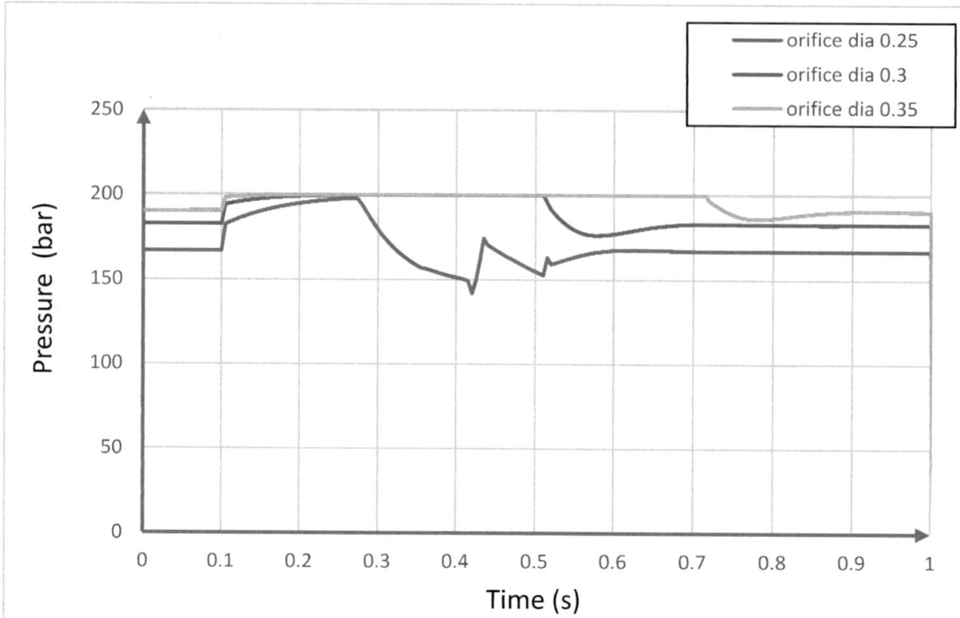

Figure 8: pressure through servo valve, input displacement (0.3 m) with varying the orifice diameter (0.25, 0.3, and 0.35) mm. right side.

6. Conclusion

The electro hydraulic servo valve is a very sensitive hydraulic control unit as it is affected by the very low sized impurities due to its tiny orifices' configuration.

It is deduced that the diameter of servo valve orifice effects on the transient time response of the system as if its orifice diameter is exposed to any small change it effects on the transient time and also effects on the system overshoot. That is due to the variation in the pressure behaviour due to the change in the orifice area.

Nomenclature

Q	flow rate
C_d	discharge coefficient
A_0	orifice area
d_f	flapper nozzle diameter.
P_s	feeding pressure.
P_T	returns pressure.
X_f	flapper displacement.
X_i	initial flapper nozzle opening.
ρ	oil density.
A_s	spool cross-sectional area.
B	bulk modulus.
V_o	initial volume of oil in the spool side.
V_3	volume of the flapper valve return chamber.
L	actuator stroke.
V_g	geometric volume.
η_v	volumetric efficiency.
D_n	nozzle diameter.
m	spool mass.
D_o	orifice diameter.
J	moment of inertia

T_l	torque due to flapper displacement limiter
f_v	damping coefficient
K_t	stiffness of flexure tube
T_F	feedback torque
T_p	torque due to the pressure forces.

References

[1] M. A. S. Aboelela, M. E. S. M. Essa, and M. A. M. Hassan, 2018 "Modeling and identification of hydraulic servo systems," *Int. J. Model. Simul.*, vol. **38**, no. 3, pp. 139–49.

[2] Gordić, Dušan, Milun Babić, and Nebojša Jovičić. 2004, "Modelling of spool position feedback servovalves." *International Journal of Fluid Power* **5.1** pp 37-51.

[3] Eryilmaz, Bora, and Bruce H. Wilson. 2000, "Combining leakage and orifice flows in a hydraulic servovalve model." *J. Dyn. Sys., Meas., Control* **122** pp: 576-79.

[4] Sohl, Garett A., and James E. Bobrow. 1999, "Experiments and simulations on the nonlinear control of a hydraulic servosystem." *IEEE transactions on control systems technology* **7.2** pp: 238-47.

[5] Gordić, D., et al. 2011, "Spool valve leakage behaviour." *Archives of Civil and Mechanical Engineering* **11.4** pp: 859-66.

[6] Tran, Ngoc Hai, et al. 2018, "Modeling of Position Control for Hydraulic Cylinder Using Servo Valve." *Asian Conference on Intelligent Information and Database Systems. Springer, Cham.*

[7] Fadel, M. Z., M. G. Rabie, and A. M. Youssef. 2019, "Motion control of an aircraft electro-hydraulic servo actuator." *IOP Conference Series: Materials Science and Engineering*. Vol. **610**. No. 1. IOP Publishing.

[8] G. Shen, Z. Zhu, J. Zhao, W. Zhu, Y. Tang, and X. Li, 2017 "Real-time tracking control of electro-hydraulic force servo systems using offline feedback control and adaptive control," *ISA Trans.*, vol. **67**, pp. 356–370.

[9] V. I. Forental, M. V. Forental, and F. M. Nazarov, "Investigation of dynamic characteristics of the hydraulic drive with proportional control," Procedia Eng., vol. **129**, pp. 695–701, 2015.

[10] M. E. S. M. Essa, M. A. S. Aboelela, M. A. Moustafa Hassan, and S. M. Abdrabbo, 2019 "Control of hardware implementation of hydraulic servo application based on adaptive neuro fuzzy inference system," ICENCO 2018 - *14th Int. Comput. Eng. Conf.* Secur. Smart Soc., pp. 168–73.

[11] M. G. Rabie, Fluid power engineering vol. 28: McGraw-Hill New York, NY, USA, 2009

IOP Conf. Series: Materials Science and Engineering 1172 (2021) 012038 doi:10.1088/1757-899X/1172/1/012038

Modeling, simulation and controller design for a typical bent axis electrohydraulic servo motor

M A Yousef[1], M El-Sanabawy[1], G Rabie[2] and R A Rateb[3]

[1] Department of Rockets, Military Technical College, Cairo, Egypt.
[2] Modern Academy for Engineering and Technology, Cairo, Egypt.
[3] Egyptian Armed Forces, Cairo, Egypt.

E-mail: mohamed_abdelwahab@mtc.edu.eg

Abstract. Bent axis electrohydraulic servo motors are one of the Electrohydraulic Servo Motors (EHSMs) family which are used in high frequency, speed, and precision applications such as aerospace applications as well as many military weapon system applications. Therefore, it is essential to understand how such motors affect the whole performance of the system which makes modeling of these motors is an important task to be achieved. Once an accurate model is obtained, an optimum controller can be applied. This paper introduces detailed mathematical modeling of a typical bent axis (EHSM). MATLAB SIMULINK package is used to simulate and control such (EHSM) using PID controller. The PID controller gains were tuned using PD-PI controller to obtain the precision response for the (EHSM). The validity of mathematical modeling was reviewed through some practical experiments.

1. Introduction

Electro-hydraulic Servo Motors (EHSMs) are used in industries where high power to weight ratios, rapid responses, and high precision are needed such as aerospace applications.(EHSMs) motors are used to convert the hydraulic power into mechanical one [1]. They combine the advantages of the hydraulic and electric powers, as they are composed of a hydraulic motor and an electro-hydraulic servo valve (EHSV) respectively[2].

The importance of such devices in recent industries makes their mathematical modeling an important task to be achieved. In the process of designing a new model for a system or improving an existing one, an exact model must be identified [3]. The output of a system can be accurately predicted with good mathematical modeling [4]. So, when the system is well defined, the mathematical model representing this system can be achieved using for example differential equations. These differential equations can be simulated using simulation programs such as the one used in[5].

The point of building a mathematical model is to identify the dynamic behavior of the system besides the ability to control such systems without overshoot or delay through a control action applied in an optimum way to ensure control stability. As it's known, mathematical modeling of a system is achieved by a huge amount of dependent differential equations that describe every component in the system. Besides the equations that express how these components are related to

The dynamic behavior of an Electro-hydraulic servo system (EHSS) was examined by Attila. The dynamic behavior was investigated using a step response for a non-linear mathematical model. The results showed that the servo valve rapidly followed the valve control signal [6]. Dechrit and Nitin

Content from this work may be used under the terms of the Creative Commons Attribution 3.0 licence. Any further distribution of this work must maintain attribution to the author(s) and the title of the work, journal citation and DOI.
Published under licence by IOP Publishing Ltd

achieved modeling and simulation of a non-linear high-speed servo system used in the operation of a mini-press machine. They built a mathematical model using MATLAB Simulink and the simulation results were tested using experimental system identification which showed the effectiveness of their model [5]. Also, J. Das et al derived a physical nonlinear model for an electrohydraulic actuator. To obtain the actuator dynamics, MATLAB / SIMULINK program was used. The model was validated through experimental work. They focused, experimentally, on the actuator friction and proportional valve characteristics to get closure of the physical model. They tested their model and then a PID controller was introduced to the model. The gain values were found to be suitable and hence providing stable results[7].

Once a good model is obtained an optimum controller can be achieved. PID controller is the most applicable controller used in the industry because of its simplicity in structure and satisfying performance in the industrial process. However, the optimum performance of the PID controller is dependent on three gain values K_P, K_I, K_D proportional, integral, and derivative gains respectively. Tuning of these gain values becomes an essential task to be achieved in order to get a precision performance of such PID controller[3][8]. T. G. Ling et al applied a PID controller to their model to achieve an accurate position tracking of an electrohydraulic actuator. They compared their simulation results with real work, and they concluded that the system with the controller achieved position tracking with high accuracy [9]. There are several types of PID tuning. R. Sen et al applied three types of tuning the PID controller to obtain a precision Positioning of a multi-axis machine[10]. The PD-PI controller is considered an extension of the classical PID controller[11]. Galal A. Hassaan achieved tuning for the classical PID controller using a PD-PI controller to a first-order time-delayed process. The PD-PI controller was compared to the classical PID controller. The results showed that it has a better response than the classical PID controller[12]. Also, Akram F. Singer used the PD-PI controller for tuning the classical PID controller. The proposed controller was applied to a third-order oscillating process. The results showed that the proposed controller has a better performance than the classical PID controller[13].

2. System description

Fig. (1) shows the construction of the considered hydraulic system.

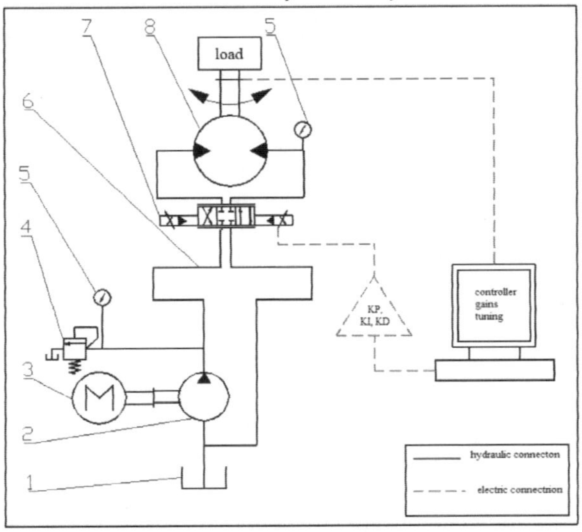

1. open tank 2. Fixed displacement pump 3. Electric motor 4. Relieve valve 5. Pressure -gauges 6. Hydraulic pipelines 7. EHSV 8. Hydraulic motor

Figure 1. System description

2.1 Hydraulic power supply.

2.1.1 construction and operation. The considered hydraulic system is illustrated in fig. (1). It consists of a hydraulic power supply to ensure a constant and controllable supply pressure (Ps). The hydraulic power supply (HPS) is composed of an electric motor (3), [3-phase power electric motor 10 HP and 380 Volt], coupled to (2) a hydraulic fixed displacement gear pump, [533/rev], immersed in (1) a tank, [250 liters]. The supply pressure Ps is limited by (4) a regulated pilot-operated relief valve.

2.1.2 mathematical modeling. The mathematical relations describing the (HPS) , from equation (1) to (20) in [14], have been used to develop a Simulink program .the calculated transient response of the system pressure is plotted Fig.(2).

2.1.3 simulation and results. The following figure shows that the exit pressure ranges from (90.5) bar in the case of throttle area equal to (Ath = 1e-6 m2) which gives a flow rate of (10) l/ min equal to the maximum flow rate of the electro-hydraulic servo valve (EHSV) to (91.4) bar at (Ath = 0) which equates to zero servo valve flow rate. This result shows that the maximum pressure variation is within (1.2%). Therefore, in this study, the HPS pressure is assumed to be constant.

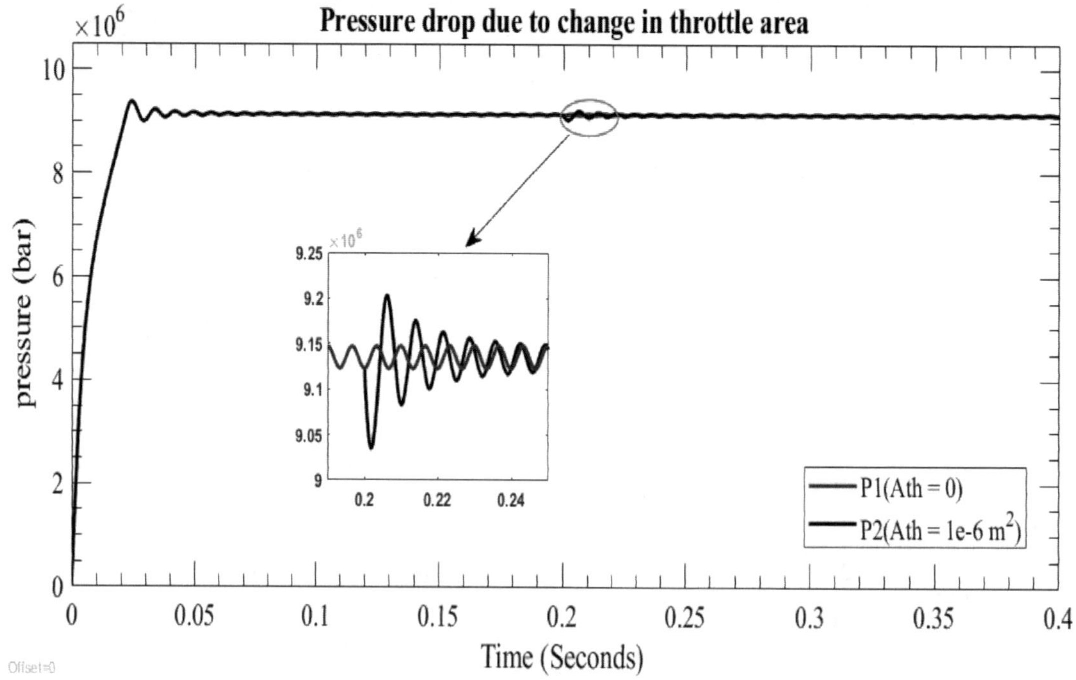

Figure 2. HPS exit pressure at throttle area (Ath= 0 & 1e-6 m2).

2.2 Electro-hydraulic servo valve.

2.2.1 Construction and operation. The (HPS) fluid is controlled by a two-stage electro-hydraulic servo valve (EHSV) of type [MOOG] numbered in fig. (1) by (7). The construction drawing of two-stage EHSV identified by complete parameters is shown in fig. (3). The typical two-stage EHSV receives the pressurized hydraulic fluid that comes from a hydraulic pump. It then transmits the fluid to a bi-directional hydraulic motor at a pressure that is relative to an electrical signal which it receives. Servo valves can guarantee an accurate control of velocity, position, force, and pressure with good damping characteristics.

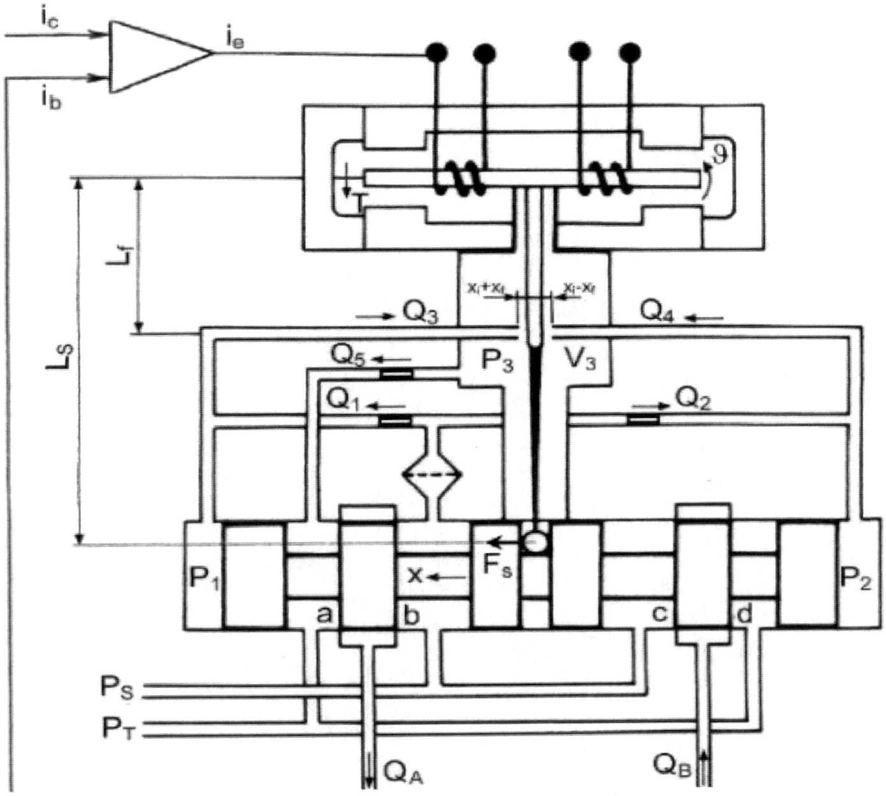

Figure 3. Schematic drawing for two-stage EHSV

2.2.2 Mathematical modeling. The mathematical modeling of the EHSV was developed through equations 1 to 18[2]. The author has partially contribution on the modeling of the two-stage EHSV. The mathematical modeling was developed as follows.

- *Torque motor*

The function of the electromagnetic torque motor is to convert the electric input which is a current of low value within 10 mA into mechanical torque. Neglecting the magnetic hysteresis effect, the equation of the torque motor can be expressed as flows:

$$T = K_i i_e + K_\theta \theta \tag{1}$$

- *Armature dynamics*

The rotating armature movement is illustrated in equations (2) & (3)

$$T = J\frac{d^2\theta}{dt^2} + f_\vartheta \frac{d\theta}{dt} + K_T\theta + T_L + T_P + T_F \tag{2}$$

$$T_p = \frac{\pi}{4} d_f^2 (P_2 - P_1) L_f \tag{3}$$

- *Feedback Torque*

The magnitude of the feedback torque depends on the displacement of the spool and the rotation of the flapper angle which can be described from equation (4) to equation (6)

$$T_F = F_S L_S \tag{4}$$

$$F_S = K_S(L_S\theta + x) \tag{5}$$
$$T_F = F_S L_S = K_S(L_S\theta + x)L_S \tag{6}$$

- *Flapper Position Limiter*

The flapper displacement is restricted by the jet nozzles. Once the flapper limiter reaching one side of the nozzle, a counter torque is produced which can be expressed in equation (7):

$$T_L = \begin{cases} 0 & |x_f| < x_{fL} \\ R_f \dfrac{d\theta}{dt} - (|x_f| - x_{fL})K_L L_f sign(\theta) & |x_f| > x_{fL} \end{cases} \tag{7}$$

- *Flow rates and the restriction areas of the flapper valve*

1. *Flow rates*

The equations of the flow rates can be expressed in equations (8) to (13).

$$Q_1 = C_d A_o \sqrt{\frac{2}{\rho}(P_s - P_1)} = C_{12}\sqrt{(P_s - P_1)} \tag{8}$$

$$Q_2 = C_d A_o \sqrt{\frac{2}{\rho}(P_s - P_2)} = C_{12}\sqrt{P_s - P_2} \tag{9}$$

$$Q_3 = C_d \pi d_f (x_i + x_f) \sqrt{\frac{2}{\rho}(P_1 - P_3)} = C_{34}(x_i + x_f)\sqrt{(P_1 - P_3)} \tag{10}$$

$$Q_4 = C_d \pi d_f (x_i - x_f) \sqrt{\frac{2}{\rho}(P_2 - P_3)} = C_{34}(x_i - x_f)\sqrt{(P_2 - P_3)} \tag{11}$$

$$Q_5 = C_d A_s \sqrt{\frac{2}{\rho}(P_3 - P_T)} = C_5\sqrt{(P_3 - P_T)} \tag{12}$$

$$x_f = L_f\theta \tag{13}$$

2. *Restricting areas*

The modeling of the restricting areas of the flapper valve are shown in equations (14) & (15).

$$\left.\begin{aligned} A_a = A_c &= \omega c \\ A_b = A_d &= \omega\sqrt{(x^2 + c^2)} \end{aligned}\right\} \quad For\ x \geq 0 \tag{14}$$

$$\left.\begin{aligned} A_a = A_c &= \omega\sqrt{(x^2 + c^2)} \\ A_b = A_d &= \omega c \end{aligned}\right\} \quad For\ x \leq 0 \tag{15}$$

3. *Continuity equations applied to flapper valve chambers.*

If the continuity equation is applied to the flapper champers, equation (16) to (18) will be introduced.

$$Q_{ba} = Q_1 - Q_3 + A_s\frac{dx}{dt} = \frac{d}{dt}\left(\frac{V_o - A_s x}{B}P_1\right) \tag{16}$$

$$Q_{dc} = Q_2 - Q_4 - A_s\frac{dx}{dt} = \frac{d}{dt}\left(\frac{V_o + A_s x}{B}P_2\right) \tag{17}$$

$$Q_3 + Q_4 - Q_5 = \frac{d}{dt}\left(\frac{V_3}{B}P_3\right) \tag{18}$$

These mathematical relations from (1) to (18) were used to develop a computer simulated program (Simulink). Fig. (4,5) show where these numerical values where found.

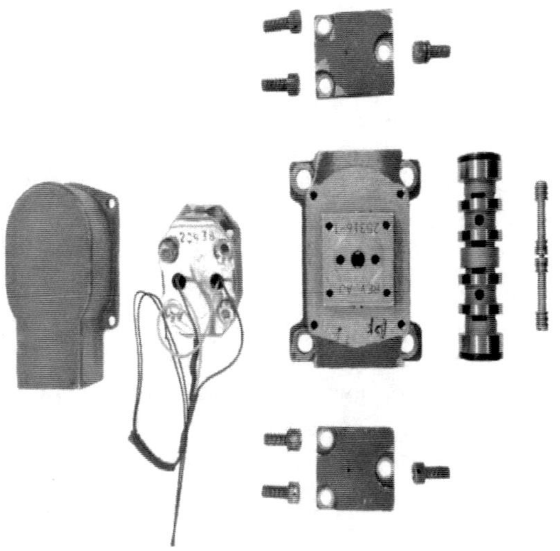

Figure 4. Disassembled two-stage EHSV

Figure 5. Typical two-stage EHSV

The servo valve is a complicated device that contains a higher order of nonlinear response. To perform an accurate mathematical model, knowledge of a big number of internal parameters is required. These parameters are adjusted together by the manufacture to tune the response of the valve and this information is not available for the users[6]. The previous section illustrates the detailed mathematical modeling of the EHSV.

2.2.3 Simulation and results. The following figure shows the pressure response (P1, P2) at the two sides of the spool. The pressure difference $\Delta P = (P1 - P2)$ is responsible for the movement of the spool. This spool movement generates a flow rate (Q_A, Q_B) which in turn rotates the hydraulic motor.

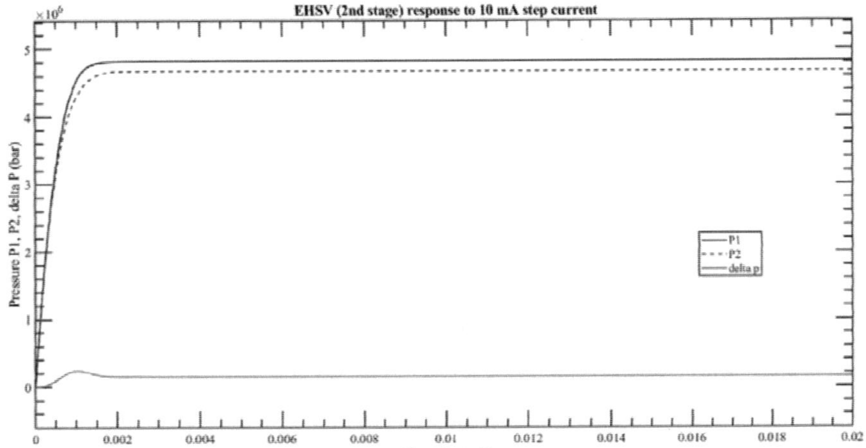

Figure 6. EHSV response to step 10 mA electric current

2.3 *Hydraulic motor*

2.3.1 Construction and operation. The hydraulic motor is a bent axis responsible for converting the hydraulic fluid pressure into mechanical torque and the hydraulic flow into angular displacement. The hydraulic circuit is provided by pressure gages at the exist of the (HPS) and the entrance of the hydraulic motor as illustrated in fig. (1). The hydraulic motor in concern is illustrated in the fig. (7) consists of a front and rear casing (1,2) port plate (3), cylinder block (4), piston head (5), piston rod assembly (6), bearing set (7) leakage port (8) and output shaft (9). The input flow rate with high pressure (Q_{in}) is supplied to the port plate (3) which transmits the oil to the cylinder block and the piston chamber (4,5). The high-pressure line connected to the piston chambers causes pressure force. This pressure force is resolved at the end of the connecting rod into two components, axial and tangential force. The axial force is sustained by the bearings while the tangential force causes the driving of the hydraulic motor shaft against the loading torque. Hence, the loading torque value depends on the supply (inlet) pressure P_s and hydraulic motor speed is based on the supply (inlet) flow rate Q_{in}. The intake and the drain of the oil flow of the hydraulic motor are controlled by the port plate (3).

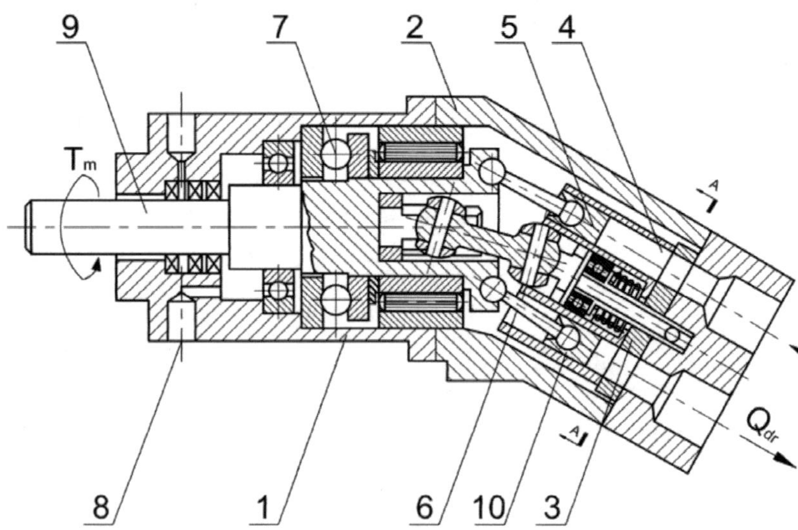

Figure 7. Schematic drawing for the hydraulic motor.

2.3.2 *Mathematical modeling*

The inlet (intake) and outlet (drain) flow rates can be calculated from the following equations:

$$Q_{ini}= C_d A_{ini}\sqrt{\frac{2}{\rho}(P_s\text{-}P_{ci})} \tag{19}$$

$$Q_{dri}= C_d A_{dri}\sqrt{\frac{2}{\rho}(P_{ci}\text{-}P_T)} \tag{20}$$

$$Q_{in}=\sum_{1}^{i} Q_{ini} \tag{21}$$

$$Q_{dr}=\sum_{1}^{i} Q_{dri} \tag{22}$$

The following equation results from applying the continuity equation to the individual cylindrical chamber of the hydraulic motor taking into consideration the volume variation of the chamber and internal leakage.

$$Q_{ini}-Q_{dri}-\frac{dx_i}{dt}A_S-\frac{P_{ci}}{R_i}=\frac{V_{ci}+X_iA_S}{B}\frac{dP_{ci}}{dt} \tag{23}$$

$$A_S=\frac{\pi}{4}d_8^2 \tag{24}$$

$$V_{ci}=A_S L_C \tag{25}$$

The piston displacement can be expressed as a function of rotational angle as follows:

$$X_i=\frac{Y_i}{\sin\propto}=\frac{d_7(1-\cos\theta_{mi})}{2\sin\propto} \tag{26}$$

The torque on one piston of the hydraulic motor is calculated as follows:

$$T_{mi}=\frac{d_7}{2}P_{ci}A_S\sin\propto\sin\theta_{mi} \tag{27}$$

Hence, the total torque (motor torque) can be expressed as:

$$T_m=\sum_{1}^{i} T_{mi} \tag{28}$$

And then the equation of motion of port plate under the action of motor torque can be expressed as follows:

$$T_m=J\ddot{\theta}_{mi}+f\dot{\theta}_{mi}+T_L \tag{29}$$

Equations from (18) to (29) were used to develop a Simulink program for the hydraulic motor[16]. Fig. (8) shows where these numerical values were found.

Figure 8. Disassembled hydraulic motor.

2.3.3 Simulation and results. The hydraulic motor was connected directly to the HPS and the response of the hydraulic motor angular velocity was illustrated in fig. (9).

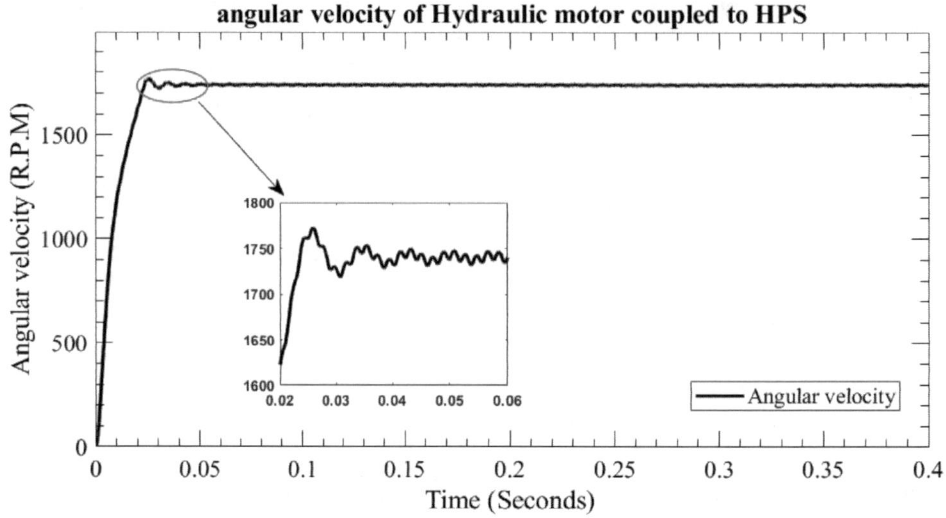

Figure 9. Hydraulic motor angular velocity response to step throttle area Ath = 1e-6 m2.

2.4 Electro-hydraulic servo motor (EHSM)

2.4.1 Construction and operation. The motor understudy shown in fig. (10) is a bent axis electrohydraulic servo motor used in the rocket launcher system. The function of this motor is the movement of the rocket system in vertical and longitudinal planes. Once the guidance commands reach the control panel of the rocket launcher, they are automatically transformed into electrical signals to the EHSV which in turn rotate the hydraulic motor at an angle that corresponds to the guidance signal. It composes 9 pistons to reduce the pulsation of the supplied pressure. The operating ranges of frequency and power for this motor are much higher compared with other applications. The operating pressure is 3000 psig i.e about 200 bar.

Figure 10. Typical electro-hydraulic servo motor.

2.4.2 Mathematical modeling. The EHSV was simulated through equation (1) to (18), moreover, the hydraulic motor was simulated through equations from (19) to (29). The continuity equations describing the room (chamber) between the EHSV and the hydraulic motor are as follows:

$$\sum_{1}^{i} Q_{ini} - Q_A = \frac{d}{dt}\left(\frac{V_o}{B} P_{ba}\right) \tag{30}$$

$$Q_{ba} - \sum_{1}^{i} Q_B = \frac{d}{dt}\left(\frac{V_o}{B} P_{dc}\right) \tag{31}$$

Fig. (11) shows the the schematic drawing for the EHSM.

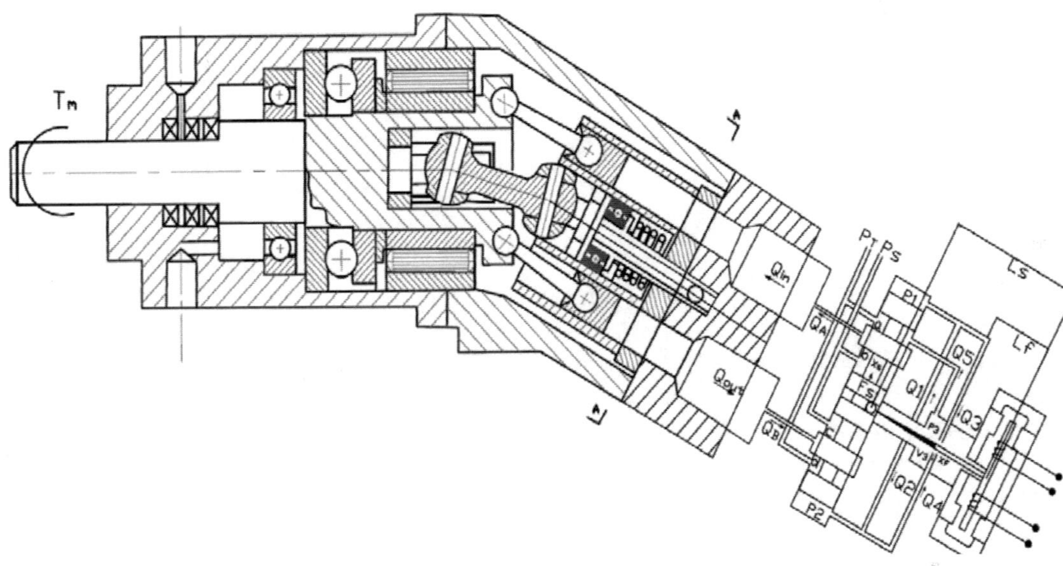

Figure 11. Schematic drawing for the EHSM

The hydraulic motor was mathematically modeled through equations from (1) to equation (31). MATLAB Simulink program was used to simulate these equations. Figure (12) shows the constructed program block veiw.

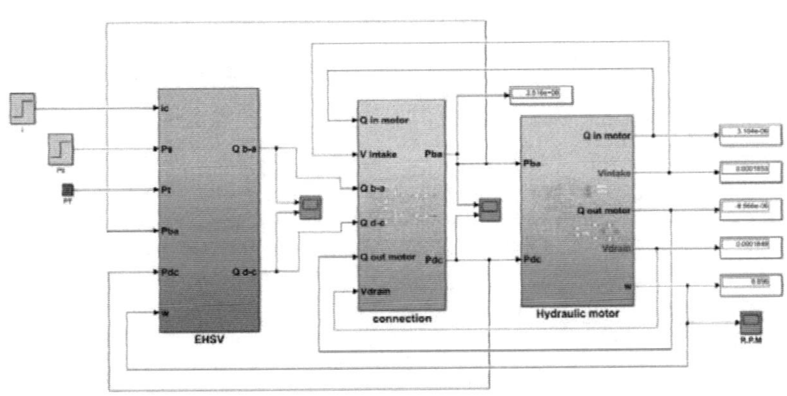

Figure 12. Simulink Program Block View.

2.4.3 Simulation and results. Figure (13) shows the dynamic behavior response of the EHSM to 10 mA electric current input to the EHSV.

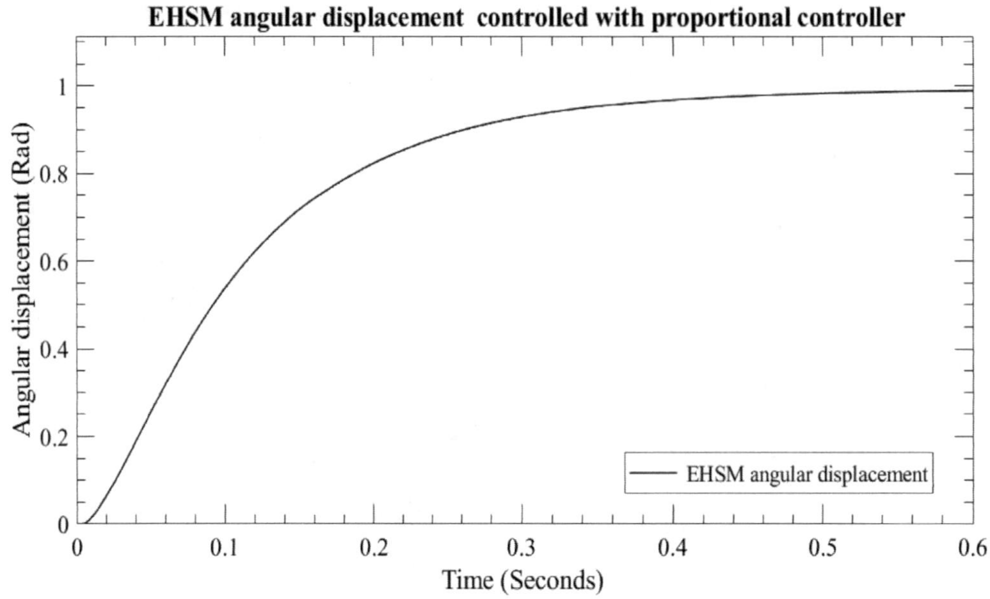

Figure 13. EHSM angular displacement response to 10 mA electric current input.

3. PID controller design

MATLAB Simulink program is also used for designing the PID controller with preliminary gain parameters (K_P, K_I, and K_D) based on trial and error of the model. The best values of controller gains were found to be Kp = 21, Ki = 5 , and Kd = 0. Fig.(14) shows the model response for the controlled EHSM compared to the EHSM.

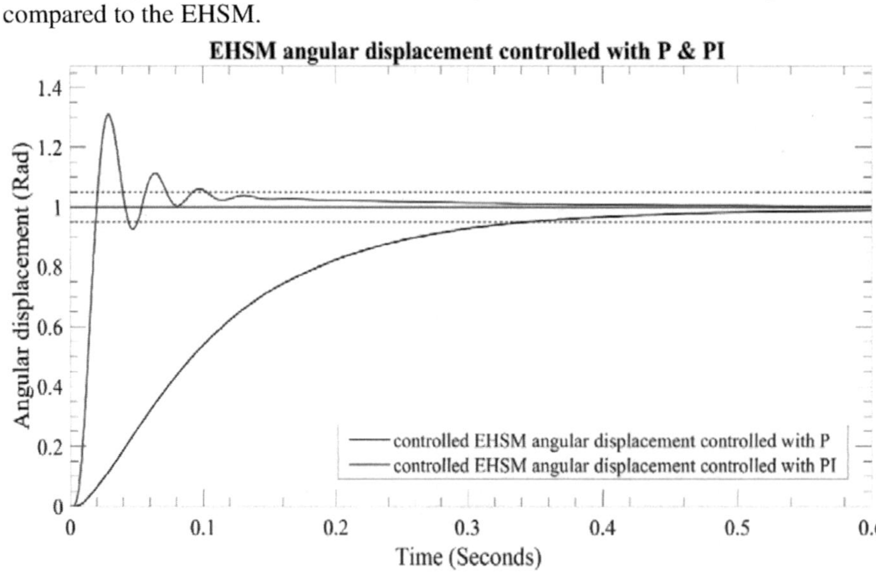

Figure 14. EHSM angular displacement response to 10 mA electric current step input.

4. PD-PI controller design

The proposed controller is a proportional- derivative (PD) proportional - integral (PI) controller. The two parts of the controller (PD-PI) are connected in series as shown in figure (15). The system error is the input of the first part and its output is the input of the second part [11], [12], [13], [17].

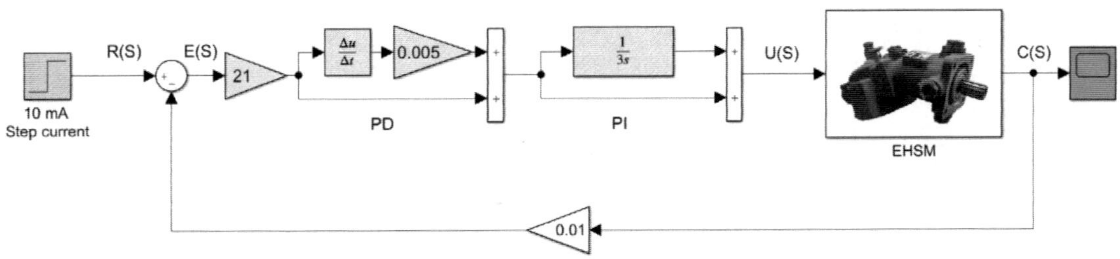

Figure 15. Construction of PD-PI controller.

The following figure shows the response of the PD-PI controller with respect to the classical PID controller.

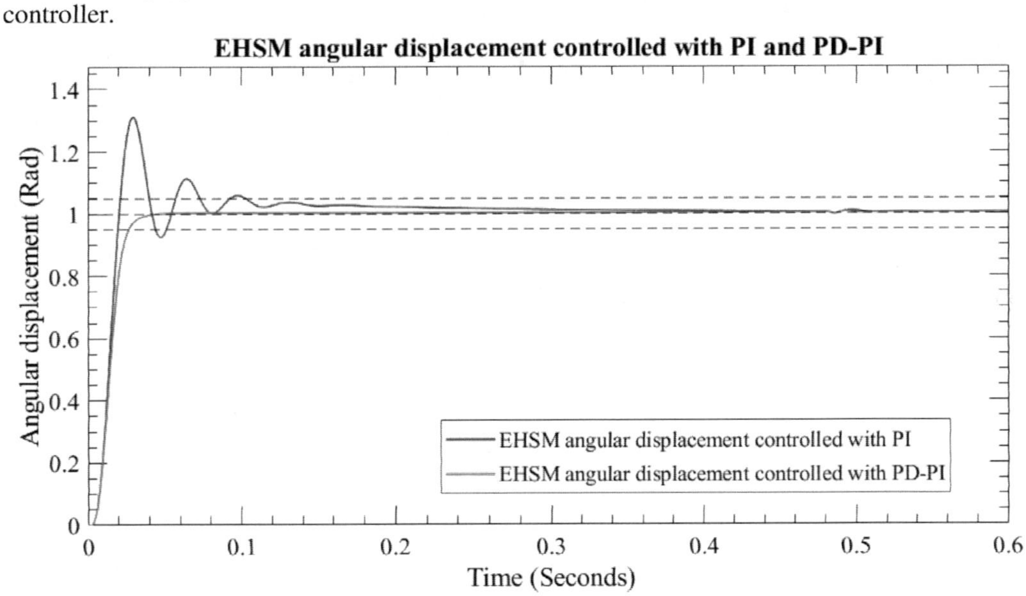

Figure 15. EHSM angular displacement controlled with PID and PD-PI.

5. Validation

The following figure shows that the more pressure you apply, the higher slope you will get. This can be explained by the nonlinearity of the model. As the model pressure is increased, the model becomes more sensitive to the input signal which increases the velocity with a small value of current.

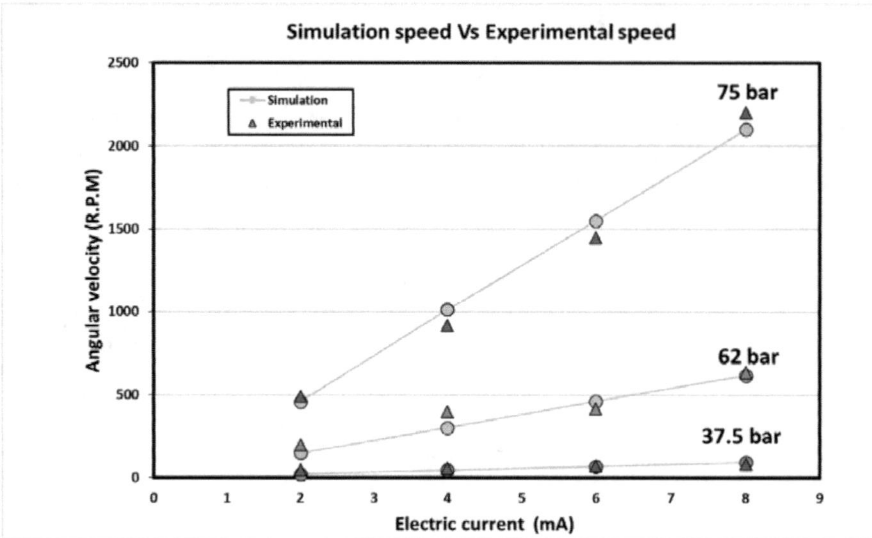

Figure. 17 Experimental and simulation results of the EHSM for different supply pressures and input current.

6. Conclusion

A complete modeling for a typical bent axis electro-hydraulic servo motor was done. MATLAB/ Simulink program was used to achieve the model. A new generation for the classical PID controller (PD-PI) was applied to the model. The model was validated through some expremental work.

Nomenclature

B	Bulk modulus of oil	=	1.5e9	(Pa)
c	Radial clearance			(m)
d_5	Diameter of return orifice N5 d5(m)	=	0.0006	(m)
d_f	flapper diameter	=	0.0005	(m)
d_{fn}	Hydraulic amplifier nozzles N1 & N2 diameter	=	0.0005	(m)
d_s	Spool diameter	=	0.0046	(m)
F_s	Spool Damping coefficient	=	2	(Ns/m)
f_{th}	Armature damping coefficient	=	0.002	(Nms/rad)
J	Moment of inertia of the rotor	=	5e-7	(Nms^2)
K_i	Current gain	=	0.556	
K_{Lf}	Flapper seat equivalent stiffness	=	5e6	(N/m)
K_s	Feedback spring stiffness	=	900	(N/m)
K_T	Flexible tube rotational stiffness	=	10	(N.m/rad)
K_{th}	Armature rotational angle torque gain	=	9.45e-4	
L_f	Flapper length	=	0.009	(m)
L_s	Mechanical feedback spring length	=	0.03	(m)
M_s	Spool mass	=	0.02	(kg)
P_{c1} & P_{c2}	Hydraulic motor pressures			(Pa)
P_T	Return Pressure	=	0	(Pa)

Q1	Flow rate in the left orifice			(m3/s)
Q2	Flow rate in the right orifice			(m3/s)
Q3	Left flapper nozzle flow rate			(m3/s)
Q4	Right flapper nozzle flow rate			(m3/s)
Q5	Flapper valve drain flow rate			(m3/s)
R_f	Equivalent flapper seat material damping coefficient	=	5000	(Nsm)
V_3	Initial Volume of oil in the return chamber	=	5e-6	(m^3)
V_o	Initial volume of oil in spool side chamber	=	2e-6	(m^3)
Xi	Flapper limiting displacement	=	30e-6	(m)
π	Pi	=	3.141 59	
ρ	Oil density	=	867	(Kg/m^3)
ω	Ports width on the valve sleeve			(m)

References

[1] Merritt H *1967. Hydraulic Control Systems.* New York: Wiley.
[2] Rabie M *2009. Fluid Power Engineering*. New York: McGraw-Hill.
[3] Ogata K *2016. Modern Control Engineering.* (Delhi): Pearson.
[4] Ljung L *1999 Wiley Encyclopaedia of Electrical and Electronics Engineering* 29 3026
[5] Maneetham D and Afzulpurkar N 2010 Modeling, Simulation and Control of High Speed 4
[6] Kővári A 2015 Mathematical Model and Simulation of Electrohydraulic Servo Systems *Tudomány Hete a Dunaújvárosi Főiskolán* 2009.
[7] Santosh J, Mishra K et al 2017 Nonlinear Modeling And PID Control Through Experimental Characterization for an Electrohydraulic Actuation System *J. Brazilian Soc. Mech. Sci. Eng.* **39** 1177.
[8] Dorf R and Bishop R *2011 Modern control systems* 12th ed New Jersey: Pearson.
[9] Ling T, Rahmat M et al 2012 System Identification and Control of an Electro- Hydraulic Actuator System *2012 IEEE 8th International Colloquium on Signal Processing and its Applications.*
[10] Sen R, Pati C et al 2014 Comparison Between Three Tuning Methods of PID Control for High Precision Positioning Stage *J. Metrol. Soc. India* **30**
[11] Atherton D 2016 PI-PD, an Extension of Proportional-Integral-Derivative Control *Measurement and Control* **49** 161 (United Kingdom).
[12] Hassaan G 2014 Tuning of a PD-controller used with Second Order Processes *International Journal of Engineering Research & Technology (IJERT)***3** 120.
[13] Singer A, Hassaan G, et al 2020 Tuning Of a PD-PI Controller Used with a Third Order Process *International Journal of Engineering Research & Technology (IJERT)* **9** 6.
[14] Rabie M 1991 Dynamic Performance of Pilot operated Hydraulic Relief Valves *International Conference on Aerospace Sciences & Aviation Technolog* **4** 135 (cairo).
[15] Automatisés S, Fadel M, et al 2019 Modeling, Simulation and Control of a Fly-by-wire Flight Control System Using Classical PID and Modified PI-D Controllers **52** 267.
[16] Sallam A, Rabie M et al 2009 Dynamic Behaviour of a Hydraulic Motor Controlled by a Poppet Type Directional Control Valve International Conference on *Aerospace Sciences & Aviation Technolog* **13** 1 (Cairo).
[17] Singer A, Hassaan G, et al 2020 Robustness of PI-PD Controller Used with Third Order Processes *International Journal of Engineering and Techniques* **6** 4.

IOP Conf. Series: Materials Science and Engineering 1172 (2021) 012039 doi:10.1088/1757-899X/1172/1/012039

Obstacle avoidance for multi-UAV path planning based on particle swarm optimization

E N Mobarez [1], A Sarhan [2], and M M Ashry [1]

[1]Department of Optoelectronics and Control,
[2]Aircraft electrical equipment's, Military Technical College, Cairo, Egypt.

E-mail: mmaashry@mt c.edu.eg

Abstract. Traffic Collision Avoidance System aims to help aircraft to avoid collision with any object or other aircraft .One of the functions of this system is that it avoids threatening UAV to collide, it also addresses each threat separately with the best collision avoidance and the best suitable horizontal separation with other aircraft in the optimal path. In this paper the flight path planning for UAVs was designed to avoid obstacles depending on how the particle swarm was improved. Optimization problems are improved by using swarm dynamics (evolutionary computational technology). This is by describing avoiding obstacles and adapt the path planning for UAVs. The concept of concurrent restructuring has been integrated into path planning to stay away from both static obstacles. This optimization technique designed to decrease processing time and the shortest route of the path planning.

1. Introduction

Multi-UAV is designed to perform various tasks in a specific environment [1], [2] and [3]. The working environment consists of different types of obstacles. UAVs can depart from their location to any destination points without pass through any obstacles. UAVs must efficiently return their tracks based on environmental change due to dynamic obstacles.

Particle Swarm Optimization is stochastic optimization based on the social behavior of some species to determine the required position in a swarm activity. It gathers local and global search techniques by balancing exploitation and exploration. In PSO, the workspace is initiate with stochastic particles in which each particle performs a possibility resolution to the problem. Each particle of the swarm randomly seeking in a multidimensional area. These particles update themselves with the optimum solutions of their experience as well as the social information collected from other particles. The optimum solutions of the particles are assess by the fitness function of their current position. The swarm modificate its location dynamically throughout the optimization operation to obtain the best solution. The velocity of each particle initiate randomly. The location of the Particles changed by updates their velocities. The velocity update is effected by three factors are particles current motion, particles previous experience and the influence of the whole swarm.

This paper is organized with an introduction in the first section. Second section represent the overview of Particle Swarm Optimization initialization parameters. Third section illustrate the effect of the PSO parameters, the Particles continuously update their positions and velocities, until they have accomplish the desired location. Each particle update its location with respect to its previous experience and experience of rest of the particles (Neighbors). Section four how to update the Fitness Value. Section

Content from this work may be used under the terms of the Creative Commons Attribution 3.0 licence. Any further distribution of this work must maintain attribution to the author(s) and the title of the work, journal citation and DOI.
Published under licence by IOP Publishing Ltd

five represent obstacle Avoidance for Multi-UAV path planning based on PSO. The summary is comes in the last section.

2. Particle Swarm Optimization initialization parameters.

Particles continuously update their positions and velocities, until they have accomplish the desired location. Each particle update its location with respect to its previous experience and experience of rest of the particles (Neighbors) [5].

Each individual particle is consist of three vectors:

- The X- record the particle current location in the search area.
- The P- (P-best) records the position of the superior settling found far via the particle.
- The V- Particle gradient (direction) if it is not disturbed

Each particle are kept as part and member of the group (swarm) during the operation in PSO

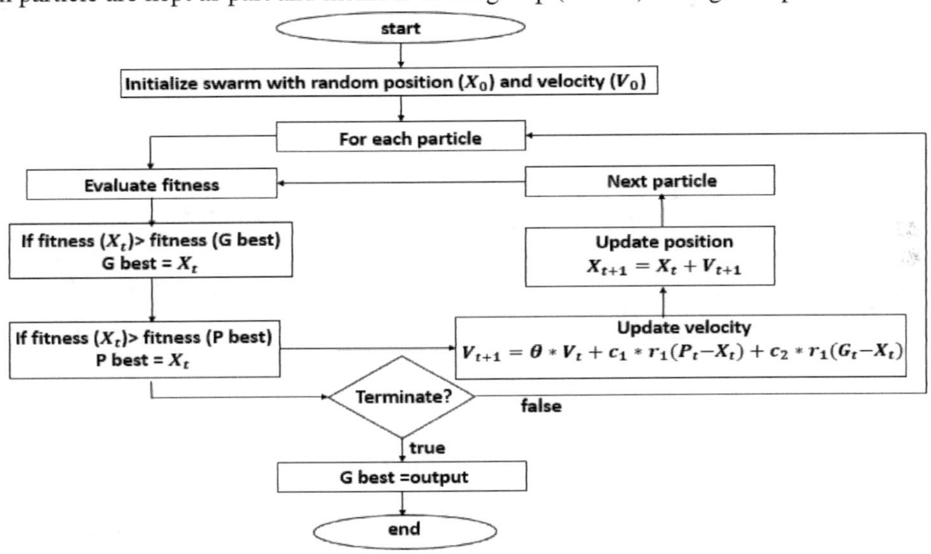

Figure 1. PSO Algorithm Schema [6].

Initialize location x_i and velocity $v_i$of n particles through the objective function $f(x), x = (x_1, \dots\dots, x_d)^T$[8]. And find G^* from min $\{f(x_1), \dots\dots, f(x_n)\}$ at $t = 0$, generate new velocity V_{t+1} using equation (2) and calculate new locations X_{t+1}, then evaluate the Objective function of a new locations X_{t+1} to find the present superior for each particle, and then find the current (global best) G^* as shown in figure (1) [5].

The main objective of PSO is to accelerate each particle to the superior position (P best) and the best global position (G best) obtained by any particle, with stochastic weighted (angles of particle direction), this is accomplished by addition of the V-vector to the X-vector to get another new X_{new} as in following equation (1). As shown in figure (2) [6].

$$X_{i\ t+1} = X_{i\ t} + V_{i\ t+1} * \Delta t \tag{1}$$

Then the new velocity is calculated by using the following equation (2):

Update velocity

$$V_{i\ t+1} = W * \underbrace{V_{i\ t}}_{} + \quad \underbrace{c_1 * r_1(P_{Best} - X_{i\ t})}_{} + \quad \underbrace{c_2 * r_2(G_{Best} - X_{i\ t})}_{} \tag{2}$$

(current motion) Inertia weight	(cognitive/Particle memory influence) particle best solution (Distance to the particle best)	(Social/swarm influence) Global best solution Distance to the global best-all particle best location

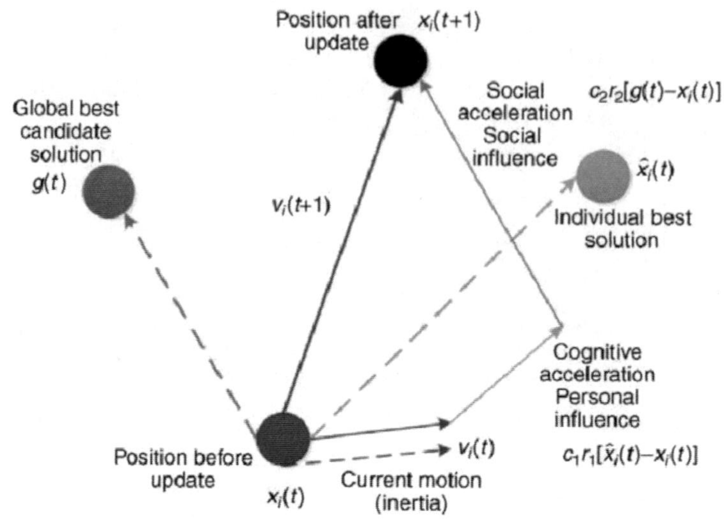

Figure 2. Particles motion.

3. The effect of the PSO parameters.

Where W an weight of inertia function (angle of particle travel) can be taken as a constant from 0.4 to 0.9 this to cause the particles converge more quickly i.e. Large inertia weight making easy for conduct global research, while small inertia facilities provide local search (controls exploits and explores of particles), $r_1$and r_2 are two stochastic vectors, and the values takes from 0:1 each entry. The parameters c_1 and c_2 are the learning parameters or Acceleration coefficients effect on the personal and global leaders on the seeking operation [5], which can be taken from 0 to 2. Kennedy and Eberhart suggest to use values of $c_1 = c_2 = 2$ also, Shi and Eberhart [5] proposed to use the value of 'W' from 0.8 to 1.4 as shown in table (1) [6].

The location of the Particles changed by updates their velocities. The velocity update is effected by three factors are particles current motion, particles previous experience and the influence of the whole swarm. Figures (3) and (4) represent the different factors that impact particles motion in PSO [6]. These figures shows the effectiveness of the updating parameters on the optimization mechanism of the particles. Figure (3) show the effect of parameter inertia weight W from [0.9: 0.4], when parameter W decreases. The exploration and exploitation process increases and converge looking for global best solution (hand) as shown in figure. Figure (4) show the effect of parameter Acceleration coefficients c_1 andc_2, the small cognition parameter c_1 and large social parameter c_2 making quickly to conduct and converge to global position. The Large cognition parameter c_1 and small social parameter c_2 making slower and not converge to global position, because no information between particles.

Table. (1) PSO parameters. [6]

W	inertia weight factor range from (0.4 : 1.4)	
c_1	cognition parameter (self-confidence range 1.5 : 2)	(learning parameters or acceleration constant)
c_2	social parameter (swarm confidence range 2 :2.5)	
r_1	independent random variables uniformly distributed (r_1 and r_2 are two stochastic vectors, and the values takes from 0:1 each entry)	
r_2		
P Best	is the personal best position achieved by the particle	
G Best	Is the globally best position achieved by the swarm	

Simulation of PSO [C1=2, C2= 2]

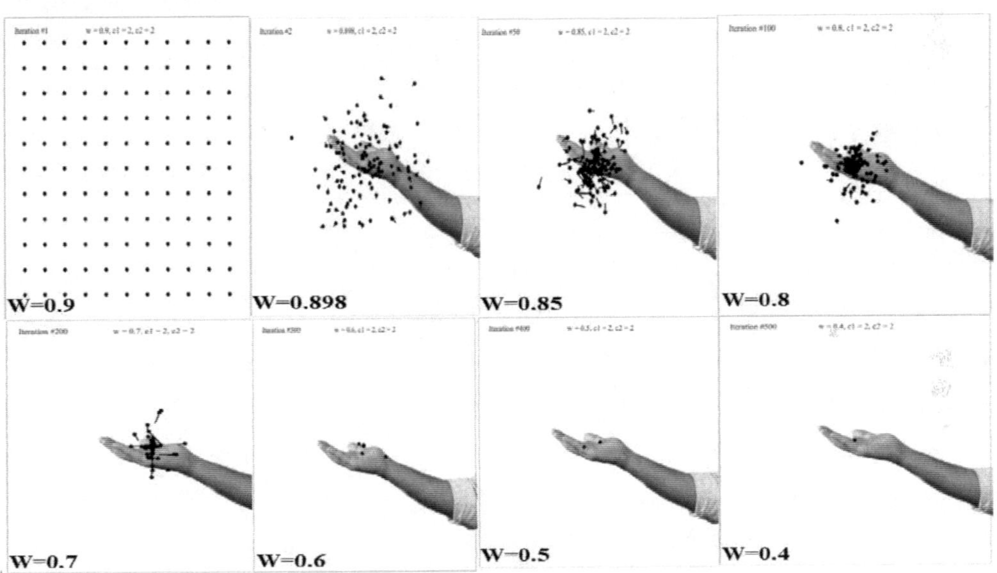

Figure 3. Effect of parameters inertia weight *W*.

Optimization is fulfilled utilize an objective function defined by the Euclidean distance between the beginning point (x1, y1) and the end point (x2, y2) as in equation (3).

$$\text{Distance} = \sqrt{(x_1 - x_2)^2 + (y_1 - y_2)^2} \quad (3)$$

In PSO, the algorithm assess the fitness function for all the particles and the solution is optimized by iterative enhancement of the particles positions [6].

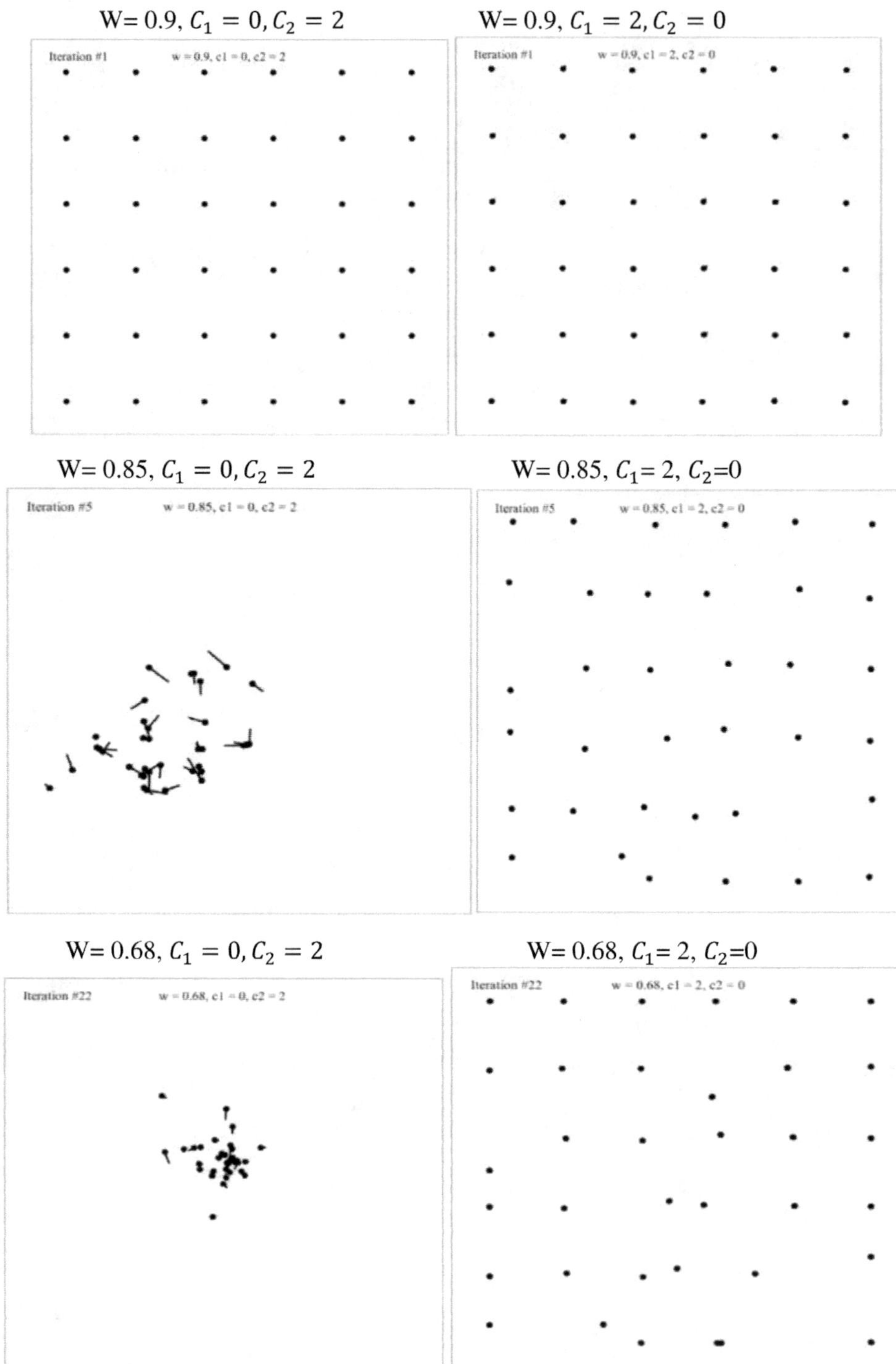

Figure 4. Effect of parameters cognition and social parameterC_1,C_2.

4. Update the Fitness Value.

In this section, the path planning tack place to avoid any kind of obstacles. At first, to keep a collision free course all the particles swarm must find out the obstacle's positions, then they determine the displacement between the particles and the obstacles [6]. A collision tack place if the condition of the equation (4) occurred:

$$D_{obs}^{a} \leq 0 \tag{4}$$

WhereD_{obs}^{a} describes the distance between obstacles and UAVs, then it considered as alert for collision. Here, the negative value of D_{obs}^{a} means that the location of the particles are inside the impact area. The particle swarm fitness values through the collision area is proposed by a larger positive functional value. As the optimization is considered as an issue reduction, the particles with larger fitness vales are not considered as the superior fitness function at all. So the trajectory within the impact area is averted.

So as to escape maneuver the particles from the impact area, the values of the parameters (ω, c1, c2) in equation (2) are increased. Thus, the particles velocities are also increased and they leave the collision area [6].

5. Path Planning by Obstacle Avoidance

A 2D surroundings is considered as a workspace. For obstacle avoidance, we considered various patterns of working surroundings. The map area of the surroundings is considered as a circle with different diameters of obstacles. The blue circles represent the static obstacles as shown in figure (5). Black line represents the trajectory (flight path) of UAVs. Yellow Square represent of initial position Green star represent of final position.

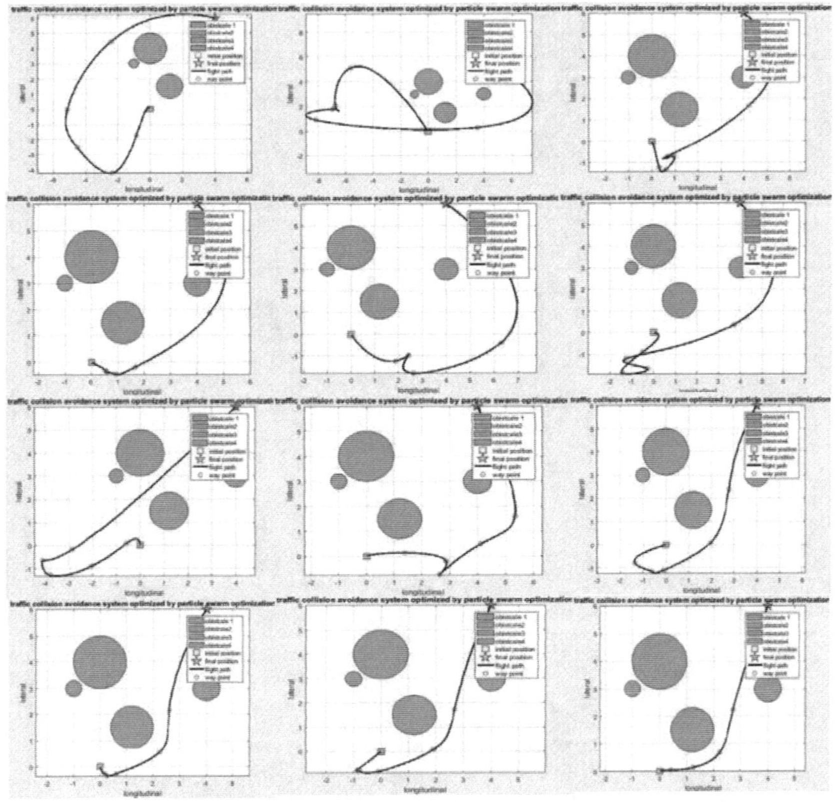

Figure 5. Sample of iteration steps to optimal solution.

Figure (5) shows that a brief or sample of the iteration where the particle swarm optimization looking for the optimal path from source (starting point) to the destination (end point) Including avoiding collisions with obstructions, the figure consist of a number of iteration starting from initial iteration upper-left one and it ends at lower-right one.

The figure (6) shows a relationship between the best cost and iteration steps where it was found that the cost was reduced with the aging of a number of iterations that got an optimal solution.

The cost of the final iterations for converging to the optimal path is shown in figure (6). This is a best cost function to achieve to the destination points without pass through any obstacles.

After getting optimal path from initial point to the destination, convert it to the heading angle from currently way point to next one as a reference command to the leader UAV as shown in figure (7).

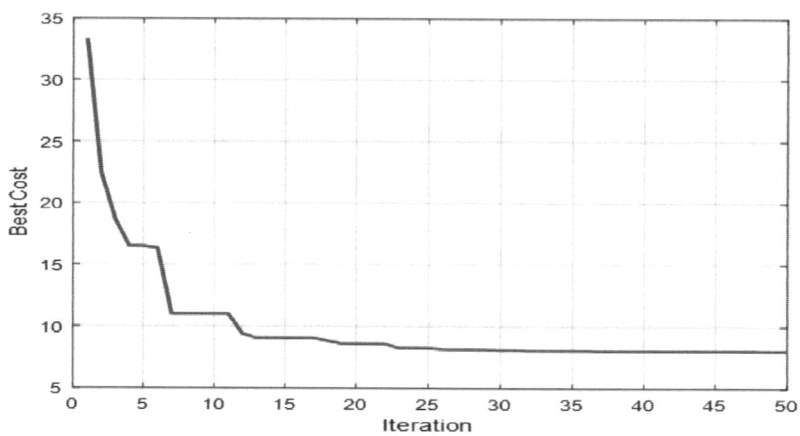

Figure 6. Best cost according to iterations.

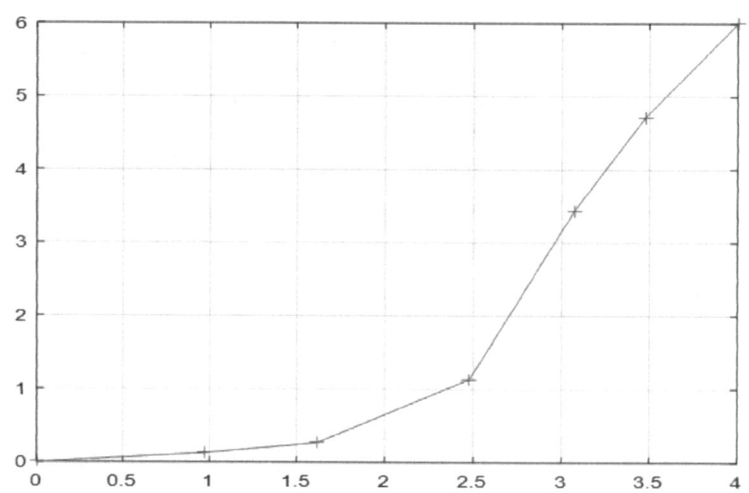

Figure 7. Way point on flight path.

In this section using the PSO to optimize the way points and best flight path to keep the UAVs in safe from obstacles through number of iteration for reach to best cost and optimum solution, and then convert the final flight path to heading angles, which using it as a reference command to leader UAV, so the followers UAV follow currently heading angle through leader UAV path as shown in figure (8).

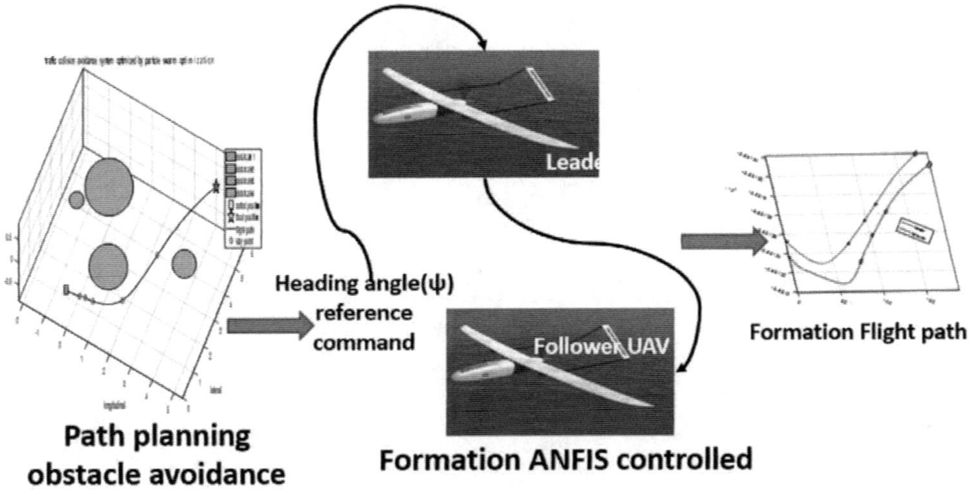

Figure 8. Traffic collision avoidance system.

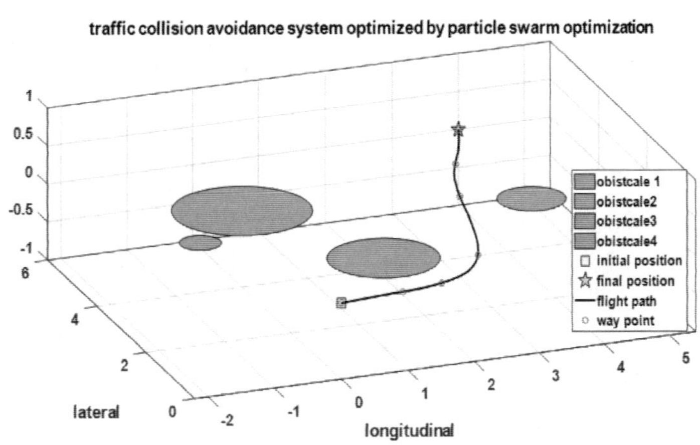

Figure 9. Optimal solution for UAV flight using PSO.

The figure (9) show the path planning of the final optimal solution to keep the UAV in safe from obstacles through number of iteration for reach to best cost and optimum solution.

The figure (10) illustrate that implementation of formation flight for two UAVs to the path planning, as shown in this figure the formation execute the heading according to the optimal path obtained from particle swarm optimization.

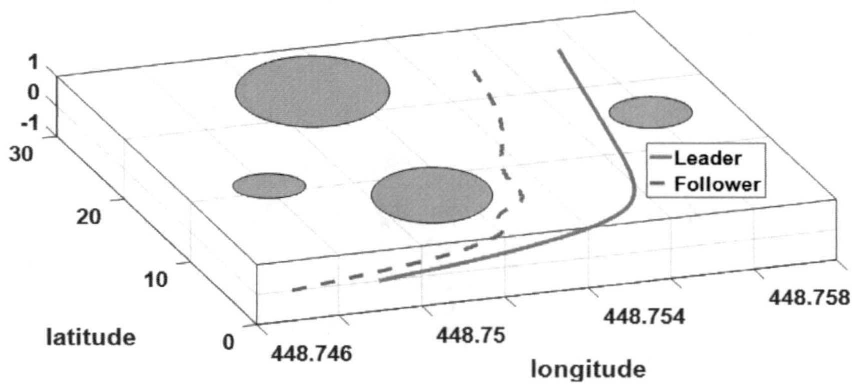

Figure 10. Path planning of multi-UAV.

6. Conclusion

In this paper the flight path planning for UAVs to avoid obstacles depending on how the particle swarm is designed. Optimization problems are improved by using swarm dynamics (evolutionary computational technology). This is by describing avoiding obstacles and adapt the path planning for UAVs. The concept of concurrent restructuring has been integrated into path planning to stay away from both static and dynamic obstacles. This optimization technique designed to decrease processing time and the shortest route of the path planning.

References

[1] Mobarez, E.N., Sarhan, A. and Mohamed, A.M., 2019, September. Modeling of fixed wing UAV and design of multivariable flight controller using PID tuned by local optimal control. In *IOP Conference Series: Materials Science and Engineering* (**Vol. 610, No**. 1, p. 012016). IOP Publishing.

[2] Mobarez, E.N., Sarhan, A. and Ashry, M.M., 2019, December. Classical and Intelligent Multivariable Controllers for Aerosonde UAV. In *2019 Ninth International Conference on Intelligent Computing and Information Systems (ICICIS)* (pp. 350-355). IEEE.

[3] Mobarez, E.N., Sarhan, A. and Ashry, M.M., 2019, December. Robust PID Flight Controller for Ultrastick-25e UAV. In *2019 15th International Computer Engineering Conference (ICENCO)* (pp. 150-156). IEEE.

[4] Mobarez, E.N., Sarhan, A. and Ashry, M.M., 2019, December. Formation Flight of Fixed Wing UAV Based on Adaptive Neuro Fuzzy Inference System. In *2019 Ninth International Conference on Intelligent Computing and Information Systems (ICICIS)* (pp. 356-361). IEEE.

[5] Wang, D., Wang, H. and Liu, L., 2016. Unknown environment exploration of multi-robot system with the FORDPSO. *Swarm and Evolutionary Computation*, ***26***, pp.157-174.

[6] Zhang, Y., Wang, S. and Ji, G., 2015. A comprehensive survey on particle swarm optimization algorithm and its applications. *Mathematical Problems in Engineering*, *2015*.

[7] Sarhan, A. and Ashry, M., 2013. Self-tuned PID controller for the Aerosonde UAV autopilot. *International Journal of Engineering Research & Technology (IJERT)*, **2(12),** pp.2278-0181.

IOP Conf. Series: Materials Science and Engineering 1172 (2021) 012040 doi:10.1088/1757-899X/1172/1/012040

PID tuning approaches for quadrotors unmanned aerial vehicles

M Mahfouz [1], A Taiomour [2], M M Ashry [1] and G Elnashar [1]

[1]Department of Optoelectronics and Control, Military Technical College,
[2]Aircraft Armament Department, Military Technical College,

E-mail: m.mahfouz.trc@gmail.com,

Abstract. In this paper, a comparative study between different PID tunning techniques is presented. The proposed techniques are applied to solve the formation configuration problem for a cooperative team of unmanned vehicles. The formation problem for the cooperative team is divided into two levels of control, one is the backstepping control technique for the stabilization of the team members positions as a higher controller. Simultaneously, PID controller receives the desired position to stabilize the attitude control as a lower controller to track the desired planning trajectories. The main contribution of this paper is the comparison between the different control approaches in tunning the PID gains to stabilize attitude control for the leader quadrotor. Simulation results present the assessment of the proposed PID control technique compared with different PID tuning approaches such as local optimal control, fraction order, Ziegler–Nichols and genetic algorithm. Moreover, disturbance rejection and white noise attenuation criterions are inspected to evaluate the ability of the proposed controllers to preserve the stability of the system.

1. Introduction

In past few decades, unmanned aerial vehicles (UAVs) have become much popular and attracted more interest from scientist and researchers due to their potential use in military and civilian utilizations [1]. Nowadays, the vertical take-off and landing (VTOL) with four-rotors aircraft called quadrotor is considered as one of the most important UAVs. This sort of airplanes is characterized with unrevealed useful features such as small size, ultimate maneuverability capabilities, simplicity of design, and control. Quadrotor has the footprint of taking-off vertically from any place and landing vertically at any place regardless the hardness of that places. Quadrotors UAVs have the advantage of being able to accomplish specific and difficult tasks that cannot be done by fixed-wing aircrafts due to its unique characteristics.

There are various space applications where this capacity would be desired, for example, discharging raised territories and transporting merchandise to regions where street arrival isn't accessible. VTOL planes also provide direct entry to buildings or zones making them a quick type of transportation between regions, particularly those which are raised or have restricted admittance [2].

Conventionally, quadrotors usage is increased rapidly in the aerospace industry. It is fact that most of the usual payloads could be lifted using one or two rotors. However, Quadrotors have unique features that make it appealing. Naturally, one of these features is the higher and diverse capability for payload. The plainness of their control system is also an advantage; just by modifying the speed of every rotor separately.

Content from this work may be used under the terms of the Creative Commons Attribution 3.0 licence. Any further distribution of this work must maintain attribution to the author(s) and the title of the work, journal citation and DOI.
Published under licence by IOP Publishing Ltd

The quadrotor is a vastly nonlinear, multi-variables, strongly coupled, under-actuated, sensitive to interference and essentially an unstable system (6 DOF, 3 translational and 3 rotational motions, with only 4 actuators), which acts as fundamental foundation for design of control plan. This makes control design is very challenging. currently, there are numerous methods to control the quadrotor, the literature [3], [4], [5], [6], and [7] respectively focused on PID, sliding mode, backstepping, fuzzy neural network, and H-infinity approaches. The adaptive neuro-fuzzy inference system (ANFIS) controller to the quadrotors is discussed in [8] and [9].

It is believed that multiple cooperative quadrotors are more stable and adaptable to fulfill the desired demands than single quadrotor [10]. Unmistakably, the general ability of the cooperative quadrotors can upgrade execution, essential time range, task allocation, system proficiency and the safety to accomplish the desired tasks. Complex tasks cannot be achieved efficiently by a single quadrotor, such as finding objects in massive areas like in searching and relief missions, basic payloads restrictions, etc. Because of these regards, collaboration of numerous UAVs has recently gotten a lot of consideration from the scientist and researchers [11]. Quick advancements on computers, actuators, sensors fusion, and communications technologies have all an impact in developing the novel epoch of collaboration methodologies for cooperative quadrotors.

State of art of the proposed controller designed based on combining couple of robust controllers that can deal with highly non-linear systems effectively [10]. The proposed controller is divided into couple of parts working together. Backstepping (BS) controller is used to stabilize the position control as a higher controller. Simultaneously, PID controller receives the desired position to stabilize the attitude control as a lower controller to track the desired planning trajectories.

The main contribution of this paper is using Local Optimal Control (LOC), and Fractional Order PID (FOPID) approach combined with backstepping controller in the proposed cascaded controller for first time in cooperative quadrotors. LOC and FOPID are used to tune PID gains to stabilize attitude control for cooperative quadrotors.

BS controller is used mostly as a higher controller [12]. BS controller is an effective control strategy for extremely nonlinear frameworks [13]. There are many benefits of utilizing BS controller such as rapid convergence rate, stability, capability to manage nonlinear frameworks using recursive operation of Lyapunov function, robustness regardless presence of uncertainties, and simplicity [5].

Multivariate control systems were given much attention because of the growing complexity of control systems operations [14].

Numerous researchers are broadly utilizing fractional order control (FOC) to accomplish better robust performance in several control systems. FOC approach has accomplished further remarkable outcomes on quadrotors particularly enhancing the stability of its frameworks.

The superiority of fractional order can be demonstrated in the ability of fractional calculus to configure the gain margin and the phase margin in the frequency domain of the quadrotor framework freely. Fractional order PID (FOPID) is developed based on FOC theory and traditional PID theory. This controller expands the resolution of tuning parameters from the triple parameters of PID,$(k_P, k_I, K_D)$and to five parameters by adding the FOC parameters λ and μ.

LOC is convenient for controlling non-linear systems with multivariable parameters as Single Input Single Output (SISO) systems [15]. LOC is characterized as well with robust performance. The proposed BS-PID controller enhanced the overall performance and stability of the underlying cooperative quadrotors system.

This paper is organized as follows:

Section 2 presents a short description and mathematical model of the quadrotor. The controller's approach is described in section 3. Section 4 shows the simulation results of trajectory tracking using the proposed controllers. Finally, conclusion is introduced at final section in section 5.

2. Quadrotor control approaches

2.1. Backstepping control design

BS approach design has been formerly studied in [13]. BS approach is a preferred control technique over various approaches. The controller utilizes a recursive algorithm, where the controller has divided its mechanism into several steps guaranteeing the stability of each step gradually. The inputs of BS controller are directly received as a desired mission plan and a streaming of information sent by the onboard quadrotor sensors. BS utilizes numerous coefficients during the computation. These coefficients represent the quadrotor dynamics and states. The yield of the controller approach is the separately sent pulse width modulation (PWM) signal for every rotor of the quadruple rotors.

The proposed controller utilizes BS in the higher control loop of positioning the underlying framework, so, BS is responsible of defining the updated positions of every quadrotors of the cooperative squad. The altitude control and the position control for i^{th} quadrotors (U_{1i}, u_{xi}, u_{yi}) is assumed as:

$$\begin{aligned} U_{1i} &= \frac{m_i}{\cos x_{1i} \cos x_{3i}}(g_i + g_{7i} - \alpha_{7i}(g_{8i} + \alpha_{7i} g_{7i}) - \alpha_{8i} g_{8i}) \\ u_{xi} &= \frac{m_i}{U_{1i}}(g_{9i} - \alpha_{9i}(g_{10i} + \alpha_{9i} g_{9i}) - \alpha_{10i} g_{10i}) \\ u_{yi} &= \frac{m_i}{U_{1i}}(g_{11i} - \alpha_{11i}(g_{12i} + \alpha_{11i} g_{11i}) - \alpha_{12i} g_{12i}) \end{aligned} \quad (1)$$

2.2. PID Control Design

PID controllers is considered as an ordinary controller. PID is generally used because of design simplicity and reduced number of tunable parameters in the production controlled systems [16], and [17]. Upwards of 90 percent of the commonly deployed controllers are routinely based on PID controllers. PID controller is designed in the proposed controller as a lower controller to control the attitude channels of each member of cooperative quadrotors. The popular mathematical equation of PID controller is formed as [18]:

$$u(t) = K_P + K_I \int_0^t e(t)dt + K_D \frac{d}{dt} e(t) \quad (2)$$

A scheme block diagram for PID controller is depicted in Figure 1.

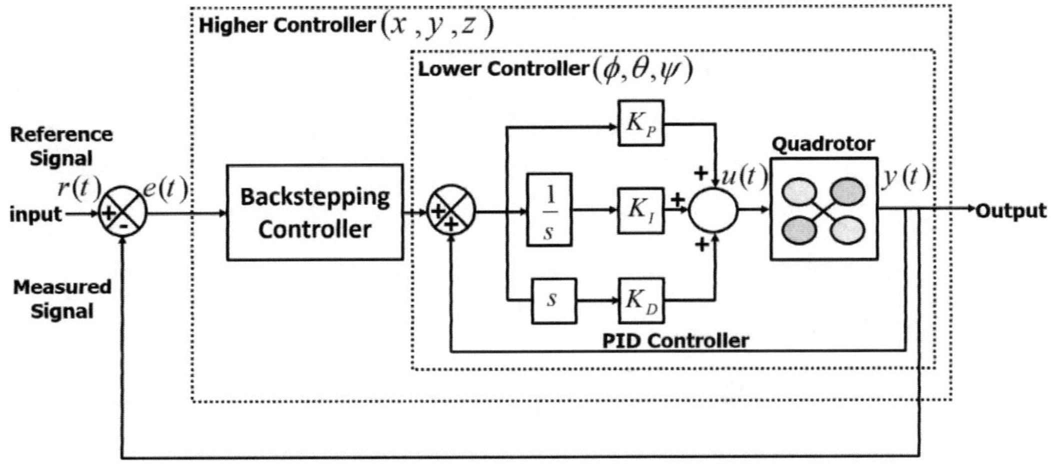

Figure 1. A scheme block diagram for proposed PID controller.

The quadrotor model should be reduced in order to make a simple reverse model to be invoked in the control techniques. That can be done by removing gyroscopic terms from the dynamic equations [19]. So, the simplified mathematical equation can be reformulated as:

$$\begin{pmatrix} \ddot{x}_{1i} \\ \ddot{x}_{3i} \\ \ddot{x}_{5i} \end{pmatrix} = \begin{pmatrix} \ddot{\phi}_i \\ \ddot{\theta}_i \\ \ddot{\psi}_i \end{pmatrix} = \begin{pmatrix} \frac{l_i}{I_{xxi}} U_{2i} \\ \frac{l_i}{I_{yyi}} U_{3i} \\ \frac{1}{I_{zzi}} U_{4i} \end{pmatrix} \tag{3}$$

Converting (3) to Laplace model and adding the rotor dynamics yields:

$$\begin{cases} \phi_i(s) = \dfrac{B_i^2 b_i l_i}{s^2 (s+A_i)^2 I_{xxi}} \left(u_{4i}^2(s) - u_{2i}^2(s) \right) \\ \theta_i(s) = \dfrac{B_i^2 b_i l_i}{s^2 (s+A_i)^2 I_{yyi}} \left(u_{3i}^2(s) - u_{1i}^2(s) \right) \\ \psi_i(s) = \dfrac{A_i^2}{s^2 (s+A_i)^2 I_{zzi}} \sum_{n=1}^{4} (-1)^{n+1} u_{ni}^2(s) \end{cases} \tag{4}$$

where, the parameters A_i and B_i symbolize the rotor dynamics.
Inserting the control inputs instead of the motor inputs in (4) yields:

$$\begin{cases} \phi_i(s) = \dfrac{A_i^2 l_i}{s^2 (s+A_i)^2 I_{xxi}} U_{2i} \\ \theta_i(s) = \dfrac{A_i^2 l_i}{s^2 (s+A_i)^2 I_{yyi}} U_{3i} \\ \psi_i(s) = \dfrac{A_i^2}{s^2 (s+A_i)^2 I_{zzi}} U_{4i} \end{cases} \tag{5}$$

Assume that the required angles are $x_{2i} = \dot{\phi}_i, x_{4i} = \dot{\theta}_i, x_{6i} = \dot{\psi}_i$, and $x_{1id} = x_{3id} = x_{5id} = 0$ if the state is hovering. So, the controller formulas can lastly be acquired as:

$$\begin{cases} U_{2i} = \dfrac{1}{b_{1i}} (K_{2i} x_{2i} - a_{1i} x_{4i} x_{6i}) + w_{2i} (x_{1id} - x_{1i}) \\ U_{3i} = \dfrac{1}{b_{2i}} (K_{3i} x_{4i} - a_{3i} x_{2i} x_{6i}) + w_{3i} (x_{3id} - x_{3i}) \\ U_{4i} = \dfrac{1}{b_{3i}} (K_{4i} x_{6i} - a_{5i} x_{4i} x_{6i}) + w_{4i} (x_{5id} - x_{5i}) \end{cases} \tag{6}$$

Various tuning approaches for PD and PID controllers are utilized to improve the overall performance and stability of the underlying cooperative quadrotors frameworks.

These tuning approaches are characterized with implementation simplicity and short execution time to exert the controller coefficients as inputs for the actuators.

2.2.1. *Tuning PID Controller via Genetic Algorithm*

GA tuning approach is characterized with its feature to be easily and promptly executed. Choice the proper estimations of the controller parameters still an essential challenge in designing the proposed PID controllers $(K_{2i},w_{2i}),(K_{3i},w_{3i}),(K_{4i},w_{4i})$. However, creating the population based on choosing reasonable GA chromosomes is the first and most critical stage. Every chromosome involves the triple PID parameters K_P, K_I, K_D with esteem limits differed rely upon the utilized objective functions and the delay.

At first, Simplex Optimization (SO) approach is utilized to obtain the ideal estimations of these controller parameters as rudimentary satisfactory parameters for GA. Then, GA acquire a fine tuning of the proposed PID parameters based on hypothetical evolutionary technique that parallels the real biology in the world. The parents' choice and genes conjunction to produce post-generation (crossover) is used as a major GA procedure. The genetic algorithm is designed to lessen multi-objective operation. This operation decreases the mean square error (MSE) amongst inputs and outputs of the system [18]. The resultant values for control parameters generated by GA are revealed in Table 1.

Table 1. **Genetic Algorithm** Controller **parameters.**

Population size	Crossover	Number of generations
200	0.8	800

The controller design strategy is devoted to the purpose of performance optimization of a specific system or subsystem [20]. The purpose of the proposed optimization technique is to achieve the stability efficiency for the cooperative quadrotors, speed up the response, and reducing overshoots as least as could be expected.

A PD controller is developed as a lower control for the underlying cooperative quadrotors for attitude and altitude channels as in (7):

$$U_{2i,3i,4i} = K_{pi_{\phi_i,\theta_i,\psi_i}}(\phi_i,\theta_i,\psi_i) + K_{Di_{\phi_i,\theta_i,\psi_i}}(\dot{\phi}_i,\dot{\theta}_i,\dot{\psi}_i) \tag{7}$$

The forthcoming control formula is utilized to obtain the required conduct of the altitude channel for the vertical location:

$$u_{1i} = \frac{r_{1i} + m_i g}{\cos\phi_i \cos\theta_i} \tag{8}$$

$$r_{1i} = -a_{1i}\dot{z}_i - a_{2i}(z_i - z_{id}) \tag{9}$$

The required altitude z_d can be determined by:

$$z_{id} = \frac{-a_{1i}\dot{z}_i - a_{2i}z_i - r_{1i}}{a_{2i}} \tag{10}$$

Where a_{1i} and a_{2i} symbolize K_{Di} and K_{pi} respectively. Considering the uncertainties and hardware constraints; the controller coefficients has been determined as demonstrated in Table 2:

Table 2. **PD controller** coefficients **tuned by GA.**

Controller Gain	Roll	Pitch	Yaw	Altitude
K_{Pi}	-14.6211	-46.6211	-64	4.7673
K_{Di}	48.7832	112.7832	42.2144	4.9794

Verifying the proposed control rule is achieved by many simulation experiments. Even with extreme initial states in the attitude channels, the controller completed multiple simulation situations successfully.

2.2.2. *Tuning PID Controller via Ziegler-Nichols*

Ziegler-Nichols approach is one of the most popular techniques in tuning PID parameters in less time [21]. Ziegler and Nichols created rules for configuring the PID gains. There are couple of techniques to tune the PID parameters using ZN. One of them dependent on open-loop technique. However, the other technique dependent on closed-loop technique. Closed-loop technique is utilized, in this paper, to regulate the gains of the PD parameters. Table 3. summarized the PD controller coefficients tuned by ZN as follow.

Table 3. **PD controller** coefficients **tuned by ZN.**

Controller Gain	Roll	Pitch	Yaw	Altitude
K_{Pi}	44.32	-50.62	42.21	21.69
K_{Di}	-9.787	52.78	-64	15.72

2.2.3. *Tuning PID Controller via Fractional Order*

The BS-PID tuned by FOPID controller is divided into couple of parts working together. Backstepping controller is used to stabilize the position control as a higher controller. Simultaneously, PID tuned by FOPID controller receives the desired position to stabilize the attitude control as a lower controller to track the desired planning trajectories as depicted in Figure 2.

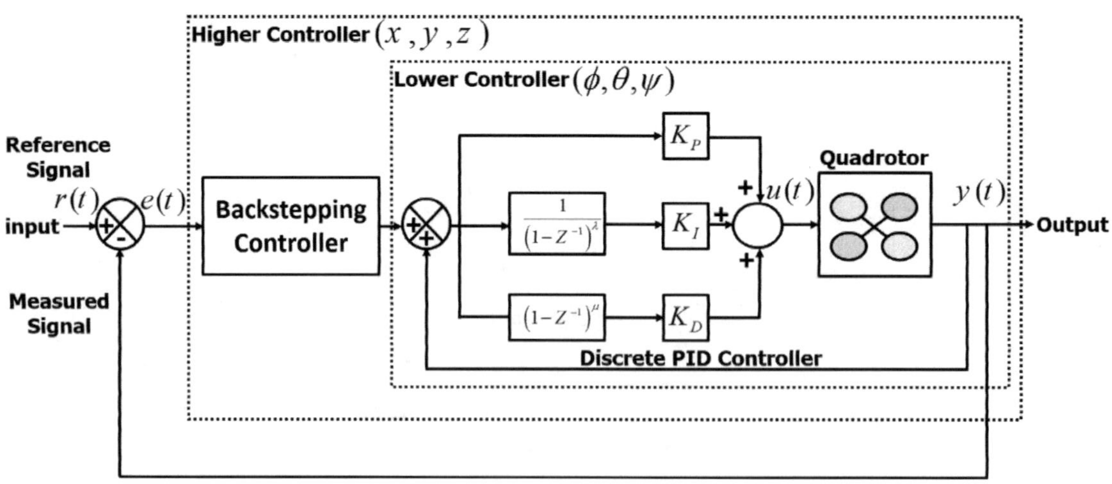

Figure 2. FOPID block diagram for 2nd order model of quadrotor framework

The general form of digital PID controller for i^{th} quadrotor can be defined as [15]:

$$G_i(z) = \frac{u_i(z)}{e_i(z)} = K_{Pi} + \frac{K_{Ii}}{1-z^{-1}} + K_{Di}(1-z^{-1}) \tag{11}$$

By multiplying equation (11) with $(1-z^{-1})$:

$$(1-z^{-1})u_i(z)=K_{Pi}e_i(z)(1-z^{-1})+K_{Ii}e_i(z)+K_{Di}e_i(z)(1-z^{-1})^2 \quad (12)$$

$$u_i(z)-u_i(z)z^{-1}=K_{Pi}(e_i(z)-e_i(z)z^{-1})+K_{Ii}e_i(z)+K_{Di}(e_i(z)-2e_i(z)^{-1}-e_i(z)z^{-2}) \quad (13)$$

Suppose that $u\left(nT\right)$ as a discrete time, and use it in (13):

$$u_i(n)-u_i(n-1)=K_{Pi}\left(e_i(n)-(e_i(n-1)\right)+K_{Ii}e_i(n)+K_{Di}\left(e_i(n)-2e_i(n-1)+e_n(n-2)\right) \quad (14)$$

So, the digital form of fractional order $PI^{\lambda}D^{\mu}$ is formulated by:

$$G_i(z)=\frac{u_i(z)}{e_i(z)}=K_{Pi}+K_{Ii}\frac{(1+z^{-1})}{(1-z^{-1})}\sum_{m=0}^{\infty}f_{mi}(1-\lambda_i)z^{-m}+K_{Di}\sum_{m=0}^{\infty}f_{mi}(\mu_i)z^{-m} \quad (15)$$

The digital form of fractional order $PI^{\lambda}D^{\mu}$ is reformulated for R^{th} order by:

$$G_i(z)=\frac{u_i(z)}{e_i(z)}=K_{Pi}+K_{Ii}\frac{(1+z^{-1})}{(1-z^{-1})}\sum_{m=0}^{R}f_{mi}(1-\lambda_i)z^{-m}+K_{Di}\sum_{m=0}^{R}f_{mi}(\mu_i)z^{-m} \quad (16)$$

The estimation of R order of the system should be chosen carefully respecting the utilized hardware limitation to satisfy the desired requirement without wasting memory in unnecessary bigger calculations with increasing of the supposed order [22].

The general form of fractional order $PI^{\lambda}D^{\mu}$ for i^{th} quadrotor is formulated by:

$$G_i(z)=\frac{u_i(z)}{e_i(z)}=K_{Pi}+\frac{K_{Ii}}{S^{\lambda_i}}+K_{Di}S^{\mu_i} \quad (17)$$

where the tunable parameters of FOC λ_i and $\mu_i \geq 0$.

The time domain differential form of fractional order $PI^{\lambda}D^{\mu}$ can be formulated as [23]:

$$u_i(t)=K_{Pi}e_i(t)+K_{Ii}D_t^{-\lambda_i}e_i(t)+K_{Di}D_t^{-\mu_i}e_i(t) \quad (18)$$

The block diagram of fractional order $PI^{\lambda}D^{\mu}$ can be shown in Figure 3.

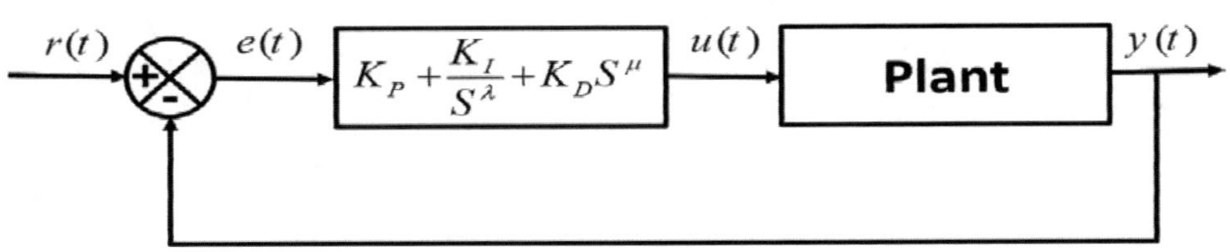

Figure 3. The block diagram of fractional order $PI^{\lambda}D^{\mu}$

The PID approach coefficients tuned by FOPID are summarized in Table 4.

Table 4. PID controller coefficients tuned by FOPID

Controller Gain	Roll	Pitch	Yaw
K_{Pi}	27.32	7.429	24.24
K_{Ii}	4.41	1.126	11.56
K_{Di}	0.73	2.85	15.86
λ_i	0.06	-0.72	0.15
μ_i	0.62	0.97	0.75

2.2.4. Tuning PID Controller via Local Optimal Control

LOC is used to tune PID gains in the planned cascaded controller to stabilize attitude control for cooperative quadrotors.

Figure 3 depicts a block diagram for the underlying platform using digital PID controller.

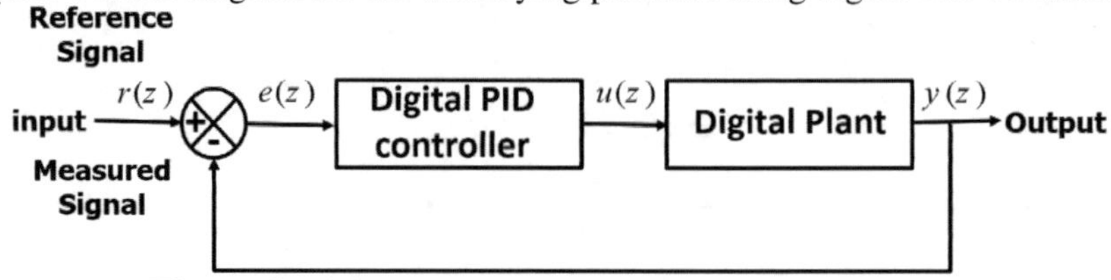

Figure 4. Digital PID controller schematic diagram.

The error in relation to the controller inputs and outputs can be defined as:

$$e_i(z) = r_i(z) - y_i(z) \tag{19}$$

The general form of digital PID approach is definable as [15]:

$$G_i(z) = \frac{u_i(z)}{e_i(z)} = K_{Pi} + \frac{K_{Ii}}{1-z^{-1}} + K_{Di}(1-z^{-1}) \tag{20}$$

$\delta u_i(n)$ is acquired by as:

$$\delta u_i(n) = u_i(n) - u_i(n-1) \tag{21}$$

From (20; $\delta u_i(n)$ is obtained by:

$$\delta u_i(n) = \left(K_{Pi} + K_{Ii} + K_{Di}\right)e_i(n) - \left(K_{Pi} + 2K_{Di}\right)e_i(n-1) + K_{Di}e_i(n-2) \tag{22}$$

The transfer function to regulate digital PID approach coefficients utilizing LOC is formed by:

$$G_i(z) = \frac{y_i(z)}{u_i(z)} = \frac{\beta_i z}{z^2 + \alpha_{1i} z + \alpha_{2i}} \tag{23}$$

α_{1i} , α_{2i} and β_i are the coefficients at the second order form of the linearized transfer function. So, the system model can be represented as:

$$y_i(n) = \alpha_{1i} y_i(n-1) + \alpha_{2i} y_i(n-2) + \beta_i u_i(n-1) \tag{24}$$

Figure 4. depicts LOC block diagram.

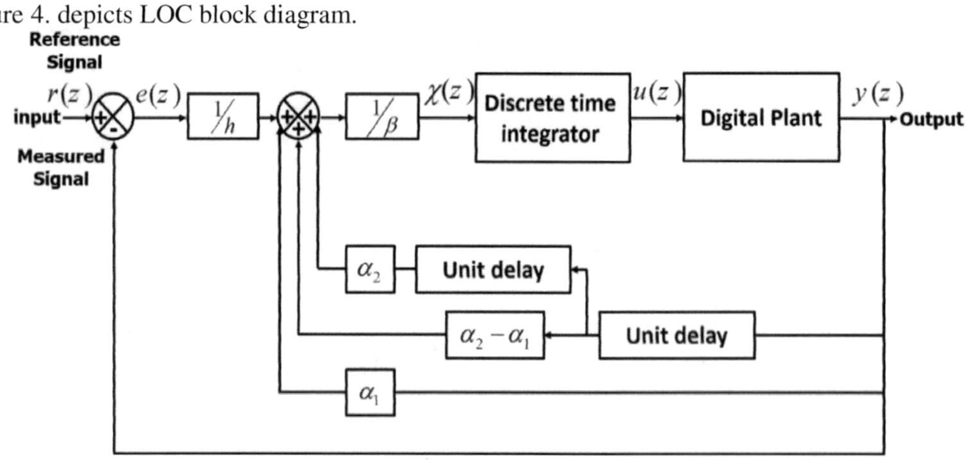

Figure 5. **LOC block diagram for 2nd order model.**

The error is obtained by:

$$e_i(n) = r_i(n) - y_i(n) \tag{25}$$

$\chi_i(n)$ can be obtained by:

$$\chi_i(n) = \frac{1}{b_i}\left[\frac{1}{h_i}e_i(n) + \alpha_{1i}y_i(n) + (\alpha_{2i} - \alpha_{1i})y_i(n-1) + \alpha_{2i}y_i(n-2)\right] \tag{26}$$

In step input case represented by $r_i(n)$:

$$r_i(n) = r_i(n-1) = r_i(n-2) \tag{27}$$

So, $\chi_i(n)$ can be obtained by:

$$\chi_i(n) = \frac{1}{\beta_i}\left[\frac{1}{h_i}e_i(n) + \alpha_{1i}y(n) + (\alpha_{2i} - \alpha_{1i})y_i(n-1) + \alpha_{2i}y_i(n-2) + \alpha_{1i}r_i(n-1) - \alpha_{1i}r_i(n)\alpha_{2i}r(n-2) - \alpha_{2i}r_i(n-1)\right] \tag{28}$$

From (26) and (27):

$$\chi_i(n) = \frac{1}{\beta_i}\left[\frac{1}{h_i}e_i(n) + \alpha_{1i}e(n) - (\alpha_{2i} - \alpha_{1i})e_i(n-1) + \alpha_{2i}e_i(n-2)\right] \tag{29}$$

Assuming τ_{si} as a time sample; $u_i(z)$ can be obtained by:

$$u_i(z) = \frac{\tau_{si}}{1 - z^{-1}} \tag{30}$$

Assuming that:

$$\delta u_i(n) = \tau_{si}\chi_i(n) \tag{31}$$

Substituting (31) in (29) yields:

$$\delta u_i(i) = \frac{\tau_{si}}{\beta_i}\left[\frac{1}{h_i}e_i(n) + \alpha_{1i}e_i(n) - (\alpha_{2i} - \alpha_{1i})e_i(n-1) + \alpha_{2i}e_i(n-2)\right] \tag{32}$$

By equalizing coefficients $e_i(n)$, $e_i(n-1)$, $e_i(n-2)$ with (22) and (32) yields:

$$K_{Pi} + K_{Ii} + K_{Di} = \frac{\tau_{si}}{\beta_i h_i} - \frac{\alpha_{1i}\tau_{si}}{\beta_i} \tag{33}$$

$$K_{Pi} + 2K_{Di} = \frac{(\alpha_{2i} - \alpha_{1i})\tau_{si}}{\beta_i} \tag{34}$$

$$K_{Di} = \frac{\alpha_{2i}\tau_{si}}{\beta_i} \tag{35}$$

From (33), (34), and (35); LOC coefficients for the digital PID can be obtained as:

$$K_{Pi} = \frac{-\tau_{si}(\alpha_{1i} + \alpha_{2i})}{\beta_i} \tag{36}$$

$$K_{Ii} = \frac{\tau_{si}}{\beta_i h_i} \tag{37}$$

$$K_{Di} = \frac{\alpha_{2i}\tau_{si}}{\beta_i} \tag{38}$$

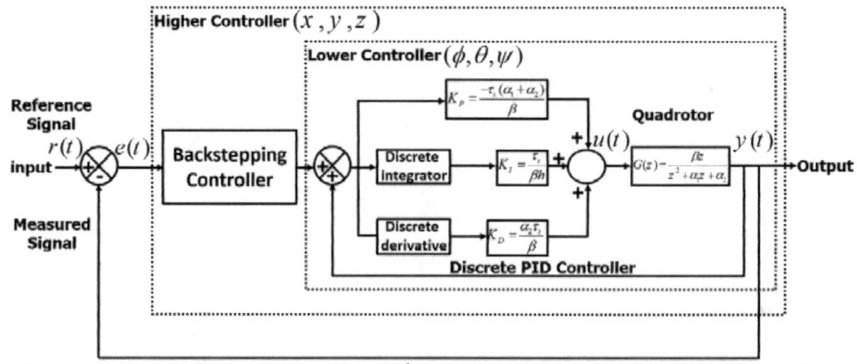

Figure 6. LOC block diagram for 2nd order model of quadrotor framework

PID coefficients tuned by LOC are dependent upon controller coefficients values and optimizable coefficient h_i. Figure 5. depicts PID approach coefficients of (36), (37), and (38) that utilized to optimize the 2nd order transfer function in (23).

From (11), the transfer function of Roll control is formed as:

$$G_i(z) = \frac{0.0002268z}{z^2 - 9.0329z + 2.8367} \tag{39}$$

The transfer function of Pitch control is formed as:

$$G_i(z) = \frac{0.0000533z}{z^2 - 1.355z + 0.39125} \tag{40}$$

The transfer function of Yaw control is formed as:

$$G_i(z) = \frac{0.0000865z}{z^2 - 2.5z + 0.5} \tag{41}$$

The PID controller coefficients tuned by LOC are summarized in in Table 3.

Table 5. PID controller coefficients tuned by LOC.

Controller Gain	Roll	Pitch	Yaw
K_{Pi}	27.32	18.08	23.12
K_{Ii}	4.41	18.76	11.56
K_{Di}	12.51	7.34	5.78

3. Simulation results and comparative study

Several simulations trials on MATLAB Simulink are performed using the entire quadrotor model to modulate the coefficients of the proposed controllers. The functions of the controllers are tested in stabilizing the orientation angles in the lower control loop to any of quadrotor of the underlying cooperative quadrotors. The isolating ranges among the cooperative quadrotors and speed are obliged at control approaches to keep the planned formation configuration. The comparative criterion between the offered controllers includes designed trajectory tracking, wind disturbance rejection, and noise cancellation.

3.1. Free flight scenario:

The underlying cooperative quadrotors team is examined in trajectory tracking in ideal case of free flight. The study of the tunning approaches is applied on the leader quadrotor, however the results on applying the tunning approaches on the other team members fulfill the desired formation requirements.

Figures 4-7 display the trajectory tracking of leader quadrotor controlled by BS-PID controller tuned by LOC, FOPID, BS-PD controllers tuned by ZN and GA, and in altitude and attitude (roll, pitch, and yaw) orientations respectively.

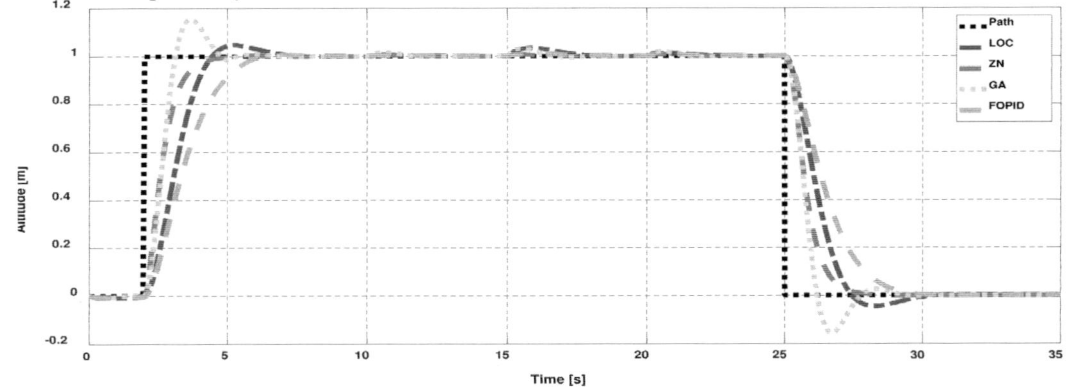

Figure 7. Trajectory tracking of BS-PID controller tuned by LOC, FOPID, BS-PD controllers tuned by ZN and GA for a single quadrotor in ***Z***-orientation.

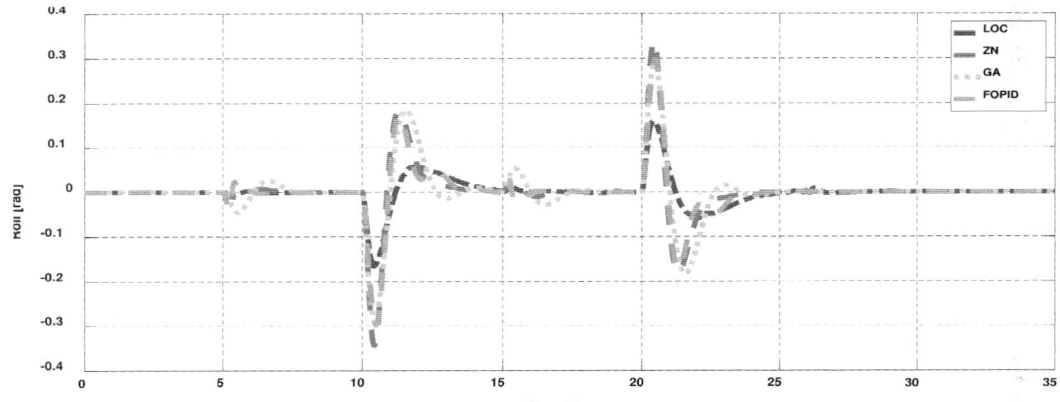

Figure 8. Roll control for BS-PID controller tuned by LOC, FOPID, BS-PD controllers tuned by ZN and GA.

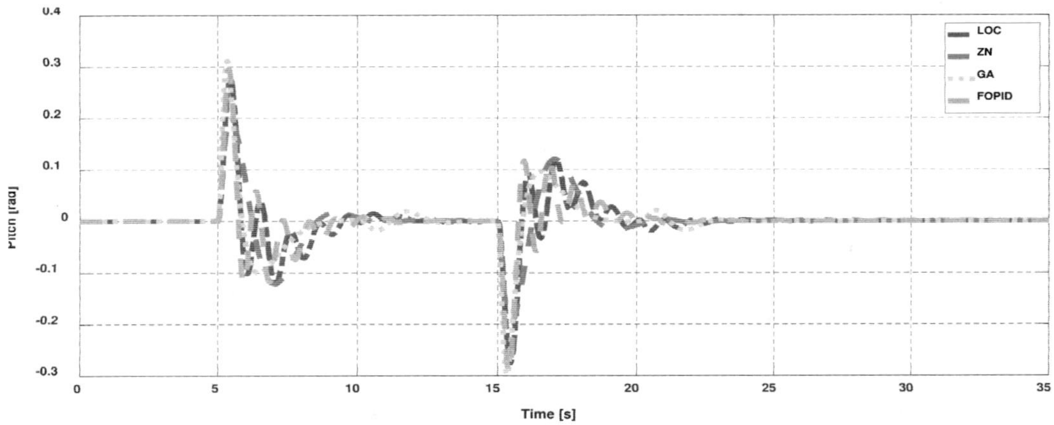

Figure 9. Pitch control for BS-PID controller tuned by LOC, FOPID, BS-PD controllers tuned by ZN and GA.

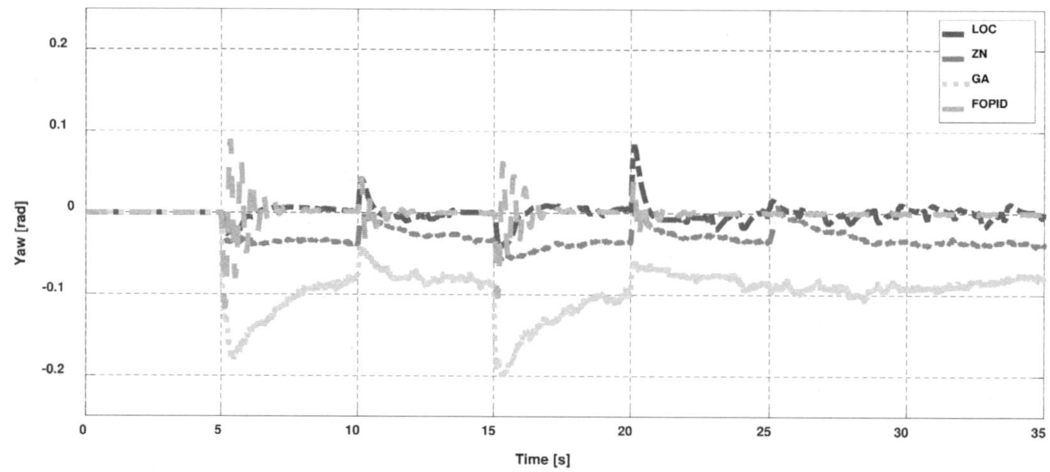

Figure 10. Yaw control for BS-PID controller tuned by LOC, FOPID, BS-PD controllers tuned by ZN and GA.

Figures 4-7 show successful of the trajectory tracking of a single quadrotor from the underlying cooperative quadrotors controlled by BS-PID controller tuned by LOC, FOPID, BS-PD controllers tuned by ZN and GA, and in altitude and attitude orientations respectively. The obtained trajectory of each approach is compared with other approaches. The analysis of the acquired results indicated that the output response of FOPID method is better than that of BS-PD tuned by ZN and GA.

3.2. Wind disturbance rejection and noise attenuation:

Stability and robustness evaluation of BS-PID controller tuned by LOC, FOPID and BS-PD controllers tuned by ZN and GA is realized. These controllers are studied and compared in view of stability scope based on the ability of these controllers on disturbance rejection and noise cancelation.

Figures 8-11 display the trajectory tracking of single quadrotor controlled by BS-PID controller tuned by LOC, FOPID and BS-PD controllers tuned by ZN and GA under effect of wind disturbance in altitude and attitude orientations, respectively.

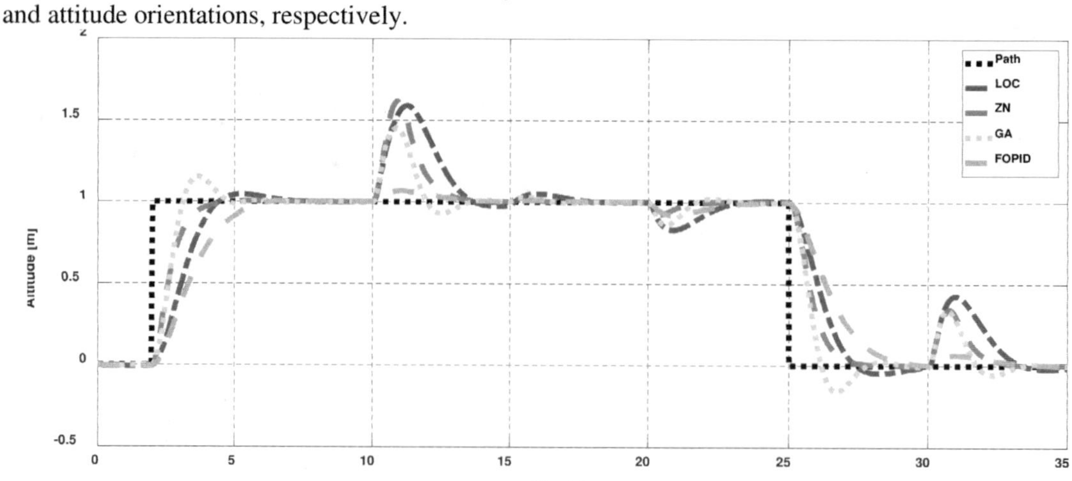

Figure 11. Trajectory tracking of BS-PID controller tuned by LOC, FOPID, BS-PD controllers tuned by ZN and GA for a single quadrotor in ***Z***-orientation under effect of wind disturbance.

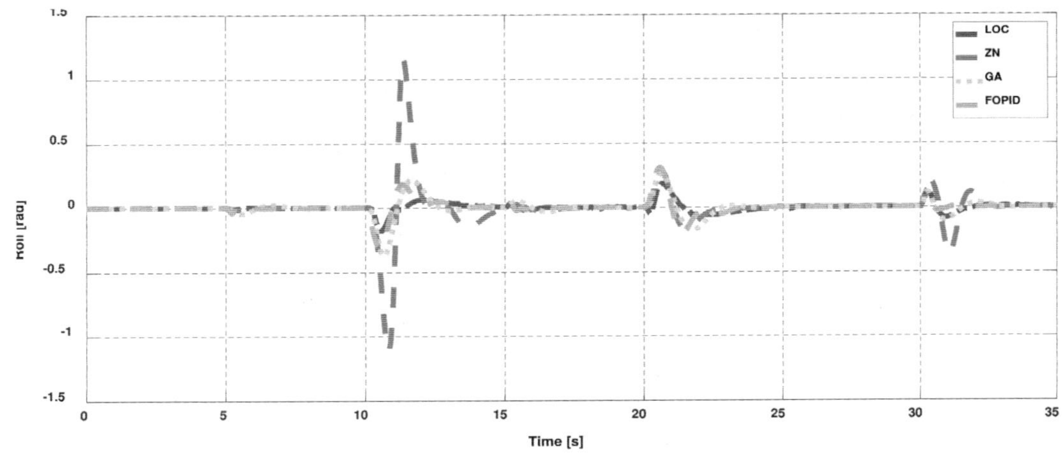

Figure 12. Roll control for BS-PID controller tuned by LOC, FOPID, BS-PD controllers tuned by ZN and GA under effect of wind disturbance.

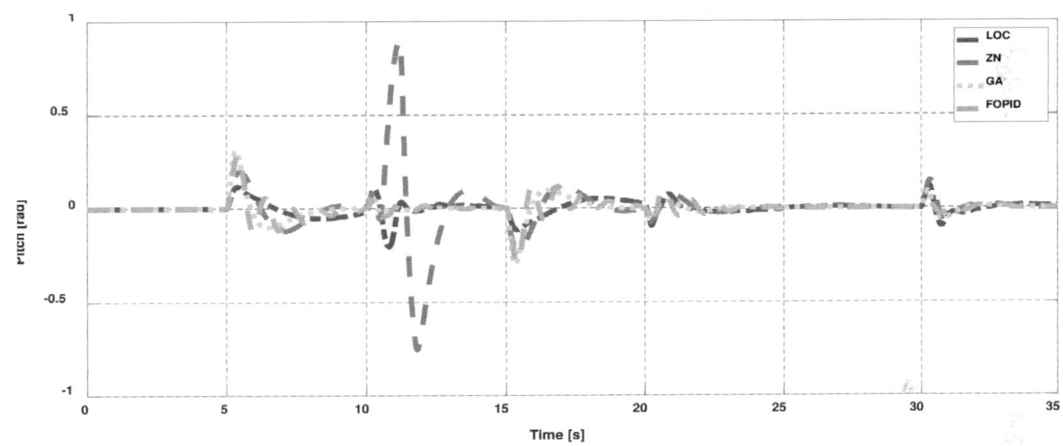

Figure 13. Pitch control for BS-PID controller tuned by LOC, FOPID, BS-PD controllers tuned by ZN and GA under effect of wind disturbance.

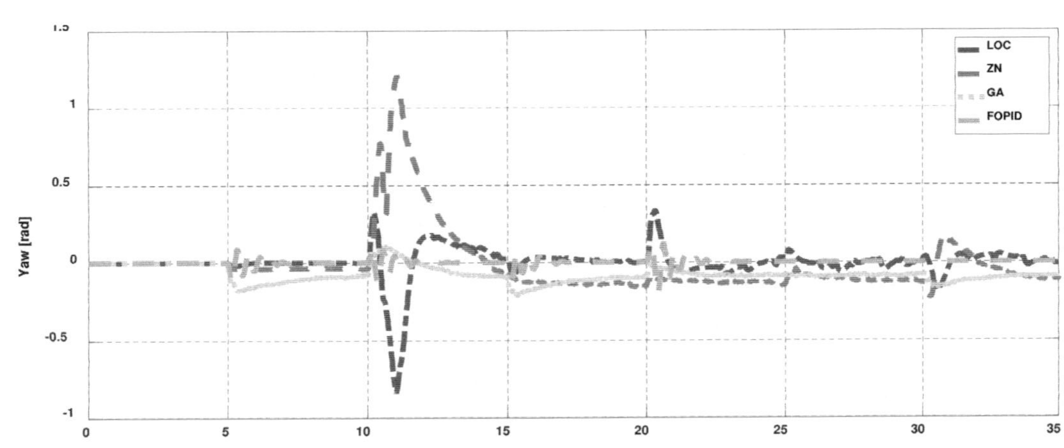

Figure 14. Yaw control for BS-PID controller tuned by LOC, FOPID, BS-PD controllers tuned by ZN and GA under effect of wind disturbance.

The obtained results show the ability of the underlying controllers to stabilize and control the leader quadrotor under effect of sudden wind disturbances with superiority of BS-PID tuned by FOPID method.

Figures 12-15 display the trajectory tracking of single quadrotor controlled by BS-PID controller tuned by LOC, FOPID and BS-PD controllers tuned by ZN and GA under effect of white noise in in altitude and attitude orientations, respectively. The obtained results show the ability of the underlying controllers to stabilize and control the quadrotor under effect of white noise with superiority of BS-PID tuned by FOPID method.

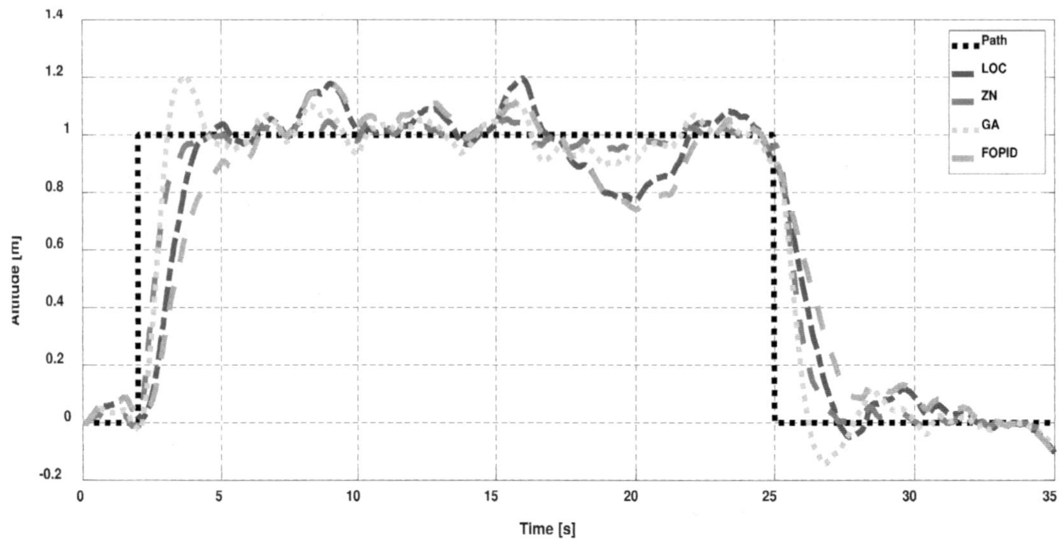

Figure 15. Trajectory tracking of BS-PID controller tuned by LOC, FOPID, BS-PD controllers tuned by ZN and GA for a single quadrotor in ***Z***-orientation under effect of noise.

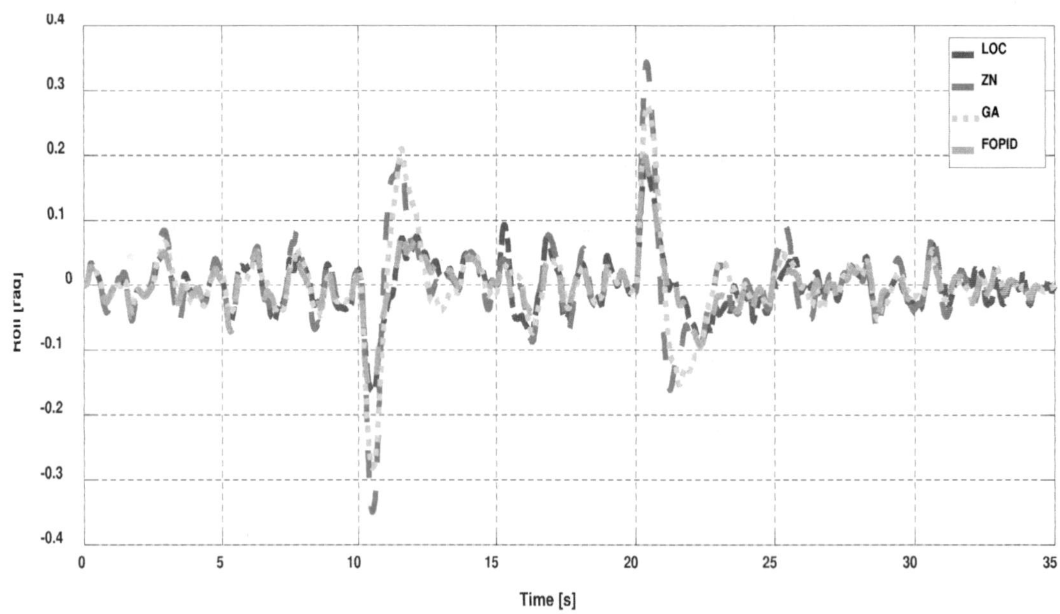

Figure 16. Roll control for BS-PID controller tuned by LOC, FOPID, BS-PD controllers tuned by ZN and GA under effect of noise.

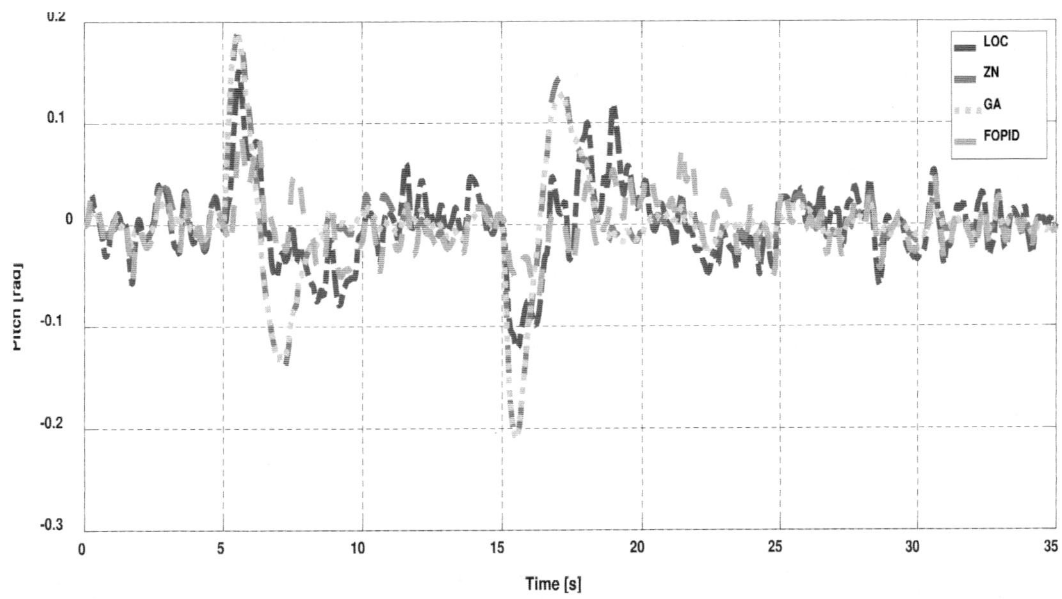

Figure 17. Pitch control for BS-PID controller tuned by LOC, FOPID, BS-PD controllers tuned by ZN and GA under effect of noise.

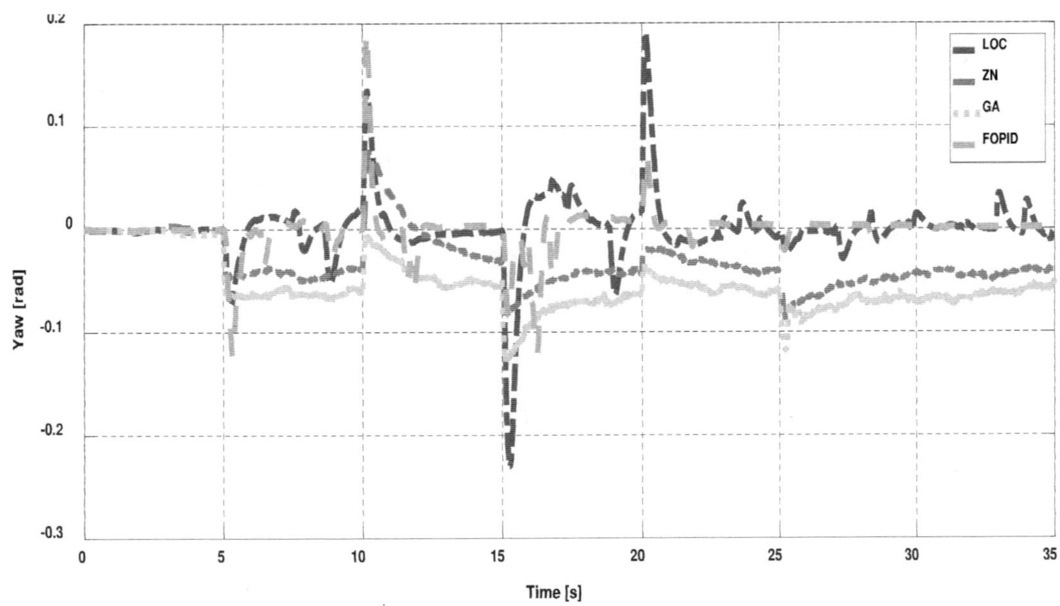

Figure 18. Yaw control for BS-PID controller tuned by LOC, FOPID, BS-PD controllers tuned by ZN and GA under effect of noise.

Figures 16 shows the (X-Y) plane, the leader succeeded in tracking the desired path along the (X-Y) plane. Moreover, follower (1) and follower (2) respect the desired separating distances 40m and 30m along the X-axis and Y-axis, respectively. Follower (3) and follower (4) respect the desired separating distances 20m and 60m along the X-axis and Y-axis, respectively. Figure 17 displays the trajectory tracking of cooperative quadrotors controlled by BS-PID controller tuned by LOC, FOPID controller in XYZ orientations.

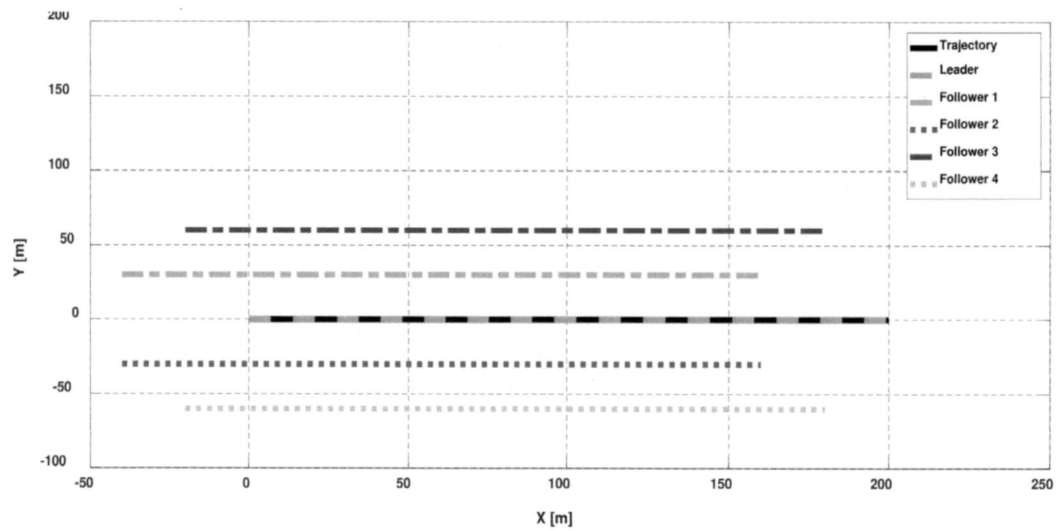

Figure 19. Trajectory tracking of BS-PID controller tuned by LOC, FOPID for cooperative quadrotors in XY-orientation.

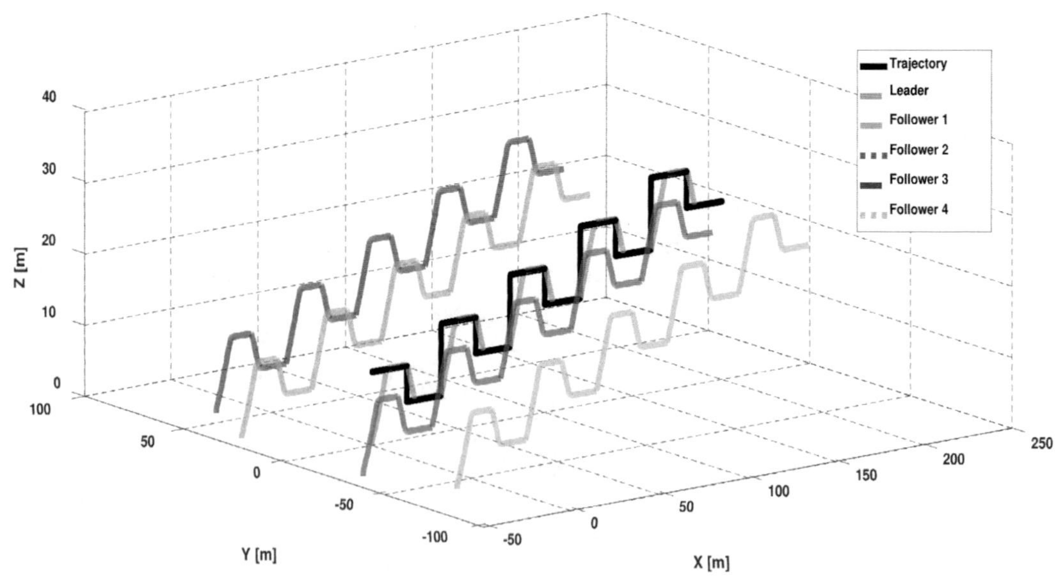

Figure 20. Trajectory tracking of BS-PID controller tuned by LOC, FOPID for cooperative quadrotors in XYZ-orientation.

4. Conclusion

In this paper, BS-PID controller tuned by LOC, FOPID, BS-PD controllers tuned by ZN and GA are conceived for dissolving formation configuration issues for cooperative quadrotors UAVs. A complete mathematical model for the underlying tunning approaches is presented. The simulation outcomes demonstrate a perfect stable flight attained by the proposed controller in the appearance and disappearance of applied disturbances and white noises. A comparison between the proposed controllers in different cases and orientations are studied. The outcomes demonstrate the superiority of BS-PID controller tuned by LOC, FOPID compared with BS-PID controller tuned by LOC, BS-PD controllers tuned by ZN and GA in the appearance and disappearance of applied disturbances and white noises.

References

[1] Hou Z, Wang W, Zhang G and Han C 2017 A survey on the formation control of multiple quadrotors. In: *Ubiquitous Robots and Ambient Intelligence (URAI), 2017 14th International Conference on*: IEEE) pp 219-25

[2] Mahfouz M, Hafez A T, Ashry M M and Elnashar G 2019 Cyclic Leader-Follower Strategy For Cooperative Unmanned Aerial Vehicles. In: *2019 IEEE International Conference of Vehicular Electronics and Safety (ICVES)*: IEEE) pp 1-6

[3] Gonzalez-Vazquez S and Moreno-Valenzuela J 2010 A New Nonlinear PI/PID Controller for Quadrotor Posture Regulation. In: *2010 IEEE Electronics, Robotics and Automotive Mechanics Conference,* pp 642-7

[4] Xu R and Ozguner U 2006 Sliding mode control of a quadrotor helicopter. In: *Decision and Control, 2006 45th IEEE Conference on*: IEEE) pp 4957-62

[5] Madani T and Benallegue A 2006 Backstepping control for a quadrotor helicopter. In: *Intelligent Robots and Systems, 2006 IEEE/RSJ International Conference on*: IEEE) pp 3255-60

[6] Wang Y, Chenxie Y, Tan J, Wang C, Wang Y and Zhang Y 2015 Fuzzy radial basis function neural network PID control system for a quadrotor UAV based on particle swarm optimization. In: *Information and Automation, 2015 IEEE International Conference on*: IEEE) pp 2580-5

[7] Wong T, Khan R and Lee D 2014 Model linearization and H∞, controller design for a quadrotor unmanned air vehicle: Simulation study. In: *Control Automation Robotics & Vision (ICARCV), 2014 13th International Conference on*: IEEE) pp 1490-5

[8] Mahfouz M, Ashry M and Elnashar G 2013 Design and control of quad-rotor helicopters based on adaptive neuro-fuzzy inference system *International Journal of Engineering Research Technology (IJERT)* **2** 479-85

[9] Mahfouz M and Kader S A 2015 Quadrotor unmanned aerial vehicle controller design and synthesis. In: *Computer Engineering & Systems (ICCES), 2015 Tenth International Conference on*: IEEE) pp 60-6

[10] Mahfouz M, Hafez A T, Ashry M M and Elnashar G 2020 Formation Reconfiguration Based on Backstepping-PID Controller for Collaborative Quadrotors. In: *2020 12th International Conference on Electrical Engineering (ICEENG)*: IEEE) pp 396-401

[11] Mahfouz M, Hafez A T, Azzam A, Ashry M M and Elnashar G 2020 Formation Reconfiguration in Dynamic Obstacle Loaded Environment via Backstepping Approach. In: *2020 12th International Conference on Electrical Engineering (ICEENG)*: IEEE) pp 441-6

[12] Rego B S, Adorno B V and Raffo G V 2016 Formation Backstepping Control Based on the Cooperative Dual Task-Space Framework: A Case Study on Unmanned Aerial Vehicles. In: *Robotics Symposium and IV Brazilian Robotics Symposium (LARS/SBR), 2016 XIII Latin American*: IEEE) pp 163-8

[13] Mahfouz M, Hafez A T, Ashry M and Elnashar G 2018 *2018 AIAA SPACE and Astronautics Forum and Exposition*: American Institute of Aeronautics and Astronautics)

[14] Ashry M, Abou-Zayed U and Breikin T 2008 Comparative robustness study of multivariable controllers for a gas turbine engine. In: *IASTED International Conference on Modelling, Identification, and Control,*

[15] Ashry M M, Kamalova Z Z and Breikin T V 2008 Tuning of digital PID controller parameters using local optimal control. In: *2008 16th Mediterranean Conference on Control and Automation*: IEEE) pp 587-92

[16] Habashi A, Ashry M, Mabrouk M H and Elnashar G 2018 Comparative Study Among Different Control Techniques For Stabilized Platform *Engineering Science and Military Technologies* **2**

[17] Habashi A, Ashry M, Mabrouk M and Elnashar G 2015 Controller Design for Line of Sight Stabilization System *International Journal of Engineering Research and Technology*

[18] Ashry M, Abou-Zayed U and Breikin T 2008 Control of multivariable systems using modified local optimal controller *IFAC Proceedings Volumes* **41** 8767-72

[19] Mahfouz M, Taiomour A, Ashry M M and Elnashar G 2019 PID vs. Backstepping Control for Cooperative Quadrotors Unmanned Aerial Vehicles. In: *IOP Conference Series: Materials Science and Engineering*: IOP Publishing) p 012057

[20] Onsy A, Sharaf A, Ashry M and Eldemerdash S 2013 Effect of Slow-Active Suspension Controller Design on the Performance of Anti-Lock Brake System. In: *15 th International Conference on ASAT-15, Military Technical College, Kobry Elkobbah, Cairo, Egypt,*

[21] Khodja M A, Tadjine M, Boucherit M S and Benzaoui M 2017 Experimental dynamics identification and control of a quadcopter. In: *2017 6th International Conference on Systems and Control (ICSC)*: IEEE) pp 498-502

[22] Mobarez E N, Sarhan A and Ashry M 2019 Fractional order PID Based on a Single Artificial Neural Network Algorithm for Fixed wing UAVs. In: *2019 15th International Computer Engineering Conference (ICENCO)*: IEEE) pp 1-7

[23] Mobarez E N, Sarhan A and Ashry M M 2018 Comparative robustness study of multivariable controller of fixed wing Ultrastick25-e UAV. In: *2018 14th International Computer Engineering Conference (ICENCO)*: IEEE) pp 146-50

IOP Conf. Series: Materials Science and Engineering 1172 (2021) 012041 doi:10.1088/1757-899X/1172/1/012041

Formation configuration of unmanned cooperative quadrotors via PID tuning approaches

M Mahfouz [1], A Taiomour [2], M M Ashry [1] and G Elnashar [1]

[1]Department of Optoelectronics and Control, Military Technical College,
[2]Aircraft Armament Department, Military Technical College,

E-mail: m.mahfouz.trc@gmail.com,

Abstract. Formation configuration is one of the major intrinsic strategies used in cooperative Unmanned Air Vehicles field. In this paper, Backstepping-PID control technique for cooperative quadrotors unmanned aerial vehicles are developed to solve the formation problem. The proposed controller is divided into couple of parts working together. Backstepping controller is used to stabilize the position control as a higher controller. Simultaneously, PID controller receives the desired position to stabilize the attitude control as a lower controller to track the desired planning trajectories. The main contribution of this paper is using Fraction Order Approach, and Local Optimal Approach to refine the PID lower controller gains. The tunning of the PID gains through the proposed PID tuning approaches guarantee the stabilization of the attitude control for all the team members. Simulation results present the success of the proposed PID tuning approaches in solving the formation problem for cooperative unmanned quadrotors tracking a desired path. Moreover, the simulation results present the ability of the proposed approaches to handle disturbance rejection and noise attenuation while preserving the stability of the system.

1. Introduction

Deploying cooperative unmanned aerial vehicles (UAVs) is extremely relevant for wide areas of implementations in either military and civilian utilizations [1]. Currently, one of the most important UAVs is the vertical take-off and landing (VTOL) with four-rotors aircraft. It is well known as quadrotor. This kind of aircrafts is characterized with unrevealed useful features such as small size, ultimate maneuverability capabilities, simplicity of design, and control. Quadrotor has the footprint of taking-off vertically from any place and landing vertically at any place regardless the hardness of that places. Quadrotors UAVs have the advantage of being able to accomplish specific and difficult tasks that cannot be done by fixed-wing aircrafts due to its unique characteristics.

There are various space applications where this capacity would be desired, for example, discharging raised territories and transporting merchandise to regions where street arrival isn't accessible. VTOL planes also provide direct entry to buildings or zones making them a quick type of transportation between regions, particularly those which are raised or have restricted admittance [2].

Conventionally, quadrotors usage is increased rapidly in the aerospace industry. It is fact that most of the usual payloads could be lifted using one or two rotors. However, Quadrotors have unique features that make it appealing. Naturally, one of these features is the higher and diverse capability for payload. The plainness of their control system is also an advantage; just by modifying the speed of every rotor

Content from this work may be used under the terms of the Creative Commons Attribution 3.0 licence. Any further distribution of this work must maintain attribution to the author(s) and the title of the work, journal citation and DOI.
Published under licence by IOP Publishing Ltd

separately.
The quadrotor is a vastly nonlinear, multi-variables, strongly coupled, under-actuated, sensitive to interference and essentially an unstable system (6 DOF, 3 translational and 3 rotational motions, with only 4 actuators) [3], which acts as fundamental foundation for design of control plan [4]. This makes control design is very challenging. currently, there are numerous methods to control the quadrotor, the literature [5], [6], [7], [8], and [9] respectively focused on PID, sliding mode, backstepping, fuzzy neural network, and H-infinity approaches. The adaptive neuro-fuzzy inference system (ANFIS) controller to the quadrotors is discussed in [10] and [11].
It is believed that multiple cooperative quadrotors are more stable and adaptable to fulfill the desired demands than single quadrotor [12]. Unmistakably, the general ability of the cooperative quadrotors can upgrade execution, essential time range, task allocation, system proficiency and the safety to accomplish the desired tasks.
Complex tasks cannot be achieved efficiently by a single quadrotor, such as finding objects in massive areas like in searching and relief missions, basic payloads restrictions, etc. Because of these regards, collaboration of numerous UAVs has recently gotten a lot of consideration from the scientist and researchers [13].
Quick advancements on computers, actuators, sensors fusion, and communications technologies have all an impact in developing the novel epoch of collaboration methodologies for cooperative quadrotors. State of art of the proposed controller designed based on combining couple of robust controllers that can deal with highly non-linear systems effectively.
The proposed controller is divided into couple of parts working together. Backstepping (BS) controller is used to stabilize the position control as a higher controller. Simultaneously, PID controller receives the desired position to stabilize the attitude control as a lower controller to track the desired planning trajectories.
BS controller is used mostly as a higher controller [14]. BS controller is an effective control strategy for extremely nonlinear frameworks. The rapid convergence rate, the stability, the capability to manage nonlinear frameworks using recursive operation of Lyapunov function, the robustness regardless presence of uncertainties, and the simplicity are the benefits of BS controller [7]. The proposed BS-PID controller enhanced the overall performance and stability of the underlying cooperative quadrotors system. Whereas, Local Optimal Approach (LOA) is convenient for controlling non-linear systems with multivariable parameters as Single Input Single Output (SISO) systems [15]. Local Optimal PID (LOPID) is developed based on LOA theory and traditional PID theory. LOA is characterized with robust performance [16].
Multivariate control systems were given much attention because of the growing complexity of control systems operations [17]. Numerous researchers are broadly utilizing Fractional Order Approach (FOA) to accomplish better robust performance in several control systems. FOA approach has accomplished further remarkable outcomes on quadrotors particularly enhancing the stability of its frameworks.
The superiority of FOA can be demonstrated in the ability of fractional calculus to configure the gain margin and the phase margin in the frequency domain of the quadrotor framework freely [18]. Fractional order PID (FOPID) is developed based on FOA theory and traditional PID theory. This controller expands the resolution of tuning parameters from the triple parameters of PID k_P, k_D, and k_I to five parameters by adding the FOC parameters λ and μ. Theses tunable parameters make the controller more precise, robust and flexible [19].
The main contribution of this paper is using LOPID and FOPID approaches with BS controller in the proposed cascaded controllers for first time in cooperative quadrotors. LOPID and FOPID are used to tune PID gains to stabilize attitude control for cooperative quadrotors.

This paper is organized as follows:

Section 2 presents a short description and mathematical model of the quadrotor. The controller's approach is described in section 3. Section 4 shows the simulation results of trajectory tracking using the proposed controllers. Finally, conclusion is introduced at final section in section 5.

2. Quadrotor control approaches

2.1. Backstepping control design

BS approach design has been formerly studied in [20]. BS approach is a preferred control technique over various approaches. The controller utilizes a recursive operation. It divides the controller mechanism into serial stages with respect of keeping the stability of every single stage progressively. The inputs of BS controller are directly received as a desired mission plan and a streaming of information sent by the onboard quadrotor sensors.

BS utilizes numerous coefficients during the computation. These coefficients represent the quadrotor dynamics and states. The yield of the controller approach is the separately sent pulse width modulation (PWM) signal for every rotor of the quadruple rotors.

The proposed controller utilizes BS in the higher control loop of positioning the underlying framework, so, BS is responsible of defining the updated positions of every quadrotors of the cooperative squad. The altitude control and the position control for i^{th} quadrotor (U_{1i}, u_{xi}, u_{yi}) is assumed as:

$$\begin{aligned} U_{1i} &= \frac{m_i}{\cos x_{1i} \cos x_{3i}}(g_i + g_{7i} - \alpha_{7i}(g_{8i} + \alpha_{7i} g_{7i}) - \alpha_{8i} g_{8i}) \\ u_{xi} &= \frac{m_i}{U_{1i}}(g_{9i} - \alpha_{9i}(g_{10i} + \alpha_{9i} g_{9i}) - \alpha_{10i} g_{10i}) \\ u_{yi} &= \frac{m_i}{U_{1i}}(g_{11i} - \alpha_{11i}(g_{12i} + \alpha_{11i} g_{11i}) - \alpha_{12i} g_{12i}) \end{aligned} \tag{1}$$

2.2. PID Control Design

PID controllers is considered as an optimal control. PID are generally used because of design simplicity and reduced number of tunable parameters in the production controlled systems [21], and [22]. Upwards of 90 percent of the commonly deployed controllers are routinely based on PID controllers. PID controller is designed in the proposed controller as a lower controller to control the attitude channels of each member of cooperative quadrotors. The popular mathematical equation of PID controller is formed as [16]:

$$u(t) = K_P + K_I \int_0^t e(t)dt + K_D \frac{d}{dt} e(t) \tag{2}$$

The quadrotor model should be reduced in order to make a simple reverse model to be invoked in the control techniques. That can be done by removing gyroscopic terms the dynamic. The simplified mathematical equation can be reformulated as:

$$\begin{pmatrix} \ddot{x}_{1i} \\ \ddot{x}_{3i} \\ \ddot{x}_{5i} \end{pmatrix} = \begin{pmatrix} \ddot{\phi}_i \\ \ddot{\theta}_i \\ \ddot{\psi}_i \end{pmatrix} = \begin{pmatrix} \frac{l_i}{I_{xxi}} U_{2i} \\ \frac{l_i}{I_{yyi}} U_{3i} \\ \frac{1}{I_{zzi}} U_{4i} \end{pmatrix} \tag{3}$$

Converting (3) to Laplace model and adding the rotor dynamics yields:

$$\begin{cases} \phi_i(s) = \dfrac{B_i^2 b_i l_i}{s^2(s+A_i)^2 I_{xxi}}\left(u_{4i}^2(s) - u_{2i}^2(s)\right) \\ \theta_i(s) = \dfrac{B_i^2 b_i l_i}{s^2(s+A_i)^2 I_{yyi}}\left(u_{3i}^2(s) - u_{1i}^2(s)\right) \\ \psi_i(s) = \dfrac{A_i^2}{s^2(s+A_i)^2 I_{zzi}}\sum_{n=1}^{4}(-1)^{n+1} u_{ni}^2(s) \end{cases} \tag{4}$$

where, the parameters A_i and B_i symbolize the rotor dynamics.
Inserting the control inputs instead of the motor inputs in (4) yields:

$$\begin{cases} \phi_i(s) = \dfrac{A_i^2 l_i}{s^2(s+A_i)^2 I_{xxi}} U_{2i} \\ \theta_i(s) = \dfrac{A_i^2 l_i}{s^2(s+A_i)^2 I_{yyi}} U_{3i} \\ \psi_i(s) = \dfrac{A_i^2}{s^2(s+A_i)^2 I_{zzi}} U_{4i} \end{cases} \tag{5}$$

Assume that the required angles are $x_{2i} = \dot{\phi}_i, x_{4i} = \dot{\theta}_i, x_{6i} = \dot{\psi}_i$, and $x_{1id} = x_{3id} = x_{5id} = 0$ if the state is hovering. So, the controller formulas can lastly be acquired as:

$$\begin{cases} U_{2i} = \dfrac{1}{b_{1i}}(K_{2i}x_{2i} - a_{1i}x_{4i}x_{6i}) + w_{2i}(x_{1id} - x_{1i}) \\ U_{3i} = \dfrac{1}{b_{2i}}(K_{3i}x_{4i} - a_{3i}x_{2i}x_{6i}) + w_{3i}(x_{3id} - x_{3i}) \\ U_{4i} = \dfrac{1}{b_{3i}}(K_{4i}x_{6i} - a_{5i}x_{4i}x_{6i}) + w_{4i}(x_{5id} - x_{5i}) \end{cases} \tag{6}$$

The BS-PID tuned by LOPID and FOPID approaches are utilized to improve the overall performance and stability of the underlying cooperative quadrotors frameworks. These tuning approaches are characterized with implementation simplicity and short execution time to exert the controller coefficients as inputs for the actuators.

The proposed controllers are divided into couple of parts working together. Backstepping controller is used to stabilize the position control as a higher controller. Simultaneously, PID tuned by LOPID and FOPID controllers receive the desired position to stabilize the attitude control as a lower controller to track the desired planning trajectories as depicted in Figures 1 and 2 respectively.

The BS-PID tuned by LOPID and FOPID approaches are studied in [23].

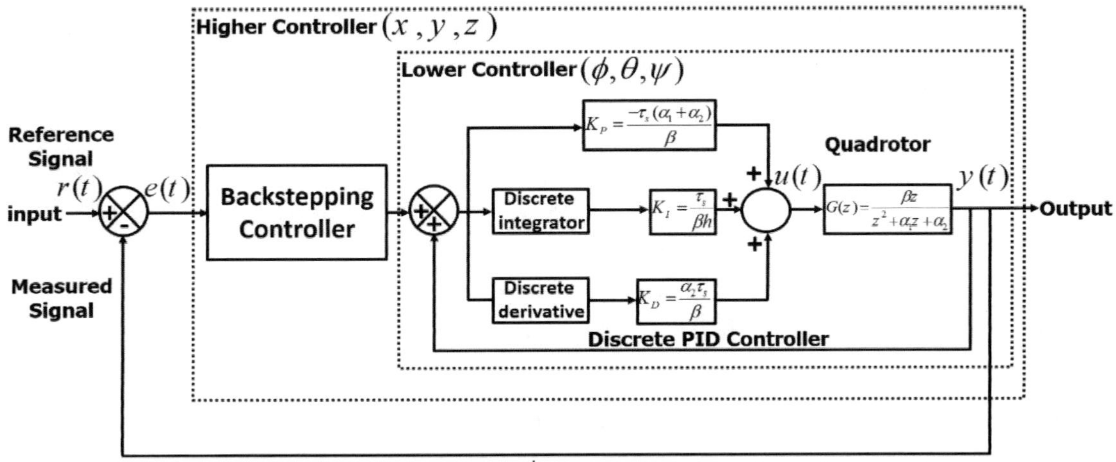

Figure 1. LOPID block diagram for 2nd order model of quadrotor framework.

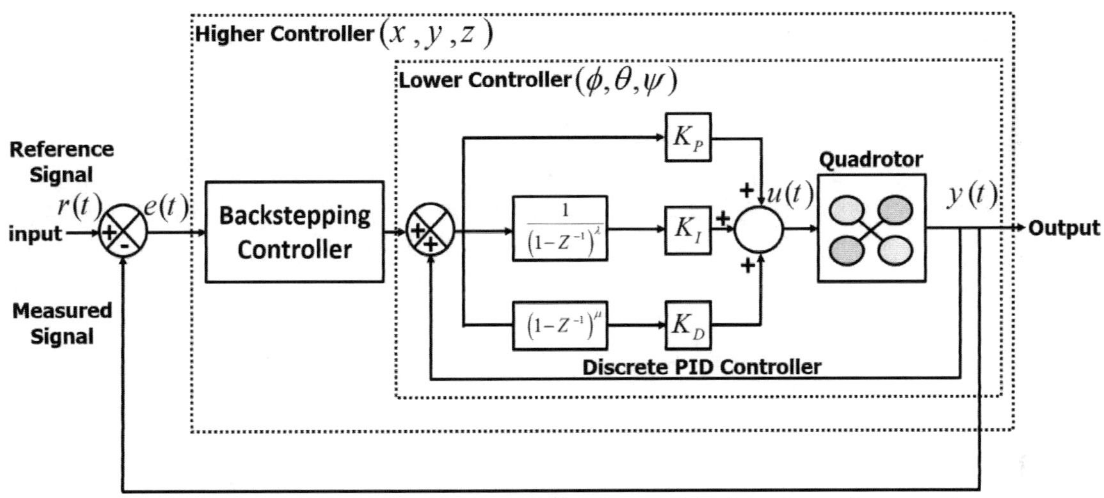

Figure 2. FOPID block diagram for 2nd order model of quadrotor framework.

3. Simulation results and comparative study

Several simulations trials on MATLAB Simulink are performed using the entire quadrotor model to modulate LOPID and FOPID control coefficients.

The controller's function of LOPID and FOPID are tested in stabilizing the orientation angles in the lower control loop to any of quadrotor of the underlying cooperative quadrotors. The isolating ranges among the cooperative quadrotors and speed are obliged at control approach to keep the planned formation configuration.

The comparative criterion between the offered controller includes designed trajectory tracking, wind disturbance rejection, and noise cancellation.

3.1. Local Optimal PID Approach

The underlying quadrotors are examined in path tracking in an ideal case of free flight and in the absence of applied noise and wind disturbance.

The team is composed from a leader and four followers tracking a desired path along the three axes. Figures 3-8 display the trajectory tracking of cooperative quadrotors controlled by BS-PID controller tuned by LOPID controller in X, Y, and Z orientations with their velocities, respectively.

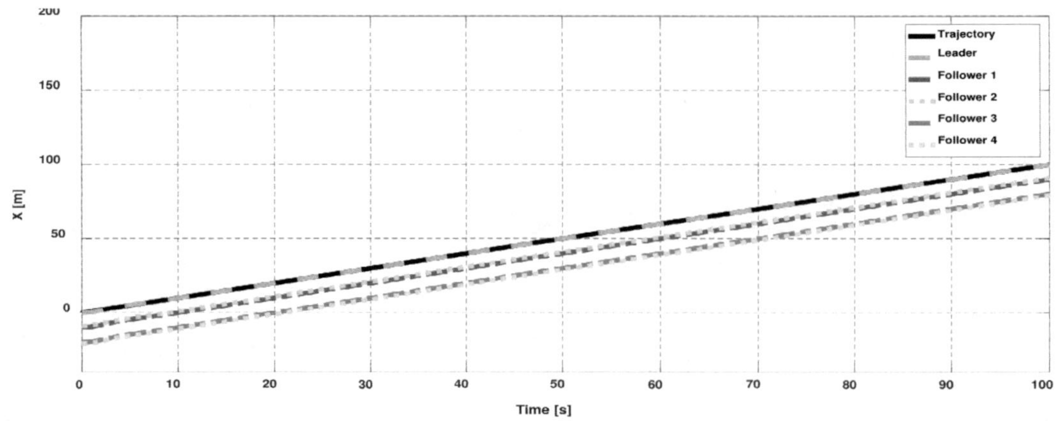

Figure 3. Trajectory tracking of BS-PID controller tuned by LOPID for cooperative quadrotors in ***X***-orientation.

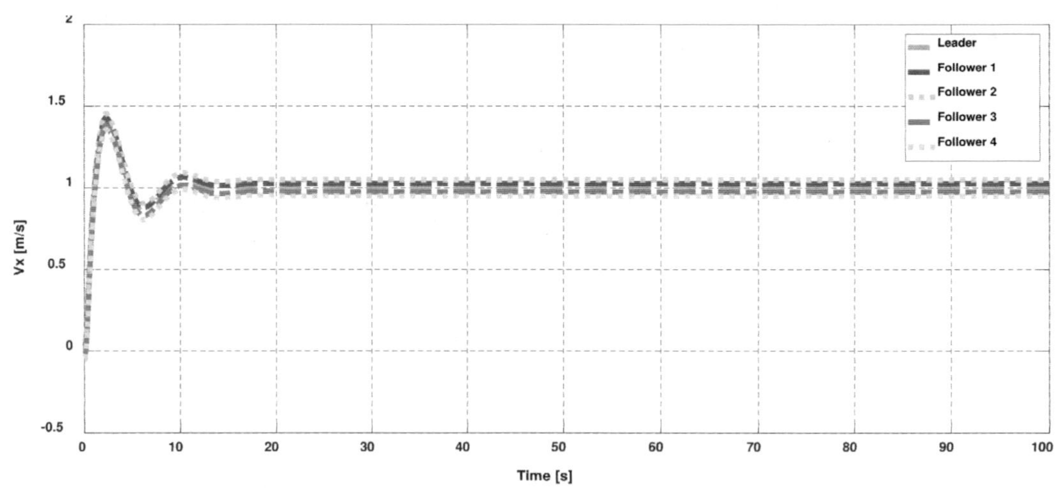

Figure 4. Velocity of cooperative quadrotors in ***X***-orientation.

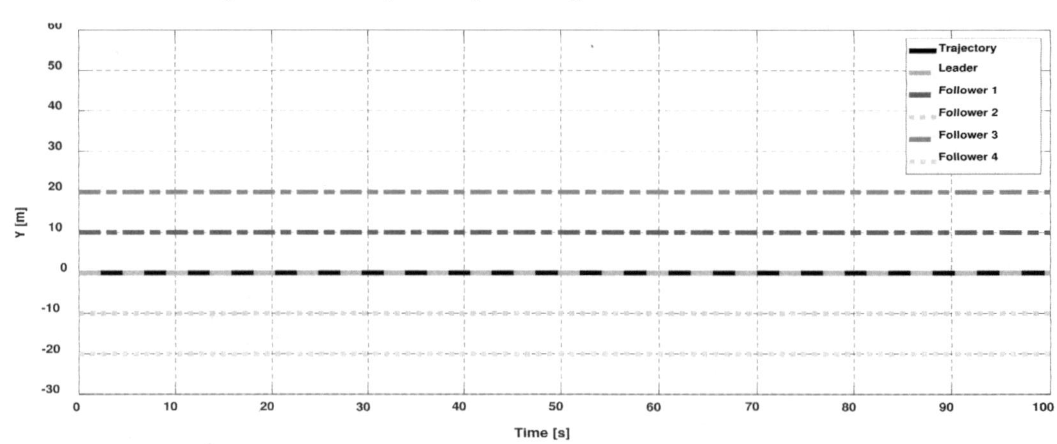

Figure 5. Trajectory tracking of BS-PID controller tuned by LOPID for cooperative quadrotors in ***Y***-orientation.

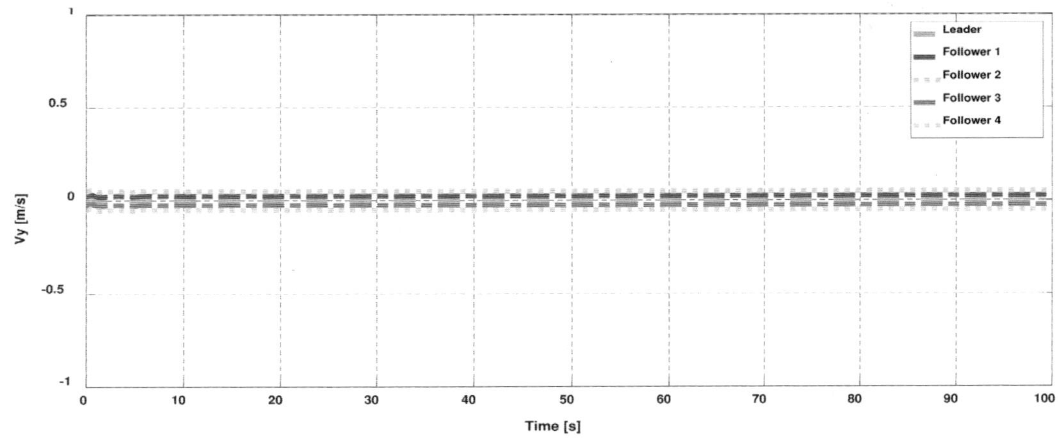

Figure 6. Velocity of cooperative quadrotors in ***Y***-orientation.

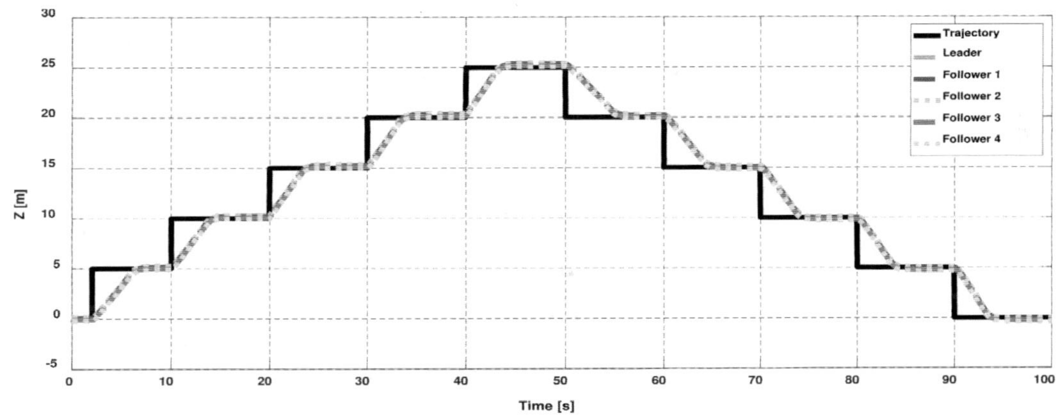

Figure 7. Trajectory tracking of BS-PID controller tuned by LOPID for cooperative quadrotors in ***Z***-orientation.

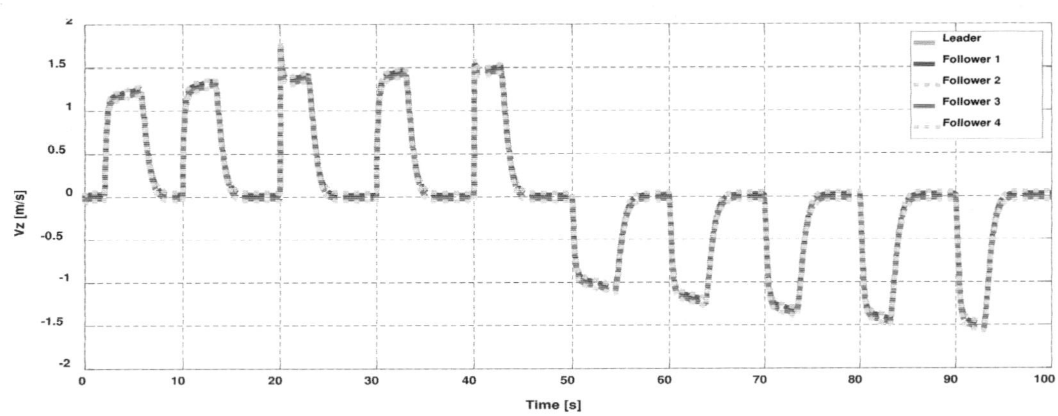

Figure 8. Velocity of cooperative quadrotors in ***Z***-orientation.

Figures 3-8 show successful of the trajectory tracking for a team of cooperative quadrotors controlled by BS-PID controller tuned by LOPID, and their velocities respect the designed constraints 2 m/sec. Figure 9 shows the $(X - Y)$ plane, the leader succeeded in tracking the desired path along the $(X - Y)$ plane. Moreover, follower (1) and follower (2) respect the desired separating distances 10m along the

X-axis and Y-axis. Follower (3) and follower (4) respect the desired separating distances 20m along the X-axis and Y-axis. Figure 10 displays the trajectory tracking of cooperative quadrotors controlled by BS-PID controller tuned by LOPID in XYZ orientations.

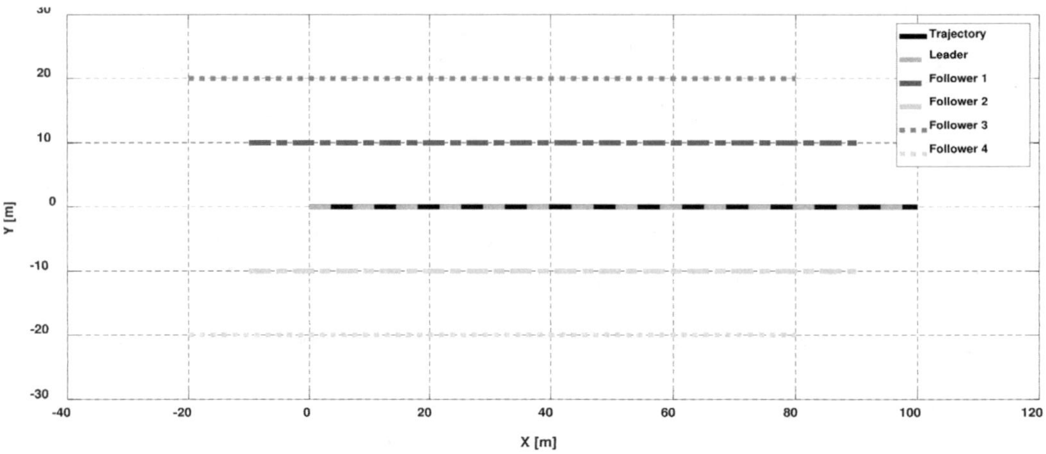

Figure 9. Trajectory tracking of BS-PID controller tuned by LOPID for cooperative quadrotors in ***XY***-orientation.

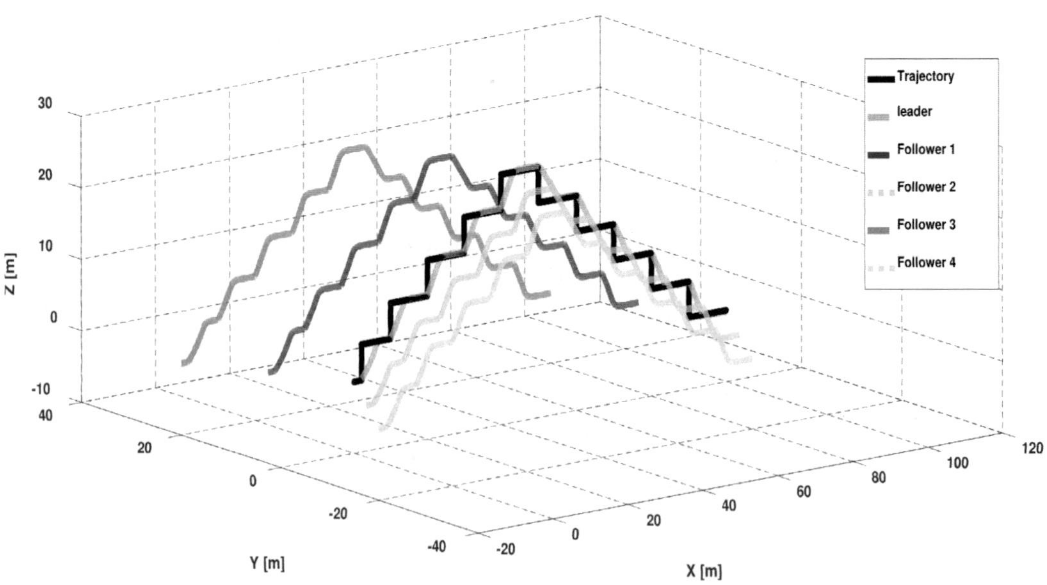

Figure 10. Trajectory tracking of BS-PID controller tuned by LOPID for cooperative quadrotors in ***XYZ***-orientation.

Figures 3-10 show successful of the trajectory tracking for a team of cooperative quadrotors controlled by BS-PID controller tuned by LOPID, and their velocities respect the designed constraints 2 m/sec. Figures 11-13 display the roll, pitch, and yaw channels of cooperative quadrotors controlled by BS-PID controller tuned by LOPID controller, respectively. All the quadrotors respect the desired angles during the free flight.

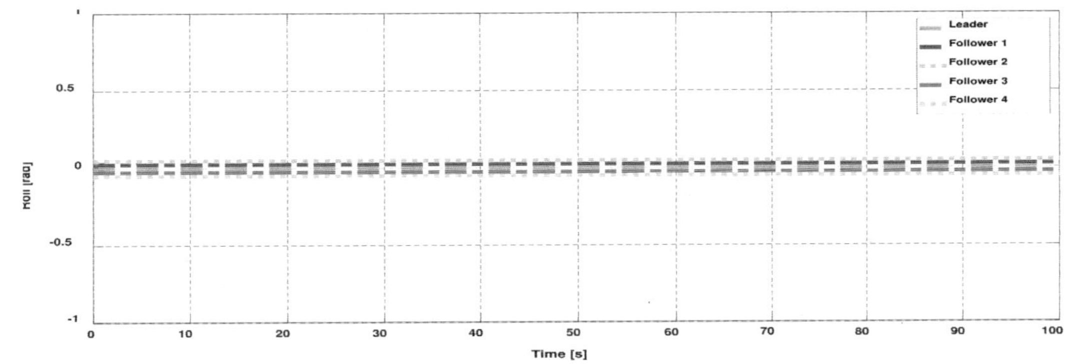

Figure 11. Roll control of BS-PID controller tuned by LOPID for cooperative quadrotors.

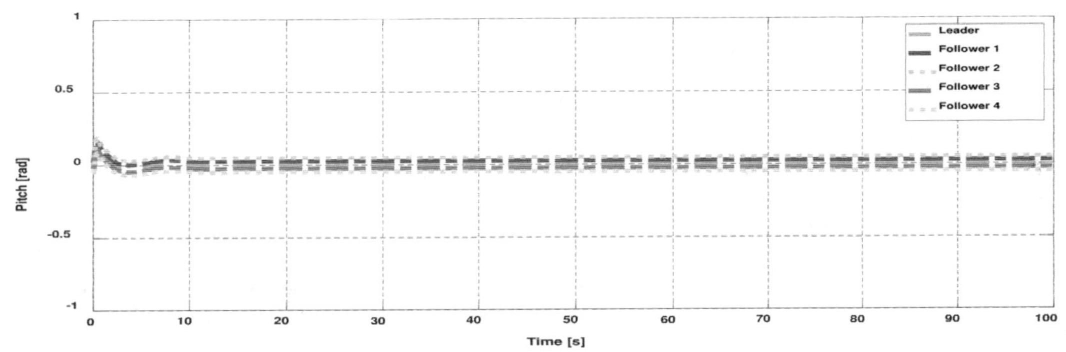

Figure 12. Pitch control of BS-PID controller tuned by LOPID for cooperative quadrotors.

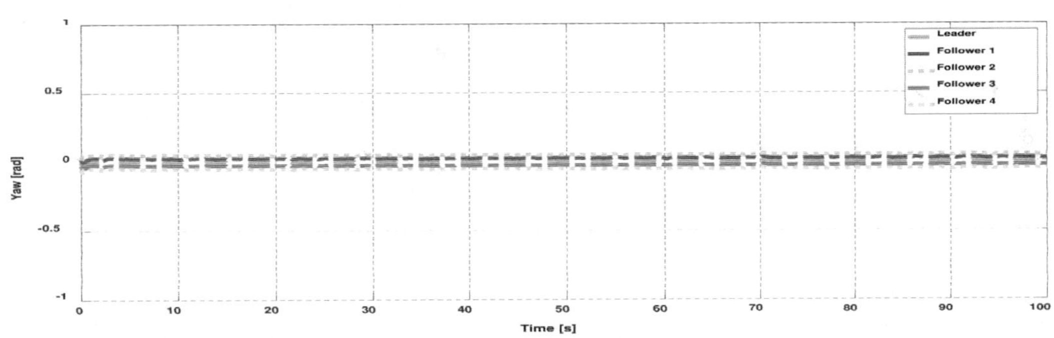

Figure 13. Yaw control of BS-PID controller tuned by LOPID for cooperative quadrotors.

Figures 11-13 show successful of the trajectory tracking of cooperative quadrotors controlled by BS-PID controller tuned by LOPID.

3.2. Fractional Order PID Approach

The underlying quadrotors are examined in path tracking in an ideal case of free flight and in the absence of applied noise and wind disturbance.

The team is composed from a leader and four followers tracking a desired path along the three axes. Figures 14-19 display the trajectory tracking of cooperative quadrotors controlled by BS-PID controller tuned by FOPID controller in X, Y, and Z orientations with their velocities, respectively.

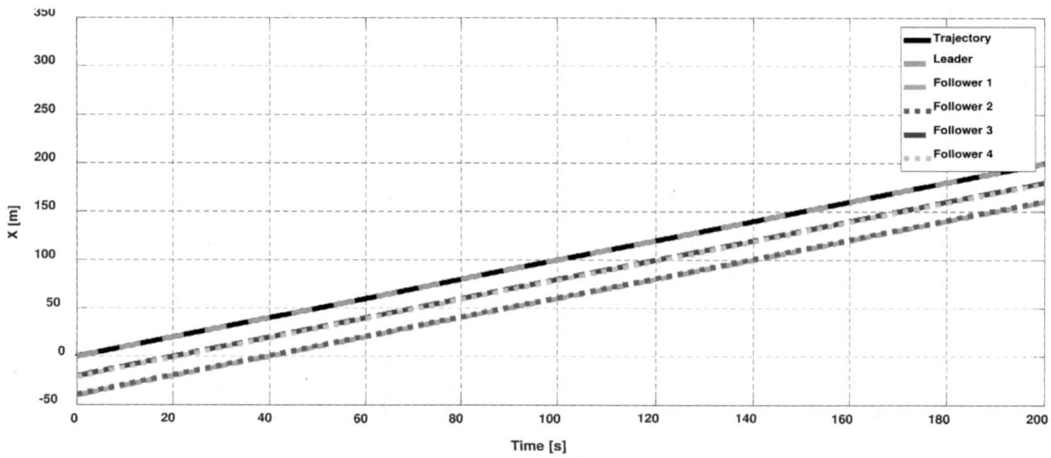

Figure 14. Trajectory tracking of BS-PID controller tuned by FOPID for cooperative quadrotors in ***X***-orientation.

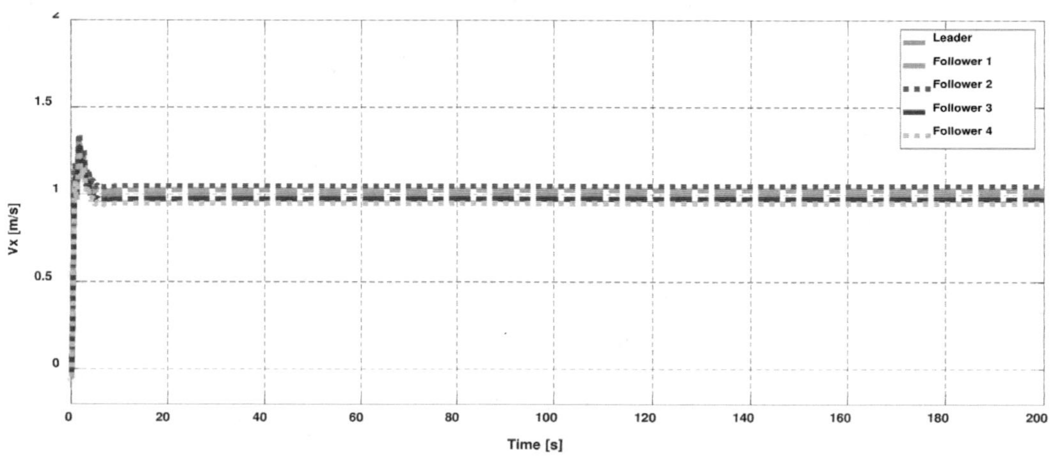

Figure 15. Velocity of cooperative quadrotors in ***X***-orientation.

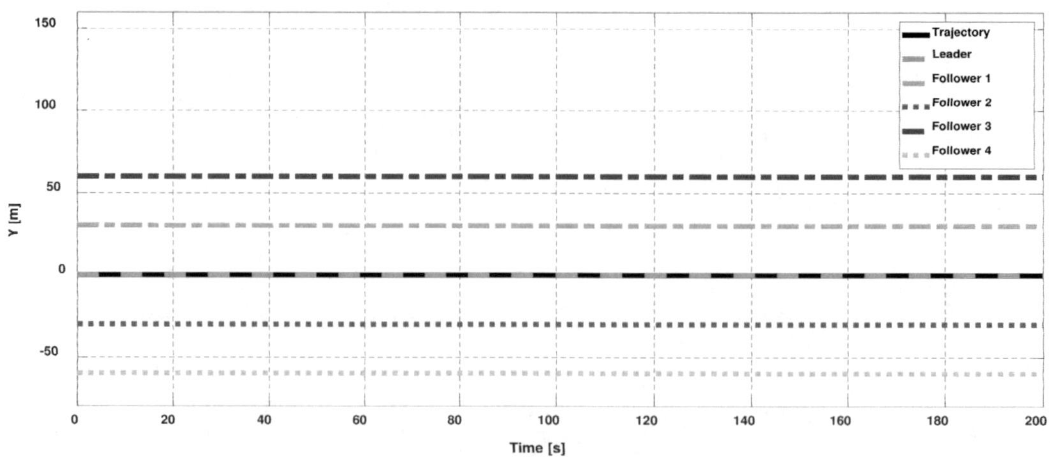

Figure 16. Trajectory tracking of BS-PID controller tuned by FOPID for cooperative quadrotors in ***Y***-orientation.

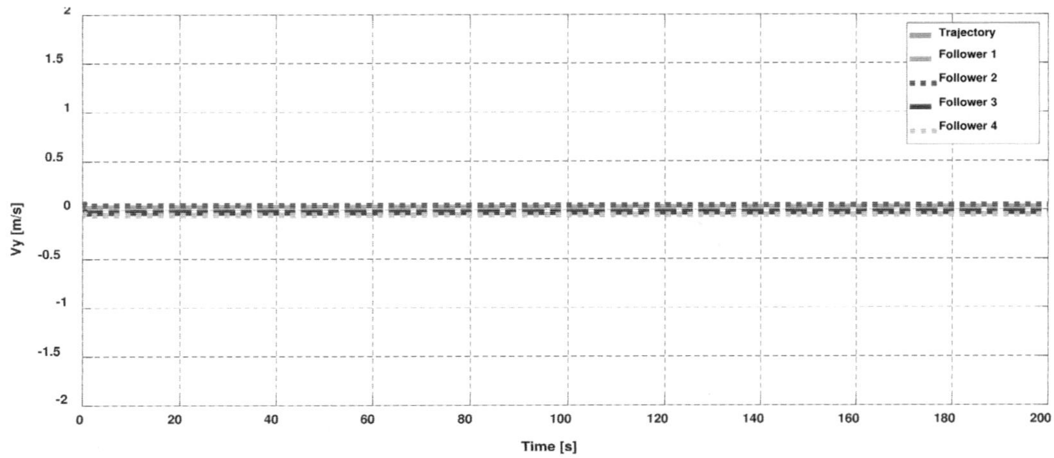

Figure 17. Velocity of cooperative quadrotors in ***Y***-orientation.

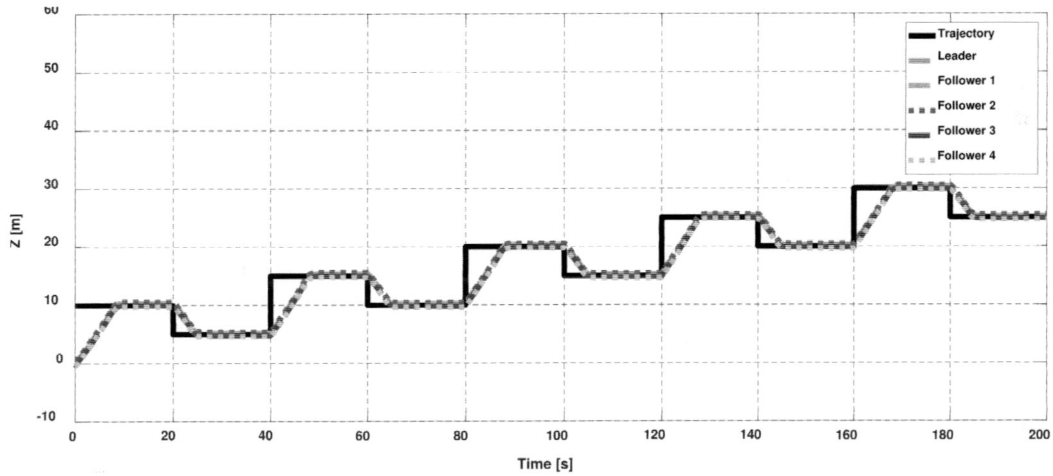

Figure 18. Trajectory tracking of BS-PID controller tuned by FOPID for cooperative quadrotors in ***Z***-orientation.

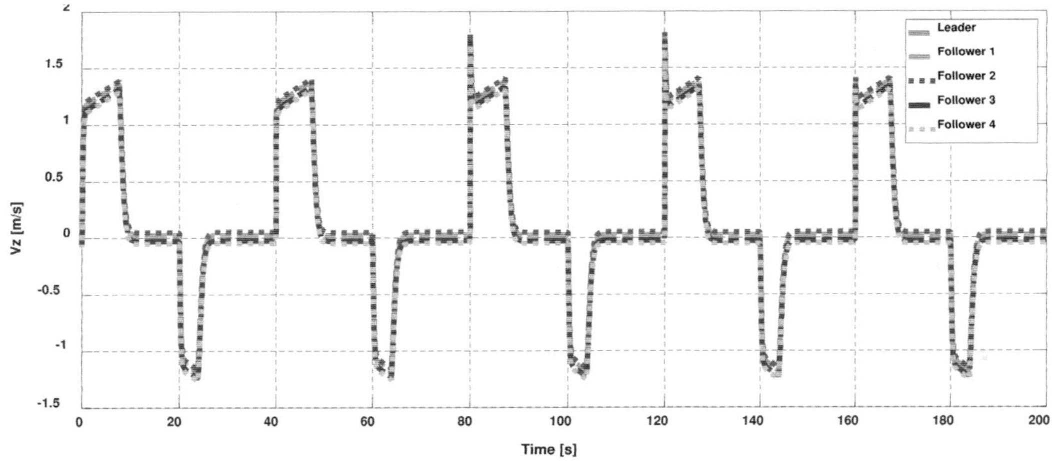

Figure 19. Velocity of cooperative quadrotors in ***Z***-orientation.

Figures 14-19 show successful of the trajectory tracking for a team of cooperative quadrotors controlled by BS-PID controller tuned by FOPID, and their velocities respect the designed constraints 2 m/sec.

Figures 20 shows the $(X - Y)$ plane, the leader succeeded in tracking the desired path along the $(X - Y)$ plane. Moreover, follower (1) and follower (2) respect the desired separating distances 40m and 30m along the X-axis and Y-axis, respectively. Follower (3) and follower (4) respect the desired separating distances 20m and 60m along the X-axis and Y-axis, respectively. Figure 21 displays the trajectory tracking of cooperative quadrotors controlled by BS-PID controller tuned by FOPID in XYZ orientations.

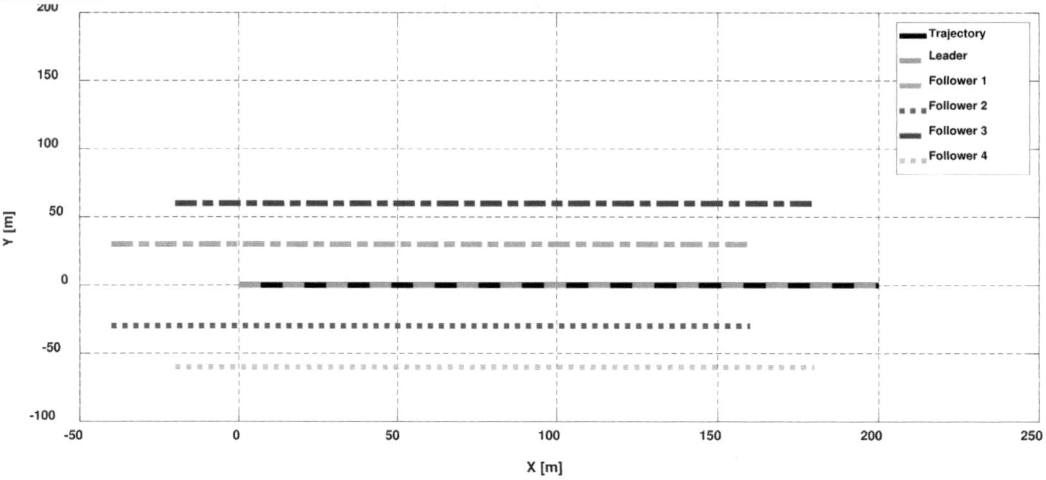

Figure 20. Trajectory tracking of BS-PID controller tuned by FOPID for cooperative quadrotors in $\boldsymbol{XY}$-orientation.

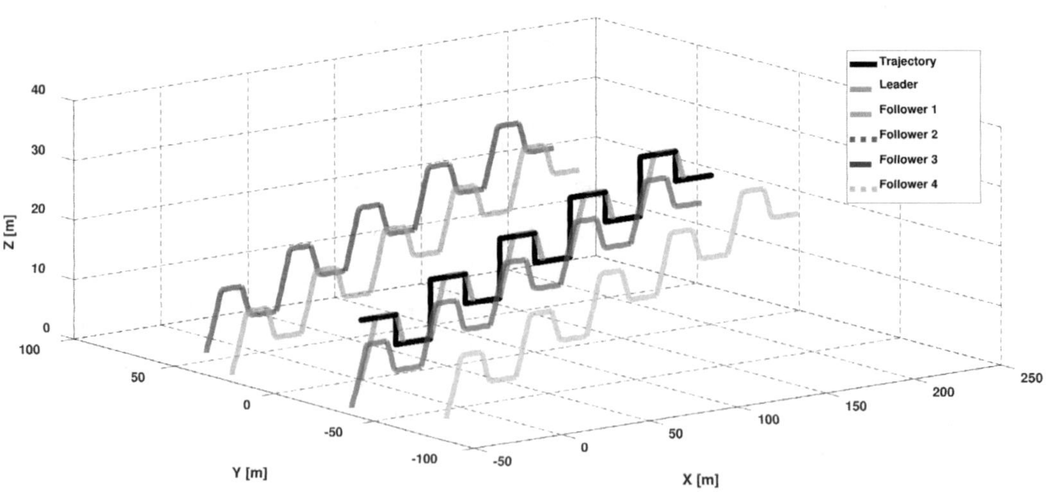

Figure 21. Trajectory tracking of BS-PID controller tuned by FOPID for cooperative quadrotors in $\boldsymbol{XYZ}$-orientation.

Figures 22-24 display the roll, pitch, and yaw channels of cooperative quadrotors controlled by BS-PID controller tuned by FOPID controller, respectively. All the quadrotors respect the desired angles during the free flight.

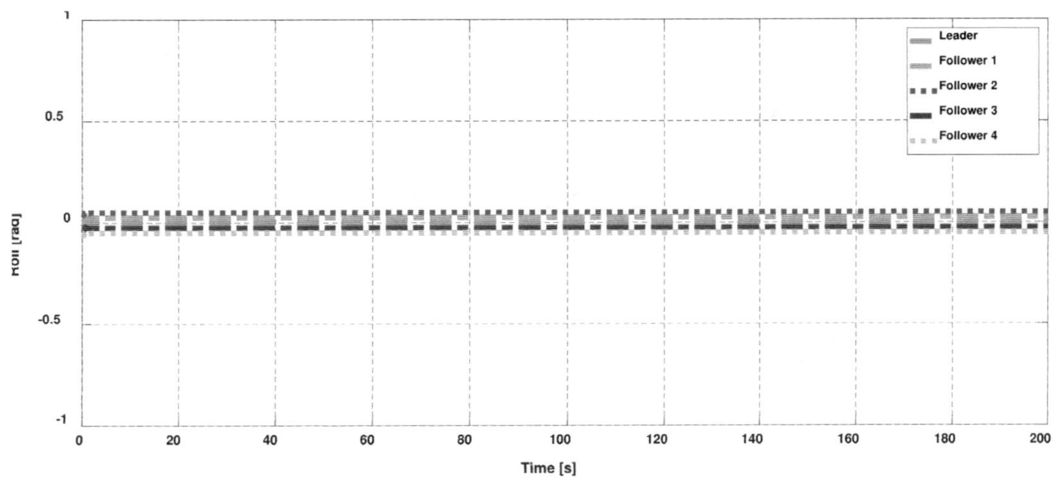

Figure 22. Roll control of BS-PID controller tuned by FOPID for cooperative quadrotors.

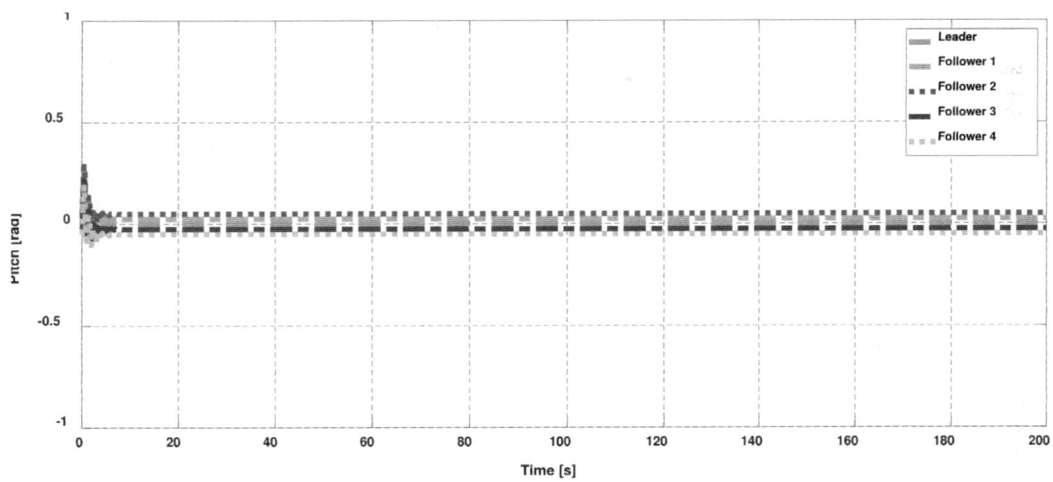

Figure 23. Pitch control of BS-PID controller tuned by FOPID for cooperative quadrotors.

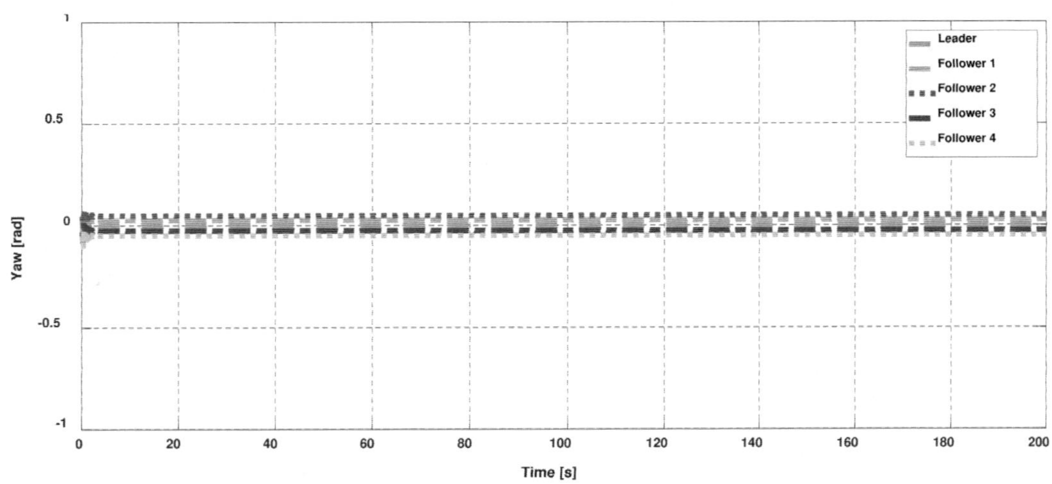

Figure 24. Yaw control of BS-PID controller tuned by FOPID for cooperative quadrotors.

3.3. Wind disturbance rejection and noise attenuation scenario:

Stability and robustness evaluation of BS-PID controller tuned by LOPID and FOPID controller are realized. The proposed controllers are checked on the leader of the cooperative quadrotors in different trajectories. The proposed controllers are checked and compared together in the presence of wind disturbance and noise to show their robustness.

Figures 25-28 display the trajectory tracking of the leader of the cooperative quadrotors controlled by BS-PID controller under the effect of wind disturbance in altitude and attitude orientations, respectively. The wind disturbances are represented as a sudden pulse at a periodic time of 10 sec in different direction to show the ability of the introduced approach to reject it. The obtained results show the ability of the underlying controllers to stabilize and control the leader quadrotor under effect of wind disturbances.

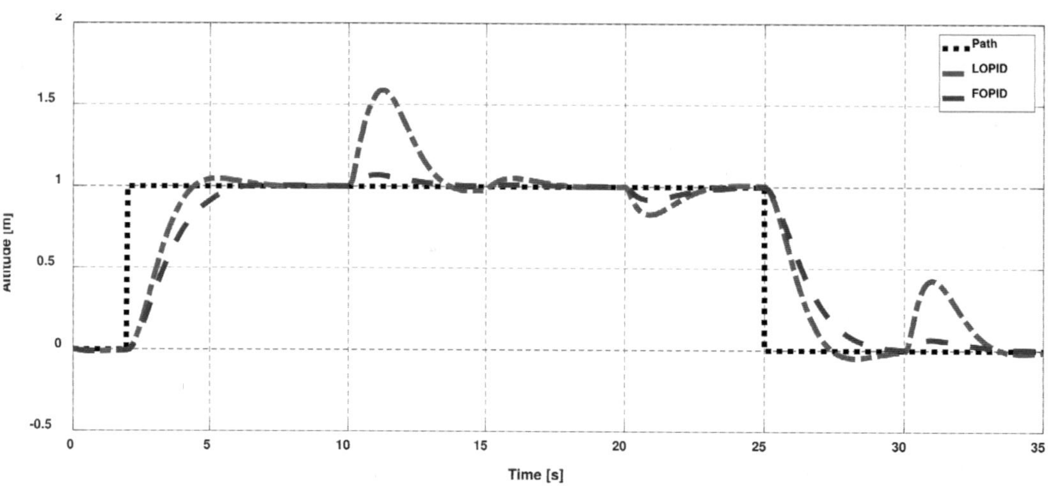

Figure 25. Trajectory tracking of BS-PID controller tuned by LOPID and FOPID for the leader of the cooperative quadrotors in ***Z***-orientation under effect of wind disturbance.

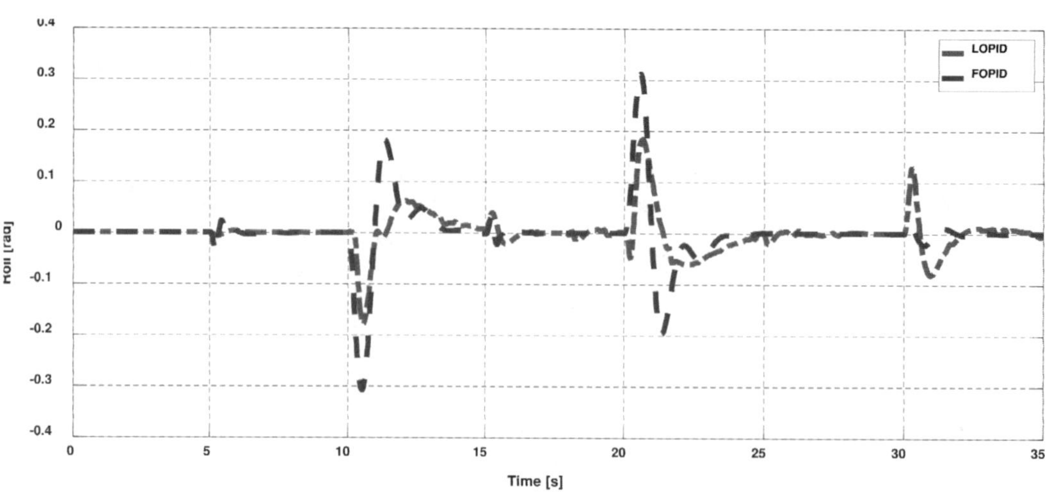

Figure 26. Roll control for BS-PID controller tuned by LOPID and FOPID under effect of wind disturbance.

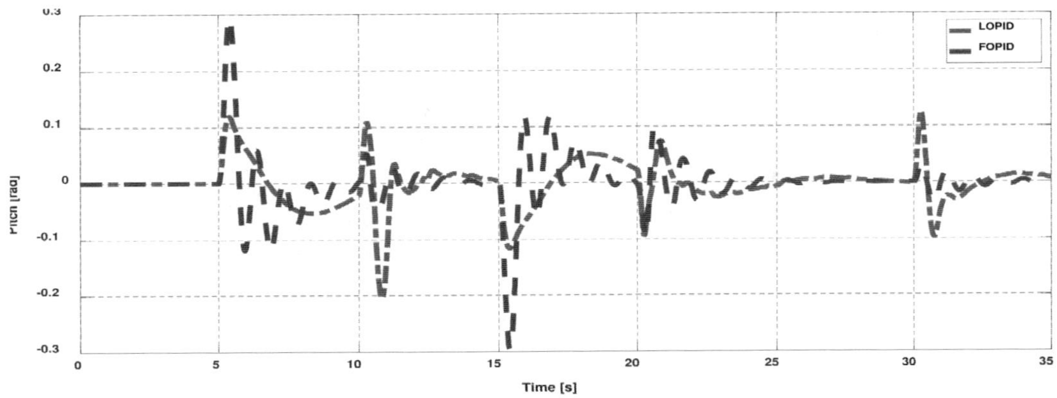

Figure 27. Pitch control for BS-PID controller tuned by LOPID and FOPID under effect of wind disturbance.

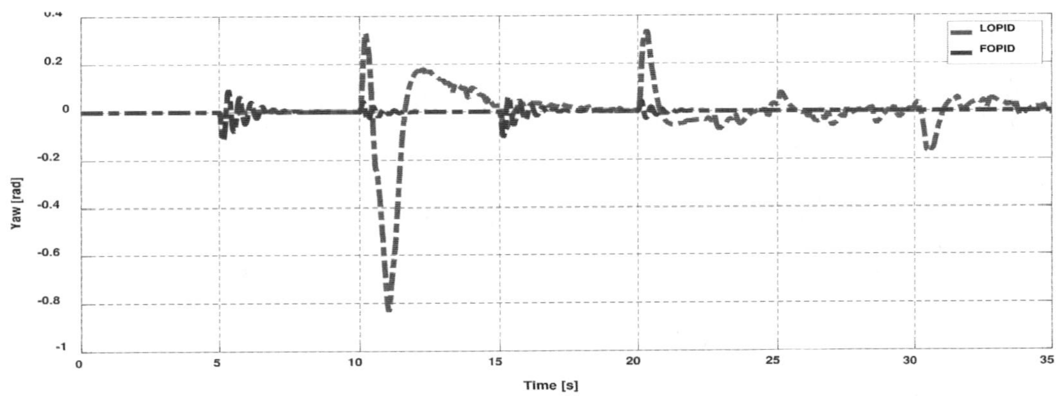

Figure 28. Yaw control for BS-PID controller tuned by LOPID and FOPID under effect of wind disturbance.

Figures 29-32 display the trajectory tracking of the leader of the cooperative quadrotors controlled by BS-PID controller under the effect of sudden noise in altitude and attitude orientations, respectively. The obtained results show the ability of the underlying controllers to stabilize and control the leader quadrotor under effect of white noise.

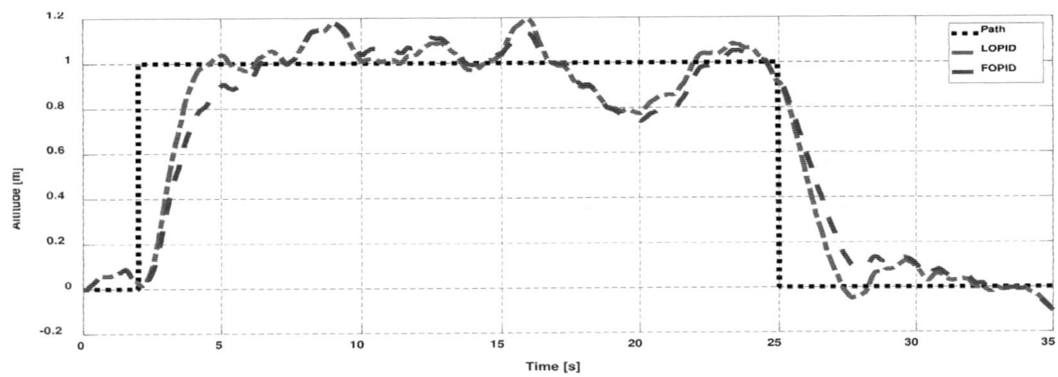

Figure 29. Trajectory tracking of BS-PID controller tuned by LOPID and FOPID for the leader of the cooperative quadrotors in ***Z***-orientation under effect of noise.

IOP Conf. Series: Materials Science and Engineering 1172 (2021) 012041 doi:10.1088/1757-899X/1172/1/012041

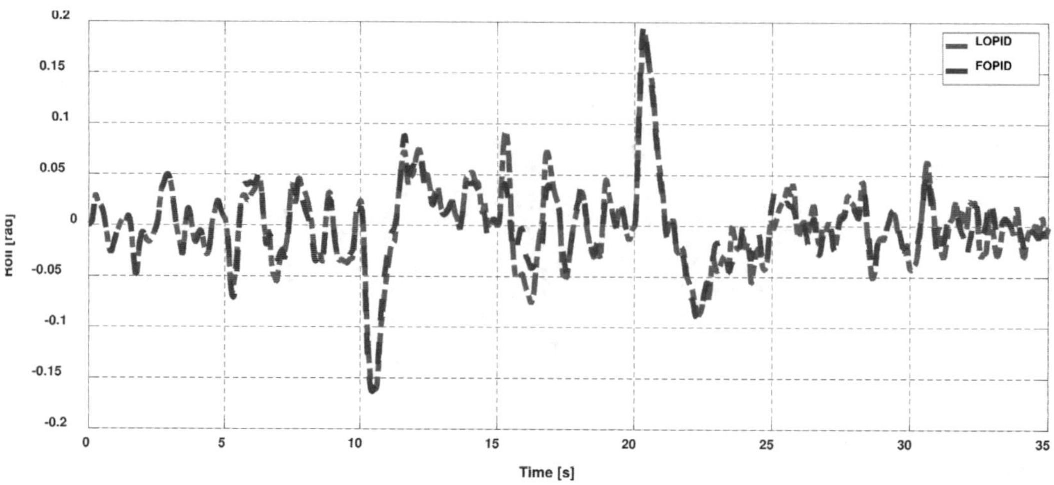

Figure 30. Roll control for BS-PID controller tuned by LOPID and FOPID under effect of noise.

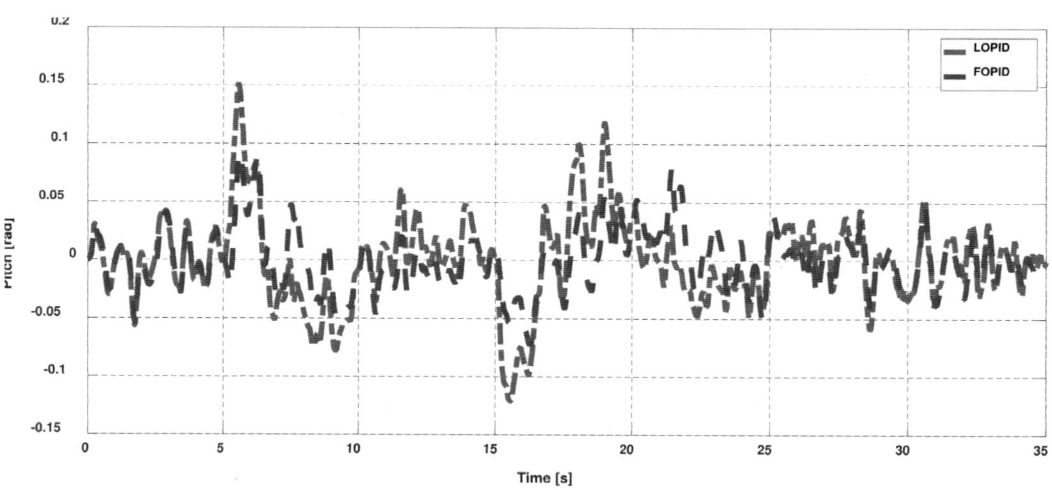

Figure 31. Pitch control for BS-PID controller tuned by LOPID and FOPID under effect of noise.

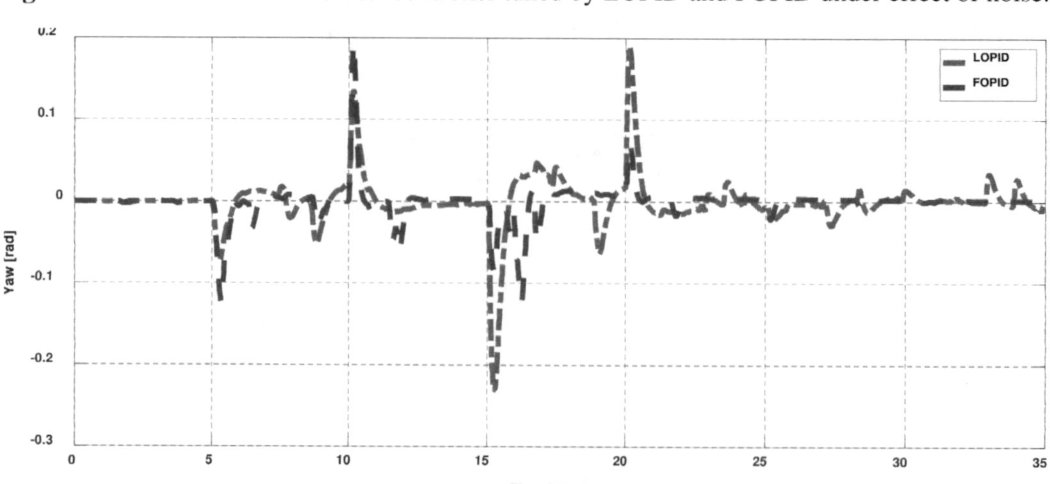

Figure 32. Yaw control for BS-PID controller tuned by LOPID and FOPID under effect of noise.

4. Conclusion

In this paper, BS-PID controller tuned by LOPID and FOPID is conceived for dissolving formation configuration issue for cooperative quadrotors UAVs. Couple of scenarios were presented to show the ability of the introduced approaches in solving the formation problem. The simulation outcomes demonstrate a perfect stable flight attained by the proposed controllers in the appearance and disappearance of applied disturbances and white noises. The simulation results show the superiority of FOPID over LOPID. Future work, the proposed approaches will be applied to solve the formation problem for cooperative unmanned vehicles in an obstacle loaded environment.

References

[1] Hou Z, Wang W, Zhang G and Han C 2017 A survey on the formation control of multiple quadrotors. In: *Ubiquitous Robots and Ambient Intelligence (URAI), 2017 14th International Conference on*: IEEE) pp 219-25

[2] Mahfouz M, Hafez A T, Ashry M M and Elnashar G 2019 Cyclic Leader-Follower Strategy For Cooperative Unmanned Aerial Vehicles. In: *2019 IEEE International Conference of Vehicular Electronics and Safety (ICVES)*: IEEE) pp 1-6

[3] Norouzi Ghazbi S, Aghli Y, Alimohammadi M and Akbari A 2016 QUADROTORS UNMANNED AERIAL VEHICLES: A REVIEW *International Journal on Smart Sensing & Intelligent Systems* **9**

[4] Zhang X, Li X, Wang K and Lu Y 2014 A Survey of Modelling and Identification of Quadrotor Robot *Abstract and Applied Analysis* **2014** 16

[5] Gonzalez-Vazquez S and Moreno-Valenzuela J 2010 A New Nonlinear PI/PID Controller for Quadrotor Posture Regulation. In: *2010 IEEE Electronics, Robotics and Automotive Mechanics Conference,* pp 642-7

[6] Xu R and Ozguner U 2006 Sliding mode control of a quadrotor helicopter. In: *Decision and Control, 2006 45th IEEE Conference on*: IEEE) pp 4957-62

[7] Madani T and Benallegue A 2006 Backstepping control for a quadrotor helicopter. In: *Intelligent Robots and Systems, 2006 IEEE/RSJ International Conference on*: IEEE) pp 3255-60

[8] Wang Y, Chenxie Y, Tan J, Wang C, Wang Y and Zhang Y 2015 Fuzzy radial basis function neural network PID control system for a quadrotor UAV based on particle swarm optimization. In: *Information and Automation, 2015 IEEE International Conference on*: IEEE) pp 2580-5

[9] Wong T, Khan R and Lee D 2014 Model linearization and H∞, controller design for a quadrotor unmanned air vehicle: Simulation study. In: *Control Automation Robotics & Vision (ICARCV), 2014 13th International Conference on*: IEEE) pp 1490-5

[10] Mahfouz M, Ashry M and Elnashar G 2013 Design and control of quad-rotor helicopters based on adaptive neuro-fuzzy inference system *International Journal of Engineering Research Technology (IJERT)* **2** 479-85

[11] Mahfouz M and Kader S A 2015 Quadrotor unmanned aerial vehicle controller design and synthesis. In: *Computer Engineering & Systems (ICCES), 2015 Tenth International Conference on*: IEEE) pp 60-6

[12] Mahfouz M, Hafez A T, Ashry M M and Elnashar G 2020 Formation Reconfiguration Based on Backstepping-PID Controller for Collaborative Quadrotors. In: *2020 12th International Conference on Electrical Engineering (ICEENG)*: IEEE) pp 396-401

[13] Mahfouz M, Hafez A T, Azzam A, Ashry M M and Elnashar G 2020 Formation Reconfiguration in Dynamic Obstacle Loaded Environment via Backstepping Approach. In: *2020 12th International Conference on Electrical Engineering (ICEENG)*: IEEE) pp 441-6

[14] Fethalla N, Saad M, Michalska H and Ghommam J 2017 Robust observer-based backstepping controller for a quadrotor UAV. In: *Electrical and Computer Engineering (CCECE), 2017 IEEE 30th Canadian Conference on*: IEEE) pp 1-4

[15] Ashry M M, Kamalova Z Z and Breikin T V 2008 Tuning of digital PID controller parameters using local optimal control. In: *2008 16th Mediterranean Conference on Control and*

Automation: IEEE) pp 587-92
[16] Ashry M, Abou-Zayed U and Breikin T 2008 Comparative robustness study of multivariable controllers for a gas turbine engine. In: *IASTED International Conference on Modelling, Identification, and Control,*
[17] Ashry M, Abou-Zayed U and Breikin T 2008 Control of multivariable systems using modified local optimal controller *IFAC Proceedings Volumes* **41** 8767-72
[18] Cajo R, Mac T T, Plaza D, Copot C, De Keyser R and Ionescu C 2019 A survey on fractional order control techniques for unmanned aerial and ground vehicles *IEEE Access* **7** 66864-78
[19] Wang C, Li J, Liu Y and Li M 2014 Research on UAV attitude control based on fractional order proportional integral controllers. In: *Proceedings of the 33rd Chinese Control Conference*: IEEE) pp 3552-7
[20] Mahfouz M, Hafez A T, Ashry M and Elnashar G 2018 *2018 AIAA SPACE and Astronautics Forum and Exposition*: American Institute of Aeronautics and Astronautics)
[21] Habashi A, Ashry M, Mabrouk M H and Elnashar G 2018 Comparative Study Among Different Control Techniques For Stabilized Platform *Engineering Science and Military Technologies* **2**
[22] Habashi A, Ashry M, Mabrouk M and Elnashar G 2015 Controller Design for Line of Sight Stabilization System *International Journal of Engineering Research and Technology*
[23] Mahfouz M, Taiomour A, Ashry M M and Elnashar G 2021 PID tuning approaches for quadrotors unmanned aerial vehicles. In: *The 19th International Conference on Aerospace Sciences & Aviation Technology (ASAT-19)*: IOP Publishing)

IOP Conf. Series: Materials Science and Engineering 1172 (2021) 012042 doi:10.1088/1757-899X/1172/1/012042

Impact of nozzle profile on ballistic performance and structural loads

M Saleh[1], S Saleh[1], A Zayed[2] and M Y M Ahmed[1]
[1] Department of Rockets, Military Technical College, Cairo, Egypt.
[2] Nahda University, Egypt.

E-mail: saleh_uchiha@ymail.com

Abstract. A nozzle is a device that is designed to regulate the direction and characteristics of the combustion gas products of jet engines. So, the nozzle performance has a significant impact on the mission achievement. This paper is concerned with the internal ballistics of the nozzle aiming to estimate pressure and thermal loads on its walls. Computational fluid dynamics is applied to analyse the effect of changing nozzle internal profile on the resulting thrust, flow energy losses, and nozzle wall structure. Area ratios at the inlet, critical, and exit sections are considered as constraints for the examined design. Two different sets with 4 different profiles for each are investigated. The results show thrust, entropy losses across the nozzle and the static pressure and temperature at nozzle wall. Bell shape profiles produce better performance compared to other profiles. Changing the internal profile of the nozzle causes significant change in pressure and temperature loads acting on nozzle wall structure.

1. Introduction

A nozzle is a device that is used to convert pressure and thermal energy, generated in the combustion chamber of the engine, into a high velocity jet flow of temperature and pressure at its exit. The converging diverging nozzle contour includes a convergent part, throat, and a divergent part. Each part affects the flow in a different way and must be considered in the analysis of the nozzle performance [1].

Nozzle contours can be categorized into two major types, conical nozzles, and bell nozzles. The conical nozzle profile has a fixed diverging angle and so can be easily produced for any acceptable expansion ratio. The conical profile are presently used for short nozzles when simple design is desired over performance [2], [3].

To enhance the efficiency by turning the flow momentum axially, contour or bell nozzles were introduced. Compared to the conical nozzle, they are generally shorter and lighter. The design of non-linear contour is however difficult and incorrect design prompts the undesirable occurrence of shocks within the nozzle.

There are other several derivatives of the two main nozzle profiles including dual bell, double bell, multi divergent angles, and arc-based nozzle.

Thrust is a key feature performance merit of the nozzle. A higher thrust reflects a better nozzle design. It is also important that losses in flow energy are reduced. However, in many cases, the ability of nozzle structure to withstand structural and thermal loads is of a special interest. Hence., understanding both ballistic performance and structural loads is sought. A nozzle profile is a key feature that is believed to decide both factors.

Content from this work may be used under the terms of the Creative Commons Attribution 3.0 licence. Any further distribution of this work must maintain attribution to the author(s) and the title of the work, journal citation and DOI.
Published under licence by IOP Publishing Ltd

The method of characteristics (MOC) has been a leading technique for designing contour nozzles. It is based on the theory for centered expansion waves for supersonic flow [4] and was applied in early attempts for designing thrust optimized nozzles by Guderley and Hantsch [5].

Rao developed a much simpler technique for nozzle contouring that was broadly used in 1960's [2], [6]. Rao and Shmyglevsky developed a way to alter MOC method in order to generate an optimum nozzle which was much shorter; the 'bell nozzle'. This approach used a combination of Lagrangian multipliers and MOC while maintaining the length of the nozzle as constant constrain. A distinctive contour that can be produced to maximize thrust for a specified length was described by Rao [7] often referred to thrust optimized contour (TOC). Rao developed the approximation through a list of contour points using skewed parabola; thrust optimized parabola (TOP).

The geometric parameters of bell profile according to Rao approximation are shown in fig. 1 below.

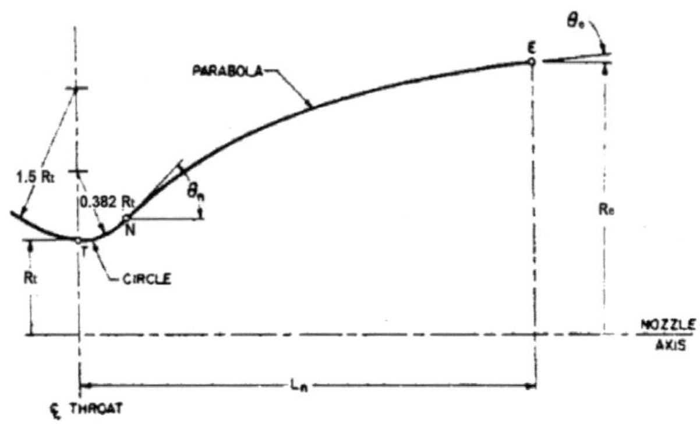

Figure 1. Sketch of a nozzle showing nomenclature and construction of parabola [7].

Ashwood et al. [8] studied the benefit of using a profiled divergent part and the effect of divergence angle on the thrust. The results show that the profiled nozzle gives about 2 percent more thrust than the conical one. Madhu et al. [9] numerically simulated expansion through different profiles for rocket nozzle. They found that the contoured nozzles produce higher exit velocity and higher degree of flow separation compared to conical nozzles. Dumitrescu et al. [10] presented a study and a design of a Laval nozzle, and obtained its influence on the performances. They concluded that the design parameters play an important role in determining the maximum attainable performance. Ömer, and Günaltay [11] investigated the design and flow differences of two different nozzle geometries and their structural effects on two different materials. They stated that bell nozzle obtains a slightly better performance by reaching a higher maximum Mach number and hence pressure inside the nozzle is decreased and thereby decreasing the structural effects on the nozzle walls.

Thies and Jordan [12] studied the effect of changing the nozzle configuration at entrance, throat and exit on thrust and specific impulse efficiencies. They found that a generous throat contour radius and a smooth exit cone increase both nozzle throat and thrust efficiencies. Joe [13] developed a method for designing compressed truncated perfect nozzles and developed a procedure for predicting the performance of such nozzles. The results show that in all cases the Rao nozzle has higher performance than compressed truncated perfect nozzles. Sreenath and Mubarak [14] studied and analyzed four different types of bell nozzles and one dual bell nozzle numerically. They stated that the dual bell nozzle has better overall performance than the single bell-shaped nozzle. At low altitudes, a vehicle could save 25-30% more fuel by using a dual bell nozzle. Mubarak [15] designed a double parabolic nozzle and compared its performance with conical and bell nozzle designs and concluded that the double parabolic nozzle achieved higher values of discharge and thrust coefficients than the conventional conical and bell nozzles. Singh et al. [16] developed a way to identify the optimum nozzle geometry that maximizes critical pressure ratio while minimizing pressure drop across the nozzle, they found that a lower diverging angle and no elongated throat resulted in a higher critical pressure ratio. Schomberg et al. [17]

IOP Conf. Series: Materials Science and Engineering 1172 (2021) 012042 doi:10.1088/1757-899X/1172/1/012042

compared a total of ten arc-based nozzle contours numerically to an existing thrust-optimized design. The results show that an increase in thrust of 0.25% could be achieved in an equivalent arc-based nozzle compared to the existing design. The analysis of the arc-based nozzles indicated that the contour angles had the greatest effect on thrust.

The nozzle is exposed to several thermal and mechanical loads acting on the internal wall as compressive stresses while the throat is the most exposed region to high temperature and pressure. Cozart [18] analyzed the stress distribution of a 3-D braided conical nozzle under mechanical and thermal loading conditions and it was concluded that radial stress is mildly compressive through most of the nozzle structure and the nozzle responds to internal pressure and thermal loading by tensile hoop stress distribution. Gomaa et al. [19] studied the thermo-structural response of a conical nozzle made of both steel and composite material. They addressed the loads acting on the nozzle due to combustion gases temperature and pressure. Wall temperature and pressure were obtained from experimental results. Gong et al. [20] investigated the thermo-structural response of submerged nozzles. They found that the stresses increased clearly in case of combining both thermal and mechanical loadings much more than in the case of each one separately. The hoop stress at throat insert was found to increase at first and then decrease with the time.

It's clear that the previous studies were almost totally devoted to understanding and optimizing nozzle ballistic performance in terms of thrust. On the other hand, the studies on thermal and structural stresses on nozzles did not take into full consideration enhancing the ballistic performance.

The aim of the present paper is to compare a variety of nozzle profiles. Namely, conical, bell, and modified bell. The objective is to estimate the structural and thermal loads acting on nozzle walls. These loads are derived from simulating the internal flow for given chamber pressure and expansion ratio. In addition, the impact of nozzle profile on its performance (in terms of thrust and losses) is addressed.

2. Case study and Methodology

2.1. Case study

A total of eight nozzle profiles are examined, they are grouped into two sets. Set 1 is characterized by a profiled convergent section with inlet-to-throat area and length-to-throat radius ratios of 45.4 and 8.87, respectively. Nozzles of set 1 are attached to a chamber of 73.2 bar, 3475.8 K, and 0.8 bar for combustion pressure and temperature and exit pressure, respectively. Set 2 is characterized by a conical convergent section with inlet-to-throat area and length-to-throat radius ratios of 5.23 and 2.59 , respectively. Nozzles of set 2 are attached to a chamber of 24.13 bar, 2964.7 K, and 1 bar for combustion pressure and temperature and exit pressure, respectively. For each set, four divergent profiles are used namely: conical, 3-angle conical, conventional bell, and modified bell, fig.2. The conical profile has the angle of 12° for set 1 and 15.3° for set 2 while the 3-angle profiles has the angles of 23° , 12° , and 4.1° for set 1 and 28° , 15° , and 7° for set 2 respectively at 30%, 30%, and 30% of divergent lengths.

The bell profiles are constructed based on Rao approximation for the method of characteristics [7]. A MATLAB code is developed to generate the bell nozzle contours using quadratic Bézier curve equations for the given data namely:

α, Equivalent conical nozzle half angle,
Rt, Throat diameter,
Re, Exit diameter,
θ_n, Angle of bell contour start point, and
θ_e, Angle of bell contour end point.

Fig. 3 shows the values of θ_n and θ_e for different expansion ratios ε.

The conventional bell profile is based on fig. 3 for the present cases, $\varepsilon = 4.59$, and hence $\theta_e = 9.5°$, and $\theta_n = 19.5°$ for set 1 $\varepsilon = 4$, and hence $\theta_e = 9.2°$, and $\theta_n = 19°$ for set 2.

At low area ratios, values of both angles θ_n, θ_e are relatively close to each other as shown in fig. 3. which generates a bell contour that is relatively similar to the conical contour.

A modified bell is proposed in the present study by putting θ_e equal zero to produce axial flow at the exi

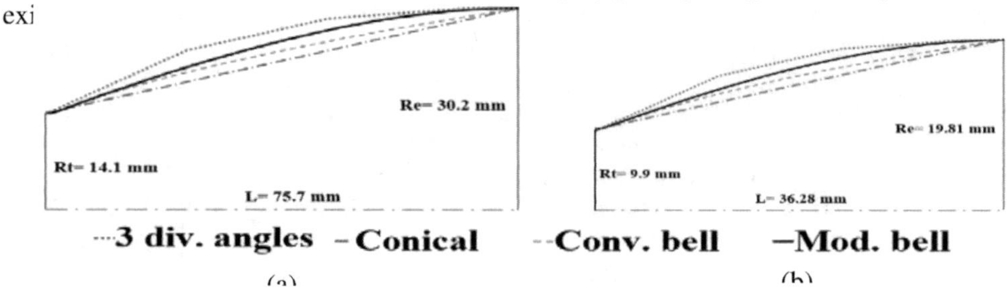

Figure 2. Sketches of the four divergent profiles for set1 and set 2, respectively.

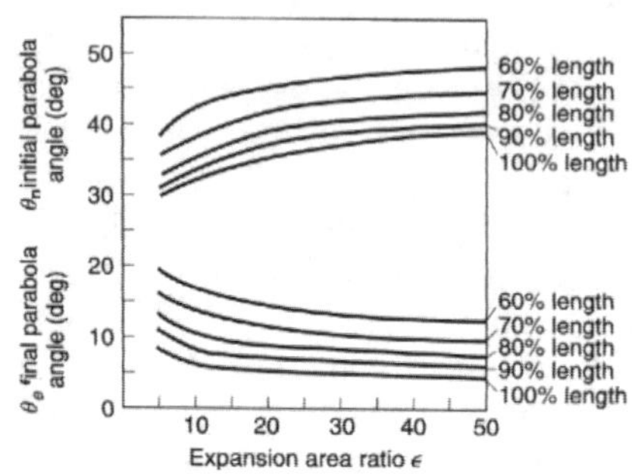

Figure 3. The initial angle θ_n and the exit angle θ_e for bell nozzles as functions of the nozzle area ratio [21].

The goal of study is to address the thermal and structural loads on the nozzle walls as well as its ballistic performance. Thrust of nozzle is adopted as a measure of nozzle ballistic performance. As for losses, entropy change across the whole nozzle is taken as the measure. The ratio of thrust to losses is introduced here as a measure of ballistic performance indicator (PI) of the nozzle profile. Flow static pressure and adiabatic wall temperature are adopted as measure for structural and thermal stresses on the nozzle, respectively.

2.2. *Flow simulation setup*

A 2D axisymmetric domain is constructed for each case. The boundary definition is illustrated in fig. 4. At the inlet, a pressure inlet is defined where absolute and total pressures are set to chamber pressure. At the exit, pressure outlet boundary is defined where total temperature is set equal to chamber temperature. The nozzle walls are set as adiabatic to nullify heat transferred no-slip condition is applied.

A commercial CFD solver is used. The second-order discretization scheme is applied for a density-based solver. Since focus is on nozzle profile, air as an ideal gas is used to represent combustion gas products.

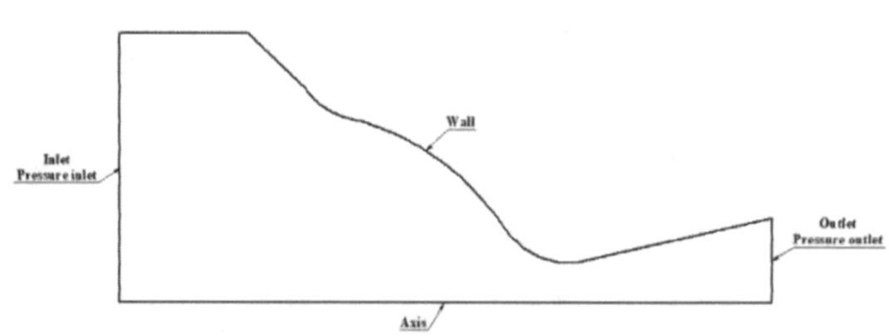

Figure 4. The boundary definition.

2.3. *Meshing*

Table. 1 Number of divisions and the bias factor for each edge.

Mesh No.	Sizing Type	V1.. V9	H1, H2	H3, H4	H5, H6	H7, H8	H9, H10	H11, H12	H13, H14	H15, H16	Nodes Number	Elements Number
1	No. of divisions	100	50	30	20	120	50	10	7	150	44238	43700
	Bias Factor	10	—	—	—	3	—	—	—	5		
2	No. of divisions	150	50	30	20	120	75	15	10	225	82446	81750
	Bias Factor	15	—	—	—	3	—	—	—	5		
3	No. of divisions	225	50	30	20	120	115	25	15	335	160686	159750
	Bias Factor	15	—	—	—	7	—	—	—	7		
4	No. of divisions	340	50	30	20	120	175	40	25	500	327701	326400
	Bias Factor	15	—	—	—	15	—	—	—	7		
5	No. of divisions	550	50	30	20	120	270	70	50	750	749911	748000
	Bias Factor	15	—	—	—	24	—	—	—	12		

The fluid domain is made up of linear hexahedral elements that have been meshed. Since the edges are simple, the mesh can be transformed to structured meshing with Mapped Face Meshing. The number of divisions and the bias factor are varied on the inlet, exit, nozzle walls, nozzle axis and other edges, as seen in fig. 5.

A structured grid is generated for each case. The grid is made clustered at the walls, and nozzle throat to capture key flow details with high resolution. A grid sensitivity check is made for five different forms of meshes obtained by adjusting the number of divisions (see fig. 6, and table 1).

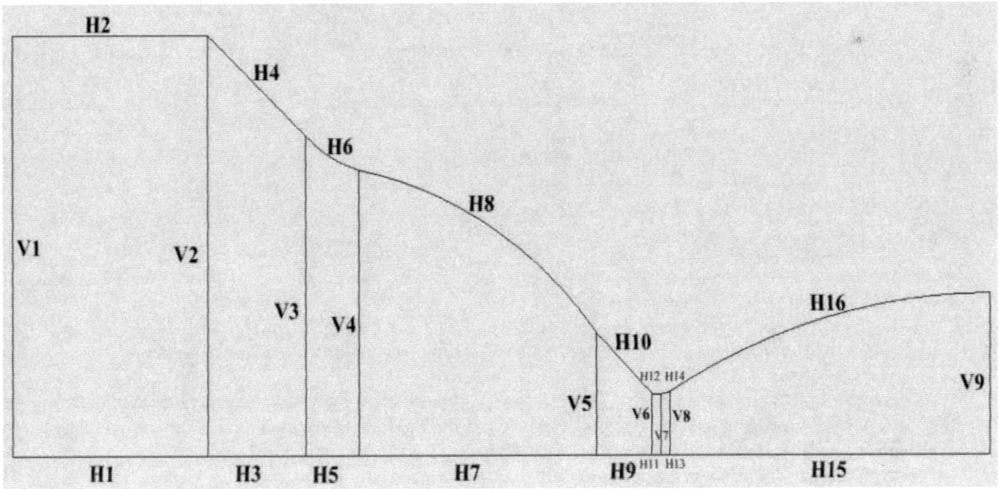

Figure 5. Nozzle edges and their nomenclature.

Mesh 5 has been selected as the most appropriate mesh due to the results convergence.

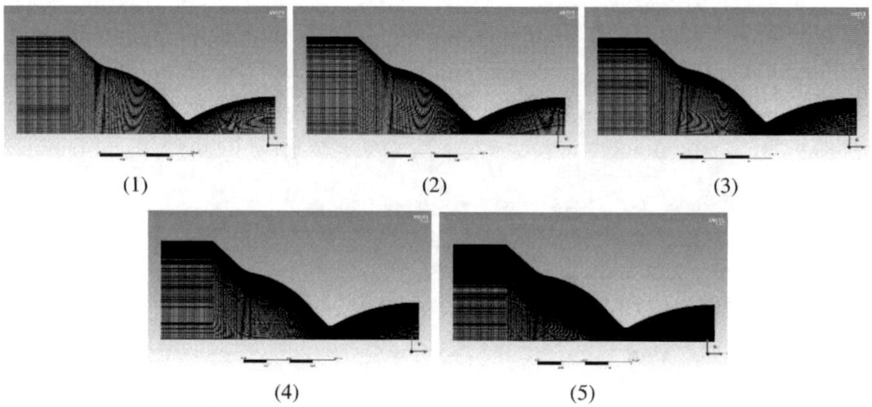

Figure 6. Mesh variation for mesh sensitivity analysis.

3. Results and Discussion.

3.1. Validation of the flow solver

A validation is done for commercial CFD package by comparing experimental results in NASA technical report [22] to these of numerical simulation. For validation, a conical nozzle with 15º half angle and area ratio 8 is chosen while the inlet-to-exit pressure ratio is varied. Shown in fig.7 the error is less than 8%.

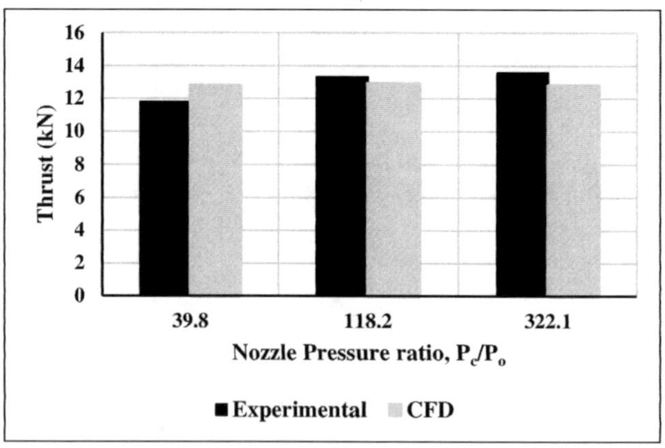

Figure 7. Results of validation for CFD solver.

3.2. Nozzle ballistic performance.

Fig. 8 shows the ballistic performance for both sets 1, and 2 for the different profiles. Thrust (F) at outlet section and the difference between flow's entropy (ΔS) at outlet and inlet sections and the ratio F/ΔS are all displayed in the figure.

It is shown that the 3-angles profile has the lowest performance and the highest losses while the conventional bell profile has the highest performance but not the lowest losses. The bell profile yields a slight improvement in nozzle PI compared to the conical one.

Through the modification in the conventional bell profile (assuming angle of bell contour end point θ_e equal to zero) lower losses values attained which leads to a higher PI as shown in modified bell profile.

The variation in Mach number contours with internal profile change is illustrated in fig. 9. It can be shown that due to simple modification for conventional bell profile, the position of the shock wave is shifted from inside the nozzle to outside which is desirable as far as ballistic performance is concerned fig. 9. c, and d.

3.3. Nozzle structural loads.

Fig. 10. Shows the contours for wall pressure and adiabatic temperature for the nozzle profiles examined. It is shown that the 3-angles and conventional bell profiles have the lowest static pressure and adiabatic wall temperature compared to those in the conical profile. The 3-angles profile is not considered as a potential candidate profile due to its poor ballistic performance. So, and as expected, the modified bell profile that has the highest PI also has the lowest thermal and mechanical loads hence the lowest thermo-mechanical stresses acting on the nozzle wall structure.

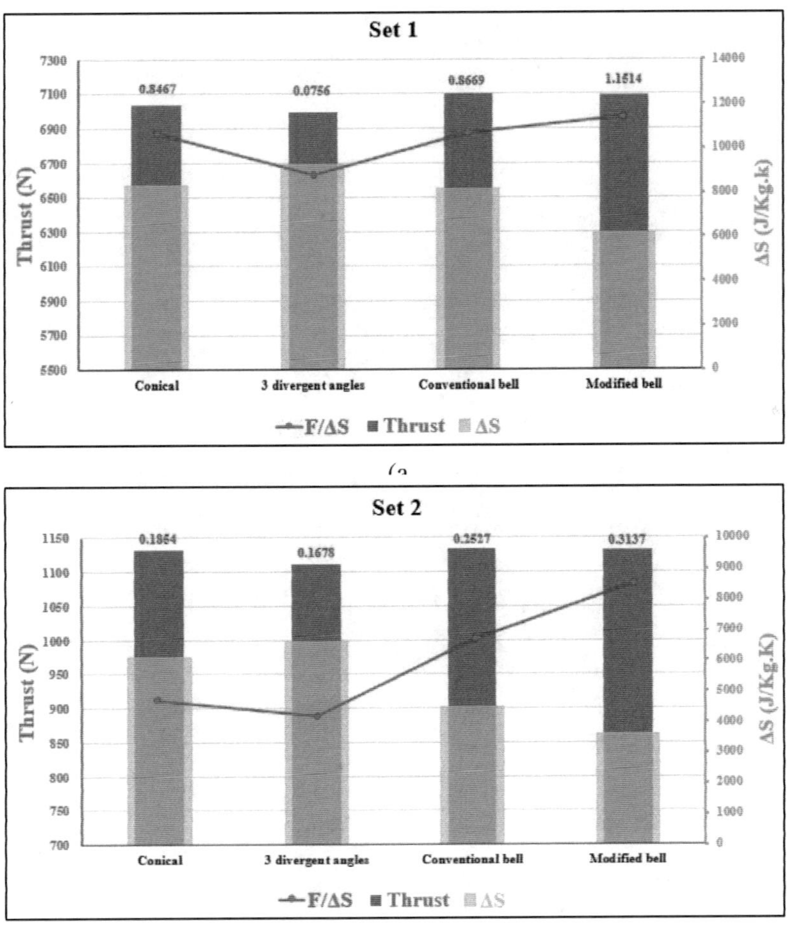

Figure 8. Ballistic performance parameters of nozzle profiles of (a) set 1, and (b) set 2.

(a)

(b)

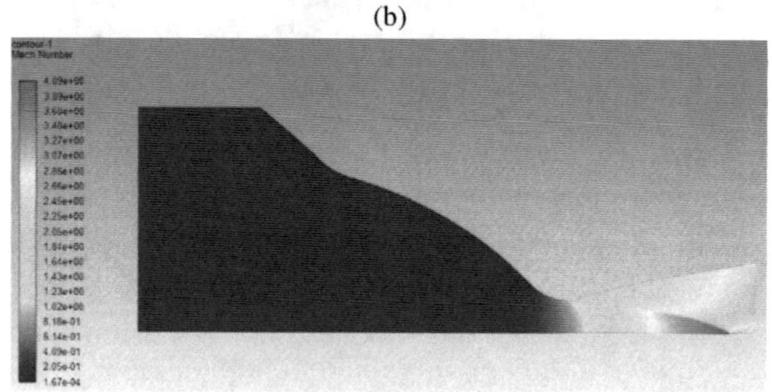

(c)

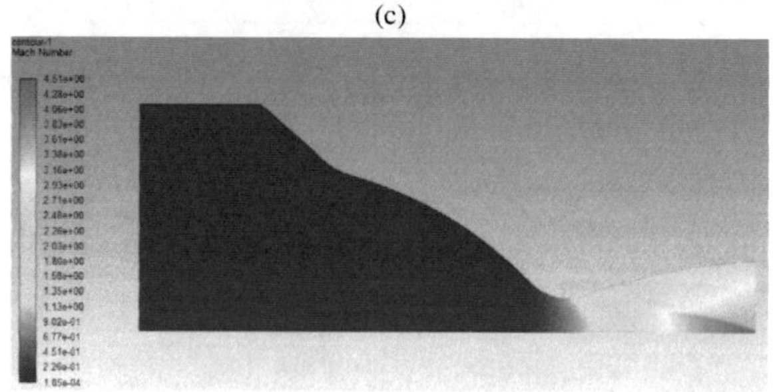

(d)

Figure 9. Mach number Contours of set 1 for (a) Conical, (b) 3-angles, (c) conventional bell, and (d) modified bell profiles.

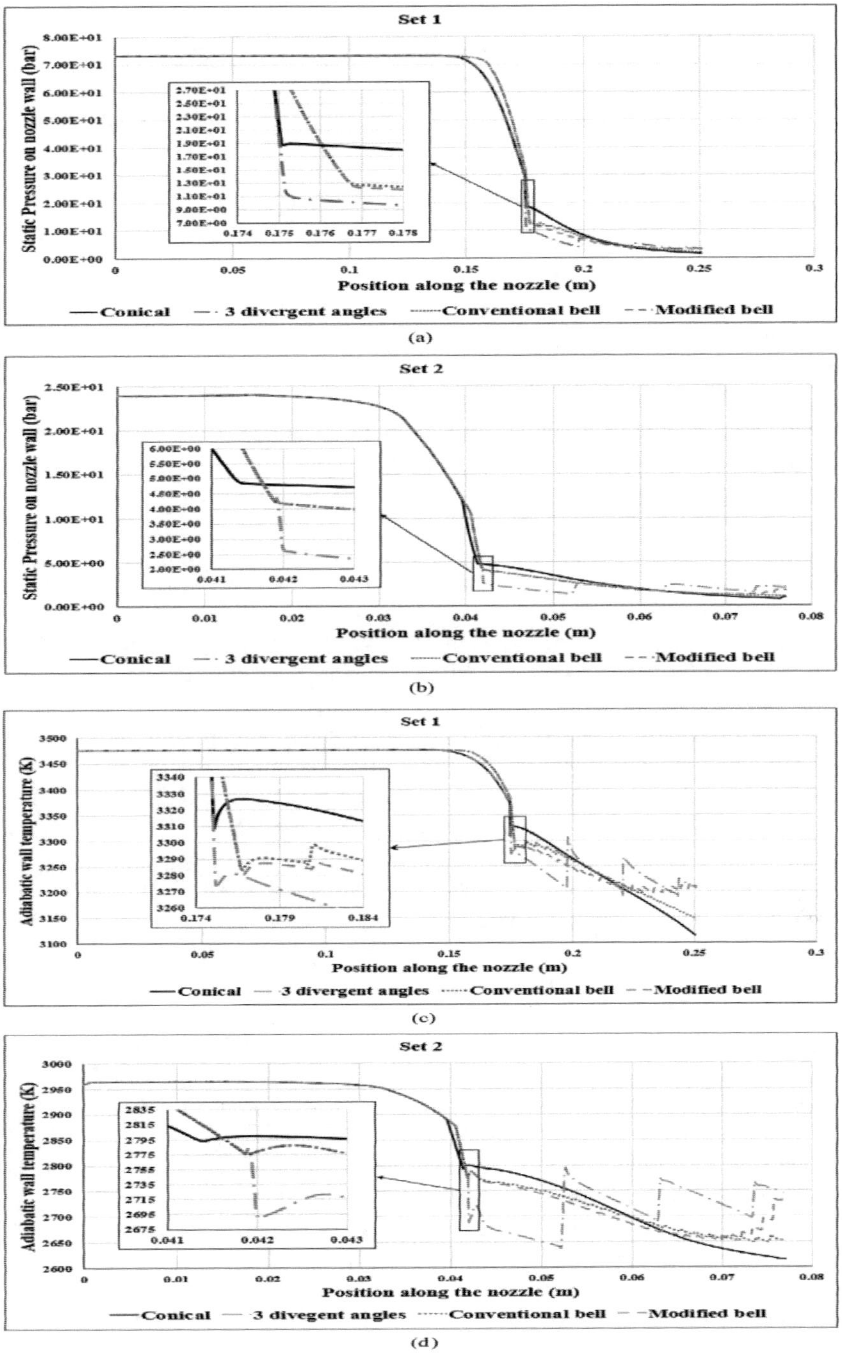

Figure 10. (a, and b) static pressure on nozzle walls for set 1, and 2, respectively, and (c, and d) adiabatic wall temperature for set 1, and 2, respectively.

4. Conclusion

The objective of the present paper was to investigate the impact of nozzle profile on both ballistic performance and structural loads. Eight different profiles were examined including conical, 3-angles, conventional bell, and a proposed modified version of it. The following can be stated as conclusions of the present research.

- Bell contoured nozzles has better PI than the conical contoured nozzles.
- Three-angle profile has the lowest PI because of its sharp edges that exist along the divergent part.
- Regarding the performance indicator results, bell contoured nozzles have also the lowest acting pressures and temperatures which produce the lowest thermo-mechanical stresses on the nozzle wall structure.
- Rao method is more effective with long nozzles that have large area ratios. In short nozzles (as in the present cases), the bell contours act as truncated below its full profile which is needed to generate an axial flow at exit. By modifying the profile to produce an axial thrust at exit, both ballistic and structural loads are improved.

References

[1] Singh R 2015 Optimal Design and Numerical Analysis of Axisymmetric Nozzles (Australia: *School of Mechanical and Manufacturing Engineering*, Faculty of Engineering, UNSW).

[2] Young R B 2012 Automated Nozzle Design Through Axis-Symmetric Method of Characteristics Coupled With Chemical Kinetics (Atlanta, Georgia: *48th AIAA/ASME/SAE/ASEE Joint Propulsion Con. & Exhibit*).

[3] Östlund J 2002 Flow Processes in Rocket Engine Nozzles With Focus on Flow Separation and Side-Loads (Stockholm: *Royal Institute of Technology Department of Mechanics*).

[4] Cozart A and Shivakumar K 1999 Stress Analyses of a 3-D Braided Composite Ablative Nozzle (St. Louis, Missouri: *40th AIAA/ASCE/AHS/ASC Structures, Structural Dynamics, and Materials Con.*) **12** 15.

[5] Gomaa A R *et al* 2011 Thermo-Structural Analysis of Combined-Wall Nozzle (Cairo Egypt: *14th International Con. on Aerospace Sciences & Aviation Technology ASAT-14).*

[6] Gong J L *et al* 2020 Study of Coupling Thermal and Mechanical Effect on Submerged Nozzle in Solid Rocket Motor (*Journal of Physics Con. Series*) **1507** 8.

[7] Ashwood P F and Higgins D G 1957 The Influence of Design Pressure Ratio and Divergence Angle on The Thrust of C D Propelling Nozzles (London: *Aeronautical Research Council C.P.*) **325**.

[8] Madhu B P *et al* 2017 Cfd Analysis of Convergent Divergent and Contour Nozzle (*Int. Journal of Mechanical Engineering and Technology*) **8** 8 670–677.

[9] Dumitrescu O *et al* 2018 Development of a Laval Nozzle For a Cold Gas Propulsion System (*IOP Conf. Series Materials Science and Engineering*).

[10] Ömer M C and Günaltay A 2019 Static Structural and Cfd Analysis of Rocket Nozzles Made From Molybdenum and Titanium (Ankara Turkey: *10th Ankara Int. Aerospace Con. , METU*) **18** 20.

[11] Thies C E and Jordan F W 1981 Effects of Nozzle Configuration and Defects on Motor Efficiency (Colorado Springs CO U.S.A.: *AIAA/SAE/ASME 17th Joint Propulsion Con.*).

[12] Joe H D 1987 Design of Compressed Truncated Perfect Nozzles (Monterey, CA: *Joint Propulsion Con.*).

[13] Sreenath K R and Mubarak A K 2016 Design and Analysis of Contour Bell Nozzle and Comparison With Dual Bell Nozzle (*Int. Journal of Research and Engineering*) **3** 6 52-56.

[14] Mubarak A K and Tide P S 2018 Design of a Double Parabolic Supersonic Nozzle and Performance Evaluation By Experimental and Numerical Methods (*Aircraft Engineering and Aerospace Technology*) **91** 1 145–156.

[15] Singh J *et al* 2019 Effect of Nozzle Geometry on Critical-Subcritical Flow Transitions (Heliyon) **5** 2.
[16] Schomberg K *et al* 2019 Design of an Arc-Based Thrust-Optimized Nozzle Contour (*the european con. for aerospace sciences*).
[17] Anderson J D 2003 Modern Compressible Flow (Maryland: *McGraw-Hill Publishing*).
[18] Guderley G 1959 Rao's Method For The Computation of Exhaust Nozzles (*Zeitschrift für Flugwissenschaften*) **12** 345-350.
[19] Rao G 1958 Exhaust Nozzle Contour For Optimum Thrust (*Jet Propulsion*) **28** 6 377-382.
[20] Rao G 1960 Approximation of Optimum Thrust Nozzle Contour (*Jet Propulsion*) **30** 561.
[21] Sutton G P and Biblarz O 2010 Rocket Propulsion Elements, 8th Ed. (New Jersey, USA: *John Wiley & Sons*).
[22] Harry E B *et al* 1961 Experimental Study of Effects of Geometric Variables on Performance of Conical Rocket-Engine Exhaust Nozzles (Washington: NASA, *National Aeronautics and Space Administration*).

Design, test and build of a monopropellant thruster using 85% hydrogen peroxide

A Deif[1], A El-S Makled[2], M K Khalil[3], and Y Elshaar[3]

[1] Space Technology Centre, Cairo, Egypt.
[2] Faculty of Engineering, Zagazig University, Zagazig, Egypt.
[3] Military Technical College, Cairo, Egypt.

E-mail: Deifmoh@yahoo.com

Abstract. The objective of this research is to design, build, and test about a 5N Hydrogen Peroxide(H_2O_2) monopropellant thruster (MPT). It is utilized in remote sensing satellites for attitude control and orbit manoeuvres. The MPT uses high test peroxide (HTP) of 85 % concentration. Firstly, H_2O_2 ≈85% concertation by weight is prepared in the laboratory. A distillation and filtration units are built. The distillation and filtration processes are performed. Next, the design of the monopropellant thruster is done based on the developed mathematical model using NASA CEA rocket performance code. The test facility is developed which consists of the thruster, the feeding system, static test stand and data acquisition system with measuring sensors. An experimental test stand is designed and fabricated with Pendulum thrust mechanism for measurements of thrust. Finally, the silver catalyst is prepared and packed inside the MPT chamber where silver screens of high purity 99.96 % are used. The 10-firing tests are conducted under atmospheric conditions. The firings performed without heating are not completely successful. The analysis of the results shows that the thruster has a thrust range from 3.8-4.2 N. The performance of thruster starts to decay after consuming 6 kg of stabilized H_2O_2. The specific impulse (I_s) is evaluated to be ≈ 93-97s at decomposition pressure of ≈10 bars and mass flow rate ≈4.18 g/s. The performance evaluation is judged to be successful. However, using the whole potential of the 85 % concentrated H_2O_2, is expected to increase (I_s) up to ≈111.5s.

1. Introduction

High test hydrogen peroxide (HTP) has been used as a monopropellant and an oxidizer with propulsion systems many years ago. It has a long heritage in the aerospace propulsion as, it has been used in many manned and unmanned space systems since 1930s up to the present time [1].A lot of work has been done by NASA on H_2O_2 decomposition as a MPT [2].

A need for a higher I_s gradually led to the replacement of H_2O_2 with Hydrazine(N_2H_4). Moreover, the development of a superior catalyst bed for N_2H_4 (Shell 405) made N_2H_4 a better option. Also, dinitrogen tetroxide (N_2O_4) replaced H_2O_2 for use as a storable oxidizer. In contrast, the Russian Soyuz launch vehicle, which has been working in space for over forty years, still depends on H_2O_2 as a working medium for its gas generator to run the turbine pump and the reaction control system (RCS) thrusters used for the descent phase of the Soyuz spacecraft re-entry capsule [3],[4]. However, recently, environment concerns and significant cost saving associated with the drastic simplification of the health and safety protection procedures necessary during propellant production, storage and handling have led to a renewed interest in H_2O_2 for use in propulsion systems for low and medium thrust rocket engines

Content from this work may be used under the terms of the Creative Commons Attribution 3.0 licence. Any further distribution of this work must maintain attribution to the author(s) and the title of the work, journal citation and DOI.
Published under licence by IOP Publishing Ltd

as it is ITAR-free and environment-friendly [1],[5]. The nominal propulsive performance of H_2O_2 as a monopropellant is about 20% lower than N_2H_4, but the volume specific impulse achievable with 90% H_2O_2 is higher than for most other propellants due to its high density. This is particularly useful for systems with significant aerodynamic drag losses or firm volume constraints, as is often the case for small satellites [6].

The most important knowhow challenge in the field of H_2O_2 MPT is the development of active, dependable, startable and durable catalytic beds which have fast, and reproducible performance. Such developed catalyst beds must be resistive to poisoning by stabilizers and impurities exist in the propellant which slow down self-decomposition of H_2O_2 such as pyrophosphate ions. In addition, such thrusters need to be capable of sustaining the large number of operation cycles imposed by typical mission profiles in small satellite applications and preferably should not require pre-heating for efficient operation. Nowadays the most popular catalyst material for H_2O_2 is metallic silver [2]. The most significant drawback of the catalytic silver is the temperature limitation and poisoning [7]. As the H_2O_2 flows through silver mesh screens of 20x20 catalyst bed, the decomposition takes place in an exothermic reaction producing oxygen and water steam.

2. Thruster Design

The sequence of the design is performed according to the flowchart in Figure 1.The NASA CEA rocket performance code [8] is used to calculate the thruster theoretical performance, where a mathematical model is developed to carry out the design of the thruster [9].

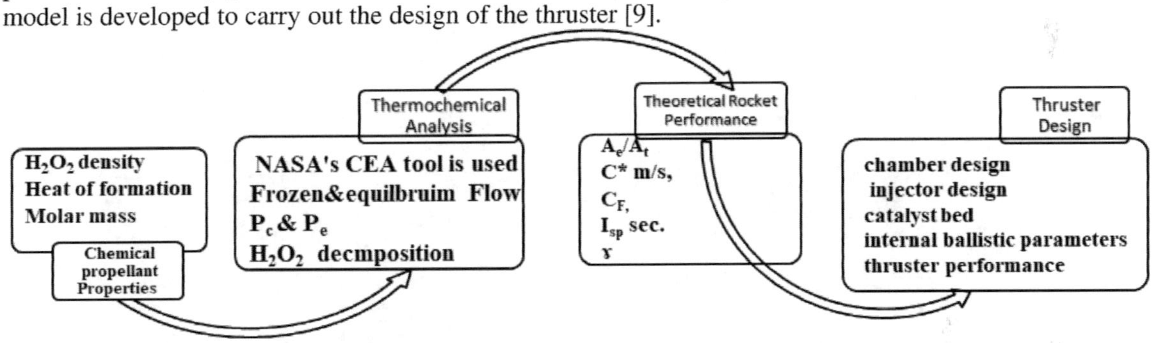

Figure 1.Thruster design flow chart

All the thruster design outcome parameters are listed in Table 1

Table 1. Design parameters of MPT.

	parameter	value
Propellant	H_2O_2	≈85%
	Mass flow rate, g/s	≈4.20
Catalytic bed	Silver mesh material 20x20 purity	99.96%
	Inner diameter, mm	14.60
	Length/diameter ratio	4.0
	Loading factor, $kg/s.m^2$	50.0
	Total number of packed screens	112
Nozzle part	Convergent-divergent nozzle shape	
	Throat diameter, mm	2.25
	Exit diameter, mm	3.32
	Nozzle half angle, deg	14.0°
Injection and retainer plate	Opening area, %	50
	Diameter, mm	14.6
	Orifice diameter, mm	1.50
	Number of orifices	23
	Thickness, mm	2.0

Thruster performance	Thrust range, N	≤ 5
	Specific impulse, s	≈ 111.49
	Characteristic velocity, m/s	≈ 884.50
	Chamber temperature, K	≈ 907.30
	Thrust coefficient	≈ 1.256
	Mass flow rate, g/s	≈ 4.19

3. Experimental Work

The concentration process and firing test are performed in chemistry and jet propulsion labs respectively in Military Technical College. The MPT is fabricated using stainless steel 316 in the AOI Engine Factory based on the mathematical model developed by Ah. Deif et al. [9]. It uses HTP of 85% concentration by weight. Commercial low concentration H_2O_2 solution 50% bought from the local market is concentrated by distillation of water due to the main impediment in getting the HTP from the local market in Egypt.

3.1 H_2O_2 distillation Unit:

A 50 litres of an industrial H_2O_2 of type HYPROX 500 manufactured in a multinational company called Evonik industries of concentration 50 % by weight is bought from the local market in Egypt. It is decided to concentrate the commercial HYPROX 500 to reach the desirable concentration but without removing all the undesirable stabilizers. A distillation unit has been built as shown in Figure 2. consists of:

1. Two flasks of 1-liter volume.
2. A glass gas trap
3. Condenser
4. Manometer
5. Vacuum pump
6. Thermometer
7. Workstation with ventilation, power source, and water source
8. Hot plate
9. Graduated cylinder
10. Sensitive weight scale
11. Water hose

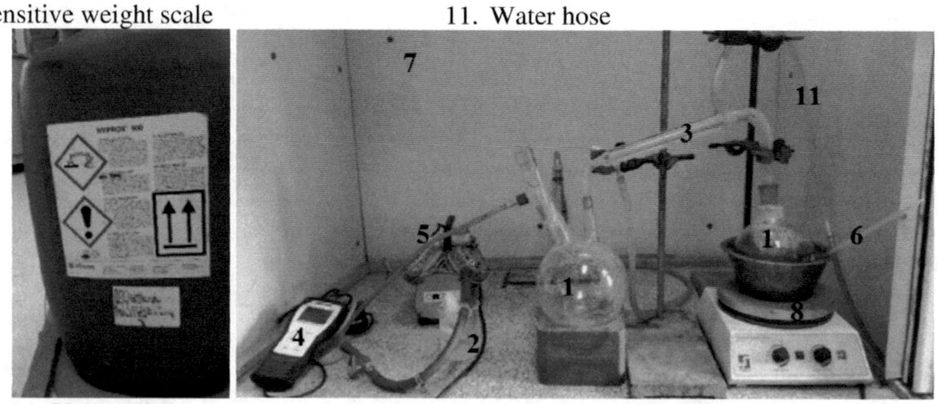

(a)A 50-liter H_2O_2 drum (b) distillation Unit Set-up

Figure 2. H_2O_2 concentration process

3.1.1 Sequence of distillation. The distillation unit is designed and built in the Department of Chemistry (Military Technical College). All the safety precautions during work are taken based on H_2O_2 handbook. The sequence of distillation is performed as follows:

1. A low concentration H_2O_2 solution is stored in the flask over the hot plate.
2. Evacuation of the distillation unit to a pressure value of about 0.133322 bar by the vacuum pump which is continuously checked by the manometer.
3. The H_2O_2 solution flask heating temperature is about 70°C which is monitored by the thermometer.
4. The H_2O_2 solution begins to boil and the contained water in the solutions starts to evaporate and condense in the adjacent flask. Consequently, the H_2O_2 concentration keeps increasing with time.

5. The glass trap is connected to the vacuum line to prevent H_2O_2 vapor from flowing to the pump.
6. Cold water is flowing in the condenser to condensate the water vapor.
7. The concentration of the boiling H_2O_2 flask can be determined from the known H_2O_2 solution starting concentration and volume, and the condensed water volume in the adjacent flask.
8. When the needed concentration is obtained in H_2O_2 flask, the hot plate is turned off to stop heating. The unit is left to cool down to the atmospheric temperature, then the vacuum is released. the concentrated H_2O_2-solution is collected.
9. The density of H_2O_2 is determined precisely by weighing the condensed volume of the concentrated H_2O_2 using the graduated cylinder and the scale to check the mass. The density is calculated by dividing the mass by the volume If the H_2O_2 density is obtained at 1.36 gr/cm^3 and 25°C then, it corresponds to H_2O_2 of 85 % concentration at ambient temperature based on Evonik calculation model [10], the concentrated H_2O_2 still has some stabilizers which slow down the process of self-decomposition which has negative effect on the decomposition efficiency and the catalysts life time.
10. An 8 liters of concentrated H_2O_2 is produced, stored in a ventilated container, and saved in the refrigerator to slow down the self-decomposition rate.

3.2 Filtration Process:

The concentrated H_2O_2 underwent filtration process to get rid of the stabilizers using G3, G4 sintered discs which are categorized into 5 grades G1, G2, G3, G4 and G5. The grades are determined by maximum pore size, which is determined by measuring pressure at which the first air bubble penetrates the filter under specific circumstances.

Table 2. porosity grades and their general use. [11]

Porosity Grade	Pore Size (Microns)	General use
1	90-150	Filtration of coarse materials / precipitates, gas dispersion, gas washing, extractor bed, and support for other filter materials.
2	40-90	Filtration of medium precipitates gas disperations and gas washing.
3	15-40	Filtration of fine grain precipitates, Analytical work with medium precipitates, and Mercury filtration.
4	5-15	Analytical work with fine and very fine precipitates, Non-return mercury valves.
5	1-2	Bacteriological filtration.

IOP Conf. Series: Materials Science and Engineering 1172 (2021) 012043 doi:10.1088/1757-899X/1172/1/012043

3.2.1Sequence of filtration. Filtration sequence is performed as follows:

1. The sintered disc (1) is mounted on a vacuum flask with a neck and a vacuum line is connected to a connector socket as shown in Figure 3.
2. The vacuum pump evacuates the flask (2) through a trap (4) to create a pressure differential between the two sides of the disc to let H_2O_2 flow through it.
3. A trap is connected through the line to prevent the filtered H_2O_2 to enter the vacuum pump (3).
4. Firstly, H_2O_2 is filtered using G3 sintered disc of pore size 15-40 microns to filter out the impurities bigger than this size,.
5. Secondly, a finer filtration process is performed using G4 sintered disc of pore size 5-15 microns to avoid blockage of the sintered disc which was harder and longer process because, it takes longer time to let H_2O_2 pass through the filter.
6. It is necessary to clean the sintered disc with distilled water to avoid blockage of the flow.
7. The filtered H_2O_2 will be collected and saved in ventilated bottle in a refrigerator.

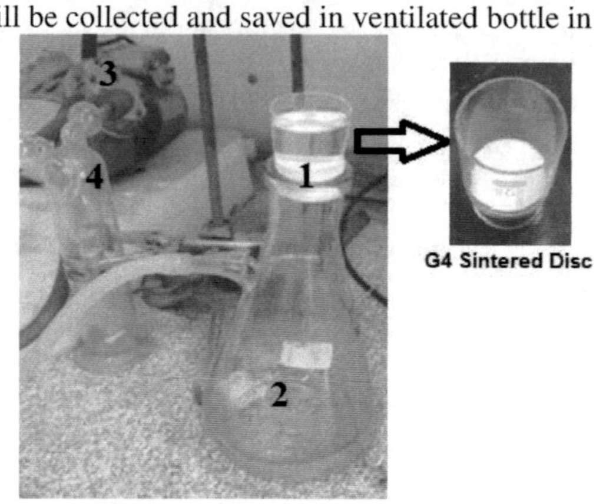

Figure 3. Filtration process set-up

3.3 Test Facility

The experimental testing of the 5N MPT prototype is done using H_2O_2 test facility. It consists of MPT, a feeding system, a custom-made test bench and data measuring system designed for delivering H_2O_2 to the MPT at the desired mass flow rate and pressure as shown in Figure 4.

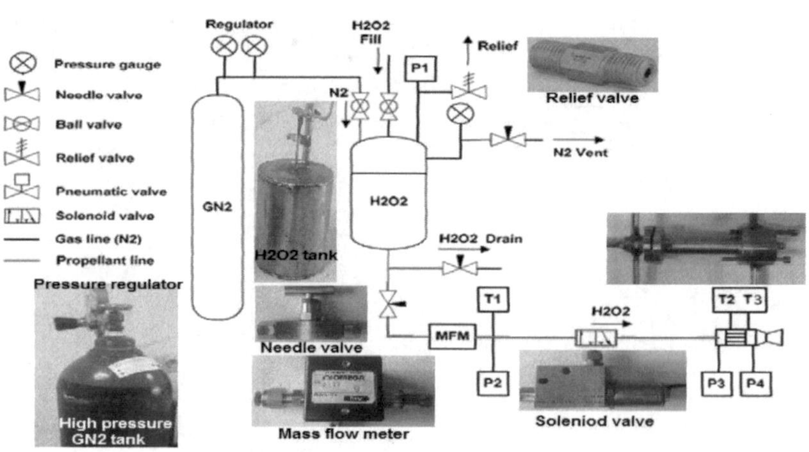

Figure 4. Thruster feeding system

3.3.1 The thruster. The MPT is made of stainless steel 316 to withstand the thermal and mechanical stresses generated during firing based on mathematical model developed by Ah. Deif et al [9]. It is manufactured on modular base where it consists of 3 parts: the injector (1) where the liquid hydrogen is entered and injected, the cylindrical parts which holds the catalyst bed where the hydrogen peroxide is decomposed by flowing through it, and the nozzle (3) where the gaseous products are accelerated into the critical section. The injector plate is designed to distribute the propellant on the surface of the catalyst, the retainer (4) is designed to block the catalyst to go out from the chamber, and an anti-channeling disc to redirect the flow into the core of the catalyst Figure 5.

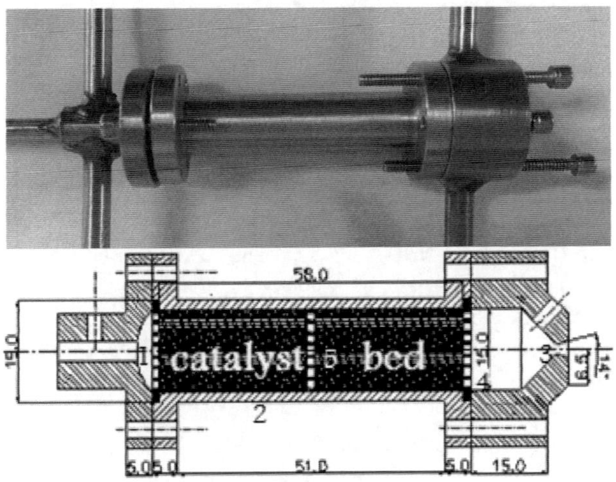

Figure 5. Thruster design

3.3.2 The feeding system. The feeding system consists of a high purity nitrogen high pressure vessel connected to pressure regulator to regulate the exit pressure. It is used as a pressurizing gas to pressurize the propellant feeding line, and purging/cleaning the pipes from propellant residuals when needed as shown in Figure 4.The H_2O_2 is stored in a stainless-steel tank of ≈3.5 liters volume and wall thickness of 3 mm is designed and manufactured to withstand the high pressure up to 25 bar, noticing that high concentration H_2O_2 is compatible with the stainless-steel 316 [12]. The pressure in the tank is monitored by a dial gauge. The tank is connected to Stainless Steel 1-Piece Poppet relief Valve with adjustable Pressure (10.4 to 24.2 bar) to release any overpressure may happen if it exceeds 15 bar. A fill valve is connected to the tank to fill the tank with H_2O_2. A manually operated ball valve is used to feed the compressed H_2O_2. A vent is connected to the circuit to make sure no air or nitrogen bubbles in the flow of the circuit. The mass flow rate of H_2O_2 is regulated by a Needle Valve. The compressed H_2O_2 flows in the thruster using electrically switching on solenoid valve remotely by a command from the Ni-user software interface, Figure 4.

3.3.3 The test Bench. The MPT is installed, by properly designed interfaces on an assembly which is composed of two L-shape beams. The MPT (1) is installed by two small brackets (2) on the small L-shape (3) by screws which in turn is hanged freely to a bigger and rigid L-shape beam (4) by two flexures (5) made of (0.5) mm thickness spring steel sheets. The thrust force generated during the operation of the thruster is transferred to the small L-shape beam (3) which, in turn transfers the thrust to the load cell through a stud screwed to it and is touched to the load cell. The compensation of the cables and connections loads to assure zero loads before of the firing test is done by tightening or untightening the screw. The bigger L-shape structure is fixed to the test bench considering the horizontal position of the MPT axis and alignment with the load cell axis. A series of longitudinal and transverse holes (6)are also incorporated into the design of the big L shaped beams to allow the alignment and adjustment in both directions A Swagelok NPT connection (7) and 5 mm diameter Teflon tube (8)are used for feeding the H_2O_2 to the MPT which is perpendicular to the thruster axis to eliminate the reaction of the fuel flow and fitted without applying any further forces in the thrust direction as shown in Figure 6.

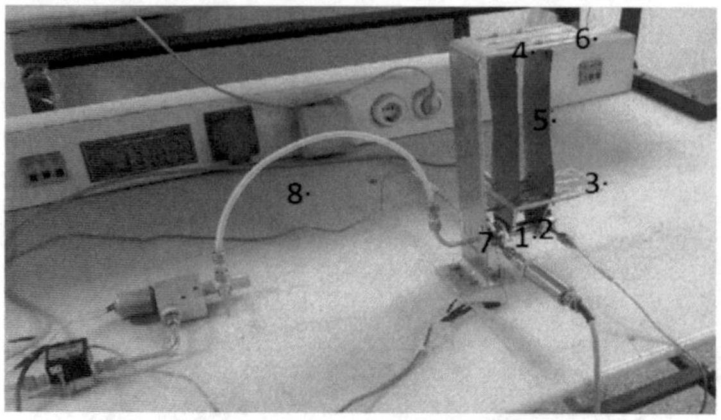

Figure 6. The H_2O_2 MPT Test Bench

3.3.4 Measurement tools and Data Acquisition System. The measurements collected during testing can be classified into two different types:

A) Measurements for detecting HTP feeding system status:

- The tank output pressure, which is associated with HTP mass flow. Additionally, irregular pressure rise is typically a sign of early decomposition of the stored HTP which is monitored by a pressure gauge.
- The propellant flow rate is measured by an omega FLR10009TH. A DC source of 12 output volt is used to provide the flow meter with the required excitation voltage.

B) Measurements for evaluating the performance of the MPT prototype:

- The two-pressure transducer used for static pressure measurements are produced by HBM, type P3MB-20 are selected for its high precision (accuracy is 1% FSO). The first one is installed in the injection part upstream while the second is installed downstream before the convergent part of the nozzle Figure 9(a). The data coming from the pressure transducers are acquired and transferred to a laptop by National Instruments acquisition board number 9237 which supplies the excitation voltage needed by the pressure transducer.
- Temperature measurements are performed using K-type thermocouples with 0.5 mm diameter and 150 mm Figure 9(b). The data coming from thermocouples are collected and carried to a laptop using a National Instruments acquisition board number NI 9219 which supplies the excitation voltage needed by the thermocouples.

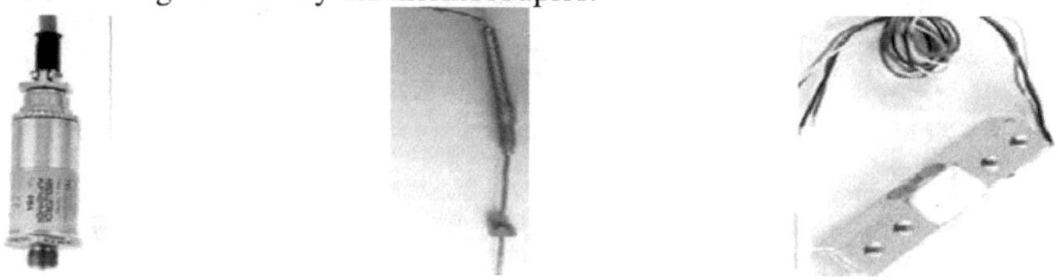

(a)Pressure transducer **(b) Thermocouple** **(c)Load cell**

Figure 7. Measuring tools

- The thrust force is measured by a load cell manufactured by 3132_0 - Micro Load Cell (0-780g) - CZL616C. This Single Point Load Cell is used in small jewelry scales, Figure 7(c).
- The data coming from the loadcell are collected and carried to a laptop using a National Instruments acquisition board number NI 9219.

- A 24 volts power source provides the solenoid valve with the required electric operating power. The solenoid valve is controlled by a laptop using a National Instrument relay card NI9481 to start the firing, Figure 8.

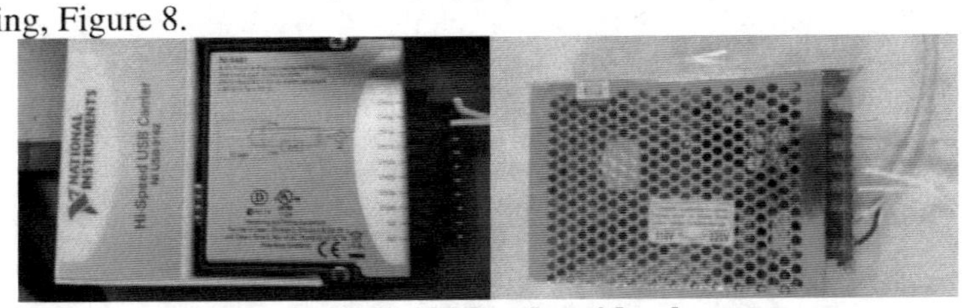

Figure 8.NI relay card NI9481& solenoid valve power source

3.3.5 Measuring tools calibration system

- Thermocouples are calibrated in the lab using a cup filled with ice cubes and cold water to fill the gaps, and another cup filled boiled water Figure 9(a).
- Two-pressure transducer are calibrated in the lab using fluke RPM4-E-PWT electronic deadweight tester and the reading matches perfectly the test report included with the transducer Figure 9(b).
- load cell is calibrated using different standard weights Figure 9(c).
- The flow meter is calibrated using a 500ml graduated cylinder and a timer at different flow rates. The data coming from flow meter is collected and carried to a laptop using a National Instruments acquisition board number NI usb-6009 as shown in Figure 9(d).

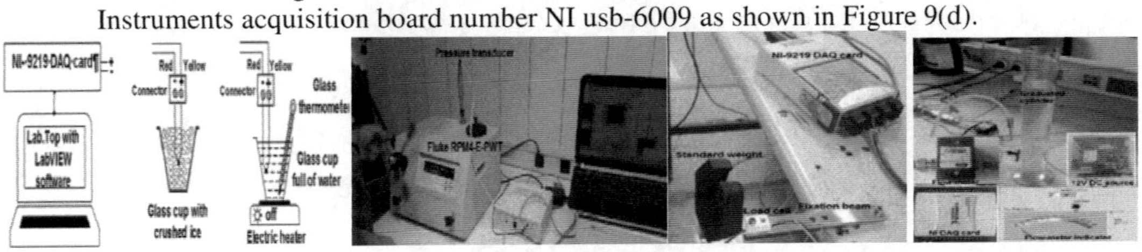

(a) Thermocouple (b) Pressure transducer (c) Load cell (d)Flowmeter

Figure 9. Measuring tools calibration system

3.4 Silver catalyst bed procurement and preparation:

The silver mesh should be cut into discs of 15 mm to be packed in the thruster. A punch tool is designed, machined and heat treated to cut the silver screen sheet as shown in Figure 10(a) The silver sheet is mounted on acrylic sheet and hammered using the punching tool to cut the silver discs as shown in Figure 10(b),(c). The discs are compacted in the thruster using a hydraulic press to increase the packing factor of the silver screens in the thruster.

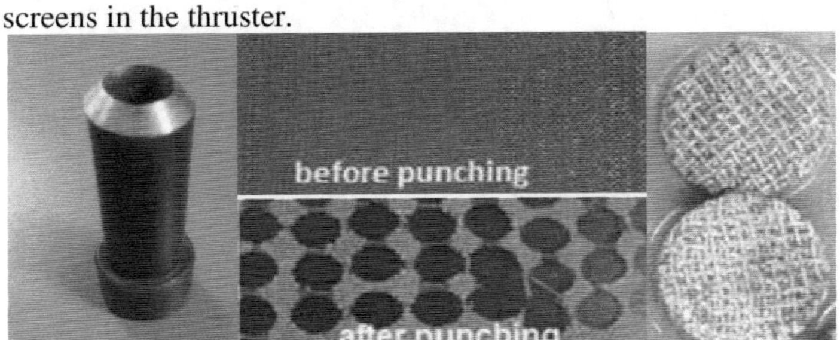

(a) punching tool **(b) Silver mesh sheet** **(c) sliver discs**

Figure 10. Silver catalyst preparation

3.5 Firing Test:

Initial testing begins with leak tightness of the system without the thruster connected, the cracking pressure of the relief valve was set to 15 bar. The system is pressurized with nitrogen gas to ≈14 bar and checked for leakage. The third test is performed with water to check the functionality of the feeding system elements, the functionality of data acquisition system, adjustment volume flow rate using the flowmeter and the needle valve by the user front panel of NI control software as shown in Figure 11.

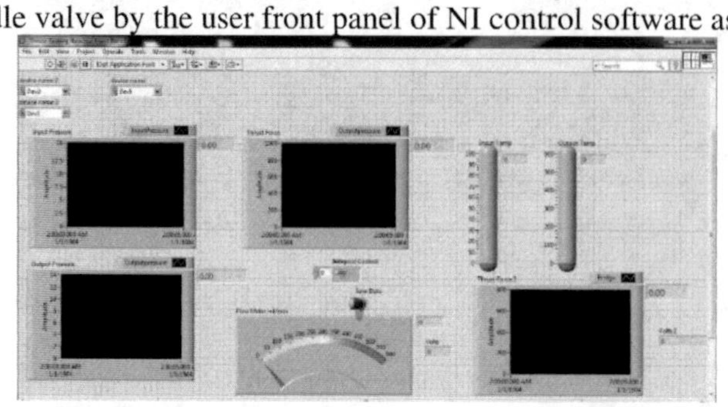

Figure 11. Thruster performance user front panel

Propellant is filled into the tank through filling valve. Pressure regulator is set to the required propellant tank pressure. Nitrogen supply is opened, and it enters the gas pressurization tank of 3.3L volume. Once the propellant tank pressure is stabilized, solenoid valve is switched on to initiate the engine operation. The thruster is fired until the solenoid valve is switched off by the relay signal from Ni control software based on the required pulse time. The thruster is purged by deionized water after firing to remove the residual H_2O_2 in the thruster and catalyst bed.

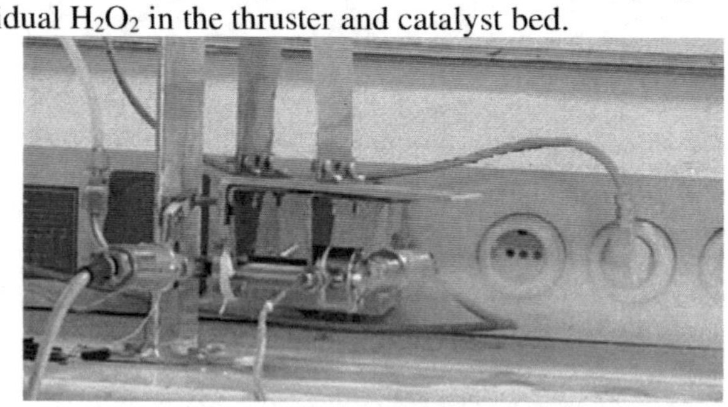

Figure 12. Thruster hot firing

4. **Results and Discussion**

The firing test of the thruster with and without heating is investigated

4.1 Firing without catalyst preheating:

The initial trials are performed without heating to fire the thruster, the initial input catalyst bed temperature is 25C°, injecting pressure 10 bar, and the silver mesh is compacted to 10 MPa in the decomposition chamber of 15 mm diameter and 58 mm length. The results are not successful. It is shown in Figure 13.The firing is comprised of injecting the concentrated H_2O_2 for a duration of ≈ 13 s. The Chamber input pressure is≈ 6.59 bar and the output pressure is ≈ 6.24 bar, where the difference is 0.36 bar represents the pressure drop due to the catalyst bed.

It is noticed that the decomposition is not complete and partial quenching takes part. The chamber input & output pressures are fluctuating, it takes ≈ 7.0.s as a rise time to reach the decomposition pressure. A white smoke is released, and an acidic odor was smelled in the decomposition products because, H_2O_2 is not decomposed completely. The thrust force reaches ≈2N which is much lower than the theoretical designed value. The possible reasons for the H_2O_2 is not decomposed at the catalyst bed could be the temperature of the catalyst and H_2O_2. At the time of the test the atmospheric temperature is

≈ 25°C. Willis [13] stated that the most noticeable effect on the initial ignition time delay is the thruster body temperature and majority of his testing were done at catalyst bed temperature of 200°C.The second possibility might be the inadequate surface contact of the catalyst material with the H_2O_2 especially the compaction pressure is low.10 MPa. Moreover, the used silver mesh 20x20 has high free volume ratio of 0.7 after compaction, knowingly that silver mesh 40x40 has less free volume fraction which increase the effective surface area used to the decompose the H_2O_2 flowing through the silver and gives better performance. however, it is more expensive and has higher pressure drop through the catalyst Runkle et al. [2] stated that mesh silver gauzes 40x40 has better performance than 20x20 mesh silver gauzes. Long M.R. et al [14],[15] revealed that catalytic beds with high packing density lead to more complete decomposition and higher produced temperatures with respect to low packing density ones. Another reason might be the contamination in the H_2O_2 which was not removed by the filtration process because soluble stabilizers cannot be removed by such filtration process especially nitric acids.

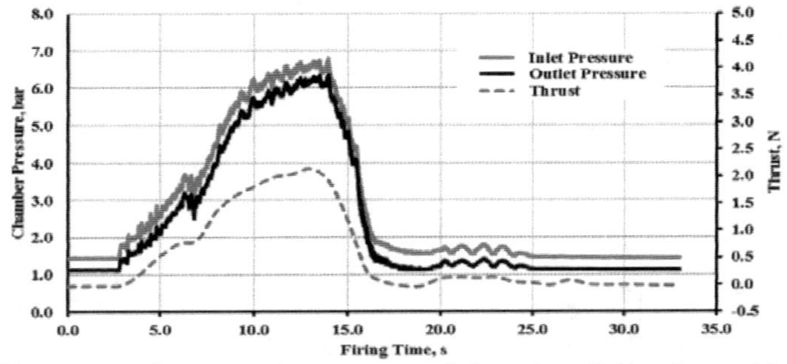

Figure 13. Thruster performance (pressures and thrust) vs firing time without preheating

4.2 Firing with catalyst preheating and increased packing factor:

This run is conducted after two taken corrective actions which are heating including propellant pulsing to pre-heat the silver screens before firing and increasing the compaction pressure from 10MPa to 12MPa to increase the contact area to enhance the decomposition process. It is better to increase the compaction pressure more than 12MPa, but the lack of silver screens won't allow to increase it as pure silver screens are not available in the local market. The results show that firing test is successful, Figure 14.

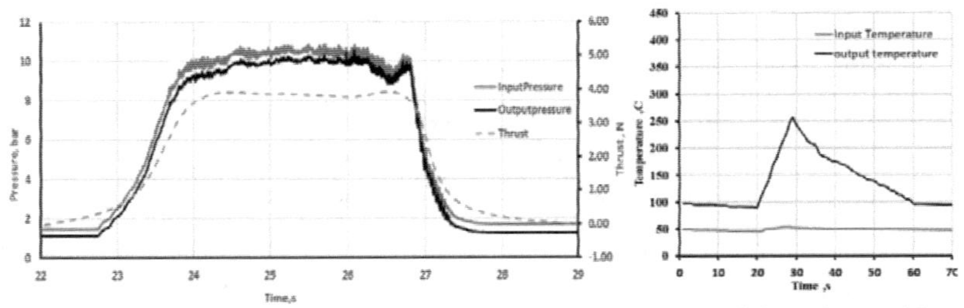

Figure 14. Thruster performance (pressures, thrust& Temp.) vs firing time with preheating

The firing is comprised of injecting the concentrated H_2O_2 for ≈ 5s, the chart shows the decomposition chamber input pressure rises to pressure 10.3 bar and the input temperature is 50°C. While the output pressure is around 10.0 bar the difference is 0.3 bar which represents the pressure drop due to the catalyst bed and the output temperature rises to 256°C and starts to drop after the end of the firing. The graph also shows the pressure is somehow stable and the H_2O_2 almost decomposes into steam and oxygen. A complete decomposition of the monopropellant is mostly obtained. The thrust reaches the value of 3.8N and the average mass flow rate is around 4.20 g/s. So, the experimental specific impulse (I_s) is calculated as in equation(1).

$$I_s = \frac{\int F dt}{mg} \qquad (1)$$

where F is the measured thrust, g = 9.8065 m/s^2 is the standard gravity acceleration. As a result, an average experimental $I_s \approx 93$s is obtained where the value calculated by the NASA CEA code is 111.4s. Therefore, the experimental result is acceptable, because several losses happen in the decomposition process (silver contact area, silver purity, starting temperature, heat losses through the thruster body, and H_2O_2 Purity), as well as, the nozzle is not adiabatic.

The H_2O_2 pulses heat the catalyst bed itself during pre-firing, but the fuel is still at the room temperature. It is noticed from the pressure time chart that it takes almost 0.65 s as a rise time to reach the working pressure and produce the required thrust. To decease the rise time of the operating pressure and accelerate the decomposition process, the catalyst bed need to be heated to higher temperature around the thruster casing before firing and also to decrease the heat losses and the dead volume should be minimized as Y. A. Chan [5] revealed that increasing the temperature to 150°C provides adequate energy to decompose the catalyst bed and the pressure rises to maximum in 100ms after opening the solenoid valve. Moreover, silver compaction pressure needs to be increased and the silver screens need to be pickled by samarium nitrate [12] but it is very expensive and not available in the local market. The H_2O_2 purity needs to be enhanced to avoid the poisoning of the catalyst bed and decay of the catalyst performance. It is noticed that temperature of the decomposition products is not high because the duration of the firing is short to heat the catalyst especially the thruster body is not insulated.

This third run is conducted with heating including propellant pulsing to pre-heat the silver screens before firing. The thruster initial input temperature is about 50°C and the output initial temperature is ≈120°C after heating which is higher than the previous two runs. It is shown in Figure 14 that the chamber input pressure rises to 11 bar and the input temperature is 50°C after firing, the output pressure and temperature are around 10.6 bar and 268°c respectively. The pressure difference is 0.4 bar which represents the pressure drop due to the catalyst. The thrust reaches the value of 4.1 N and the average mass flow rate is about 4.19 g/s so the experimental I_{sp} is =97s. It is noticed that the pressure and thrust increases due to the increase in the initial temperature at the same mass flow rate which agrees with [3], [9].

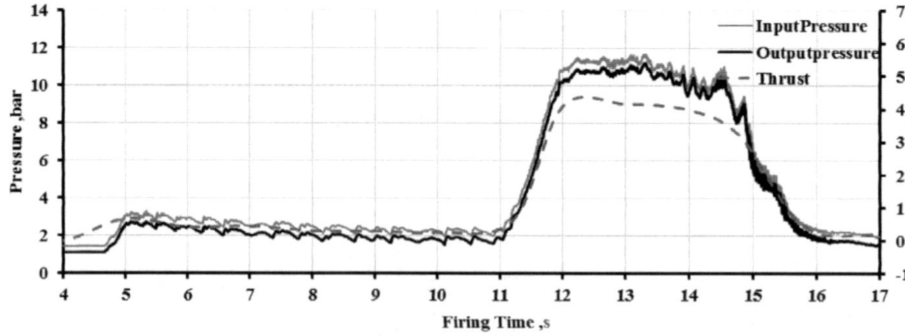

Figure 15. Thruster performance (pressures and thrust) vs firing time with heating

In the last firing of the thruster, the initial input temperature is 48 °C, the initial output temperature is 90 °C, the injecting pressure is 10 bar, The results show they are not successful, it is shown in Figure 16. the firing is comprised of injecting the concentrated H_2O_2 for a duration of about 12s, but no decomposition takes part and a white smoke was released, and an acidic odour was detected in the combustion products because, the H_2O_2 is not decomposed. The main reasons for the H_2O_2 is not decomposed by catalyst bed could be the poisoning effect which insulates and stops the reaction to take part especially the H_2O_2 which is used is in this run is not filtered so the contaminations and stabilizers especially chromium chloride and sodium stannate which agrees with the result given by [16] are not removed and stick on the surface of the silver mesh as shown in Figure 16 which prevents the contact with H_2O_2. Whitehead [12] highlights the necessity of decreasing the contaminations during preparation of concentrated H_2O_2.

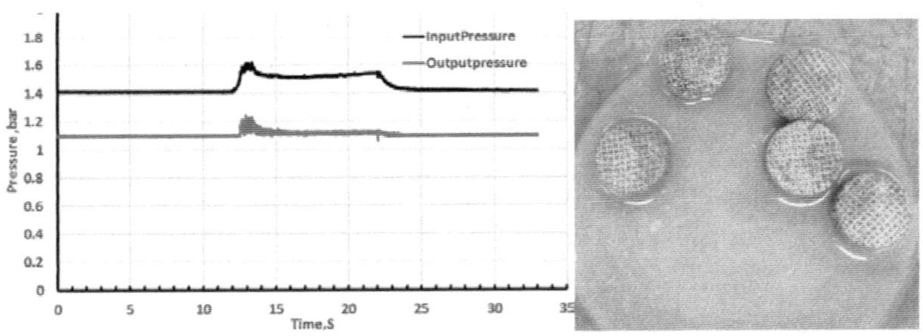

Figure 16. Poor decomposition due to silver poisoning

5. **Conclusion**:

The main conclusions of the present work can be summarized as:

- Building a distillation unit and concentrating the H_2O_2 successfully using the available commercial 50 % H_2O_2 to the reach the required 85 % concentration
- Building a filtration unit and removal of the solid stabilizers using mechanical filtration is achieved which enhances the service life of the catalyst bed.
- Design and development of the test facility capable of performing the MPT firing successfully and safely. The test facility includes the feeding system which capable of providing the MPT with required operating conditions. The test bench can measure all the thruster performance parameters precisely and remotely using developed automated front panel interface.
- Design, manufacture, and test of 5 newton H_2O_2 monopropellant thruster which can be utilized in remote sensing satellites for attitude control and orbit maneuver. However, the actual measured thrust reaches the value of ≈ 4.1 N, the average mass flow rate is ≈ 4.19 g/s and the thruster experimental I_{sp} is =97s. which is an accepted performance due to the prementioned difficulties in the present work.

References

[1] Marsland E F, Roberts G, Gibbon D. and Ryan C.2019 Development of a low-cost 0.1N high test peroxide thruster using additive manufacturing *AIAA 2019-* 4227 Indianapolis, USA.

[2] Runckel F, Willis C M and Salters L B 1967 *Investigation of Catalyst Beds for 98-Percent-Concentration Hydrogen Peroxide* NASA TN D-1808, Langley research center Hampton, Virginia.

[3] Pasini A, Torre L, Romeo L, Cervone A, Musker A and Saccoccia G 2007. *43rd AIAA/ASME/SAE/ASEE Joint Propulsion Conf. & Exhibit.* 5465 Cincinnati OH.

[4] Company, "https://corporate.evonik.com/en," June 2018. [Online]. Available: *https://active-oxygens.evonik.com/en/application-areas/aeronautics* [Accessed 1 August 2018].

[5] Chan Y. A, Liu H J, Tseng K C and Kuo T C 2013 *Int. J of Aerospace and Mechanical Engineering* vol **7**.No7.

[6] Cervone A, Romeo L, Torre L, d'Agostino L, Calderazzo F and Musker A. 2006: Development of green hydrogen peroxide monopropellant rocket engines and testing of advanced catalytic beds *3rd Int. Conf. on Green Propellants for Space Propulsion and 9th Int. Hydrogen Peroxide Propulsion Conf.* Poitiers, France.

[7] Wernimont E and Mullens P 2000 *36th AIAA/ASME/SAE/ASEE Joint Propulsion Conf. and Exhibit.* 3555 Las Vegas, NV, U.S.A.

[8] Gordon S. and McBride B. J 1971*Computer program for calculation of complex chemical equilibrium compositions, rocket performance, incident and reflected shocks, and chapman-jouguet detonations* NASA Glenn Research Center Cleveland.

[9] Deif Ah, El-Zahaby A, Makled Ah.El-S and Khalil M k 2015 *16th Int. Conf. on Aerospace Sciences & Aviation Technology* ASAT-144-PP Cairo, Egypt.

[10] Schumb W C, Satterfield C N and Wentworth R L *Evonik company*[Online]. Available: *https://active-oxygens.evonik.com/en/products/hydrogen-peroxide/general-information/calculations* [Accessed 17 may 2018].
[11] B.J. Scientific Company [Online]. Available: *http://bjscientific.com/sintered-ware.html.* [Accessed 12 May 2018].
[12] Whitehead J C 1998 *12th Annual AIAA/USU Conf. on Small Satellite* 98-VIII*1.
[13] Willis C. M 1960 *The Effect of Catalyst-Bed Arrangement on Thrust Buildup and Decay Time for a 90% Hydrogen Peroxide Control Rocket* NASA TND–516.
[14] Long M. and Rusek J 2000. *36th AIAA/ASME/SAE/ASEE Joint Propulsion Conf. and Exhibit* 3683 Las Vegas, NV, U.S.A.
[15] Amri R, Gibbon D, Rezoug T.2013 *Aerospace Science and Technology* vol 25(1) pp 266–272.
[16] Palmer M. J. 2014 *Experimental Evaluation of Hydrogen Peroxide Catalysts For Monopropellant Attitude Control Thrusters* University of Southampton, Southampton, England.

IOP Conf. Series: Materials Science and Engineering 1172 (2021) 012044 doi:10.1088/1757-899X/1172/1/012044

Parametric study of solid propellant slotted grain

M El-Naggar[1], H Belal[2] and H M Abdalla[3]

[1] M.Sc. Student, Rocket Department, Military Technical College, Egypt

[2] Independent Researcher, AIAA Member, Egypt

[3] Associate Professor. Rocket Department, Military Technical College, Egypt

Email: khatab80@hotmail.com

Abstract. One of the goals in solid rocket motor design is to have as large volumetric loading as possible keeping the basic requirements unaffected. Slotted grain can achieve this goal as it has the advantages of sliver-free and no stress-concentration regions that occur in other internal burning grains as star grain and wagon wheel grain. It has the disadvantage of exposing the motor wall to hot gases. In this paper, the geometry of slotted grain is discussed and the effect of design parameters (e.g., number of slots, dimensions of the slot, etc.) of slotted grain on grain burn back is explained. Also, a comparison between results and experimental data is performed.

1. Introduction

Volumetric loading is defined as the ratio of propellant volume to chamber inner volume. One of the goals in solid rocket motor design is to have as large volumetric loading as possible keeping the basic requirement unaffected. These basic parameters are the blocking factor, sliver area, and progressivity or neutrality parameters [1, 2]. There are various types of slots, for example, the fin slot or radial slots used in the second stages of French ballistic missiles (SEP P4.0) (figure 1) [3,4]. and in space motors like in the Indian SLV-3 Apogee (figure 2) [5] and FW-4D motor (figure 3) [6]. Other examples are the through-web axial slot, slots connecting to the inner and outer surfaces and extending to a part of the grain length as in SA-3 Goa sustainer motor [7].

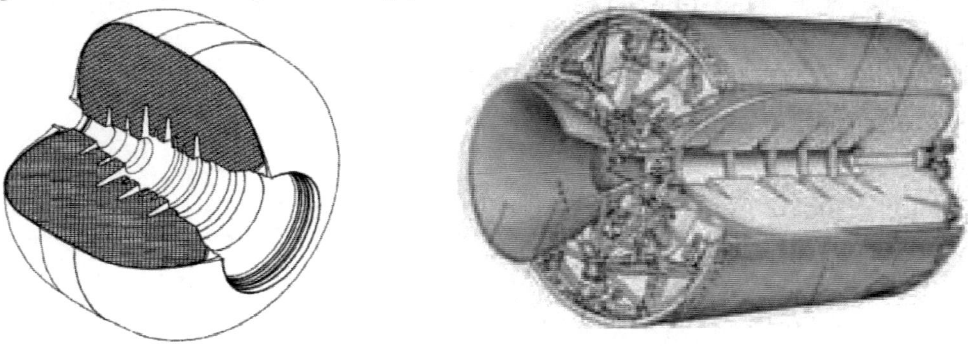

Figure 1 Second stages of French ballistic missiles (SEP P4.0) [3,4]

Content from this work may be used under the terms of the Creative Commons Attribution 3.0 licence. Any further distribution of this work must maintain attribution to the author(s) and the title of the work, journal citation and DOI.
Published under licence by IOP Publishing Ltd

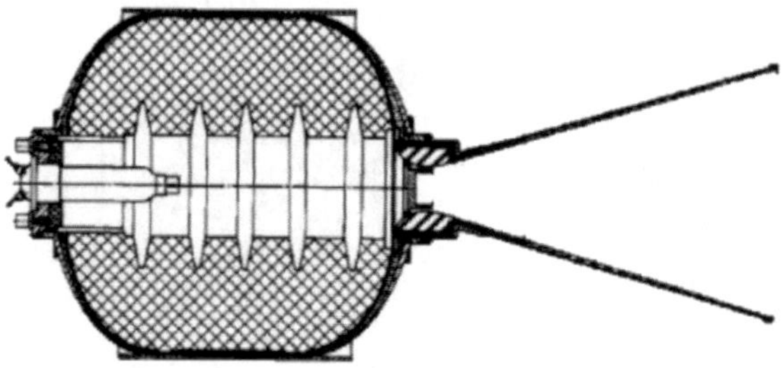

Figure 2 SLV-3 Stage 4 motor [6]

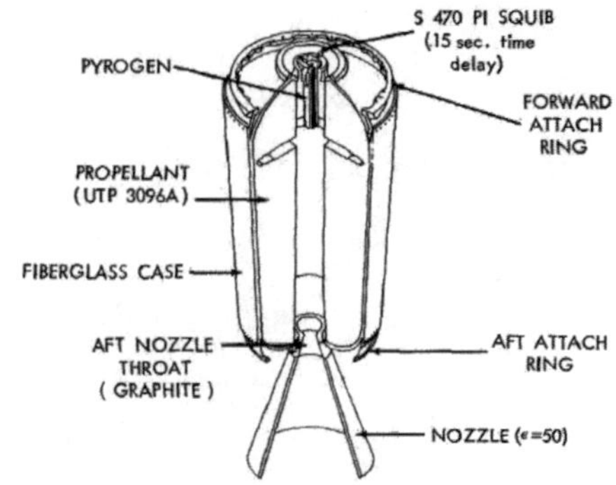

Figure 3 FW-4D Motor [7]

Through-web axial slot grain (in this research, it will be shortened as slotted grain) has the advantages of being free of sliver and free of stress-concentration regions that occur in other internal burning grains (e.g., star grain and wagon wheel grain). It has a larger volumetric loading compared to other internal-external burning grains with a disadvantage of exposing the motor wall to hot gases. Zayed [8] studied grains with through-web slots and showed that there are two important parameters: the length ratio $\lambda = \frac{L_1}{L}$, and the number of slots (figure.4)

b. 3 Slots Grain

a. 4 Slots Grain

Figure 4 Configuration of slotted grains [8]

Stone [9] derived a set of equations that describe the evolution of burning perimeter and port area at a given burn distance in terms of the design parameters shown in figure.5. Also, the slotted grains find a wide application in base-bleeding artillery projectiles [10]. It was found that increasing the number of slots, can increase the range of the base bleed artillery.

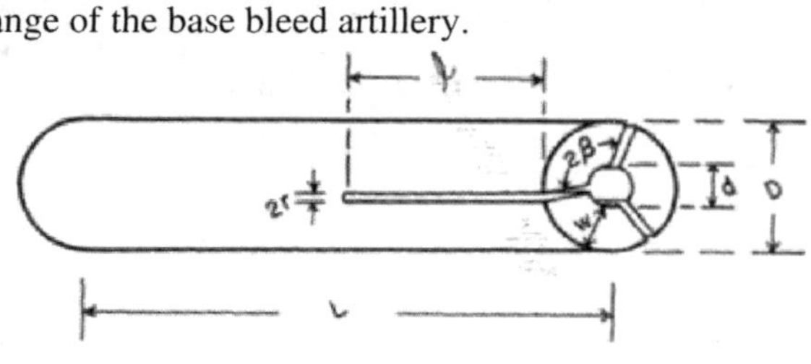

Figure 5 Slotted tube grain design [9]

2. Slotted grain: baseline case

2.1. Geometry

The baseline grain is the solid propellant grain used in the second stage of ex-USSR air-defense missile SA-3 Goa [11]. The basic dimensions of that grain are shown in figures .6 and 7. Solid propellant charge 5b29 is made as a cylindrical single-perforated grain with six radial slots near its front end face. The outer surface and the tail end face of the grain are inhibited. The grain is loaded into the engine body through its tail portion with the nozzle fairing removed; the grain is fixed in the body against axial and radial movement employing supporters.

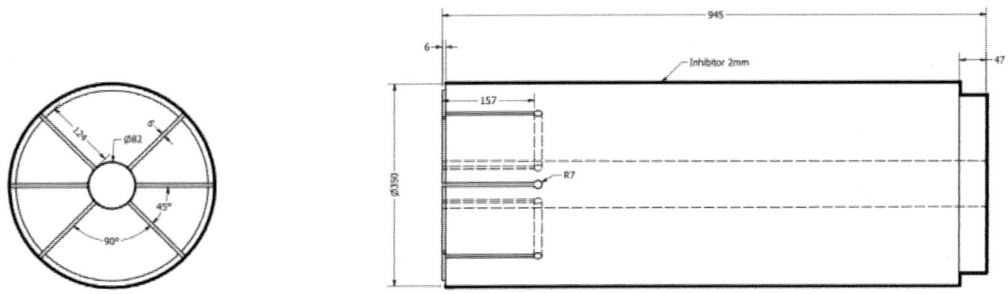

Figure 6 Grain of Sustainer of SA-3 Goa

Figure 7 3D CAD of the slotted grain of SAM-3

2.2. *Grain Burnback*

Due to the complexity of the analytical solution of the equations presented in [9], in this research drafting technique is utilized for studying burnback analysis of slotted grain. The drafting technique is the oldest technique used in grain burnback analysis, where the evolution of the burning surface is evaluated by combining basic shapes to describe the grain initial geometry and the burning surface is assumed to propagate normal to itself [1]. In this case, CAD software is used to propagate the initial surfaces till burnout. The main drawbacks of this method are that it is discrete, requires high human-computer interaction which may lead to errors, and it is also a time-consuming method. However, in this case, where the grain is 3D and complicated analytical analysis, the drafting technique is a good choice for the preliminary study of slotted grain. The steps of this burnback are shown in figure. 8

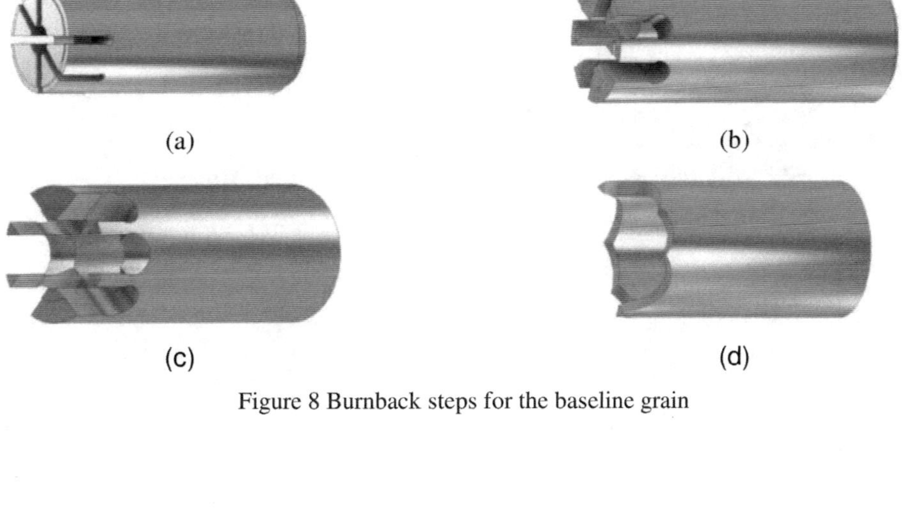

Figure 8 Burnback steps for the baseline grain

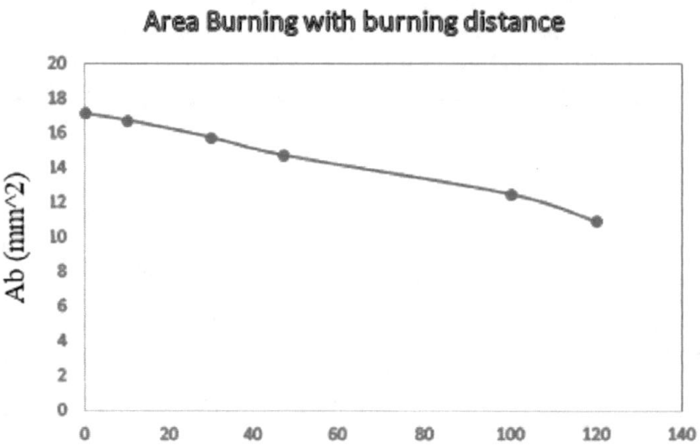

Figure 9 Burnback results for the baseline grain

The results (figure. 9) show that the baseline grain has regressive behavior which is expected as from a tactical viewpoint, the main function of the sustainer motor is to maintain the missile speed constant, with propellant depletion, it requires less thrust to keep the same velocity.

3. Experimental Validation

To predict the pressure-time curve a 0-D internal ballistic prediction module (IBPM) was applied. The basic equation for this module according to [1] is

$$V_c \frac{dP_c}{dt} = \rho_{sp} RT_c A_b a P_c{}^n - \Gamma P_c A_{cr} \sqrt{RT_c} \quad (1)$$

where V_c= chamber free volume, Pc = combustion pressure (stagnation pressure), ρ_{sp} = solid propellant density, R = gas constant, T_c = combustion temperature, A_b = burning area, and A_{cr}= nozzle critical area. Comparison between the experimental results and the analytical results for the baseline grain is shown in figure. 10, the comparison shows a good agreement with the progressive behavior captured. However, there is a deviation between the experimental and analytical with the error increase near the end of burning. This error can be attributed to uncertainties in burning rate law or nozzle erosion as discussed in a previous article for the same research group. These uncertainties can be evaluated using an optimization module [12].

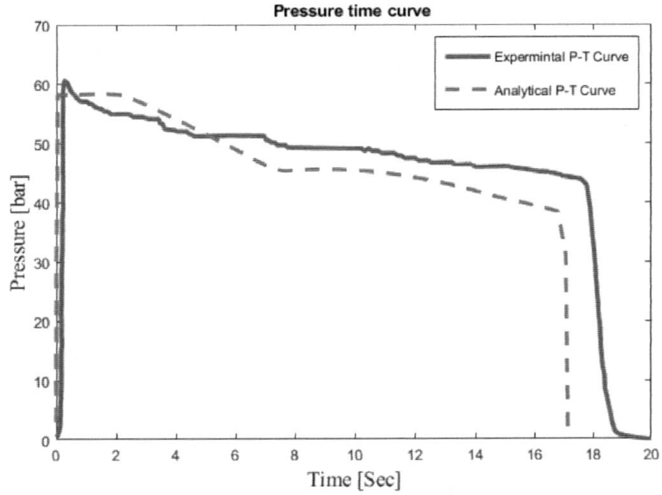

Figure 10 Pressure time curve for sustainer of SAM-3

4. Parametric study

In order to evaluate the importance of each geometric parameter, a parametric study using a single parameter at a time variation is pursued considering the basic grain as shown in figure.6, There will be three groups, each group has one variable parameter keeping all other parameters constant

- **Group A**: Effect of slot length relative to total grain length.
- **Group B**: Effect of the internal radius of grain
- **Group C**: Effect of slot width.
- **Group D**: Effect of the number of slots.

4.1 Group A: Effect of slot length relative to total grain length

In this group, the variation of percentage of slotted part is investigated, with 3 more cases other than the baseline case; the slotted length is 25%, 50%, and 75% of total grain length, with values of slotted length in table 1.

Table 1 Changes in lengths of slotted grain

$Case\ A$	$Lsg1 = 0.25\ Lg$	$Lsg2 = 0.5\ Lg$	$Lsg3 = 0.75\ Lg$
	236.25	472.5	708.75

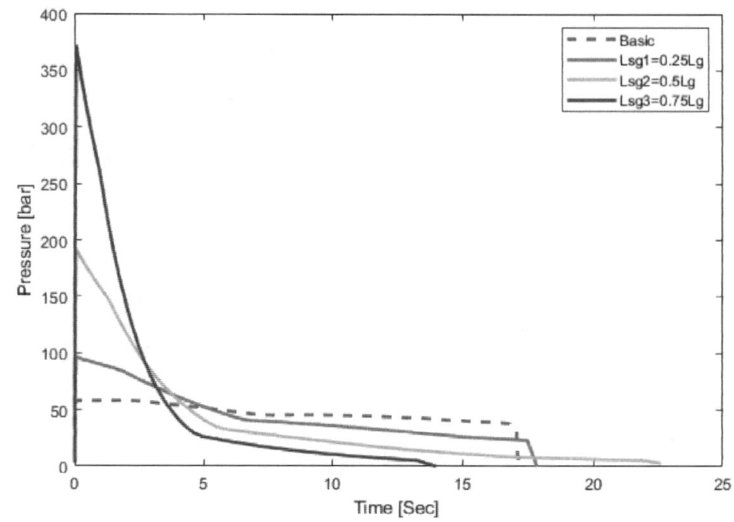

Figure 11 Pressure time curve for cases in group A

From figure.11, it is obvious that increasing the slotted length increases regressivity in the grain burnback analysis by increasing the initial burning surface and decreasing the final burning area with a shorter burning time.

4.2 Group B: Effect of the internal radius of grain

In this group, the variation of the inner radius of the grain is investigated, with 2 more cases other than the baseline case, with values of slotted length as in table 2.

Table 2 Changes in the internal radius of slotted grain

Case B	r_{in1}(mm)	r_{in2}((base radius)	r_{in3}
	25	41	60

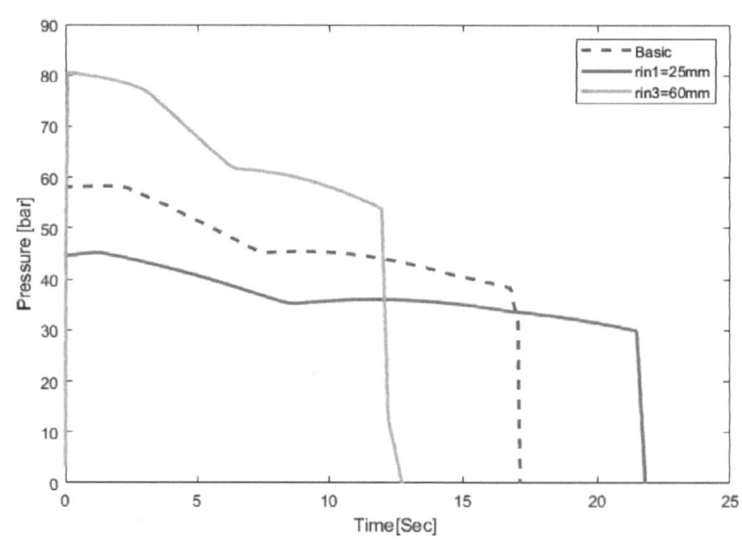

Figure 12 Pressure time curve for cases in group B

From figure.12, it is obvious that changing the inner radius does not affect the regressive behavior as both initial and final burning surfaces increase. However, increase the inner radius, increases the burning area and combustion pressure and decreases the burning time.

4.3 Group C: Effect of slot width

In this group, the variation of slot width is investigated, with 3 more cases other than the baseline case, with values of slotted length as in table 3.

Table 3 changes in the width of the slotted grain

$Case\ D$	b $(width\ of\ slots)$	$b2$	$b3$	$b4$
	3 (Basic)	4	8	10

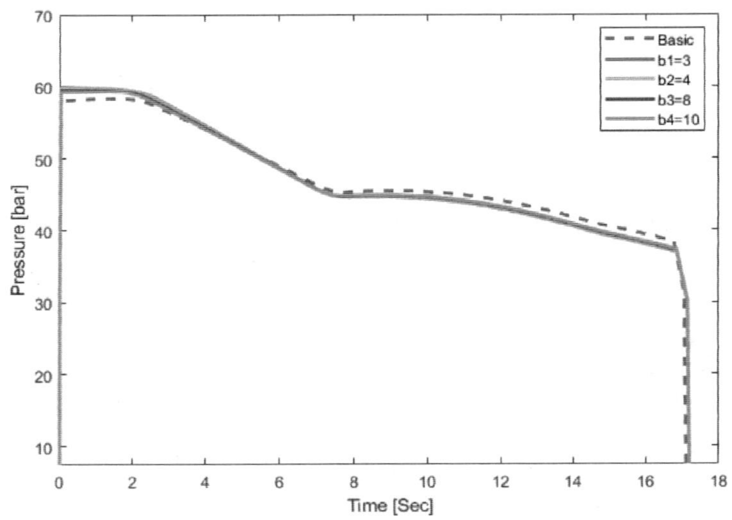

Figure 13 Pressure time curve for cases in group C

From figure.13, it is obvious that changing the slot width has a very minor effect on grain burnback.

4.4. Group D: Effect of the number of slots.

In this group, the variation of the number of slots is investigated, with 3 more cases other than the baseline case, with values of slotted length as in table 4.

Table 4 changes in the number of slotted grain

$Case\ D$	n $(number\ of\ slots)$	$n1$	$n2$	$n3$
	3 (Basic)	4	5	6

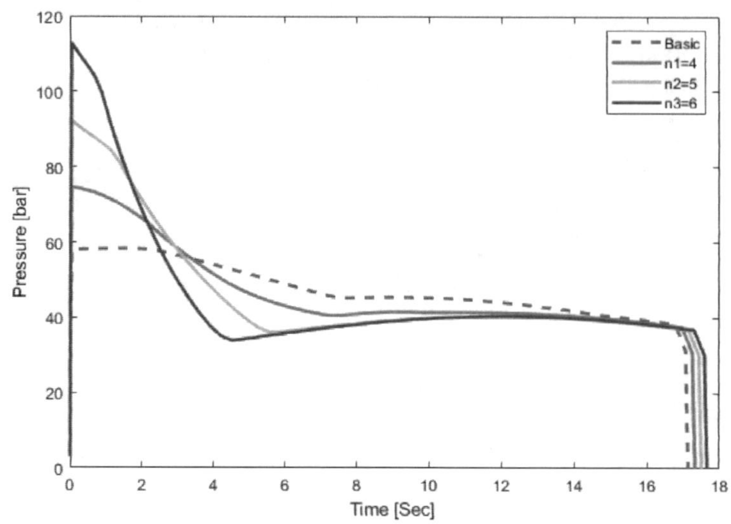

Figure 14 Pressure time curve for case D

From figure.14, it is obvious that changing the number of slots has a large effect on grain burnback, increasing the number of slots increases the regressive behavior of the grain with increased initial burning surface and decreased final burning area.

5. Conclusion and future work

Through-web axial slot grain (slotted grain) has the advantages of sliver free and no stress-concentration regions that occur in other internal burning grains as star grain and wagon wheel grain. Moreover, it has larger volumetric loading compared to other internal-external burning grains. It has the drawback of exposing the motor wall to hot gases. The drafting technique was used to study the effect of geometric parameters on the performance of slotted grain. First, the technique was validated against experimental data, then a variation of one single parameter at a time technique was applied to find the effect of each parameter. It was clear that the width of the slot has a minor effect, while slot length and number of slots have a great effect on the regressive behavior.

While regressive behavior is required for some missile applications especially in the sustainer phase, this objective may compromise other performance parameters. So, a multidisciplinary optimization module should be applied to achieve the requisite multi-objective optimization.

References

[1] Barrere M, et al. 1960 *Rocket propulsion* Elsevier Publishing Company New York
[2] Sutton, G. and Biblarz .2001 Rocket propulsion elements 7th edition, John Wiley & Sons, Inc New York
[3] Brügge N. 2021 Diamant missile Description, http://www.b14643.de/Spacerockets_1/West_Europe/Diamant/Description/Frame.htm (retrieved Jan, 28th, 2021)
[4] Davenas A 1995 History of the development of solid rocket propellant in France *Journal of Propulsion and Power* **11**, No. 2, 285,.
[5] Nagappa, R et al.1989 ISRO's solid rocket motors Acta *Astronautica.* **19**, No. 8, 681
[6] Daniel D 1969 *Flight Performance Of FW-4D solid-propellant Rocket Motor On Delta 50 (Aimp-E)* NASA TN D-4987.
[7] Anon. 2021 Guided Missile 5B27y Technical Description http://historykpvo.narod2.ru/ (retrieved Jan, 28th, 2021)
[8] Zayed, A 2005 Design Optimization and Geometric Analysis of The Slotted Solid Propellant Grains *Proceeding of the 11th ASAT Conference*, 17-19 May 2005, 441-462

[9] Stone, M.1961 Slotted tube grain design," *ARS Journal* **31**, No. 2 223,
[10] Jaramaz S and Injac M 1991 Effect of Grain Characteristics on Range of Artillery Projectiles with Base Bleed *1st International Symposium on Special Topics in Chemical Propulsion* 143
[11] Helmy A.1981 Performance analysis of SA-3 missile second stage *17th Joint Propulsion Conference July 27-29*, 1981 Colorado Springs, Colorado, AIAA-1981-1514
[12] El-Naggar M. et al 2020 IOP Conf. Ser.: Mater. Sci. Eng. 973 012001

IOP Conf. Series: Materials Science and Engineering 1172 (2021) 012045 doi:10.1088/1757-899X/1172/1/012045

Application of the level set method in solid propellant grain burnback with dogbone grain as a case study

H Adel[1] and H Belal[2]

[1] M.Sc. Student, Rocket Department, Military Technical College, Egypt
[2] Independent Researcher, AIAA Member, Egypt

Email: khatab80@hotmail.com

Abstract. Solid propellant grain burnback is a crucial step internal ballistic module. There are different methods to predict grain burnback; the oldest method is the drafting technique where the burning surface is tracked using manual drawing applying surface regression rules. Then there is the analytical method, where a geometric analysis of the surface is performed and analytical expressions for the burning surface segments are derived. These expressions are evaluated at different values of burnt distance to find the evolution of burning surface and port areas with time. A level set method is a different approach which, in some aspects, may be considered a drafting technique where the burning surface segments are tracked. From another point of view, it is a numerical method that tracks the burning surface with no need of trying to derive analytical expressions for burning segments. Dogbone grain is known for its structural superiority; however, researches dealing with dogbone grain design parameters are scarce. In this paper, the level set method is introduced and the main steps for its implementation are elaborated. Validation of the level set method is performed against the drafting technique using CAD and analytical expressions for the well-known star grain. Finally, dogbone grain geometry with different configurations is introduced and a parametric study for the governing parameters that affect dogbone grain burnback is performed.

1. Introduction

Burnback analysis is an important step in predicting the ballistic performance of a solid-propellant rocket motor. There are several methods to perform the burnback analysis; the oldest method is the drafting technique where the simulation of the burning is done by combining basic shapes to describe the grain initial geometry and the burning surface is assumed to propagate normal to itself [1]. The famous Solid Performance Program (SPP™) is based on the drafting technique [2]. The main drawbacks of this method are that it is discrete, requires high human-computer interaction which may lead to errors, and it is also a time-consuming method.

Another way to perform the burnback analysis is the analytical method where the main idea is to describe the burning surface or interface through mathematical relations. Unlike the drafting technique, the analytical method is fast and continuous in representing the burning progression [3]. The disadvantage of the analytical method is that the relations are dedicated to a single grain configuration, so for some complex configurations, it might be difficult to deduce the proper mathematical relations.

The other method which is the scope of this research is one of the numerical interface tracking methods or namely the level set method. The level set method is first introduced in 1988 by Sethian and Osher [4] and offers a numerical technique for tracking the motion of a curve with time.

Content from this work may be used under the terms of the Creative Commons Attribution 3.0 licence. Any further distribution of this work must maintain attribution to the author(s) and the title of the work, journal citation and DOI.
Published under licence by IOP Publishing Ltd

IOP Conf. Series: Materials Science and Engineering 1172 (2021) 012045 doi:10.1088/1757-899X/1172/1/012045

The level set method has so many applications in different fields from physics, chemistry, fluid mechanics, image processing, materials sciences, and fabrication of microelectronic components [4]. Also, it has many applications in rockets such as burning rate modeling [5], sloshing in liquid propellant tanks [6], droplet collision in dense spray applications [7], and G-equation for flame front modeling [8].

Unlike the analytical method, the level set method affords a high ability in dealing with complex shapes as it can easily work with the sharp edges, cusps, and corners; as well as the change in the topology of the interface. This enables to perform the burnback analysis for any grain configuration such as dogbone grain; which is an internal burning grain known for its structural superiority [9]. Its geometry can be represented typically as a special case of the wagon wheel grain. This configuration is commonly used when high initial burning area or chamber pressures are desired [10]. However, the current work aims to implement a procedure to perform a level-set-based burnback analysis and apply this procedure to different grain geometries to validate it. Finally, a parametric study for a dogbone grain is performed. As to the author's knowledge, the researches dealing with dogbone grain design parameters are scarce.

2. Level Set Methodology

2.1 Level set basics

There are two approaches in studying moving interfaces or fluids, the Eulerian approach and the Lagrangian approach, which are completely different. "*Lagrangian approach follows the path of each particle of fluid as it moves, Eulerian approach sees which particles pass through each point it sits*". The Lagrangian approach is more direct as it tracks the flow, so if we mark a set of equally spaced points on it at the beginning, some difficulties will appear later in the mesh such as the spacing could be too narrow or too wide, the initial curve can intersect itself or can split apart. The level set method cures all these difficulties by using the Eulerian approach where the coordinates are fixed and it only captures the interface implicitly, when expanding or tangling the interface it is translated into changes in the level set function '$\emptyset$' not in the mesh [11].

This remarkable feature made it able to deal with different types of cusps in the propellant grain configurations. For example, as shown in figure 1; the burning progression does not affect the convex cusp and it remains as it was; while the concave one becomes an arc of a circle and its center is the cusp vertex. This can cause some difficulties in using the drafting technique for the grain burnback analysis, since one has to create a new surface for the arc, unlike the level set method which deals with this easily as this topological change can be handled by converting the 2D interface into a 3D surface called the level set function '$\emptyset$' as shown in figure 2. This function is constructed numerically and on intersecting it at zero level one gets the initial interface which is star grain.

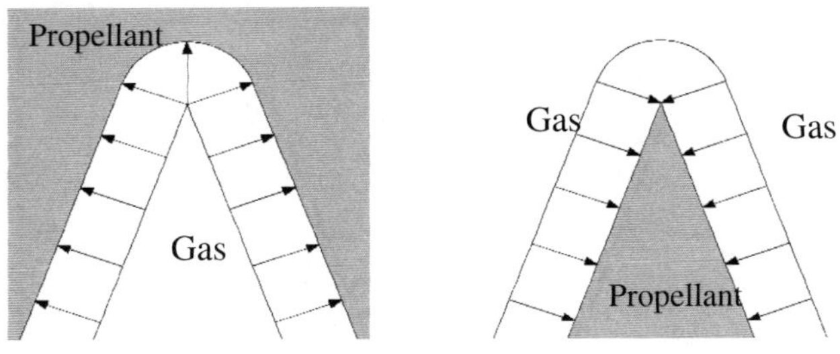

Figure 1 Effect of burning on a convex cusp (right) and concave cusp (left).

For further explanation, consider a simple 2D circular interface shown in figure (3) such that:

$$\emptyset(x,y) = x^2 + y^2 = c \quad (1)$$

When 'c' is positive the level sets are circles and if it is negative there are no level sets. To create the function '$\emptyset$' one has to introduce a third dimension 'z' such that:

$$z = \emptyset(x, y) = x^2 + y^2 = c \tag{2}$$

This will result in a 3D surface shown also in figure 3. That is exactly what happens in any other interface.

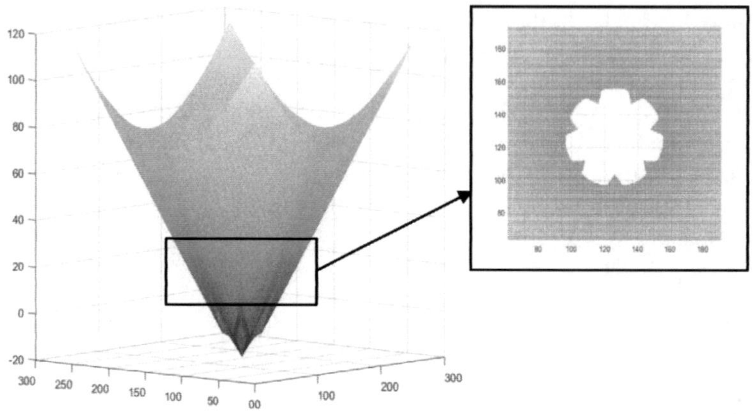

Figure 2 Representation of the function '$\emptyset$' for star grain configuration.

The characteristic feature of the level set method consists of representing the interface implicitly, by using the zero-level set of a time-dependent function '$\emptyset$' [12]. During the formation of the level set method a principle of *"once a particle burns, it remains burnt"*. This means that when two interfaces pass by the same point whatever one arrives first; the particle located there will be ignited. Thus, the function $\emptyset$ is created as an indicator function of the propagating interface, from which the boundary can be determined.

The first interface is represented by Eqn. 3:

$$\emptyset[\vec{x}(t), t] = 0 \tag{3}$$

which expresses that at any time the interface points are defined by $\emptyset$ function for all x not just for the points of the boundary. Figure (3) is a schematic representation of the level set method for an interface (circle) propagating with a velocity F (which in grain burnback analysis is equivalent to burning rate) normal to the boundary and to the outward direction:

The velocity F (burning rate) can be represented as:

$$\frac{d\vec{x}}{dt} \cdot \vec{n} = F \tag{4}$$

Differentiating $\emptyset[\vec{x}(t), t] = 0$ with respect to time:

$$\frac{\partial \emptyset}{\partial t} + \nabla\emptyset \cdot \frac{\partial \vec{x}}{\partial t} = 0 \tag{5}$$

Because $\emptyset$ is constant at a certain level set, $\nabla\emptyset$, is normal to the interface. So the normal direction is calculated using Eqn. 4:

$$\vec{n} = \frac{\nabla\emptyset}{|\nabla\emptyset|} \tag{6}$$

Substituting, the level set equation is obtained:

$$\frac{\partial \emptyset}{\partial t} + \mathrm{F} \cdot |\nabla\emptyset| = 0 \tag{7}$$

with the initial condition given by

$$\emptyset[\vec{x}, t = 0] \tag{8}$$

The initial condition involves knowing all the values of $\emptyset$ for all x at t=0. So, Starting from this initial condition, the interface can be propagated by solving the level set equation for a given velocity F, the previous equations are implemented in a MATLAB code, with an explanation of the procedure as discussed in the next section

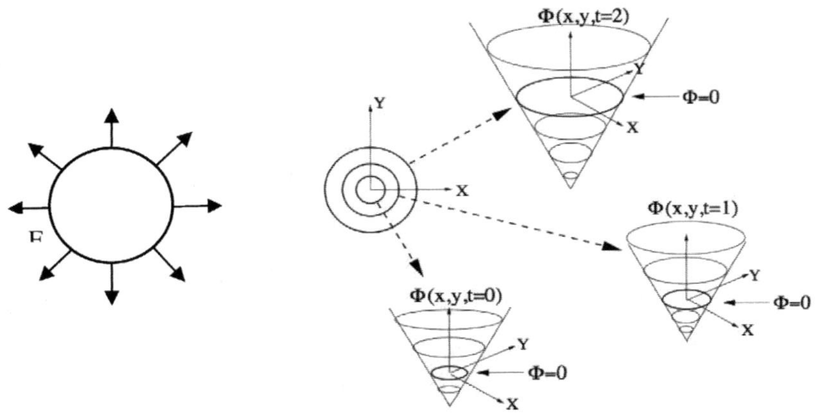

Figure 3 Level set representation for a circular interface

2.2 *Level set procedure*

The procedure starts with drawing the 2D grain configuration using CAD software (in this case SOLIDWORKS), then this 2D grain configuration –which is basically a line describing grain initial burning area- is discretized into XYZ points and exported into a text file. This text file is received by a MATLAB code where the basic equations of the level set method are implemented (as introduced in the previous section), allowing the calculation to start. This MATLAB code initializes values of Ø and adds the background mesh where the calculation will be performed. A block diagram for the procedures is shown in figure 4.

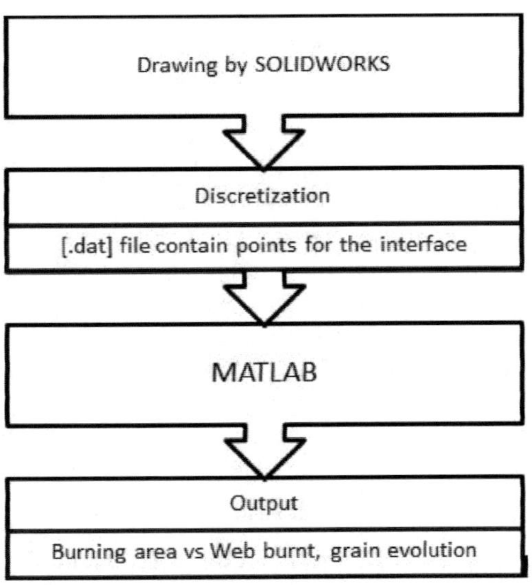

Figure 4 Block diagram for the procedures of level set code

3. Validation and mesh size effect

To validate the method, a standard 7-point star grain, figure 5, with zero cusp radius was selected with a total length of 1.6 m and a web thickness of 33.5 mm [13]. The progress of the burning area against the web burnt was represented by the 3 aforementioned techniques.

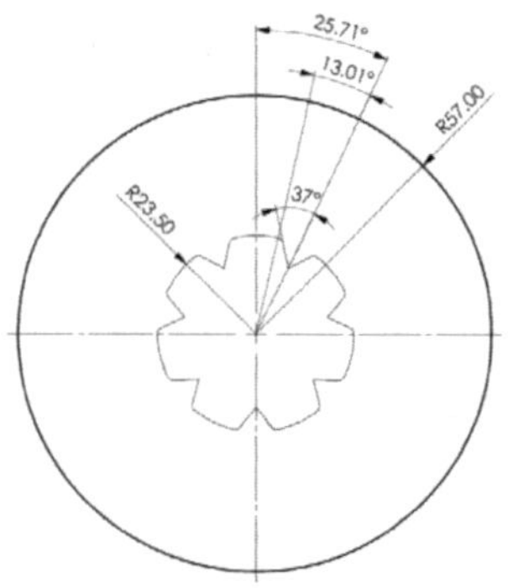

No of star points [N]	7
Star point angle [θ]	74°
Angle fraction [ε]	0.5058
Grain inner radius [R_{in}]	23.5 mm
Fillet radius [f]	1.6 mm
Web thickness [w]	33.5 mm
Length of star grain L_g	1604.1 mm

Figure 5 Geometric parameters for the used star grain

First, the drafting technique is applied to the star grain with data as in Table 1. The grain is drawn using CAD software with several parallel offsets as sketched in figure 6.

Figure 6 Drafting technique for the selected star grain

Second, the analytical method is applied to the same grain using star grain equations for different phases of burning into a MATLAB code written by Hashish et al. [13]. The inputs are the star grain parameters (N, f, θ, w, L, R), and the burning area vs. the web burnt is calculated. Finally, the level set method is applied using the procedure illustrated in figure 4. For this case (star grain); it was first drawn and represented by SOLIDWORKS and then it was imported as an IGS file format into a module which is used for discretization of the model by meshing its outlines for a 2D section into an XYZ file that contains model meshing points to be recognized by the MATLAB code. The code creates the initial burning surface and then propagates it normal to itself to represent the grain regression. The comparison between the three methods is shown in figure 7.

As perceived from these figures, the results from drafting and analytical methods are nearly the same with a very narrow error that could be minimized by increasing the number of the offsets in the drafting method at the expense of consuming more time and effort. However, for the level set method, there is an error in comparison with the other methods. This error is a characteristic of any numerical method due to discretization and local truncation error in solving the level-set equation (Eqn. 7). This error can be minimized by decreasing the cell size of the background mesh at which the level-set equations are solved (figure 8 a-c). A comparison between calculation times for solving this case using different mesh sizes is shown in Table 1, with the root mean square error (RMSE) in each case.

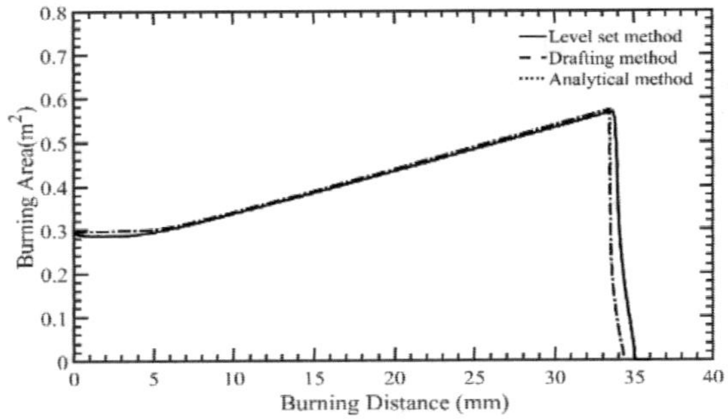

Figure 7 Validation results for the star grain

From figure 8c and Table 1, the maximum error with using mesh size of 250x250 is less than 4 % and RMSE is ~1.5 % compared to the analytical method, which may be considered as an accepted error margin without consuming much more time especially in cases of studying parameters effects (as in this paper)-, while denser mesh size (1000x1000 or 2000x2000) may be used in cases of experimental validation to minimize the discrepancy between prediction and experimental results.

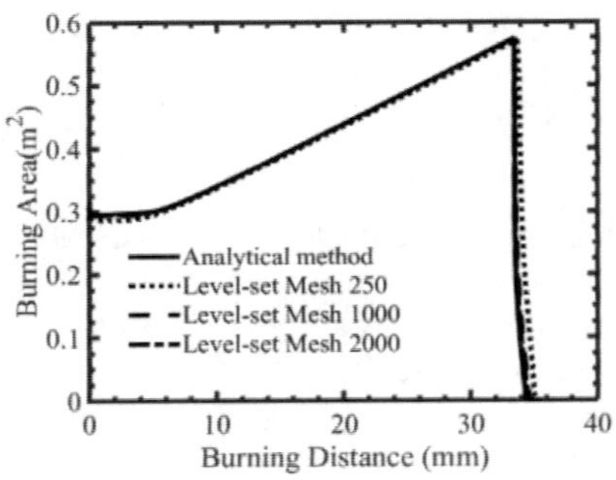

a. Effect of mesh size

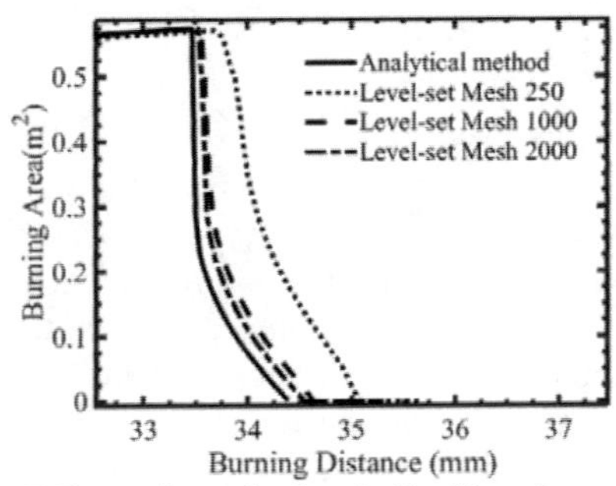

b. Effect of mesh size (tail-off region zoomed)

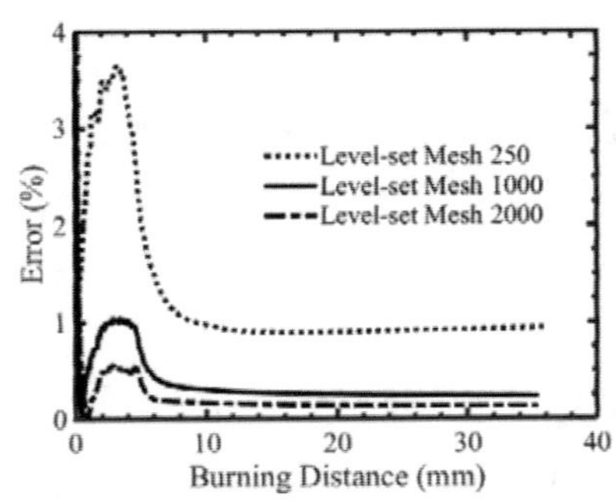

c. Error % for different mesh size-discarding error for the tail-off region-

Figure 8 Effect of mesh size on level-set accuracy

Table 1 Calculation time for different mesh sizes

Mesh size	Time (sec)	RMSE (%)
250x250	20	1.5
500x500	84	0.8
1000x1000	280	0.46
2000x2000	1148	0.36

IOP Conf. Series: Materials Science and Engineering 1172 (2021) 012045 doi:10.1088/1757-899X/1172/1/012045

4. Dogbone grain

Dogbone configuration is selected to test the applicability of the level-set method as actually -to the author's knowledge- published sources have neither studied the dogbone grain nor provided the analytical relations that can be used to find the effect of different design parameters. The dogbone is an internal burning grain that has been developed and it is known by its structural reliability, it can be defined by 8 parameters: (R_{in}, e, θ, h, L_b, L_a, N, R) as shown in figure 9a, with the baseline case, (shown in figure 9b).

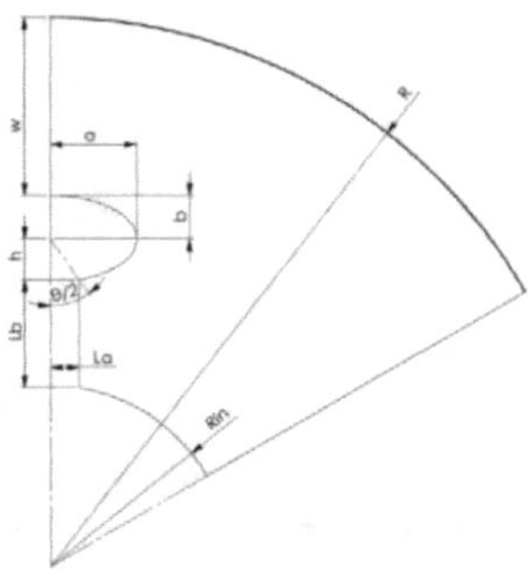

(a) Dogbone grain basic parameters

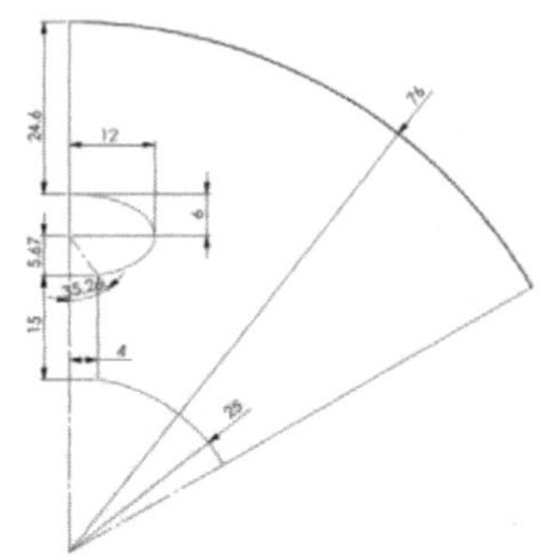

(b) Baseline dogbone grain

Figure 9 Dogbone grain configuration

The baseline case study is a 3-arm dogbone grain (figure 10a), with all geometric data presented in Table 2. First, a drafting technique was executed for the baseline case, where different burning phases are shown in figure 10b. Phase one ends when the arm vanishes; this phase is slightly progressive as the decrease in the burning area due to diminishing of the arm is less than its increase due to the growth of the elliptical part size and the inner radius. Phase two is also progressive and it ends when the web disappears. In the third phase, the tail-off starts, with a regressive behavior as the area of burning decreases. The last phase is the sliver phase where the burning area decreases sharply near the burnout until the propellant grain is completely burned.

Table 2 Baseline case Parameters

Rin	e (a/b)	θ	h	Lb	La	N	R
25	2	70.53°	5.66 mm	15 mm	4 mm	3	76 mm

a- 3D Cross-section for baseline case

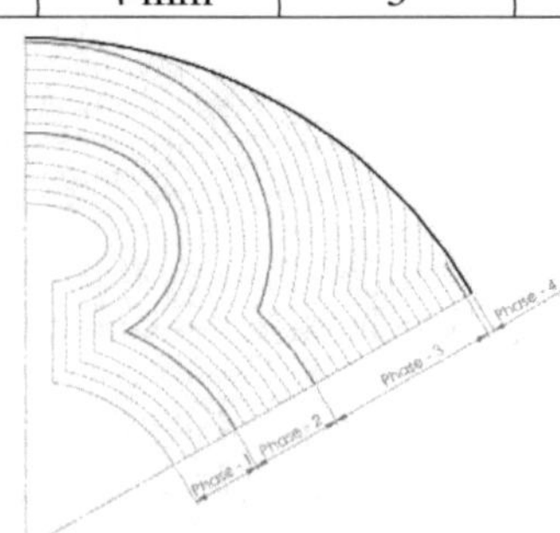

b- Burning phases representation

Figure 10 Dogbone grain

The following is a comparison between the drafting technique and the LS method (figure 11).
As one can see here the error (figure 12) is less than 3% which is also an acceptable error. This proves that the level set method is a reliable technique and can be used in this study.

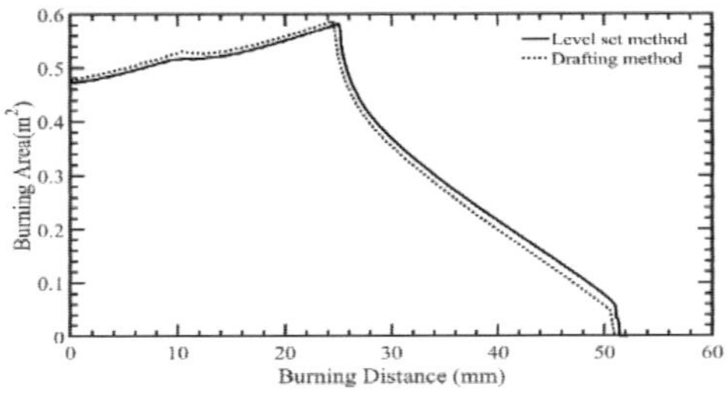

Figure 11 Results for baseline case using drafting technique and LS method.

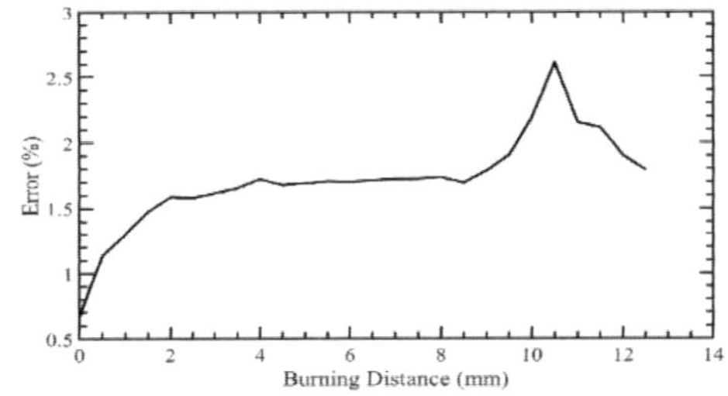

Figure 12 Error percentage for baseline grain.

5. Parametric study for dogbone grain

In order to conclude the effect of each geometric parameter on grain burnback, a parametric study is performed where a single parameter variation at a time method is applied for the following parameters: (R_{in}, e, θ, h, L_b, L_a, and N). Each parameter was changed 3 times to study its effect on the burning area. Table 3 presents a total of 16 studied cases. To compare between different cases three well-known parameters were introduced for each case. These parameters are the blocking factor (K_I) which is the ratio between the burning area to the port area, the clamping factor (J) which is the ratio between the critical area to the port area, and the filling coefficient ($K_{F.C.}$) which is the ratio between the propellant area to chamber cross-section area.

5.1 Effect of changing inner radius Rin':

The effect of changing the inner radius 'R_{in}' was studied by applying 2 different values for it (35 mm and 15 mm) and comparing it with the baseline case#1 (25 mm). For the 3 different configurations, the results of performance parameters and the grain evolution during burning are presented in (figures.13, 14) respectively. Increasing the inner radius obviously decreases the filling coefficient, increases the blocking factor (initial burning perimeter), and increases the sliver phase.

Table 3 Parametric Study Parameters

Case No.	Rin (mm)	L_b (mm)	L_a (mm)	a/b	θ °	h (mm)	N
Case#1	25	15	4	2	70.53	5.66	3
Case#2	35	15	4	2	70.53	5.66	3
Case#3	15	15	4	2	70.53	5.66	3
Case#4	25	10	4	2	70.53	5.66	3
Case#5	25	20	4	2	70.53	5.66	3
Case#6	25	15	6	2	70.53	5.66	3
Case#7	25	15	2	2	70.53	5.66	3
Case#8	25	15	4	2	70.53	5.66	5
Case#9	25	15	4	2	70.53	5.66	6
Case#10	25	15	4	1	56.14	7.50	3
Case#11	25	15	4	0.5	48.19	8.94	3
Case#12	25	15	4	0.5	180	0	3
Case#13	25	15	4	1	180	0	3
Case#14	25	15	4	2	180	0	3
Case#15	25	15	5.37	2	90	5.37	3
Case#16	25	15	5.37	2	270	-5.37	3

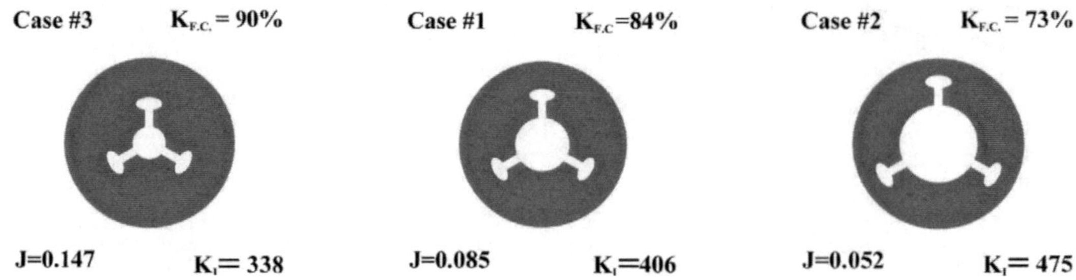

Figure 13 Effect of the inner radius on performance parameters.

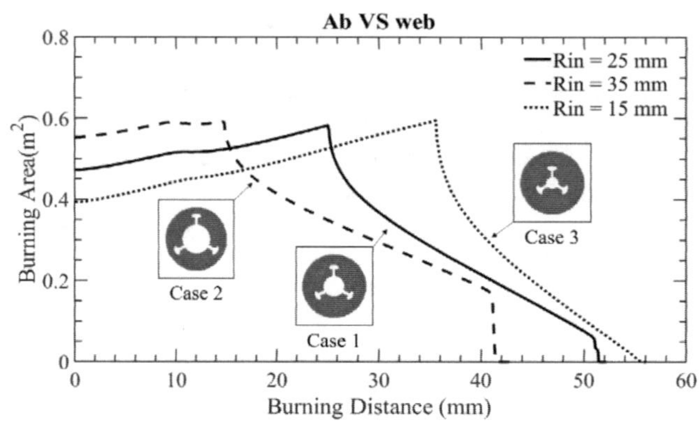

Figure 14 Effect of the inner radius on grain burnback

5.2 *Effect of changing arm length 'Lb'*

The effect of changing the thickness of the arm 'L_b' was studied by applying 2 different values for it, (20 mm, and 10mm) and comparing it with the baseline case#1 (15 mm). For the 3 different configurations, the results of performance parameters, and the grain evolution during burning are shown in (figures 15, 16) respectively. Increasing the arm length does not affect the filling coefficient, but increases the blocking factor (initial burning perimeter) and the sliver phase. From grain burnback, it is noticed that the start of the sliver phase is earlier as arm length is increased. The final burning phases are almost the same.

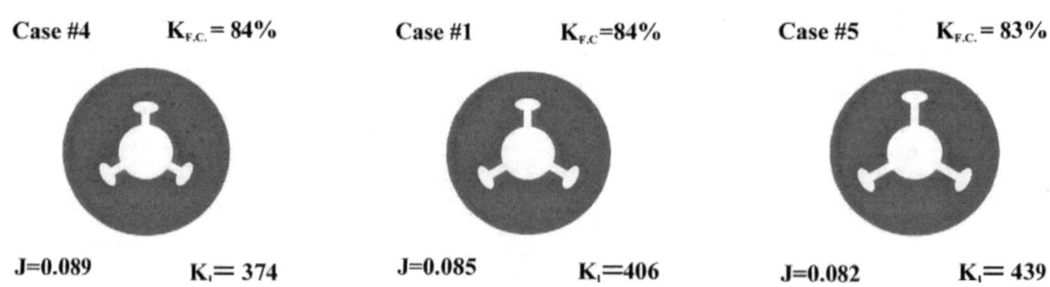

Figure 15 Effect of arm length on performance parameters.

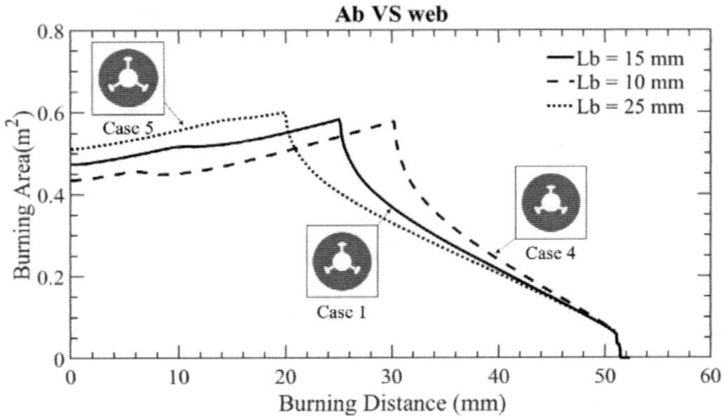

Figure 16 Effect of arm length on grain burnback

5.3 *Effect of changing arm thickness 'La'*

The effect of changing the thickness of the arm 'L_a' was studied by applying 2 different values for it, (2 mm, and 6 mm) and comparing it with the baseline case#1, (4 mm). For the 3 different configurations, the results of performance parameters, and the grain evolution during burning are shown in (figures.17, 18) respectively. Increasing the arm thickness decreases the filling coefficient and the blocking factor, but the total change can be observed only in the initial burning area; the other changes are quite small.

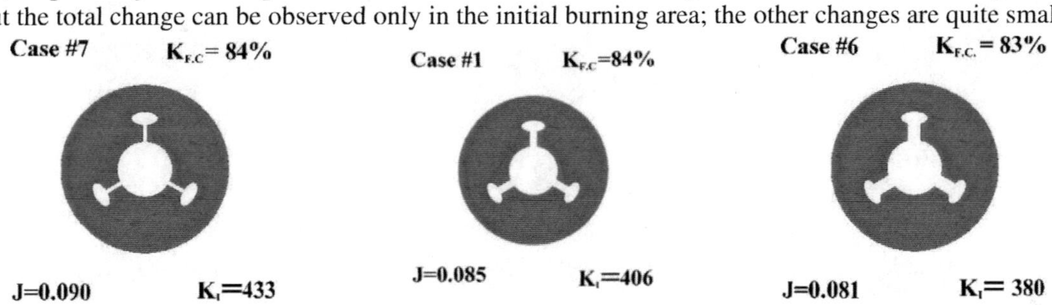

Figure 17 Effect of arm width on performance parameters.

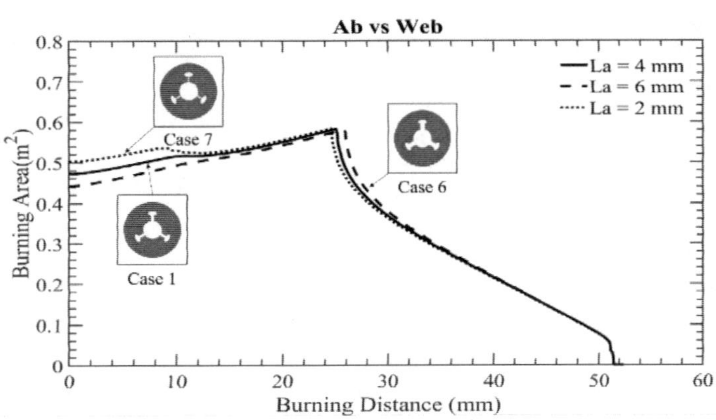

Figure 18 Effect of arm width on grain burnback

5.4 *Effect of changing the number of arms 'N'*

The effect of changing the number of arms 'N' was studied by applying 2 different values for it (5 and 6) and comparing it with the baseline case#1. For the 3 different configurations, the results of performance parameters, and the grain evolution during burning are shown in (figures.19, 20) respectively. Increasing the number of arms made an obvious effect on the behavior of burning as shown. It decreases the filling coefficient and visibly increases the blocking factor but the major effect can be observed in the burning area evolution with burned distance; a dual thrust behavior is observed and the sliver area is reduced.

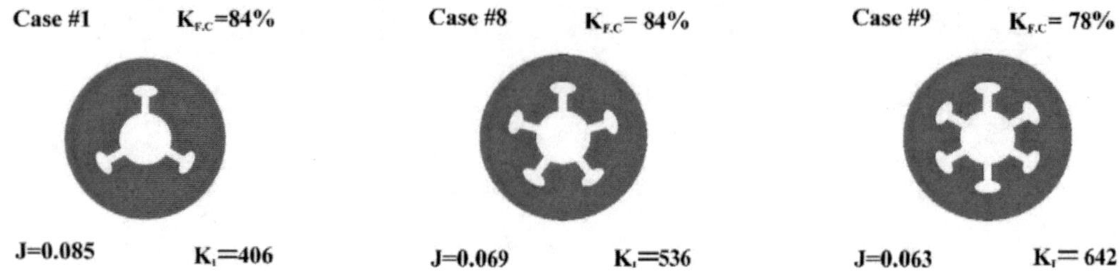

Figure 19 Effect of number arms on performance parameters.

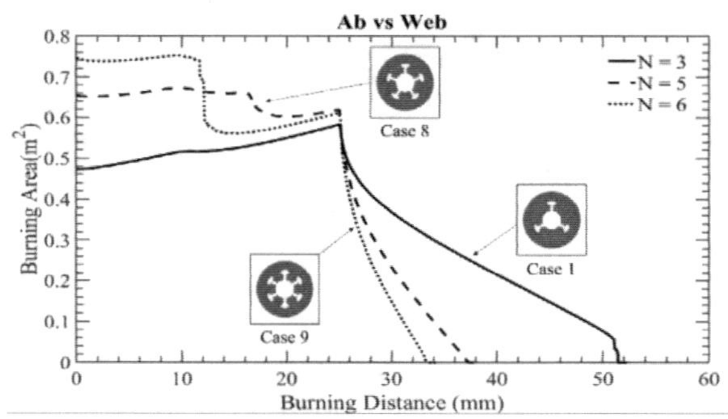

Figure 20 Effect of number of arms on grain burnback

For further understanding of the dual thrust behavior, the drafting technique for the 3 aforementioned cases was performed as shown in figures 21a-c. The blue dashed line is the sector border for each case; it was found that the surfaces became much closer and a detached sliver is formed as shown in figure 22. This indicates that the dual thrust behavior shown in figure 20 is caused by a sudden vanishing of the surface with the formation of a detached sliver, not a smooth decrease like star grains or sliver-less jump like wagon wheel grain.

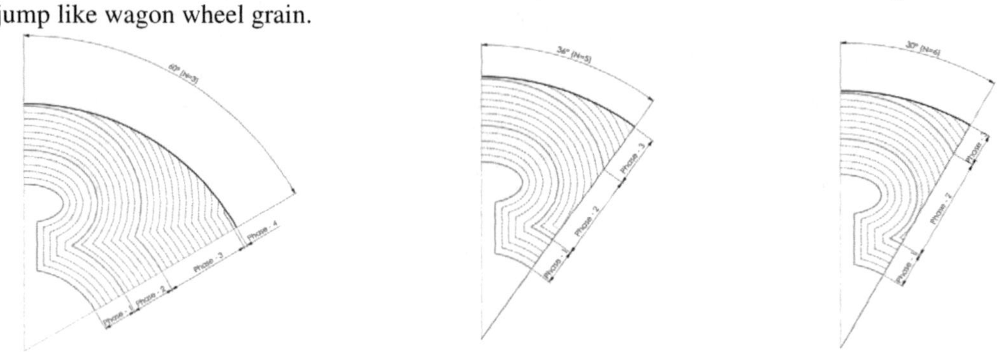

a. Case #1 b.Case #8 c. Case #9

Figure 21 Drafting technique for cases #1, #8 and #9: Explanation of dual thrust behavior

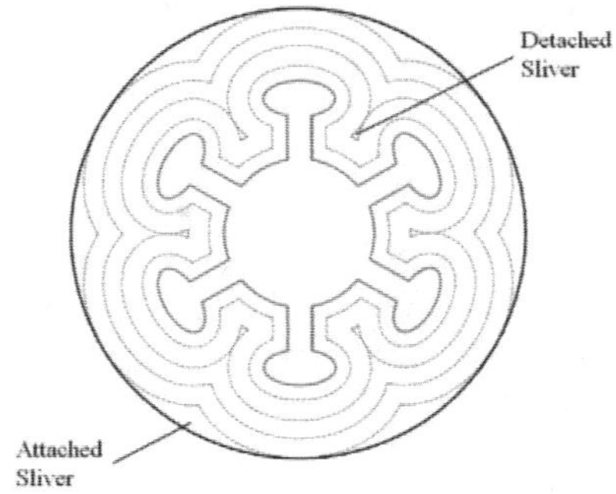

Figure 22 Dogbone grain sliver illustration

5.5 *Effect of changing ratio (a/b)*

The effect of changing the ellipse shape or (a/b) ratio was studied by applying 2 different values for it (2 and 0.5) and comparing it with the baseline case#1. For the 3 different configurations, the results of performance parameters, and the grain evolution during burning are shown in (figures. 23, 24) respectively.

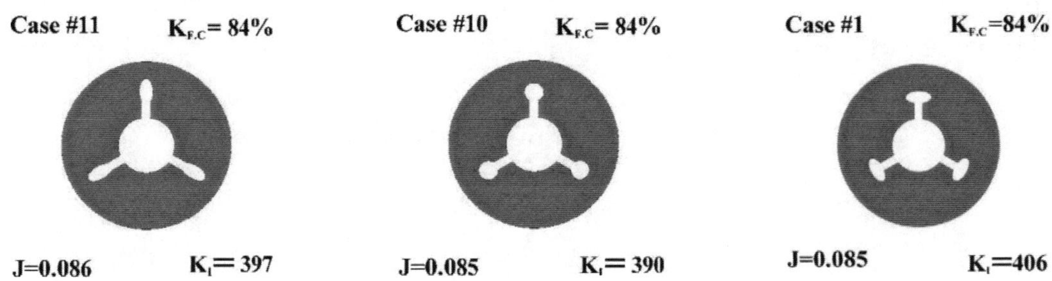

Figure 23 Effect of ellipse ratio a/b on performance parameters.

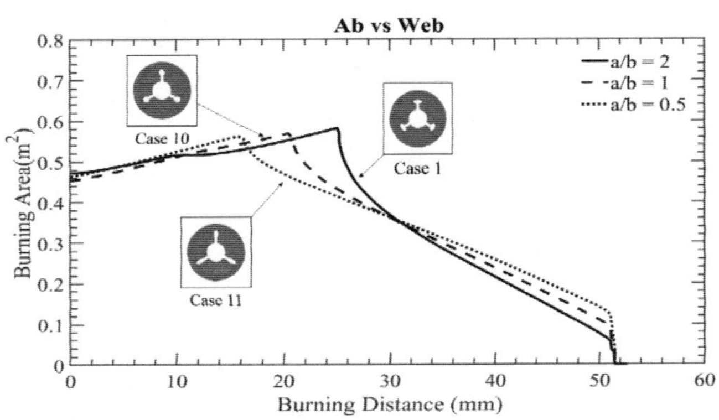

Figure 24 Effect of the ellipse ratio or (a/b) ratio on grain burnback

Changing the ratio (a/b) quietly affects the blocking factor and the filling coefficient. For a/b=0.5, the sliver area increases obviously, and the progressive phase decreases. For a/b=1 (circular tip) change can be observed in the sliver area too, and the progressive phase decreases a little.

5.6 *Effect of tip center shape*

The change in the tip center and shape was studied by assuming 3 cases as follows. The first one is case #12 as shown below. It is a special case of case #11 where a=La, b=2a, and the center h=0. When a=b=La and the center h=0, the grain configuration became finocyl. Finally, case#14 is a special case of case#1 where a=La=2b and radius h=0. For the 3 different configurations, the results of performance parameters, and the grain evolution during burning are shown in figures 25, 26 respectively.

As shown in the figure when a=La, b=2a, and the center h=0, the blocking factor is maximum and the filling is minimum, the sliver area is also the minimum. For the finocyl configuration, the sliver area increased but the progressive phase of burning is also increased and the blocking factor is decreased. Finally, when a=La=2b and radius h =0, the progressive phase is maximum and the sliver area is also increased, the blocking factor change is negligible.

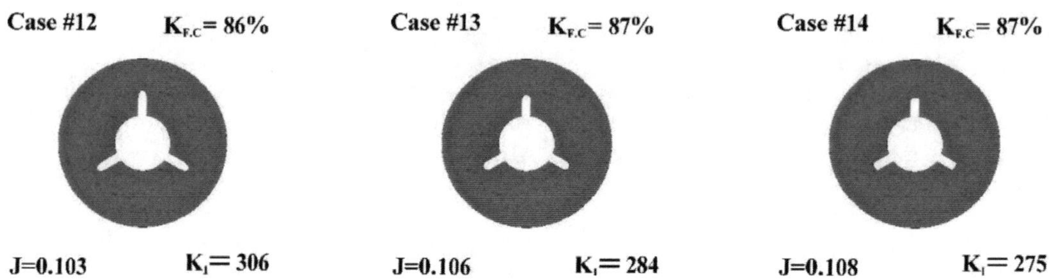

Figure 25 Effect of tip center shape on performance parameters.

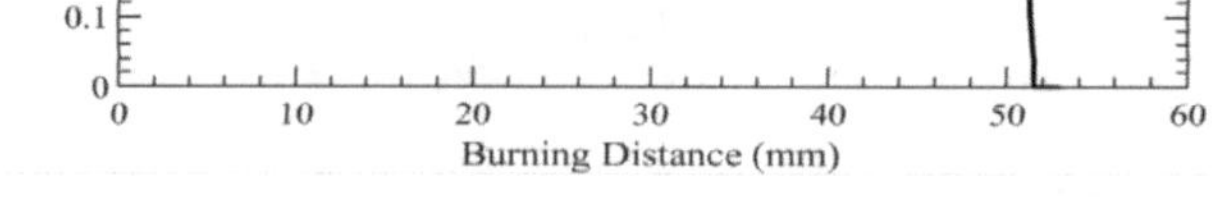

Figure 26 Effect of tip center shape on grain burnback.

5.7 *Effect of ellipse angle 'θ'*

The ellipse angle 'θ' effect was studied by applying 2 different values for it (270°and 90°). The two cases were compared together and the results of performance parameters are shown in figure 27. The grain evolution during burning is shown in figure 28, which indicates that the ellipse center angle changes the entire configuration. When θ=270, the blocking factor decreases, the sliver area decreases obviously, the progressive phase increases, and the filling coefficient increases.

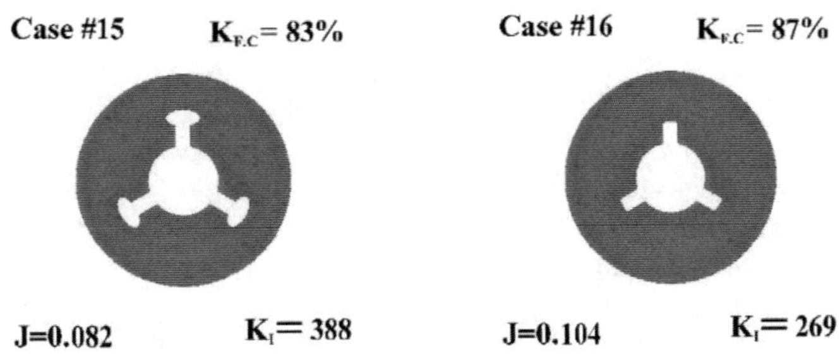

Figure 27 Effect of ellipse angle on performance parameters.

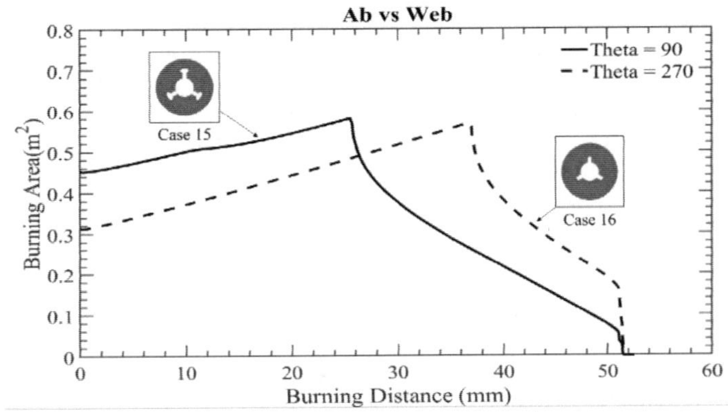

Figure 28 Effect of ellipse angle 'θ' on grain burnback

6. Conclusion and Future Work

The level set method and its application in evaluating grain burnback are introduced. Validation of the procedure is performed by comparing level set results for both star and dogbone grains with other validated techniques, namely drafting and analytical techniques, with an acceptable error. The level set method has the advantage of the ability to deal with complex geometry in a rapid way suitable throughout the preliminary design phase of a rocket motor. A parametric study for dogbone grain was performed, which showed that the dogbone grain is capable of providing a dual thrust profile. It is shown that the most important parameters are the number of arms, arm length, and inner radius, while the other parameters show minor effects on performance and grain burnback.

However, in order to get useful results from the proposed procedure, it needs to be automated to plug in an optimization module to get the most favorable design parameters according to the required mission. Furthermore, as the present methodology deals only with 2D grains, it could be modified to be usable in axisymmetric and 3D grains.

References

[1] Barrere M, et al. 1960 *Rocket propulsion* Elsevier Publishing Company New York
[2] Peterson, E, et al., 1968 Generalized coordinate grain design, and internal ballistics evaluation program, *ICRPG/AIAA 3rd Solid Propulsion Conference* - Atlantic City, New Jersey/ AIAA Paper, No. 1968-490.
[3] Ceyhun T and Melike N, March 2019, Internal ballistic modeling of a solid rocket motor by analytical burnback analysis *Journal of Spacecraft and Rockets,*. **56**, No. 2, 498.
[4] Stanley O and James A S, 1988, Fronts propagating with curvature-dependent speed: algorithms based on Hamilton-Jacobi formulations. *J. Comput. Phys.*, **79** No.1:12.
[5] Wang X and Jackson T L, 11-14 July 2004, Modeling of aluminized composite solid propellants, *40th AIAA/ASME/SAE/ASEE Joint Propulsion Conference and Exhibit,* Fort Lauderdale,

Florida, AIAA-2004-4041.
[6] Tadashi W, 2011, Numerical simulation of liquid sloshing using arbitrary Lagrangian-Eulerian level set method, Int. J. of Multiphysics **5** No.4, 339.
[7] Kwon T, 2003, *Simulating collisions droplets with walls and films using a level set method*, Ph.D. Dissertation, Purdue University, West lafayette, IN,USA.
[8] Norbert P, 2010 *Combustion theory* June 28th - July 2nd, 2010, *CEFRC* Summer School Princeton University.New Jersey, USA
[9] B*rooks W. 1972 Solid propellant grain design and internal ballistics" NASA Report SP-8076,*
[10] Roy H, et al., 2003, A review of analytical methods for solid rocket motor grain analysis *39th AIAA/ASME/SAE/ASEE Joint Propulsion Conference and Exhibit*, 20-23 July 2003, Huntsville, Alabama, AIAA-2003-4506.
[11] Arnau P L, January 2013*, Study of grain burnback and performance of solid rocket motors*, Graduation Project*, ETSEIAT-* Universitat Politècnica de Catalunya, Spain
[12] Gilbert Strang.2006 *18.086 Mathematical Methods for Engineers II.* Spring 2006. Massachusetts Institute of Technology: MIT OpenCourseWare, https://ocw.mit.edu.
[13] Hashish A. April 2018, "Hybrid optimization of star grain performance prediction tool", *18th Int. AMME Conference,* Military Technical College, Egypt.

IOP Conf. Series: Materials Science and Engineering 1172 (2021) 012046 doi:10.1088/1757-899X/1172/1/012046

Experimental investigation of recirculation zone characteristics of a pre-filming airblast injector

N Darwish[1], K P Shanmugadas[2]and S R Chakravarthy[2]

[1] University of Science and Technology at Zewail City, Giza, Egypt
[2] National Centre for Combustion Research and Development & Department of Aerospace Engineering, Indian Institute of Technology Madras, Chennai, India

Email: s-mohammednady@zewailcity.edu.eg

Abstract. The toroidal flow at the recirculation zone has a vital role in combustion process as it helps in mixing hot combustion products with incoming fresh air and fuel which increases combustion efficiency. In the present work, characteristics of recirculation zone are investigated using a pre-filming airblast injector. Particle Image Velocimetry is used to characterize the swirl flow field generated by the airblast injector. Moreover, olive oil is used as a tracer to be captured by a high-speed camera. Different flow rates are investigated in order for finding out the effect of varying flow rates on the characteristics of the recirculation zone. Results for recirculation zone shape, velocity field and shear strength at the primary zone of combustion are represented. Additionally, results show that recirculation zone is almost symmetrical with increasing trend in shear strength with increasing flow rates.

1. Introduction

An increasing demand for better combustion performance from the perspective of decreasing pollutants emissions and increasing combustion efficiency was raised in the recent decades in order to fulfil the firm regulatory requirements [1], [2]. Most of gas turbine systems utilize swirl injectors that create Central Toroidal Recirculation Zone (CTRZ) in the primary combustion region downstream the exit of the injector [3]. The CTRZ plays a vital role in flame stabilization and increasing combustion efficiency as it affords an appropriate forward flow speed that matches with flame speed which aids in flame holding [4]. Moreover, CTRZ creates a region of high shear and strong turbulence that promotes rapid mixing and heat transfer between hot combustion products and fresh incoming air and fuel which improves the combustion efficiency [5]. In addition to this, a flow characteristic that is of significant importance of a swirl injector is the vortex breakdown [6]. It can be identified as a sudden change in the centre of a thin vortex and in most cases it develops into a recirculating spiral pattern forming the CTRZ [7]. Also, it is characterized by a low static pressure in the central core downstream the injector resulting in recirculating the flow [8]. Besides, the flow within the core of the CTRZ has a reverse axial velocity component with no tangential or radial velocity components [9]. Reddy [10] reported a decrease in the tangential velocity along the flow path and the flow transitioned to an axial flow with vanishing tangential and radial components.

Numerical and experimental investigations by Lilley and Gupta [11] (1984) led to identify a term, Swirl Number (SN) for measuring the strength of swirl. SN is the ratio between the axial flux of angular

Content from this work may be used under the terms of the Creative Commons Attribution 3.0 licence. Any further distribution of this work must maintain attribution to the author(s) and the title of the work, journal citation and DOI.
Published under licence by IOP Publishing Ltd

IOP Conf. Series: Materials Science and Engineering 1172 (2021) 012046 doi:10.1088/1757-899X/1172/1/012046

momentum to the axial flux of axial momentum, studied at the exit of the injector. Additionally, Chigier & Beer [12] (1972) concluded that when the swirl number is around 0.5 ~ 0.6, the swirling motion starts to induce radial and axial pressure gradients in the flow field and when the adverse pressure gradient at the centre of the axis is large enough, CTRZ is formed. Furthermore, Tangirala [13] examined the effect the of swirl numbers by investigating the advantages and drawbacks of high swirl operations. The main result was that when increasing the SN, the mixing and flame stability is enhanced for values approximately till 1 and beyond that value turbulence levels and flame stability are reduced. The main drawback of excessive swirl is that it causes excessive wall heating due to forcing the flame to move upstream getting closer to the wall. Besides, Lu X [14] (2005) carried out a large-eddy simulation analysis to study the vortex breakdown in a coaxial swirl injector. It was concluded that for low swirl number, SN=0.3, vortex breakdown was not observed, however, for SN=0.5, a vortex breakdown was observed resulting in forming a CTRZ.

Chieger & Beer [12] (1972) investigated the types of devices for achieving swirling flows and categorized them as; mechanically rotating vanes with tangential entry of fluid into an axial flow tube called radial swirler and angled vanes in axial flow tube called axial swirler. Moreover, Candel [15] (2014) found that the flame dynamics and the topology of the spray flow field are mainly affected by the size of the CTRZ which is influenced by some geometrical parameters. In addition to, Kilik [16] stated that an enhancement of the size of the CTRZ is achieved by increasing the vane outlet angle. Moreover, a stronger and larger CTRZ is achieved by increasing the curvature of the vanes which minimizes the pressure loss in the swirler compared to a flat axial swirler having the same vanes outlet angle. Also, Beer and Syred [17] (1974) concluded that swirls through enhancing the flow mixing on CTRZ boundaries can reduce the combustion length.

Hadef and Lenze [18] (2008) examined the effect on spray characteristics for different relative swirl rotations of flow, co and counter swirlers. They reported that flame stability was enhanced using counter rotating swirler due to the finer atomization achieved by it. Additionally, Wang [19] reported that the counter-rotating configuration is more preferable than the co-rotating one as the counter-rotating arrangement helps in the formation of fine droplets. Moreover, a drawback of the co-rotating arrangement is that it produces a large recirculation zone leading to a more susceptibility to flame oscillation. Besides, Merkle [20] reported that in the case of a counter-rotating arrangement, more mass flow rate was recirculated in addition to a shorter axial length of the CTRZ in comparing with the co-rotating arrangement.

Fuel atomizers are devices used for spreading liquid fuel to thinnest possible sheet and to atomize it, using the aerodynamic forces generated from swirling flows, into fine droplets [19]. Most modern gas turbine combustors achieve fuel atomization either using an airblast atomizer or using a pressure atomizer [21]. In addition, airblast atomizers are more preferable to be used when a reduction in soot formation and low pollutant emissions are desirable [22]. Furthermore, the atomization process consists of a primary break up and a secondary breakup. During the primary breakup shear force plays a vital role in breaking liquid films into liquid ligaments while through the secondary breakup surface tension works on breaking liquid ligaments into droplets [23]. Additionally, Lefebvre and Rizkalla [24], [25] studied the effects of both liquid and air properties on atomization quality and reported that atomization quality decreases with increasing liquid-air mass ration and with increasing liquid viscosity as liquid films get thicker. In contrast, increasing air velocity and air density accomplishes better atomization.

2. Experimental Procedures

In the present work, Particle Image Velocimetry (PIV) is used to investigate the swirl flow field generated by a pre-filming airblast injector fitted into a vertical chamber to study characteristics such as shear regions, CTRZ shape and velocity components and the effect of varying only inlet air mass flow rates on these characteristics. Olive oil is used as the tracer particle and SN for inner and outer swirlers are 0.6 and 1.6 respectively.

2.1. Pre-filming airblast injector

The injector used in the present work is a piloted pre-filming airblast injector. The geometry of a pre-filming airblast injector is in figure 1 and more information can be found in [26]. The injector consists of primary and secondary swirlers and a curved pre-filmer called the venture. A dual orifice nozzle is used as the pilot nozzle. The dual orifice fuel nozzle has two fuel circuits: a central primary nozzle that forms a narrow hollow cone spray and a secondary orifice surrounding the nozzle at the centre. The central nozzle creates a primary spray, which is atomized by the primary and secondary air swirl. The secondary orifice spray and part of the primary spray are made to impinge on the venture and proceed with pre-filming airblast atomization [26].

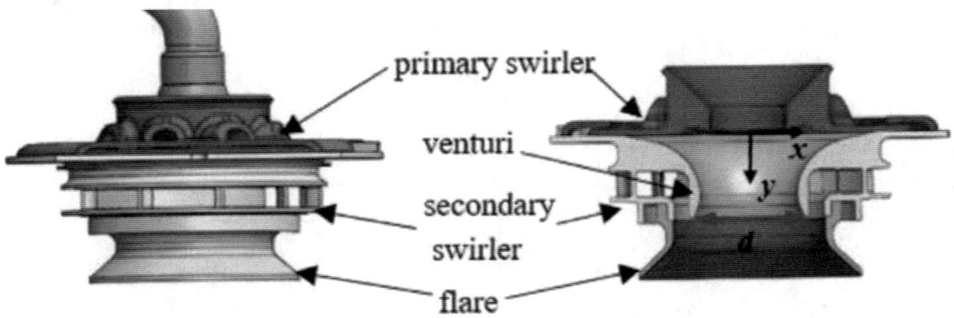

Figure 1. Geometry of a pre-filming airblast injector [26]

2.2. Test conditions

Six different air mass flow rates at room temperature are investigated as tabulated in table 1 along with the bulk velocity at injector exit and Reynolds number associated with each case. Mass flow rates are normalized to the highest used value which is 1512 standard litre per minute (SLPM) for case 6. In addition, bulk velocities are normalized to the highest value which is 9 m/s for case 6.

Table 1. Flow characteristics for different test cases.

	Air mass flow rate	Bulk velocity	Reynolds number
Case 1	0.48	0.48	$2.14*10^4$
Case 2	0.58	0.58	$2.58*10^4$
Case 3	0.64	0.64	$2.845*10^4$
Case 4	0.72	0.72	$3.215*10^4$
Case 5	0.83	0.83	$3.71*10^4$
Case 6	1	1	$4.5*10^4$

2.3. Particle Image Velocimetry and tracer particles

PIV is used to capture the flow field of the seeded air flow downstream the injector. It is based on adding tracer particles to a fluid flow and considering the principle that tracer particles follow fluid flow particles so its motion can be considered as the fluid flow motion. Thereafter, photographing the system twice within a short time interval to capture the displacement moved by the tracer particles. After that by comparing the positions of the same tracer particle at the two different instants of time, the displacement and velocity vectors of the fluid flow can be calculated. The difference between the two shots of laser could be around 40 microsecond.

Moreover, the tracer particles should be small enough to follow the flow and at the same time have correct density and good dispersion of light. A balance has to be achieved between these sides during the experiment [27]. A commonly used tracer particles are oil particles which are prepared by injecting air at high pressures into the oil via a number of nozzles. Each nozzle consists of a tube with a number of small holes in them. Hence, each nozzle produces micro-bubbles with micro-droplets inside. When these bubbles reach the oil surface they burst releasing these droplets. Internal baffles are used to remove the larger unwanted particles. Density of seeding is determined by the air flow and number of nozzles that's why seeding density can be controlled by altering the air flow to the nozzles or by turning one or more of the nozzles on or off [28].

For the present experiment, an ND: YLF Litron laser is used synchronized with one Photron high speed camera with adapters. The acquired pictures are processed using the commercial software Davis. More than 6000 pictures are used for processing of each measurement. An example of similar setup arrangement is shown in figure 2 noting that for the present experiment only one camera is used. Furthermore, Olive oil was used for this experiment for its good characteristics that meet the mentioned requirements [29].

Figure 2. Example of an experimental setup arrangement

3. Results and discussion

Through the current section, results for measurements are represented and for each characteristic a comparisons between results for different flow rates are provided. Figure 3 shows the Seeded air flow downstream the injector.

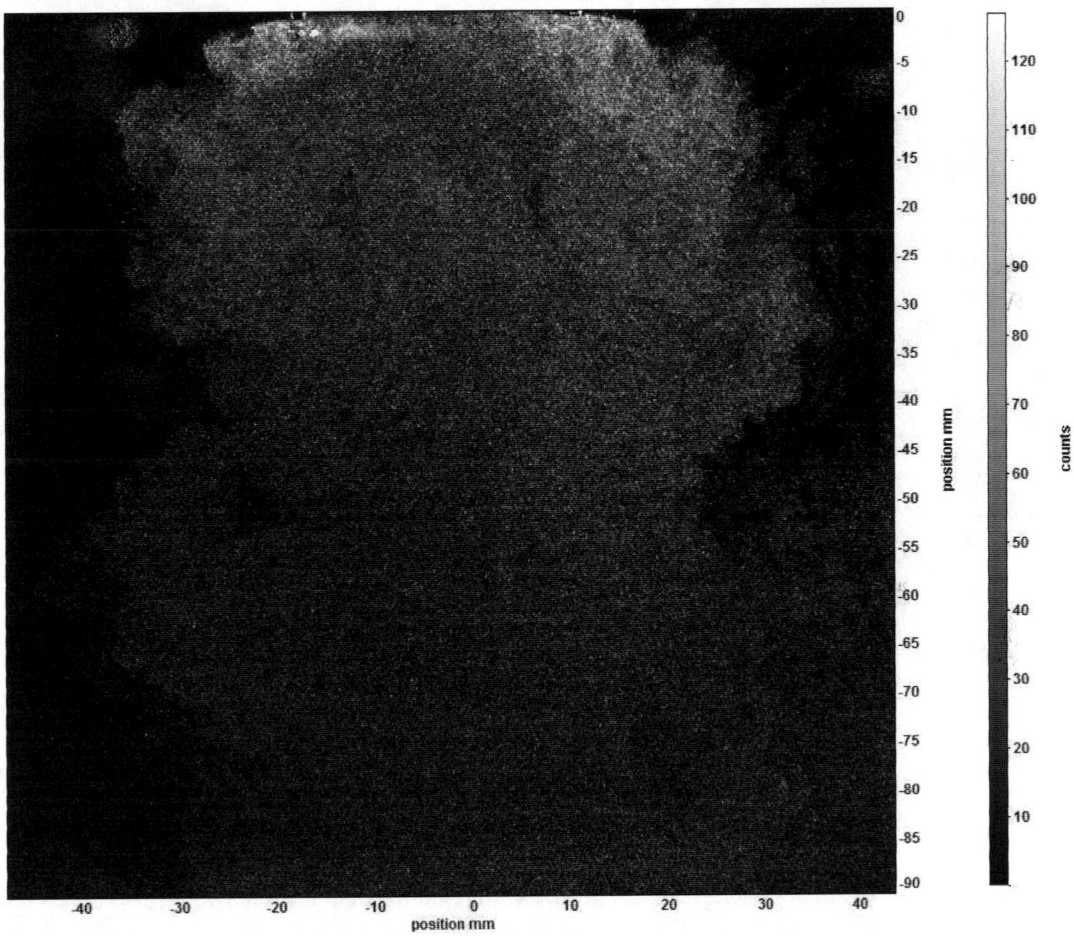

Figure 3. Seeded air flow downstream the injector.

3.1. Recirculation zone shape

From the results, CTRZ shape and dimensions for different flow rates are similar. In figure 4, symmetric configuration around a central vertical axis is clear. Width of the zone is approximately 34 mm while the height is approximately 37 mm. Moreover, for the different flow rates, results show almost independency on the air mass flow rate.

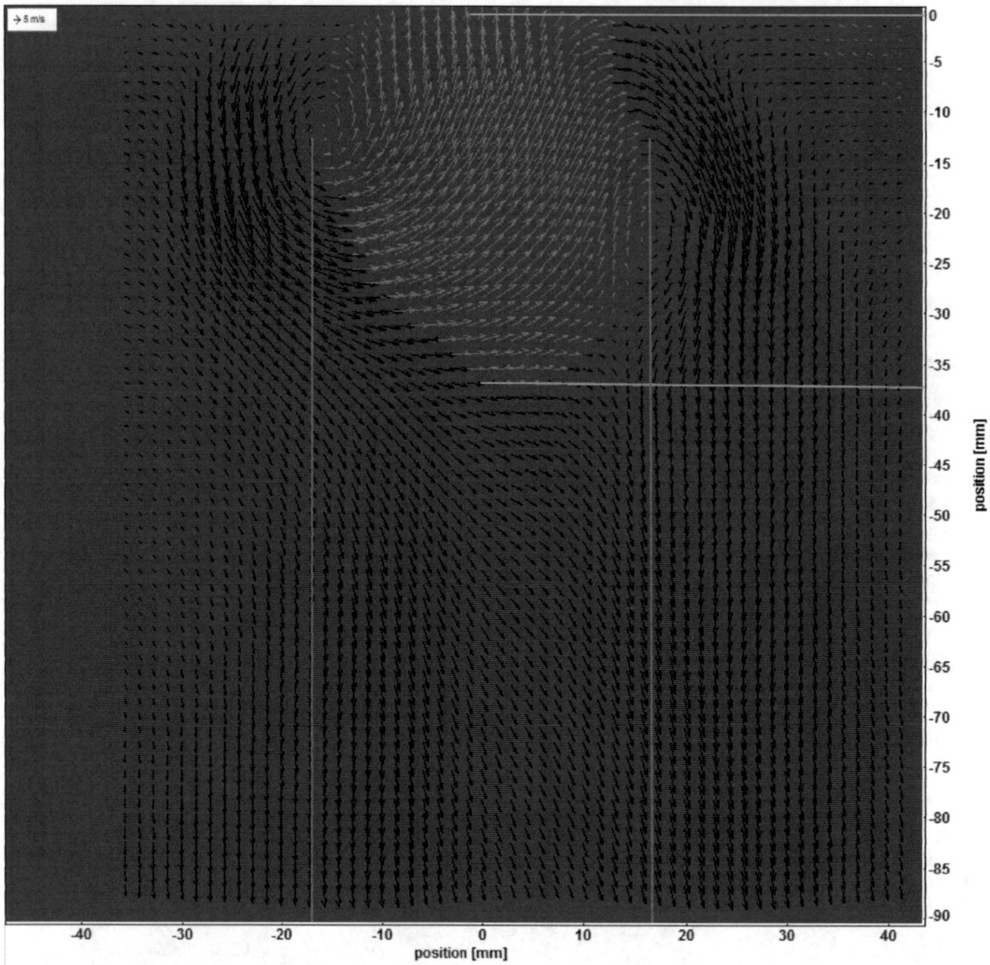

Figure 4. Recirculation zone for test case number 5.

3.2. Velocity field and shear strength

Velocity fields and shear strength for different test cases are shown in figure 5 and figure 6. As shown in the figures, velocity fields are similar in characteristics for the different test cases with change in velocity magnitude. Boundaries of the CTRZ have maximum shear strengths along with the largest velocity magnitudes. In addition, results show that increasing air mass flow rate increases the shear strength magnitude. CTRZs are indicated in the following figures by white circles while the boundaries of the CTRZs are indicated by white arrows.

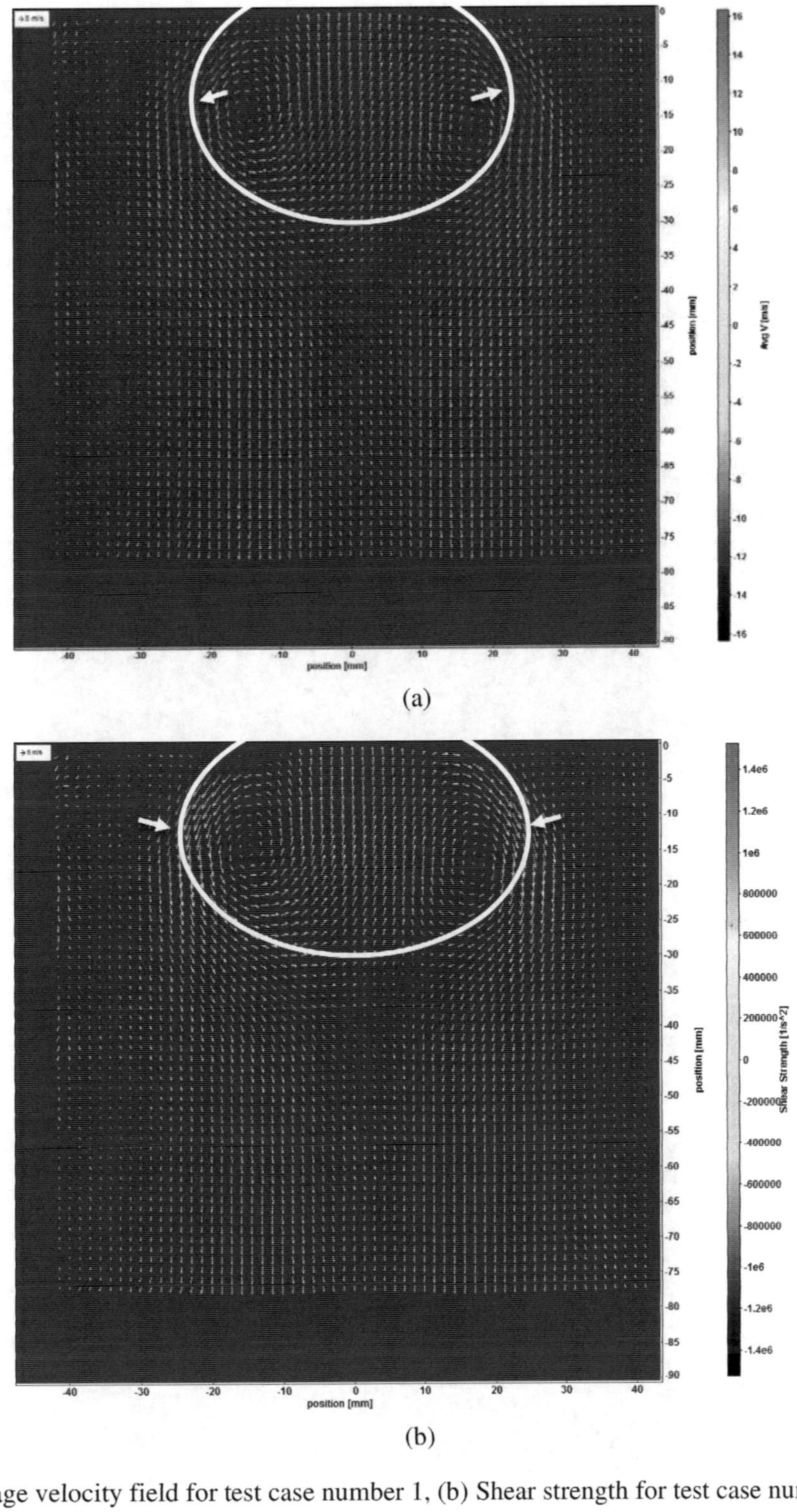

(a)

(b)

Figure 5. (a) Average velocity field for test case number 1, (b) Shear strength for test case number1.

IOP Conf. Series: Materials Science and Engineering 1172 (2021) 012046 doi:10.1088/1757-899X/1172/1/012046

The core region in figure 5 (a) has a velocity magnitude of 8 m/s while the maximum velocity is at the boundaries of the recirculation zone which is 16 m/s. The maximum shear strength in figure 5 (b) is $1.4*10^6$ ($1/s^2$) which is also at the boundaries of the recirculation zone.

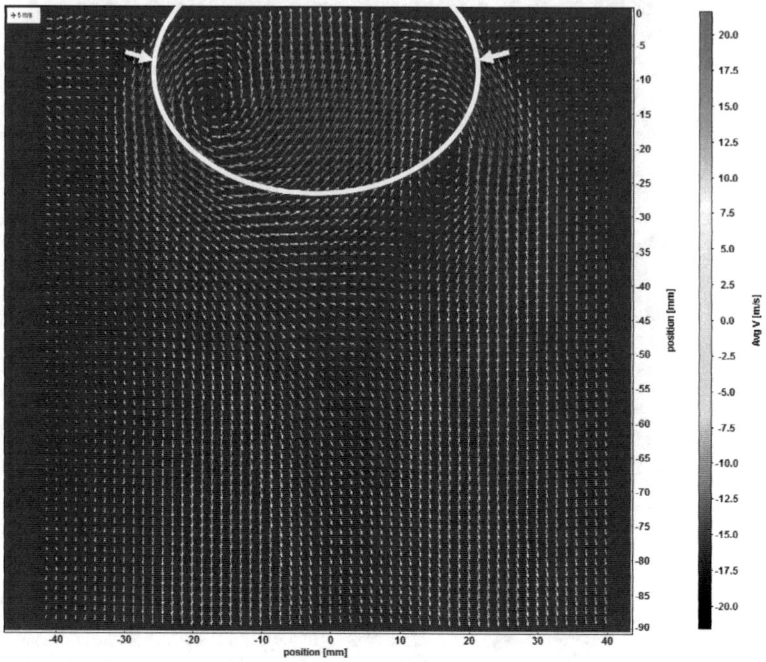

(a)

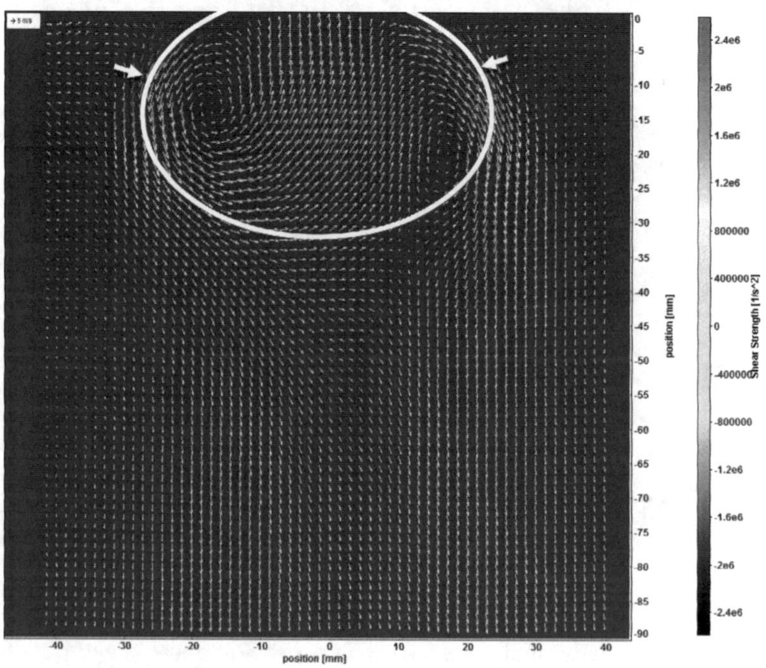

(b)

Figure 6. (a) Average velocity field for test case number 3, (b) Shear strength for test case number 3.

The core region in figure 6 (a) has a velocity magnitude of 14 m/s while the maximum velocity is at the boundaries of the recirculation zone which is 22 m/s. The maximum shear strength in figure 6 (b) is $2.4*10^6$ ($1/s^2$) which is also at the boundaries of the recirculation zone. The other test cases have similar results with same profile shapes but with differences in magnitudes.

3.3. V_y along a horizontal line

V_y along a horizontal line has similar profile shape for different flow rates with change in maximum velocity magnitudes at the core and at the boundaries of the CTRZ. Recirculation zone width can also be noticed as the region between points that cross the zero velocity line which has the same width similar to those results in the subsection 3.1.

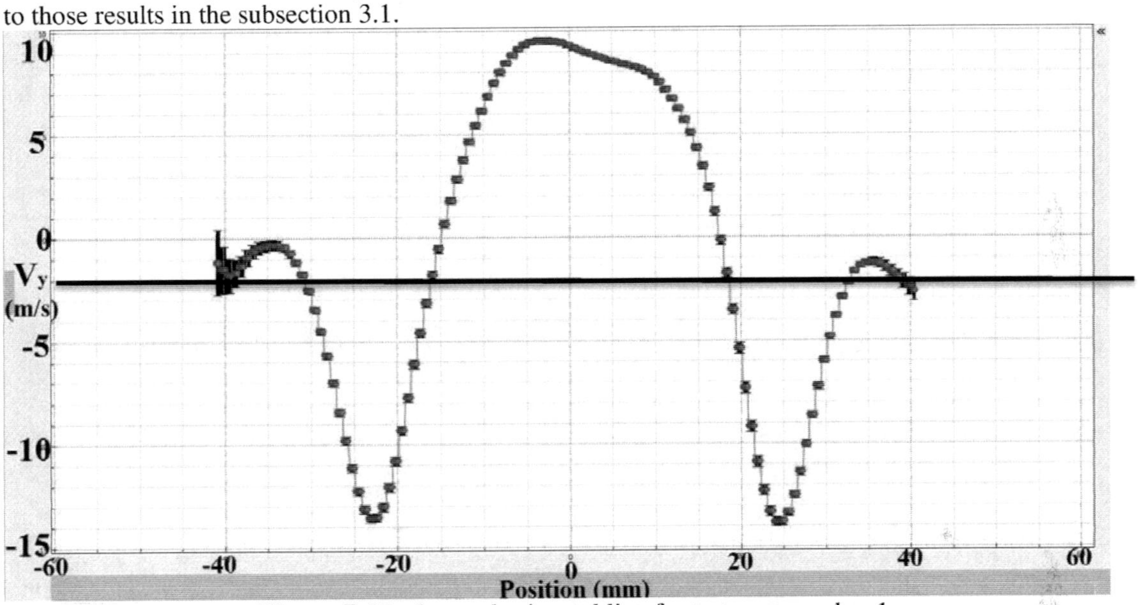

Figure 7. V_y along a horizontal line for test case number 1.

For the test case number 1 as shown in figure 7, the core has maximum velocity magnitude of 10 m/s while the CTRZ boundaries have magnitude of 14 m/s. Additionally, for the test case number 3, the core region has a maximum velocity magnitude of 14 m/s while the CTRZ boundaries have a magnitude of 19 m/s. The width of the CTRZ for both cases is around 34 mm. The other test cases have similar results with same profile shapes with same widths but with differences in velocity magnitudes.

4. Conclusion

Recirculation zone is formed by a toriodal flow in the region immediately downstream the injector. That flow has a significant effect in increasing combustion efficiency by mixing the fresh air and fuel downstream the injector with combustion products. Experimental investigation for recirculation zone characteristics is performed using a pre-filming airblast injector. Moreover, PIV is used for imaging the flow downstream the injector and the captured date is processed by a commercial software Davis. Olive oil is used as a tracer for its good characteristics in following the flow and in scattering the light. Results are represented for recirculation zone shape and geometrical characteristics, velocity field and shear regions. The shape of the CTRZ is found to be symmetrical about the central axis having the same dimensions for different test cases. Furthermore, the boundaries of the CTRZ have the maximum velocity magnitudes and maximum shear strength values. Vy along a horizontal axis is found to have the same shape profile for various test cases with difference in maximum velocity magnitudes.

References

[1] Bahr D W 1996 *Unsteady combustion* (Dordrecht: Springer)
[2] Correa S M 1993 *Combustion science and technology*, **87(1-6)** pp 329-62
[3] Lefebvre A H 1999 *Gas Turbine Combustion, 2nd edition* (Philadelphia: Taylor & Francis)
[4] Han Y M, Seol W S, Lee D S, Yagodkin V I and Jeungm I S 2001 *Journal of Engineering for Gas Turbines and Power* **123(1)** pp 33-40
[5] Winterfeld G, Eickhoff H E and Depooter K 1990 *Design of modern gas turbine combustors* (San Diego: Academic Press)
[6] Hall M G 1972 *Annual review of fluid mechanics* **4(1)** pp 195-218
[7] Leibovich S 1978 *Annual review of fluid mechanics* **10(1)** pp 221-46
[8] Leibovich S 1984 *AIAA journal* **22(9)** pp 1192-1206
[9] Lucca-Negro O and O'doherty T 2001 *Progress in energy and combustion science*, **27(4)** pp 431-81
[10] Reddy A P, Sujith R I, and Chakravarthy S R 2006 *Journal of Propulsion and Power* **22(4)** pp 800-08
[11] Gupta A K, Lilley D G and Syred N 1984 *Swirl Flows* (Kent: Abacus Press)
[12] Beer J and Chigier N 1972 *Combustion Aerodynamics* (London: Applied Science Publishers LTD)
[13] Tangirala V, Chen R H and Driscoll J F 1987 *Combustion Science and Technology* **51(1-3)** pp 75-95
[14] Lu X, Wang S, Sung H G, Hsieh S Y and Yang V 2005 *Journal of Fluid Mechanics* **527** pp 171-95
[15] Candel S, Durox D, Schuller T, Bourgouin J and Moeck J P 2014 *Annual Review of Fluid Mechanics* **46 (1)** pp 147-73
[16] Kilik E 1985 *23rd Aerospace Sciences Meeting* pp 187
[17] Syred N and Beer J M 1974 *Combustion and flame* **23(2)** pp 143-201
[18] Hadef R and Lenze B 2008 *Chemical Engineering and Processing: Process Intensification* **47(12)** pp 2209-2217
[19] Wang S, Yang V, Hsiao G, Hsieh S and Mongia H C 2007 *Journal of Fluid Mechanics* **583** pp 99-122
[20] Merkle K, Haessler H, Büchner H and Zarzalis N 2003 *International Journal of Heat and Fluid Flow* **24(4)** pp 529-37
[21] Benjamin M A 2000 *Atomization and Sprays* **10** pp 427-38
[22] Lefebvre A H and McDonell V G 2017 *Atomization and sprays* (Boca Raton, FL: CRC Press)
[23] Engelbert C, Hardalupas Y and Whitelaw J H 1995 *Proceedings of the Royal Society of London Series A: Mathematical and Physical Sciences* **451** pp 189-229
[24] Rizkalla A and Lefebvre A H 1975 *Journal of Engineering for Gas Turbines and Power* **97(3)** pp 316-20
[25] Rizk N K and Lefebvre A H 1980 *Journal of Engineering for Gas Turbines and Power* **102** pp 706-10
[26] Shanmugadas K, Manuprasad E, Chiranthan R and Chakravarthy S 2020 *Proceedings of Combustion Institute* **000** pp 1-8
[27] Raffel M, Willert C, Werely S and Kompenhans J 2007 *Particle Image Velocimetry A Practical Guide* (Berlin Heidelberg, New York: Springer)
[28] Hanson R K 1988 *Symposium (International) on Combustion* **21(1)** pp 1677-91
[29] Wu Q, Ye Q and Meng G 2012 *Journal of Mechanical Engineering Science* **227(9)** pp 1927-37

Acknowledgment

I wish to acknowledge the National Centre for Combustion Research and Development (NCCRD) for giving me access to its facilities and for the experimental support and guidance provided to me by NCCRD team.

AUTHOR INDEX